Exploring Data
in Engineering, the Sciences and Medicine

Exploring Data
in Engineering, the Sciences and Medicine

Ronald K. Pearson

OXFORD
UNIVERSITY PRESS

OXFORD

UNIVERSITY PRESS

*Oxford University Press, Inc., publishes works that
further Oxford University's objective of excellence
in research, scholarship, and education.*

Oxford   New York
Auckland   Cape Town   Dar es Salaam   Hong Kong
Karachi   Kuala Lumpur   Madrid   Melbourne   Mexico
City   Nairobi   New Delhi   Shanghai   Taipei   Toronto

With offices in
Argentina   Austria   Brazil   Chile   Czech Republic
France   Greece   Guatemala   Hungary   Italy
Japan   Poland   Portugal   Singapore   South Korea
Switzerland   Thailand   Turkey   Ukraine   Vietnam

Published by Oxford University Press, Inc.
198 Madison Avenue, New York, New York 10016

www.oup.com

Oxford is a registered trademark of Oxford University Press

Library of Congress Cataloging-in-Publication Data

Pearson, Ronald K., 1952-
  Exploring data in engineering, the sciences and medicine / Ronald
K. Pearson.
     p. cm.
  ISBN 978-0-19-508965-3 (hardcover : alk. paper) 1. Mathematical
statistics--Textbooks. I. Title.

  QA276.12.P396 2011
  001.4'22--dc22

                                           2010031400

1 3 5 7 9 8 6 4 2
Printed in the United States of America on acid-free paper

Contents

Exploring Data
in Engineering, the Sciences and Medicine

Chapter 1

The Art of Analyzing Data

Sections and topics:

The art of *exploratory data analysis* involves the application of mathematical and graphical techniques to a collection of data records in order to better understand their source (e.g., the plant species, material samples, or medical patients the dataset describes). These techniques are usually applied either to new, vaguely understood data collections, or to well established (and often large) collections approached with new, often vaguely formulated questions. Although there is considerable overlap between the techniques of exploratory data analysis and those of *confirmatory data analysis*, where we are attempting to answer a specific question (e.g., "is species A more resistant to root rot than species B?"), the underlying philosophies are quite different. Traditional introductory statistics texts generally emphasize confirmatory analysis methods—which give better (i.e., stronger, more concise) results when they are applicable—but in practice, confirmatory analysis can only take place once we understand our data source well enough to have precise hypotheses to confirm or reject. This book provides a detailed introduction to the techniques of exploratory data analysis that can lead us to the level of understanding required to formulate these hypotheses and set up a careful confirmatory analysis to examine them in detail.

Key points of this chapter:

1. This book introduces the subject of *exploratory analysis* of (often very large) collections of data, with a primary focus on *quantitative data* (i.e., numbers). Other data types are also discussed in varying levels of detail (e.g., categorical variables like "type of material," "species of animal," or "color"), but this book does *not* discuss the highly specialized analysis methods that have been developed for other data types like text or images.

2. An important distinction between the data analysis problems discussed in this book and traditional math problems (e.g., "how much is 2 plus 3?" or "what is the square root of π?") is that data analysis problems almost never have a unique "correct answer": different results may be obtained by different methods, and the appropriateness of the results depends on the reasonableness of the working assumptions on which the different methods are based. For this reason, *validation* of the results is even more important in data analysis than it is for traditional math problems.

3. For all intents and purposes, real data observations are *never* "exactly correct," and this is one of the reasons that different methods can give very different results for the same data analysis problem. In particular, practical data analysis methods must make working assumptions about the nature and magnitude of this data uncertainty, and different methods make different assumptions, which can lead to different end results.

4. Classical statistics books generally emphasize *confirmatory data analysis* where we are given a working hypothesis we wish to examine: e.g., drug X is more effective against insomnia than drug Y, or polymeric material A has greater tensile strength when it is made using additive B than without it. Confirmatory analysis is primarily concerned with designing experiments to collect data as efficiently and informatively as possible, and in applying formal statistical tests of our hypothesis.

5. As the title implies, the focus of this book is on *exploratory data analysis*, where we typically start with an existing dataset (very often collected by someone else), some possibly vague (and also possibly incorrect) notions about the source of the data, and a desire to improve our understanding.

6. Not surprisingly, the methods used in exploratory analysis differ in both detail and orientation from those used in confirmatory analysis. The methods discussed in this book are generally of one of three types: *descriptive characterizations* like means, medians, standard deviations, or correlations coefficients; *graphical characterizations* like boxplots, scatterplots, and quantile-quantile plots; and *data models* like linear or logistic regression models that predict one variable from one or more other variables.

7. This chapter gives a general overview of some of these methods and their motivations; the rest of the book describes these and related methods in more detail, illustrating their use with a variety of examples.

1.1 What this book is about

In his book *Glut*, Alex Wright notes that the total volume of recorded information produced every year—including documents, e-mail, television programs, web pages, books, and everything else—has been estimated to exceed 5 exabytes (5×10^{18} bytes), more than the total number of words *spoken* in human history [298, p. 6]. While no single individual is faced with the task of making sense of *all* of these data records, Wright's observation does highlight the fact that very large collections of data of all kinds are becoming increasingly available to be analyzed, interpreted, and understood. To provide a framework for approaching these problems, this book introduces the basic notions of *exploratory data analysis*. A more precise definition of this term is given in Sec. 1.6, but Diaconis [89] provides the following general description:

> We look at numbers or graphs and try to find patterns. We pursue leads suggested by background information, imagination, patterns perceived, and experience with other data analyses.

This quote implicitly assumes that we are dealing, at least primarily, with *quantitative data*, and this important working assumption represents one of the main criteria defining the scope of this book. In particular, while one of the examples discussed in Chapter 3 is based on text data, examining disputed authorship attribution in the *Federalist Papers*, this book does not consider any of the growing number of techniques developed specifically for the purpose of text analysis. These techniques typically construct quantitative characterizations of text, which are then subjected to quantitative analysis, including possibly some of the characterizations described in this book, but questions of how to define and compute these quantitative text characterizations are beyond the scope of this book. Similarly, the notions of *qualitative data analysis* used in the social sciences to make sense of nonquantitative data also lie beyond the scope of this book; for an introduction to these ideas, refer to the book by Miles and Huberman [209].

Another important working assumption concerns the size of the datasets considered here. Specifically, many of the datasets considered in this book are large by the standards of traditional statistics texts (e.g., "microscale" datasets that can be completely described by a table occupying one page or less) but small by the standards of large-scale data mining applications (e.g., "macroscale" datasets large enough that loading the complete dataset into a computer's main memory for analysis is not possible). The reason for this focus on "intermediate-sized" or "mesoscale" datasets is threefold. First, specialized techniques are required to deal with datasets so large that they cannot be put into a single data object for analysis purposes, and this is undesirable in an introductory data analysis text. Further, many of these large dataset techniques are extensions or variations of the techniques described here, and it is important to understand these basic notions before attempting to deal with the added complications introduced by huge datasets. The second reason for considering intermediate-sized datasets here is that certain historically significant notions that have evolved into simple "rules of thumb" on the basis of experience with small datasets break down in the intermediate case. A case in point is the idea of considering at most a fixed number of possible outliers in a dataset discussed in Chapter 7: this may be feasible when dealing with a set of ten or so data observations, but not when considering a few thousand observations. Finally, the third reason for considering intermediate-sized datasets is that the exploratory analysis of these datasets is extremely important in practice.

1.1.1 Useful data characterizations

The data characterizations described in this book can generally be divided into three types: descriptive numerical summaries, graphical characterizations, and data models. Descriptive numerical summaries are real numbers or collections of real numbers with simple interpretations that tell us something about one or more of the variables under consideration. Typical examples include means, medians, standard deviations, ranges (e.g., minimum and maximum values), and correlation coefficients between pairs of variables. The principal graphical characterizations discussed in this book include simple scatterplots that show how one variable relates to another, boxplots that show how the range of a variable changes between discrete subgroups (e.g., how the range of polymer viscosities differs between one polymerization reactor design and another, or how age range compares between diabetic and nondiabetic patients), and a variety of other tools like nonparametric density estimates or quantile-quantile plots that tell us something about how the observed values of a variable are distributed over its range of possible values. The data models considered here can generally be viewed as extensions of or variations on the two-variable regression models discussed in detail in Chapter 5, where the idea is to predict the value of one variable from another; in the more general cases considered in other chapters of this book, the objective is to predict either the value or some other characterization of one variable from one or more other variables. All of these characterizations are aimed at giving us useful insights into the behavior of the variables included in the dataset, either individually or collectively.

It is also important to note that many of these different ideas and techniques are closely related. For example, the mean of a sequence of numbers represents a simple, widely used method for combining multiple observations or estimates of what should be "the same" value, and idea strongly advocated by Gauss at the beginning of the nineteenth century [271]. In addition, this descriptive data characterization may be derived as the constant value that best approximates the data sequence, under the *least squares error criterion*, historically the most important criterion used for fitting regression models to datasets. Similarly, the standard deviation of the data sequence represents a measure of the approximation error associated with the mean value under the least squares error criterion. Alternatives to the mean, such as the median, may be derived as best-fit constant approximations under other, non-least squares fitting error criteria. Further, these results are all intimately related to specific probability distributions, corresponding to *maximum likelihood estimates* of parameters in data models under specific random error models. As an important example, the mean and standard deviation completely characterize the Gausisan data model, and Gaussian quantile-quantile plots represent a useful graphical tool for assessing the reasonableness of this data model. All of these ideas and their interrelationships are discussed in detail throughout the rest of this book, to provide a useful introduction to a range of specific data analysis techniques, illustrate their similarities and differences, and provide practical guidance in selecting the appropriate techniques for analyzing a particular dataset.

More specifically, the primary objective of this book is to introduce a collection of data analysis techniques that satisfy the following criteria:

1. they are useful in a wide variety of different applications;

2. implementations are widely available at little or no additional cost;

3. they can be explained in a few pages to a reader at the intermediate to advanced undergraduate level.

With respect to the second of these criteria, note that all of the data analysis methods described in this book are either available as built-in procedures or easily implemented using built-in procedures in the open source statistical software package *R* (see Appendix 1 for a brief discussion of this package and how to obtain it, free of charge). Further, many of these techniques are also widely available (or, again, easily implemented) on other platforms, ranging from spreadsheet packages like Microsoft *Excel* to scripting languages like *Python*. In addition, this book attempts to introduce the basic theory necessary to understand how to choose between the various method options available, and to illustrate via example why it is sometimes important to choose carefully.

In connection with this last point, a word of caution is in order: the simultaneous availability of almost unlimited data and powerful data analysis tools makes it important to understand what the analysis tools do and how to apply them, a point that Hosmer and Lemeshow note in the preface of their book, *Applied Logistic Regression* [148, p. ix]:

> As is well-recognized in the statistical community, the inherent danger of this easy-to-use software is that investigators are using a very powerful tool about which they may have only limited understanding.

The situation is somewhat analogous to that of being given unlimited access to a jet fighter aircraft without the benefit of any instruction in how to fly it; while the consequences are arguably less catastrophic, they can still lead to a good deal of frustration and embarrassment. The intent of this book is to give the reader the background to make effective use of some of the most important and widely known techniques available in software packages like *Excel* and *R*. Pursued carefully, this type of data analysis can have profoundly useful and far-reaching consequences. Conversely, as with many other technological advances, these techniques can be put to extremely dubious uses. To mention only one, the use of large-scale exploratory analysis or *data mining* techniques to find "hidden codes" in biblical texts has given support to those who disparage exploratory data analysis as "unprincipled data snooping" by those with a clear hypothesis in mind and who are looking for "proof" in datasets, often by repeating a questionable analysis with modified data until they get the answer they are looking for. These objections do raise a legitimate concern: it is possible to find a lot of different patterns in a dataset, especially a large one; to avoid being misled, it is important to proceed carefully and test the validity of our results as we go along. The quesions of what validity tests are available, how to perform them, and how to interpret the results are important issues discussed throughout this book.

1.1.2 Ohm's law

The following example is based on a small (i.e., "microscale") dataset whose analysis ultimately had a historically significant impact; it is introduced here because it illustrates a number of the key points made in this chapter and emphasized in subsequent chapters of this book.

In the first half of the nineteenth century, Georg Ohm conducted a number of experiments into the electrical behavior of metallic conductors. Fig. 1.1 shows the results of one set of experiments conducted in 1826 relating the length in inches of eight copper conductors to a measure of the magnetic field generated by the current flowing through each one in response to an applied voltage [98, p. 470]. These magnetic field intensities were measured with a torsion balance and were directly proportional to the current I flowing in the conductor, which Ohm had connected to a stable source of constant voltage V. A regular pattern of variation is evident in Fig. 1.1: the measured torsion T decreases monotonically as the sample length ℓ increases.

Ohm found that these values could be described quite well by the following equation:

$$T = \frac{a}{b + \ell},\tag{1.1}$$

where a and b were constants having the values $a = 6800$ and $b = 20.25$. A plot of the observed torsions against their predictions from Eq. (1.1) is shown in Fig. 1.2, from which it is clear that the agreement between the equation and the data values is excellent. In modern terminology, the constant a was proportional to the source voltage V and b was proportional to the external resistance R_s of the leads connecting Ohm's voltage source to the copper wire samples. Rearranging Eq. (1.1) then leads to the following expression:

$$a = (b + \ell)T \;\Rightarrow\; V = I(R + R_s),\tag{1.2}$$

where R represents the resistance of the conductor under consideration, a value proportional to the length ℓ. Expressed in terms of voltage and current, this result has come to be known as *Ohm's law* and today represents one of the fundamental principles used in the design and analysis of electronic circuits in

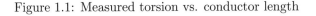

Figure 1.1: Measured torsion vs. conductor length

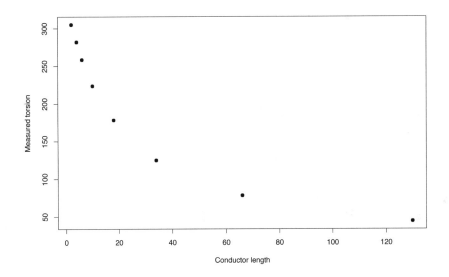

physics and electrical engineering. One of the key points of this example is that Ohm's law evolved as a consequence of the careful analysis of empirical data.

Another one of the key points of this example is that the analysis of the data (i.e., the curve-fitting process used to estimate the model parameters a and b), the interpretation of the results (i.e., the plot shown in Fig. 1.2), and the modern interpretation and application of the results all rely on transformations of the variables in which the original problem was formulated. In particular, recall that Ohm's experiment was described in terms of magnetic field intensities and conductor lengths, while the variables included in the dataset considered here were measured values of a torsion T (proportional to magnetic field strength) and the conductor's length ℓ. Also, note that the hypothesized relationship between T and ℓ given in Eq. (1.1) can be rewritten as:

$$1/T = \alpha + \beta\ell, \qquad (1.3)$$

where $\alpha = b/a$ and $\beta = 1/a$ represent the intercept and slope, respectively, of a line relating ℓ to the transformed variable $1/T$. The problem of fitting straight lines in the plane—i.e., of estimating the values of the constants α and β that give the best fit to the observed $(\ell, 1/T)$ data pairs—is one with a long history. Many methods have been proposed, and a number of these are described in detail in Chapter 5; the key point here is that the transformation from Eq. (1.1) to Eq. (1.3) makes the problem much easier to solve, although it can have undesirable side consequences, an issue discussed further in Chapter 12. Also, note that once

Figure 1.2: Measured vs. predicted torsion

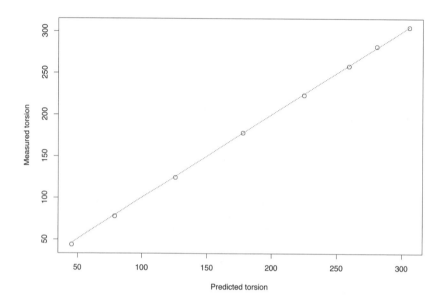

we have estimated α and β, we can compute the coefficients a and b appearing in Eq. (1.1) as $a = 1/\beta$ and $b = \alpha/\beta$ and, from these values, compute the predicted torsion according to Eq. (1.1) needed to construct Fig. 1.2. Finally, note that the results of this data model—specifically, the observed linear relationship between the model parameter a and the measured torsion T described by Eq. (1.2)— lay the foundation for Ohm's law as it is used today, expressing the linear relationship between the applied voltage (proportional to a) and the current (linearly proportional to magnetic field intensity and thus to the torsion T) in a resistor of resistance R (linearly proportional to the length ℓ of the conductor).

This example will be revisited several times in this chapter because it nicely illustrates a number of other important practical points. Before leaving the present discussion, however, it is worth saying something about the effort required to obtain the data shown in Fig. 1.1. Voltaic batteries were available in 1826 when Ohm performed these experiments, but they exhibited significant voltage fluctuations with time due to chemical changes and were therefore unsuitable. Instead, Ohm exploited the discovery by Seebeck in 1821 of the *thermoelectric effect*: a junction between two dissimilar metals generates a small voltage whose value depends on the temperature and the composition of the two metals. In practice, thermoelectric voltages tend to be small but stable, provided the temperature is kept constant. Also, real circuits necessarily involve at least two such junctions; in Ohm's experiments, he kept one junction in boiling water and the other in ice since these choices provided excellent temperature regulation. The care and effort required to make these measurements in the early nineteenth century partially explain why the dataset considered here was so small: the magnitudes of both the data collection effort and the subsequent analysis effort involved in generating and characterizing a "mesoscale" dataset of even a few thousand observations would have been unthinkable at the time. Another closely related issue is that Ohm was personally involved in all aspects of the collection, analysis, and interpretation of this dataset. In the nineteenth century, this situation was typical, but it is not today when large datasets are commonly collected by one group, analyzed by another, and the results interpreted by a third. In this setting, clear communication between these groups is extremely important since poor communication can lead to problems like that of *disguised missing data* and others discussed in Chapter 2.

1.2 How much can we learn from data?

In analyzing observed data, it is important to recognize that there are a number of inherent limitations on what we can hope to learn. One of these limitations arises from the unavoidable uncertainty present in virtually all measurement data, a point discussed further in Sec. 1.3 and in greater detail in Chapter 2. A second limitation is the fact that we have only a finite amount of this uncertain data on which to base our analysis. Beyond these two considerations, there are additional limitations that would hold even in the absence of uncertainty and even if we had an infinite amount of data. Sec. 1.2.1 illustrates this point with a simple example, and Sec. 1.2.2 briefly considers its main practical consequence: the necessity of imposing *some* working assumptions to guide us in our analysis.

1.2.1 Can one hear the shape of a drum?

In 1966, M. Kac published a paper titled, "Can One Hear the Shape of a Drum?" that posed the following problem [166]. Consider the sound waves emanating from a drum after it has been struck. Mathematically, these sound waves can be well approximated by the *Fourier series* [212], which decompose oscillating signals of complex shape into sums of sines and cosines at different frequencies. Physically, these frequencies are determined by the shape and size of the vibrating drum head. In mathematical terms, this statement represents a *forward problem*—the vibrations of a drum are described approximately by the two-dimensional wave equation:

$$\frac{\partial^2 v}{\partial t^2} = c^2 \left(\frac{\partial^2 v}{\partial x^2} + \frac{\partial^2 v}{\partial y^2} \right), \tag{1.4}$$

subject to the boundary conditions that define the shape and size of the vibrating membrane. Given this formulation, we can uniquely compute the sound waves eminating from the drum. In particular, the drum head shape determines the relationship between the intensities of the different frequency components in the drum sound. The question posed by Kac [166] was the corresponding *inverse problem*: if we analyze these sound waves accurately, would the results be sufficiently informative to reconstruct the shape of the drum head?

The title of a recent paper by Gordon *et al.* [121] gives the answer: "You Cannot Hear the Shape of a Drum." These authors show that different boundary conditions (i.e., different drum shapes) can give rise to *exactly the same* vibrations. Chapman [59] extends these results, describing a simple construction based on paper folding that leads to drum heads with arbitrarily complicated shapes that are distinct but carefully related and that would generate the same sound waves. Here, then, is a fairly simple practical problem for which the philosophy of "letting the data speak for itself" is inadequate, *even in the limit of an infinite amount of perfectly accurate data.* The fundamental difficulty illustrated in this example is that the solution of the inverse problem of interest is not unique. Unfortunately, this difficulty is generic, as a variety of other examples in this book demonstrate.

1.2.2   The role of assumptions

The key point of the preceeding example is that data analysis cannot be performed in a vacuum—we need educated prior assumptions to guide us in asking the right kinds of analytical questions. In particular, it is easy to formulate questions to which there is no unique answer, as Kac's question illustrates. One alternative is to ask less ambitious questions: it *is* possible, for example, to determine the area and perimeter of the drum head from an analysis of the sound waves [59, 239]. Conversely, it is often not obvious from the outset whether a particular analytical question is too ambitious or not (note, e.g., that a quarter of a century lapsed between Kac's published question and the answer given by Gordon *et al.*). Another alternative is to impose explicit *constraints* on the problem, an approach that can guarantee a unique solution in favorable cases, or lead to inconsistencies (i.e., lack of any solution) in unfavorable cases. For example, although we cannot answer the general question, "what shape is the drum?" we can ask—and answer—the more specific question, "is this drum elliptical in shape?" If it is, certain relations will exist between the Fourier series coefficients of the emanated sound waves [212] and conformance or non-conformance with these relations may be tested. In practice, we would begin by determining the shape parameters for the elliptical drum head that best fit the observed sound waves (a typical data analysis problem that usually has a unique answer), and then assessing the reasonableness of the result. That is, if the drum head is rectangular rather than elliptical in shape, the relative amplitudes of the different frequency components will not correspond to those generated by any ellipse, and careful analysis of our results should reveal that the predicted intensities for our "best fit elliptical drum head" are not in adequate agreement with the observed intensities. An important question here is what constitutes "adequate" agreement, a question that is examined repeatedly in subsequent chapters of this book, in connection with a variety of different data analysis problems. For now, the following two points are worth noting.

First is the need for some sort of working assumptions as the basis for any type of practical data analysis. Initially, it is desirable not to make these assumptions too specific in order not to bias our results too badly. For example, given any finite sequence of real numbers $\{x_k\}$, it is always possible to compute the arithmetic mean $\bar{x}$ and the standard deviation $\hat{\sigma}$ using the simple expressions given in Sec. 1.7.4. In addition, *if $\{x_k\}$ can be reasonably approximated by a sequence of statistically independent Gaussian random variables*, these two numbers provide a complete characterization of the data sequence. This idea—commonly imposed as a working assumption, perhaps tacitly—forms the basis for many classical statistical analysis procedures. Although enormously popular historically, this working assumption is often *not* satisfied in practice, a point discussed at length in Chapter 7. Alternatively, we can start with the *much* weaker assumption that "$\{x_k\}$ may be approximated by *some* random variable sequence." This alternative assumption then leads us to exploratory analyses using techniques like the nonparametric density estimates or Q-Q plots introduced in Chapter 8 to first assess the reasonableness of classical normality assumptions before allowing our results to depend heavily on them. The real point here is that, since assumptions are unavoidable, it is important to recognize this need from the outset and make them as explicit as possible.

The second key point is that working assumptions can be wrong. Consequently, it is useful to check these assumptions where possible and even more important to be willing to consider alternatives, even when these alternatives are at odds with our initial expectations and beliefs. A famous example that illustrates this point is the *Literary Digest* prediction of the outcome of the 1936 U.S. presidential election [153, p. 20]. Based on a survey of 10 million voters, whose names were drawn from magazine subscription lists, club membership rosters, and telephone directories, the magazine predicted that Landon would defeat Roosevelt by a substantial margin. The same sampling strategy had allowed the magazine to correctly predict the outcome of the 1932 election. When the 1936 election was actually held, however, the outcome was precisely the opposite of the *Literary Digest* predictions: Roosevelt defeated Landon by the greatest electoral landslide in U.S. history (523 votes to 8; the question of how well or poorly electoral and popular votes agree is examined in Chapter 10). The difficulty with the *Literary Digest* predictions was not that the sample size of 10 million was too small, but rather that it was not representative of the overall U.S. population at the time: 1936 was in the middle of the Great Depression, and most people could not afford magazines, club memberships, or telephones. Not surprisingly, this "unsampled majority" voted quite differently from *Literary Digest's* minority sample of voters who could afford such things.

1.3 Numerical mathematics versus data analysis

An important practical aspect of data analysis is that, in general, no "correct answer" exists to the questions we pose. This situation stands in marked contrast to the problems considered in elementary arithmetic or numerical mathematics (e.g., "what is the probability of observing a sample drawn from a Gaussian distribution that lies more than 3 standard deviations from the mean?"), which have precise, unambiguous solutions, at least in principle. Due to computational considerations, we may have to settle for approximations of these exact solutions (e.g., the value of π rounded off to the nearest 10, 100, or 1,000 digits), but we can generally obtain approximations as accurate as we like if we are willing to compute long enough. In contrast, questions like "what is the relationship between an animal's body weight and their brain weight?" involve at least two additional complications. The first is that these questions are not directly mathematical (e.g., what do we mean by "relationship" between two variables?), and the second is that observed data values almost always involve some degree of uncertainty (e.g., how accurately are the body weights and brain weights measured?). Both of these points are extremely important in practice and are emphasized throughout this book. Questions of the first type—e.g., "how do we characterize a relationship between variables?"—are essentially questions of what analysis method to choose, and they generally have more than one feasible answer; these different methods can yield results that either agree reasonably well or disagree strongly. Questions of this type are necessarily quite problem-specific and are addressed throughout the rest of this book. Questions of the second type—e.g., "how accurate are the available data observations?"—are concerned with how we measure and characterize data uncertainty, and the following paragraphs briefly examine some of the consequences of this uncertainty. More detailed discussions of the nature and sources of the uncertainty in real data are given in Chapter 2.

1.3.1   Numbers, arithmetic, and roundoff errors

The simplest numerical problems are those involving *integer arithmetic*, for which it is possible both in principle and in practice to obtain *exact* answers. The prototype problem here is the simple question, "how much is $2 + 2$?" As we all learn at a young age, this answer is 4 (at least in decimal arithmetic), and this answer is exact: any other answer is simply wrong. In somewhat more mathematical terms, integer arithmetic is based on the fact that the integers form a *ring* [157, p. 87], which is a collection of mathematical entities satisfying the following six conditions:

1. associativity of addition: $(a + b) + c = a + (b + c)$
2. commutativity of addition: $a + b = b + a$
3. existence of an additive identity: $a + 0 = 0 + a = a$
4. existence of an additive inverse: $a + (-a) = (-a) + a = 0$
5. associativity of multiplication: $(ab)c = a(bc)$
6. existence of a multiplicative identity: $a \cdot 1 = 1 \cdot a = a$

It follows from these results that any combination of integers formed via addition, subtraction, or multiplication yields a well-defined result that is also an integer. Conversely, note that this conclusion does *not* extend to division, since the quotient of two integers is, in general, not another integer. Consequently, so long as we are content with addition, subtraction, and multiplication, exact answers are possible, but once we extend beyond this extremely narrow set of mathematical operations, we lose the ability to compute exact answers.

To deal with division, it is necessary to consider either *rational numbers—* fractions of the form n/m where n and m are integers—or *real numbers*. Computationally, we typically adopt the practical compromise of considering *floating point numbers*, which are equivalent to *pairs* of integers, the *exponent* E and the *mantissa* M. In a decimal floating point number, for example, this combination yields a real number of the form:

$$x = M \times 10^E. \tag{1.5}$$

Adopting this representation, it is possible to represent an enormous range of real numbers quite efficiently, and to develop approximate procedures for all four of the standard arithmetic operations: addition, subtraction, multiplication, and division. For example, with only three digits for the mantissa and one digit for the exponent, it is possible to represent numbers ranging in magnitude from 0.01×10^0 to 9.99×10^9, spanning a range of 12 orders of magnitude. In contrast, the same number of digits (4) used in an integer representation span a range of only four orders of magnitude, from 1 to 9999.

The price we pay for this wider range of values and the ability to perform division operations is that the floating point representation is an approximate one that exhibits certain unpleasant subtleties. For example, the exact value of the product of 1.22×10^{-7} and $3.07 \times 10^{+4}$ is 3.7454×10^{-3}. In contrast, the result based on a three-digit mantissa would either be *truncated* to 3.74×10^{-3} or *rounded off* to 3.75×10^{-3}, depending on the details of the floating point multiplication routine used. In practice, such computations generally use a larger mantissa and are based on a binary representation instead of a decimal one, but the key point is that the results of such computations are no longer exact, as in integer arithmetic, but involve some degree of approximation. This approximation can be controlled by making the mantissa M and the exponent E sufficiently large, but an important practical issue is that long sequences of computations can, in unfavorable circumstances, result in roundoff errors that grow at each stage until the final result is dominated by these accumulated errors. The art of designing useful numerical algorithms lies in avoiding these difficulties and is a subject to which much attention has been paid, particularly since the advent of large-scale digital computers. For more detailed discussions of these ideas, refer to any standard numerical analysis text; two representative references are the books by Acton [2] and by Golub and van Loan [118].

A more subtle feature of floating point arithmetic is that floating point operations are not associative. This undesirable characteristic is a consequence of the fact that, if a is too small relative to b, the sum $a+b$ simply rounds off to b in floating point arithmetic. As a specific example, consider the sum of 1.00×10^0 and 1.00×10^{-3}: the exact answer is 1.001, but in our three-digit floating point arithmetic, this result rounds off to $1.00 = 1.00 \times 10^0$. As a consequence, for a sufficiently small floating point number ϵ, the sum of $+1$, -1, and ϵ depends strongly on how they are grouped: $(1+[-1])+\epsilon = \epsilon$, whereas $1+([-1]+\epsilon) = 0$. These difficulties are well understood in numerical analysis and are the basis for advice like, "avoid computing small differences of large numbers."

1.3.2   Computing mathematical functions

As we move to more complex computations, new sources of inaccuracy enter in the form of approximation errors. That is, the basic operations of addition, subtraction, multiplication, and division are subject only to the roundoff errors just described. Similarly, if we evaluate polynomials like $f(x) = a+bx+cx^2$, we may incur some accumulation of roundoff errors, but if we are careful, these errors need not be severe. However, once we attempt to evaluate more complicated functions like square roots, exponentials, or arctangents, we will generally incur some form of *approximation error*, arising from the fact that these functions cannot be computed exactly from the four basic operations of arithmetic. For many classes of mathematical functions, this approximation error can be made as small as we like, at least over a specified range of values for the function's arguments. Specifically, it follows from the Weierstrass approximation theorem [174, p. 334] that any continuous function can be approximated arbitrarily accurately on any closed, bounded interval by a polynomial of sufficiently high order. As a practical matter, less complex *rational approximations* can usually be obtained, approximations given by the ratio of two polynomials.

As a specific example, the *error function* is defined by the definite integral [1, p. 297]:

$$\text{erf}(z) = \frac{2}{\sqrt{\pi}} \int_0^z e^{-t^2}\,dt. \tag{1.6}$$

The importance of this function in statistics and data analysis applications lies in the fact that the probability of observing a sample x drawn from a standard normal distribution (i.e., a Gaussian distribution with mean zero and standard deviation 1) is given by [162, p. 81]:

$$\mathcal{P}\{x > z\} = \frac{1 + \text{erf}(z/\sqrt{2})}{2}. \tag{1.7}$$

Unfortunately, the definite interval defining the error function cannot be expressed exactly in terms of elementary functions, so it must be approximated in terms of such functions. Not surprisingly, given the historical importance of the error function, many different computational approximations have been developed. Abramowitz and Stegun give several, the simplest being [1, p. 298]:

$$\text{erf}(z) = 1 - \frac{1}{[1 + a_1 z + a_2 z^2 + a_3 z^3 + a_4 z^4]^4} + \epsilon(z), \tag{1.8}$$

where a_1 through a_4 are the constants:

$$\begin{aligned} a_1 &= 0.278393 & a_2 &= 0.230389 \\ a_3 &= 0.000972 & a_4 &= 0.078108, \end{aligned} \tag{1.9}$$

and $\epsilon(z)$ is the approximation error, which satisfies $|\epsilon(z)| \leq 5 \times 10^{-4}$ for all $z \geq 0$. More accurate approximations are possible through the use of more complex expressions of the same basic form. The key point here is that the computations required to evaluate these approximations are the four basic arithmetic operations of addition, subtraction, multiplication, and division.

1.3.3 Data analysis and uncertainty

In contrast to the function approximation problems just described, data analysis problems introduce a fundamentally different type of complication: the data values $\{x_k\}$ on which we must base our computations exhibit some degree of *inherent uncertainty*. Some of the reasons for this uncertainty are discussed in Chapter 2, which includes a brief discussion of the the problems of making measurements on physical systems. The basic difficulty is that all real measurement systems exhibit inaccuracies arising from a variety of sources. Further, one of the key points of Chapter 2 is that these inaccuracies are generally *much* larger than the roundoff errors arising from numerical calculations and these inaccuracies *cannot* generally be made "as small as we like." For example, the *S-plus* statistical software package [272] displays numbers in a floating point decimal format with a seven-digit mantissa, corresponding to a roundoff error on the order of 1 part in 10^7. In contrast, typical measurement uncertainties generally range from a fraction of a percent (i.e., a few parts in 10^3) to a few percent. Further, even in the best possible circumstances (e.g., measurements of fundamental physical constants like the mass of the electron under the incredibly careful conditions maintained at national standards laboratories), accuracies are very rarely better than one part in 10^6. In contrast, if we require higher accuracies in numerical computations, we can always use multiple precision arithmetic, which amounts to increasing the size of the mantissa in the floating point computations. While this approach may be complicated and unpleasant, it is always possible in principle. Conversely, real-world measurement accuracies are ultimately limited by factors beyond our control, a point discussed further in Chapter 2. Hence, it is necessary to account for these dominant uncertainties in analyzing real-world data; the next section considers the question of how we can do this in practice.

1.4 Dealing with uncertain data

Most of the data characterizations described in this book effectively partition the observed data into two components: a *predictable component* describing the fundamental behavior of interest, and an *unpredictable component* describing the uncertainty itself. The following paragraphs describe this decomposition in more detail, emphasizing the unpredictable component; the predictable component is discussed in Sec. 1.5.

1.4.1   Additive uncertainty models

The easiest way of decomposing an observed data sequence into predictable and unpredictable components is to construct an *additive uncertainty model* of the following general form. For simplicity, this discussion considers a sequence $\{y_k\}$ of N observations of a single real variable y, but the basic ideas described here extend readily to other settings and several of these will be described in detail later in this book. Given this sequence of observations of the dependent variable y, the additive uncertainty model takes the following form:

$$y_k = \hat{y}_k + e_k, \quad \text{for } k = 1, 2, \ldots, N, \tag{1.10}$$

where $\hat{y}_k$ is the predicted value of y_k generated by a mathematical data model $\mathcal{M}$, and e_k is the prediction error associated with $\hat{y}_k$ (i.e., the difference between the observed value y_k and the predicted value $\hat{y}_k$). The model $\mathcal{M}$ generally depends on one or more observed explanatory variables x_i and one or more model parameters θ_j that are initially unknown. In this case, the task of data modeling reduces to one of deciding what structure the mathematical model $\mathcal{M}$ should have, and determining the unknown parameter values θ_j for which $\mathcal{M}$ best fits the data (i.e., for which the prediction error sequence $\{e_k\}$ is "most reasonable"). These tasks can be accomplished in more than one way, and the selection of an appropriate data model structure and fitting method is an important part of the art of data analysis, as are the subsequent tasks of assessing the quality of the model once we have obtained it and interpreting the results in the context of the original real-world problem that generated the data.

A concrete example to illustrate these ideas is useful here. Chapter 5 is devoted to the problem of fitting straight lines to a set of points in the plane, a problem that is both a prototype for many data analysis problems and one that arises frequently in practice, possibly after suitably reformulating the original problem (the Ohm's law problem discussed in Sec. 1.1.2 is a case in point). The basic problem is to determine the single line in the x, y plane that "best fits" a given set of points defined by the set of coordinate pairs $\{(x_k, y_k)\}$, and the additive uncertainty model for this problem has the following form:

$$y_k = ax_k + b + e_k. \tag{1.11}$$

The prediction model $\mathcal{M}$ here is $\hat{y}_k = ax_k + b$, which involves the single explanatory variable x and the two parameters a and b that represent the slope and the y-axis intercept of the line defined by this model. The prediction error term $e_k = y_k - \hat{y}_k$ gives a measure of how well the line defined by the parameters a and b fits the data point (x_k, y_k). In particular, note that if $e_k = 0$, it follows that $y_k = ax_k + b$, implying that the line with slope a and intercept b passes exactly through the point (x_k, y_k). In fact, this error term is a measure of the vertical distance the point lies from the line, with positive values indicating that the point lies above the line and negative values indicating that the point lies below the line.

Note that in this example, the prediction error e_k depends on both the data values (x_k, y_k) and the model parameters a and b. In fact, this situation holds generally and provides a basis for developing model fitting procedures. That is, since the prediction errors $\{e_k\}$ are defined as the differences between the observed data values $\{y_k\}$ and their corresponding predictions $\{\hat{y}_k\}$, it follows that, for a given dataset $\mathcal{D}$ and a given form for the model $\mathcal{M}$, the prediction error sequence $\{e_k\}$ is determined by the set of model parameters $\{\theta_i\}$ (e.g., the slope a and intercept b in the line-fitting problem considered above). The key consequence of this observation is that we can choose these model parameters to make the prediction errors "as acceptable as possible." Different interpretations of the phrase in quotation marks lead to different procedures for fitting data models and interpreting the results.

1.4.2   The minimum uncertainty model

Conceptually, the simplest way of making the prediction errors $\{e_k\}$ "as acceptable as possible" is to make them "as small as possible." To do this requires a method for measuring the size of a sequence of numbers like $\{e_k\}$, and several such measures are available. Given a particular size measure, we can then define an optimization-based model-fitting strategy: we choose as a "best fit" the model $\mathcal{M}$ defined by the chosen structure and the set $\{\theta_i\}$ of parameters that minimize the size of the resulting error sequence $\{e_k\}$, as defined by the chosen size measure. Different choices of size measure lead to different optimization problems, with different strengths and weaknesses. In addition, many of the most popular size measures are closely related to specific cases of the random variable problem formulation discussed in Sec. 1.4.3.

One of the most popular size measures for an error sequence $\{e_k\}$ is the sum of squared errors:

$$J = \sum_{k=1}^{N} e_k^2, \tag{1.12}$$

where N is the length of the error sequence. Note that J is a nonnegative quantity for any sequence $\{e_k\}$ and that $J = 0$ if and only if e_k is identically zero for all k, corresponding to perfect agreement between the model predictions $\{\hat{y}_k\}$ and the observed data values $\{y_k\}$. Further, note that if J and J' are computed from two different sequences $\{e_k\}$ and $\{e_k'\}$ and if the magnitude of the errors $\{e_k'\}$ is consistently larger than the magnitude of the errors $\{e_k\}$ (i.e., if $|e_k'| > |e_k|$ for all k), then $J' > J$. Also, note that multiplying the sequence $\{e_k\}$ by some positive constant λ causes J to be multiplied by λ^2. The key point is that J exhibits many properties that make it a natural choice for comparing the sizes of different sequences $\{e_k\}$ and $\{e_k'\}$. Further, minimizing this particular size measure J often leads to mathematically well-behaved problems (e.g., they frequently have unique solutions that are relatively easy to compute). For this reason, fitting data to models by minimizing the squared error criterion J has been quite popular historically and is known as the *method of least squares*.

This book gives detailed discussions of many specific data analysis procedures based on this general idea, emphasizing both their considerable advantages and their sometimes glaring disadvantages. Because of these disadvantages, alternative procedures have become the focus of much research in the statistics community, and some of these ideas will also be considered in this book, particularly some of those that have led to practical computational procedures like the *M-estimators* discussed in Chapter 7.

1.4.3   The random variable model

Another popular basis for data fitting models is to model the prediction errors $\{e_k\}$ as a sequence of *random variables*. A detailed introduction to random variables is given in Chapters 3 and 4, but it is useful to give a few introductory comments here since random data models are extremely important in practice, leading to a wide variety of useful data analysis procedures. In its simplest form, the random variable model assumes that $\{e_k\}$ is a sequence of statistically independent samples, all drawn from a common probability distribution. Given the parameters of this distribution, we can assess the "reasonableness" of each individual prediction error e_k by considering the probability of drawing it from the assumed distribution. Under the assumption that the errors are statistically independent, it is easy to combine the probabilities associated with each individual observation, giving us a measure of the likelihood of observing a given error sequence $\{e_k\}$. Since this error sequence is determined by the unknown model parameters $\{\theta_i\}$, we can associated this likelihood with the model parameters. This process leads to the class of *maximum likelihood methods* for fitting the prediction model $\mathcal{M}$ to the dataset $\mathcal{D}$. That is, given the form of the model $\mathcal{M}$ and the dataset $\mathcal{D}$, the maximum likelihood method selects as "best fit" the model parameters $\{\theta_i\}$ that maximize their likelihood, given the probability distribution assumed for the prediction errors $\{e_k\}$.

As with the case of different size measures in the minimum uncertainty model described in Sec. 1.4.2, different choices of probability distribution are possible in the random variable approach, and these lead to different data-fitting methods with different advantages and disadvantages. In fact, maximum likelihood formulations often lead to problems whose solutions are identical with particular cases of the minimum uncertainty model. An important example is the case of Gaussian maximum likelihood. Indeed, one of the most popular random variable models assumes that $\{e_k\}$ is a statistically independent sequence of random samples from a Gaussian distribution with mean μ and variance σ^2. Two examples are shown in Fig. 1.3. The upper left plot shows $N = 100$ samples drawn from the *standard normal distribution*, defined as the Gaussian distribution with mean $\mu = 0$ and standard deviation $\sigma = 1$, while the upper right plot shows the corresponding probability density function, exhibiting the familiar "bell-shaped curve" proposed to describe many different phenomena. The lower left plot shows $N = 400$ samples drawn from the $N(1, 0.4)$ Gaussian distribution, with $\mu = 1$ and $\sigma = 0.4$, while the lower right plot shows the corresponding probability density function.

A detailed introduction to the Gaussian distribution is given in Chapters 2 and 4, but the key point here is that this distribution is completely specified

Figure 1.3: Two Gaussian random error models: the standard normal model, $N(0,1)$ (top), and a second example, $N(1,0.4)$ (bottom)

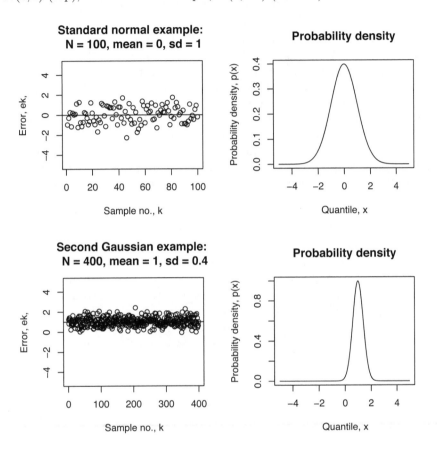

by these two parameters. Here, the mean μ specifies a "typical" value for the prediction error e_k and this is usually assumed to be zero, corresponding to an assumption that there is no *systematic* error in the model predictions $\{\hat{y}_k\}$. That is, these predictions generally exhibit some error, but *on average*, this error is zero, implying that the predictions $\hat{y}_k$ are neither consistently smaller nor consistently larger than the observed values y_k. The other parameter required to completely specify the Gaussian error distribution—the variance σ^2—is assumed to be unknown, but given a sequence of prediction errors $\{e_k\}$ drawn from this distribution, this parameter can be estimated as:

$$\widehat{\sigma^2} = \frac{1}{N-1} \sum_{k=1}^{N} e_k^2. \tag{1.13}$$

For a dataset with a fixed number of observations N, the maximum likelihood solution to the model-fitting problem under these distributional assumptions reduces to one of minimizing the estimated prediction error variance $\widehat{\sigma^2}$. Since N is a constant in this problem formulation, this problem reduces to the least squares problem described in Sec. 1.4.2. That is, since $\widehat{\sigma^2} = J/(N-1)$, it follows that minimizing $\widehat{\sigma^2}$ and minimizing J are equivalent problems.

The question of when the Gaussian assumptions just described are reasonable and when they are unreasonable is one that will be revisited a number of times for different specific problems in subsequent sections of this book. The key points to note here are, first, that these assumptions often lead to computationally simple solutions (e.g., simple analytic expressions in the best of cases), and because of this, they have been extremely popular historically. A second key point is that an advantage of random variable formulations over minimum uncertainty formulations is that the availability of an explicit prediction error distribution leads to a number of useful procedures for assessing the reasonableness of the resulting model. Several of these procedures are introduced in Chapters 9 and 15. Finally, the third key point to note is that the distributional assumptions on which a particular random variable model is based (either the Gaussian distribution or any other distribution) are only approximate, in general, and for datasets where this approximation is poor enough, the data-fitting results we obtain using it can be very poor as well. The practical importance of this problem has led to significant developments in the field of *robust statistics*, which attempts to develop analysis methods that are not strongly dependent on (possibly erroneous) distributional assumptions. A brief introduction to some of these ideas is given in Sec. 1.7.4, and a more detailed treatment is given in Chapter 7.

1.4.4 Other uncertainty models

While the minimum uncertainty model described in Sec. 1.4.2 and the random variable model described in Sec. 1.4.3 are unquestionably the most popular uncertain data models in practice, other options do exist. Some of these alternatives have theoretically attractive features but are complicated enough that they have been little used in describing and analyzing data. A case in point is Dempster-Shafer theory [263], an approach that is capable of distinguishing between *uncertainty* (where probability models are appropriate) and *ignorance* (where probability models are not appropriate); an application of Dempster-Shafer theory to robotic vision that exploits this distinction has been described by Hughes and Murphy [154]. A detailed treatment of more general alternatives in this direction is given by Walley [287], including a brief discussion of fuzzy sets, described further below. First, however, it is useful to give a short introduction to the set-theoretic uncertainty model, which is conceptually simple but frequently leads to significant computational complexity.

The *set-theoretic* or *unknown-but-bounded* error model assumes that the prediction error e_k can assume any value in a specified set S_k. Since the prediction errors are determined by the unknown model parameters $\{\theta_i\}$, the requirement that $e_k \in S_k$ implies that the parameter vector θ, defined as the vector whose components are the model parameters $\{\theta_i\}$, must lie in a corresponding parameter set P_k. Imposing this requirement for each of the N prediction errors e_k then leads to an overall constraint that the parameter vector θ lie in the intersection of these individual parameter sets P_k. That is, given a model structure $\mathcal{M}$, a dataset $\mathcal{D}$, and a collection of assumed error bounding sets $\{S_k\}$, the set-theoretic modeling approach constructs the set P of admissible parameter vectors, given by:

$$P = \cap_{k=1}^{N} P_k. \tag{1.14}$$

In the simplest case, all of the error uncertainty sets S_k are the same, analogous to the assumption in the random variable model that the prediction errors $\{e_k\}$ have a common distribution. Similarly, the assumption of no systematic errors that corresponds to assuming a zero mean error distribution in the random variable model is typically translated to an assumption that the common uncertainty set S for the prediction errors is a symmetric interval around zero, of the form $S = [-a, a]$ for some constant a.

Despite the appealing simplicity of the set-theoretic formulation, it suffers from two important limitations in practice. The first was mentioned above: the computational complexity can become impractical even for simple problems. Specifically, the set-theoretic uncertainty model leads to parameter estimation problems that involve simultaneous sets of inequalities (a detailed example is given in Chapter 5, where the set-theoretic formulation of the line-fitting problem is considered). In the simplest cases, the solution of these inequalities—the parameter uncertainty set P defined in Eq. (1.14) above—is a *polytope* [305]. The best-known examples of polytopes are *polygons*—figures defined by line segments connecting verticies with all interior angles less than $180°$—a class that includes triangles, squares, squares, etc. The computational difficulty with the set-theoretic parameter estimation problem is that the number of vertices (i.e., "corner points" in the p-dimensional space of the parameter vector θ) can grow rapidly with the size of the dataset. As a result, in unfortunate cases, it is not possible to define the admissible parameter set P much more simply than through the set of N simultaneous inequalities that defines it. To address this problem, much effort has been made to define useful *bounding sets* B, which are sets of simpler structure (and thus amenable to simpler mathematical descriptions) than P, but which contain P (i.e., $P \subset B$) so that $\theta \in P$ implies $\theta \in B$. In general, this increases computational complexity and introduces conservatisim since $\theta \in B$ does not guarantee that $\theta \in P$. For a more detailed treatment of these issues, refer to the survey article by Combettes [67], the books by Schweppe or Walter [261, 288] and the references cited there.

The second basic difficulty with the set-theoretic uncertainty formulation is that, like the minimum uncertainty formulation discussed in Sec. 1.4.2, it provides limited basis for subsequent model validation and refinement, relative to those available in the random variable formulation. For both of these reasons—the complexity of the exact solution and the limited range of model validation and refinement tools—the set-theoretic formulation has not achieved the popularity as a data analysis approach enjoyed by random variable methods. Consequently, the set-theoretic formulation will not be considered further in this book, beyond the illustrative example on fitting straight lines presented in Chapter 5.

Finally, it is worth briefly describing *fuzzy sets*, which have been advocated as the basis for a nonprobabilistic formulation of uncertain problems [96, 301, 306]. The essential idea is to allow uncertainty in set membership: in an ordinary or *crisp* set S, an element x either belongs ($x \in S$) or it does not ($x \notin S$). In a fuzzy set F, elements are assigned a continuous *grade of membership* $\mu(x)$ that lies between 0 and 1, usually with the following interpretation:

1. $\mu(x) = 1$ means that x belongs to the set F, with certainty;

2. $\mu(x) = 0$ means that x does not belong to the set F, with certainty;

3. $0 < \mu(x) < 1$ means that x may or may not belong to the set F, with increasing likelihood as $\mu(x)$ increases.

Probably the most popular application of fuzzy sets is in dealing with linguistic uncertainties by providing a way of partially quantifying descriptors like "very," "somewhat," or "slightly." The mathematics of set theory has been extended to fuzzy sets, but this can be done in several different ways, adding to the complexity of working with fuzzy sets. For example, the intersection of two fuzzy sets can be defined in at least three distinct ways [96]. Fuzzy sets do make an appearance in the data mining literature, and there has been some work in the area of data analysis based on fuzzy sets [167, 306]. Like the set-theoretic approach described above, however, these methods have not yet achieved the practical utility of those based on the random variable uncertainty model and will therefore not be considered further here; for those interested in a detailed introduction, Zimmermann's book is recommended [306].

1.5   What is a good data model?

Several of the chapters in this book are concerned with detailed procedures for constructing different types of data models, frequently—but not always—based on the additive decomposition introduced in Sec. 1.4.1. As the preceeding discussions have indicated, this process typically involves selecting both a structure for the data model and an uncertainty description. Both of these choices influence many aspects of the resulting data model that we obtain for a given dataset, including its complexity, its interpretation, and its accuracy in approximating the original dataset. While different data modeling situations lead naturally to different evaluation criteria for the resulting models, one key aspect of data models that is worth discussing in more general terms is model complexity.

1.5.1   Empirical versus fundamental models

In considering the problem of how to construct a mathematical data model, it is useful to begin by considering two distinct philosophies of model-building:

empirical modeling and fundamental modeling. In its strictest interpretation, the empirical philosophy attempts to build a model based *only* on the available data, and Kac's drum head question discussed in Sec. 1.2.1 illustrates that this interpretation is too extreme to be practical. In particular, one of the key points of that discussion was the necessity for working assumptions to give us a data analysis problem that is even solvable in principle. In practice, additional— and generally stronger—working assumptions are usually imposed to give us a problem that is more easily solved. Part of the art of data analysis, then, lies in choosing these working assumptions wisely, so that we obtain a mathematical problem that satisfies two key criteria: first, it can be solved with reasonable effort, and second, the solution bears a useful relationship to the real-world situation that yielded the data we are analyzing.

Alternatively, a *fundamental model* is a mathematical description of the physical system from which the dataset was obtained. In its strictest form, fundamental modelling proceeds by first identifying the fundamental phenomena present in the physical system and then describing the behavior of these phenomena in terms of the generally accepted fundamental principles governing that behavior. Thus, in applications like chemical or mechanical engineering, fundamental models are generally developed from the principles of heat, mass, and momentum transfer, whereas fundamental models of electronic systems are generally developed on the basis of Kirchoff's laws and the constitutive relations for the various circuit components. Even here, however, the practical situation is less straightforward than it might seem, as the following example illustrates.

It was noted in Sec. 1.1.2 that Ohm's law has come to be one of the fundamental principles on which electronic circuit design and analysis is based. Specifically, Ohm's law states that the voltage drop across an *ideal resistor* is proportional to the current flowing through it. In practice, although a great deal of manufacturing effort goes into producing resistors that are good, well approximated by Ohm's law over a useful range of currents and voltages, no real resistor is perfect. Thus, the ideal resistor characterization represents only an approximate mathematical model to describe the characteristics of real resistors, arguably about the simplest electronic component there is. Deviations from Ohm's law—and the practical consequences of these deviations—depend on a number of application-specific factors. For example, at sufficiently high voltages, nonlinear effects set in and voltage is no longer simply proportional to current. In fact, this behavior is exploited in the *varistor*, an electronic component whose resistance is significantly lower at high voltages than at low voltages, behavior that can be useful in transient suppression for inductive loads [221, p. 811]. At low voltages, the random thermal noise generated by resistors can be a significant factor [221, p. 198]. Further, the characteristics of a real resistor are frequency dependent: at very low frequencies, the effects of "flicker noise" or "popcorn noise," which exhibit an inverse dependence on frequency (e.g., the noise voltage V_n varies with the frequency f as $1/f^n$ for some exponent $n \geq 1$), dominate, while effects like thermal noise become dominant at high frequencies. Also, so-called "parasitic effects" like the finite inductance of the lead wires used to connect the resistor into an electronic circuit and the capacitance between adjacent conductors, either within the resistor itself or associated with its mounting configuration, become extremely important at high frequencies. To make matters worse, the relative importance of these effects depends on the construction details of the resistor and on extremely important secondary considerations like operating temperature on which all resistor characteristics depend strongly. The key point is that even an apparently simple electronic circuit like the Wheatstone bridge, composed of a fixed voltage source, two fixed resistors, and one variable resistor, present more serious fundamental modeling problems than we might initially expect. For a glimpse at the gory details that become vitally important in developing and analyzing Wheatstone bridges for very precise measurements (i.e., "accuracies approaching a part in 10^9"), refer to the book by Kibble and Rayner [173]. Because of the rapid increase in analytical complexity of fundamental models as we attempt to make them more precise, even these models are necessarily approximate, as the following discussion emphasizes.

1.5.2 The principle of zig-zag-and-swirl

A rather colorful, if obscure, first-principles extremist was Alfred William Lawson. Born in London in 1869, Lawson grew up in the U.S. and achieved a sufficiently dubious prominence to rate a chapter in Martin Gardner's book, *Fads and Fallacies in the Name of Science* [111]. Lawson's multifaceted career included nineteen years in professional baseball, beginning at age 19 as the pitcher for a team in Goshen, Indiana. In 1904, his Utopian fantasy novel *Born Again* appeared, a book that Gardner characterizes as "the worst work of fiction ever published" (one memorable line in the novel is "What strange phenomena is this, I soliloquized?" Shortly after uttering it, the protagonist finds a woman who has been asleep for more than 4,000 years and wakes her with his kiss). From there, Lawson went on to be a mover and shaker in the early aviation industry, first founding two popular aviation magazines while heavier-than-air flight was still in its infancy (he is credited with introducing the term "aircraft" into widespread use with his second magazine [165, p. 61]), and then founding an aircraft company that obtained the first U.S. airmail contract before falling into receivership. During the Great Depression, Lawson founded a group called the Direct Credits Society, which advocated the "Lawson Money System." This unique economic system included a form of "valueless money" that was not redeemable for anything. But in many ways, Lawson's crowning achievement was his purchase in 1942 of the bankrupt University of Des Moines, subsequently rechristened the "Des Moines University of Lawsonomy."

Lawson defined Lawsonomy as "the knowledge of Life and everything pertaining thereto" and he argued that "the basic principles of physics were unknown until established by Lawson." One of these basic principles was the replacement of Newtonian mechanics and the other generally accepted laws of physics with the dual notions of "Suction and Pressure." Indeed, the applicability of these ideas went well beyond the limits of physics. For example, these forces provided an explanation of the brain's function in terms of two microscopic societies, constantly waging war in our heads: the Menorgs were the "mental organizers" responsible for constructive thought, while the Disorgs were the "mental disorganizers" responsbile for all manner of destructive activities. Less than universal acceptance of these insights by his contemporaries led Lawson to once remark, "when I look into the vastness of space and see the marvelous workings of its contents . . . I sometimes think that I was born ten or twenty thousand years ahead of time."

One of the fundamental tenents of Lawsonomy was the Principle of Zig-Zag-and-Swirl, for which Lawson offered the following illustrative example. Consider a germ, moving across a blood corpuscle in the body of a man who is walking down the aisle of a flying airplane. Viewed locally, the germ's path may appear to be fairly regular and uncomplicated. But, the corpuscle is moving in the blood stream, which is circulating throughtout the man's body, who is walking down the aisle of the airplane, which is flying over the earth's surface, which is rotating and revolving around the sun, which is rushing through space. Hence, the "true" path of the germ is infinitely more complicated than it first appeared to be. The Principle of Zig-Zag-and-Swirl attempted to deal with this overwhelming complexity. Lawson's definition was equally impressive:

> "movement in which any formation moves in a multiple direction according to the movements of many increasingly greater formations, each depending upon the greater formation for direction and upon varying changes caused by counteracting influences of Suction and Pressure of different proportions."

To deal with complexity of this nature, Lawson argued that ordinary mathematics was inadequate and he advocated the development of a "Supreme Mathematics" in its place.

The difficulty is that Lawson's point is correct, as far as it stands: any attempt to be *complete* and *exact* in describing natural phenomena is doomed to failure by the ultimate complexity of life. That is, if we want to be exact—with no approximations—we must include the full effects of quantum chromodynamics, flow of the polar icecaps, possibly influential legislation pending in Azerbaijan, the effects of Jupiter's gravity, and a great deal more. Of course, we aren't usually after *exact* descriptions of the source of our data, only *sufficiently accurate* ones. It is this subtle—but important—change in the problem formulation that avoids the need for Lawson's Supreme Mathematics.

1.5.3 Ockham's razor and overfitting

The question of how complex a data model should be is frequently answered qualitatively by invoking *Ockham's razor*, also known as "the principle of economy," typically translated as [194, p. 185]:

> Entities are not to be multiplied beyond necessity.

A somewhat more intuitive translation is that "facts should be interpreted with a minimum of explanatory causes" [84, p. 433]. This advice comes from William of Ockham, a fourteenth-century English philosopher whose date of birth is uncertain, but who was educated at Oxford University, completing the requirements for a doctorate in about 1324 when he was summoned to appear before the Pope on charges of heresy [69, p. 43]. He was excommunicated in 1328 and went into exile under the protection of Emperor Ludwig of Bavaria; he died in Munich in 1349, probably as a victim of the bubonic plague or "black death" that is estimated to have killed one third of Europe [84, p. 412].

In the context of data modeling, Ockham's razor argues that we accept the simplest model that is consistent with the available data. Given the unavoidable uncertainty that infests real datasets, the question of what we mean by "consistent with the available data" is subject to interpretation, but it is easily demonstrated that "reasonable data models" are necessarily constrained in their complexity. In particular, if we allow a data model to become too complex, we risk the problem of *overfitting* in which the data model captures both the main behavior of interest and the extraneous details we wish to eliminate as noninformative approximation error.

A specific illustration of the overfitting problem is provided by the following observation: under very general conditions, almost any sequence $\{(x_k, y_k)\}$ of N points in the plane can be fit *exactly* to a polynomial of degree $N - 1$. To accomplish this objective, we need to solve the *polynomial interpolation problem*, determining the set of N polynomial coefficients $\{a_j\}$ that satisfy the following exact data matching conditions:

$$y_k = \sum_{j=0}^{N-1} a_j x_k^j, \quad k = 1, 2, \ldots, N. \tag{1.15}$$

This problem may be cast in matrix-vector notation by defining the $N \times N$ independent variable data matrix $\mathbf{X}$ as:

$$\mathbf{X} = \begin{bmatrix} 1 & x_1 & x_1^2 & \cdots & x_1^{N-1} \\ 1 & x_2 & x_2^2 & \cdots & x_2^{N-1} \\ \vdots & \vdots & \vdots & \vdots & \\ 1 & x_N & x_N^2 & \cdots & x_N^{N-1} \end{bmatrix}, \tag{1.16}$$

defining the N-vector of unknown parameters $\mathbf{a}$ as:

$$\mathbf{a} = [a_0, a_1, \ldots, a_{N-1}]^T, \tag{1.17}$$

and defining the N-vector of observed dependent variable values $\mathbf{y}$ as:

$$\mathbf{y} = [y_1, y_2, \ldots, y_N]^T. \tag{1.18}$$

The N simultaneous linear equations in Eq. (1.15) that define the polynomial interpolation problem may now be written in matrix-vector notation as:

$$\mathbf{Xa} = \mathbf{y}. \tag{1.19}$$

If the matrix $\mathbf{X}$ is nonsingular, the unknown coefficient vector $\mathbf{a}$ may be computed from the available data by premultiplying both sides of Eq. (1.19) by the inverse matrix $\mathbf{X}^{-1}$, to obtain the explicit solution:

$$\mathbf{a} = \mathbf{X}^{-1}\mathbf{y}. \tag{1.20}$$

Perhaps the best known condition for $\mathbf{X}$ to be nonsingular is that its determinant $|\mathbf{X}|$ be nonzero, and the determinant of the matrix $\mathbf{X}$ in the form shown in Eq. (1.16) is called the *Vandermonde determinant*, which can be written in the following form [39, p. 104]:

$$|\mathbf{X}| = \prod_{k=2}^{N} \prod_{j=1}^{k-1} (x_j - x_k). \tag{1.21}$$

An extremely useful consequence of this result is that it immediately yields simple conditions under which an exact fit is possible: the determinant $|\mathbf{X}|$ is nonzero if and only if $x_j \neq x_k$ for all $j \neq k$, a result equivalent to the *Haar condition* discussed further in Chapter 5 [177, p. 337].

This example demonstrates that any set of N points in the plane, $\{(x_k, y_k)\}$ can be fit *exactly* to a polynomial of degree $N - 1$, provided all of the x_k values are distinct. In the context of pure empirical modeling, this result means we can achieve a perfect fit, but in practice it is difficult to imagine circumstances under which this polynomial represents a useful data model. For one thing, this model achieves no simplification of the data: we have replaced a set of N data points with the N coefficients necessary to define the exact fit polynomial, $\{a_0, a_1, \ldots, a_{N-1}\}$. More important, however, this polynomial model rarely gives reasonable approximations to *unobserved* data values, either at intermediate values x_j that might represent missing data observations, or at x_j values outside the range of the original dataset where we might wish to extrapolate our data model to predict as yet unobservable future responses or responses outside the range of our data collection capabilities. In particular, since any polynomial $p(x)$, of any degree higer than zero (i.e., anything other than the constant $p(x) = a_0$), approaches $\pm\infty$ as x goes to either ∞ or $-\infty$, it follows that polynomial models cannot describe any form of *bounded* response behavior, including oscillations of limited magnitude, saturation phenomena, decay phenomena, or other common situations where observed response variables remain bounded as explanatory variables (e.g., time) take on increasingly larger values. This type of undesirable behavior is often cited as a reason not to use empirical data models as a basis for extrapolation, although the real difficulty here is that the form of the data model chosen (i.e., a polynomial) is not appropriate in cases where this type of behavior is to be expected [226, sec. 1.1.4]. This point is discussed further in Chapters 12 and 15 in connection with the use of transformations and the selection of structures for empirical data models. Nevertheless, the key point of this discussion remains: models that fit the available data exactly can often be constructed, but they are rarely useful.

The Ohm's law example considered in Sec. 1.1.2 provides a practical illustration of this point. Recall that Ohm's original experiment examined the rela-

Table 1.1: Ohm's torsion vs. length data

Length	Torsion
2	305.00
4	282.00
6	258.25
10	223.50
18	178.00
34	124.75
66	78.00
130	44.00

tionship between the length of a copper conductor and a measured torsion value that reflected the magnitude of the current flowing in the conductor in response to a fixed applied voltage. Table 1.1 summarizes the data, giving the lengths ℓ of each conductor and the measured torsion T resulting from each experiment. In principle, this relationship can be described exactly by a seventh-order polynomial predicting torsion from powers of length, but attempting to fit this model in R exhibits a numerical failure. That is, although the data matrix $\mathbf{X}$ for this example is nonsingular since all of the conductor lengths are distinct, the extremely heterogeneous distribution of these lengths causes the resulting matrix to be *ill-conditioned* or "nearly singular," a problem discussed in detail in Chapter 15 in connection with the problem of *collinearity*. As a consequence, the **solve** procedure in R does not return a solution for the exact-fit polynomial parameters. Alternatively, taking only the first seven data observations does yield an exact-fit sixth-order polynomial, and the predictions generated by this data model are shown in Fig. 1.4 (the line), overlaid with the data points (the solid circles). Note that while this model exactly fits all seven of the original data points, both its extrapolation behavior and its interpolation behavior are horrible, and it is strongly inconsistent with the systematic behavior clearly evident in the original data plot discussed in Sec. 1.1.2. That is, all conductors of length greater than 66 inches are predicted to generate unphysical *negative* torsions, while conductors of lengths intermediate between 34 inches and 66 inches are predicted to generate *huge* positive torsions, an order of magnitude larger than the largest values observed in Ohm's data. Overall, the key point of this example is that it is often easy to construct a mathematical model that *exactly* describes most or all of our data points, but that is glaringly inconsistent with the clearly observable qualitative behavior of those points (i.e., the monotonic decay of the torsion values with length seen in the Ohm's law data).

Figure 1.4: Plot of the first seven of Ohm's data observations (solid circles), overlaid on the exact fit polynomial model fit to these points (solid line)

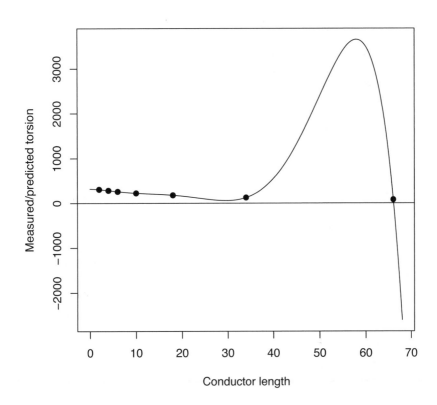

1.5.4 Wei's elephant and Einstein's advice

Often, particularly in the physical sciences, excess complexity has been avoided through an invocation of Ockham's razor dogmatic enough to be worthy of fourteenth-century popes and emperors. In particular, the complexity of empirical models has frequently been limited by fiat: no parametric model p of order greater than p^* will be entertained, with p^* commonly taken as 3. Many beginning data analysts have been brow-beaten by the Three Parameter Zealots with some version of the following tirade: "FIVE parameters?? You used FIVE parameters to fit your data??? With FOUR parameters, I can fit an elephant—with FIVE, I can make it wave it's trunk!" In a paper that should be required reading for anyone who even thinks about fitting data—or criticizing other people's efforts at doing so—Wei accepted this challenge [291]. Specifically, he took a cave drawing of an elephant, plotted it on polar graph paper, and fit the resulting curve to a series of sines and cosines, using standard regression methods like those discussed in Chapter 13. The adjustable parameters in this model were constants multiplying the sine and cosine terms. For the infamous FIVE parameter model, he got an egg-shaped blob that didn't even have a trunk, much less wave it. Ultimately, by including enough parameters in the model, he did get a fairly reasonable approximation to the original elephant. Here, however, "enough" was *thirty* parameters—and his elephant never did wave its trunk.

One of the reasons the three-parameter limit has been widely advocated in the past is probably a simple consequence of working with small datasets. In particular, if each individual data point represents the results of a day's or a week's intensive manual effort in setting up an experiment, keeping it running, and collecting the results, it follows that a dataset of size $N = 10$ is a significant accomplishment. Since data models should involve more than one (and preferably many more than one) observation per parameter fit, it follows that fitting a dataset of size $N = 10$ to more than about three parameters probably is a poor practice. While it is still true that in some cases, experimental results are difficult enough to obtain that we are unlikely to have more than a few results to analyze, this is increasingly the exception rather than the rule. In particular, one of the important differences between data analysis now and data analysis in the past is the availability of computers, both for data collection and for its subsequent analysis. As a consequence, it is now reasonable to consider much larger datasets and fit them to correspondingly more complex models. Thus, while it is important to bear Ockham's razor in mind as useful advice, it is important not to let it degenerate into counterproductive dogma like the rule of the Three Parameter Zealots. In particular, it is important to also remember Albert Einstein's variation of Ockham's razor:

> Everything should be as simple as possible, but no simpler.

1.6 Exploratory versus confirmatory analysis

As with the dichotomy between empirical and fundamental modeling of data sources, the analysis of observed data can be approached from two philosophically distinct perspectives: *confirmatory analysis*, which seeks to confirm or refute a stated hypothesis, and *exploratory analysis*, which has the somewhat vaguer objective of "finding interesting structure in a dataset." The following paragraphs describe each of these approaches in somewhat more detail, followed by an example that illustrates some of the risks inherent in exploratory analysis.

1.6.1 Confirmatory data analysis

Typically, introductory statistics texts have approached the problem of data analysis from a *confirmatory* perspective. In its purest form, confirmatory data analysis consists of the following three steps:

1. The analysis problem is formulated, typically in the form of a testable hypothesis (e.g., the mean of data sequence $\{x_k\}$, collected under controlled conditions A is the same as the mean of data sequence $\{y_k\}$, collected under controlled conditions B).
2. A dataset is collected, from which a test statistic is computed (e.g., a normalized difference of the computed means)
3. Based on the results of this test statistic, the hypothesis is either *rejected* (i.e., the hypothesis that the means are the same is inconsistent with the observed data) or *provisonally accepted* (i.e., the hypothesis is *not inconsistent* with the data).

Perhaps *the* illustrative example of confirmatory data analysis is the clinical trial, a highly regulated procedure designed to assess the efficacy and safety of a new drug. There, in order to comply with federal law, it is necessary to be very explicit about the hypothesis to be tested ("new drug X is a more effective treatment of Grunge's syndrome than the current standard treatment, drug Y, as assessed by the following criteria . . . ") and a detailed statistical analysis plan must be developed for analyzing the results once they are available, including a specification of how many patients need to be recruited for the study to provide the basis for unambiguous conclusions, and a detailed plan of exactly what data values are to be collected and how.

The primary advantage of this approach is that it does lead to specific answers to specific questions, taking advantage of the techniques of *experimental design* [159, ch. 5], developed to maximize the useful information content in the data to be collected. Conversely, the principal weakness of confirmatory data analysis lies in its narrow scope: it is well suited to providing specific answers to specific questions like "is the average noise level from Muffler A significantly lower than the average noise level from Muffler B?" but it is not designed to yield insights into questions like, "if so, why and if not, why not?" In particular, as Dempster [86] notes, the strict confirmatory data analysis prescription just described dictates that all of the analyst's "judgment and intelligence is exercised before the data arrives, and afterwards it only remains to compute and report the results." Accepted strictly, this prescription rules out graphical analysis of the raw data or any analysis of model prediction errors, *except as a basis for rejecting the analysis altogether*. In this context, notions like prescreening the data for outliers or transforming it to better conform with the modeling assumptions are unacceptable: their necessity implies either the analysis has not been set up correctly or the data has not been collected with sufficient care. In either case, the strict interpretation of the procedure just described requires that the analysis be terminated: if the analytical procedure was deemed inadequate, a new one must be proposed; in any case, new data must be collected, either to conform to the new analytical procedure or to overcome deficiencies in the orignial data collection.

In practice, confirmatory data analysis is generally not viewed this strictly. Instead, at least a limited amount of examination of the data is encouraged, to check the validity of the analytical working assumptions before the analytical calculations are performed. In particular, notions like outlier detection and either removal or replacement and transformation of the data to improve its consistency with working assumptions are often advocated in practice. The practice that is *not* advocated is repeating the analysis until we get the answer we really want. That is, as Diaconis [89] notes, since there is normally *some* probability that random fluctuations in the available data will *appear* to support a given explanation, if we repeatedly reanalyze the *same* data with a lot of different techniques, *one* of these analyses will often "confirm" the results we were originally expecting. If repeated reanalysis of the same dataset *does* turn up something interesting, the confirmatory analysis philosophy dictates that the favorable analysis procedure that uncovered it be repeated with *new* data, collected specifically with that analysis in mind.

1.6.2 Exploratory data analysis

As noted at the beginning of this chapter, the primary focus of this book is not on confirmatory data analysis, but rather on *exploratory data analysis*. This alternative analysis philosophy is appropriate in the common situation where we do not yet know enough to formulate a precise hypothesis and pursue the path of confirmatory data analysis. Indeed, as noted in Sec. 1.6.1, confirmatory data analysis offers no help in formulating hypotheses to test, only in setting up and conducting careful tests of hypotheses once they have been formulated.

Another important limitation of confirmatory data analysis is that strict adherence to this philosophy prohibits the analysis of *historical* or *observational* data, meaning any data that was not collected under carefully controlled conditions to serve as the basis for a specific confirmatory analysis. Increasingly, vast quantities of historical data are available in many different fields due to the growing use of computer data acquisition and archiving. In contrast to strict confirmatory data analysis, the exploratory data analysis philosophy *encourages* the analysis of historical data, precisely to find interesting hypotheses for further investigation. The principal difference between exploratory data analysis and Diaconis's "unrestricted data snooping" examples [89] is in the *intent* of the analysis: in exploratory analysis, we don't know how to explain or describe the data and we are looking for possibilities. In the worst of the "data snooping" examples, hypotheses are very clear in advance and the analyst persists in seeking "proof" of this assertion, *ignoring more plausible alternative explanations*. This issue is important because, as Diaconis [89] notes in the conclusion of his discussion, "it is an empirical fact that using the tools of exploratory data analysis to root about in large rich collections of data leads to useful results."

1.6.3 A cautionary example: the killer potato

Before giving a more detailed introduction to the mechanics of exploratory data analysis, it is important to emphasize that this type of analysis has two distinct aspects. First, there is the mathematical aspect of computing numerical summaries of one sort or another from the available data. The second aspect of

Figure 1.5: Reported U.S. homicides vs. potato production, 1900 to 1997

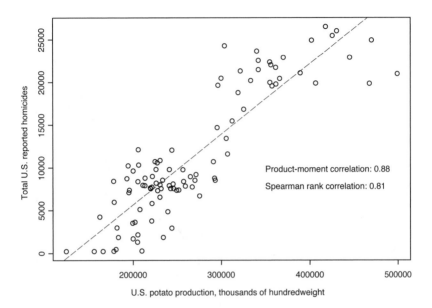

U.S. potato production, thousands of hundredweight

exploratory data analysis is that of interpreting and explaining what we have found. One of the points often emphasized in introductory treatments of data analysis is the distinction between *association*—the existence of mathematical evidence suggesting that two or more variables are related—and *causality*—the interpretation that changes in one variable are responsible for changes in the other. This distinction is important, and the following example provides a nice illustration of the point.

Fig. 1.5 shows a plot of the total reported U.S. homicides against the U.S. potato production in thousands of hundredweight, from 1900 to 1997. The source of these numbers is *Historical Statistics of the United States, Earliest Times to the Present: Millennial Edition* [55], based on table Da768-773 (Alan L. Olmstead and Paul W. Rhode, "Irish potatoes and sweet potatoes—acrage, production, and price: 1866-1998 [Annual]") and table Ec190-198 (Douglas Eckberg, "Reported homicides and homicide rates, by sex and mode of death: 1900-1997"). It is clear from Fig. 1.5 that both of these variables tend to increase together, a conclusion emphasized by the reference line in the plot, which was obtained by fitting the data using the method of least squares discussed in Chapter 5. Quantitative evidence in support of this conclusion is provided by the large product-moment correlation value of 0.88 and the large Spearman rank correlation value of 0.81, two numerical association measures discussed in detail

Figure 1.6: Common axis plot showing the increase in U.S. reported homicides, potato production, and total population, from 1800 to 2000

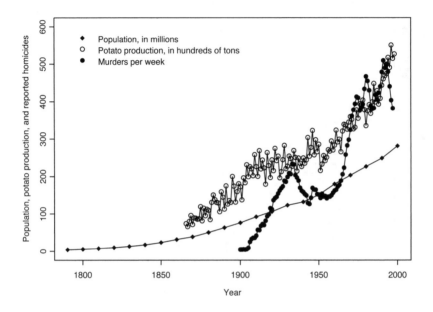

in Chapter 10. The key point here is that these characterizations—the visual appearance of the plot, the qualitative consistency with the reference line, and the large correlation values—provide evidence in support of an association between U.S. potato production and reported homocides over essentially all of the twentieth century.

We are now left with the question of interpretation: what does this association mean? A naive interpretation in terms of causality would lead us to the conclusion that potatoes are dangerous: either potato farmers are dangerous people or the consumption of potatoes induces homicidal rages. Since both of these conclusions seem pretty unlikely, it is reasonable to examine other possibilities. One possibility is suggested by Fig. 1.6, which shows both of these variables, expressed in different units, plotted together with the total U.S. population (from table Aa22-35, Michael R. Haines, "Selected population characteristics—median age, sex ratio, annual growth rate, and number, by race, urban residence, and nativity: 1790-2000" [55]). It is clear from this plot that all three of these variables are generally increasing with time, suggesting an alternative explanation: both the increase in reported homicides and the increase in potato production are consequences of the substantial growth in U.S. population during the twentieth century.

Besides emphasizing the important difference between the mathematical aspect of exploratory data analysis (e.g., characterizing the association between two variables) and the interpretation of these results (in particular, that association *is not* sufficient to imply causality), this example illustrates another key point. Specifically, a strong association between two variables can reflect the influence of one or more other variables that were not included in the original analysis. In the practice of confirmatory data analysis discussed in Sec. 1.6.1, the possibility of associations induced by omitted variables provides one of the motivations for *randomization*, where experimental units (e.g., patients in a clinical trial) are randomly assigned to "treatment groups," each exposed to different experimental conditions (e.g., one group given an experimental drug, the other group assigned to a standard treatment). The motivation for randomization is to avoid systematic differences between treatment groups with respect to variables not controlled in the experimental design. This approach can be very effective when it is applicable, but in exploratory data analysis it is generally not applicable: the available dataset has already been collected, and we must deal with any systematic differences that may exist between variables not included in our analysis. Of course, if we begin to suspect another variable may be partially responsible for our results (e.g., the observed association between reported homicides and potato production), we can attempt to include it in our analysis. In some cases, this is simple (e.g., look at *per capita* homicide rates versus *per capita* potato production), while in other cases, a certain amount of cleverness is required. Some excellent examples of this kind of cleverness are the judicious choices of data subsets chosen for comparison in the book *Freakonomics*, coauthored by economist Steven Levitt [182]; even better examples are provided by his technical papers, which include more details. Some of these ideas are discussed further here in Chapter 15.

1.7 The four R's of exploratory data analysis

As the preceeding discussions have emphasized, exploratory data analysis can lead us badly astray if it is practiced without sufficient care. To provide useful guidance in formulating a careful analysis, Velleman and Hoaglin [282] offer "the four R's of exploratory data analysis":

1. revelation;
2. residuals;
3. reexpression;
4. resistance.

Revelation refers, basically, to *looking* at the data to see what it reveals, and a variety of useful techniques for doing this are discussed throughout this book. The key to success lies in selecting an *informative view*, a task that frequently requires multiple attempts. *Residuals* are the prediction errors $e(k) = \hat{y}(k) - y(k)$ associated with any tentative data model: given any sequence $\{\hat{y}(k)\}$ of model predictions, the residuals $\{e(k)\}$ describe whatever structure may be present in the data that is not accounted for by the model. Analysis of model residuals is considered in various chapters, in conjunction with specific model structures appropriate to different types of data analysis. *Reexpression* refers to the use of data transformations to emphasize important structure that is inherent in the data, but which may not be initially evident. For example, developing a fundamental model in a coordinate system that coincides with the natural symmetry of the problem (e.g., cylindrical or spherical vs. rectangular) can simplify both the structure of the model and its interpretation. The same is true of exploratory data analysis, as the example discussed in Sec. 1.7.3 illustrates: a plot of the dataset in its original form is not particularly informative, but applying a logarithmic transformation to both variables offers strong evidence for a relationship between these variables; a detailed discussion of the use of transformations in data analysis is presented in Chapter 12. Finally, *resistance* refers to limited sensitivity to small changes in the data being analyzed. Here, "small changes" can arise either from "small changes in most of the individual data values" or "large changes in a small fraction of the total data set." This notion is closely related to the concept of *robustness*, and both of these ideas are discussed at some length in Chapter 7. The key point here is that all four of these notions are fundamental to the task of searching for *inherent structure* that is *representative of the bulk of the data.*

1.7.1   The first R: revelation

A number of useful tools for looking at data—particularly for looking at moderately large datasets—are described in this book. A key consideration is the view taken: which variables do we examine, and how do we examine them? For the case of a single, real-valued variable, a particularly useful tool is the *boxplot*, as the following example illustrates.

Fig. 1.7 shows plots of four industrial pressure data sequences, each obtained from the same chemical reactor but at different times under different operating conditions. Differences between these pressure sequences are apparent from these plots, and a detailed discussion of these differences is presented in Chapter 8. Boxplots for these four these four pressure sequences are shown in Fig. 1.8, based on the following five-number data summary [280, p. 32]:

1. the *sample minimum* $x_{0.00}$: $x_i < x_{0.00}$ for 0% of the sample;
2. the *lower quartile* $x_{0.25}$: $x_i \leq x_{0.25}$ for 25% of the sample;
3. the *sample median* $x_{0.50}$: $x_i \leq x_{0.50}$ for 50% of the sample;
4. the *upper quartile* $x_{0.75}$: $x_i \leq x_{0.75}$ for 75% of the sample;
5. the *sample maximum* $x_{1.00}$: $x_i \leq x_{1.00}$ for 100% of the sample.

These values are easily computed and often give an extremely useful summary of the data; *boxplots* are obtained by simply plotting these numbers as a sequence of five horizontal lines.

The basic structure of the boxplot is perhaps best seen in the plot labeled "P2" (for pressure sequence 2) in Fig. 1.8: the horizontal lines at the top and bottom of the plot represent the value of the sample maximum and the sample

Figure 1.7: Four industrial pressure data sequences

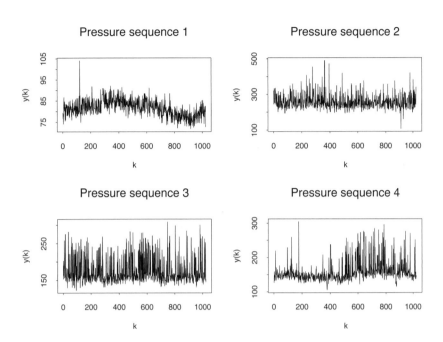

minimum, respectively, whereas the white line in the center of the plot represents the sample median. The black box in the center extends from the lower quartile at the bottom of the box to the upper quartile at the top of the box; hence, the black portion of the box plot contains 50% of the observed data values. Note that this figure emphasizes differences between these four sequences that are not obvious from the raw data plots in Fig. 1.7. For example, pressure sequence 1 (P1) exhibits both the smallest median pressure value and the smallest range of variation, whereas pressure sequence 2 (P2) exhibits the largest median value and the widest range of variation of the four pressure sequences. It is also clear from these boxplots that all four of these pressure sequences exhibit a somewhat asymmetric distribution, since the sample maximum lies farther away from the sample median than the sample minimum does. This asymmetry is not pronounced in pressure sequences 1 and 2, but it is quite pronounced in pressure sequences 3 and 4. Also, these boxplots suggest that pressure sequences 3 and 4 are much more similar to each other in distribution than any of these sequences are to P1 or P2.

Before leaving this example, two points are worth noting, both of which will be revisited in subsequent chapters. First, examination of the original data plots in Fig. 1.7 reveals "spikes," "anomalous data values," or *outliers*, a topic discussed in more detail in Chapters 2 and 7. These anomalous values are not well represented by the boxplots shown in Fig. 1.8, but there is a simple extension of

Figure 1.8: Boxplot comparison of the pressure sequences

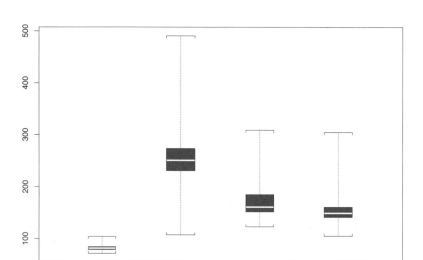

the basic boxplot described here that represents outliers in a dataset much more reasonably. This extension is discussed in Chapter 8, where the four pressure datasets are revisited. The second point is that the pressure data sequences are not well described by irregular fluctuations around a constant mean value, but they exhibit evidence of relatively smooth variation over time. Smoothing methods for estimating these long-term trends are discussed in Chapter 8, and removal of these trends allows us to focus on the residuals, as discussed in the next section. The key point here is that characterization techniques like boxplots are often more useful in examining residuals after the dominant gross structure of the data has been removed than they are in examining the raw data itself. This is also true of a number of other simple data characterization techniques described in this book.

1.7.2   The second R: residuals

As noted above, *residuals* are defined as the difference between the observed values of a data sequence $\{y_k\}$ and its predictions, $\{\hat{y}_k\}$, obtained from some mathematical model that attempts to describe or approximate the data. An important point is that this definition holds for any type of model, including straight-line fits, more complex regression models, or dynamic models that predict data values from their neighbors in the sequence. The following example

Figure 1.9: Four data plots: a physical property data sequence (upper left), two *lowess* smoothing results computed from this sequence (upper right), and the residuals associated with both of these smoothing results (lower two plots)

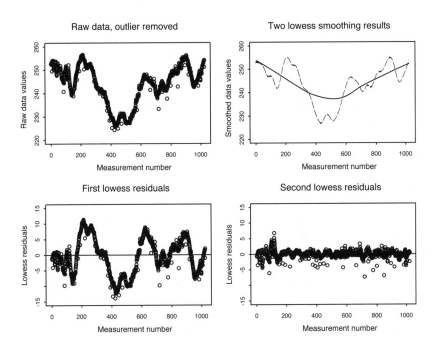

illustrates the utility of examining residuals in connection with predictions generated by the *lowess smoother*, a member of the class of *scatterplot smoothers* discussed in Chapter 8.

The upper left plot in Fig. 1.9 shows a sequence $\{y_k\}$ of 1,024 physical property measurements made once per hour for a period of approximately six weeks. This example is closely related to one discussed in Sec. 1.7.4 and, like that example, exhibits a single gross measurement error at hour 291. In the plot of the data shown here, this anomalous zero value has been replaced with the median value of the data sequence, a strategy discussed further in Chapter 7. The key point here is that this cleaned data sequence exhibits a relatively smooth, long-term variation, in contrast to the erratic, short-term variation seen in the industrial pressure data sequences discussed in Sec. 1.7.1.

The *lowess smoother* is a computational procedure discussed in Chapter 8 that attempts to estimate the long-term trend in a data sequence, based on the idea that it should vary smoothly. As a consequence, it is necessary to specify a *smoothing parameter f* that specifies "how smooth" the result should be. Two predictions $\{\hat{y}_k\}$ obtained from the raw data sequence $\{y_k\}$ are shown in the upper right plot in Fig. 1.9, for two different choices of f. The solid line represents the prediction sequence obtained with the *S-Plus* default value for the lowess smoother, which is $f = 2/3 \simeq 0.67$. This smoothing parameter value is quite large and the resulting prediction sequence is quite smooth as a result. The dashed curve in the upper right plot corresponds to the prediction sequence $\{\hat{y}_k\}$ obtained by the lowess smoother with the much smaller smoothing parameter $f = 0.05$. Comparing these two smoothing results with the raw data sequence shown in the upper left plot, it is clear that, while the results for $f = 0.67$ are much smoother than those for $f = 0.05$, these latter results do a much better job of capturing the long-term variation of the data sequence.

This point may be seen more clearly by comparing the residuals for the two smoothers, shown in the bottom two plots in Fig. 1.9. The lower left plot shows the residuals $e_k = y_k - \hat{y}_k$ from the first lowess smoothing result, with the default smoothing parameter $f = 0.67$, while the lower right shows the residuals from the lowess smoothing result with $f = 0.05$. Other than being roughly centered about zero, the first sequence of residuals looks almost like the original data sequence, showing that very little of the long-term variation seen in the data is accounted for by the trend estimate obtained for $f = 0.67$. In contrast, the residuals from the trend estimate obtained for $f = 0.05$ are much smaller in magnitude and exhibit much less regular variation than the sequence obtained for $f = 0.67$. In fact, these observations are two of the hallmarks of a good data model: the residuals are small in magnitude and free of evident structure. Of course, these are not the only criteria for a good data model, and it is worth emphasizing that the long-term trend models generated by scatterplot smoothers like the lowess procedure are not describable by a simple mathematical prediction model, making them quite unsuitable for some applications. The key point here is that, even with a relatively complicated prediction model like the lowess smoothing result, the examination of the residuals associated with it can be extremely informative.

1.7.3 The third R: reexpression

The utility of the scatterplot, obtained by plotting one variable against another, has been illustrated several times in this chapter. The basic motivation behind scatterplots is to discover relationships between two different attributes of some particular data source, an idea that works better in some cases than in others. The points of the following example are, first, that the right transformation applied to these variables can sometimes reveal structure that was not at all apparent in the original scatterplot, and second, that the association measures introduced in Chapter 10 can sometimes suggest when to consider such a transformation.

Fig. 1.10 shows a scatterplot of the brain weight in grams against the body weight in kilograms for 64 different animal species, similar to an example discussed by Rousseeuw and Leroy [252]. In contrast to the killer potato example discussed in Sec. 1.6.3, where the scatterplot was strongly suggestive of a relationship between the two variables, no such relationship is evident from the brain weight vs. body weight plot in Fig. 1.10. Two other characterizations examined

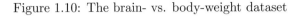

Figure 1.10: The brain- vs. body-weight dataset

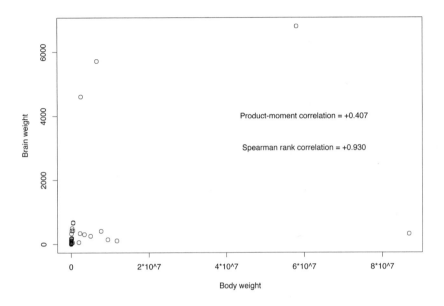

in the killer potato example were the product-moment correlation coefficient and the Spearman rank correlation coefficient, both defined and discussed in detail in Chapter 10. The product-moment correlation coefficient gives a measure of the tendency for two variables to exhibit a linear relationship, with a value of +1 indicating that the points fall exactly on a straight line in the plane with a positive slope. In contrast, the Spearman rank correlation coefficient gives a measure of the tendency for large values of one variable to be associated with large values of the other, with a value of +1 indicating that the points fall exactly on a monotonically increasing curve of otherwise arbitrary shape. For the brain weight vs. body weight example considered here, the product-moment correlation coefficient is +0.407, suggesting the possibility of a weak linear association, but the much larger Spearman rank correlation value of +0.930 suggests the existence of a fairly strong monotone relationship that is not evident from the plot of the raw data shown in Fig. 1.10.

Fig. 1.11 shows the result of applying a logarithmic transformation to both variables, plotting the log of brain weight against the log of body weight. The appearance of this plot is much more strongly suggestive of a linear relationship between these transformed variables than is the original plot in Fig. 1.10. In fact, the product-moment correlation coefficient for the transformed variables is +0.911, even larger than that seen in the killer potato example. Note that since the Spearman rank correlation is a measure of monotone association, it

Figure 1.11: The log brain- vs. log body-weight dataset

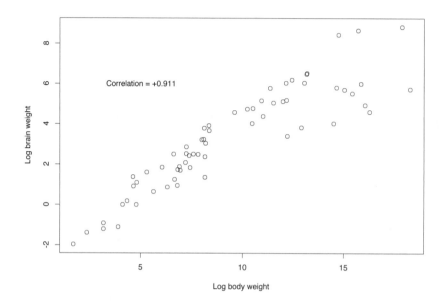

Log body weight

is invariant to monotone transformations like the logarithms used here, just as the product-moment correlation coefficient is invariant to linear rescalings of the variables. The key point here is that the large difference between the product-moment correlation value and the Spearman rank correlation value seen in the original data representation suggests the possibility that a monotone transformation may reveal a relationship that was not apparent in the original representation.

A more detailed discussion of transformations is given in Chapter 12, but the point of this example is to emphasize that they can give us a very different view of a dataset than the original dataset does. In cases like the one considered here, this modified view can be *much* simpler and more comprehensible than our first view of the original, untransformed data values. Another motivation for applying transformations is that they often permit the use of simple techniques like the linear fits discussed in Chapter 5 to datasets that are not originally suitable to such analysis. This point is illustrated in the Ohm's law dataset considered in Sec. 1.1.2: the original torsion vs. length dataset exhibits a clear, regular relationship but not a linear one. Applying a simple transformation, however, reduces this relationship to a linear one and permits the use of the techniques discussed in Chapter 5. Conversely, the use of transformations introduces two new considerations, both of which can be quite important. First, if we adopt the random variable model described in Chapter 4 for the data sequence $\{x_k\}$, the application of a transformation generally changes the distribution of this data sequence, and this change can have important practical consequences. The second important practical consideration is that the use of transformations generally enhances some data features and suppresses others: if the features enhanced are the ones we wish to focus on and the features suppressed are those we wish to ignore, the transformation can greatly facilitate our analysis. Conversely, transformations can suppress important features and enhance unimportant or even nonexistent ones to the extent that the results of our analysis are highly misleading. Some cautionary examples and discussions are included in Chapter 12 to help in preventing these difficulties.

1.7.4 The fourth R: resistance

Resistance refers to the sensitivity of data analysis results to violations of working assumptions, an important issue since, as noted in several of the previous sections, these assumptions are unavoidable in practice. In favorable cases where these working assumptions are approximately satisfied, our analysis can be expected to lead to reasonable results and conclusions. Sometimes, however, gross violations of one or more of these working assumptions can occur, and in unfavorable cases, these violations can completely negate the results of our analysis. Further, unless we are aware of this possibility, there is a danger of uncritically accepting these results as correct when they are not.

Data homogeneity is one of the classic working assumptions on which many data analysis procedures are based: whatever random variation is inherent in the data, it is described by a single probability distribution. The notion of outliers or anomalous data values that violate this working assumption has been mentioned previously and, in fact, forms one of the major themes of this book. The primary reasons for this emphasis are, first, that outliers occur frequently in real datasets, and second, that the unsuspected presence of even a few outliers in a dataset can *profoundly* distort the results of an otherwise reasonable analysis. As a simple but typical example, consider the dataset shown in the upper plot in Fig. 1.12. This dataset consist of $1,024$ physical property measurements taken from a different part of the same manufacturing process as the physical property dataset discussed in Sec. 1.7.2. As in that example, measurements were taken once per hour for a period of approximately six weeks. At hour $k = 291$, the measurement system failed temporarily, so no value was recorded for this data point. The results obtained from this data collection system subsequently underwent further computer processing that included storage in an intermediate file, which was ultimately read by a FORTRAN program. Because of the way READ statements work in the FORTRAN language, this "blank" data value was converted to *zero*, a value lying approximately 30 standard deviations from the mean of this data sequence. Replacing this zero value by the median of the dataset results in the cleaned dataset shown in the lower plot in Fig. 1.12.

Various aspects of this dataset and several related datasets will be considered in some detail in subsequent chapters of this book, but initially it is instructive to examine the influence of this single isolated outlier on a simple and repre-

Figure 1.12: Original and cleaned physical property datasets

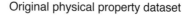

Original physical property dataset

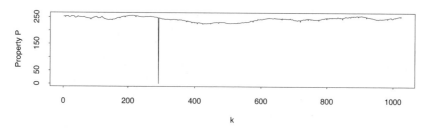

Dataset with missing data outlier removed

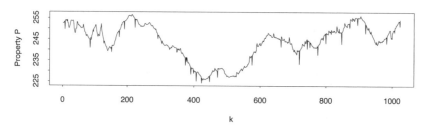

sentative data characterization: estimation of the standard deviation, $\hat{\sigma}$. The standard deviation is one of the general family of *moment estimators* discussed at length in Chapter 6, and it is easily computed from the following formula:

$$\hat{\sigma} = \sqrt{\frac{1}{N-1} \sum_{k=1}^{N} [x_k - \bar{x}]^2},$$

where $\bar{x}$ is the arithmetic mean of the data sequence:

$$\bar{x} = \frac{1}{N} \sum_{k=1}^{N} x_k.$$

Table 1.2 compares the standard deviations estimated from the original and cleaned data sequences shown in Fig. 1.12 using these two expressions. These results are given in the second column of the table, and they clearly illustrate the influence of this single anomalous data point: the estimated standard deviation increases by approximately 37%. Although it is true that the outlying data value lies far from the mean, it is important to emphasize that this change in $\hat{\sigma}$ is due to a single point in a dataset of size $N = 1024$, representing a *contamination level* of approximately 0.1%. Further, it is also worth emphasizing that this

Table 1.2: Comparison of two scale estimates

Dataset	Standard deviation $\hat{\sigma}$	MADM scale estimate S
Original	11.16	7.46
Cleaned	8.16	7.45

dataset is a real one, and although the outlier is glaringly obvious once we examine the dataset, we *will not* find this data point if we do not look for it. This point is important because there is a strong and growing practical motivation to automate data collection and analysis as much as possible to permit treatment of very high volumes of data, as in *data mining* applications. Consequently, one of the topics considered at length in Chapter 7 is the problem of automated outlier detection.

One difficulty with many automated outlier detection procedures is, ironically, their tendency to fail in the presence of multiple outliers. One practical way around this difficulty is through the use of *robust* or *resistant* techniques that are relatively insensitive to the presence of outliers in the dataset. One example of such a procedure is the *MADM scale estimate S* introduced in Chapter 7 as a resistant alternative to the standard deviation. Discussion of the definition and evaluation of this scale estimate will be deferred until Chapter 7, but here we note the following two key points. First, the MADM scale estimate S is easily computed, making it a practical alternative to $\hat{\sigma}$, and second, it is *much* less sensitive to outliers than $\hat{\sigma}$. This second point is demonstrated in Table 1.2, which also gives the MADM scale estimates computed from the original and cleaned physical property datasets considered here. Here, replacement of the single outlying data point with the sample median results in a change in S of approximately 0.1%, in marked contrast to the 37% change seen in $\hat{\sigma}$.

In the context of exploratory data analysis, the key points of this example are, first, that robust and nonrobust estimators exist for many quantities of interest (e.g., the median vs. the mean as measures of a variable's "typical value," the MADM scale vs. the more familiar standard deviation as measures of spread in this example, etc.). This point is important in part because many of the most familiar characterizations are also the most sensitive to outliers, as in the example considered here. The second point of this example is that significant insights can often be obtained by using *both* of these assessments—one robust and one nonrobust—and comparing the results: if they are similar, this is a suggestion (but not a proof) that outliers are not a major concern; if they are not similar, this may be an indication of outliers lurking in the data, a point considered further in Chapter 7. Alternatively, as the brain weight vs. body weight example discussed in Sec. 1.7.3 illustrated, the primary source of these differences may lie elsewhere: as results presented in Chapter 15 demonstrate, outliers are present in this dataset, but the primary reason for the difference between the product-moment and Spearman rank correlations was the strong nonlinear relationship between the original variables, and not the presence of these outliers.

1.8   Working with real datasets

Many of the examples discussed in this book are based on real datasets, generally collected under noncontrolled conditions. The following two examples illustrate some of the important practical issues that arise in working with such datasets. Specifically, Sec. 1.8.1 introduces the notion of missing data in connection with a historically important example from the early days of astronomy; sources of missing data are discussed further in Chapter 2 and strategies for dealing with this problem are discussed in Chapter 16. The second example, presented in Sec. 1.8.2 describes the preliminary characterization of a moderately large clinical dataset that characterizes patients diagnosed with chronic fatigue syndrome. A more complete discussion of this example is presented in Chapter 8, which describes a variety of techniques for initially characterizing a dataset.

1.8.1   A missing data example: the asteroid belt

In the late eighteenth century, Johannes Titius and Johann Bode each independently observed that the distances d_n between the seven known planets in the solar system and the sun could be described by the *Titus-Bode law* [176, p. 1]:

$$d_n = 0.4 + 0.3N_n, \tag{1.22}$$

where $N_1 = 0$ and $N_n = 2^{n-2}$ for $n \geq 2$ and the distances are expressed in astronomical units AU, equal to the distance from Earth to the sun or approximately 93 million miles (1.5×10^8 kilometers). In fact, these predictions are accurate to about 5% or better for the first four planets (Mercury, Venus, Earth, and Mars) and are even more accurate for the next three planets (Jupiter, Saturn, and Uranus), *provided the indices for these planets are taken as $n = 6$, 7, and 8 instead of* 5, 6, *and* 7. This observation of a "gap" in the sequence motivated many eighteenth-century astronomers to look for a "missing planet" between Mars and Jupiter at approximately 2.8 AU. In fact, Giuseppe Piazzi discovered the asteroid Ceres at 2.77 AU in 1800. The asteroid moved into the daytime sky and became unobservable, but Carl Friedrich Gauss developed a procedure for predicting planetary orbits and accurately predicted where Ceres would reappear in the "Titus-Bode gap" during the winter of 1801-1802. Fig. 1.13 shows a plot of the seven planetary distances *versus* the Titus-Bode law predictions, including the location of the asteroid Ceres. The dashed reference line is included as a visual aid to illustrate the accuracy of the Titus-Bode predictions.

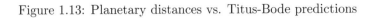

Figure 1.13: Planetary distances vs. Titus-Bode predictions

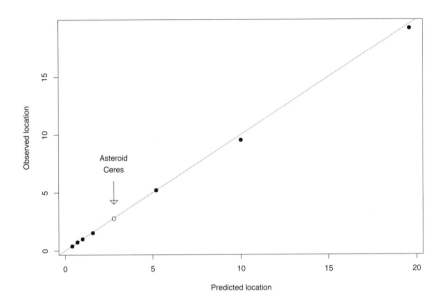

Between 1802 and 1807, three more asteroids (or *minor planets* as they are sometimes called [49, 176]) were discovered in the Titus-Bode gap: Pallas, Juno, and Vesta. In fact, there are more than 10,000 such asteroids large enough to be seen from earth at present, along with many other smaller ones. Vesta is one of the largest members of what has come to be called the asteroid belt; it is approximately 576 kilometers in diameter, located at a mean distance of 2.362 mean earth radii from the sun. In addition, it is interesting to note that the densities of the planets appear to exhibit a somewhat regular pattern of variation, although nothing like the regularity of their distances. This pattern is shown in Fig. 1.14, which shows the estimated planetary densities plotted against their mean distances from the sun for the eight now officially accepted planets of the solar system, along with the recently demoted astronomical object formerly known as the planet Pluto. In addition, the density of Vesta (approximately 2.9 g/cm^3) is also shown in this plot, further emphasizing the regularity of the density/distance relationship.

One possible explanation for the asteroid belt offered soon after the discovery of the second asteroid, Pallas, was that a planet had previously existed in the Titus-Bode gap and it had exploded or disintegrated. This view does not appear to be generally supported today (see, e.g., the discussion of Kowal [176, p. 58]); instead, the prevailing view seems to be more the opposite—that a planet between Mars and Jupiter *failed to coalesce* from the planetesimals that now form

Figure 1.14: Estimated density vs. log planetary distance

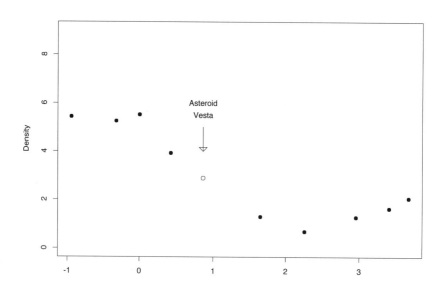

the asteroid belt. One possible cause of this failure is the enormous gravitational influence of Jupiter, which is about 340 times more massive than Earth. While it is not nearly as well known as Ohm's law, Titus and Bode's observation of the pattern of planetary spacings continues to motivate new investigations and lead to new discoveries [9, 49, 176].

1.8.2   A large dataset: chronic fatigue syndrome

Chronic fatigue syndrome (CFS) has been described by Wagner *et al.* as [286]:

> an incapacitating illness defined by disabling chronic fatigue and characteristic accompanying symptoms [107]. CFS has no confirmatory physical signs or characteristic laboratory abnormalities and its etiology and pathophysiology remain unknown [7].

A dataset that will be examined further in Chapter 8 summarizes clinical data from 227 patients in the Witchita CFS Surveillance Study [243], including some who were ultimately diagnosed with CFS, others who were not, and still others who had some of the typical symptoms of CFS but were ultimately deemed not to have the disease. This dataset was one made available for analysis by the organizers of the 2006 Critical Assessment of Microarray Data Analysis

(CAMDA) meeting; at the time of this writing, it may be obtained from the following Web site:

http://www.camda.duke.edu/camda06/datasets

Each record in this dataset corresponds to one of the 227 patients, and it consists of 87 fields, including demographic data like gender, age, and race, along with responses to three quesionnaires: the Medical Outcomes Survey Short-Form-36 (SF-36), the Multidimensional Fatigue Inventory (MFI), and the CDC Symptom Inventory (CDC SI). Altogether, these 87 variables cover most of the different variable types discussed in Chapter 2, including real-valued variables like patient age, binary diagnosis variables (e.g., answers to questions like, "have you been diagnosed with asthma within the last two years?"), and categorical variables like the Intake Classification, that classified patients into one of the following five groups:

1. CFS: patients who had been classified as CFS at least once during the four-year Witchita study;

2. NF: "nonfatigued" patients who had not been classified as CFS during the Witchita study and who were matched with the CFS patients on the basis of gender, race, ethnicity, age, and body mass index (BMI);

3. ISF: patients with unexplained chronic fatigue symptoms that did not meet the defining criteria for CFS, classified as "insufficient symptoms for fatigue";

4. CFS-MMDm: patients who met the CFS criteria, but who also exhibited melancholic depression, considered exclusionary for CFS;

5. ISF-MMDm: patients who met the ISF criteria, but who also exhibited melancholic depression.

One of the key reasons for considering this dataset is that it is a real example of a *mixed* dataset with many different kinds of variables, typical of the large datasets that are becoming increasingly common in many fields. It is no accident that this example is a clinical dataset, since mixed datasets are extremely common in clinical applications, but the problems posed by this dataset arise more and more often in all fields.

In the more detailed analysis of this dataset given in Chapter 8, preliminary dataset characterizations like the following will be examined:

1. How many records N (i.e., observations) are present, and how many fields M (i.e., attributes) does each record have?

2. What types of variables are each of these attributes (e.g., real-valued, binary, categorical, etc.)?

3. What are the ranges of any real-valued variables?

4. How many levels do each of the categorical variables have?

5. What fraction of the observations are missing?

6. Do any of the observations appear to be *anomalous* (i.e., *outliers*)?

7. Is there evidence to suggest that either the missing observations or the anomalous observations exhibit unusual patterns of occurrance?

8. How homogeneously or heterogeneously distributed are these variables?

Answers to these and other similar questions, along with any available information about their real-world interpretations, provide the starting point for our analysis of a dataset, so it is worth taking the time to do this initial data survey carefully and summarize the results for future reference. This is particularly true in the common situation where the data analyst was not involved in the data collection, so that some of the information about the dataset is either incorrect or subject to misinterpretation. This point is discussed further in Chapter 2 where the notion of *disguised missing data* is discussed.

The key point of this example is to give an indication of the scope of the preliminary assessment problem for an unfamiliar dataset. This example is not a particularly large one by present standards, with only 227 records, but it is complex enough (i.e., we have 87 potential analysis variables to select from in characterizing it) that the task of understanding what we have to analyze is nontrivial and will necessarily take some time.

1.9   Software considerations

When Ohm was analyzing his data, the only real computational aids available were his brain and his pencil and paper. Today, most data analysis is done using a computer—indeed, for large datasets, this is the only feasible option. As a consequence, most data analysis is now ultimately based on some kind of analysis software written, tested, and debugged by someone other than the data analyst. Of course, in analyzing data we often construct our own special-purpose procedures tailored to our particular problem, but generally these procedures rest on other procedures that someone else wrote: unless our basic interest lies in developing nonparametric smoothing procedures, for example, we are much more likely to use built-in smoothing procedures than to develop our own.

All of the computational procedures used in analyzing the data described in this book were based either on the commercial package *S-Plus*, or its open-source near-equivalent, *R*. A brief discussion of the mechanics of obtaining the *R* software package is given in Appendix 1 at the end of this book, which also describes how to obtain the specific routines used to generate most of the results presented here. Finally, it should be noted that most of the analysis procedures discussed in this book are described in enough detail that they can be implemented on many other computational platforms, and a number of them are widely available as built-in procedures.

1.10   Organization of the rest of this book

Beyond this introductory first chapter, the rest of this book consists of 15 additional chapters, each devoted to a specific topic or group of closely related topics. Each of these subsequent chapters starts with a brief overview that introduces the topics it addresses in broad terms, emphasizing the basic notions covered and their relations to the material covered in other chapters. Of necessity, some of this material is presented sequentially, with later discussions building on ideas presented in earlier ones, but to the extent possible discussions of different topics are intended to stand alone, and any essential connections with earlier material are noted explicitly.

Chapter 2 considers the data on which all data analysis ultimately rests, describing the different types of data considered in this book (e.g., real-valued variables like temperature or pressure, integer-valued count variables, or categorical variables like the chronic fatigue syndrome classifications discussed in Sec. 1.8.2). In addition, Chapter 2 also briefly examines the question of what we can measure and how well (e.g., what level of inherent uncertainty should we expect and why?) and describes a number of practically important data anomalies like outliers and missing data. Chapters 3 and 4 then consider the characterization of categorical and real-valued variables, respectively, primarily based on the random variable uncertainty model. The categorical case is treated first because it involves discrete random variables, which are conceptually simpler than the continuous random variables used in characterizing real-valued variables and because it provides useful background on which the continuous random variable discussion builds.

Chapter 5 presents a detailed discussion of a simple but extremely important application: fitting a set of points representing two variables to a straight line in the plane. By focusing on a single example, this chapter introduces a variety of important ideas in a common framework that emphasizes their similarities and differences. In addition, the techniques presented in Chapter 5 are extremely useful in practice, as various examples discussed throughout this book demonstrate. Chapter 6 gives a brief introduction to estimation theory, focusing on those aspects of the theory that are useful in understanding the data analysis problems and techniques discussed in this book. Chapter 7 builds on some of this material in treating the topic of outliers and their impact, especially on "standard" characterizations like the mean and standard deviation. In addition, Chapter 7 also introduces a number of general ideas and specific techniques that are useful in developing analysis procedures with much lower outlier sensitivities than these standard methods.

Chapter 8 presents a variety of useful techniques for characterizing a dataset, emphasizing the practical utility of generating a simple overall summary (e.g., how many variables are included? What kinds? What fraction of each are missing?). Other techniques presented in Chapter 8 are useful in characterizing individual real-valued variables, including both quantile-quantile plots and nonparametric density estimates to give an idea of how the data values are distributed for a given variable.

As noted in the discussion of exploratory versus confirmatory data analysis presented earlier, a key component of confirmatory data analysis is the use of formal *hypothesis tests*. In fact, hypothesis testing is also extremely useful in exploratory data analysis, particularly in connection with model validation. Chapter 9 gives a brief introduction to hypothesis testing, concentrating on the basic notions and classical model validation techniques (e.g., we estimated a value of 0.02 for a particular model parameter: is this value large enough that the corresponding model term is important, or could we just as well simplify the model by setting this coefficient to zero?). These techniques are popular and are often simple, but their Achilles heel is their underlying distributional basis: a probability distribution must be assumed, and when this assumption is reasonable, the results obtained may be useful, but when it is not reasonable, the results may not be useful. Alternatives are available in cases where these distributional assumptions are not reasonable, and several of these are discussed in Chapter 14, including resampling methods like the *bootstrap*, subsampling methods, and permutation methods. The essential idea of these methods is to repeatedly and randomly vary the data used to obtain a result and see what happens: when we vary the dataset only a little, do the results change a little or a lot? Conversely, if we modify the dataset in a way that destroys the relationship we think we have found, how much does our characterization change?

Chapter 10 considers the problem of quantifying the degree of association between two variables, providing detailed discussions of both the product-moment correlation coefficient and the Spearman rank correlation coefficient used in both the killer potato and the body weight vs. brain weight examples discussed in this chapter. As the brain weight vs. body weight example demonstrated, large differences between the product-moment and Spearman rank correlation values for two variables may suggest the use of transformations to reexpress one or both variables. Chapter 12 presents a reasonably detailed treatment of transformations, including discussions of what types of transformations are commonly used and why, some further illustrations of their advantages, and some cautionary examples to demonstrate that they can also have unexpected (and thus potentially damaging) side effects.

Chapters 11, 13, and 15 give a reasonably detailed treatment of regression analysis in various contexts, interspersed with treatments of the closely related topics of transformation (Chapter 12) and characterizing analysis results (Chapter 14). Specifically, Chapter 11 considers the basic linear regression problem where a real-valued response variable is predicted from a linear combination of one or more prediction variables. These techniques extend those presented in Chapter 5 for fitting straight lines to more general problems that involve an arbitrary number of prediction variables, including simple transformations and combinations (e.g., products of two or more variables). Chapter 13 considers a variety of important mixed cases, including *logistic regression* where a binary outcome is predicted from real-valued independent variables, *analysis of variance (ANOVA) models*, where a real response variable is predicted from one or more categorical variables, and *Poisson regression models*, used to model the dependence of count data on other variables. Finally, Chapter 15 presents a reasonably detailed treatment of the problem of validating and refining a regression model, including techniques like *deletion diagnostics* to determine whether a few unusual observations exhibit an undue influence on the final regression model.

Finally, Chapter 16 is concerned with the treatment of missing data. As emphasized in Chapter 2, this is a significant issue in many applications, especially in cases where a large dataset has been assembled from a number of smaller ones. The basic difficulty is that data observations we would like to have for our analysis are simply unavailable, forcing us to decide how to proceed: do we omit incomplete records, omit incomplete variables, or do something else? Various approaches are possible, and the best one to choose is often a fairly strong function of the data analysis problem we are considering. Chapter 16 presents a number of different techniques, discusses their strengths and weaknesses in different circumstances, and offers some suggestions for specific cases.

Chapter 2

Data: Types, Uncertainty and Quality

Sections and topics:

In many cases, most or all of the data we have available to analyze consists of observations of real variables; examples include time, temperature, pressure, area, density, concentration, and many others. Consequently, many data analysis books focus entirely on real variables, and this book will focus primarily on them, but it is important to recognize that we will sometimes encounter other data types that require different analysis methods. The treatment of these other data types will be considered in various places throughout the rest of this book, including cases involving multiple data types. Overall, the objective of this chapter is threefold, as suggested by the title: first, to describe the different data types that we may encounter, particularly in large datasets and/or medical applications; second, to describe the types of uncertainty we are likely to encounter in real-world data, with brief descriptions of sources and basic characterizations of this uncertainty; and third, to introduce some important data quality issues like outliers and missing data that we must be aware of in analyzing real-world datasets.

Key points of this chapter:

1. In analyzing data, it is useful to regard a dataset as a rectangular array or table, with each row corresponding to a single data source (e.g., an individual chemical compound, material sample, manufacturing lot, patient, etc.) and each column corresponding to a single characterization of these data sources (e.g., melting point, tensile strength, yield, or tumor stage).

2. Variables can be real-valued like temperature or pressure, integers like counts, *ordinal* (nonnumeric but with an obvious order, like "never," "rarely," "sometimes," "often," or "always"), or *nominal* (nonnumeric with no obvious ordering, like "compound type," "color," or "texture").

3. *Metadata* describes the contents of a data file, ideally defining all variables, specifying representations for missing data, and providing pointers to additional sources of information. In practice, metadata is often less complete than we would like, and it can be wrong in important respects.

4. The things we can measure and the ways we can measure them are limited only by our imagination, as examples presented in this chapter illustrate. For real-valued quantities like temperature or pressure, the result of the measurement is never perfect, but typical measurement uncertainties may be usefully described as random variables and characterized by their means and standard deviations, closely related to *accuracy* and *precision*.

5. Some entries in a dataset are highly inconsistent with other entries that we expect to be comparable: these *outliers* may be due to gross measurement errors (e.g., sensor failures, analogous to photographs taken with the lens cap still on the camera), but they can also be legitimate surprises. As noted in Chapter 1, the detection and appropriate treatment of outliers is important, a point discussed further here and refined in a variety of different contexts in subsequent chapters of this book.

6. While not generally as widely recognized as outliers, many other types of data anomalies also exist, and this chapter introduces several of them, including *inliers*, a variety of *file merge and manipulation errors*, the problem of *duplicate records*, and several types of *noninformative variables*.

7. Another important class of data anomalies is *missing data,* corresponding to entries that *should* be present in a dataset but aren't. The fact that there is no universal standard for representing missing data causes significant practical complications in dealing with it, as does the necessity of modifying logical constructs when basing decisions on variables that may have missing values (e.g., "if statements" or record subset selection procedures). Chapter 16 presents a detailed treatment of handling missing data in data analysis, but this chapter introduces some important aspects of the problem, including the possibility of *disguised missing data*.

Table 2.1: Structure of a typical dataset **X**

Observation i	Variable 1	Variable 2	$\cdots$ $\cdots$	Variable N_c
1	x_{11}	x_{12}	$\cdots$	x_{1N_c}
2	x_{21}	x_{22}	$\cdots$	x_{2N_c}
$\vdots$	$\vdots$	$\vdots$	$\vdots$	$\vdots$
N_r	$x_{N_r 1}$	$x_{N_r 2}$	$\cdots$	$x_{N_r N_c}$

2.1 The structure of datasets

Datasets exhibit an extremely wide range of types, and one of the great advances in storing data for efficient retrieval is the *relational database* [81]. As the name suggests, the advantage of relational databases is that they permit efficient searches to find collections of related variables, where the relations between variables can be rather complex (e.g., "give me all records for female patients admitted to Hospital X between January 1, 1993, and October 17, 1995, and treated for one of the following seven cardiac conditions ...”). While the software packages that implement relational databases do have some built-in computational capabilities, these are limited enough that they either do not support most of the data analysis tasks considered in this book, or do so only with great effort. Consequently, it is assumed here that even if the data values of interest are originally stored in a relational database, they have been extracted and are available in the form of a data table like that shown in Table 2.1, corresponding to a *flat file* with N_r records and N_c fields per record. Each of these rows or records will be assumed to represent a collection of observations characterizing a single real-world entity, one for each of N_c distinct variables, which can include experimental conditions (independent variables), experimental responses (dependent variables), or both. The values stored in this data table will often be designated x_{ij}, where i is a row index, ranging from 1 to N_r, and j is a column index, ranging from 1 to N_c. The value x_{ij} will generally be referred to as the ith observation of the jth variable. As noted in the introduction to this chapter, the observation x_{ij} is typically a real number, but this need not be the case, since any variable can be any one of the data types discussed in Sec. 2.2.

Associated with the data table $\mathbf{X} = \{x_{ij}\}$ is a set $\mathcal{M}$ of *metadata* that describes it. Ideally, this description should be complete, fully describing all of the variables involved, giving their names, their ranges of admissible values, codes used for missing data values, and literature references giving more complete descriptions of the variables and how they were measured. In practice, the metadata $\mathcal{M}$ is almost never this complete, is sometimes partly or completely missing (e.g., some variables may be named while others are not), and is sometimes incorrect, a point discussed further in Secs. 2.3 and 2.6. In traditional statistics texts, the datasets considered were typically small enough to be represented as a single table of numbers on a page, and the associated metadata consisted of a few paragraphs of accompanying text. Even in these cases, the metadata can be incorrect, a point discussed further in Sec. 2.3, where it is illustrated with an extremely well-known example. Most of the datasets considered in this book are large enough that it is neither feasible nor desirable to represent the complete table X of data observations in the text, and for very large datasets, even a complete list of variable names can occupy many pages. Consequently, even when it is present and consistent, the available metadata is seldom as complete as we would like, a point discussed further in Sec. 2.3.

2.2 Data types

As noted above, in many data analysis problems we will be concerned primarily or exclusively with *real variables* like time, temperature, or pressure that can assume any real value, possibly over a restricted range. So long as we are not near these range limits, it is usually simplest to treat these variables as real numbers that can, in principle, take any value from $-\infty$ to $+\infty$. There are cases, however, where it is important to take explicit account of these range limits. For example, some variables (e.g., absolute temperatures, pressures, and weights) can only assume positive values, while others like probabilities are restricted to lie in the interval $[0, 1]$. These restrictions may be regarded as *fundamental* since values that violate them are meaningless and represent errors, either in data collection or in subsequent analysis steps. In contrast, there are other cases where these restrictions are *field-specific:* values of the variable that lie outside some application-specific range are unrealistic, but these limits may be strongly application dependent.

As a case in point, consider the admissible range for a percentage: if it describes the partitioning of a fixed whole, valid numbers must be nonnegative and sum to 100% (e.g., percentages of male patients, female patients, and patients with unspecified gender). In contrast, if the percentage describes a change in some variable (e.g., the increase in population or grocery prices in a given city between one year and the next), the numbers can be positive or negative and values greater than 100% are certainly possible. There are, of course, exceptions to almost every rule, and it is important to be aware of them. For example, "fuming sulfuric acid" is pure sulfuric acid with sulfur trioxide dissolved in it; since sulfur trioxide and water react to form sulfuric acid, adding just the right amount of water to fuming sulfuric acid dilutes it down to pure—100%—sulfuric acid. Thus, fuming sulfuric acid may be regarded as sulfuric acid with a concentration greater than 100%, depending on how much sulfur trioxide is dissolved in it. The key points of this example are, first, that there are circumstances where the expected limitations on a variable can be violated, but second, that to accept these violations as real anomalies rather than data recording errors, it is important to have an explanation like the one just given for concentrations greater than 100%.

Another important type of variable that arises frequently in data analysis is the class of *count variables,* nonnegative integers that result from counting discrete events or objects. If the counts are large (e.g., national populations), it may be reasonable to treat them approximately as real variables, but in some cases this view leads to undesirable artifacts. A classic example is the "typical" family with 2.5 children, a specification with the paradoxical side effect that no typical families actually exist. Some of the important theoretical differences between count variables and real variables are discussed in Chapter 4 and the practical consequences of these differences are illustrated in various places throughout the rest of this book.

An important characteristic of both real variables and count variables is that they are *ordered:* given two values, it is possible to say one is larger than the other, smaller than the other, or equal to the other. Also, the notion of *distance* is well defined for both of these variable types: given two values of a real variable, say, x_1 and x_2, the absolute difference $|x_1 - x_2|$ gives us a measure of how far apart these two values are, and the same holds for count variables. Variables that are ordered but for which such distances are either not well defined or not meaningful are called *ordinal.* An example is the *Karnofsky score*, a ten-level index of general health summarized in Table 2.2. The values for this variable are *coded* as percentages, with lower scores corresponding to poorer states of health, but it is not possible to say that a "distance of 40 Karnofsky percentage points" has any particular meaning: while any decline is bad, it is clear from the interpretations given in the table that a decline from a Karnofsky score of 50% to one of 10% is *much* worse than a decline from 100% to 60%.

The least structured data type we will encounter is the *nominal variable,* which assigns a classification but neither order nor distance. Examples here include characteristics like color, species, shape, religion or political affiliation, race, or nationality. A specific example that will be revisited in subsequent chapters is the *UCI mushroom dataset* from the University of California at Irvine's Machine Learning Archive, available at the following Web site:

> http://www.ics.uci.edu/ mlearn/MLRepository.html

The mushroom dataset consists of 22 nominal variables like cap type and gill color that describe 8,124 mushrooms from *The Audobon Society Field Guide to North American Mushrooms* [185], together with a binary classification of whether each mushroom is edible or not. Nominal variables arise frequently in the social sciences and medicine; while they are less common in the physical sciences and in some engineering disciplines, they do arise there as well (e.g., chemical structure, compound class, manufacturing process type, crystal structure, etc.). Because they possess no inherent mathematical structure, these variables are commonly used either as a basis for comparison (e.g., comparing average melting points across different classes of compounds) or converted to count variables. A variety of useful nominal variable characterizations is presented in Chapter 3.

Table 2.2: The Karnofsky scale, used to assess patient health

Score	Interpretation
100%	Normal, no complaints, no signs of disease
90%	Capable of normal activity, few symptoms or signs of disease
80%	Normal activity with some difficulty, some symptoms or signs
70%	Caring for self, not capable of normal activity or work
60%	Requiring some help, can take care of most personal requirements
50%	Requires help often, requires frequent medical care
40%	Disabled, requires special care and help
30%	Severely disabled, hospital admission indicated but no risk of death
20%	Very ill, urgently requiring admission, requires supportive measures of treatment
10%	Moribund, rapidly progressive fatal disease processes
0%	Death

An important special case of the nominal variable is the *binary variable*, which assumes either one of two possible values. One example is the edible vs. nonedible classification of mushrooms in the UCI mushroom dataset, and binary variables arise frequently in medical data (e.g., malignant vs. benign, male vs. female, survived vs. died, etc.). A particularly important class of medical binary variables are diagnoses: the patient has the disease or does not. In fact, analogous classifications occur in many other applications: a given manufacturing batch produced acceptable yield or not, a particular chemical compound or multicompound formulation exhibited certain desired physical properties or not (e.g., adequate tensile strength, corrosion resistance, etc.), knocking out a specific gene produced a particular mutation or not, etc. The special structure of binary variables is often exploitable, leading to special analysis techniques like *logistic regression,* discussed in Chapter 14.

Finally, it is important to note that the variable types just described are only the simplest examples, and do not represent all possibilities. Other data types include text, graphs, and more complex structures like XML (Extensible Markup Language) documents. Specialized techniques have been developed to analyze some of these other data types, but to keep the scope of this book reasonable and achieve a useful degree of unity among the topics treated, none of these other data types will be examined here, except for special cases where they lead to simple quantitative characterizations (e.g., word counts, word lengths, etc.). For those interested in exploring these rapidly developing specialized analysis approaches, the recent literature in the fields of data mining, artificial intelligence, and machine learning is recommended.

2.3   Metadata

As the discussion in Sec. 2.1 implied, metadata is essentially "data about data," but this definition is too broad to be completely satisfactory. For example, metadata can be developed for different purposes, and what is fully adequate for one purpose may be completely inadequate for another. As a case in point, Bentley [35] identifies four distinct contexts for metadata: a data administration context concerned with how records are organized in a database, a data warehouse back-end context concerned with details of the "extract-clean-transform-load" process, an application development context concerned with how specific applications access the data, and a data warehouse front-end context concerned with the real-world meaning of the data. Another extremely significant complication is the fact that metadata, by its nature, cannot conform to a rigid format specification except within a narrow application area. As a specific example, an ISO (International Standards Organization) standard has been developed to facilitate the exchange of geographic data over the Internet [184]. This standard (ISO 19115) "divides the description of resources into six levels: dataset series, dataset, feature type, feature instance, attribute type and attribute instance." It is also noted that "the comprehensive set of 19115 metadata entity set information consists of 11 entities (UML classes), with more than 300 either optional or mandatory elements (UML attributes)," raising another important point about metadata: even when a complete, standardized specification is possible, it is often quite complicated. As a consequence, even when "ideal metadata" is achievable in principle, it is rarely available in practice.

Although his definition of "good metadata" is specific to business applications, the guidelines that Bentley offers in table 4 of his paper [35] provide the basis for a metadata framework that is very broadly applicable. Based on these guidelines, Table 2.3 lists 20 metadata components recommended for any analysis dataset. Those attributes marked "(DS)" in the *Variables* column are dataset attributes that need only appear once in the metadata for a given dataset; all other attributes should appear once per variable, as appropriate. Specifically, those metadata elements marked "All" should be present for all variables included in the dataset, while those marked "Cat." are only appropriate for categorical variables (including nominal, ordinal, or binary), those marked "Real" are only appropriate to real-valued variables, and those marked "R/C" are appropriate to both real-valued and integer-valued count variables. Note that constrained real-valued variables (e.g., variables that can only be positive or that must fall between 0 and 1) should be treated as real-valued variables, with the appropriate restrictions included in metadata component 17 ("Admissible range of values"), and similarly for bounded count variables. Finally, the abbreviation "Calc." appearing in the *Variables* column in metadata component 11 refers to *calculated variables* that are computed from other variables in the dataset.

The variable type appearing in metadata component 5 can be any of the variable types discussed in Sec. 2.2, including "other" types like text strings. The variable definition appearing in component 6 should be sufficient to identify the

Table 2.3: Proposed metadata framework for analysis datasets

No.	Metadata component	Variables
1	Dataset name	(DS)
2	Dataset description	(DS)
3	Dataset size (number of included variables)	(DS)
4	Variable name	All
5	Variable type (nominal, ordinal, real, etc.)	All
6	Variable definition, brief	All
7	Variable description, expanded	All
8	Number of observations	All
9	Missing value codes and interpretations	All
10	Number of missing observations	All
11	Defining formulae or procedures	Calc.
12	Information on any cleaning or transformations applied	All
13	Cautions, tips, and hints on usage	All
14	List of valid values and their definitions	Cat.
15	Frequency distribution, including missing	Cat.
16	Units of measure	Real
17	Admissible range of values	R/C
18	Descriptive statistics: mean, standard deviation, MAD scale estimate, and 5-number summary	R/C
19	Fraction of ESD outliers, specified threshold (see Ch. 7)	R/C
20	Fraction of Hampel outliers, specified threshold (see Ch. 7)	Real

variable for someone with field-specific knowledge, while the variable description appearing in component 7 should provide enough information for someone without field-specific knowledge but involved in analyzing or attempting to use the data to understand what the variable is. Ideally, this description would include citations of generally available references to provide this type of background information. Missing data is an extremely important issue, discussed further in Sec. 2.6 and forming the basis for Chapter 16; a key point discussed further in Sec. 2.6 is that there is no "standard" code for missing data. Therefore, to avoid problems like that of *disguised missing data*, it is important to include the code used to designate missing values for all variables in the dataset.

To be useful, metadata components 11 and 12 need to be reasonably detailed when they are applicable, including either mathematical defining expressions for the calculated variables or transformations involved, or references to such detailed definitions. Similarly, any data cleaning procedures (e.g., outlier detection and replacement procedures like those discussed in Chapter 7) or missing data imputation strategies employed (see Chapter 16 for a detailed discussion of this topic) should be described in enough detail that they could be repeated by the analyst, either to consistently incorporate new data or to perform simulation studies to evaluate their impact on the available data.

As this discussion demonstrates, the range of data types that can appear as metadata elements is vast, and useful metadata is typically highly heterogeneous. In the simplest cases, these metadata components are easily interpreted numerical quantities (e.g., the number of data observations), but in other cases (e.g., "variable description, expanded" or "cautions, tips, and hints on usage"), useful metadata corresponds to what has been called *thick description* in the social sciences [113, 249]. In particular, Geertz defines the aim of ethnography as "to render obscure matters intelligibly by providing them with an informing context" [113, p. 152], a statement adopted by Rosenbaum and Silber as their preferred definition of thick description, "as it relates to the statistical analysis of large quantities of data" [249]. The key points here are, first, that no "standard format" is possible for thick description and, second, that a description may be "thick" for one group of users and "thin" for another. Both of these factors contribute to the common problem of inadequate metadata, which can lead to a number of specific data quality problems discussed later in this chapter.

Another critical aspect of metadata is its source. That is, it is important to distinguish between *local metadata*, which can be computed from a given dataset, and *external metadata* that comes from elsewhere. The reason this distinction is important is that some metadata components may be difficult or impossible to construct from the dataset alone. Examples include the problems of disguised missing data discussed in Sec. 2.6.2 and of incorrect measurement units discussed in Sec. 2.7.3; as both of these examples demonstrate, incomplete or incorrect local metadata can cause significant data analysis errors.

Finally, to put the metadata problem in perspective, it is useful to note that since metadata does represent data, it is subject to the same data quality issues as the dataset with which it is associated. That is, metadata components can be slightly or grossly inaccurate, missing, or misaligned (i.e., accurate for one variable but erroneously ascribed to another). Conversely, since metadata available for an analysis dataset is generally much less regularly structured than the data contained in the dataset, the detection of metadata quality problems is generally more challenging than the detection of data quality problems. An example illustrating both of these issues—the possibility of metadata being incorrect, and the potential difficulty of uncovering metadata errors—is given by Dodge [92], who investigates the origin of a small dataset that has been used as an example in at least 90 papers and books on linear regression. This dataset may be viewed as a table of real numbers with 21 rows and 4 columns and is commonly known as the *Brownlee stack loss dataset*, described in a book by Brownlee in 1960 [48] as follows:

> It was obtained from 21 days of operation of a plant for the oxidation of ammonia to nitric acid. The first column represents the rate of operation of the plant. The nitric oxides produced are ab-

sorbed in a countercurrent absorption tower. The third variable is the concentration of acid circulating, minus 50, times 10: that is, 89 corresponds to 58.9 per cent acid. The second column is the temperature of cooling water circulated through coils in the absorption tower. The dependent variable y is 10 times the percentage of the ingoing ammonia to the plant that escapes from the absorption column unabsorbed; that is, an (inverse) measure of the overall efficiency of the plant.

As subsequent authors adopted this dataset as an example, modifications of this description emerged, with many regarding it as a sequence of measurements made on *successive days of operation*. Dodge presents 60 different regression models fit to this dataset and 26 different assessments of outliers detected in it by different means, which range from zero to nine in number with only 5 of the original 21 observations *not* declared to be an outlier by at least one of these analyses. Dodge ultimately traces the the dataset to a pamphlet issued by the Ministry of Supply in England in 1946 that includes a nine-element portion of a data table for what appears to be the same process. The information included in this pamphlet suggests strongly that the observations in the stack loss dataset are *not* measurements made on successive days, but include both gaps in the date sequence and repeated measurements made on the same day. Since the successive observation working assumption formed the basis for some of the outlier detection strategies noted above, this apparent error in the metadata brings these outlier classifications strongly into question.

2.4 What can we measure and how well?

As the discussion in Sec. 2.4.1 illustrates, the variety of things that can be measured in the real world is enormous, as is the range of ways they can be measured. For measurements that yield real values, quality is typically expressed in terms of accuracy and precision, and these terms are defined and discussed briefly in Sec. 2.4.2. Since measurements of the fundamental physical constants like Planck's constant do not have to deal with the effects of time variation, they represent a "best case" for making high-quality measurements; further, because of their fundamental importance in physics, a great deal of very specialized effort has gone into making these measurements over the years. Thus, measurements of these constants may be taken as representative of the best possible measurement quality, both in terms of what is achievable and in terms of the effort required to achieve it. A brief summary of these points is given in Sec. 2.4.3. Finally, to give an idea of the quality of more typical measurements, Sec. 2.4.4 briefly describes several examples of high-quality laboratory measurements described in the scientific literature. The key points of these discussions are, first, that we can measure almost anything we can think of, and second, that the uncertainty in these measurements is essentially always large compared with the numerical roundoff errors discussed in Chapter 1.

2.4.1   What can we measure?

As noted above, the range of things we can measure is almost unimaginably vast, a point that the following examples illustrate. Conceptually, the simplest type of measurement is that of counting discrete objects, so this discussion begins with the apparently simple question, "what is the population of Peoria?"

How many people live in Peoria?

Peoria is a U.S. city in central Illinois, established in the early part of the nineteenth century, and the 1991 *World Almanac* [144] provides two population estimates: based on the 1970 Census, the population was 126,963, and the population reported in the 1980 Census was 124,160. Similarly, the *International Geographic Encyclopedia and Atlas* [270] quotes a 1978 estimated population of 126,000 and other sources yield different numbers, all about the same order of magnitude. What, then, is the *true* population of Peoria, and why don't all of these sources simply give it?

In fact, large manual data collection and recording efforts like determining the population of a city face formidible obstacles and the outcome can be the subject of considerable debate. For example, the September, 1993 issue of the *Journal of the American Statistical Association* devoted a special section to the undercount in the 1990 U.S. Census, defined as the difference between the true population size and the count obtained by the Census Bureau. This quantity was estimated at about 5.4% in 1940, declined to an estimated 1.2% in 1980, and increased again to about 1.8% in 1990 [259]. The undercount is the subject of much concern because the Census results are used to determine the representation in the U.S. House of Representatives, decide the allocation of federal funds to state and local governments, and influence the planning of other surveys. Further, note that the objective of the Census is not only to determine the total U.S. population but, more importantly, to assess its *distribution* with respect to region, race, age, sex, and other factors (e.g., home ownership).

Various approaches have been used to estimate the undercount, including a follow-up study called a postenumeration survey, or PES [145], and demographic analyses using other data sources like Medicare enrollments [246]. Reading through either of these specific accounts makes clear the magnitude of the effort required, both in the original Census and in estimating the undercount. For example, the PES subdivided the U.S. population into 1,392 distinct groups based on geographic region, race, population density, housing tenure, age, and sex. One of the objectives of the PES was to estimate the percentage of erroneous census enumerations—e.g., people who got counted twice, assigned to the wrong group, etc. Of course, since the world changes constantly and the PES was an attempt to resample a portion of the original Census, some discrepancies were unavoidable—people who moved between Census Day and the PES, for example, might be recorded at the wrong address in the PES. In addition, the PES excluded people in jails or nursing homes, military personnel on ships, and the homeless.

Returning to the original example, we are faced with a closely related question: how do we define "Peoria"? For example, *The World Almanac* gives population estimates for both Peoria and Peoria Heights, Illinois: do we include both places in our definition of Peoria or not? Even more fundamentally, how do we define a "resident of Peoria"? The PES excluded people in jails or nursing homes, military personnel on ships, and the homeless—do we follow suit, or do we include some or all of these groups? Similarly, the population estimates given in *The World Almanac* are based on the U.S. Census Bureau's numbers for "urban population," and a three-paragraph definition of this term is given there. Although it is not stated in these three paragraphs whether the groups just listed are included or excluded, it is noted that "persons living in the rural portions of extended cities" are excluded, and one of the three paragraphs is devoted to a definition of this term. Overall, the point of this example is that questions that can be phrased simply may be extremely difficult to answer, even in cases where this answer is an integer and the problem *seems* straightforward.

Is Pluto a planet?

In the preface of his book *Is Pluto a Planet?* [292], David Weintraub notes that he has taught an introductory astronomy course at Vanderbilt University for 14 years and his students frequently pose the title question of his book. Indeed, there have been newspaper articles discussing arguments by various astronomical organizations to downgrade Pluto to some lesser category. The key point of Weintraub's book is extremely relevant here and is summed up by the following quote [292, p. 2]:

> The question *Is Pluto a planet?* illustrates a difficult challenge common to all areas of research and thought: how do we draw the lines we use to categorize objects and ideas?

Indeed, this was one of the issues that made the preceeding example less straightforward than it seemed like it ought to be (i.e., "how do we define a resident of Peoria?"). In Weintraub's case, the essential question is how exactly we define a planet, and he devotes an entire book to exploring it from historical, cultural, and physical perspectives. It becomes clear from these discussions that the appropriate definition of a planet remains a somewhat controversial issue within the astronomy community. Weintraub's own preference is based on the following three *physical* defining criteria for planethood:

1. it is small enough never to have generated energy through nuclear fusion (thus distinguishing planets from stars or brown dwarfs);

2. it is large enough that its internal gravitational forces make it essentially spherical in shape;

3. it has a primary orbit around a star.

Under these three criteria, Weintraub argues that our solar system has 24 planets: the eight noncontroversial planets (Mercury, Venus, Earth, Mars, Jupiter,

Saturn, Uranus, and Neptune), the controversial planet Pluto, and 15 other objects that are usually labeled as members of the asteroid belt between Mars and Jupiter or of the Kuiper Belt in the outer reaches of the solar system.

The key point here is that even for something as *conceptually* simple as binary classification (planet? nonplanet?), the answer depends strongly on the definition on which the classification is based. As the next four examples illustrate, closely related issues arise—along with new ones—when we attempt the more complex task of making real-valued measurements.

Measuring line tension

Line tension is a quantity describing boundary effects in thin liquid films and layers that has been proposed to explain a wide variety of physical phenomena (e.g., stability of liquid crystals, adhesion effects, spreading and wetting phenomena, etc.). More than one definition is possible, but one is an excess force κ that acts along the boundary line of a thin liquid film, tending to shorten this line if $\kappa > 0$ or lengthen it if $\kappa < 0$ [22]. In addition, many different measurement approaches are possible, including measurements of the force required to rupture an aqueous film by a glass sphere of sufficiently small diameter (less than 50 micrometers), measurements of the variation of the contact angle between two air bubbles in water, or measurement of the contact angle between a fluid droplet and a solid surface. Overall, these measurements are sufficiently challenging that there is considerable debate over both the sign and the relative magnitude of line tension, with reported results on identical systems made by identical methods differing by as much as five orders of magnitude (see, e.g., fig. 5 of Babak [22]). Part of this difficulty is due to the small values expected for line tension, ranging from 10^{-13} to 10^{-10} newton, with *reported* values lying between $\lesssim 10^{-12}$ and $\sim 10^{-6}$ newton. Some of this variability arises from the fact that line tension can be defined in more than one way, analogous to the Peoria and Pluto problems just discussed: line tension is a thermodynamic excess quantity, defined with respect to a specified reference model. Since this model is necessarily an idealization of any real physical system under consideration, different reference models can lead to line tensions differing by more than three orders of magnitude [22].

The key point of this example is that line tension is a quantity that is inherently much more complicated in its definition and principles of measurement than the population of Peoria, but it suffers from precisely the same kinds of difficulties. For line tension, these difficulties are mangified enormously by the complexity and small magnitude of the phenomena involved. For example, Babak [22] describes the determination of line tension for a system of air bubbles in an aqueous solution of sodium dodecyl sulfate and sodium chloride. For sufficiently high sodium chloride concentrations, contact angles between two air bubbles of different radius were observed to be approximately 10 degrees, from which the line tension was computed to be negative, with a magtitude of $\sim 10^{-9}$ newton. Lowering the sodium chloride concentration by about 10% reduced the contact angle to approximately 8 degrees, from which a *positive* line tension was computed, of about the same order of magnitude. Most of the physical quantities we encounter in data analysis are not *so* sensitive to experimental conditions and can therefore be measured with greater certainty, but this example is included to emphasize the potential severity of measurement uncertainty under difficult circumstances.

Thermophysical properties of refrigerants

Following the international ban on the use of chlorofluorocarbons (CFCs), a number of alternative refrigerants have been developed, including the hydrofluorocarbon 1,1,2,2-tetrafluoroethane, denoted R134a. To modify or redesign refrigeration equipment to use these alternative refrigerants, it is necessary to have reasonable measurements of their thermophysical properties, including their thermal conductivity, thermal diffusivity, and viscosity. Prior to 1992, measurements reported by different laboratories differed by 15–30% for viscosity and up to 25% for thermal conductivity [18]. These observations led a group of university and national standards laboratories to perform an extensive set of comparative measurements, all based on samples of the substance R134a provided by a single manufacturer from a single production run. In addition, it was noted that even trace amounts of water could significantly alter sample viscosity, so careful measurements of water content were included as part of the sample preparation. Also, gas chromatography was used to determine the presence of organic contaminants, with the largest contaminant determined to be a chemically related compound, designated R134 and present at a concentration of 850 parts per million (water concentration was measured at 6 parts per million).

In this comparison study, sample viscosity was measured by five different methods (oscillating-disk, vibrating-wire, falling-ball, capillary flow and falling-body viscometers). The largest observed differences in vapor-phase viscosity measurements was 2.5% between the oscillating-disk and vibrating-wire viscometers; the largest differences in liquid viscosity was 6% between the falling-ball and vibrating-wire measurements, and it was noted that the reasons for these differences were not fully understood. For comparison, the largest differences between falling-body and vibrating-wire liquid viscosity measurements were less than 3%. Thermal conductivities were also measured several different ways, including a steady-state hot wire system, five transient hot-wire systems, and computation from thermal diffusivity measurements made via light scattering. Most of these results were found to agree within ±2%.

One of the primary conclusions reached in this comparison study was the importance (and difficulty) of maintaining adequate sample purity [18]:

> It is not enough to obtain a high-purity sample from a reputable supplier and then to make measurements assuming that this high purity is maintained throughout the experiments, in particular, for materials such as polar refrigerants which are also very good solvents. Even the materials of the supply cylinder are potential sources of contamination that should not be overlooked.

It was further noted that the materials of construction for the measurement apparatus were also possible sources of sample contamination, especially elastomeric seals and any sources of residual water, both of which are described as sources of "dramatic errors."

Microwave properties of rice wevils

Rice wevils (*Sitophilus oryzae*) are insects that feed on rice and other grains and are a potential source of damage to commercial wheat crops. Concerns about chemical pesticides in feed grains provides a motivation for exploring alternative approaches to pest control, and one option is microwave heating. To investigate the feasibility of this approach, it is necessary to know the dielectric properties of both the insects and the grain they infest, motivating the permittivity and loss factor measurements described by Nelson *et al.* [217], who report measurement results over a frequency range of 200 megahertz (MHz) to 20 gigahertz (GHz) and a temperature range from 10°C to 65°C. Detailed descriptions are given of sample preparation:

> The rice weevils used for these measurements were reared in hard red winter wheat, *Triticum aestivum L.*, in standard quart Mason jars . . . ,

and the sample holder:

> To provide for confinement of the insect samples and control of sample temperature, a stainless steel sample cup with a depth of 19.05 mm (0.750 in) and an inside diameter of 18.95 mm (0.746 in) was constructed.

Because bulk density strongly influences dielectric property measurements, this sample holder was designed to be able to adjust and measure bulk density by adjusting the volume of the sample holder and measuring the sample weight. From these results, it was confirmed that the cube root of the measured dielectric constant varies approximately linearly with bulk density, a result known to hold for other samples besides rice weevils [216]. The authors also made the following observation [217]:

> Although the mosture content of insects is not ordinarily of interest, water has an important influence on permittivity, so moisture content of the insects was determined by drying triplicate samples of about 1 g for 16 h in a forced-air oven at 105°C.

Ulitmately, moisture contents were estimated to be ∼48% by weight. Uncorrected permittivity and loss factor results were described as "highly variable," in part because the insect dimensions were of the same order as the diameter of the coaxial probe used in making the measurements. Consequently, the dielectric behavior of the overall sample (the air-insect mixture) reflected this spatial homogeneity. Correction for bulk density partially addressed this problem and results were ultimately presented showing a general decrease in dielectric permittivity with increasing frequency and an estimated dielectric loss factor that decreases with increasing frequency to about 3 GHz, above which the loss factor increases slowly with increasing frequency. Error bars representing deviations of ± 1 standard deviation are shown in the plots of these results, corresponding to $\sim 10\%$ variation. In addition, the permittivities exhibit a strong temperature dependence, increasing $\sim 50\%$ between 15°C and 65°C. The loss factor exhibits an even more pronounced temperature dependence at the low end of the frequency range, but this dependence diminishes sharply with increasing frequency, becoming comparable to the measurement variability above ~ 5 GHz.

Assessing temperature before they invented thermometers

The preceeding example illustrated that what we can measure is limited only by our imagination, and the following example illustrates that this is also true of *how* we choose to make measurements. Atkinson *et al.* [20] examined climate changes in Britain during the previous 22,000 years, a period going back to well before traditional measurements (e.g., mercury thermometers) were available. To overcome this difficulty, the authors note that fossil records of beetle remains *are* available over this time period and that beetles are [21, p. 851]:

> extremely diverse, identifiable to species level in many fossil occurrences, and have undergone very few extinctions and almost no morphological evolution for the past several hundred millenia.

In addition, the authors note that present-day species can be grouped according to their climatic preferences (e.g., warm vs. temperate vs. cool summers), and they further note both that species with similar climatic preferences tend to occur together within fossil assemblages and that the simultaneous occurrances of incompatible species are "remarkably rare." Consequently, the authors propose the use of these fossil records as a means of inferring climate changes.

More specifically, Atkinson *et al.* [20] begin by constructing *climatic envelopes* for each species of beetle considered, corresponding to two-dimensional regions in the plane of maximum temperature T_{max} (defined as the mean temperature of the warmest month) and temperature range T_{range} (defined as the difference between the warmest and coldest months) for which each species of beetle is known to survive today. Intersecting the envelopes for two different species leads to a smaller set in the (T_{range}, T_{max}) plane called the *mutual climatic range*. By studying beetle fossil samples that have been radiocarbon-dated over the last 22,000 years, the authors constructed a record of estimated temperatures over this period from the mutual climatic ranges of 350 species of beetles, obtaining what they believed to be "the most reliable yardstick yet produced of climatic change on the oceanic margin of north-west Europe."

2.4.2   Accuracy and precision

Accuracy and precision are the two primary ways of characterizing the quality of
measurements of a real variable. To define these terms, consider the following
measurement scenario: we have a measurement system (e.g., a thermocouple
for temperature measurements, together with its necessary interface electron-
ics and a computer data collection system), and we use it to make repeated
measurements of a real variable, called the *measurand*, at a sequence of distinct
measurement times $\{t_k\}$. The output of this measurement system is a sequence
$\{y(k)\}$ of numbers that should be equal to $m(t_k)$. If we assume the true value of
the measurand is a constant, say m^*, over the duration of our measurement sequence, it follows that $m(t_k) = m^*$ for all k, and a perfect measurement system
will generate the output $y(k) = m^*$ for all k. The *accuracy* of the measurement
system is defined as the average difference between the output $y(k)$ and the true
value of the measurand, m^*.

The four plots in Fig. 2.1 represent sequences of repeated position measurements $\{(x(k), y(k))\}$ for a point in the plane. The correct value for the
coordinates of this point is $(6, 6)$, indicated with a solid circle in all four plots.
Here, these position measurements are *accurate* in the left-hand plots in Fig. 2.1
because they cluster around the correct location $(6, 6)$. Conversely, the measurements shown in the right-hand plots in Fig. 2.1 are *inaccurate* because they do
not cluster around the point $(6, 6)$. In contrast, *precision* is a measure of the
range or *spread* of the fluctuations of the measurement sequence about their
nominal value, *regardless of the accuracy of this nominal value*. Generally, we
will refer to this nominal value as a *location* parameter, and any measure of
spread we choose for these fluctuations will be called a *scale parameter*. In
these terms, measurements precise *precise* if this scale parameter is small, and
imprecise if this scale parameter is large. This notion is illustrated in the upper two plots Fig. 2.1, where the precise measurement sequences $(x(k), y(k))$
fluctuate only slightly around their location parameters. In contrast, the lower
two plots in Fig. 2.1 illustrate imprecise measurement sequences because the
magnitude of these fluctuations is large.

It is important to remember that accuracy and precision are independent
quantities. This point is illustrated in Fig. 2.1, which show measurement sequences that are both accurate and precise (upper left), precise but not accurate (upper right), accurate but not precise (lower left), and neither accurate
nor precise (lower right). One reason this distinction is important is that if the
measurand's true value m^* has not changed over a sequence of measurements,
we can estimate precision from these measurements, but we cannot estimate
accuracy without a reference standard capable of maintaining a *known* measur-
and value m^*. In the context of the problem of estimating this reference value
from data, *precision* is closely related to *estimator variance*, and *accuracy* is
closely related to *estimator bias*. These points are discussed further in Chapter
4. Specific examples of measurement accuracy and precision are discussed in
later sections of this chapter.

Figure 2.1: Four examples of measurement accuracy and precision

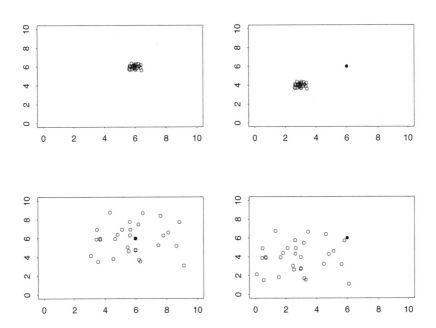

Finally, in cases where estimates of both accuracy and precision are available—or in cases where we are willing to assume values for accuracy and/or precision—it is frequently useful to combine them into a single overall uncertainty measure. This is most commonly done by considering the *root mean square (RMS) error*, defined as follows. Given an estimate δ of the accuracy (i.e., average or systematic error) and an estimate σ of the precision (i.e., the standard deviation of the random error), the RMS error is:

$$e_{\mathrm{RMS}} = \sqrt{\sigma^2 + \delta^2}. \tag{2.1}$$

In a number of the examples discussed later in this chapter (e.g., the Newtonian gravitational constant measurements of Gundlach and Merkowitz [126] discussed in Sec. 2.4.3), measurement uncertainties are expressed as RMS errors.

2.4.3   The limits of measurement quality

In his book *The Fundamental Physical Constants and the Frontier of Measurement* [233], B.W. Petley gives a detailed view of the high-precision measurements made at national standards laboratories, especially those related to measurements of the fundamental physical constants like the mass of the electron m_e, the Rydberg constant R_∞, and the Newtonian constant of gravitation G. Because this discussion is more than 20 years old, the measurement uncertainties he quotes for the fundamental constants are no longer correct, but the current values and the measurement techniques required to obtain them may be viewed as the latest installment of the fascinating history he presents as of 1985. At the end of his book, Petley gives a graph showing the estimated measurement uncertainties that he expected to see by the year 2000 for eleven fundamental physical constants. The best result was that for the Rydberg constant R_∞, with a predicted uncertainty of $\sim 10^{-10}$, and the poorest was that for the Newtonian gravitational constant G, with a uncertainty a little better than 10^{-4}. Currently, the National Institute of Standards and Technology (NIST) lists recommended values for fundamental constants on their Web site:

```
http://physics.nist.gov/constants
```

which gives the uncertainty for R_∞ as 6.6×10^{-12}, an order of magnitude better than Petley's prediction, and the uncertainty for G given as 1×10^{-4}, slightly worse than Petley's prediction. Since 1×10^{-4} corresponds to 0.1%, comparable to the uncertainty in some of the better temperature measurements discussed in Sec. 2.4.4 below, it is reasonable to ask why the uncertainty in G is as large as it is.

Gundlach and Merkowitz [126] describe their own results in estimating the Newtonian gravitational constant, ultimately arriving at an uncertainty of 13.7 parts per million (1.37×10^{-5}), an order of magnitude better than the current NIST recommended value. They note that G was first measured by Cavendish more than 200 years ago, and that the measurement uncertainty has only improved by an order of magnitude per century. One reason the uncertainty in G is so large, relative to the other fundamental physical constants, is that there is no known material to shield against gravitational forces, in contrast to electric fields, which can be effectively excluded by metal enclosures (Faraday cages). In an earlier publication, Gundlach et al. [125] noted that a contributing factor to the large uncertainty in the current recommended value for G is due to a systematic bias caused by torsion fiber inelasticity in the torsion pendulum on which previous measurements had been based. To address this problem, the authors developed a modified technique for measuring G that was designed to be insensitive to inelasticity effects. Data collection involved measurements of the angular acceleration of a torsional pendulum made once per second over a period of approximately three days. The experimental setup involved precisely machined and calibrated attractor masses, a torsional pendulum in a vacuum chamber with a geometry and dimensions chosen to cancel certain known sources of measurement bias. The largest sources of systematic error were uncertainty in the diagonal separation of the attractor masses (which was less than 1.0 micrometers) and temperature uncertainty (which was less than 0.1 degrees Kelvin). In their preliminary experiments, Gundlach et al. [125] noted that "gravitational fluctuations from human activity in the vicinity of the test setup were the dominant noise source." To overcome this difficulty, the final measurements were made in a temperature-stabilized, partially underground room ("former cyclotron cave") on a "massive platform 3.5 m above the floor."

2.4.4 More typical measurements

The measurements just described represent extremes in both achievable measurement quality and required measurement effort. More typically, measurement quality is lower in part because it is not usually practical to exert the effort required to achieve ultimate measurement accuracy and in part because the conditions under which the measurements are made cannot be as carefully controlled as in national standards laboratories. The following four examples describe a range of less extreme measurement scenarios and the accuracies achievable in them. It is important to emphasize that these summaries are not in any way presented as criticisms: they are intended to give an idea of the achievable quality of *good* measurements. Poor measurements can be substantially worse; also, sometimes measurements are so hard to make that even the best achievable measurement quality is much poorer than that described here.

Temperature measurements

Temperature is one of the most important and widespread measurements in the manufacturing industries, and it is also extremely important in physics, chemistry, and biology. The *Comité Consultatif de Thermométrie* (CCT) is the group within the French national standards organization *Bureau International des Poids et Mesures* that is concerned with maintaining the International Temperature Scale (ITS). To extend the utility of the 1990 ITS, this group published a report [237] that provides detailed recommendations for those "who wish to approximate the ITS using simpler techniques." These recommendations cover many different measurement techniques for use under many different circumstances, and the following rough estimates are given in the introduction for the associated measurement uncertainties in different temperature ranges:

> ± 0.05K below 100K
> ± 0.2K near and above room temperature
> ± 0.5K up to 1000°C
> ± 1K to ± 2K above this temperature,

although it is noted that substantially better accuracies can be achieved if sufficient care is taken. In addition, this document provides a useful summary of the typical measurement uncertainties for 21 different measurement devices, including nine different types of standard thermocouples. The smallest measurement uncertainty listed in this summary is $\Delta T/T < 2 \times 10^{-4}$ for a germanium electrical resistance thermometer in the temperature range between 1 Kelvin and 100 Kelvin. Conversely, the largest measurement uncertainties listed in this summary were 3–10 Kelvin for a tungsten/rhenium thermocouple for use between 1000°C and 2400°C and 5 Kelvin for radiative temperature measurements made above 1000°C.

The CCT report is divided into two parts: part 1 describes techniques for approximating the ITS of 1990 using relatively high-accuracy techniques that are generally complicated, inconvenient, and expensive, whereas part 2 describes techniques that are widely used in practice because they are simpler, more convenient, and less expensive. Germanium resistance thermometers are typical of the methods discussed in part 1; these devices are based on the known temperature dependence of the electrical resistance of germanium and are made by connecting a single crystal of doped germanium to four small-diameter gold leads. These devices are used in two ranges: zone 1, corresponding to temperatures from 1K to 10K, and zone 2, corresponding to temperatures from 10K to 100K. This division arises due to the differences in physical mechanisms responsible for the electrical conductivity in the germanium crystal. In addition, this electrical conductivity is strongly dependent on impurities in the germanium, which is typically *doped* with known impurities to achieve better (i.e., smoother) resistance/temperature characteristics. Conversely, as the CCT report notes [237, p. 42], the sensitivity of germanium's electrical resistivity to these dopants is great enough that manufacturing tolerances cannot be made tight enough to produce interchangeable germanium thermometers. Typical sensitivity of these devices is on the order of 1 microvolt per millikelvin. Consequently, these devices are rather sensitive to external electromagnetic interference, particularly in the frequency range from 30 to 300MHz and the temperature range from 70K to 300K, where induced relative errors on the order of 30% have been observed [302]. In addition, measured temperatures depend on both the voltage and frequency of the applied excitation signal: voltages must be kept small enough that the power dissipated by the germanium resistance thermometer itself is less than 1×10^{-6} watt to avoid temperature errors due to self-heating of the thermometer [237, p. 46]. Similarly, alternating-current measurements yield smaller temperatures than direct-current measurements due to Peltier heating and cooling at the contacts between the germanium crystal and the gold leads attached for current and voltage measurements. These effects result in temperature errors on the order of 0.2K at 100K, decreasing rapidly as the temperature is lowered down to ~0.1 millikelvin at $T = 10$K. Finally, it is important to note that most of the heat exchange between the germanium thermometer and its surroundings occurs via the electrical connection leads [237, p. 44], so the accuracy of the resulting temperature measurement depends strongly on how well the whole assembly (sensor and leads) is anchored thermally to the measurement environment and insulated from undesirable external influences.

Part 2 of the CCT report discusses the performance that can be expected from a variety of temperature sensors that are in extremely wide use in both laboratory and industrial settings. A typical example is the industrial platinum resistance thermometer (IPRT), which can be accurate to within ±0.05K between −180C and 0C and within about ±0.01K between 0C and 420C, *provided* sufficient care is taken in assembling them into protective sheaths, annealling them, and calibrating them. Note that this last step is usually done by measuring the response at known comparison points (e.g., melting points of metals like tin or zinc) and fitting the results to a polynomial of order 3 to 8. The CCT report also notes that mechanical shocks, vibration, high pressure, and other hostile conditions encountered in industrial environments means that IPRTs do not conform to the strain-free mounting requirements imposed in standard platinum resistance thermometers (SPRTs) [237, p. 88]. Consequently, IPRTs exhibit inherently lower reproducibilities than SPRTs, although the CCT report notes [237, p. 129]:

> Nevertheless, it is possible to select IPRTs that are only an order of magnitude less reproducible.

Again, it is important to note that this measurement quality comes only after the careful installation and calibration procedure described above.

In typical industrial practice, base-metal thermocouples comprise one of the most important classes of temperature sensors, and the CCT report briefly discusses nine of these sensors, eight of which have standard one-letter type designations: T, J, K, E, N, R, S, and B. Of these devices, it is noted that the type T thermocouple (rated for use from $-200C$ to $350C$) is the most accurate, typically on the order of $0.1C$, but with special calibration, accuracies on the order of $0.01C$ may be approached up to $200C$. By comparison, the CCT report rates the performance of the type J thermocouple—noted to be the most common in industrial use—as "fair to poor," corresponding to $\sim \pm 0.1C$ to $\sim \pm 0.5C$ below $300C$ and $\sim \pm 1C$ to $\sim \pm 3C$ above $300C$; further, it is noted that accuracy suffers above $500C$ due to oxidation of the iron thermoelement.

Physical properties of pentane-alcohol mixtures

Motivated by their widespread industrial use, Sastry and Valand [257] present detailed measurement results for a series of pentane-alcohol mixtures. The results illustrate the level of accuracy achievable under very carefully controlled laboratory conditions, perhaps not quite as rigorous as those of national standards laboratories, but also much better than average, representing a probable upper bound on "typical" measurement accuracy. For example, the purity of all starting materials was verified to be better than 99.9% on a molar basis via high-pressure liquid chromatography (HPLC), mass measurements were made on a single pan analytical balance accurate to ± 0.02 milligrams, and computed mole fractions were quoted as accurate to within ± 0.0001. Two different viscometers were used to cover the viscosity range of all of the mixtures considered, and the instruments used to measure viscosity, density, and permmitivity were calibrated prior to making this series of measurements; also, the measurement temperature was regulated at $298.15 \pm 0.01K$. Note that this regulation accuracy is an order of magnitude better than the "typical" values described above for this temperature range.

Reported accuracies were $\pm 0.0001 \ g/cm^3$ for densities, $\pm 0.002 \ mPas$ for viscosities, ± 0.001 dimensionless units for relative permittivities (ϵ/ϵ_0 where ϵ_0 is the permittivity of free space), and ± 0.0002 dimensionless units for refractive index, which may be viewed as the permittivity measured at visible light frequencies (specifically, in this case, the frequency of the sodium D line). In terms of relative errors, these measurement accuracies correspond to $\sim 1 - 2 \times 10^{-4}$ for the densities, $\sim 1 \times 10^{-4}$ to $\sim 1 \times 10^{-2}$ for the viscosities, $\sim 1 \times 10^{-4}$ to $\sim 1 \times 10^{-3}$ for the permittivities, and $\sim 1 \times 10^{-4}$ for the refractive index values. Again, it is worth emphasizing that these reported accuracies are impressive and should not be taken as "typical," but rather as a rough upper bound on what is possible under very carefully controlled conditions, although less extreme than the pristine conditions than those described in Sec. 2.4.3. In particular, note that these results are generally one to three orders of magnitude better than the certified accuracies of the calibration standards for specific heat measurements described by Cezairliyan [57]. More typical measurement accuracies are described in the following two examples.

Thermal properties of powders

Glatzmaier and Ramirez [116] describe a technique for studying the thermal properties of unconsolidated (e.g., powdery) materials, consisting of a sample holder that is thermally isolated from the ambient environment by a vacuum barrier, a platinum wire heater, and a Wheatstone bridge system for measuring the wire temperature. This measurement scheme is an inferential one in which the wire heats the sample, causing a temperature rise. Monitoring the temperature rise of the wire as a function of time, it is then possible to estimate both the thermal diffusivity of the sample and its thermal conductivity. More specifically, for a wire heater of sufficiently small radius r generating a heat input q per unit length, the relationship between the temperature rise ΔT of the sample and the measurement time t is given by:

$$\Delta T = \frac{q}{4\pi K} \ln\left(\frac{4\alpha t}{r^2 C}\right), \tag{2.2}$$

where C is a known constant, K is the thermal conductivity of the sample, and α is its thermal diffusivity. Plotting ΔT against $\ln t$ then yields a straight line of slope A and y-axis intercept B, from which the desired physical properties may be computed via:

$$K = \frac{q}{4\pi A} \quad \text{and} \quad \alpha = \frac{Cr^2 \exp[B/A]}{4}. \tag{2.3}$$

The authors note that the parameters A and B were obtained from a least squares fit of ΔT to $\ln t$ (i.e., using the techniques discussed in Chapter 5).

The temperature rise ΔT was measured by first calibrating the platinum wire's temperature as a function of its resistance R, fitting measured temperatures and resistances to a fourth-order polynomial of the form:

$$T = T_0 + T_1 R + T_2 R^2 + T_3 R^3 + T_4 R^4, \tag{2.4}$$

over the temperature range 0C$\leq T \leq$ 250C. In operation, wire temperatures were not directly measurable, but were inferred from measurements of the wire resistance R, obtained with an unbalanced Wheatstone bridge operating at constant source voltage to keep the electrical power dissipation q constant; this value is given by $q = V^2/R\ell$, where V is the voltage drop across the platinum wire, R is the resistance from which the wire temperature is inferred, and ℓ is the length of the wire. It is noted that for a 10C temperature rise ΔT, the change in q is less than 0.03% in this measurement scheme.

The data acquisition system used in these measurements incorporated a 12-bit analog-to-digital converter, implying a voltage resolution of about 0.025% of full scale. Overall measurement accuracy and precision was assessed by comparing results obtained for 99.5% glycerin with published values. Based on these comparisons, the accuracy in thermal conductivity measurements was estimated at about 1% with a precision of 0.4%, while the accuracy in diffusivity measurements was estimated at about 6% with a precision of 4.5%. It is important to note that these numbers represent significantly larger uncertainties than either the converter resolution or the national standards laboratory results discussed in Sec. 2.4.3. This point is not raised as a criticism—expectations based on either converter resolution or ultimate achievable measurement accuracies are unreasonable for carefully collected laboratory data. The results described here are much more representative; in fact, they are comparable with the accuracies reported for most of the heat capacity results discussed by Cezairliyan [57].

Characterizing asphalt

Talpe *et al.* [276] present a detailed description of an undergraduate physics laboratory experiment to measure thermal conductivity and heat capacity in asphalt samples. As in the previous example, these quantities are inferred from the slope and intercept of straight lines fitted to tranformed temperature rise vs. time data values. The paper gives a fairly detailed description of a relatively simple setup for making experiments at room temperature, in air, and without requirements for precision machining. The heat source used in this experiment is an electrical resistance heater constructed from copper, and temperature measurements are made using a 1N914 switching diode operating at constant current to provide inexpensive, rapid measurements over a wide temperature range. Significant consideration is given to the errors inherent in the approximations on which the thermal conductivity and heat capacity esitmates are based. For example, it is noted that sample response times less than about 15 minutes or more than a few hours are inconvenient for laboratory measurements; given rough estimates of material properties (specifically, thermal diffusivity), this operational constraint translates into limits on the sample length. Further restrictions are imposed to limit measurement errors due to unwanted radiative heat losses.

The experimental results reported by Talpe *et al.* [276] gave a final sample temperature rise as $19.5 \pm 1K$. Estimated thermal conductivity was given as $3.3 \pm 0.4 mW/Kcm$ and the value of ρC where ρ is the sample density and C is the heat capacity were given as $\rho C = 2.3 \pm 0.2 J/Kcm$. Note that these measurement uncertainties correspond to $\sim$5% for the temperature and $\sim$10%

for the physical properties. It was noted that these physical property values are consistent with those reported elsewhere for asphalt, but it was also noted that:

> ashpalt is not a well-defined product, so these values give only the order of magnitude one should expect.

In many respects, this example may be the most representative of those considered here in terms of "nominal" measurement environments: significant care was taken in developing the measurment system, minimizing unwanted interference effects (e.g., radiative heat losses), and quantifying the limitations of the analytical approximations involved. In addition, this measurement system was built from reliable components, but not from extremely expensive or complicated custom-built components (e.g., the 1N914 switching diode as a temperature sensor vs. the hydrogen-annealed, custom-wound transformer cores described by Kibble and Rayner for making impedance measurements that are accurate to a few parts in 10^9 [173]).

2.5   Variations: normal and anomalous

One of the key points of the discussion in Sec. 2.4 was that even extremely high-quality measurements of real-valued variables like time, temperature, or pressure cannot be measured with unlimited accuracy or precision. Thus, we should not expect repeated measurements of the same variable to have exactly the same value (in fact, if it does, this may be an indication of measurement system failure, a point discussed further in Sec. 2.7.2). This observation leads to the question of how measurement uncertainty should be treated, a topic introduced in Chapter 1. The following discussion extends this introduction, giving more detailed discussions of both the "normal variations" we should expect to see in data sequences and some of the causes of "anomalous variation" that we should always be on the lookout for.

2.5.1   Normal variations and "noise"

One of the files included in the UCI Machine Learning Archive is the Pima Indians diabetes dataset, which gives values for eight clinical characterizations of women from the Pima Indians tribe. One of these characterizations is body mass index (BMI), defined as the weight in kilograms divided by the square of the height in meters. Fig. 2.2 shows these BMI values plotted against a sample number that identifies each patient. In addition, a solid horizontal line is included at the median BMI value for these 768 patients, along with horizontal dashed lines at the upper and lower outlier detection limits of the *Hampel identifier* (the detection threshold value here is $t = 3$; see Chapter 7 for a discussion). The key point of this example is that the variation we see between these dashed lines is representative of the *normal variation* we expect to see in most datasets. In this particular example, the scatter we see around the median value

Figure 2.2: Recorded body mass index values from the Pima Indians diabetes dataset

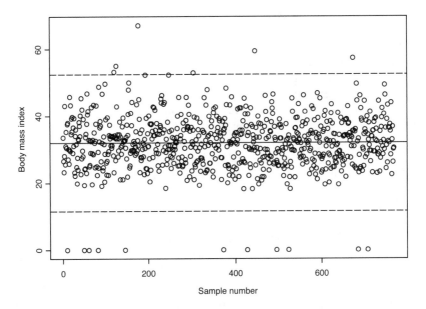

corresponds to the variability in BMI between patients. In the measurement examples considered in Sec. 2.4, the primary source of variation was measurement error, and in the general case we can expect to see variations due to a combination of effects: variations between measurement units (e.g., between patients in a medical study, between material samples in a physical property characterization, etc.), variations due to measurement errors, and even variations due to changes in the same measurement unit over time.

While it is not always reasonable, the most popular working assumption imposed to describe this natural variation or "noise" in real-valued data sequences is the zero-mean Gaussian noise model. This data model is discussed further in Chapter 4, and it is popular because it has a number of mathematically convenient characteristics. First, the assumption that the measurement error sequence has zero mean is a statement of accuracy, invoking a belief that there are no *systematic* measurement errors in the data sequence under consideration. Second, the assumption that the data sequence is Gaussian means that it is completely characterized by its second moment properties. In many data analysis applications, it is also assumed that the random variation in the data sequence is well described by a sequence of *statistically independent, identically distributed (i.i.d.)* random variables. This assumption is extremely convenient, extremely restrictive, and extremely popular in practice. The concept of statistical independence is discussed in detail in Chapter 8, and the notion of an i.i.d. sequence is discussed further in Chapter 10. For now, it is enough to note that this assumption is appropriate for describing highly irregular variation between successive elements of a data sequence, *provided there is no reason to expect these successive elements to be vary together.* Roughly, the i.i.d. model corresponds to an assumption that different measurement units have the same random basic characteristics but are distinct, so their random variations are independent and not coordinated in any way.

The points falling outside the horizontal dashed lines in Fig. 2.2 seem to fall outside the normal range of variation seen in the rest of the data and may thus be regarded as outliers in the sense introduced in Chapter 1 and discussed further below. Indeed, the eleven points at the bottom of this plot all exhibit the *same* value (zero), a highly suspicious finding discussed in detail in Sec. 2.6.2. As the next section discusses, anomalous variation can arise from a number of sources, and when detected, this variation is usually worth investigating at least enough to assure ourselves that it is not interesting or important.

2.5.2   Outliers: gross errors and legitimate surprises

One of the points emphasized in Chapter 1 was that practical data analysis necessarily rests on certain working assumptions. In favorable cases where these assumptions are approximately satisfied, our analysis can be expected to lead to reasonable results and conclusions. Conversely, gross violations of one or more of these assumptions can, in unfavorable cases, completely negate our results. Unless we are aware of this possibility, there is a danger of uncritically accepting these results as correct when they are not. Consequently, one of the most important things to bring to any analysis of real-world data is an open mind, as the following examples illustrate.

A standard working assumption behind many if not most data analyses is that of *homogeneity.* This means that every data point is assumed to be *representative* of the underlying physical, chemical, biological, medical, or other phenomenon of interest, and that there is only *one* such phenomenon present. In practice, it is often true that *most* of the dataset has been generated by the mechanism of interest, but that a small fraction (frequently on the order of a few percent) has been generated by some *other* mechanism that is fundamentally different in nature. One of the most common sources of these *anomalies* or *outliers* is some form of gross measurement error (e.g., measurement system malfunction, saturation, or power failure), but other mechanisims are possible and some of them may prove more interesting than the one that generated the majority of the data.

The term *outlier* can be defined in various ways, but this book adopts the following informal working definition, essentially the same as that given by Barnett and Lewis [26, p. 4]:

> An outlier is a data value x_i in a dataset $\mathcal{D}$ that is inconsistent with the *nominal behavior* exhibited by most of the data values in $\mathcal{D}$.

Depending on the exact interpretation of some of the terms in this definition, it can describe a number of very different situations. In the simplest case, the data values $\{x_k\}$ are real numbers and the nominal behavior of the dataset is described by a range $[x_-, x_+]$ in which these numbers are expected to lie. The working definition given above then partitions $\{x_k\}$ into two classes: nominal points are those for which $x_- \leq x_k \leq x_+$, and outliers are those points for which $x_k < x_-$ or $x_k > x_+$. In more complicated cases, the nominal data values are defined by some more complex structure (e.g., the points lying near some line in the plane), and outliers represent those points that violate this general pattern.

Fig. 2.3 shows plots of four real-valued data sequences, each containing one or more visually obvious points that may be regarded as outliers in the sense just described. The upper two plots show two of the four industrial reactor pressure sequences introduced in Chapter 1, each of length $N = 1024$ and measured under different operating conditions. The upper left plot illustrates what is often assumed to be the most common outlier scenario: all but one of these data points lie between $x_- \simeq 71.7$ and $x_+ \simeq 92.2$, but the single observation at $k = 122$ exhibits the anomalously large value $y(k) \simeq 104.1$. One popular but frequently ineffective procedure for detecting outliers is according to their distance from the mean, normalized by the standard deviation, an approach discussed in some detail in Chapter 7; here, the value at $k = 122$ lies approximately 5.7 standard deviations above the mean and would be declared an outlier under this procedure. Careful examination of this plot also shows the presence of a slow, systematic variation with time that may be removed using the scatterplot smoothers discussed in Chapter 8; if this trend is removed and the *residuals* (i.e., the fluctuations about this trend) are examined, the point $k = 122$ appears even more extreme, lying approximately 8 standard deviations above the mean.

The second data sequence, shown in the upper right plot in Fig. 2.3, illustrates an important subtle variation on the first example. Specifically, note that most of the data points in this sequence lie between $x_- \simeq 193.9$ and $x_+ \simeq 490.8$, but the data points at $k = 913$ and $k = 930$ exhibit the unusually small values $x_k \simeq 107.5$ and $x_k \simeq 161.4$, respectively. Further, close examination of this data sequence reveals a pronounced asymmetry, with large fluctuations *above* the mean value $\bar{x}_N \simeq 258.4$ being much more common than large fluctuations *below* this mean value. This observation is important for two reasons: first, the violation of this asymmetry by the data points at $k = 913$ and $k = 930$ is visually obvious and leads us to declare these points as at least "suspicious," and second, this asymmetry represents a significant complication in the development of practical procedures for automated outlier detection, a point discussed further in Chapter 7. *The key point of both of these examples is that "outliers" are context-dependent.* That is, even within a closely related family of data sequences, it is generally not possible to formulate simple rules for unambiguously declaring a point to be an outlier. In this particular example, it is known that all of these data points are *valid* in the sense of being real pressure values and *not* instrument failures or other gross measurement errors. Hence, the anomalous

Figure 2.3: Four measurement data sequences

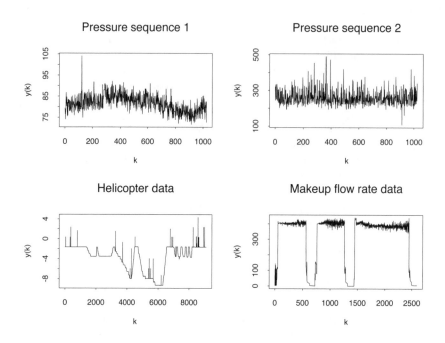

variations seen in these datasets have some unknown but possibly significant *physical* explanation, a point considered further at the end of this discussion.

The plot shown in the lower left in Fig. 2.3 is a sequences of collective pitch commands used to control a small autonomous helicopter described by Chapuis *et al.* [60]. The nominal variation in this sequence is the relatively smooth sequence of steps seen in most of the data, and the outliers correspond to the isolated spikes that occur throughout the data sequence, caused by noise in the triggering circuits of the pulsewidth modulators that generated the data. This data sequence contains more outliers than the two pressure sequences, but it is nine times as long and in fact exhibits roughly the same *contamination level* as the two pressure sequences. This important characterization is defined as the fraction of outliers in the dataset, and here it is $\sim 0.1\%$ to $\sim 0.2\%$, representative of lightly contaminated datasets. In contrast, the data sequence shown in the lower right plot in Fig. 2.3 exhibits a much higher contamination level: nearly 80% of the data values lie between $x_- \simeq 330$ and $x_+ \simeq 440$, but the remaining $\sim 20\%$ exhibit values between 0 and ~ 160. This dataset consists of approximately 2,500 regularly sampled flow rate values from a manufacturing process with a recycle stream: solvent is evaporated from the reaction products in a distillation column and is recycled back to the reactor. Since the separation in this distillation column is not perfect, some of the solvent leaves the distillation column with the product and must therefore be "made up" from another source; the values plotted in Fig. 2.3 represent this makeup flow rate. Here, the anomalous values clustered around zero correspond to times when the process was shut down—i.e., no product was being produced—so no solvent was being fed to the reactor. In fact, if we examine these numbers carefully, we find that they are not *exactly* zero, but they are negligibly small compared to the nominal value of about 400, reflecting a small *systematic* flow rate measurement error. Also, before and after these shutdown episodes are *transitions* to and from the nominal operation episodes. This example is important because, historically, much effort has been devoted to the detection of a single isolated outlier in a larger dataset, and the question of what fraction of outliers should be considered remains somewhat controversial [152, p. 144]. This point is revisited in Chapter 7, but the key observation here is that datasets containing ~20% outliers or more do arise in practice; another such example is discussed by Davies and Gather [83], who consider a dataset of 2,001 measurements with ~400 outliers arising from a known physical cause.

Outliers need not be "gross measurement errors"

At the end of the nineteenth century, Lord Rayleigh performed a series of experiments to determine the density of nitrogen gas, in which he noted a discrepancy between densities obtained by extracting nitrogen from air versus those obtained by extracting nitrogen from other nitrogen-containing compounds [280, p. 49]. The motivation for this series of experiments was a conjecture that all atomic weights should be multiples of that of hydrogen, but Rayleigh found that nitrogen densities obtained by removing oxygen, carbon dioxide, and water from air were consistently 0.5% higher than those for nitrogen obtained from ammonia [123, p. 1042]. In fact, this difference was due to the presence of the inert gas argon in air, and careful analysis of this small discrepancy led Rayleigh to the discovery of this previously unknown element, for which he was awarded the Nobel Prize in Physics in 1904.

In a similar vein, the development of the commercial polymer Teflon (a registered trademark of the DuPont company) also followed from the successful investigation of a small anomaly. On April 6, 1938, Roy Plunkett and Jack Rebok were performing some experiments that used cylinders of the gas tetrafluoroethylene (TFE), one of which appeared to be empty since no gas flowed out when the valve was opened. Because the cylinder weighed too much to be empty, Plunkett sawed it in half and found a solid material inside. Physical property tests on this previously unknown polymer of TFE showed an unusually high corrosion resistance and other remarkable characteristics, including much greater thermal stability than any other plastic known at the time. Ultimately, this discovery formed the basis for the Teflon business [149, p. 157].

The key point of both of these examples is one that will be repeated elsewhere in discussions of outliers: even in cases where some subset of the observed data is clearly anomalous (i.e., inconsistent with most of the other data values and/or our expectations), it can sometimes be more beneficial to understand the nature of the anomaly than to proceed with the original analysis. Consequently, it is important not to fall into the habit of regarding the term *outlier* as synonymous with "bad data points" or "gross measurement errors," unworthy of further examination.

2.5.3   Inliers: a subtle data anomaly

In the case of a single variable, the characteristic of outliers is that they are extreme, lying unusually far from the "center" of the data. As discussed in detail in Chapter 7, this characterization leads to a number of outlier detection procedures, based on different ways of defining the center of the data and distance from it. It was emphasized in the preceeding discussion of outliers that, while they can be erroneous, they need not be. In contrast, an *inlier* is usually defined as "a data value that lies in the interior of a statistical distribution and is in error" [88, 296]. Common sources of inliers are duplicate records, discussed in Sec. 2.7.4, and disguised missing data, discussed in Sec. 2.6.2, but other sources include some of the file merge errors discussed in Sec. 2.7.3, especially the case of correct data values attributed to the wrong data source (e.g., the wrong patient in medical data or the wrong material sample in physical or chemical assays), or data values recorded in the wrong units (e.g., temperatures recorded in degrees Farenheit instead of Celsius, or vice versa).

Because they are not extreme values in a data sequence, inliers cannot be detected using simple distance-based procedures like the outlier detection methods discussed in Chapter 7. Indeed, well-hidden inliers may be very difficult to detect at all, although they do sometimes have characteristic features that can be exploited to find them. For example, because they are erroneous, inliers will often violate the relationships that exist between the variable containing them and other variables in a dataset [296]. In these cases, scatterplots of the inlier-contaminated variable against other, related variables will convert the inliers into visually obvious bivariate outliers like those discussed in Chapter 11.

Another common feature of inliers is that they frequently correspond to a single data value that appears unusually often in a dataset. A case in point is the plot in Fig. 2.4 of triceps skinfold thickness (TSF) values from the Pima Indians diabetes dataset discussed in Sec. 2.5.1. This case represents one of the examples of disguised missing data discussed in Sec. 2.6.2, where missing data observations have been coded with the value zero. Because almost 30% of these data values are missing, the resulting group of zeros is visually obvious in Fig. 2.4, forming a band along the bottom of the plot. To emphasize the point made above that inliers cannot be detected using distance-based methods, the Hampel identifier limits for outlier detection discussed in Chapter 7 are included as dashed horizontal lines in Fig. 2.4, and the median value is indicated as a solid horizontal line. While one point lies outside these outlier detection limits, the inliers do not: these anomalous zero values actually lie closer to the median than many of the nominal data values do.

In fact, the pattern of repeated inliers seen in Fig. 2.4 often appears as a consequence of disguised missing data. Specifically, in cases where the same inlying data value is used to represent missing data values, this value often

Figure 2.4: Recorded triceps skinfold thickness values from the Pima Indians diabetes dataset

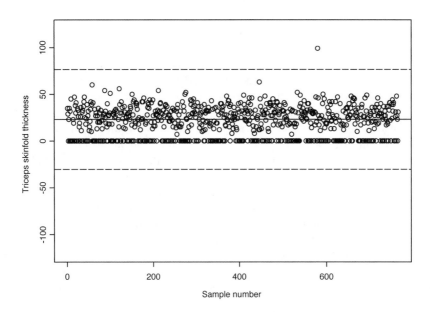

appears more frequently than most or all of the other data values in the data sequence. In these cases, the quantile-quantile plots discussed in Chapter 8 can be very effective in detecting the presence of inliers. Similarly, if these repeated values appear frequently enough, they can sometimes be detected as spurious peaks in density estimates; a case in point is the reporting latency analysis example discussed in Sec. 2.6.2, where these peaks are clearly evident.

2.6 Missing data

The problem of missing data is an extremely common one in data analysis, about which much has been written in the statistics literature. Essentially, the problem is that an observation x_{ij} that should be present in a data table X is absent. Unfortunately, there is no general agreement on how to best represent missing data, leading to a variety of practical complications, including the problem of *disguised missing data* discussed in Sec. 2.6.2. Thus, the following discussion begins with a brief treatment of popular representations and some of their consequences, in Sec. 2.6.1. These discussions are followed by a brief examination of some of the main causes of missing data in Sec. 2.6.3, laying the groundwork for a discussion of the important distinction between *ignorable missing data* and *nonignorable missing data* in Sec. 2.6.4. Essentially, the practical distinction is this: ignorable missing data corresponds to the omission of representative data and leads to an increase in the variability of our results (i.e., a loss of precision), while nonignorable missing data corresponds to selective omission of data, which can lead to large biases (i.e., gross inacurracies).

2.6.1   The problem of coding missing data

An important source of practical difficulties associated with missing data is that there is no universally accepted standard for coding missing values. This is particularly a problem with numeric data, which presents us with two basic options, each with its own advantages and disadvantages. The first option is to use a special symbol for missing data values, and this option is widely adopted: in the *S-Plus* and *R* software packages, missing numeric values are represented as "NA," while in the *SAS* software package, missing values are represented as "." and in the *Minitab* software package the default representation is "*" but users can also define their own symbols. The second option is the use of special numeric codes for missing values, and this is commonly done by adopting values that are clearly impossible for the variable under consideration. As the examples discussed in Sec. 2.6.2 illustrate, zero is a popular code for missing data values, but negative values are sometimes adopted to code missing values for variables that must be positive (e.g., counts, ages, absolute temperatures, etc.), and extremely large values are sometimes also used (e.g., +9999 for variables lying in a finite range like 0.0 to 1.0 or 1 to 100).

The difficulty with the first of these options is that the introduction of non-numeric values into numerical variables requires the computational procedures applied to these variables to be modified to handle these special values. In simple cases, this is not difficult: the logarithm or square of a missing value can simply be coded as a missing value, as can the sum, difference, product, or quotient of a missing value with a nonmissing value. In characterizing a collection of variables, however, while this option is consistent it is frequently not what we really want. Thus, for the minimum, maximum or mean of a sequence $\{x_k\}$, we are typically interested in the corresponding characterization of the nonmissing portion of this sequence, and not in a "missing" final result. As a consequence, analysis routines that deal with possibly incomplete data sequences frequently provide multiple options for dealing with missing data: the default option may be to return a missing value, but another option might be to omit the missing values and return the characterization of the non-missing portion of the data.

Another, more subtle issue that arises in dealing with special codes for missing data is how to interpret the logical operations that frequently arise in data analysis. As a specific example, consider the interpretation of a selection criterion like "$x_k > 1.0$," used to select all records $\mathcal{R}_k$ for which x_k has a value greater than 1.0: if x_k is missing, is this record selected or not? A common approach to dealing with problems of this type—one adopted in both the *S-Plus* and *R* packages—is the use of *three-valued logic,* in which a logical expression can be "TRUE," "FALSE," or a third value, "NULL" or "UNKNOWN." In his book on relational databases, C.J. Date argues strongly against the adoption of three-valued logic because it can greatly complicate the correct specification of database queries [81]. In particular, note that if A has either the logical value "TRUE" or "FALSE," then the logical conditions $A = A$ and $A \neq A$ have the logical values "TRUE" and "FALSE," respectively, while if A has the logical value "NULL," then both of these conditions have the value "NULL" in three-valued logic. As a more analysis-oriented example, note that if x_k can be missing and three-valued logic is adopted, then the comparison $x_k > 1.0$ has three possible outcomes: "TRUE" if x_k is a nonmissing value larger than 1.0, "FALSE" if x_k is a nonmissing value smaller than or equal to 1.0, and "NULL" if x_k is missing. If, in developing an analysis procedure that involves selection conditions like this one, we forget to include the third possibility explicitly, it is possible to get the wrong answer. Specifically, note that since $x_k > 1.0$ and $x_k \leq 1.0$ are *not* logical compliments in three-valued logic, it is not enough to test for the first condition and assign all records failing this test to the second group, as this will assign both those intended cases ($x_k \leq 1.0$) and all cases where x_k is missing to the second class. A recent example illustrating incorrect handling of missing values in statistical analyses of medical data for diabetic patients in the *R*, *S-Plus* and *Minitab* software packages is described by Cordeiro [70], where default treatments of missing data values led to incorrect sample sizes, altering the conclusions of the study.

The difficulties just described provide one motivation for the use of numeric codings for missing data values, but as the discussion in Sec. 2.6.2 illustrates, this practice can lead to the problem of disguised missing data, with arguably worse consequences. A second motivation for the use of numerical codings for missing data is that it permits the use of *multiple codes*, appropriate to circumstances where missing values can be of different types. For example, Little and Rubin [188, p. 3] note that separate codes might be used for different types of non-response in survey data, to distinguish responses like "don't know," "out of legitimate range," and "refused to answer." Similarly, Date notes that some have argued in favor of multiple "NULL" values in databases to represent "value unknown," "value not applicable," "value does not exist," "value undefined," or "value not supplied" [81, p. 585], potentially leading to logic involving more than three levels. A specific example where multiple types of missing data are distinguised is the vegetation index data described on the following Web site:

http://islscp2.sesda.com/ISLSCP2_1/html_pages/groups/veg
/fasir_ndvi_monthly_xdeg.html

This vegetation index is nominally a nonnegative quantity, but negative values are used to distinguish three types of missing data: -99 indicates measurements made over bodies of water, -88 indicates missing vegetation data over land areas, and -77 indicates measurements made over regions of permanent ice.

Overall, the key point of this discussion is to note that missing data values represent a significant practical complication in data analysis. At the very least, the problem of incomplete data records needs to be recognized because failure to do so can lead to blatantly incorrect analysis results. For this reason, while the use of numerical codes is simpler in some respects than the use of special nonnumeric codes, it is also inherently more dangerous since it can lead to the problem of disguised missing data discussed next. *Ultimately, whatever coding we adopt for missing data values, it is the responsibility of the data analyst to ensure that these values are handled appropriately, whether by omission or any of the other techniques discussed in Chapter 16.* The essential point here is the importance of identifying missing values for what they are and structuring our subsequent analysis procedures to handle them explicitly, either through the careful use of built-in three-valued logic in packages like *S-Plus* or through the use of explicit exception-handling logic in our analysis procedures. For example, in the vegetation index example mentioned above, we might choose to omit all records with measured values -99 (bodies of water) or -77 (regions of permanent ice) from our analysis, but adopt one of the missing data imputation procedures discussed in Chapter 16 for records with measured values of -88 (missing vegetation measurements over land areas).

2.6.2 Disguised missing data

The problem of *disguised missing data* arises when the metadata $\mathcal{M}$ associated with a dataset $\mathcal{D}$ is either absent or erroneously indicates that there are no missing observations when in fact some fields do have missing values. Since, as emphasized in the preceeding discussion, there is no standard code for missing data, it is possible that what initially appear to be "normal" data values are in fact codes for missing data. The following two examples illustrate this point.

The BMI data example from the Pima Indians diabetes dataset discussed in Sec. 2.5.1 exhibited a range of variation from 0 to 70. The Centers for Disease Control establish guidelines for BMI, with values less than 18.5 being classed as underweight and values greater than 40 being classed as "extreme obesity." The 1998 *Guiness Book of World Records* lists the world's heaviest man at $1,400$ pounds, corresponding to a BMI in excess of 150, so the maximum value seen in the Pima Indians dataset is plausible. Conversely, Ferro-Luzzi and James have proposed a classification of "extreme wasting malnutrition" for BMI values less than 10, noting that adults with BMI values this low are "at a high risk of imminent death" [103]. Thus, the minimum recorded value of zero is not at all feasible and is, in fact, a code included to indicate that the BMI values for these patients were not recorded.

These BMI values are shown in the lower right plot in Fig. 2.5, along with plots of three other variables from the Pima Indians diabetes dataset: diastolic blood pressure (upper left), the triceps skinfold thickness (TSF) discussed in Sec. 2.5.3 (upper right), and serum insulin concentration (lower left). The metadata accompanying the Pima Indians diabetes dataset in the UCI Machine Learning Archive indicates that there are no missing data values, and many analyses of this dataset appear to have accepted this description without question. One of those who analyzed this dataset, however, was a physician who noted that five of the eight variables in the dataset exhibit biologically implausible zero values [42], including the four plotted in Fig. 2.5. In fact, these

Figure 2.5: Recorded values for four different variables from the Pima Indians diabetes dataset

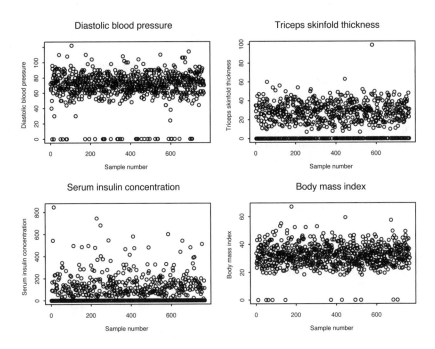

cases illustrate the all too common practice of using zeros to represent missing observations in a variable that assumes only positive values, and the fraction of these missing observations ranges from 0.7% to 48.7% for the five variables with missing data. Failure to recognize this undocumented coding of missing data can have serious analytical consequences: Breault describes the results of approximately 70 previous classification analyses of the Pima Indians diabetes dataset, most of which treated the dataset as complete [42]. The objective of these analyses was to classify patients as diabetic or nondiabetic (a binary variable included in the dataset) on the basis of the other variables. Since 500 of the 768 patients characterized in this dataset were nondiabetic, simply classifying *everyone* as nondiabetic achieves a classification accuracy of 65.1%, and several of the previous results described by Breault [42] do barely better than this trivial classification rule. It is not difficult to show that the results of many analyses of this dataset change substantially when the zero values are viewed as missing data and either omitted from subsequent analysis or handled as missing values in some of the other ways described in Chapter 16 [229].

The second example considered here illustrates the point made earlier that inliers—again corresponding to disguised missing data—can appear as spurious peaks in estimated probability distributions. The U.S. Food and Drug Administration's (FDA) Adverse Event Reporting System (AERS) database documents reports of adverse reactions to prescription drugs [279]. This database is assembled from many different sources, and each report included lists the drug or drugs the patient was taking, the adverse reaction or reactions they experienced, and a certain amount of additional information like gender, age, and the date on which the adverse reaction was reported to have occurred. Updates are released quarterly and the following example considers the $\sim$50,000 reports included in the first quarter 2002 update of this database. Define the *latency* of a report as the time in months between the reported adverse event date for a given report and the end of the quarter when the AERS update containing the report was released. This number gives us an idea of how long it takes an incident report to move through the FDA's processing system and it is highly analogous to the residence times considered in chemical engineering applications. In both cases, we expect to see a distribution of values, generally with a single peak at some positive value at approximately the average time required to work through the system and possibly a slowly decaying right tail, corresponding to difficult cases that take longer to process.

Fig. 2.6 shows two plots of the fraction of records exhibiting latency values between zero (i.e., reports filed in the last month of the quarter but processed quickly enough to be included in the quarterly update) and 120 months (10 years), for the AERS reports included in the third quarter 2002 update that list adverse event dates ($\sim$23% of these values are explicitly coded as missing). The left-hand plot shows the unmodified latency values, while the right-hand plot shows the results we obtain if we remove all records listing an adverse event date of "January 1" in any year. Overall, these figures show exactly what we expect, based on the above description: a dominant peak at 2 months, with a slowly decaying right tail. In the left-hand plot, however, we also see a number of unexpected secondary peaks of decaying amplitude that are spaced 12 months apart. After some investigation, it was found that if all records listing an event date of "January 1" in any given year were removed from the dataset, these secondary peaks disappeared, as seen in the right-hand plot. In fact, what appears to be happening is that the event date "January 1" is frequently being used as a surrogate for "event date unknown," causing the extra peaks seen in the latency distribution shown in Fig. 2.6 [229]. In addition, a close examination of the right-hand plot shows a small secondary peak at 14 months in the right-hand plot, exactly 12 months behind the main peak at 2 months. A possible explanation for this extra peak is data recording errors, where the event date is recorded one year too early, causing an image of the main peak to appear shifted 12 units to the right, exactly as seen in the right-hand plot in Fig. 2.6. Further, note that this peak is also evident in the left-hand plot and is of about the same magnitude, suggesting it is not significantly affected by the removal of the January 1 event dates.

Disguised missing data can manifest itself as outliers in a dataset, as in the Pima Indians BMI example considered in Sec. 2.5.1, but it more commonly manifests itself as inliers, as in the Pima Indians TSF example discussed in Sec. 2.5.3 or the AERS report latency example just described. Thus, while it is sometimes possible to detect disguised missing data using the outlier detection

Figure 2.6: Estimated report latency values for the third quarter 2002 AERS data. The left-hand plot shows the results computed from the event dates as they appear in the dataset, while the right-hand plot shows the results after removing all "January 1" event dates

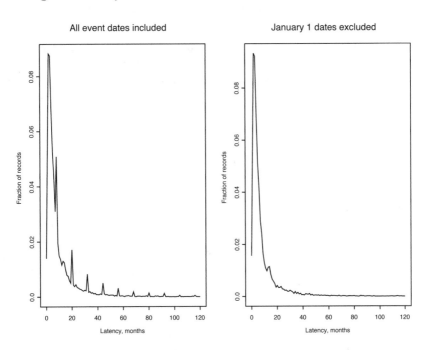

procedures described in Chapter 7, this is generally not the case, as in the TSF example. Fortunately, in cases where disguised missing data is present in high concentration and coded with a single data value, as in the four examples shown in Fig. 2.5, it is frequently possible to detect its presence either from a careful examination of a scatterplot of the data, or better from the quantile-quantile plots discussed in Chapter 8. In more subtle cases like the AERS adverse event report dates, other types of preliminary data analysis are needed to detect the presence of disguised missing data.

2.6.3 Causes of missing data

It was noted earlier that a contributing factor to the lack of a standard representation for missing data is the fact that missing data may be classified into more than one type: observations that were not acquired for whatever reason, responses to questions for which there are no appropriate answers (e.g., pregnancy-related medical questions for men or prostate-related questions for women), or responses to questions that are out of the allowable range. As a specific example of this last type, Adriaans and Zantinge offer the following example [6, p. 84]:

> Recently, a colleague rented a car in the USA. Since he was Dutch, his post code did not fit into the fields of the computer program. The car hire representative suggested that she use the Zip code of the rental office instead.

The key point here is that each of these different types of missing data represents a different class of missing data sources. While all of these sources may be viewed as instances of a more general failure to acquire all of the data intended, it is important to distinguish different failure mechanisms, both because it may be possible to acquire more data in some cases (or possibly influence the design of subsequent data collection systems) and because different treatments may be appropriate to different types of missing data. For example, the discovery of significant numbers of observations missing because the measured quantity is inherently inapplicable (e.g., prostate-specific antigen (PSA) scores for women) may suggest partitioning the original dataset into subsets, with separate analyses for each (e.g., analyzing men and women separately, including PSA scores as a prediction variable for men and number of times pregnant as a prediction variable for women, but not vice versa).

Another extremely important source of missing data lies in the fact that large datasets are often obtained by combining smaller subsets, sometimes collected by different individuals or organizations, and sometimes with different analytical intentions at the time of collection. As a result, these different subsets need not all contain the same variables. Clearly, *some* of these variables need to be the same or there is no basis for combining them, but any variable that was not included in all of the original subsets will exhibit missing data in the combined dataset, and the fraction of this missing data can be quite large. As a specific example, Kooiman *et al.* [175] describe a social science dataset obtained by combining a complete demographic dataset, containing age, gender and other information, with three other datasets. One of these was from a labor survey that included some health and education information, another was from a health survey with some education information in common with the labor survey, and the third was an economic survey that contained salary information for a subset of the individuals covered in the demographic dataset. While there was no missing data in the original demographic dataset, the authors show that the fraction of missing data from the composite dataset varied between 50% and 90%, depending on which combination of variables was considered for analysis.

A detailed discussion of the options for treating missing data is given in Chapter 16, with specializations appropriate for the different types of data analysis considered in this book. The key points here are, first, that missing data does arise frequently enough in practice that it cannot be ignored, and second, that it usually forces us to make explicit choices in how to proceed with the analysis. Can't we simply omit the incomplete records and analyze what we have left? Maybe, but it depends on a number of factors. One is the fraction of missing data, which can be quite large in some situations, as the example discussed by Kooiman *et al.* illustrates. Another factor is the type of analysis considered: a small fraction of missing data can be handled by simple omission in computing descriptive statistics like the mean or standard deviation, but the situation is much more complicated in applications like dynamic data characterization where the assumption of regularly spaced data samples is important. A third factor is the question of ignorability, discussed next: if the missing observations are systematically different from the available observations, simple omission can seriously distort our analysis results.

2.6.4 Ignorable versus nonignorable missing data

The distinction between *ignorable missing data* and *nonignorable missing data* will be discussed further in Chapter 16, but it is important enough to warrant a few words here. Many data analysis problems may be viewed as computations of one or more numerical characterizations from a collection of N uncertain data values. If we model the data values as random variables—the most common way of modeling data uncertainty—it follows that any characterization computed from these data values is also a random quantity. In particular, this characterization can, like the data on which it is based, be at least partially described in terms of its mean and standard deviation. As discussed in greater detail in Chapter 6, an estimator whose mean (specifically, its expected value) is equal to the correct value we are seeking is called *unbiased*. In practical terms, unbiased estimators do not exhibit *systematic errors*. Similarly, as with real-valued measurements, a data characterization is *precise* if it exhibits a small standard deviation. One of the points made in Chapter 6 is that many (indeed, probably most) of the data characterizations we are likely to use in practice exhibit a standard deviation that decays to zero in the limit of infinitely large samples. The practical corollary to this result is that, generally speaking, the larger our data sample the better.

Given a dataset $\mathcal{D}$, if we randomly omit some fraction of the data observations and compute our characterization from the remaining observations, we reduce the effective sample size and so reduce the precision (i.e., increase the standard deviation) of the computed result. This situation corresponds to the case of *ignorable missing data*, where the effect of simply omitting the incomplete records from our analysis causes a loss of precision. A much more serious problem is that of *nonignorable missing data*, corresponding to a case where the missing data values are not omitted randomly, but are omitted *systematically* in a way that can introduce serious biases into our data characterization.

As a simple example, suppose we have a sample of N Gaussian random variables with zero mean and unit standard deviation. If we compute the mean of the complete sample, the result is a random variable that again has a mean of zero but a standard deviation of $1/\sqrt{N}$. If we randomly omit half of these data values and recompute the mean, the result is again a random variable with a mean of zero but a standard deviation of $\sqrt{2/N}$, about 40% larger than that of the result computed from the complete dataset. Instead, suppose we omit the *smallest* half of the data values, corresponding to most of the negative

values in the data sample. The distribution of these values is approximately the
half-normal distribution that describes the absolute value of a Gaussian random
variable [162, p. 417], which has a mean of $\sqrt{2/\pi} \simeq 0.798$. Thus, in addition
to incurring the same ~40% increase in the standard deviation we had before,
we now have a substantial bias in our estimate of the mean, *which does not go
away as our sample size increases.*

The key point of this example is to emphasize that, even in cases where omitting missing values still leaves us with a large enough dataset to analyze, this
strategy is only reasonable in the face of ignorable missing data. With *systematically missing data*, simple omission can lead to badly biased results since the
remaining data sample may not be representative of the data source, effectively
leading to problems of selection bias like the *Literary Digest* example discussed
in Chapter 1. For this reason, it is a good idea to characterize incomplete data
records as well as possible on the basis of the other available fields in the data
records, to see whether there is evidence of systematic differences between the
complete and the incomplete portions of the dataset.

2.7   Other data anomalies

It has been noted that both outliers and missing data are the subject of large,
specialized literatures. The problem of inliers (Sec. 2.5.3) is more subtle and
less widely discussed, as are the issues of coarse quantization (Sec. 2.7.1), noninformative variables (Sec. 2.7.2), file merge and manipulation errors (Sec. 2.7.3),
duplicated records (Sec. 2.7.4), and categorical data errors (Sec. 2.7.5).

2.7.1   Coarse quantization

Mathematically, real numbers exhibit infinite precision: no matter how close
two *distinct* real numbers, x_1 and x_2, are to each other, there are always an
uncountably infinite number of distinct real values between them. This infinite precision is not characteristic of real measurement data, however, nor is it
characteristic of the numerical variables stored in computer memory or external
databases. Thus, as discussed in Chapter 1, the data values we have available
for analysis are always *quantized* to some finite level of resolution. In favorable
cases, this level of quantization is fine enough not to make any significant difference: even if we represent the Gaussian random numbers generated by our
computer's random number generator to only six significant figures, the likelihood of obtaining exactly the same number more than once in a sample of
size N is extremely small unless N is really huge. Conversely, there are circumstances in which the actual measurements of quantities like temperature,
pressure, voltage, or anything else that we might model conceptually as real-
valued are *coarsely quantized,* meaning that the probability of observing two
data samples with exactly the same value is nonnegligible. As the example of
the MAD scale estimator discussed in Chapter 7 illustrates, coarse quantization
can have some important practical consequences. The basic difficulty is that,

for real-valued random variables, *ties*—where two or more observations have exactly the same value—have zero probability; as a consequence, the continuous random variable models that provide the foundation for the overwhelming majority of real-valued data analysis are inadequate for dealing with coarsely quantized data. Special analysis techniques may therefore be required, like the Poisson regression models discussed in Chapter 14 for dealing with count data.

Historically, coarse quantization in computer-based data collection systems often resulted from limitations of either memory or data converters. For example, rounding temperatures to the nearest 0.1 degree meant that they occupied less space in an ASCII data file than temperatures recorded more precisely. Similarly, when analog-to-digital converters with more than 8-bit resolution were prohibitively expensive for many applications, the 256-level quantization artifacts introduced by these converters were clearly evident in the raw data. While both of these limitations are much less significant today, it is important to be aware of them since the resulting coarse data quantizations may persist in *legacy databases*, built completely or partly from data collected during this earlier era. Further, there are those who advocate deliberate coarse quantization of uncertain data to avoid giving the impression of unwarranted measurement accuracy. A case in point is Jacob's advice concerning published population data, where he recommends rounding counts to the nearest 1,000 [156, p. 15].

2.7.2 Noninformative variables

It is important to recognize that not all fields present in a dataset are useful in any given data analysis. Some may be unrelated to the analysis questions, although this may not be obvious at the outset, becoming clear only as the analysis progresses. There are cases, however, where the variables are *inherently* noninformative, and the following discussion describes three such situations.

Constant fields

Data analysis is inherently concerned with explaining the variation seen in some response variable as a function of the variation seen in one or more explanatory variables. In cases where the response variable does not change at all, there is no relationship to explain: the best we can say is that, within the joint range of variation exhibited by the potential explanatory variables, they have no effect on the observed response variable. This situation arises rarely in practice, and if it does arise, it is prudent to carefully review the measurement/data collection process to make sure that the observed results do not reflect a consistent measurement system failure. For example, a partial power failure in an electronic data collection system can result in a constant sequence of measurements that means something like, "the preamplifier is not working" or "the range of the input sensor has been exceeded."

The case of a potential explanatory variable x that assumes a constant value $x_k = c$ for every record in a dataset arises more commonly and may simply reflect that, for the portion of the data under consideration, this particular experimental condition has been kept constant. Conversely, it is important to note that if $x_k = c$ for all records under consideration in the dataset, the variable x has no power to explain variations in any response variable. Certainly, if several different datasets are considered—each with their own distinct but constant value for x—this variable may be useful in explaining differences between the datasets, but within the context of a given data analysis, if $x_k = c$ for all records, the variable x is inherently noninformative. In practice, this means that constant fields should be excluded from consideration, other than to possibly note their constant value in the summary of the analysis results.

A similar situation arises in the case of potential explanatory variables that are *mostly constant,* with two important exceptions. The first is that "almost-constant" variables can play an extremely significant role in dynamic analysis. The reason is that in dynamic analysis, the value of a response variable y_k can depend strongly on previous values y_{k-j} for $j > 0$, so the influence of isolated "spikes" in an input variable x_k can persist. Conversely, in the common case where x_k and x_j are assumed statistically independent for $k \neq j$, if $x_k = c$ for the vast majority of records, it follows from the discussion in the preceeding paragraph that x can only be explanatory for a small fraction of the data values. Probably the best analysis strategy in these situations is to, at least initially, omit the records where $x_k \neq c$ and analyze the remainder of the dataset, excluding x from the class of candidate explanatory variables.

The second exception to this line of reasoning is that if there are *several* "almost constant" variables x^i whose nonconstant regions do not overlap, it may be best to combine them into a single categorical variable. As a specific example, suppose we have a dataset containing $N = 1,000$ records and there are 25 potential explanatory variables, each of which has $x_k^i = 0$ for 960 records and $x_k^i \neq 0$ for 40 records. If the nonzero regions for these variables are mutually exclusive, each record may be regarded as having one and only one of these 25 variables "active." In this case, combining these 25 individual variables into a single categorical (i.e., nominal or ordinal) variable with 25 possible values may yield a highly informative new explanatory variable. Note that while situations like this one are unlikely to arise by chance, it is sometimes useful in analysis to recode an M-level categorical variable as M mutually exclusive binary variables, leading to exactly the situation just described. A specific illustration of this idea is presented in Chapter 14, in connection with the use of categorical explanatory variables in logistic regression analysis.

Mostly or completely missing fields

As in the case of a constant field in a dataset, a field that is entirely missing in a dataset has no explanatory variable. In cases where we are analyzing a subset of a larger dataset, it may be possible to impute these missing values from other fields that are available (see Chapter 16 for a detailed discussion of imputing missing data values), but even when this is possible, relying *completely* on imputed data values seems risky, so it is probably better to drop the field from the analysis altogether. Conversely, note that if all or most of the missing values of a particular variable appear in the portion of the dataset that is of greatest interest, this may be an indication of nonignorable missing data.

As in the case of "mostly constant" explanatory variables, "mostly missing" explanatory variables are also of limited utility. Again, a possible exception would be a collection of variables, each of which was individually "mostly missing" but that taken together were "mostly present." As in the case of the collection of M "mostly constant" but jointly informative variables discussed above, it might be possible to, convert these "mostly missing" variables into a single categorical variable with one possible value for each of the original variables. This would, however, suggest a highly structured pattern of missing data, and some effort should be made to understand the origin of this pattern before fully embracing this analysis alternative.

Record keys

Another class of variables that is not useful in data analysis is the class of *record keys*. It was noted in Sec. 2.1 that relational databases represent an extremely useful development in managing data. Conceptually, a relational database may be viewed as a collection of data tables, linked through special records that uniquely identify each record in the table. This structure permits the efficient construction of many-to-many mappings between tables and is directly responsible for the utility of the relational database. For analysis purposes, however, these key variables are simply surrogate record indices, informationally equivalent to the observation number i. Thus, they contain no additional information that is useful in data analysis.

2.7.3   File merge and manipulation errors

Frequently, large datasets are obtained by combining smaller ones, either whole or in part. It is important to recognize that this combination process involves three components, any or all of which can be the source of errors:

1. the data contained in the files to be combined;

2. the procedures used to combine the files, which may be manual operations, software, or a combination of both;

3. the metadata associated with each of these files.

To see some of the things that can go wrong in this process, consider the simple example of a file $\mathbf{Z}$ constructed from two other files, $\mathbf{X}$ and $\mathbf{Y}$. For concreteness, suppose $\mathbf{Z}$ is a data table with N_r rows and N_c columns and consider Z_{ij}, the ith observation of the jth variable in the data table $\{Z_{ij}\}$. In the simplest case, suppose the data tables $\{X_{ij}\}$ and $\{Y_{ij}\}$ are each of the same size as $\{Z_{ij}\}$ (i.e., N_r rows by N_c columns), and that Z_{ij} is some simple mathematical function of the corresponding observations in the $\mathbf{X}$ and $\mathbf{Y}$ tables:

$$Z_{ij} = f(X_{ij}, Y_{ij}). \tag{2.5}$$

Here, the function $f(\cdot, \cdot)$ can be any function of two variables, including sums, averages, maxima, minima, or anything else that may be appropriate. While there are exceptions, the following statements are *usually* true:

1. data errors in X_{ij} or Y_{ij} result in data errors in Z_{ij};

2. missing data in X_{ij} or Y_{ij} resulting in missing data in Z_{ij}.

In this simple example, the function $f(\cdot, \cdot)$ is implemented by the procedure that combines files, so procedural errors (whether manual or in software) can cause Z_{ij} to be incorrect even if X_{ij} and Y_{ij} are both correct. Finally, the role of the metadata is more subtle and metadata errors can be more insidious. In particular, there are usually implicit (and, all too often, undocumented) relationships between the file combination procedure and the metadata. For example, note the implicit assumption in Eq. (2.5) that the ith record in the three datasets **X**, **Y**, and **Z** all refer to the same data source (e.g., the same material sample, the same patient, the same time period, etc.). Similarly, note the implicit assumption that the jth variable in the output file **Z** is obtained by combining the jth variable in the input files **X** and **Y**. The reasonableness of both of these assumptions is strongly dependent on the metadata describing the three files. Thus, errors in the metadata associated with these files (e.g., incorrect labeling of records or variables) can lead to errors in the output file entries $\{Z_{ij}\}$, as can incomplete metadata. A dramatic example of the consequences of incomplete metadata is the loss in 1999 of the $125 million Mars Climate Orbiter because certain variables were supplied in one set of units and *assumed* to be in a different set of units when they were used to compute midcourse corrections.

A significant contributing factor to file merge errors is the fact that the software used to build large datasets from smaller ones is often a "one off" procedure written by a busy researcher (or their student) *that is used once and never tested.* A published example illustrating this point is the analysis of the Project Normal microarray dataset presented at the 2002 Critical Assessment of Microarray Data Analysis (CAMDA) meeting by Stivers *et al.* [273]. The basis for this analysis was a dataset developed to show the baseline (i.e., "normal") variation in gene expression levels in mouse liver, kidney, and testis samples, each yielding 24 microarray results. For convenience, the results for each organ type were combined into a single dataset, and Stivers *et al.* made a preliminary examination of these three datasets. Each individual microarray gives intensity values for a sample channel and a reference channel, and the authors found that while the sample channel results were clearly distinguishable for the three organs—as expected—but so were most of the reference channel results, which should have been nearly identical. On further examination, described in detail in their paper [273], the authors ultimately concluded that a record misalignment had occurred, in which approximately 36% of the gene expression values in the testis dataset were misidentified (i.e., correct expression measurements for gene X were assigned erroneously to gene Y). One of the conclusions of the paper is the following:

We recognize that most microarray data sets presently extant have neither been stored nor transported under ideal conditions. The data sets from Project Normal are a case in point. The data sets posted to the Web site were not primary data. Primary data are obtained by scanning one microarray at a time. The data used here were assembled into packages, probably using an ad hoc database, spreadsheet, or PERL script. Under these conditions, it is remarkably easy for the row order to be changed accidently, for example, by sorting rows based on the values in an arbitrarily chosen column.

Unfortunately, while these errors are easy to introduce into a large composite dataset, their subsequent detection sometimes requires near-heroic efforts.

2.7.4 Duplicate records

As discussed further in Chapter 3, the analysis of categorical data frequently involves counting records and applying various quantitative characterizations to the results. Similarly, logistic regression analysis discussed in Chapter 14 involves estimating the probability of a particular binary outcome under different circumstances, again based on counts of data records satisfying certain selection criteria. In either of these cases, the presence of *duplicate records*—i.e., duplicate rows in the data table X introduced in Sec. 2.1—will cause these counts to be incorrect, potentially leading to large errors in our analysis and its conclusions.

The presence of duplicate records can also cause errors in the analyses of real variables, as the following simple example illustrates. Before presenting this example, however, it is important to emphasize the difference between duplicate records—repetitions of what should be the same data record—and *replications* or *repeated measurements*. The difference is that duplicate records represent errors in a dataset—the inclusion of records that should not be present—while replications are distinct measurements of the same quantity, included either to permit assessment of measurement variability (as in the case of repeated physical or chemical analyses made by the same method on the same material sample, or repeated biological assays made on a single tissue sample) or to assess variability over time (as in the case of the same clinical measurement repeated at subsequent doctor visits for the same patient). The critical distinction is that deliberately repeated measurements provide a basis for assessing some quantity of interest—either measurement uncertainty or time evolution in the examples just presented—while accidentally duplicated records only degrade the quality of our results.

As a simple illustration of the impact of duplicate records, suppose $\{x_k\}$ is a sequence of N valid data observations that has been recorded as a sequence $\{y_j\}$ of M data records with some degree of duplication. That is, the recorded values $\{y_j\}$ may be subdivided into N groups, with each group containing the single data value x_k, repeated d_k times, for $k = 1, 2, \ldots, N$. The length of the data sequence $\{y_j\}$ is then given by:

$$M = \sum_{k=1}^{N} d_k \geq N. \tag{2.6}$$

Denote the averages of the two sequences by $\bar{x}$ for $\{x_k\}$ and $\bar{x}_D$ for $\{y_j\}$ to emphasize that $\{y_j\}$ should be the same sequence as $\{x_k\}$ but contains duplicates. These averages are given by:

$$\bar{x} = \frac{1}{N} \sum_{k=1}^{N} x_k, \text{ and}$$

$$\bar{x}_D = \frac{1}{M} \sum_{j=1}^{M} y_j = \frac{\sum_{k=1}^{N} d_k x_k}{\sum_{k=1}^{N} d_k}. \tag{2.7}$$

Thus, the effect of duplication on the average is to replace the correct sample average $\bar{x}$ with a *weighted average* that may be expressed as:

$$\bar{x}_D = \sum_{k=1}^{N} w_k x_k, \quad w_k = \frac{d_k}{\sum_{k=1}^{N} d_k}. \tag{2.8}$$

Note that the weights w_k are all positive and sum to 1, reflecting the relative influence of the data point x_k in the overall average. In the unweighted case, corresponding to $d_k = 1$ for all k and $M = N$ (i.e., no duplication), these weights are all equal: $w_k = 1/N$ for all k. It should be clear from Eq. (2.8) that duplication can increase the computed mean $\bar{x}_D$ relative to the correct value $\bar{x}$ if records with larger values of x_k are duplicated, and it can decrease $\bar{x}_D$ relative to $\bar{x}$ if records with smaller values are duplicated. It is not difficult to show that record duplication can have analogous effects on standard deviation estimates or other simple data characterizations like the skewness and kurtosis measures discussed in Chapter 4. In other cases, the effects can be more serious. As a specific and important example, duplicate records can cause the Harr condition discussed in Chapter 5 to fail, so that apparently well-formulated regression problems can fail to have solutions.

Finally, another important practical point about duplicate records is that they are frequently not *exact duplicates* but only *approximate duplicates*, making them more difficult to detect. For example, if some of the fields used to construct the dataset are text fields, alternate spellings, misspellings, or abbreviations can all refer to the same data source, leading to a duplication of records. If this text field is included in the final dataset, these fields will be distinct, so a simple search for rows i and i' in a data table $\{X_{ij}\}$ for which all columns are identical (i.e., $X_{i'j} = X_{ij}$ for $j = 1, 2, \ldots, N_c$) will not identify these duplicates. Also, in cases where some data observations are missing, record pairs with exactly the same values in the fields that are complete in both records may or may not represent duplicate records. This problem of detecting approximate duplicates in a data file is considered further in Chapter 16 as part of the general treatment of missing data in analysis datasets.

2.7.5   Categorical data errors

The notion of additive errors discussed in Sec. 2.5 to describe either measurement uncertainty or outlying observations is not applicable to categorical variables, since simple arithmetic operations like sums or products are not defined for these variables. Still, the recorded values for categorical variables in a dataset can be incorrect, just as in the case of real-valued variables. In a categorical data sequence $\{c_k\}$ where most of the observed values are correct, the following error models are useful:

1. *random substitutions*, where the observed value of c_k represents a random variable drawn from the distribution of possible values for the categorical variable;

2. *random deletions*, where the observation c_k is missing but c_j is replaced with c_{j+1} for all $j \geq k$;

3. *random insertions*, where the observation c_k represents a random substitution as described above, but c_j is replaced by c_{j-1} for all $j \geq k+1$;

4. *random transpositions*, where observations c_k and c_{k+1} are interchanged.

Note that random deletions correspond to a combination of missing data and record mislaignment, and therefore change the length of the data sequence; in particular, the last value c_N in the sequence is not defined in the random deletion model described above. Similarly, random insertions also involve record misalignment and, in the definition given here, lead to an increase in the length of the data sequence $\{c_k\}$. To apply these models as descriptions of categorical data errors for a specific dataset $\mathbf{X}$, it is necessary to specify a value for c_N for a random deletion, and to exclude the extra observation c_{N+1} created by the random insertion model.

These error models are useful for two reasons. First, they provide a basis for generating simulated categorical data errors that can be used to characterize analysis procedures, answering questions like "are random substitutions, insertions, deletions, or transpositions more damaging in the analysis procedure I am considering?" Second, these nominal variable error models are related to the different error generation mechanisms described in the preceeding sections, so if we detect errors of a particular type, we may be able to determine their origin. In particular, random substitutions correspond essentially to individual measurement errors, while the other three error types correspond to specific types of record misalignments.

2.8   A few concluding observations

In view of the range of data anomalies discussed in this chapter, it is important to emphasize that the main message is *not* "data quality is usually so bad that there is really no point in analyzing it at all." Instead, the main message is that, while we can expect most of the data contained in most datasets to be reasonably accurate, we can't expect it to be perfect. In broad terms, we should generally expect real-valued variables to be no more accurate than a few percent, but this should not cause great errors in our analysis results. The aspects of data that we need to be more concerned with are the detection and appropriate treatment of outliers—whether they represent gross data errors or not—along with the detection and treatment of other data anomalies like coarse quantization, duplicate records, or noninformative variables. Similarly, recognition that missing data is a possibility—even disguised missing data—is important since it requires special handling, a subject discussed in detail in Chapter 16.

To protect ourselves from the adverse consequences of these data anomalies, it is a good idea to do two things. First, we should carefully examine any dataset we intend to analyze before beginning our analysis, with an eye toward detecting any anomalies that may be present. Detailed recommendations on how to do this are presented in Chapter 8, but the basic idea is to pose and attempt to answer questions like the following:

1. How many variables are included in the dataset, and what are they? (E.g., what do we think each variable means and where can we turn for verification or additional information?)

2. What are the ranges of possible values for these variables, and how are they distributed over these ranges? (E.g., if they can have values between 0 and 100, do they appear to be integers or real values? Are they approximately evenly distributed over this range, or are they concentrated into a narrow portion of it? In particular, do they all or almost all have the *same* value?)

3. How many records do we have, and what fractions of missing values are associated with each variable? (Do these counts of records and missing fields agree with any metadata that we obtained with the data file? Do they seem reasonable or are they unexpected in any way? How do they compare with record counts for other related analyses we might have done previously or might be considering as a basis for evaluating our results?)

4. What do we know (or are we assuming) about how the different variables are related? If we have certain expectations, are they supported by the appropriate scatterplots, boxplots, or other graphical displays?

The objective of this preliminary examination of the data is to determine whether there are any unusual features in the dataset that might warrant special analytical treatment or might be capable of strongly influencing our results.

The second thing we should do is, after we have obtained our analysis results, examine them carefully to see how well they correspond to our expectations. In doing this, we should compare any estimated quantities with prior estimates, with any reference ranges that might be available, or with related work by ourselves or others. The key objective is to detect anything unusual in our results and understand them as well as possible before making them available for others to consider, argue with, or attempt to extend. The mathematician

George Polya emphasizes the importance of checking your work in his little book *How to Solve It*, noting that [236, p. 59]:

> Some students are not disturbed at all when the find 16,130 feet for the length of the boat and 8 years, 2 months for the age of the captain who is, by the way, known to be a grandfather.

Conversely, as in the case of outliers, it is important to keep in mind that unusual results can arise either from errors (in the analysis, in the data on which it is based, or in both) or from legitimate surprises. Indeed, this second possibility provides a strong motivation for exploratory data analysis.

Chapter 3

Characterizing Categorical Variables

Sections and topics:

This chapter presents a detailed introduction to some aspects of the analysis of categorical data variables, considering the characterization of one variable at a time. This introduction focuses primarily nominal rather than ordinal variables since the incorporation of order information, in those cases where it is present, uses different techniques that will be introduced in later chapters. Further, it has been noted that in the preliminary analysis of categorical variables (also frequently called *factors* or *factor variables*), the type of the factor (i.e., nominal vs. ordinal) is usually of secondary importance, with primary interest focusing on whether the variable has an effect in the analysis (i.e., whether it is a reasonable candidate for an explanatory variable) [266, p. 53]. In these problems, the primary quantitative information we have to work with is derived count data: the number of times a categorical variable takes on each of its possible values. The characterization techniques considered in this chapter are therefore mainly based on counts. The sole exception is the Zipf distribution, defined on the basis of *ranks*, which are only appropriate to ordinal variables.

Key points of this chapter:

1. This chapter is concerned with the univariate characterization of *categorical variables,* corresponding to the *nominal* and *ordinal* data types introduced in Chapter 2.

2. In practice, observations of these variables exhibit a finite number of *possible values,* also called *levels,* although it is sometimes useful to consider models involving an infinite number of *potential levels.* The Zipf distribution characterization discussed in Sec. 3.7 represents a case in point.

3. Most of the categorical variable characterizations considered here involve counts of the number of observations that exhibit each of the possible levels; for a fixed sequence of observations, these counts may be converted into *relative frequencies* or *estimated probabilities.*

4. The advantage of representing categorical data in terms of estimated probabilities is that it allows us to take advantage of an enormous body of literature on *discrete probability theory,* providing the basis for a wide range of characterization techniques.

5. Random variable distributions introduced in this chapter include those like the binomial and hypergeometric distributions that are appropriate to describing binary events like coin tossing, along with *count distributions* like the Poisson and negative binomial distribution that are useful in characterizing the number of times an event occurs in a sequence of observations. Both of these types of distributions will be considered further in later chapters, where more complex categorical variable characterizations are considered (e.g., the logistic and Poisson regression models discussed in Chapter 13).

6. Specific characterizations included here include *entropy,* which may be normalized to obtain an *interestingness measure* or *heterogeneity measure* that tells us how uniformly or heterogeneously a categorical variable is distributed over its range of possible values. In addition, this chapter introduces three other heterogeneity measures and compares their performance for a simple categorical data example.

7. An unusual discrete distribution that has recently become quite popular is the *Zipf distribution,* which seems to provide—at least to lowest order—a useful description of the behavior of a wide range of ordinal phenomena in terms of their ranks. Examples include word frequencies in long documents, city sizes, market share in business, and various characterizations of the Internet. A particularly interesting characteristic of this distribution is that its mean and variance are both generally infinite, rendering many popular data characterizations useless in these settings.

3.1 Three categorical data examples

As noted in Chapter 2, nominal or categorical variables arise frequently in medicine and the social sciences, and while they are less common in the physical sciences, where real numbers abound, they do sometimes arise there as well. In contrast to real-valued variables, subject to an infinite variety of mathematical analyses, our mathematical options for categorical variables are mainly limited to counting the number of times each possible level occurs. The following three examples show some of the ways categorical variables arise in real-world datasets and will provide the basis for illustrating several of the analysis methods introduced in subsequent sections of this chapter and in later chapters.

3.1.1 The UCI mushroom dataset

The UCI Machine Learning Archive is a collection of more than 100 datasets, typically fairly large ones, that are available to be downloaded and analyzed, generally with no restriction other than an appropriate acknowledgement. The archive may be accessed at:

```
http://www.ics.uci.edu/~mlearn/MLRepository.html
```

Each dataset includes a brief description of the data included and its source. The *mushroom dataset* nicely illustrates a number of issues that arise in dealing with categorical variables and has been widely studied in the machine learning and data mining literature [117]. This dataset contains values for 22 categorical variables measured for each of 8,124 different mushrooms, along with their classification as either edible or nonedible. The original source of these data values is *The Audubon Society Field Guide to North American Mushrooms* [185], and one motivation for assembling and analyzing this dataset noted in the information available from the UCI Machine Learning Archive is that there is no simple rule for determining the edibility of mushrooms.

One of the variables included in this dataset is "gill color," which can assume any one of the following 12 values, each encoded by the single letter indicated in parentheses: black (k), brown (n), buff (b), chocolate (h), gray (g), green (r), orange (o), pink (p), purple (u), red (e), white (w), yellow (y). One of the key differences between a categorical variable like this one and real- or integer-valued variables like temperatures or population totals is that categorical variables cannot be plotted. It is possible to encode a categorical variable like gill color into a sequence of numbers, say, from 1 to 12, but this encoding is of limited utility, since numerical characterizations based on these encoded sequences may be as strongly influenced by the choice of encoding scheme as they are by the actual characteristics of the categorical variable. Further, an enormous number of different encodings is possible, with no obvious way of selecting a "best" one: for the 12-level gill color example, there are $12! \simeq 4.79 \times 10^8$ possible codings. To see the impact of these different choices, consider "mean gill color" based on the following two encodings: first, alphabetically by the color name, and second, in increasing order of the number of ocurrances of each color. For the alphabetical ordering, this mean value is 5.73, representing a color "between" black (code 5) and brown (code 6). In contrast, the mean for the occurrance-based ordering is 9.30, representing a color that is "between" brown (code 9) and white (code 10). These results emphasize both the sensitivity of the numerical results obtained to the codings used for categorical variables and the difficulty of interpreting the results: what does "a color between brown and white" mean, exactly? For these reasons, this book makes no use of numerical codings of categorical variables, relying instead on the counts derived from these variables: the number of times each possible value of the variable is observed.

This focus on counts raises the following important point. Given only a sequence of catagorical data values, with no external information or *metadata* to help us interpret them, the counts derived from a sequence $\{c_k\}$ of N observations of a categorical variable c are necessarily positive integers, lying between 1 and N, where N is only possible if c_k has the same value for all k. A characterization of this type—derived solely from counts of observed values, will be called an *internal categorization* of the variable in the discussions that follow. Alternatively, if we have metadata available that tells us something about a categorical variable c, it is possible to have zero counts for values that are possible in principle but absent from the dataset at hand. An example from the mushroom dataset is "gill spacing," a variable that can assume, according to the metadata provided with the dataset, three possible values: close (c), crowded (w), or distant (d). This third possible value or *level* of the variable is absent from the mushroom dataset: no mushrooms with gill spacing "distant (d)" are included in the dataset. Thus, if we treat this variable as internally categorized and simply count the number of times each level appears, we conclude it has two possible values. Conversely, if we treat this variable as *externally categorized*, we define its levels based on the metadata, and it becomes a three-level categorical variable, with one of the levels not observed in the dataset. This distinction is important in practice, especially when comparing subsets of a given dataset: adopting an external categorization means that the subsets of a categorical variable are always comparable since they always involve the same number of levels. In contrast, adopting internal categorizations based on different subsets can lead to the same variable having different numbers of levels, which can make certain types of comparisons impossible. For this reason, whenever possible, this book uses external categorizations of categorical variables, based on the metadata that—at least in favorable cases—helps us interpret the data and our subsequent analysis results. An important distinction between internal and external categorizations also arises in connection with the range of admissible counts: the counts for each level for an externally categorized variable can range from 0 to N, and a count of N for one level does not imply that the variable only has one level, but only that the counts are 0 for all other levels.

3.1.2 Who wrote the *Federalist Papers*?

Although smaller both in terms of the number of data sets included and their sizes than the UCI Machine Learning Archive introduced in the previous example, another very useful source of analysis datasets is the book *Data* by Andrews and Herzberg [12]. This collection includes a number of classic datasets that have been studied extensively in the statistics literature, including both the one considered here and that considered in the next section. The example considered here involves the analysis of word frequencies in text documents, and while it was noted in Chapter 1 that specialized techniques not considered in this book have been developed for text analysis, this example is considered here because words do represent an important categorical data subtype. The point of the following discussion is to illustrate the point noted earlier that categorical data analysis frequently reduces to characterizations of count data.

The *Federalist Papers* is a collection of short essays written in 1787 and 1788 by Alexander Hamilton, John Jay, and James Madison encouraging citizens of the state of New York to vote for ratification of the U.S. Constitution. Many of these essays were published in newspapers under pseudonyms, and the question of which author actually wrote several of these essays has been the subject of some debate. Mosteller and Wallace [214] devoted an entire book to the problem of inferring authorship from the word distributions in these essays, and the two tables of word counts presented by Andrews and Herzberg [12, pp. 423–425] are taken from that book. The underlying idea behind this analysis was that the distribution of certain "marker words" should be characteristic of an author; while both Madison and Hamilton later claimed to have written 12 of the essays, Mosteller and Wallace concluded that Madison's word count distributions better matched those of the 12 disputed essays than Hamilton's did.

Table 3.1 lists the numbers of times the word "an" appears in blocks of approximately 200 words of running text from documents known to have been written by either Hamilton or Madison. These counts are based on 247 blocks of exactly 200 words each for Hamilton and 262 blocks that are also 200 words long in most cases for Madison, but occasionally differ by one or two words. In their discussion of this example, Andrews and Herzberg note that these word counts were initially expected to have a Poisson distribution, but this was not found to be the case. As the following example illustrates, the Poisson distribution is often assumed for rare events, and it may be argued that any individual word should be a rare event in a much larger text like the essays considered here. This example is revisited in Chapter 8 in connection with assessing the reasonableness of the Poisson distribution in describing count data.

3.1.3   Horse-kick deaths in the Prussian army

One of the classic illustrations of the Poisson model for rare events is its use to characterize the number of Prussian soldiers killed by horse kicks between 1875 and 1894. In 1898, Bortkewitsch published a paper titled "The law of small numbers" (*"Das Gesetz der kleinen Zahlen"*) that included the number of soldiers killed each year by horse kicks in 14 different Prussian army corps [12]. Bortkewitsch concluded that these numbers were in reasonable agreement with

Table 3.1: Frequencies of occurrance of the word "an" in documents known to have been written by Hamilton and Madison, based on blocks of approximately 200 words of running text

Count:	0	1	2	3	4	5	6
Hamilton	77	89	46	21	9	4	1
Madison	122	77	40	14	8	0	1

the following discrete random variable model:

$$\text{Prob } \{k \text{ killed each year}\} \quad = \quad e^{-\lambda}\frac{\lambda^k}{k!}. \tag{3.1}$$

Since then, this particular example has frequently been used to illustrate that rare events are often well approximated by the Poisson distribution.

In fact, there is a lot of statistical theory behind this observation [179, sec. 2.4]. Just as the Gaussian distribution arises naturally as the limiting distribution for averages of most sequences of independent, identically distributed random variables, an idea discussed in detail in Chapter 6, the Poisson distribution arises naturally as the limiting distribution for *high-level exceedences* for most such sequences. This idea is illustrated in Fig. 3.1, which shows 100 random variables drawn from the gamma distribution discussed in Chapter 4 with parameters $\alpha = 3$, $\beta = 1$, and $\xi = 0$. The horizontal line corresponds to $x = 7$ and represents a positive threshold roughly 4 standard deviations above the mean. The exceedences are marked with solid circles, and these marked points exhibit the general character of the Poisson point process model discussed in the next paragraph. Specifically, if we regard each threshold crossing as a discrete event (i.e., if we ignore *how much* the threshold is exceeded and only note that it has been exceeded), this model provides a reasonable approximation of the resulting sequence of randomly ocurring discrete events.

A classic application of the Poisson point process model is as a description of the decay of radioactive isotopes, resulting in the emission of particles that may be detected by a Gieger counter [124, p. 228]. The number of particles $N(t)$ observed in the time interval $(0, t]$ is an example of a *counting process*, that satisfies the following two conditions:

1. $N(0) = 0$ and $N(t)$ assumes a nonnegative integer value for all $t > 0$;
2. if $s < t$, then $N(s) \le N(t)$.

The *Poisson point process model* assumes that $N(t)$ has the Poisson distribution defined in Eq. (3.1) with the dimensionless constant λ appearing in that

Figure 3.1: Gamma distribution exceedences

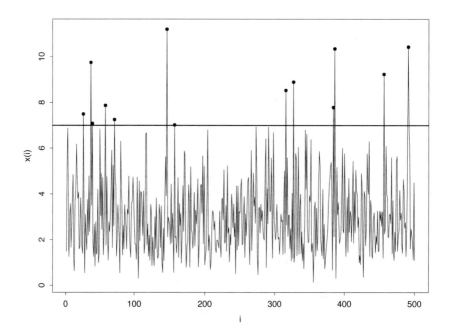

equation replaced by λt, where λ here has the units of a rate or frequency (i.e., reciprocal units of time); specifically, $N(t)$ is described by the discrete probability distribution:

$$\mathcal{P}\left\{N(t) = k\right\} \;=\; \frac{(\lambda t)^k}{k!} e^{-\lambda t}. \tag{3.2}$$

The parameter λ is commonly called the *interarrival rate*, for the following reason: set $T_0 = 0$ and define T_n as the time at which the nth event occurs. The time differences $x_n = T_n - T_{n-1}$ for $n \geq 1$ then represent a *statistically independent* sequence of real-valued random variables called *interarrival times*. One of the characteristic features of the Poisson point process model is that these interarrival times exhibit the exponential distribution:

$$p(x) = \lambda e^{-\lambda x}. \tag{3.3}$$

This result is closely related to the fact that, for the Poisson point process model, if $s < t$, the number of events occuring in the interval $(s, t]$ is independent of the number of events that occurred previously, in the interval $[0, s]$ [124, p. 228]. In addition, both of these results are closely related to the *lack of memory property* of the exponential distribution: if x is exponentially distributed then

$$\mathcal{P}\left\{x > v + w \mid x > v\right\} \;=\; \mathcal{P}\left\{x > w\right\}, \tag{3.4}$$

where $\mathcal{P}\{A|B\}$ denotes the conditional probability of event A, given that event B has occurred (see Chapter 10 for a discussion of conditional probabilities). The key point here is that the Poisson point process model describes *discrete events* that occur at *random times* and that the times between these events are statistically independent and exponentially distributed.

A detailed discussion of the Poisson point process model and its many applications and extensions is beyond the scope of this book. For a more thorough introduction, refer to the book by Grimmett and Stirzaker [124], or for more advanced treatments, refer to the books by Daley and Vere-Jones [77] or Aldous [10]. The exponential distribution is discussed further in Chapter 4.

3.2 Discrete random variables

Two key points emphasized throughout this book are, first, that real-world datasets are almost always subject to some degree of uncertainty, and second, that random variable models can often provide useful characterizations of this uncertainty. This is true both for *discrete variables* like the counts considered in the preceeding examples, and for *continuous variables* like velocities, concentrations, or areas. Since the mathematics required to describe discrete random variables is somewhat simpler than that required for continuous random variables, the discrete case is introduced first, laying the foundation for the discussion of continuous random variables given in Chapter 4.

3.2.1 The discrete random variable model

The discrete random variable model concerns *discrete events* like the toss of a coin, the roll of a die, or the drawing of a colored ball from an urn containing many such balls. It is common to consider sequences of such events, with each individual repetition in the sequence called a *trial* [43, p. 1]. In simple cases like the three just noted, each trial exhibits one of a finite number of possible *outcomes*: "heads" or "tails" for the coin toss, the integers 1 through 6 for a standard six-sided die, or the color of the ball drawn for the urn problem. For each trial, the specific outcome observed is traditionally called an *event*, and the defining feature of the discrete random variable model is that these events are characterized by *probabilities*.

More specifically, if we label each possible outcome with an index $i = 1, 2, \ldots, N$, the *relative likelihood of observing outcome i* is described by the *probability p_i*. A more detailed introduction to the discrete random variable model is presented in Sec. 3.2.2, but the following observation is useful in clarifying the general notion: probabilities must have numerical values between 0 (corresponding to an event that has no likelihood of occurring) and 1 (corresponding to an event that is certain to occur). In the case of the standard six-sided die, the impossible event "the number 7 is observed" has zero probability, while the event "a number between 1 and 6 is observed" has probability 1 since such an outcome is certain. For a *fair die*, each of these six possible outcomes has equal probability, $p_i = 1/6$, and this observation illustrates another feature of the discrete random variable model: each trial must exhibit one of the possible outcomes, so if there are N possible, mutually exclusive outcomes, the following *normalization condition* must hold:

$$\sum_{i=1}^{N} p_i = 1, \tag{3.5}$$

meaning that, whatever the probabilities p_i of the individual outcomes, the probability of observing *one* of them is 1. It is also possible to consider experimental formulations for which the number of possible outcomes is countably infinite, and this extension does not introduce serious complications: as in the finite case, individual probabilities must satisfy $0 \leq p_i \leq 1$ and the normalization condition (3.5) must hold for the infinite sum obtained as $N \to \infty$. In fact, any such sequence of numbers $\{p_i\}$, finite or infinite, may be interpreted as a collection of probabilities.

It is important to conclude this introductory discussion of random variable models with a critical observation: random variable models are *mathematical models, not physical entities.* Consequently, this book adopts the axiomatic view of probability theory, as described informally by Bremaud [43]. That is, discrete probabilities p_i are real numbers between 0 and 1, while continuous probabilities are defined in terms of the real-valued cumulative distribution functions or probability density functions discussed in Chapter 4. In particular, it is not meaningful to assign "physical significance" to probabilities, any more than it is meaningful to assign such significance to other mathematical constructs like real numbers, sets, functions, derivatives, or integrals. Indeed, all of these mathematical constructs—including random variable models—are useful in *approximately describing the behavior of physical systems*, but it is important to emphasize the lessons of Lawsonomy discussed in Chapter 1: in constructing these descriptions, we are necessarily engaged in a process of approximation. Consequently, such descriptions can never be regarded as "correct," only as characterizations that are either "good enough to be useful" or "too poor to be useful." As a practical matter, random variable models have been found to be useful approximations in a wide variety of circumstances, and it is for this reason that random variable models appear in most of the chapters in this book.

A recent paper by Diaconis *et al.* [90] provides a nice illustration of this point. It was noted at the beginning of this discussion that coin tosses are classic discrete probability examples, appearing frequently in introductions to probability theory. Conversely Diaconis *et al.* point out that tosses involving *physical coins in the real world* are governed by mechanics rather than the theory of random variables. In fact, they constructed a mechanical coin tossing machine and they note that the result—i.e., whether the coin lands "heads up" or "tails up"—depends strongly on the initial conditions. Indeed, they are able to adjust the initial conditions so that a coin that is initally "heads" *always* lands "heads." More generally, they show that the coin is always biased toward its initial position: if it starts out "heads," it is more likely to land "heads."

The amount of this bias, however, depends on many things and can be made quite small. The authors cite earlier work by Keller [171], who conducted:

> a study of the physics assuming that the coin spins about an axis through its plane. Then, the initial upward velocity and the rate of spin determine the final outcome. Keller showed that in the limit of large initial velocity and large rate of spin, a vigorous flip, caught in the hand without bouncing, lands heads up half the time.

Ultimately, Diaconis, *et al.* conclude that:

> Keller's analysis gives a good approximation for tossed coins. To detect the departures of the order of magnitude we have found would require 250,000 tosses. The classical assumptions of independence with probability 1/2 are pretty solid.

Conversely, the key point of this example is that the validity of this standard random variable model—i.e., the idea that a flipped coin is equally likely to land either "heads" or "tails"—depends *strongly* on working assumptions. In particular, for a practically achievable range of initial conditions, the coin never turns over, always landing with the initial face up. The authors note that "magicians and gamblers can carry out such controlled flips which appear visually indistinguishable from normal flips." This example illustrates that random variable models can be extremely accurate when they are appropriate, and extremely misleading when they are inappropriate.

3.2.2 Events and probabilities

The axiomatic view of discrete probability adopted in this book is similar to that presented in more detail by Bremaud [43]. There, a *probability space* is defined as a collection of three things:

1. a *sample space* Ω, defined as the set of all possible experimental outcomes;
2. a collection $\mathcal{F}$ of *events*, each corresponding to a subset of Ω;
3. a (probability) function $\mathcal{P}$ mapping events into the interval $[0, 1]$.

To qualify as a valid collection of events, $\mathcal{F}$ must satisfy three conditions:

i. the collection $\mathcal{F}$ must include the sample space Ω;
ii. if A is an event in $\mathcal{F}$, then the complement $\bar{A}$ (i.e., the set of all elements of Ω not contained in A) must also be an event in $\mathcal{F}$;
iii. if $\{A_n\}$ is a sequence of events in $\mathcal{F}$, then $\mathcal{F}$ must also contain the union:

$$\bigcup_{n=1}^{\infty} A_n \in \mathcal{F}.$$

The function $\mathcal{P}$ defines the *probability* of each event in Ω, and it must satisfy the following two conditions:

(a) the probability of the event Ω is 1, i.e. $\mathcal{P}\{\Omega\} = 1$;

(b) any sequence $\{A_n\}$ of *disjoint* events (i.e., subsets of Ω satisfying the condition $A_n \cap A_m = \emptyset$) must exhibit the property of σ-*additivity*:

$$\mathcal{P}\left\{\bigcup_{n=1}^{\infty} A_n\right\} = \sum_{n=1}^{\infty} \mathcal{P}\{A_n\}.$$

Taken together, these axioms lead to an extremely flexible mathematical framework that can be used to derive many useful results and characterizations of valid probability mappings $\mathcal{P}$. For example, note that the collection $\{A_n\}$ is called a *partition* of Ω if it is disjoint, and its union is equal to the set Ω. Denoting $\mathcal{P}\{A_n\}$ by p_n and combining probability conditions *a* and *b* then yields the normalization condition for discrete probabilities (3.5) presented in Sec. 3.2.1. Specializing this result to the case of the two-set partition defined by an event A and its complement $\bar{A}$, we also have:

$$\mathcal{P}\{\bar{A}\} = 1 - \mathcal{P}\{A\}. \tag{3.6}$$

It follows as a further corollary of this observation that $\mathcal{P}\{\emptyset\} = 0$. Many other useful results can be obtained fairly simply from these axioms, and the following discussion summarizes a few of them. For more detailed discussions, including derivations, refer to the introductory probability text by Bremaud [43] or that by Grimmett and Stirzaker [124].

First, suppose A and B are two arbitrary events in $\mathcal{F}$; it follows from these axioms that the events $A \cup B$ (corresponding to the occurrence of either A or B separately or both together) and $A \cap B$ (corresponding to both events A and B together) must also belong to $\mathcal{F}$. Further, it also follows that the probabilities of these events are related by:

$$\mathcal{P}\{A \cup B\} = \mathcal{P}\{A\} + \mathcal{P}\{B\} - \mathcal{P}\{A \cap B\}. \tag{3.7}$$

This result may be extended to arbitrary finite unions to obtain the *inclusion-exclusion formula*:

$$\mathcal{P}\left\{\bigcup_{i=1}^{n} A_i\right\} = \sum_{i=1}^{n} \mathcal{P}\{A_i\} - \sum_{i=1}^{n}\sum_{j>i} \mathcal{P}\{A_i \cap A_j\}$$

$$+ \cdots + (-1)^n \mathcal{P}\left\{\bigcap_{k=1}^{n} A_k\right\}. \tag{3.8}$$

Closely related are *Boole's inequalities*, which give bounds on the union and intersection probabilities of *arbitrary* (i.e., not necessarily disjoint) events A_i:

$$\mathcal{P}\left\{\bigcup_{i=1}^{n} A_i\right\} \leq \sum_{i=1}^{n} \mathcal{P}\{A_i\} \quad \text{and} \quad \mathcal{P}\left\{\bigcap_{i=1}^{n} A_i\right\} \geq 1 - \sum_{i=1}^{n} \mathcal{P}\{\bar{A}_i\}, \tag{3.9}$$

where $\bar{A}_i$ denotes the complement of the event A_i. The first of these inequalities is also known as a *sub-σ-additivity* condition, representing a weakened form of the σ-additivity axiom imposed on the probability function $\mathcal{P}$ for disjoint events. Finally, it is also worth noting that as a consequence of σ-additivity, the function $\mathcal{P}$ is necessarily monotone increasing, meaning that:

$$A \subset B \;\Rightarrow\; \mathcal{P}\{A\} \leq \mathcal{P}\{B\}. \tag{3.10}$$

This observation has the important consequence that the cumulative distribution functions on which continuous random variable models are based are also necessarily increasing.

Another useful notion that can be discussed in terms of the axiomatic probability model presented here is that of *conditional probability*, an extremely important topic in many applications. Because of its intimate connection with data analysis problems involving relations between variables, discussion of this concept is deferred until Chapter 10, where it is introduced in connection with joint distributions and association measures between variables.

3.3 Three important distributions

The discussions presented in Sec. 3.2 introduced the basic notion of a discrete random variable, emphasized that it is a mathematical notion, and gave a brief introduction to the mathematical framework that underlies it. The following discussions consider three important discrete random variable models, introducing the mechanics of working with these models and giving an indication of where they are useful.

3.3.1 Urn models and the binomial distribution

Like coin flips, another classic problem appearing in introductory probability texts is the *urn problem*, typically phrased something like:

> A large urn contains N balls, of which M are white and the remainder, $N - M$, are black. Three balls are drawn from the urn, with replacement (i.e., after each ball is drawn from the urn, its color is noted and it is returned to the urn before the next ball is drawn). What is the probability that all three balls are white?

In fact, problems of this sort arise frequently in a wide variety of practical applications [161], but they are popular in probability books because they are relatively easy to state, solve, and interpret.

To illustrate this point, consider the problem just stated. First, note that the problem only makes sense if $0 \leq M \leq N$, since otherwise the urn will contain a negative number of white or black balls. Because these situations are physically impossible, we assign them zero probability. Further, the two limiting cases are easy to solve: if $M = 0$, the urn contains no white balls, so the event that we draw three successive white balls is again impossible, implying a probability of zero. Similarly, if $N = N$, the urn contains *only* white balls, so the probability of drawing three successive white balls is necessarily 1 (i.e., the event is certain).

For intermediate values of M, note that $p = M/N$ is a number strictly between 0 and 1 representing the fraction of white balls in the urn, and $q = 1 - p$ represents the fraction of black balls in the urn. If we say each ball is drawn *at random*, we mean that the probability of choosing a white ball is equal to the fraction p of white balls in the urn. In particular, under such a random selection scheme, we are not permitted to *choose* the color we want on each drawing, nor is there any hidden mechanism at work that causes one color to be chosen preferentially over the other. This point is important, because *physical* sampling may involve unknown factors; for example, if our mechanism for withdrawing balls involved using ferromagnetic tongs and white balls were magnetized but black balls were not, the likelihood of drawing a white ball would *not* simply depend on the number of white balls in the urn but would be enhanced due to the force of attraction between the white balls and the tongs.

Under the random sampling model, then, when we draw from the urn with replacement, the numbers of white and black balls in the urn remains constant, so the probability of drawing a white ball each time is simply p. Further, the random sampling model assumes successive samples are *statistically independent*, meaning that the color of the rth ball we draw from the urn is not influenced at all by the color of any previous or future balls drawn. Under these assumptions, the probability of drawing three successive white balls is $p \cdot p \cdot p = p^3$. The notion of statistical independence is important and is discussed further, first in Chapter 10 and then again in Chapter 16, where alternative assumptions are discussed in some detail. More generally, the distribution of the number of white balls obtained in n successive draws from an urn with replacement is given by the *binomial distribution*:

$$p_k = \binom{n}{k} p^k (1-p)^{n-k}. \tag{3.11}$$

The special case $n = 1$—corresponding to a single trial (e.g., a coin flip)—is often called a *Bernoulli trial*, in honor of James Bernoulli, who derived this distribution around the end of the seventeenth century; further discussion of the history of this distribution and typical applications is given in Johnson *et al.* [164]. Alternatively, if balls are drawn from the urn *without replacement*, the number of white and black balls in the urn changes with each drawing, and this must be taken into account in answering questions like the one posed here; such problems lead to the *hypergeometric distribution* discussed in the next section.

To verify that the binomial distribution satisfies the normalization condition for discrete probabilities, consider the *binomial expansion*:

$$(x+y)^n \; = \; \sum_{k=0}^{n} \binom{n}{k} x^k y^{n-k}. \tag{3.12}$$

Substituting Eq. (3.11) for p_k into the normalization sum then yields:

$$\sum_{k=0}^{\infty} p_k = \sum_{k=0}^{n} \binom{n}{k} p^k (1-p)^{n-k} = [p + (1-p)]^n \; = \; 1, \tag{3.13}$$

establishing that $\{p_k\}$ indeed satisfies the normalization condition. Further, if $f(k)$ is any real-valued function of k, its *expected* value is defined by:

$$E\{f(k)\} = \sum_{k=0}^{\infty} f(k)p_k. \tag{3.14}$$

Intuitively, this expression may be interpreted as follows. Suppose $\{p_k\}$ describes an experiment whose possible outcomes are indexed by k, and each outcome occurs with probability p_k. Further, suppose $f(k)$ represents some associated response of the experiment that is observed if the outcome is k. The expected value of $f(k)$ then represents a weighted average of these responses over all possible outcomes, each weighted by its probability of occurance. As a special case, note that if $p_j = 1$ for some j (implying $p_k = 0$ for all $k \neq j$), the outcome of the experiment is certain—i.e., outcome j will be observed—and the expected value of $f(k)$ is equal to $f(j)$. At the other extreme, suppose the response $f(k)$ is the same for all possible experimental outcomes—i.e., $f(k) = f_0$ for all k. In this case, it follows from the normalization condition that $E\{f(k)\} = f_0$. In intermediate cases, where more than one outcome k is possible and the different outcomes have different observed responses $f(k)$, the expected value represents a reasonable compromise between the possible values $f(k)$ that accounts for the relative likelihood of the different possible outcomes k.

A particularly important application of this idea is the following: the rth *moment* of k is denoted m_r and defined as the expected value of $f(k) = k^r$, for any positive integer r. For $r \geq 2$, it is often more convenient to work with *central moments* μ_r, defined as the expected value of $f(k) = (k - m_1)^r$; i.e.:

$$\mu_r = \sum_{k=0}^{\infty} (k - m_1)^r p_k. \tag{3.15}$$

Moments of the binomial distribution may be derived by means similar to the normalization result just presented, yielding [164, p. 107]:

$$
\begin{aligned}
m_1 &= np, \\
\mu_2 &= np(1-p), \\
\mu_3 &= np(1-p)(1-2p), \\
\mu_4 &= 3n^2p^2(1-p)^2 + np(1-p)(1-6p+6p^2). \tag{3.16}
\end{aligned}
$$

General recursion relations are available for higher moments [164, p. 108], along with many other specific results for this distribution. Intuitively, m_1 represents a measure of the average outcome of the sequence of trials considered, μ_2 represents a measure of variability about this average, and higher moments provide additional information about the shape of the distribution. These interpretations and related notions are even more important for the case of continuous random variables, a topic discussed in detail in Chapter 4.

Finally, a very useful property of this distribution is the following. If x_1 and x_2 are independent random variables with binomial distributions characterized by the same probability p but arbitrary counts n_1 and n_2, the sum $x_1 + x_2$ also has a binomial distribution with probability p and count $n_1 + n_2$. Intuitively, this is reasonable since x_1 may be regarded as the number of successes observed in n_1 independent trials and x_2 as the number of successes observed in n_2 independent trials, where each trial has probability p of success: $x_1 + x_2$ is then simply the number of successes in $n_1 + n_2$ independent trials, each with probability p. As subsequent discussions illustrate, similar forms of additive behavior are often seen in distributions appropriate to count data, but this is not always the case. As a specific example, this result does *not* hold for either the hypergeometric distribution described in the next section or the discrete uniform distribution described after that.

3.3.2　The hypergeometric distribution

The *hypergeometric distribution* is closely related to the binomial distribution introduced in Sec. 3.3.1 and arises naturally in problems involving *random sampling without replacement*. Specifically, consider an urn containing N balls, and compare the following two experiments. In the first experiment, we draw a sample of n balls, removing them successively but *with replacement:* we reach into the urn, extract the first ball, note its color, number, or whatever distinguishing feature is of interest, *and then we put it back*. We then repeat this process until we have withdrawn and examined n balls. Because there are always the same number of balls in the urn when we make our selection, the probability of drawing a white ball (or an "even-numbered ball," a "molecule of chirality d," etc.) is the same each time. Hence, if there were pN white balls and $N - pN$ black balls in the urn, the probability of drawing a white ball each time is simply p, and we have just described the experiment discussed in Sec. 3.3.1 that leads to the binomial distribution.

For the second experiment, suppose we *do not* replace the ball each time we select and remove one from the urn: the probability of drawing a white ball then *changes* on each drawing as we change the populations of white and black balls in the urn. In particular, if we draw from an urn *without replacement*, the probability of drawing k white balls in n trials is given by the *hypergeometric distribution:*

$$p_k = \binom{n}{k}\binom{N-n}{Np-k} \Big/ \binom{N}{Np}, \tag{3.17}$$

for k between the limits $k^- \leq k \leq k^+$, where:

$$k^- = \max\{0, n - N + Np\}, \qquad k^+ = \min\{n, Np\}. \tag{3.18}$$

Here, the lower bound k^- reflects the fact that if the fraction pN of white balls in the urn is sufficiently large (specifically, if $Np > N - n$), we are *guaranteed* to draw at least k^- white balls since there are only $n - k^-$ black balls in the urn. Similarly, the upper bound k^+ in Eq. (3.18) reflects the fact that the number of white balls drawn from the urn cannot exceed either the total sample size,

Figure 3.2: Four hypergeometric probability distributions

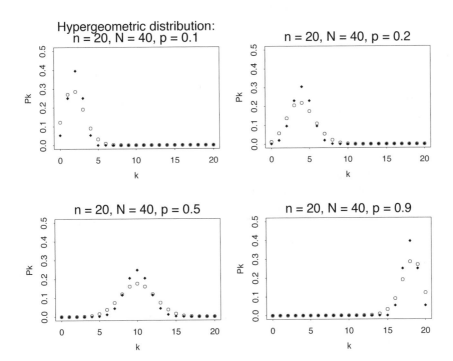

n, or the total number of white balls in the urn, Np. In the extreme case $p = 1$ (i.e., the urn is filled entirely with white balls), it follows that $k^- = k^+ = n$, implying that any sample drawn must consist of all white balls. Finally, note that if n, the number of balls drawn, is negligibly small compared to the total numbers of both white and black balls in the urn (Np and $N(1-p)$, respectively), the probability of drawing either color does not change significantly whether we sample with or without replacement. In particular, as $N \to \infty$ for any fixed p in the range $0 < p < 1$, these probabilities do not change at all and the situation becomes equivalent to that of sampling *with replacement*. Consequently, the hypergeometric probabilities p_k defined in Eq. (3.17) approach the binomial probabilities p_k defined in Eq. (3.11) as $N \to \infty$. As a practical matter, Johnson *et al.* note that it is often adequate to use the binomial approximation for p_k, provided $n < 0.1N$ [164, p. 257].

To ilustrate the differences between the binomial and hypergeometric distributions when these differences are significant, Fig. 3.2 shows plots of the probabilities p_k for both distributions, for $p = 0.1$, 0.2, 0.5, and 0.9. Specifically, the upper left plot shows the distribution for $p = 0.1$, the upper right shows the case $p = 0.2$, the lower left shows the case $p = 0.5$, and the lower right plot shows the case $p = 0.9$. The binomial probabilities are indicated by open circles, whereas the hypergeometric probabilities are indicated by the solid diamonds. In this example, the hypergeometric probabilities are computed from Eq. (3.17) with $N = 2n = 40$, implying $k^- = 0$ if $0 \le p \le 0.5$, but $k^- = (2p - 1)n$ for $p > 0.5$; conversely, $k^+ = n$ for $p \ge 0.5$, but $k^+ = 2pn < n$ for $0 \le p < 0.5$. Here, note that p_k is identically zero for $k > 4$ when $p = 0.1$, for $k > 8$ when $p = 0.2$ and for $k < 16$ when $p = 0.9$.

The name "hypergeometric distribution" is a consequence of the connections that exist between this distribution and the hypergeometric function [1]. The mean of the hypergeometric distribution is the same as that for the binomial distribution; i.e.:

$$E\{k\} \;\; = \;\; np. \tag{3.19}$$

This observation is interesting because it implies that if we just look at the *average behavior* of either of the urn models just considered (i.e., drawing balls with or without replacement), we cannot distinguish them. The variance, however, does reflect the difference between these two cases: specifically, the variance of the hypergeometric distribution is given by:

$$\sigma^2 \;\; = \;\; np(1 - p)\left[\frac{N - n}{N - 1}\right], \tag{3.20}$$

as compared with $\sigma^2 = np(1 - p)$ for the binomial distribution (Eq. (3.16)). The difference between these variance expressions is the term in brackets in Eq. (3.20), a quantity that is strictly less than 1. Note that this term approaches zero as $n \to N$, reflecting the fact that the sample drawn then represents a significant portion of the total population in the urn; hence, the sample very effectively characterizes the urn's contents, and the variability in the number k of white balls drawn about its expected value $np \simeq Np$ (the total number of white balls in the urn) goes to zero. At the other extreme, if n is any fixed finite value and $N \to \infty$, we recover the binomial result, in agreement with the arguments presented in the preceeding discussion.

Finally, it is worth noting that a number of extensions of the hypergeometric distribution have been proposed and investigated by various authors. One particularly interesting example is the *Polya-Eggenberger distribution*, which describes sampling with an *aftereffect*. There, rather than simply withdrawing n balls from an urn one at a time, we replace each ball after it is drawn, *together with c more of the same color*. If we take $c = 0$, we recover the binomial distribution for successive drawings with replacement, whereas if we take $c = -1$ (i.e., we first replace the ball we removed and then remove another, "adding -1 more" and canceling the original replacement), we recover the hypergeometric distribution. Conversely, taking $c = +1$ leads to a situation in which sampling is *contagious* in the sense that observing a white ball on the kth drawing increases the probability of observing white balls on successive trials. This distribution, along with many other variations on the hypergeometric distribution, are discussed further in the books by Johnson and Kotz [161], Johnson *et al.* [164], and the references cited there.

3.3.3 The discrete uniform distribution

The third discrete distribution considered here is the *discrete uniform distribution*, which is a generalization of the distribution of outcomes for a standard six-sided die. There, all six outcomes are equally likely, so the probability of observing any one, say, outcome i, is $p_i = 1/6$. In the general case, the discrete uniform distribution applies to settings where there are M possible outcomes, each of which is again equally likely, implying that $p_i = 1/M$ for $i = 1, 2, \ldots, M$. In cases where we have no reason to believe any one of M possible outcomes is inherently more likely than any other, it follows by the *principle of insufficient reason*, essentially a corollary of Ockham's razor, that the discrete uniform distribution is the preferred choice for a random variable model. Conversely, it is important to emphasize that one of the expectations of exploratory data analysis is that we will learn what was not initially obvious, so we may well develop specific reasons for assuming that all outcomes are *not* equally likely. Still, the discrete uniform distribution is simple, sometimes directly applicable in data analysis, and extremely useful as a reference distribution for characterizing heterogeneity, a point discussed further in Sec. 3.5.

It is a simple matter to derive the mean $m_1 = E\{x\}$ and variance $\mu_2 = E\{(x - m_1)^2\}$ for the discrete uniform distribution (Exercise 4) from the following extremely useful finite sum results [122, p. 1]:

$$\sum_{k=1}^{M} k = \frac{M(M+1)}{2},$$

$$\sum_{k=1}^{M} k^2 = \frac{M(M+1)(2M+1)}{6}. \tag{3.21}$$

The resulting mean and variance expressions are:

$$m_1 = \frac{M+1}{2},$$

$$\mu_2 = \frac{M^2-1}{12}. \tag{3.22}$$

As the following discussion illustrates, the discrete uniform distribution may be regarded in a certain well-defined sense as a "least presumptive" data distribution, giving equal weight to all possible outcomes in the absence of any compelling reason to favor any subset over the rest. As a related application, data samples generated under the discrete uniform distribution will provide a useful frame of reference in interpreting the heterogeneity assessments considered in Sec. 3.5.

3.4 Entropy

The argument made in the preceeding discussion of the discrete uniform distribution, appealing to the principle of insufficient reason, may be formalized by introducing the notion of *entropy*. That is, one approach to defining a "natural" probability distribution for a particular application (e.g., counts of rare events) is to seek a distribution that satisfies various constraints satisfied by the data (e.g., agreement of the mean and standard deviation computed from the data with that for the distribution). In addition, we can ask for a probability distribution that is *least presumptive* in the sense of requiring the fewest *additional* assumptions. This philosophy leads to the *method of maximum entropy*, in which we adopt *entropy* as a measure of unpresumptiveness that we can associate with any probability distribution.

Although entropy can be defined in more than one way, probably the most common definition of entropy is the *Shannon entropy* [71, p. 14]:

$$S = -E\{\ln p\} = -\sum_{i=1}^{M} p_i \, ln \, p_i, \qquad (3.23)$$

where M is the number of distinct values for which p_i is defined. In fact, the Shannon entropy is usually defined based on the binary logarithm $\log_2$ instead of the natural logarithm $\ln$, but this choice makes little difference in practice, since the effect is to simply rescale S, multiplying it by a constant that is independent of n or the probabilities $\{p_i\}$. Also, note that so long as $0 < p_i < 1$ for all i, the entropy S is well defined and easily seen to be positive. If $p_j = 1$ for some j, this poses no difficulty since $p_j \ln p_j = 0$ for this case, but if $M > 1$, it follows from the normalization condition that $p_i = 0$ for all $i \neq j$. This does pose a difficulty, since $\ln p_i = -\infty$ for this case. It is, however, possible to resolve this difficulty by appealing to L'Hospital's rule [174, p. 245]:

$$\lim_{x \to 0} x \ln x = \lim_{x \to 0} \frac{\ln x}{1/x} = \lim_{x \to 0} \frac{1/x}{-1/x^2} = \lim_{x \to 0} -x = 0. \qquad (3.24)$$

Thus, taking $p_i \ln p_i = 0$ whenever $p_i = 0$ in Eq. (3.23) makes the entropy well defined for all valid probabilities $0 \leq p_i \leq 1$. It also follows from these results that the minimum possible value of S is 0, achievable if and only if $p_j = 1$ for some j and $p_i = 0$ for all $i \neq j$.

Since the Shannon entropy is the expected value of the negative logarithm of the probability density, it may be viewed as somewhat analogous to the moments and other characteristics of probability densities discussed in Sec. 3.3.1. That is, the entropy is a single number that tells us something about the underlying distribution. If the discrete probabilities p_i appearing in Eq. (3.23) are regarded as relative frequencies of M possible outcomes in a sufficiently long experiment, Jaynes [158] interprets the entropy S as a measure of the number of different ways the observed outcome can be realized. For example, suppose we roll a six-sided die N times where N is a large number divisible by six (e.g., six million). The number of sequences with an equal number ($N/6$) of 1's, 2's, ..., and 6's is much larger than the number of sequences with any other combination of these numbers. At the other extreme, note that there is only one sequence with *all* 1's or *all* 2's. Therefore, in the absence of any other knowledge, evidence, or beliefs (e.g., that the die is loaded to favor 1's), our *least presumptive* guess for the probability distribution that describes the die's behavior should be $p_i = 1/6$ for $i = 1, 2, \ldots, 6$. As noted earlier, this notion is often expressed as *the principle of insufficient reason*, meaning that we adopt the hypothesis that all outcomes are equally likely because we have insufficient reason to adopt any other hypothesis.

More generally, to show that the discrete uniform distribution maximizes the Shannon entropy, proceed by the method of Lagrange multipliers [104, 247], adjoining the normalization condition as the only constraint. That is, taking λ as the Lagrange multiplier associated with this condition, we replace the entropy S with:

$$S_\lambda = -\sum_{i=1}^{M} p_i \ln p_i + \lambda \left(\sum_{i=1}^{M} p_i - 1 \right). \tag{3.25}$$

Note that this quantity is equal to the entropy S if the normalization constraint is satisfied, since the extra term in Eq. (3.25) then vanishes; otherwise, for any $\lambda > 0$, it follows that $S_\lambda < S$. The maximum entropy probabilities p_i are those that satisfy the conditions:

$$\partial S_\lambda / \partial p_j = 0, \; j = 1, 2, \ldots, M, \qquad \partial S_\lambda / \partial \lambda = 0. \tag{3.26}$$

The first of these conditions yields:

$$-\ln p_j - 1 + \lambda = 0 \;\; \Rightarrow \;\; p_j = e^{\lambda - 1}, \tag{3.27}$$

for $j = 1, 2, \ldots, M$. Hence, it follows from the normalization condition (obtained from the second condition in Eq. (3.26)) that $p_j = 1/M$ for all j, as claimed. Note that the maximum entropy value associated with this discrete uniform distribution is $S^* = \ln M$, which follows directly from substituting $p_i = 1/M$ for all i into Eq. (3.23).

The practical interpretation of this result is that, for discrete random variables, the entropy may be viewed as a measure of distributional homogeneity. That is, it follows from the results just presented that $0 \leq S \leq \ln M$ for a discrete distribution with M distinct values, where S achieves its maximum value for the discrete uniform distribution, and it achieves its minimum value for distributions concentrated entirely on one of these M values. These observations motivate the discussion of heterogeneity measures presented in the next section.

3.5   Interestingness and heterogeneity

As noted, the results presented in Sec. 3.4 provide an interpretation of entropy as a measure of uniformity or homogeneity for discrete distributions. In fact, a number of measures of either homogeneity or, equivalently, heterogeneity have been proposed in a wide variety of fields over the years. In the computer science literature, these measures have been considered together under the general heading of *interestingness measures*. The following discussions present, first, some general characterizations of these measures and then focus on four specific examples, including one based on entropy.

3.5.1   Four heterogeneity measures

Hilderman and Hamilton [140] consider 13 different ways of measuring interest-ingness, characterizing them in terms of five behavioral axioms. All of these measures apply to sequences of counts that may be viewed as estimates of discrete probability distributions. Of the five axioms they considered, the following three are both extremely simple and extremely useful in interpreting the results:

A1: minimum value principle: the measure should exhibit its minimum value for the discrete uniform distribution, $p_i = 1/M$ for $i = 1, 2, \ldots, M$;

A2: maximum value principle: the measure should exhibit its maximum value for a distribution concentrated entirely on one of its possible values, i.e., $p_j = 1$ for some j and $p_i = 0$ for all $i \neq j$;

A3: permutation invariance: the numerical value of the measure should be invariant under arbitrary relabelings of the variable levels i.

The first two of these axioms are important because they establish the interpretation of the numerical value as a measure of heterogeneity: the minimum value corresponds to the least heterogeneous possible distribution (i.e., the uniform distribution), while the maximum value corresponds to the most heterogeneous possible. In Hilderman and Hamilton's paper, they deal with strictly positive counts and do not allow $p_i = 0$ for any i, but as noted in Sec. 3.1.1, allowing $p_i = 0$ is appropriate when dealing with externally categorized variables, and particularly when comparing subsets of a categeorical data sequence. For this reason, zero counts or probabilities are allowed here. Finally, the permutation-invariance axiom (A3) means that the characterization does not depend on the specific labeling considered for the levels of a categorical variable. The utility of this label independence was also noted in Sec. 3.1.1 in connection with the enormous number of possible labelings and the general lack of a reason for preferring any one over the others.

Eleven of the 13 interestingness measures considered by Hilderman and Hamilton satisfy these three axioms, and 3 of these 11 reduce to others on the list when the measures are normalized to lie between 0 and 1. Four of these remaining eight measures are considered in the following discussions [229].

The Shannon measure

The Shannon measure is based on the Shannon entropy discussed in Sec. 3.4, where it was shown that the entropy S is maximized by the discrete uniform distribution. As noted, S represents a measure of homogeneity, but it is easily renormalized to a heterogeneity measure taking the minimum value of 0 for the discrete uniform distribution, and the maximum value of 1 for the completely concentrated distribution, for which $S = 0$. Specifically, the Shannon measure of heterogeneity is given by:

$$I_{Shannon} = 1 + \frac{1}{\ln M} \sum_{i=1}^{M} p_i \ln p_i. \tag{3.28}$$

The Simpson measure

The normalized Simpson measure is defined by:

$$I_{Simpson} = \frac{M \sum_{i=1}^{M} p_i^2 - 1}{M - 1}. \tag{3.29}$$

In fact, it is not difficult to show (Exercise 5) that the normalized Simpson measure is simply equal to the standard variance estimate of the probability sequence $\{p_i\}$, which assumes its minimum possible value of zero when $p_i = 1/M$ for all i and assumes its maximum possible value of 1 when $p_j = 1$ for some j and $p_i = 0$ for all $i \neq j$.

The Gini measure

The importance of the variance as a measure of variability for real-valued data sequence is discussed at some length in Chapter 4, where several other measures of variability are also introduced. One of these alternatives to the variance is the Gini mean difference, and applying it to the probability sequence $\{p_i\}$ leads to the following interestingness measure:

$$I_{Gini} = \frac{1}{2(M-1)} \sum_{i=1}^{M} \sum_{j=1}^{M} |p_i - p_j|. \tag{3.30}$$

A practically important point about this interestingness measure is that direct computation of the double sum indicated in Eq. (3.30) is extremely inefficient for large M. A much more efficient representation, discussed further in Chapter 4, is:

$$I_{Gini} = \left(\frac{M+1}{M-1}\right) \sum_{i=1}^{M} \left(\frac{2i}{M+1} - 1\right) p_{i:M}, \tag{3.31}$$

where $p_{i:M}$ represents the ith order statistic for the sequence $\{p_i\}$: $p_{1:M}$ is the smallest of the p_i values, $p_{2:M}$ is the second smallest, and so forth, with $p_{M:M}$ the largest. Both the notion of order statistics and the derivation of this result are discussed in Chapter 4. The key point here is that Eq. (3.31) can be *much* better computationally than Eq. (3.30) even though the two expressions are *algebraically* equivalent.

The Bray measure

Finally, the fourth heterogeneity measure considered here is the Bray measure, defined as:

$$I_{Bray} = \left(\frac{M}{M-1}\right) \left[1 - \sum_{i=1}^{M} \min\left\{p_i, \frac{1}{M}\right\}\right]. \tag{3.32}$$

Figure 3.3: Comparison of four heterogeneity measures for the categorical variable "gill color" from the UCI mushroom dataset. The points (solid circles) represent the heterogeneity values computed from the gill color data, and the boxplots summarize the corresponding heterogeneity values for 1,000 independent reference samples, each uniformly distributed over 12 levels

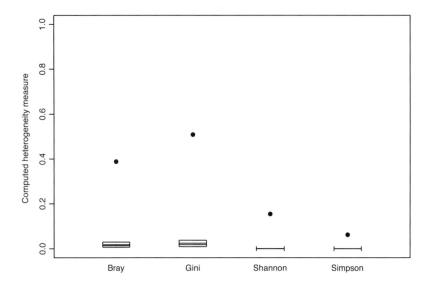

3.5.2   Application to the UCI mushroom dataset

To see how these four heterogeneity measures compare, the following discussion applies all of them to categorical variables from the UCI mushroom dataset introduced in Sec. 3.1.1. Comparisons are presented between the different measures for the same sets of variables to illustrate differences between the measures, and comparisons are presented between different variables for the same measure to illustrate differences between variables. To aid in interpreting these comparisons, the idea of using a uniformly distributed random reference to assess the *significance* of the heterogeneity characterization is introduced. This approach represents an application of some of the hypothesis testing ideas introduced in Chapter 9, but it is included here because it is both simple and useful, illustrating the role of random variability in characterizing real categorical data sequences. All of these ideas are considered further in Chapter 8 in connection with using these interestingness measures to characterize a new dataset.

Fig. 3.3 shows the results obtained when the four heterogeneity measures defined above are applied to the "gill color" variable from the UCI mushroom dataset. Specifically, the points (solid circles) in this figure represent the numerical values for each of the indicated measures computed from the gill color sequence, and these results demonstrate clearly that, while all four measures are attempting to characterize the same thing, their numerical values differ substantially. Recall from the discussion given in Sec. 3.1.1 that this categorical variable exhibits 12 distinct levels, so it is reasonable to compare these heterogeneity values with reference values computed from independent random samples that are uniformly distributed over 12 levels. The boxplots shown in Fig. 3.3 summarize the range of heterogeneity values computed from 1,000 independent reference samples based on this idea. The fact that all of the points in this plot lie well above the range of the boxplots provides support for the interpretation that the gill color sequence is heterogeneously distributed (i.e., not uniform) over its 12 possible values. In fact, these results can be used to compute quantitative significance measures for these heterogeneity values, a point discussed in detail in Chapter 9 where this example is revisited.

The ordering of the numerical heterogeneity values computed from the gill color data sequence—i.e., with the Gini measure the largest, followed by the Bray measure, then the Shannon measure, and finally the Simpson measure as smallest—seems to be more an inherent property of the heterogeneity measures themselves than a function of the variable considered, although there are some exceptions as the following example illustrates. Fig. 3.4 compares the Gini heterogeneity measure computed from eight different categorical variables from the UCI mushroom dataset. The first of these variables, *Edibility*, is a binary variable that classifies each mushroom as either "edible (e)" or "poisonous (p)." The relative fractions of these two levels are fairly close (i.e., 51.8% edible vs. 48.2% poisonous), and this is reflected in both the small value for the corresponding Gini measure and the fact that it does not exceed the most extreme of the uniform reference values. In contrast, both the "bruise classification" *Bruise* and the "stalk shape" *Stalk-sh* are also binary variables, but these are not as evenly balanced between the two possible values, so their heterogeneity measures are both larger and well above the range of the random reference values. In particular, the distribution for *Bruise* is 58.4% bruised vs. 41.6% nonbruised, while that for *Stalk-sh* is 43.3% "enlarging (e)" and 56.7% "tapering (t)." Note that the Gini heterogeneity measures for these three binary examples are consistent with the differences Δ between percentages of the two possible values: $\Delta = 3.6\%$ for *Edibility*, 16.8% for *Bruise*, and 12.6% for *Stalk-sh*.

The variables "cap surface" (*Cap-surf*) and "gill attachment" (*Gill-att*) are each categorical variables with four levels and they both appear to be much more heterogeneously distributed than the three binary variables *Edibility*, *Bruise* or *Stalk-sh*. The possible cap surface characterizations are "fibrous" (28.6%), "grooves" ($< 0.1\%$), "smooth" (31.5%), and "scaly" (39.9%), and the fact that one of these possible values is almost absent is responsible for the relatively large heterogeneity value seen here. The case of gill attachment is even more extreme. There, two of the four possible values ("descending" and "notched") are *entirely* absent, and a third value ("attached") only appears 2.6% of the time, with the fourth value ("free") overwhelmingly dominant, appearing 97.4% of the time.

Figure 3.4: Comparison of the Gini heterogeneity measures computed for the eight categorical mushroom variables listed in the text; as in Fig. 3.3, the solid points represent the heterogeneity values computed from each data variable, and the boxplots summarize the range of values computed from the corresponding uniformly distributed random references

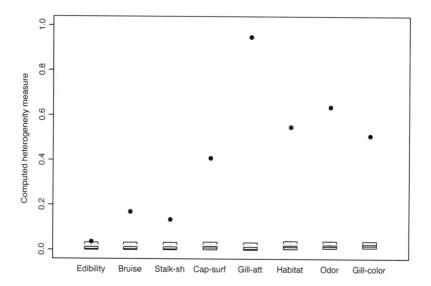

The fact that this distribution is almost entirely concentrated on one of four possible values is responsible for the very large heterogeneity value seen in Fig. 3.4 for this case. The remaining three cases shown in Fig. 3.4 are *Habitat*, a categorical variable with seven levels, *Odor*, with eight levels, and *Gill-color*, with 12 levels. All of these variables exhibit intermediate heterogeneity values between the relatively homogeneous behavior of the three binary variables considered and the nearly maximal heterogeneity of *Gill-att*.

Fig. 3.5 shows the corresponding heterogeneity results computed for the same eight variables shown in Fig. 3.4, but based on the Simpson heterogeneity measure instead of the Gini measure. Consistent with the results shown in Fig. 3.3 comparing the four measures for gill color, the numerical values for the Simpson measure are almost all substantially smaller than the corresponding Gini measure values. The sole exception is the case of gill attachment, noted in the preceeding discussion to exhibit a distribution that is almost maximally heterogeneous. Thus, it is not surprising that both the Gini and Simpson measures give this case a numerical value near 1. Also, careful comparison of figs. 3.4 and

Figure 3.5: Comparison of the Simpson heterogeneity measures computed for the eight categorical mushroom variables listed in the text; as in Fig. 3.3, the solid points represent the heterogeneity values computed from each data variable, and the boxplots summarize the range of values computed from the corresponding uniformly distributed random references

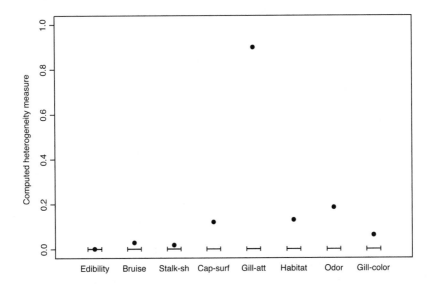

3.5 shows that the relative ordering of these heterogeneity values is identical in all cases but one. The sole exception is the relative ordering of gill color and cap surface: the Gini measure classifies gill color as more heterogeneous than is cap surface, while the Simpson measure classifies gill color as less heterogeneous.

The practical corollary of this last observation is that results do sometimes depend significantly on the analysis method used to obtain them, implying that the choice of method can be important. It follows from this observation that it is important to understand the relative strengths and weaknesses of alternative analysis methods, understanding best gained by applying them in a variety of different settings. Thus, before using a new method, it is a good idea to first apply it to cases where the *expected results* are known and carefully examine the *actual results* to see how they compare with expectations. Overall, the results presented here do not provide the basis for a compelling recommendation. While they do suggest that the Gini and Bray measures may exhibit a greater sensitivity to nonextreme heterogeneity differences than the Shannon and Simpson measures, this question should be examined further in whatever applications are considered. The specific question of how and why these heterogeneity measures differ will be revisited in Chapter 9.

3.6 Count distributions

It has been noted repeatedly in the preceeding discussions that the analysis of categorical variables often reduces to the analysis of count data. When these counts arise from binary phenomena, as in the urn model discussed in Sec. 3.3.1, the first distribution of choice is usually the binomial distribution discussed there because it is the simplest. As Johnson *et al.* note [164, p. 134], this distribution is appropriate whenever it is reasonable to assume that the probability of each of these two possible outcomes is constant and that the trials are statistically independent. Conversely, the authors note that these assumptions do not always hold, so the binomial distribution is not always adequate; a case in point was that of sampling without replacement discussed in Sec. 3.3.2, governed by the hypergeometric distribution instead of the binomial distribution.

As a specific and important case, note that the probability p and the total number of trials n completely define the distribution, determining both the mean number $m_1 = np$ of "positive" outcomes and the variance of this number, $\mu_2 = np(1 - p)$. Cases where the variance of the observed number of positive outcomes is larger than μ_2 are termed *overdispersion*, and cases where the variance is smaller than μ_2 are termed *underdispersion*. These situations also arise with other standard distributions, particularly the Poisson distribution discussed next. The key point is that, while simple distributions like the binomial and Poisson are appealing and widely regarded as "natural" in a variety of circumstances, there are practical situations where they are inadequate.

3.6.1 The Poisson distribution

The Poisson distribution was introduced in Sec. 3.1.3 as a popular model for rare events (e.g., horse-kick deaths in the Prussian army). Because it does arise frequently, it is worth giving its distributional characterizations here. Recall that this distribution is defined by Eq. (3.1), repeated here for convenience:

$$p_k = \frac{e^{-\lambda}\lambda^k}{k!}, \quad \text{for } k = 0, 1, 2, \ldots \quad (3.33)$$

Here, p_k represents the probability of observing a count of k as the outcome of an independent trial of some experiment that returns a nonnegative integer. This distribution plays a key role in building log-linear regression models for count data, a topic discussed in Chapter 13. In addition, since the Poisson distribution is often *assumed* for count data, Chapter 8 presents a graphical method for evaluating the reasonableness of this distribution for any given sample of count data. The following characterizations of the Poisson distribution are useful in these and a variety of other applications.

The moments of the Poisson distribution are easy to compute (Exercise 3), and the first four are given by [164, p. 157]:

$$
\begin{aligned}
m_1 = E\{k\} &= \lambda, \\
\mu_2 = E\{(k - m_1)^2\} &= \lambda, \\
\mu_3 = E\{(k - m_1)^3\} &= \lambda, \\
\mu_4 = E\{(k - m_1)^4\} &= 3\lambda^2 + \lambda.
\end{aligned}
\tag{3.34}
$$

In words, both the mean and the variance of the Poisson distribution are simply equal to the one parameter λ that defines the distribution. This observation is extremely useful because it provides a simple, informal assessment of the adequacy of the Poisson distribution for count data. In particular, as noted above, if the variance is too large to be consistent with the Poisson distribution, this phenomenon is often called *overdispersion*, while if the variance is too small, it is called *underdispersion*. The fact that these effects have reasonably standard names is an indication of the practical importance of the problem: while the Poisson model is appealing in its simplicity and sometimes very useful as a basis for analysis, it is not always a good approximation for count data. Recall that this point was noted in the discussion of the *Federalist Papers* word counts given in Sec. 3.1.2.

Another important characterization of the Poisson distribution is the following. Generally, if x and y are two different random variables, each having a known distribution of the same form, the distribution of the sum $x + y$ has a different form that generally cannot be written down as a simple expression in terms of the individual distributional parameters for x and y. The Poisson distribution is unusual in this respect, exhibiting the following *countable additivity property* [53, p. 4]. Assume x_i is an integer-valued random variable exhibiting a Poisson distribution with parameter λ_i, which we will denote $x_i \sim P(\lambda_i)$ for convenience. It follows that the sum (here assumed infinite, but the property holds for finite sums as well) has a Poisson distribution with parameter λ equal to the sum of the individual λ_i values, provided this sum is finite; i.e.,:

$$
\sum_{i=1}^{\infty} x_i \sim P\left(\sum_{i=1}^{\infty} \lambda_i\right), \quad \text{provided } \sum_{i=1}^{\infty} \lambda_i < \infty.
\tag{3.35}
$$

3.6.2 The negative binomial distribution

A discrete distribution that is often used as an alternative to the Poisson distribution for describing count data is the *negative binomial distribution*, defined as [164, ch. 5]:

$$
p_i = \binom{k + i - 1}{k - 1} \left(\frac{P}{1+P}\right)^i \left(1 - \frac{P}{1+P}\right)^k,
\tag{3.36}
$$

for $i = 0, 1, 2, \ldots$. The positive constants P and k define the characteristics of the distribution; in particular, the first four moments are given by [164, p. 207]:

$$
\begin{aligned}
m_1 &= kP, \\
\mu_2 &= kP(1 + P), \\
\mu_3 &= kP(1 + P)(1 + 2P), \\
\mu_4 &= 3k^2 P^2 (1 + P)^2 + kP(1 + P)(1 + 6P + 6P^2). \tag{3.37}
\end{aligned}
$$

The key point here is that, in contrast to the Poisson distribution, where the mean m_1 and the variance μ_2 are equal, for the negative binomial distribution, the variance is strictly larger than the mean, by an amount that depends on the distribution parameter P. This is a consequence of the fact that the negative binomial distribution is defined by two free parameters (i.e., P and k) vs. the single parameter λ defining the Poisson distribution. As a practical matter, since $P > 0$, this result means that the negative binomial distribution is useful in situations where overdispersion (relative to the Poisson model) arises, but not in situations involving underdispersion. Since overdispersion is generally regarded as a more serious problem in practice than underdispersion is [53, p. 79], the negative binomial distribution is a popular alternative to the Poisson distribution. This idea is discussed further in Chapter 13 in connection with regression models for count data.

3.6.3 Zero-inflated count models

The class of *finite mixture distributions* is introduced in Chapter 10, along with some of the mathematical machinery behind this extremely useful notion. One useful special case—introduced here and revisited in Chapter 10—is the class of *zero-inflated count models*, which can be used to describe one particular mechanism for overdispersion that arises frequently in practice [164, pp. 312–318]. The basic idea is that zero counts occur more frequently than expected under a standard count model (e.g., the Poisson model), which otherwise describes the data adequately. Zero-inflated models postulate *two* data generation mechanisms: one that generates only zeros and a second mechanism that corresponds to a standard count distribution. The additional parameter required to specify this model is the probability α that a given count comes from the zero-only mechanism, implying that observations come from the standard count distribution with probability $1 - \alpha$. Again, a precise formulation of this model is deferred until Chapter 10, where it is discussed further and useful characterizations are presented. The key point here is that the fact that the observed count data contains more zeros than expected under the standard count distribution means that the observed counts are overdispersed relative to that count model. Other related notions are discussed by Cameron and Trivedi [53, sec. 4.7.3].

3.7 The Zipf distribution

On May 21, 2007, the *New York Times* included a short note by Eric Pfanner titled "Zipf's Law, or the Considerable Value of Being Top Dog, as Applied to Branding." The article notes that:

> In the 1930's, the Harvard linguist George Kingsley Zipf found that "the" — the most-used English word — ocurred about twice as often as "of" (second place), about three times as often as "and" (third) and so on.

This observation defines what has come to be known as the *Zipf distribution*, and the article goes on to note, first, that the Zipf distribution appears to apply to consumer brand preferences, providing the basis for attaching monetary values to competitive rankings. This idea is discussed in more detail in the report by Riemer *et al.* [244], which shows that market share data from 70 markets conform reasonably well to a Zipf distribution. As specific examples illustrating the consequences of this result, the authors cite the following:

1. the top 25 retailers of consumer packaged goods sell 53% of the volume generated by the largest manufacturers (e.g., Kraft and General Mills), while the rest is sold through thousands of smaller retailers;

2. the top two personal computer manufacturers (HP and Dell) had 30.4% of the total market share in second quarter 2002, IBM ranked third with 6.6%, and the rest of the total was "divided among many different firms."

Johnson *et al.* [164] note that the Zipf distribution is known in linguistics as the "Zipf-Estoup law," after both Zipf and the French linguist Estoup, who observed in 1916 that word frequencies seemed to decay inversely with ranking in a long document. In fact, the Zipf distribution has been invoked as an approximate distribution for many phenomena, and this provides one motivation for introducing it here. A second motivation is more subtle and is intimately related to some of the unexpected behavior seen in these settings: commonly used Zipf distributions usually have infinite means and variances. Examples and consequences of this mathematical pathology are examined in the following discussions.

3.7.1 Definition and properties

The Zipf distribution describes an infinite sequence of objects, $i = 1, 2, \ldots$, whose relative frequency of occurrence (i.e., probability) p_i is inversely proportional to some power of their rank i. That is, the Zipf distribution has the form:

$$p_i = \frac{A}{i^\alpha}, \quad i = 1, 2, \ldots, \tag{3.38}$$

where A and α are real numbers. The normalization requirement that the probabilities sum to 1 determines A and implies that $\alpha > 1$ in order for the sum to converge. Conversely, note that the mean of the Zipf distribution is given by:

$$m_1 = \sum_{i=1}^{\infty} \frac{Ai}{i^\alpha} = \sum_{i=1}^{\infty} \frac{A}{i^{\alpha-1}}. \tag{3.39}$$

It follows as an immediate consequence that the mean m_1 is *infinite* unless $\alpha > 2$. Similarly, the variance of the distribution is derived from the *second moment* m_2, which is given by:

$$m_2 = \sum_{i=1}^{\infty} \frac{A}{i^{\alpha-2}}, \tag{3.40}$$

which is only finite if $\alpha > 3$. As the discussion in Sec. 3.7.2 illustrates, the values of α encountered in applications are rarely this large. As a consequence, the resulting Zipf distributions generally have infinite variances and frequently also have infinite means. As a practical matter, this means that moment characterizations like these are of little use in situations where Zipf distributions arise. It is important to emphasize that this situation is highly unusual, and it means that these situations are likely to exhibit a range of counterintuitive behavior.

A variation on the Zipf distribution with similar behavior is the two-parameter distribution proposed by Mandelbrot [197, p. 344]:

$$p_i = \frac{A}{(1 + Bi)^\alpha}, \tag{3.41}$$

where again A is a constant determined by the normalization condition, $\alpha > 1$ is necessary for convergence of the normalization sum, and B is another positive parameter that gives the distribution added flexibility relative to the Zipf distribution. It is not difficult to see that the same general behavior holds: Mandelbrot's distribution has finite mean m_1 only if $\alpha > 2$ and finite variance only if $\alpha > 3$.

3.7.2   Examples and consequences

As noted, the Zipf distribution has been proposed to describe a very wide variety of phenomena, ranging from linguistics to marketing. At the time of this writing, a reasonably extensive list of references related to the Zipf distribution is available at the following Web site:

http://linkage.rockefeller.edu/wli/zipf/

Included in this list are papers discussing applications of the Zipf distribution to income distributions, file size distributions, linguistics (including French, English, Chinese, speech, children's speech in German, and random text generated by monkeys), population distributions, Internet traffic, scientific productivity, scientific popularity, size distributions of businesses, debt distributions for bankrupt companies, size distributions in ecology, earthquake intensity distributions, RNA secondary structure distributions in biology, and the distribution of family names.

In many of these applications, the exponent α is approximately or exactly equal to 1, while in others it is on the order of 2, but it is rarely reported to be much larger than this, suggesting that the distributions considered are essentially always infinite variance distributions and most generally infinite mean distributions. As a consequence, a characteristic feature of the Zipf distribution is its lack of a useful "typical size" characterization. As Adamic and Huberman note [5]:

> Unlike the more familiar Gaussian distribution, a power law distribution has no "typical" scale and is hence frequently called "scale-free." A power law also gives a finite probability to very large elements, whereas the exponential tail in a Gaussian distribution makes elements much larger than the mean extremely unlikely. For example, city sizes, which are governed by a power law distribution, include a few mega cities that are orders of mangitude larger than the mean city size. On the other hand, a Gaussian, which describes for example the distribution of heights in humans, does not allow for a person who is several times taller than the average.

The authors' primary interest is in characterizations of the Internet, and they observe that the Zipf distribution is widely applicable there, noting that "there are many small elements contained within the Web, but few large ones." Plots in their paper suggest, for a given website, the Zipf distribution is applicable to the number of pages, visitors, out links, and in links. In addition, the authors argue that Internet service providers have benefitted greatly from the applicability of the Zipf distribution, noting that one of its consequences is the existence of a very small number of extremely popular sites and a low popularity of the majority of the enormous number of sites that exist. Thus, it is practical to cache a few very popular sites in order to achieve rapid response to the majority of requests.

This last example illustrates a practically important aspect of analyzing data that is approximately described by a Zipf distribution. Since, as noted, these datasets are "scale-free" and cannot be characterized by a single "typical value," it is important to characterize the "high-probability portion" of the distribution separately from the "low-probability portion." For example, suppose we are characterizing a large collection of cities, which exhibit an approximatly Zipfian size distribution. If we are attempting to characterize residents of these cities, sampling from the total pool of residents will yield a sample that is overwhelmingly represented by those from the few largest cities. In contrast, sampling from the list of cities and *then* selecting a resident from that city will yield a sample that is overwhelmingly represented by those from the enormous number of small cities in the sample. Since neither of these samples are "fully representative," it is important to recognize the distinction and characterize *both* sets of residents. In general, we can expect to find the characterizations of these two samples to be quite different. A specific application of this notion is the separate characterization of "mice" (very short transmissions) and "elephants" (very long transmissions) in Internet traffic described by Marron *et al.* [201].

3.8 Exercises

1. Show that the first moment of the binomial distribution is $m_1 = np$. [Hint—the following result may be useful:

$$k \binom{n}{k} = \frac{kn!}{k!(n-k)!} = \frac{n(n-1)!}{(k-1)!([n-1]-[k-1])!} = n \binom{n-1}{k-1}.$$

It is easily derived from the definition of the binomial coefficient and the properties of factorials.]

2. Show that the Poisson distribution satisfies the normalization requirement that $\sum_{k=0}^{\infty} p_k = 1$.

3. Derive the moments m_1 and μ_2 for the Poisson distribution.

4. Derive the mean and variance for the discrete uniform distribution.

5. The standard estimator for the variance σ_x^2 of a sequence $\{x_k\}$ of N real numbers is given in Chapter 1 and discussed further in Chapter 6. It is:

$$\hat{\sigma}^2 = \frac{1}{N-1} \sum_{k=1}^{N} (x_k - \bar{x})^2,$$

where $\bar{x}$ is the mean of the sequence. Show that the Simpson heterogeneity measure for a normalized sequence $\{p_i\}$ of M probabilities is exactly equal to $\hat{\sigma}_p{}^2$.

Chapter 4

Uncertainty in Real Variables

Sections and topics:

Chapter 3 introduced the basic notions of probability in the context of discrete events and discrete random variable models like the binomial distribution. This chapter extends these ideas to the class of continuous random variables, a concept that will play a central role throughout this book. In particular, the continuous random variable model is the most popular way of describing measurement uncertainty and other natural sources of irregular variation for real-valued variables in a dataset. Without question, the most popular continuous random variable model is that based on the Gaussian distribution, which is important for a number of both theoretical and practical reasons. Consequently, the Gaussian distribution is introduced here and a number of its properties are discussed. Conversely, one of the key points of this chapter is that, despite its popularity, the Gaussian distribution is only one of an infinite variety of possible continuous random variable models, and it is not always the most appropriate one. For this reason, a number of alternative distributions are also introduced, distributions that arise naturally in a number of applications throughout this book. In addition, this chapter also introduces moment characterizations and several other important techniques for describing and working with random variables.

Key points of this chapter:

1. This chapter introduces the basic machinery of the continuous random variable model, including *cumulative distribution functions, probability densities, expected values,* and *moments*.

2. The continuous random variable model is introduced here because it forms a useful basis for analyzing real-valued data, although this requires us to specify a particular distribution to describe our data. Historically, the most popular choice has been the *normal* or *Gaussian distribution*. While this choice remains very useful in practice, it is not always appropriate, so one of the main points of this chapter is to discuss a number of alternative distributions.

3. Three specific characterizations of continuous random variables are the *mean*, which exists for most but not all of the distributions considered in this book; the *median*, which exists for all distributions; and the *mode*, which exists for the important class of unimodal distributions. All three of these characterizations are *location parameters* that characterize the "typical value" of a random variable x.

4. To describe the range of variation of a random variable about this "typical value," the most popular *scale parameter* is the *standard deviation*. Like the mean, this characterization does not always exist, but when it does, it follows from the *Chebyshev inequality* that the probability of observing a value of x more than c times the standard deviation from the mean decays monotonically with c. In particular, for *any* finite variance distribution, the probability of observing a value more than 10 standard deviations from the mean is less than 1%. Because this bound is so broadly applicable, it is generally extremely conservative: for the popular Gaussian distribution, the probability of observing a value more than 3 standard deviations from the mean is approximately 0.3%.

5. Historically, much use has been made of the first four moments of a distribution, both in selecting the form of the distribution to use in describing a data sequence and in fitting the parameters of the distribution that best match their observed values. The standardized forms of these moments are the *mean*, along with the *standard deviation*, the *skewness*, and the *kurtosis*, and they are commonly interpreted as measures of "typical value" or "central tendency," of *scale* or *variability* about the mean, of *asymmetry* of the distribution, and of *tail behavior*, respectively. These characterizations are useful within well-defined families of distributions, but these informal interpretations have their limitations, and just as it was noted in Chapter 1 that one cannot hear the shape of a drum, the first four moments are *not* fully adequate to specify a data distribution in the absence of other constraints.

4.1   Continuous random variables

The continuous random variable model is the most popular way of describing uncertainty in real-valued measurement data. An important difference between continuous random variables and the discrete random variables introduced in Chapter 3 is that most continuous random variables have zero probability of exhibiting any *specific* value, say, $x = x_0$ but nonzero probabilities of exhibiting values within finite ranges: $x_- \leq x \leq x_+$. This difference has important consequences, both theoretical and practical: theoretically, it means that the mathematical machinery required for describing and working with continuous random variables is fundamentally different from that required for discrete random variables; practically, it means that phenomena like *ties*, where two observations have exactly the same value, cannot occur by random chance in the continuous random variable model. As we will see in Chapter 7, this difference can have extremely important implications for the utility of some data characterizations. The following discussions introduce some of the essential mathematical machinery for working with continuous random variables, covering those ideas that will come up in subsequent chapters in connection with specific data analysis problems. For a more thorough introduction to the theory of continuous random variables, refer to books like the one by Billingsley [37].

4.1.1   Distributions and densities

In this book, the following simple definition is adopted for a continuous random variable: x is a continuous random variable if it is described by a *cumulative distribution function* (CDF), usually denoted $F(x)$. This function assigns probabilities to all possible comparison events of the form $x \leq z$; because these comparisons may be viewed as discrete events, their associated probabilities may be described using the machinery of discrete event probability introduced in Chapter 3. More specifically, the cumulative distribution function $F(z)$ is defined for all real z as the probability of the event $x \leq z$:

$$F(z) = \mathcal{P}\{x \leq z\}. \tag{4.1}$$

A number of useful properties for the function $F(\cdot)$ follow immediately from this definition. For example, since discrete event probabilities must lie between 0 (for impossible events) and 1 (for certain events), it follows that $F(\cdot)$ is bounded: $0 \leq F(z) \leq 1$ for all real z. In addition, suppose $z_1 \leq z_2$, from which it follows that $x \leq z_1$ implies $x \leq z_2$; hence, the probability of the event $x \leq z_2$ must be at least as great as the probability of the event $x \leq z_1$, meaning that $F(\cdot)$ is also an increasing function: $z_1 \leq z_2$ implies $F(z_1) \leq F(z_2)$. In fact, any right-continuous, increasing function with at most a countable number of discontinuities that approaches 0 as $x \to -\infty$ and approaches 1 as $x \to \infty$ defines a continuous random variable [37, p. 261]. As a practical consequence, continuous random variable models can exhibit an enormous variety of forms, as subsequent examples illustrate.

Figure 4.1: Weibull cumulative distribution functions

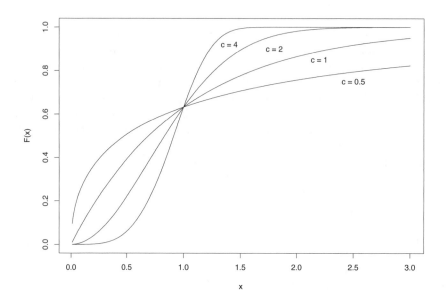

Four cumulative distribution functions are shown in Fig. 4.1, corresponding to special cases of the *Weibull distribution*, originally proposed to describe
breaking strength in material samples by the Swedish physicist Weibull in 1939,
and subsequently applied to the analysis of many different phenomena [162, ch.
21]. A contributing factor to the popularity of this random variable model is
that its CDF has the simple explicit form:

$$F(x) = \begin{cases} 1 - e^{-x^c} & x > 0 \\ 0 & x \le 0, \end{cases} \tag{4.2}$$

where $c > 0$ is a *shape parameter* that determines the qualitative character of
$F(x)$. In particular, note that for $0 < c \le 1$ (e.g., for $c = 0.5$ and 1.0 in Fig. 4.1),
this distribution function is concave, exhibiting a positive first derivative (as it
must to be an increasing function) but a negative second derivative. In contrast,
for $c > 1$ (e.g., for $c = 2$ and $c = 4$ in Fig. 4.1), the function initially exhibits
a *positive* second derivative, which ultimately becomes negative for sufficiently
large x. Important special cases of this distribution include the exponential
distribution ($c = 1$) discussed in Sec. 4.5.3 and the Rayleigh distribution ($c = 2$)
discussed in Sec. 4.5.9. Also, a more detailed discussion of the general Weibull
distribution is given in Sec. 4.5.10.

If we impose the stronger condition on $F(\cdot)$ that it be *absolutely continuous* [37, p. 434], it follows that $F(x)$ may be represented as the integral of a

probability density function p(x). Specifically, $F(x)$ and $p(x)$ are related by:

$$F(x) = \int_{-\infty}^{x} p(z)dz \Leftrightarrow p(x) = \frac{dF(x)}{dx}. \tag{4.3}$$

Although there are continuous distributions for which densities *do not* exist [43, pp. 109–117], most of the continuous random variable models in common use do exhibit densities and, in fact, densities that have a fairly simple form. More specifically, this observation holds for all of the continuous random variable models considered in this book, with the sole exception of the class of *stable random variables* introduced in Chapter 6: there, densities are known to exist, but simple expressions are only known for these densities for three special cases: the Gaussian distribution discussed in Sec. 4.2.1, the Cauchy distribution discussed in Sec. 4.5.2, and the Levy distribution discussed in Chapter 6.

For the Weibull distribution, the corresponding density function is given by:

$$p(x) = cx^{c-1}e^{-x^c}. \tag{4.4}$$

Plots of this density function for the same four values of c considered previously are shown in Fig. 4.2. For $0 < c \leq 1$, this density is *J-shaped*, meaning that it decreases from its maximum value at $x = 0$ toward zero as $x \to \infty$. Conversely, for $c > 1$, the density is *unimodal*, exhibiting a unique maximum for some unique, finite $x^* > 0$. More generally, as c increases, the density becomes more sharply peaked around this mode, which occurs at approximately $x = 1$. This general behavior is discussed further in Sec. 4.1.2.

An important consequence of the fact that $F(x) \to 1$ as $x \to \infty$ is that any *proper density p(x)* must satisfy the following normalization condition:

$$\int_{-\infty}^{\infty} p(x)dx = 1. \tag{4.5}$$

Further, $p(x)$ must be nonnegative for all x, a result that follows from the following observation. Note that the probability of the event $a \leq x \leq b$ is given by the integral:

$$\mathcal{P}\{a \leq x \leq b\} = \mathcal{P}\{x \leq b\} - \mathcal{P}\{x \leq a\} = F(b) - F(a) = \int_{a}^{b} p(x)dx. \tag{4.6}$$

Now, suppose the density function $p(x)$ were negative on any finite interval $[a, b]$; it would then follow that the probability of observing x in this interval would be negative, violating the basic axioms of discrete probability discussed in Chapter 3. Hence, for any valid probability density function, $p(x) \geq 0$ for all x.

Finally, note that valid probability density functions need not be bounded. As a specific example, note that for $0 < c < 1$, the Weibull density defined in Eq. (4.4) diverges to $+\infty$ as $x \to 0$. Despite this singularity, the integral of $p(x)$ on any interval $[0, a]$ is finite, which is all that is required for $p(x)$ to be a valid density function.

Figure 4.2: Weibull probability density functions

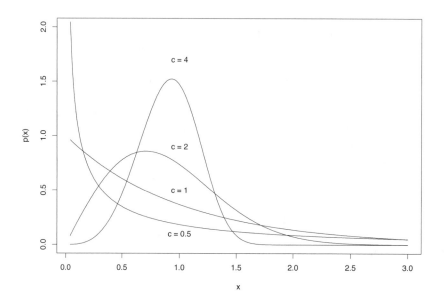

4.1.2   Location parameters: mean, median, and mode

Sec. 4.1.3 introduces the general notions of expected values and moments for continuous random variables, extending the notions introduced in Chapter 3 for discrete random variables. First, however, it is instructive to briefly consider three important characterizations that are often applied to distributions: the mean value $\bar{x}$, the median value $x^{\dagger}$, and the mode x^*. In fact, only the second of these quantities is guaranteed to exist for all continuous random variables, but there are extremely important cases (e.g., the popular Gaussian distribution) where they all exist and are equal. More generally, these three values may all be viewed as attempts to answer the question, "what is the *typical* value of x?" and are therefore sometimes called *location parameters*. Also, when these three values differ, these differences can tell us something useful about the shape of the distribution.

The *mean value* $\bar{x}$ of any distribution is the *first moment* of the density, often denoted m_1 and given by:

$$\bar{x} = m_1 = \int_{-\infty}^{\infty} xp(x)dx. \tag{4.7}$$

A density $p(x)$ is said to be *symmetric about* x_0 if it satisfies $p(x-x_0) = p(x_0-x)$ for all real x; for such a density, it follows on substituting $x = x_0+z$ into Eq. (4.7) that $\bar{x} = x_0$ (Exercise 2). The mean value exists for most popular distributions, but it need not, as in the case of the Pareto distribution discussed in Sec. 4.5.8; specifically, the Pareto distribution is characterized by a shape parameter $a > 0$, and the mean does not exist if $a \leq 1$. The *median value* $x^{\dagger}$ of any continuous distribution is the value of x for which the cumulative distribution function is equal to $1/2$:

$$F(x^{\dagger}) = \frac{1}{2} \Rightarrow \int_{-\infty}^{x^{\dagger}} p(x)dx = \frac{1}{2}. \tag{4.8}$$

Again, for a density that is symmetric about x_0, it is not difficult to show that $x^{\dagger} = x_0$ (Exercise 3). Further, if the cumulative distribution function $F(x)$ is continuous and strictly monotone, it is invertible [174, p. 181], and the median value is given explicitly by $x^{\dagger} = F^{-1}(1/2)$; even in cases where these conditions are not met, it is possible to define $x^{\dagger}$ in a reasonable way, a point discussed further in Chapter 12. Finally, recall from the previous discussion of the Weibull density example that $p(x)$ is unimodal if it exhibits a unique global maximum for some x^{*}, called the *mode* of the distribution. As with the mean and the median, if $p(x)$ is both unimodal and symmetric about x_0, it is not difficult to show that $x^{*} = x_0$ (Exercise 4), implying that all three of these location parameters are equal for these distributions. Conversely, important distributions exist for which x^{*} either does not exist (e.g., the uniform distribution) or the mode is not unique, as in the *arc-sine distribution* discussed in Sec. 4.5.1.

For the Weibull distribution, the mean value $\bar{x}$ is given by [162, p. 632]:

$$\bar{x} = \Gamma\left(\frac{1}{c} + 1\right) = \int_0^{\infty} t^{1/c} e^{-t} dt, \tag{4.9}$$

a quantity that remains well defined for all $c > 0$. In fact, $\bar{x}$ is a decreasing function of c, rapidly declining from an infinite limiting value at $c = 0$: for $c = 0.1$, $\bar{x} = 10! \simeq 3.63 \times 10^6$ vs. $\bar{x} = 2$ for $c = 0.5$ and $\bar{x} = 1$ for $c = 1$; further, $\bar{x}$ exhibits a minimum value of approximately 0.89 when $c \simeq 2.2$, and ultimately increases toward the limiting value 1 as $c \to \infty$. As noted previously, the Weibull distribution is J-shaped for $0 < c \leq 1$, implying that the mode x^{*} occurs at $x^{*} = 0$. For $c > 1$, the distribution is unimodal and x^{*} may be determined by setting the derivative of $p(x)$ to zero:

$$\frac{d}{dx}\left[cx^{c-1}c^{-x^c}\right] = 0 \Rightarrow x^{*} = \left[1 - \frac{1}{c}\right]^{1/c}. \tag{4.10}$$

Finally, because the cumulative distribution function has a simple form, it is also possible to obtain an explicit representation for the median $x^{\dagger}$ of the Weibull distribution:

$$F(x^{\dagger}) = 1 - e^{-(x^{\dagger})^c} = \frac{1}{2} \Rightarrow x^{\dagger} = [\ln 2]^{1/c}. \tag{4.11}$$

For the special case of the Rayleigh distribution, corresponding to $c = 2$ and discussed further in Sec. 4.5.9, the three location parameters are:

$$x^{*} \simeq 0.7071, \ x^{\dagger} \simeq 0.8326, \ \bar{x} \simeq 0.8862.$$

Figure 4.3: Mean, median, and mode for the Weibull density, $c = 2$

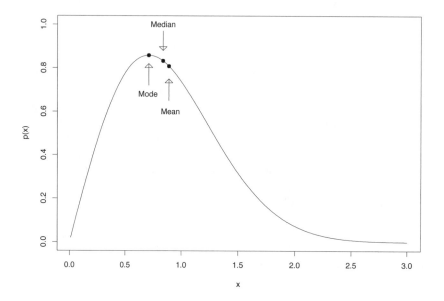

Fig. 4.3 shows where these points fall on the Weibull density function for this example, illustrating the important point that the differences between these location parameters provide information about distributional asymmetry, an idea discussed further in Chapter 6.

4.1.3   Expected values and moments

Recall that the expected value of a function $f(k)$ of the discrete random variable k was defined as the sum of the possible values for $f(k)$, each weighted by the probability p_k of observing that value. For a continuous-valued random variable x, the corresponding definition replaces the discrete probabilties p_k with the density function $p(x)$ and replaces the sum with an integral to obtain:

$$E\{f(x)\} = \int_{-\infty}^{\infty} p(x)\, f(x)\, dx. \tag{4.12}$$

Again, as in the discrete case, the moments m_n are defined as the expectations of the functions $f(x) = x^n$:

$$m_n = E\{x^n\} = \int_{-\infty}^{+\infty} x^n\, p(x)\, dx, \tag{4.13}$$

and the central moments μ_n are defined as:

$$\mu_n = E\{(x - m_1)^n\} = \int_{-\infty}^{+\infty} (x - m_1)^n \, p(x) \, dx. \qquad (4.14)$$

Using the binomial expansion discussed in Chapter 3, we obtain the following relationship between the moments m_n and the central moments μ_n:

$$\mu_n = E \left\{ \sum_{j=0}^{n} \binom{r}{j} x^j (-m_1)^{n-j} \right\} = \sum_{j=0}^{n} \binom{r}{j} m_j (-m_1)^{n-j}. \qquad (4.15)$$

In deriving this result, use has been made of the fact that $E\{c\} = c$ for any constant c (specifically, any quantity c that does not depend on x) and the fact that the expectation operator $E\{\cdot\}$ is *linear*:

$$E\{aX + bY\} = aE\{X\} + bE\{Y\}, \qquad (4.16)$$

where a and b are any constants and X and Y are quantities that depend on x. This result is extremely useful and is easily established from the definition of expected values given in Eq. (4.12). Another useful observation is that expectations are *monotonic*:

$$f(x) \leq g(x) \text{ for all } x \Rightarrow E\{f(x)\} \leq E\{g(x)\}, \qquad (4.17)$$

for all probability density functions $p(x)$. This result follows from the fact that $E\{g(x) - f(x)\}$ may be expressed as the integral of a product of nonnegative functions, which is therefore nonnegative.

Again, it is important to emphasize that moments need not exist, a point illustrated in the discussion of the mean m_1 presented in Sec. 4.1.2. If $p(x)$ is symmetric about x_0, it has been noted that the first moment is $m_1 = x_0$ and it follows more generally that the central moments μ_n may be taken as zero for all odd n. Conversely, for even n, μ_n generally increases (often dramatically) with increasing n and there are important distributions for which μ_n only exists for n sufficiently small. A case in point is the Student's t-distribution discussed in Chapter 9 and indexed by an integer parameter ν called the *degrees of freedom* of the distribution; it is a standard result that m_n does not exist for even integers $n \geq \nu$ [163, p. 365]. Conversely, it is also important to note that both *negative moments* and *fractional order moments* can be defined and are sometimes useful, particularly in connection with the class of *stable random variables* introduced in Chapter 7 [220] and in connection with reciprocals and ratios, a topic discussed further in Chapter 12.

Finally, one more extremely useful expectation is the *characteristic function* $\phi(t)$, defined as the expectation of the complex exponential e^{itx}:

$$\phi(x) = E\{e^{itx}\} = \int_{-\infty}^{\infty} e^{itx} p(x) dx. \qquad (4.18)$$

Because $|e^{itx}| \leq 1$ for all real x and t, it follows from the Cauchy-Schwartz inequality that $|\phi(x)| \leq 1$ for all proper densities $p(x)$. In fact, one of the advantages of the characteristic function is that it always exists and it uniquely determines the probability distribution [37, p. 352]. Indeed, the class of stable random variables introduced in Chapter 6 is most easily defined in terms of its characteristic function: densities have been shown to exist and even be unimodal, but general expressions for these densities are not known [256, p. 574].

4.2 How are data values distributed?

The motivation for introducing the basics of probability theory presented here is the fact that the random variable model provides an extremely useful basis for analyzing data. Essentially, we model our uncertain data variable as a random variable and use the machinery of probability theory to build useful characterization and analysis tools. Of course, a crucial aspect of this development is the assumption of a particular probability distribution to use in modeling the observed data. Historically, many—arguably most—data characterization and analysis tools have been based on the assumption that the observed data sequence $\{x_k\}$ may be usefully modelled as a sequence of *independent, identically distributed (i.i.d)* random variables with a common Gaussian distribution. The meaning and consequences of this first assumption—that $\{x_k\}$ may be viewed as an i.i.d. sequence—will be discussed further in subsequent chapters, especially Chapters 10, 11, and 16. The points of the following two discussions are, first, to say exactly what the Gaussian distribution is and why it is so popular, and second, to emphasize that this assumption is *not* always reasonable, popular though it may be.

4.2.1 The normal (Gaussian) distribution

Without question, the most popular model for measurement uncertainty in real-valued variables is the *normal* or *Gaussian* error model, which postulates that the measurement errors in a sequence of repeated measurements are statistically independent random variables drawn from the Gaussian probability distribution. In fact, a contemporary of Poincare's once observed [194, p. 155]:

> Experimentalists tend to regard it as a mathematical result that data values obey a Gaussian distribution, whereas mathematicians tend to regard it as an experimental result.

As a practical matter, *assuming* that data values obey a Gaussian distribution can be a tremendous mathematical simplification, *but this assumption may or may not be reasonable in practice*, a point discussed further in Sec. 4.2.2 and subsequently throughout this book.

The Gaussian distribution is defined by the probability density function:

$$p(x) = \frac{1}{\sqrt{2\pi}} \exp\left\{ -\frac{1}{2}\left(\frac{x-\mu}{\sigma}\right)^2 \right\}, \tag{4.19}$$

Figure 4.4: The Gaussian density function

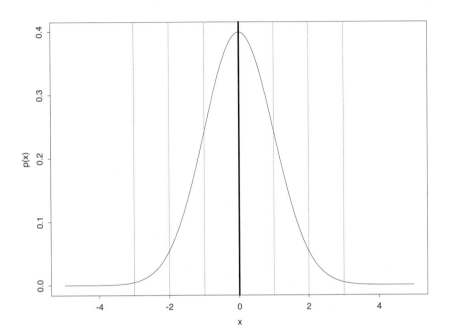

where the mean μ can assume any real value and $\sigma > 0$ is the standard deviation. This distribution is often designated $N(\mu, \sigma^2)$ and the *standard normal distribution* corresponds to the case $N(0, 1)$. A plot of this density function is shown in Fig. 4.4 for x between -5 and 5. The mean $\mu = 0$ is indicated by the heavy vertical line in the center of the plot, and vertical dashed lines indicate deviations of $\pm n$ standard deviations σ for $n = 1$, 2, and 3.

The Gaussian cumulative distribution function is often denoted $\Phi(x)$, and it may be related for $x \geq 0$ to the error function erf x [1, ch. 7] via the expression:

$$\Phi(x) = \begin{cases} \frac{1}{2} + \frac{1}{2}\text{erf}\, x, & x \geq 0 \\[2mm] \frac{1}{2} - \frac{1}{2}\text{erf}\, |x|, & x < 0. \end{cases} \tag{4.20}$$

Both the error function erf x and the Gaussian CDF $\Phi(x)$ are widely available, either in the form of tables [1], or as built-in functions in statistical software packages (e.g., the function **pnorm** in the R and *S-Plus* packages [272, 283]).

The Gaussian distribution is symmetric about the mean value μ, and its first four standardized moments are given by:

$$m_1 = \mu, \qquad \mu_2 = \sigma^2,$$
$$\gamma = 0, \qquad \kappa = 0, \tag{4.21}$$

where γ and κ are the *skewness* (normalized third moment) and *kurtosis* (normalized fourth moment) discussed in Sec. 4.3.2. More generally, all odd central moments of the Gaussian distribution are identically zero, while the even central moments are given by:

$$\mu_r \;=\; (r-1)!! \; \sigma^r, \tag{4.22}$$

where the double factorial $(r-1)!!$ is defined as [1, p. 258]:

$$(r-1)!! \;=\; (r-1)(r-3)\cdots 3\cdot 1. \tag{4.23}$$

Finally, the mean deviation for the Gaussian distribution is:

$$\nu_1 = E\{|x|\} = \sigma\sqrt{\frac{2}{\pi}} \simeq 0.7986\sigma. \tag{4.24}$$

The history of the Gaussian distribution goes back at least to 1733, when de Moivre published a pamphlet in Latin describing the Gaussian limit of the discrete binomial distribution. The name "Gaussian" reflects the influence of Carl Friedrich Gauss, who studied the distribution extensively in the early nineteenth century, publishing an early version of the Central Limit Theorem in 1816. This result has far-reaching significance and is considered in some detail in Chapter 6. Simply stated, this theorem gives broadly applicable sufficient conditions for sums of random variables to converge to a limiting Gaussian distribution. This result is one of the reasons the Gaussian distribution arises frequently in statistics. Not surprisingly, *many* results are available for the normal distribution, arguably the most widely studied of all probability distributions. This summary has given only a few of the most useful results; others are presented in connection with various problems discussed in other chapters. For more comprehensive discussions of the normal distribution, refer to the books by Johnson *et al.* [162] and Patel and Read [223].

4.2.2   Clancey's survey of data distributions

In 1947, Vernon J. Clancy published a paper in *Nature* based on his work in the chemical inspection department of the Ministry of Supply in London [64]. Altogether, he examined the results of about 50,000 chemical analyses, yielding 250 estimated distributions. Of these distributions, only about 10—15% could be reasonably approximated by the Gaussian model, with the rest exhibiting clear evidence of nonnormality. Clancy divided these cases into six categories, briefly described here.

Approximately 15% of the cases could be described adequately by a truncated normal distribution. This class of distributions is discussed in the extensive reference book on continuous univariate distributions by Johnson *et al.* [162, p. 156], and it is noted that the resulting distribution can exhibit many different forms, depending on where the truncation occurs. Many results are available for truncated Gaussian distributions, but they are generally rather complicated. For example, if the normal distribution $N(\mu, \sigma^2)$ is restricted to the range $A < x < B$, the mean of the trucated distribution is:

$$E\{x\} \;=\; \mu + \sigma \frac{p\left(\frac{\mu-A}{\sigma}\right) - p\left(\frac{B-\mu}{\sigma}\right)}{\Phi\left(\frac{B-\mu}{\sigma}\right) - \Phi\left(\frac{A-\mu}{\sigma}\right)}, \qquad (4.25)$$

where $p(x)$ is the Gaussian density function and $\Phi(x)$ is the cumulative distribution function.

About 10% of the distributions Clancy observed were symmetric and *heavy-tailed* or *leptokurtic,* a characterization discussed further in Sec. 4.3.2. To deal with these cases, Clancy proposed an alternative distribution, based on a Bessel function of the second kind, giving expressions for its first four moments. Another 10% of the distributions Clancy observed were classified as "truncated leptokurtic curves apparently derived from the above parent distribution." The details of this truncation were not specified, nor were detailed moment expressions given.

Between 20% and 25% of Clancy's distributions were *J-shaped,* exhibiting densities that decay monotonically from a maximum attained at the minimum possible value of x. For these cases, Clancy proposed the rather complicated density:

$$p(x) \;=\; \frac{1}{\sqrt{2\pi a x}} \left(\frac{\bar{x}}{x}\right)^{ax} \exp[a(x-\bar{x})], \quad x \ge \bar{x}. \qquad (4.26)$$

As simpler alternatives, note that several of the distributions discussed in Sec. 4.5 are J-shaped for the appropriate choices of parameter values, including the exponential distribution, the gamma distribution (when the shape parameter is less than 1), the Pareto distribution, and the Weibull distribution (with shape parameter less than 1). In addition, the beta distribution is J-shaped for the right choices of its two shape parameters.

Another 20—25% of the cases Clancy investigated were described as *skewed,* and he noted that transformations (presumably to normality) sometimes corrected this skewness. To deal with cases where transformation was inadequate, Clancy proposed another distribution, again based on the Bessel function of the second kind, but also involving square roots and exponentials. As in the case of Clancy's leptokurtic distributions, simple expressions are given for the first four moments of this distribution. Finally, Clancy notes that the remainder of the cases considered (less than 15% altogether) were "irregular," including a few bimodal examples that could be treated adequately as normal mixture densities (see Chapter 11 for a discussion of this class of non-Gaussian random variables).

In summarizing his work, Clancy offered three conclusions. The first was that "great caution must be exercised in applying the ordinary simple statistical methods to the data of chemical analysis," a point emphasized (not just for chemical analysis) throughout this book. Clancy's second conclusion was that "many of the distributions depart radically from the normal and are possibly more correctly to be regarded as forms of 'rare occurrence' distributions."

An introduction to these distributions (e.g., the Poisson distribution) was given in Chapter 3. Finally, Clancy's third conclusion was that "the continuous nature of the variable is, in many cases, probably superficial, and it should, more correctly, be regarded as a discontinuous variable." In fact, a similar view is expressed in the second paragraph of Chapter 12 of Johnson *et al.* [162], although they go on to argue that continuous distributions are far more convenient approximations in many cases and are therefore to be preferred, so long as the approximation is adequate. Physically, arguments in favor of an discrete distributions are generally based on *measurement quantization*, an important issue discussed in Chapter 2. Ultimately, the issue reduces to the question of what we are attempting to characterize: inherent variations in the (continuous) variable x we have attempted to measure, or the total variations in the quantized measurement of that variable? If this quantization is sufficiently coarse, it may be absolutely essential that we treat the observed data sequence as discrete. In particular, one of the inherent features of continuous random variable model is that any *specific* value x has zero probability of occurance, implying that no two observations in a sample drawn from a continuous distribution can have exactly the same value. In contrast, discrete distributions assign nonzero probabiblities to specific values of x, implying that such *ties* (i.e., exact equality between observations) *are* possible in the discrete case. The consequences of this difference are illustrated in Chapter 7, where it is shown that coarse quantization (resulting in a discrete distribution of values) can cause the otherwise very robust MADM scale estimator to fail completely.

4.3   Moment characterizations

Because they are easy to estimate, the moments introduced in Sec. 4.1.3 provide a simple basis for characterizing data sequences. This fact has led to a number of methods for both deciding what kinds of distributions are or are not appropriate as data models, and fitting these distributions to an observed data sequence. While these methods have their limitations, a point discussed further in Sec. 4.4, they are both historically important and practically useful.

4.3.1   The Markov and Chebyshev inequalities

It was noted previously that the expectation operator is both linear and monotonic. These observations may be used to prove the *Markov inequality* for nonnegative functions $f(x)$. Specifically, for any $a > 0$, it follows that:

$$\mathcal{P}\left\{f(x) \geq a\right\} \ \leq \ \frac{E\{f(x)\}}{a}. \tag{4.27}$$

For a simple proof of this result, see the discussion in Bremaud [43, pp. 92–93]. As an application of this inequality, consider the function $f(x) = |x - m_1|$, with m_1 denoting the mean of x, leading to the result:

$$\mathcal{P}\left\{|x - m_1| \geq c\right\} \ \leq \ \frac{E\{|x - m_1|\}}{c} \ \equiv \ \frac{\nu_1}{c}, \tag{4.28}$$

where ν_1 is the *mean deviation* of the distribution. To interpret this result, define $c' = c/\nu_1$ and apply Eq. (4.28) to obtain:

$$\mathcal{P}\left\{\left|\frac{x - m_1}{\nu_1}\right| \geq c'\right\} \leq \frac{1}{c'}. \qquad (4.29)$$

This expression provides an interpretation of the mean deviation ν_1 as a measure of scale: the probability of observing a value of x more than a few mean deviations from the mean value m_1 is necessarily small, provided the mean deviation is finite.

Chebyshev's inequality is an important corollary of the Markov inequality, obtained by taking $f(x) = (x - m_1)^2/\sigma^2$ in Eq. (4.27), where $\sigma^2 = \mu_2$ is the variance of x. Substituting this function into the Markov inequality yields:

$$\mathcal{P}\left\{\left(\frac{x - m_1}{\sigma}\right)^2 \geq c^2\right\} \leq \frac{1}{c^2}, \qquad (4.30)$$

since $E\{f(x)\} = 1$ for this case. Also, since $z^2 \geq c^2$ is equivalent to $|z| \geq c$ for any $c \geq 0$, it follows that:

$$\mathcal{P}\left\{z^2 \geq c^2\right\} = \mathcal{P}\left\{|z| \geq c\right\}, \qquad (4.31)$$

and as a consequence that Eq. (4.30) may be rewritten as:

$$\mathcal{P}\left\{\left|\frac{x - m_1}{\sigma}\right| \geq c\right\} \leq \frac{1}{c^2}. \qquad (4.32)$$

This result is quite similar to the one obtained for the mean deviation ν_1, and it leads to a similar interpretation: the probability of observing a value x more than a "few" standard deviations σ from the mean m_1 is "small."

Historically, much use has been made of the first four moments of the distribution, and the interpretation of these moments is considered briefly in the following subsections. It follows from the Chebyshev inequality that the first moment m_1 represents a reasonable choice for the approximate center of the distribution and the second central moment μ_2 gives an indication of how far from this central value we are likely to observe data values. Traditionally, the third and fourth moments have been taken as measures of asymmetry and (non-Gaussian) *elongation* or "tail behavior," respectively, although these interpretations cannot be taken too literally. More specifically, the traditional measures of asymmetry and elongation are the *skewness* γ and the *kurtosis* κ, defined as:

$$\gamma = \frac{E\{(x - m_1)^3\}}{\sigma^3}, \qquad \kappa = \frac{E\{(x - m_1)^4\}}{\sigma^4} - 3, \qquad (4.33)$$

where σ is the standard deviation of the random variable x. The mysterious "-3" appearing in the definition of kurtosis simply makes κ zero for the Gaussian distribution, consistent with its traditional interpretation as a measure of non-Gaussian behavior. Recent examples illustrating the use of these parameters in experimental characterization of electronic noise are described in the papers by Mantegna *et al.* [199] and [200].

Figure 4.5: Plot of a simple but pathological density function

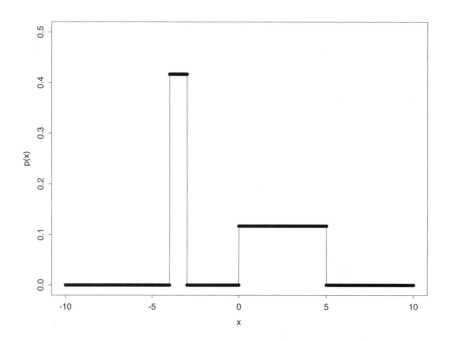

4.3.2   Skewness and kurtosis

As noted, γ is often viewed as a measure of distributional asymmetry, since it can be shown that if $p(x)$ is symmetric, then $\gamma = 0$. It is important to note, however, that the converse implication does *not* hold. As a specific example, consider the piecewise-constant density $p(x)$ shown in Fig. 4.5 and defined by:

$$p(x) = \begin{cases} 5/12 & -4 \leq x \leq -3, \\ 7/60 & 0 \leq x \leq 5, \\ 0 & \text{otherwise.} \end{cases} \tag{4.34}$$

All positive moments exist for this distribution and may be obtained by direct integration. In particular, it is not difficult to show that both the first and third moments of this distribution are zero. *Therefore, the skewness of this distribution is zero even though the density is highly asymmetric, as seen in Fig. 4.5.* In other words, $\gamma = 0$ is only a necessary condition for distributional symmetry, not a sufficient condition. A more standard example is provided by the asymmetric Weibull distribution: the skewness is equal to 2 when the shape parameter c is 1, decreases to zero for $c \simeq 3.6$, becomes slightly negative for larger c, but ultimately increases with increasing c.

Like skewness, the interpretation of kurtosis is also somewhat complicated, as discussions of kurtosis for mixture densities presented in Chapter 11 illustrate further. Within well-behaved families of distributions, it is *sometimes* reasonable to regard the kurtosis as a measure of the decay of the tail of the distribution. That is, distributions with positive kurtosis are called *leptokurtic* [40] and—at least for some symmetric, unimodal distributions—tend to decay more slowly as $|x| \to \infty$ with increasing κ. This interpretation is consistent with the following special case of the Markov inequality: consider the function $f(x) = (x - m_1)^4$, from which it follows that:

$$\mathcal{P} \left\{ \left| \frac{x - m_1}{\sigma} \right| \geq c \right\} \leq \frac{\kappa + 3}{c^4}. \tag{4.35}$$

Therefore, large values of κ would appear to be necessary for the probability of large deviations to be significant. At the other extreme, distributions with negative kurtosis are called *platykurtic* [40] and—again, within well-behaved families of distributions—tend to be associated with *light-tailed* behavior, meaning that the probability of observing large deviations is less than that for the Gaussian case. Unfortunately, these interpretations are not valid in general, as subsequent examples illustrate, especially for asymmetric distributions where the situation is rather complicated [24]. In particular, different kurtosis measures have been proposed and can yield different answers to the question "which of several distributions has heavier tails?" As a simple illustration of the difficulty with asymmetric distributions, note that the kurtosis for the symmetric Laplace distribution is $\kappa = 3$, whereas that for the asymmetric exponential distribution is $\kappa = 6$. This result is counterintuitive: since the Laplace and exponential distributions are identical on $[0, +\infty)$ (other than a factor of 2 for normalization) and the exponential distribution is identically zero on $(-\infty, 0)$, the Laplace distribution would seem to be "heavier tailed," despite its smaller kurtosis.

Despite the general difficulty of interpreting kurtosis for asymmetric distributions, certain useful results are available relating skewness and kurtosis. For example, Rohatgi and Szekely [248] show that for arbitrary distributions:

$$\kappa \geq \gamma^2 - 2, \tag{4.36}$$

and they note that this lower bound is achievable if and only if the density function $p(x)$ is concentrated on at most two points (i.e., x can assume at most two values). Further, note that the lowest possible value for this lower bound is $\kappa = -2$, achievable only for distributions for which $\gamma = 0$; in fact, it is not difficult to show that the only distribution that can achieve $\kappa = -2$ is the symmetric discrete distribution:

$$\mathcal{P}\{x\} = \begin{cases} 1/2 & x = \mu \pm \sigma, \\ 0 & \text{otherwise.} \end{cases} \tag{4.37}$$

Another particularly useful result given by Rohatgi and Szekely [248] is that *for unimodal distributions*:

$$\kappa \geq \gamma^2 - \frac{6}{5}. \tag{4.38}$$

Figure 4.6: A zero-kurtosis, bimodal density

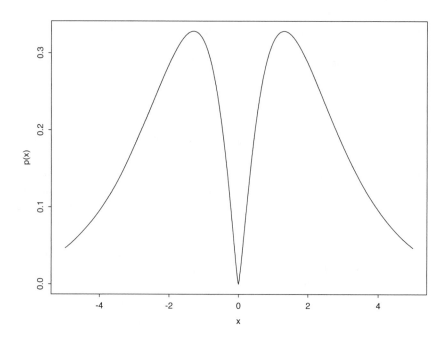

Again, specializing this result to symmetric distributions, note that this lower limit is achieved by the uniform distribution: $\kappa = -6/5$. Although this distribution is not unimodal, it does represent the limit of a symmetric, unimodal distribution on $[a, b]$ as the central peak broadens to cover the entire interval. These bounds define limits to how platykurtic a probability density function can be; in contrast, density functions can be arbitrarily leptokurtic since there is no general upper bound on κ.

Finally, it is important to note that, as in the case of the skewness results considered earlier, the *converse* implications seldom hold. For example, it follows from the above arguments that the kurtosis of a unimodal, symmetric density necessarily satisfies $\kappa \geq -6/5$. On the other hand, it *does not* follow that if $\kappa \geq -6/5$ and the density is symmetric, that it is necessarily unimodal. This point is illustrated in Fig. 4.6, which shows a *bimodal*, symmetric density for which $\kappa = 0$, discussed further in Sec. 4.4.2. This example is particularly interesting since $\kappa = 0$ for a Gaussian density, so we *might* be tempted to conclude that a density for which $\kappa = 0$ should be "near Gaussian," but the density function shown here is clearly not close to Gaussian in any useful sense. An even more disturbing example is provided by the *symmetrical Tukey lambda distribution*, discussed by Johnson *et al.* [162, p. 39] and defined as the following

Figure 4.7: Two zero-kurtosis Weibull densities

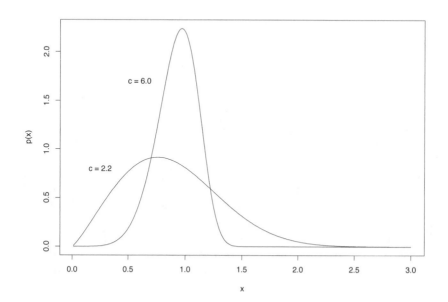

transformation of a uniformly distributed random variable U:

$$X = \frac{U^\lambda - (1 - U)^\lambda}{\lambda}. \tag{4.39}$$

This family of distributions achieves $\kappa = 0$ for *two different values of* λ, namely, $\lambda \simeq 0.135$ and $\lambda \simeq 5.20$, corresponding to two very different shaped distributions: the one with the smaller λ value looks approximately Gaussian, but the other one does not. Another similar example is provided by the Weibull distribution, where the kurtosis again vanishes for two values of the shape parameter: $c \simeq 2.2$ and $c \simeq 6.0$. Plots of these densities are shown in Fig. 4.7, from which it is clear that both of these distributions are very non-Gaussian and exhibit visually obvious asymmetries, further emphasizing the difficulty of interpreting kurtosis for asymmetric distributions.

4.3.3   The method of moments

In addition to the qualitative interpretations just discussed, the first four moments of a distribution have played an extremely important role historically in fitting distributions to observed data sequences. Specifically, easily computable estimators for the mean m_1 and the central moments μ_n are:

$$\bar{x}_N = \frac{1}{N}\sum_{k=1}^{N} x_k, \quad \text{and} \quad \hat{\mu}_n = \frac{1}{N}\sum_{k=1}^{N}[x_k - \bar{x}_N]^n. \tag{4.40}$$

Given the estimates $\hat{\mu}_2$, $\hat{\mu}_3$, and $\hat{\mu}_4$, the skewness and kurtosis may then be estimated as:

$$\hat{\gamma} = \frac{\hat{\mu}_3}{[\hat{\mu}_2]^{3/2}} \quad \text{and} \quad \hat{\kappa} = \frac{\hat{\mu}_4}{[\hat{\mu}_2]^2} - 3. \tag{4.41}$$

The *method of moments* is a technique introduced by Karl Pearson at the end of the nineteenth century for fitting data to members of a family of distributions he defined and studied. The basic method rests on two practically important facts: first, that the first four moments are sufficient to uniquely specify a member of this large family of distributions, and second, that these moments are easily computed from the available data. This basic approach has also been applied to several other families of distributions, including the Johnson family described in Sec. 4.3.5. Conversely, the method of moments also has its limitations, a point discussed further in Sec. 4.4. Still, the basic ideas are important for three reasons: their historical significance, their continuing use in some applications, and their relationship to more general data analysis problems like spectral estimation.

4.3.4 Karl Pearson's 1895 system of distributions

At the end of the nineteenth century, Karl Pearson introduced a family of probability distributions whose densities $p(x)$ were defined by the following differential equation [162, p. 16]:

$$\frac{1}{p}\frac{dp}{dx} = -\frac{a+x}{c_0 + c_1 x + c_2 x^2}. \tag{4.42}$$

This equation can be integrated to obtain an analytic solution, although the character of this solution depends strongly on the values of the coefficients a, c_0, c_1, and c_2. Imposing the conditions for a proper density function (i.e., nonnegativity and normalizability), these solutions may be divided into eight classes. The first of these classes is the Gaussian distribution, corresponding to the special case $c_1 = c_2 = 0$; this class is not numbered, as it represents a limiting case of all of the other classes. The remaining seven classes are generally referred to as type I through type VII, and the resulting distributions exhibit an extremely wide variety of shapes. Although it is well beyond the scope of this book to give a detailed discussion of this classification system, it does include a number of important families that are discussed here; for a more detailed account, see Johnson *et al.* [162, ch. 12]. The type I and type II distributions correspond to the asymmetric and symmetric cases, respectively, of the beta distribution, and the type III distributions correspond to the gamma densities, both discussed further in Sec. 4.5. Interestingly, although the type IV density function may be computed analytically, it is rather complicated, and it cannot be integrated explicitly to obtain a cumulative distribution function; probably because of its complexity, it does not correspond to any family of distributions that have come into common use. The type V distribution corresponds to a transformation of the form $f(x) = 1/x - a$ applied to a gamma distributed variable. A similar result holds for the type VI distribution, which is also known as the *beta distribution of the second kind* [163, p. 248] and may be obtained from the standard beta distribution discussed in Sec. 4.5.1 by the transformation $f(x) = x/(1 - x)$. This class of distributions includes both the F-distribution introduced in Chapter 9 and the Pareto distribution discussed in Sec. 4.5.8. Finally, the type VII distribution class consists of a family of symmetric distributions defined on the entire real line, and it includes the family of Student's t-distributions introduced in Chapter 9, of which the Cauchy distribution discussed in Sec. 4.5.2 is a member.

One advantage of the Pearson system of distributions is that the coefficients a, c_0, c_1, and c_2 may be related to the first four moments of the distribution. For any random variable z with finite mean μ and finite variance σ^2, the transformed variable $x = (z - \mu)/\sigma$ has zero mean and unit variance, so there is no loss of generality in restricting consideration to this case. There, the four coefficients in Eq. (4.42) may be related to the skewness γ and the kurtosis κ by the following expressions [162, p. 22]:

$$
\begin{aligned}
c_0 &= \frac{4\kappa - 3\gamma^2 + 12}{2(5\kappa - 6\gamma^2 + 6)}, \\
a = c_1 &= \frac{\gamma(\kappa + 6)}{2(5\kappa - 6\gamma^2 + 6)}, \\
c_2 &= \frac{2\kappa - 3\gamma^2}{2(5\kappa - 6\gamma^2 + 6)}.
\end{aligned}
\tag{4.43}
$$

One of the historically important consequences of this result is that it provides an extremely convenient and flexible basis for fitting probability distributions to data. Here, the key point is that the system of distributions defined by Eq. (4.42) includes many practically important special cases, and it is completely characterized by the first four moments. The broader question of how well an arbitrary density function $p(x)$ is characterized by its first four moments is an important one that is considered in Sec. 4.4. First, however, it is useful to discuss another family of distributions that can also be fit using the first four moments.

4.3.5 Johnson's system of distributions

Another useful family of distributions is that proposed by Johnson [160], based on three simple transformations of a standard normal (i.e., $N(0, 1)$) random variable Z. The first is the class of *lognormal* random variables, defined by:

$$
Y = \exp\{\sigma Z + \zeta\}.
\tag{4.44}
$$

This distribution is called "lognormal" because the logarithm of Y is normally distributed with mean ζ and standard deviation σ. Note that since Z is real-valued, the random variable Y is necessarily nonnegative. One advantage of this distribution is that it may be transformed to normality by the inverse transformation:

$$Z = \frac{\ln Y - \zeta}{\sigma}. \tag{4.45}$$

The idea of transformation to normality is applied frequently in practice, and it is discussed further in Chapter 12; here, it is enough to note that the idea is useful and is one of the underlying motivations for the Johnson system of distributions, since all of these distributions may be transformed *simply* to normality.

The second member of the Johnson family of distributions is the S_B family, obtained by the tranformation:

$$Y = \left[1 + e^{-(\sigma Z + \zeta)}\right]^{-1}. \tag{4.46}$$

The resulting random variable Y is distributed on the bounded interval $[0, 1]$, and it may be transformed to normality via the inverse transformation:

$$Z = \frac{1}{\sigma}\left[\ln\left(\frac{Y}{1 - Y}\right) - \zeta\right]. \tag{4.47}$$

Finally, the third member of the Johnson family of distributions is the class of S_U distributions, defined by the transformation:

$$Y = \sinh[\sigma Z + \zeta]. \tag{4.48}$$

This transformation maps the normally distributed variable $\sigma Z + \zeta$ into the real line, so Y may assume any real value. As before, the inverse transformation that takes Y to normality is simple:

$$Z = \frac{\sinh^{-1} Y - \zeta}{\sigma}. \tag{4.49}$$

Overall, the rationale for defining these three classes of distributions was to have a single family of distributions, based on simple transformations of normally distributed random variables, that was completely characterized by its first four moments. In fact, any pair of (γ, κ) values corresponding to a valid probability distribution corresponds to one distribution from the Johnson family. In addition, some interesting extensions of the Johnson family have been obtained by replacing the standard normal random variable Z with a random variable having the Laplace or logistic distribution discussed in Sec. 4.5. For a discussion of these families of random variables, see the references cited in Johnson *et al.* [162, p. 39].

4.4   Limitations of moment characterizations

As the preceeding discussions have illustrated, moment characterizations have played a historically signficiant role in both theoretical and applied statistics. Further, these characterizations and their extensions to correlation and spectral analysis continue to be extremely important in both theory and practice today. Conversely, it is important to emphasize that moment characterizations, though useful, have both fundamental and practical limitations. The following subsections discuss this point in some detail, starting with the notion of exact characterizations: if we are given all of the moments, when do these completely characterize the underlying distribution? Following that discussion, the question of approximate characterization is considered: given only the first four moments, how well can we distinguish different distributional families?

4.4.1   Exact characterizations

In general, the characterization of a distribution by a collection of its moments $\{m_n\}$ is incomplete, even if we have all positive integer moments. As a specific example, the lognormal distribution cannot be reconstructed from its moments: there exist other densities whose moments are identical [37, p. 407]. In fact, Stoyanov [275, p. 89] presents a detailed proof that if the lognormal density is multiplied by the function:

$$f(x) \quad = \quad 1 + \epsilon \sin(2\pi \ln x), \tag{4.50}$$

for any ϵ with $|\epsilon| < 1$, the resulting density exhibits the same moments as the lognormal density. The pronounced differences that exist between these densities are illustrated in Fig. 4.8, which shows two cases: $\epsilon = 0$ (i.e., the lognormal density) is shown as the solid line, and $\epsilon = 0.5$ is shown as the dotted line. Note that for $\epsilon \neq 0$, the resulting densities are generally multimodal, in contrast to the unimodal behavior of the lognormal density. Hence, it is clear that certain important qualitative information (e.g., the number of modes in the density function) is *not* generally contained in the moments.

Conversely, there are cases in which it *is* possible to completely reconstruct a density from its moments $\{m_n\}$; one sufficient condition is [63, p. 285]:

$$\limsup_{n \to \infty} \frac{m_{2n}^{1/2n}}{2n} \quad < \quad \infty. \tag{4.51}$$

This result may be used to establish that if $p(x)$ is any density taking nonzero values only on the finite interval $[a, b]$ for $a < b$, then $p(x)$ is determined by its moments. To establish this result, begin by rescaling $p(x)$ to the normalized density $\phi(z)$ defined on the unit interval $[0, 1]$. This rescaling is accomplished by the linear transformation $z = (x - a)/(b - a)$, and it follows that $p(x)$ is determined by its moments if and only if $\phi(z)$ is determined by its moments. For this normalized density, note that:

$$m_{2n} \quad = \quad \int_0^1 z^{2n}\phi(z)dz \leq \int_0^1 \phi(z)dz \; = \; 1, \tag{4.52}$$

Figure 4.8: Two distributions with identical moments

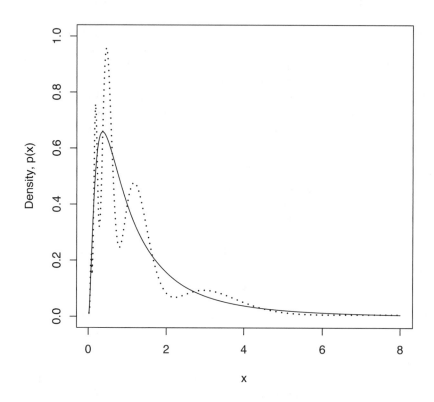

from which it follows that the limit in Eq. (4.51) is the finite value 0. *There-fore, any density function that is nonzero only on a finite interval $[a, b]$ can be reconstructed from its moments.*

4.4.2   Approximate characterizations

It is important to emphasize that the reconstructions just described require knowledge of *all* of the moments of the distribution, in general. Within specific families of distributions, it may be possible to reconstruct the density from a finite number of moments, but it is necessary to specify this family *a priori.* For example, within either the Pearson family of distributions described in Sec. 4.3.4 or the Johnson family of distributions described in Sec. 4.3.5, is possible to specify a density function uniquely from a consistent specification of the first four moments alone. Note, however, that the densities corresponding to a particular moment specification are, in general, *different* in these two families. In fact, an interesting question is *how different* the densities corresponding to the same four moments can be. Pearson *et al.* [224] consider precisely this question, comparing cumulative distribution functions whose first four moments agree, taken from the following eight families:

1. the Pearson family described in Sec. 4.3.4
2. the Johnson S_U and S_B families described in Sec. 4.3.5,
3. the Burr type XII distribution,
4. the lognormal distribution discussed in Sec. 4.3.5
5. the Weibull family discussed in Sec. 4.5.10,
6. the noncentral t-distributions,
7. the noncentral χ^2 distributions,
8. the log-χ^2 distribution.

The Burr type XII distribution is closely related to the Pareto distribution discussed in Sec. 4.5.8; in particular, if x has a Burr type XII distribution [162, p. 54] with parameters c and a, then $1 + x^c$ has the $P(I)(1, a)$ distribution discussed in Sec. 4.5.8. The noncentral t and χ^2 distributions are discussed by Johnson *et al.* [163], and the log-χ^2 distribution is obtained by replacing the unit normal distribution in the definition of the lognormal distribution with the χ^2 distribution discussed in Chapter 9.

The first four moments are sufficient to uniquely determine a member of each of these eight families, and Pearson *et al.* present detailed comparisons of the corresponding cumulative distribution functions. These comparisons are simplified by normalizing all distributions to have zero mean and unit variance, reducing the results to a comparison of the effects of the skewness γ and kurtosis κ. This summary consists of two figures and 20 tables, the most comprehensive of which have the following form. For specified values of γ (denoted as $\sqrt{\beta_1}$, another standard nomenclature) and κ (expressed as $\beta_2 = \kappa + 3$), 15 *quantiles* of the corresponding Pearson distribution are given. For a cumulative distribution function $F(x)$, the qth quantile is the value x_q such that:

$$F(x_q) = \text{Prob}\{x \le x_q\} = q. \tag{4.53}$$

The specific quantiles considered may be grouped qualitatively as:

L *lower tail:* $q = 0.0025, 0.005, 0.010, 0.025,$
M *middle:* $q = 0.05, 0.10, 0.25, 0.50, 0.75, 0.90, 0.95,$
U *upper tail:* $q = 0.975, 0.990, 0.995, 0.9975.$

Comparisons were summarized by tabulating the differences between the quantiles of the Pearson distribution and those of the other seven families.

In general, for the eight families of distributions compared, it was found that agreement was reasonable for the middle quantiles, becoming worse as we move into the tails. A typical result is summarized in Table 4.1, based on table I from Pearson *et al.* [224]. In this case, it was not possible to achieve exactly the same values of γ and κ for all five of the distributions considered, but the agreement is impressive even out into the upper and lower 1% points

Table 4.1: A comparison of five distributions

Quantile:			0.01	0.05	0.95	0.99
	γ	κ	x_q	x_q	x_q	x_q
Family						
Pearson VI	0.800	1.200	-1.80	-1.40	1.82	2.90
Johnson S_U	0.800	1.200	-1.80	-1.40	1.82	2.91
Noncentral t	0.780	1.229	-1.83	-1.42	1.81	2.89
Lognormal	0.814	1.200	-1.78	-1.40	1.83	2.91
Log-χ^2	0.780	1.188	-1.83	-1.41	1.81	2.90

($q = 0.99$ and $q = 0.01$, respectively). Conversely, the disagreement between distributions generally becomes worse as these families move away from their common Gaussian limit ($\gamma = 0$, $\kappa = 0$). In fact, the paper does not consider the U-shaped (i.e., bimodal) members of the Pearson family, corresponding to $\kappa < -6/5$, nor does it consider distributions where the difference $\kappa - \gamma^2$ is too large, noting that in these cases, "the difference between the shape of the Pearson and Johnson curves had become so large that comparisons ceased to be useful." More specifically, comparisons were restricted to the range $0.8 \leq \kappa - \gamma^2 \leq 8.0$. Finally, the paper also provided a comparison of the 5th through the 8th normalized central moments (μ_r/σ^r) for three distributions whose first four moments were identical (specifically, $\gamma = 2.00$, $\kappa = 7.8635$). These results are summarized in Table 4.2, and they illustrate both the troublesome magnitude of the higher moments and the rate at which these moments can diverge for "similar" distributions.

Perhaps a more dramatic illustration of the difference possible between distributions having the same first four moments is provided by the bimodal distribution shown in Fig. 4.6 and discussed previously. There, the first four moments exactly match those of the Gaussian distribution, but the general character of the density function shown in Fig. 4.6 is nothing like that of the Gaussian distribution. In fact, it is possible to construct entire families of distributions with zero skewness and kurtosis, similar in character to this example. Specifically, consider a density $p(x)$ of the form:

$$p(x) = \begin{cases} (1/2)\phi(-x - L) & x \leq -L, \\ 0 & -L < x < L, \\ (1/2)\phi(x - L) & x \geq L, \end{cases} \qquad (4.54)$$

where L is any nonnegative real number and $\phi(x)$ is any density function defined on the positive real line R^+. It is not difficult to show that this construction leads to a proper density (i.e., $p(x)$ is nonnegative for all x and integrates to 1 over the entire real line). It is also easy to see that the resulting density is

Table 4.2: Higher moments for three distributions

Moment r	Pearson VI μ_r/σ^r	Lognormal μ_r/σ^r	Burr μ_r/σ^r
5	71.84	69.96	75.12
6	705.2	638.9	844.9
7	10,209.3	7,859.9	18,089.4
8	235,007.3	129,791.6	2,500,459.8

symmetric about zero, but *not* unimodal for any $L > 0$: the number of modes of $p(x)$ is necessarily twice the number of modes of $\phi(x)$. It follows by symmetry that $E\{x\} = 0$ for any choice of $\phi(x)$, and $\mu_r = m_r = 0$ for all odd r. The even moments of this distribution depend on both the real number L and the moments of the density $\phi(x)$. For example, the variance of x is given by:

$$\mu_2 = f_2 + 2Lf_1 + L^2, \qquad (4.55)$$

where f_r is the rth moment of $\phi(x)$. The kurtosis is given by the following extremely messy expression:

$$\kappa = \frac{f_4 - 3f_2^2 + 4Lf_3 - 12Lf_1f_2 - 12L^2f_1^2 - 8L^3f_1 - 2L^4}{f_2^2 + 4Lf_1f_2 + 2L^2f_2 + 4L^3f_1 + 4L^2f_1^2 + L^4}. \qquad (4.56)$$

Two limiting cases yield useful insight; first, for $L = 0$, we have:

$$\kappa = \frac{f_4}{f_2^2} - 3. \qquad (4.57)$$

This expression strongly resembles the kurtosis expression for $\phi(x)$ but is *not* the same since f_2 and f_4 are moments about zero rather than central moments about the mean f_1. As a specific example, if $\phi(x)$ is the standard gamma density discussed in Sec. 4.5.4, it is characterized by a single shape parameter α and the kurtosis for $L = 0$ may be written as:

$$\kappa = -2\left[\frac{\alpha^2 - \alpha - 3}{\alpha(\alpha + 1)}\right]. \qquad (4.58)$$

For $\alpha = (1+\sqrt{13})/2 \simeq 2.30$, it follows that $\kappa = 0$, and the resulting density is the one plotted in Fig. 4.6. For α values between 0 (representing an infinite kurtosis limit) and this zero-kurtosis value, the kurtosis is positive, whereas for α larger than the zero-kurtosis value, the resulting kurtosis is negative, approaching the limit $\kappa = -2$ as $\alpha \to \infty$.

The second limiting case is $L \to \infty$, implying $\kappa \to -2$, its minimum possible value. In fact, this observation is useful because it illustrates one of the fundamental conflicts that obscures the interpretation of kurtosis. Specifically, recall that no symmetric unimodal distribution can exhibit a kurtosis κ more negative than $-6/5$. Consequently, we can expect that bimodal distributions will tend to exhibit strongly negative kurtosis values, whereas heavy-tailed distributions will tend to exhibit positive kurtosis values. In the example shown in Fig. 4.6, the gamma distribution $\phi(x)$ is somewhat heavy-tailed, with a kurtosis value of ~ 2.6, but the density $p(x)$ constructed from this distribution is bimodal; consequently, there is a competition between these two tendencies, and the overall kurtosis is exactly zero.

4.5   Some important distributions

Two examples of continuous random variables have been described in detail in this chapter: the Weibull distribution introduced initially because both its density and cumulative distribution function have simple functional forms, and the Gaussian distribution because of its enormous historical and continuing practical importance. Additional distributions that arise naturally in problems of hypothesis testing are described in Chapter 9, and the following subsections briefly summarize some other useful distributions that will appear in examples throughout the rest of this book.

4.5.1   The beta distribution

The standard beta distribution is defined on the unit interval $0 \leq x \leq 1$ by the probability density function:

$$p(x) \quad = \quad \left[\frac{\Gamma(p+q)}{\Gamma(p)\Gamma(q)} \right] x^{p-1}(1-x)^{q-1}, \tag{4.59}$$

where $\Gamma(x)$ is the gamma function, equal to the factorial $(x-1)!$ when x is a positive integer [1], and the shape parameters p and q may assume any positive values. This distribution is extremely flexible, as illustrated in figs. 4.9 and 4.10. The upper left plot in Fig. 4.9 shows the special case $p = q = 1/2$, called the *arc-sine distribution* because its cumulative distribution function is can be written explicitly as:

$$F(x) = \frac{2}{\pi} \sin^{-1} \sqrt{x}. \tag{4.60}$$

This distribution is *bimodal* or *U-shaped*, exhibiting maxima at both $x = 0$ and $x = 1$. In fact, the beta density always exhibits a maximum at $x = 0$ if $0 < p < 1$, and a maximum at $x = 1$ if $0 < q < 1$; this behavior is illustrated in the upper right and lower left plots in Fig. 4.9, which show the beta densities for $p = 1/2$, $q = 3/2$ and $p = 3/2$, $q = 1/2$, respectively. For $p = q = 1$, the beta distribution reduces to the *uniform distribution*, $p(x) = 1$ for all $0 \leq x \leq 1$, shown in the lower right plot in Fig. 4.9.

Figure 4.9: Four examples of the beta distribution

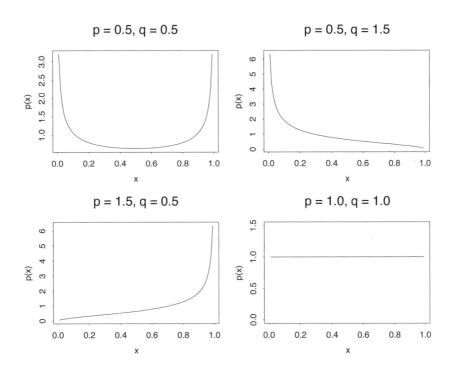

If $p > 1$ or $q > 1$, the beta density is *unimodal*, exhibiting a single peak; if $q > 1$ but $p < 1$, the density is *J-shaped* as in the upper right plot in Fig. 4.9, while if $p > 1$ but $q < 1$, the density is *reverse J-shaped* as in the lower left plots in Fig. 4.9. If $p > 1$ and $q > 1$, the mode of the distribution lies in the interior of the interval $[0, 1]$, as seen in all but the upper left density plots shown in Fig. 4.10. The mean of the beta distribution is related to the shape parameters p and q by the following simple expression:

$$m_1 = \frac{p}{p+q}, \tag{4.61}$$

from which it follows that the mean is $\bar{x} = 1/2$ if $p = q$. Further, the density is *symmetric* about this mean value in this case, as seen in the upper right plot in Fig. 4.10 for $p = q = 2$, and in the lower right for $p = q = 4$ where the result looks quite similar to the Gaussian density discussed in Sec. 4.2.1. This similarity illustrates an important feature of the symmetric beta distribution: it approaches a Gaussian limit as $p = q \to \infty$, one of many distributions to exhibit some form of Gaussian limiting behavior. Finally, the two left-hand plots in Fig. 4.10, along with the upper right and lower left plots in Fig. 4.9, illustrate the range of *asymmetric* densities that are included in the family of beta distributions when $p \neq q$.

Figure 4.10: Four more examples of the beta distribution

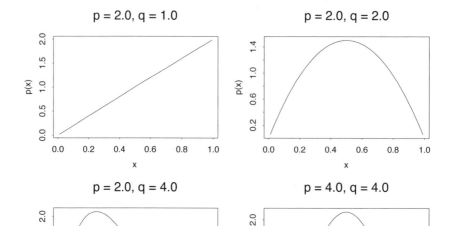

The variance is also determined by the shape parameters of the distribution, but the expression is somewhat more complicated than that for the mean:

$$\mu_2 = \frac{pq}{(p+q)^2(p+q+1)}. \tag{4.62}$$

Some interesting characterizations are available in terms of the variance [163, p. 217]: for arbitrary beta distributions, it has been shown that $\mu_2 < 1/4$; further, if the density is unimodal, $\mu_2 < 1/12$, whereas if the density is U-shaped (i.e., if $p < 1$ and $q < 1$), then $\mu_2 > 1/12$ [234]. Note that the critical value $\mu_2 = 1/12$ corresponds to the variance of the uniform distribution that separates these two types of behavior. Explicit results are also available for the skewness and kurtosis, but they are substantially messier than that for the variance:

$$\begin{aligned} \gamma &= \frac{2(p-p)\sqrt{pq^2+p^2q+pq}}{pq(p+q+2)}, \\ \kappa &= \frac{6[p^3-2p^2q-2pq^2+q^3+p^2-4pq+q^2]}{pq(p+q+2)(p+q+3)}. \end{aligned} \tag{4.63}$$

Also, the beta distribution is somewhat unusual in that its first *negative* moment exists and has the following relatively simple expression:

$$E\left\{\frac{1}{x}\right\} = \frac{p+q-1}{p-1}. \tag{4.64}$$

This result is discussed further in Chapter 12, in connection with the problem of random variable models for "denominator variables."

Finally, it is also useful to note that these moments simplify significantly when $p = q$, leading to the class of symmetric distributions designated type II in Karl Pearson's system of distributions. Specifically, for these distributions, the moments are given by:

$$m_1 = \frac{1}{2}, \qquad \mu_2 = \frac{1}{8p+4},$$
$$\gamma = 0, \qquad \kappa = \frac{-6}{2p+3}. \tag{4.65}$$

As noted earlier, when $p = 1$, we recover the uniform distribution, for which $\mu_2 = 1/12$ and $\kappa = -6/5$. As $p \to 0$, the beta distribution approaches a degenerate limit, assigning equal probabilities to $x = 0$ and $x = 1$. This degenerate distribution achieves the limiting variance $1/4$ noted above, and the limiting kurtosis $\kappa = -2$ discussed in Sec. 4.3.2. At the other extreme, as $p \to \infty$, note that $\kappa \to 0$, consistent with the Gaussian limiting behavior of the symmetric beta distribution as $p \to \infty$. Further, note that $\kappa < 0$ for all finite p, suggesting the symmetric beta distribution may be viewed as a *light-tailed* alternative to the Gaussian distribution.

4.5.2 The Cauchy distribution

Fig. 4.11 shows plots of the density functions for the standard Gaussian distribution $N(0,1)$ discussed in Sec. 4.2.1 and the *Cauchy distribution* defined on the real line by the symmetric density:

$$p(x) = \frac{1}{\pi\lambda}\left[1 + \left(\frac{x-\mu}{\lambda}\right)^2\right]^{-1}, \tag{4.66}$$

plotted for $\mu = 0$ and $\lambda = 1$. This density is *unimodal*, exhibiting a maximum or *mode* at $x = \mu$; the rate at which $p(x)$ decays away from this maximum value depends on the *scale parameter* λ, which may have any positive value. Study of this distribution goes back to Poisson and Cauchy in the nineteenth century because it provided some informative counterexamples to widely accepted statistical notions [162, p. 298]. In fact, it is the best-known member of the class of *non-Gaussian stable random variables* discussed in Chapter 6 and all characterized by infinite variance and higher-order even moments.

These characteristics are a direct consequence of the extremely *heavy tails* of the Cauchy density, which decay much more slowly than those of the Gaussian density. The implications of this slow decay may be seen clearly in Fig. 4.12,

Figure 4.11: The Cauchy and Gaussian probability density functions

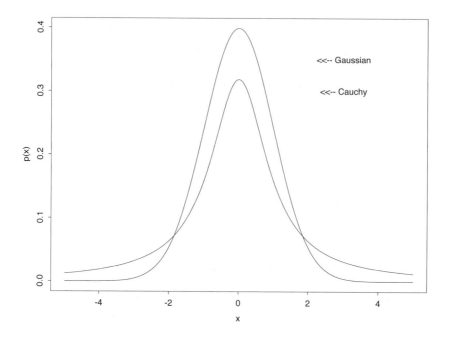

which shows boxplots of 32 random samples drawn from the Gaussian (on the left) and the Cauchy distribution (on the right). The median value for each of these samples is approximately equal to the median of the distribution ($x^\dagger = 0$), and the interquartile range—the central shaded portion of the boxplots—is roughly the same in both plots. The total range of these samples is dramatically, different, however: the Gaussian sample lies within the range $|x_k| < 2$, but the Cauchy sample exhibits values approaching ± 15.

Finally, a desirable feature of the Cauchy distribution is the simple form of its cumulative distribution function:

$$F(x) \;=\; \int_{-\infty}^{x} p(r)dr \;=\; \frac{1}{2} \;+\; \frac{1}{\pi}\tan^{-1}\left(\frac{x-\mu}{\lambda}\right), \qquad (4.67)$$

leading to the following simple result for the quantiles x_q:

$$F(x_q) = q \;\Rightarrow\; x_q = \mu + \lambda \, \tan\left[\pi\left(q - \frac{1}{2}\right)\right]. \qquad (4.68)$$

Figure 4.12: Boxplots of Cauchy and Gaussian random samples

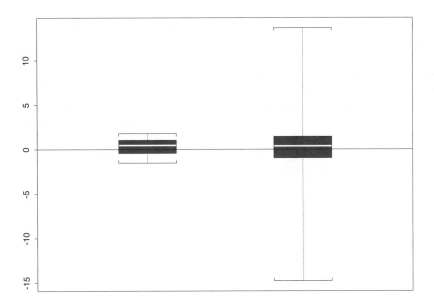

4.5.3 The exponential distribution

When the shape parameter in either the gamma distribution or the Weibull distribution is equal to 1, we recover the important special case of the *exponential distribution*, defined for all $x > \xi$ by the following density:

$$p(x) = \frac{1}{\beta} \exp\left[-\left(\frac{x-\xi}{\beta}\right)\right]. \qquad (4.69)$$

Because the form of this density is extremely simple, it can be integrated explicitly to obtain the cumulative distribution function:

$$F(x) = 1 - \exp\left[-\left(\frac{x-\xi}{\beta}\right)\right], \qquad (4.70)$$

leading to the following result for the quantiles x_q:

$$F(x_q) = q \;\Rightarrow\; x_q = \xi + \beta \ln\left(\frac{1}{1-q}\right). \qquad (4.71)$$

The first four moments of this distribution are:

$$m_1 = \xi + \beta, \qquad \mu_2 = \beta^2,$$
$$\gamma = 2, \qquad \kappa = 6. \qquad (4.72)$$

The exponential distribution is used extensively in applied statistics, both because of its extreme mathematical convenience and because it arises naturally in a number of contexts. For example, it is noted in Chapter 3 that the inter-arrival times for a Poisson point process are exponentially distributed; because the Poisson point process is often invoked as a model for *rare, discrete events*, the exponential distribution arises as a direct consequence. The exponential distribution also exhibits the *lack of memory property* discussed in Chapter 3. Further, because of the simple form of the cumulative distribution function, simple expressions may be obtained for the distributions of *order statistics* from an independent, identically distributed sequence of exponential random variables, a result discussed in Chapter 6. Finally, the exponential distribution is also interesting because maximum likelihood estimators for the parameters ξ and β behave differently from most other parameter estimation procedures, a point discussed further in Chapter 6. For a much more detailed discussion of this distribution, refer to the book by Balakrishnan and Basu [23].

4.5.4   The gamma distribution

The *gamma distribution* is the better-known name for Karl Pearson's type III family of distributions, discussed in Sec. 4.3.4 and defined for $x > \xi$ by the density function [162, p. 337]:

$$p(x) \;=\; \frac{1}{\beta\Gamma(\alpha)} \left(\frac{x-\xi}{\beta}\right)^{\alpha-1} \exp\left[-\left(\frac{x-\xi}{\beta}\right)\right]. \qquad (4.73)$$

Here, ξ is a location parameter, most commonly taken to be zero, and β is a positive scale parameter. The positive parameter α defines the shape of the density function, as illustrated in Fig. 4.13 for four values: $\alpha = 0.5$, $\alpha = 1.0$, $\alpha = 1.5$, and $\alpha = 2.0$. In all cases, these plots correspond to the *standard gamma density* with $\xi = 0$ and $\beta = 1$. For $0 < \alpha \leq 1$, the gamma density is J-shaped— i.e., decaying monotonically from a maximum at $x = 0$—whereas for $\alpha > 1$, the density is unimodal with mode $x_0 = \alpha - 1$. The critical value $\alpha = 1$ corresponds to the *exponential distribution*, an extremely important special case discussed further in Sec. 4.5.3. Another extremely important special case of the gamma distribution is the χ^2 *distributionn with ν degrees of freedom*, corresponding to $\alpha = \nu/2$, $\beta = 2$ and $\xi = 0$, discussed further in Chapters 9 and 12.

Like the Gaussian distribution, the cumulative distribution function for the gamma distribution cannot be expressed in terms of elementary functions. For $\xi = 0$, the moments m_n (i.e., the moments about zero, *not* the central moments) are given by the following ratio of gamma functions [162, p. 339]:

$$m_r \;=\; \frac{\beta^r \Gamma(\alpha + r)}{\Gamma(\alpha)}. \qquad (4.74)$$

For arbitrary ξ, the first four standardized moments are:

$$m_1 = \xi + \alpha\beta, \qquad \mu_2 = \alpha\beta^2,$$
$$\gamma = \frac{2}{\sqrt{\alpha}}, \qquad \kappa = \frac{6}{\alpha}. \qquad (4.75)$$

Figure 4.13: Four gamma densities

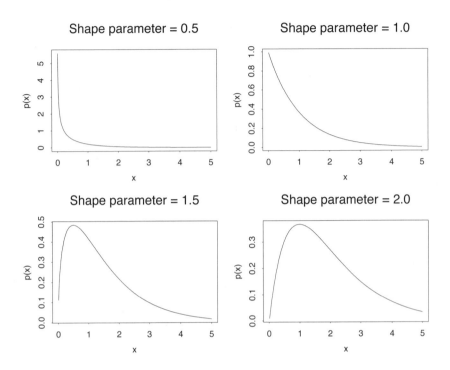

Note that $\gamma > 0$ for all α, consistent with the qualitative notion of a positively skewed distribution as one with an infinitely long right tail. Also, note that $\kappa > 0$ for all α, suggesting this right tail is heavy relative to the Gaussian decay. In both cases, these shape parameters go to zero as $\alpha \rightarrow \infty$, corresponding to the fact that the gamma distribution approaches normality in this limit. In general, the gamma distribution is an important one that appears frequently in a wide variety of applications; for a *much* more detailed discussion of the properties and applications of this distribution, refer to the 78 page chapter in the book by Johnson *et al.* [162].

4.5.5 The Laplace distribution

The *Laplace* or *double exponential* distribution is defined by the density function:

$$p(x) \quad = \quad \frac{1}{2\phi} \, e^{-|x-\theta|/\phi}, \tag{4.76}$$

where θ defines the mean of the distribution, m_1, and ϕ represents the mean deviation, $\nu_1 = E\{|x - m_1|\}$. Because $e^{-|x|}$ decays more slowly than e^{-x^2}, the Laplace distribution is heavy-tailed relative to the Gaussian distribution, but

Figure 4.14: Laplace, Gaussian, and Cauchy densities

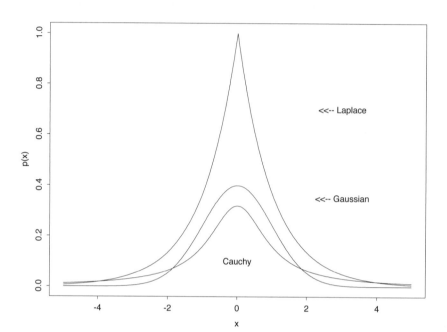

less so than the Cauchy distribution. In particular, both the variance and the kurtosis are finite for the Laplace distribution, unlike the Cauchy distribution. Hence, in terms of its tail behavior, the Laplace distribution represents a less severe nonnormal alternative than does the Cauchy distribution, although it exhibits a much more pronounced peak at $x = \theta$. Plots of all three densities— Laplace, Cauchy, and Gaussian—are shown in Fig. 4.14 for comparison. The dominant feature of the Laplace distribution is the sharp peak or cusp at $x = \theta$, where the derivative of the density is undefined.

Because the Laplace distribution is symmetric about $x = \theta$ and unimodal, this value represents the mean, median, and mode, analogous to the Gaussian and Cauchy distributions. Also, it follows from the symmetry of this distribution that all odd-order central moments are identically zero. The even-order central moments are given by the simple expression [163, p. 165]:

$$\mu_r = r! \, \phi^r. \tag{4.77}$$

The first four standardized moments for the Laplace distribution are therefore:

$$m_1 = \theta, \quad \mu_2 = 2\phi^2,$$
$$\gamma = 0, \quad \kappa = 3. \tag{4.78}$$

Like the Cauchy distribution, the cumulative distribution function for the Laplace distribution may be expressed simply:

$$F(x) \;=\; \begin{cases} \frac{1}{2}\exp\left[-\left(\frac{\theta-x}{\phi}\right)\right] & x \le \theta, \\[2mm] 1 - \frac{1}{2}\exp\left[-\left(\frac{x-\theta}{\phi}\right)\right] & x > \theta. \end{cases} \tag{4.79}$$

Consequently, the quantiles x_q of the distribution may be computed easily; specifically, the quantiles for $q \ge 0.5$ are given by

$$F(x_q) = q \;\Rightarrow\; x_q = \theta + \phi\,\ln\left[\frac{1}{2(1-q)}\right]. \tag{4.80}$$

A particularly useful special case is that of the upper and lower quartiles (i.e., x_q for $q = 0.25$ and $q = 0.75$), which are $\theta \pm \phi\ln 2 \simeq \theta \pm 0.693\phi$.

4.5.6 The logistic distribution

In marked contrast to the Laplace and Cauchy distributions discussed above, the *logistic distribution* may be regarded as a mildly non-Gaussian distribution, exhibiting slightly heavier tails than the Gaussian case, but lighter tails than the Laplace case. Specifically, the logistic distribution is defined by the probability density function [163, p. 116]:

$$p(x) \;=\; \frac{1}{4b}\,\operatorname{sech}^2\left[\frac{1}{2}\left(\frac{x-a}{b}\right)\right]. \tag{4.81}$$

The strong similarity of the logistic and Gaussian distributions is illustrated in Fig. 4.15, which shows the standard normal density and the logistic density with $a = 0$ and $b = \sqrt{3}/\pi$, yielding a distribution with zero mean and unit variance. Although it is clear that these densities are different, these differences are much less pronounced than those seen in Fig. 4.14 between the Gaussian, Cauchy, and Laplace densities.

Historically, the logistic distribution is an outgrowth of the use of a *logistic curve* to model growth, a development that dates back to the nineteenth century. In particular, the cumulative distribution function for this distribution is:

$$F(x) \;=\; \left[1 + \exp\left\{-\left(\frac{x-a}{b}\right)\right\}\right]^{-1}, \tag{4.82}$$

which is an example of a logistic curve. As with the Cauchy and Laplace distributions, the existence of a simple closed form expression for the CDF of the logistic distribution can sometimes be quite advantageous. For example, the quantiles x_q of this distribution are easily computed as:

$$F(x_q) = q \;\Rightarrow\; x_q = a + b\,\ln\left(\frac{q}{1-q}\right). \tag{4.83}$$

Figure 4.15: The logistic and Gaussian densities

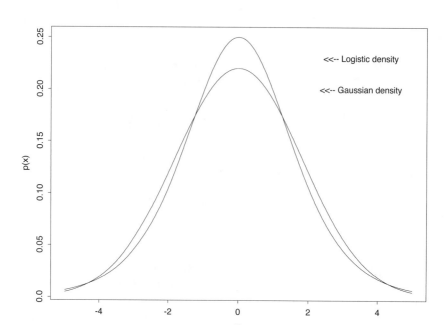

Ironically, despite the existence of this simple expression for the CDF, the *moments* of the logistic distribution are much more complex than those of the Gaussian distribution. It follows from the symmetry of the distribution that all odd-order central moments vanish, and the even-order central moments μ_r correspond to the *absolute* central moments of the same order. For $r > 1$, these are given by [163, p. 117]:

$$E\{|x - a|^r\} \;=\; 2b^r \Gamma(r + 1)[1 - 2^{-(r-1)}]\zeta(r), \qquad (4.84)$$

where $\zeta(r)$ is the Riemann zeta function, defined as:

$$\zeta(r) \;=\; \sum_{k=1}^{\infty} k^{-r}. \qquad (4.85)$$

For $r = 1$, $\zeta(r)$ corresponds to the *harmonic series*, one of the classic examples of a divergent infinite series, but for $r > 1$, this series converges and its value is known analytically for certain special values of r, including $r = 2$ and $r = 4$. Hence, it is possible to obtain the following simple, explicit expressions for the

Table 4.3: Exact probabilities and bounds

Deviation d	Gaussian probability	Logistic probability	Laplace probability	Chebyshev bound	Cauchy probability
0.5	0.6171	0.5753	0.4931	4.0000	0.7048
1.0	0.3173	0.2804	0.2431	1.0000	0.5000
1.5	0.1336	0.1235	0.1199	0.4444	0.3743
2.0	0.0455	0.0518	0.0591	0.2500	0.2952
2.5	0.0124	0.0212	0.0291	0.1600	0.2422
3.0	0.0027	0.0086	0.0144	0.1111	0.2048
3.5	0.0005	0.0035	0.0071	0.0816	0.1772

first four standard moments of the logistic distribution:

$$m_1 = a, \qquad \mu_2 = \frac{\pi^2 b^2}{3},$$
$$\gamma = 0, \qquad \kappa = 1.2. \tag{4.86}$$

This kurtosis result (which is exact) supports the observation made earlier that the logistic distribution has slightly heavier tails than does the Gaussian distribution. In fact, this kurtosis value is exactly the same as that for the Student's t-distribution with 9 degrees of freedom, and it is noted in Chapter 9 that the Student's t-distributions converge rapidly to their Gaussian limit. For $0 < r \le 1$, the absolute moments $E\{|x - a|^r\}$ exist, although they are no longer given by Eq. (4.84), since the Riemann zeta sum does not converge. Instead, these moments are given by [163, p. 117]:

$$E\{|x - a|^r\} \;=\; 2b^r \Gamma(r + 1) \sum_{j=1}^{\infty} (-1)^{j-1} j^{-r}. \tag{4.87}$$

In particular, for $r = 1$, this sum reduces to $ln\,2$, so the mean deviation is:

$$\nu_1 \;=\; 2b \ln 2 \simeq 0.7643\sigma. \tag{4.88}$$

This result again emphasizes the similarity of the logistic distribution to the Gaussian distribution, for which $\nu_1 \simeq 0.79866\sigma$.

To further illustrate the similarity between the logistic and Gaussian distributions, Table 4.3 compares the tail behavior of these two distributions with that of the Laplace and Cauchy distributions. More specifically, this table presents the exact probabilities $\mathcal{P}\{|x| > d\}$ for the Gaussian, logistic, and Laplace distributions with mean $\mu = 0$ and standard deviation $\sigma = 1$. In addition, Table

4.3 also gives the universal Chebyshev bound described in Sec. 4.3.1 for *all* distributions with $\mu = 0$ and $\sigma = 1$. Comparing these results, it is clear that the Gaussian distribution decays the most rapidly, followed by the logistic distribution, consistent with its slightly heavier tails. By comparison, the Laplace distribution decays still more slowly, illustrating the heavier tails of this distribution. These results are consistent with the standard interpretation of kurtosis as an elongation measure, since $\kappa = 0$ for the Gaussian distribution, vs. $\kappa = 1.2$ for the logistic distribution and $\kappa = 3.0$ for the Laplace distribution. Also, note how much more slowly the Chebyshev bound decays than all of these unit variance distributions. This slow decay is a direct consequence of the universality of this bound, which applies to arbitrarily heavy-tailed distributions, including the Student's t-distributions with 3 and 4 degrees of freedom (suitably scaled to have unit variance), which both exhibit infinite kurtosis. The last column in Table 4.3 shows the tail behavior of the Cauchy distribution, which decays even more slowly than the Chebyshev bound. This decay reflects the Cauchy distribution's infinite variance, which makes the Chebyshev bound inapplicable.

4.5.7 The lognormal distribution

The *lognormal distribution* is one of the three members of the Johnson family of distributions discussed in Sec. 4.3.5, obtained by assuming that the *logarithm* of the random variable x is normally distributed. Specifically, if $\ln x \sim N(\zeta, \sigma^2)$ for some real ζ and some $\sigma > 0$, we obtain the density:

$$p(x) \;=\; \frac{1}{\sqrt{2\pi}\sigma x}\, \exp\left\{-\frac{1}{2}\left(\frac{\ln x - \zeta}{\sigma}\right)^2\right\}. \tag{4.89}$$

It is also possible to extend this definition to a three-parameter distribution by assuming $\ln(x - \theta)$ is normally distributed for some nonzero parameter θ. A plot of the lognormal distribution is shown in Fig. 4.16 for the standard case $\theta = 0$, $\zeta = 0$, and $\sigma = 1$ to illustrate its general behavior. The overall shape of this distribution is somewhat similar to that of the gamma distribution with shape parameter $\alpha \sim 1.5$. For comparison, this distribution is also shown in Fig. 4.16, along with the exponential distribution (i.e., the gamma distribution with $\alpha = 1$). Close inspection of this plot reveals that the lognormal distribution exhibits a slower decay than either of these gamma densities.

The moments of the lognormal distribution are most easily expressed in terms of the quantity:

$$\omega \;=\; \exp\left[\sigma^2\right]. \tag{4.90}$$

In particular, the four standardized moments are given by:

$$\begin{aligned}
m_1 &= e^{\zeta}\omega^{1/2} = e^{\zeta + \sigma^2/2}, \\
\mu_2 &= e^{2\zeta}\omega(\omega - 1), \\
\gamma &= (\omega + 2)\sqrt{\omega - 1}, \\
\kappa &= \omega^4 + 2\omega^3 + 3\omega^2 - 6.
\end{aligned} \tag{4.91}$$

Figure 4.16: A comparison of lognormal and gamma densities

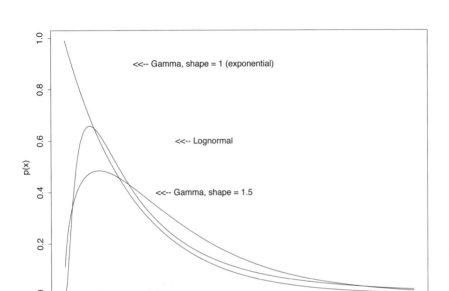

Although it may not be obvious from these expressions, it turns out that both γ and κ are positive for all ζ and σ, like the gamma distributions discussed in Sec. 4.5.4. In fact, as $\sigma \to 0$, the lognormal distribution approaches a normal limit, a result closely related to the fact that, for small x, $\ln(1+x) \simeq x$.

In addition, a number of other useful characterizations are also available for the lognormal distribution. For example, the quantiles x_q of the lognormal distribution may be computed from those η_q for the standard normal distribution by the following expression [162, p. 213]:

$$x_q = \exp\left[\zeta + \eta_q \sigma\right]. \qquad (4.92)$$

As an immediate corollary, since the median is $\eta_{0.50} = 0$ for the standard normal distribution, it follows that the median of the lognormal distribution is:

$$x^\dagger = x_{0.50} = e^\zeta. \qquad (4.93)$$

Finally, note that the lognormal distribution is always unimodal, with mode:

$$x^* = \omega^{-1} e^\zeta = e^{\zeta - \sigma^2}. \qquad (4.94)$$

Combining the expressions for the mean, median and mode, it follows that:

$$\bar{x} > x^\dagger > x^*. \qquad (4.95)$$

Figure 4.17: Four Pareto densities

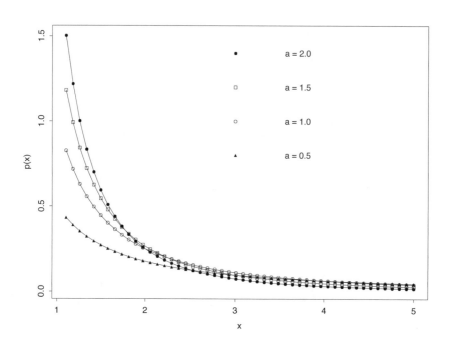

4.5.8   The Pareto distribution

The name "Pareto distribution" can refer to any one of three models proposed by Vilfredo Pareto [162, ch. 20] to describe the distribution of income over a population. Two of these distributions—the Pareto type I and Pareto type II distributions—belong to the Pearson VI family of distributions discussed in Sec. 4.3.4. In all three cases, the cumulative distribution function $F(x)$ exhibits a simple form that can be differentiated explicitly to obtain the density function. Here, consideration is restricted to the type I distribution, which will be denoted $P(I)(k, a)$ following Johnson *et al.* [162] and is defined by the cumulative distribution function:

$$F(x) \quad = \quad 1 - \left(\frac{k}{x}\right)^a, \tag{4.96}$$

for all $x \geq k > 0$, where a is a positive constant. The corresponding density is given by the simple expression, for all $x \geq k$:

$$p(x) \quad = \quad \frac{ak^a}{x^{a+1}}. \tag{4.97}$$

Plots of Pareto densities for $k = 1$ and $a = 0.5$, 1.0, 1.5, and 2.0 are shown in Fig. 4.17. This density function is J-shaped, decreasing monotonically with increasing x for $x > k$. The rth moments of x are finite only for $r < a$, and these values are easily determined due to the simplicity of the density function. In particular, the standardized four moments are [162, p. 577]:

$$m_1 = \frac{ak}{a-1}, \ a > 1,$$

$$\mu_2 = \frac{ak^2}{(a-1)(a-2)}, \ a > 2,$$

$$\gamma = 2\frac{a+1}{a-3}\sqrt{\frac{a-2}{a}}, \ a > 3,$$

$$\kappa = \frac{6(a^3 + a^2 - 6a - 2)}{a(a-3)(a-4)}, \ a > 4. \tag{4.98}$$

Note that the skewness and kurtosis are both positive for all admissible values of a, decreasing monotonically as $a \to \infty$, with the skewness approaching 2 and the kurtosis approaching 6, which are precisely the same values as those for the exponential distribution. In fact, one of the reasons the Pareto distribution is interesting is its close relationship to the exponential distribution via a simple logarithmic transformation, a point discussed further in Chapter 12. Hence, the Pareto distribution inherets some of the analytical convenience of the exponential distribution, as for example in the characterization of order statistics. Also, the Pareto distribution is sometimes a reasonable alternative to the lognormal distribution, and like the lognormal distribution, the negative moments of the Pareto distribution are well defined. This characteristic is sometimes advantageous, a point discussed further in Chapter 12 in connection with reciprocal transformations.

4.5.9 The Rayleigh distribution

The *Rayleigh distribution* is defined by the density:

$$p(x) = \frac{x}{\sigma^2}e^{-x^2/(2\sigma^2)}, \tag{4.99}$$

for $0 \le x < \infty$ and $\sigma > 0$ [162, p. 456]. This distribution belongs to the family of Weibull distributions discussed in Sec. 4.5.10 with location parameter $\xi = 0$, scale parameter $\alpha = \sqrt{2}\sigma$, and shape parameter $c = 2$. Probably the most common context in which this distribution arises is the following: if $\mathbf{v}$ is a two-component random vector with statistically independent, zero-mean Gaussian components, each having variance σ^2, then the *magnitude* of this vector has a Rayleigh distribution.

The first four normalized moments of the Rayleigh distribution all have fairly simple expressions [162, p. 459]. Specifically, the mean is proportional to the single distribution parameter σ:

$$m_1 = \sigma\sqrt{\frac{\pi}{2}} \simeq 1.25331\sigma, \tag{4.100}$$

while the variance is proportional to the square of this parameter:

$$\mu_2 = \frac{\sigma^2(4 - \pi)}{2} \simeq 0.42920\sigma^2. \tag{4.101}$$

Similarly, the skewness and kurtosis are constant, given by:

$$\gamma = \frac{2(\pi - 3)\sqrt{\pi}}{(4 - \pi)^{3/2}} \simeq 0.63111,$$

$$\kappa = \frac{32 - 3\pi^2}{(4 - \pi)^2} - 3 \simeq 0.24509. \tag{4.102}$$

It follows from the Weibull distribution results given in the next section that the mode of the Rayleigh distribution is $x^* = \sigma$ and the median is $x^\dagger = \sigma\sqrt{2 \ln 2} \simeq 1.17741\sigma$.

4.5.10 The Weibull distribution

As noted at the beginning of this chapter, the Weibull distribution is a location-scale family defined by the following density function [162, p. 629]:

$$p(x) = \frac{c}{\alpha}\left(\frac{x - \xi}{\alpha}\right)^{c-1} \exp\left[-\left(\frac{x - \xi}{\alpha}\right)^c\right], \tag{4.103}$$

for $x > \xi$. Here, ξ is a location parameter that may assume any real value, α is a positive scale parameter, and c is a positive shape parameter. Plots of the Weibull distribution for $\xi = 0$, $\alpha = 1$, and four different values of c are shown in Fig. 4.18. These plots correspond to the same four values considered for the shape parameter α in the gamma distribution, plotted in Fig. 4.13. In fact, there are certain similarities in these densities, although they become less similar as the shape parameters increase in magnitude. As with the gamma densities, the distribution is J-shaped for $0 < c \leq 1$, exhibiting a single mode at $x = \xi$. The case $c = 1$ corresponds to the *same* special case as $\alpha = 1$ in the gamma distribution family: the exponential distribution discussed in Sec. 4.5.3. For $c > 1$, the Weibull density is unimodal with mode:

$$x^* = \xi + \alpha\left(\frac{c - 1}{c}\right)^{1/c}, \tag{4.104}$$

and it has been noted that this mode approaches $\xi + \alpha$ very rapidly as $c \to \infty$ [162, p. 630]. Another important special case of the Weibull distribution is the *Rayleigh distribution* discussed in Sec. 4.5.9, obtained for $c = 2$.

The difference between the gamma and Weibull densities for $\alpha = 1.5$ and $c = 1.5$, respectively, is shown in Fig. 4.19. In addition, the lognormal density is also shown in this plot for further comparison. All three of these densities exhibit similar unimodal shapes with modes at roughly the same values, and as noted in Sec. 4.5.7, the lognormal density exhibits a slower asymptotic decay

Figure 4.18: Four different Weibull densities

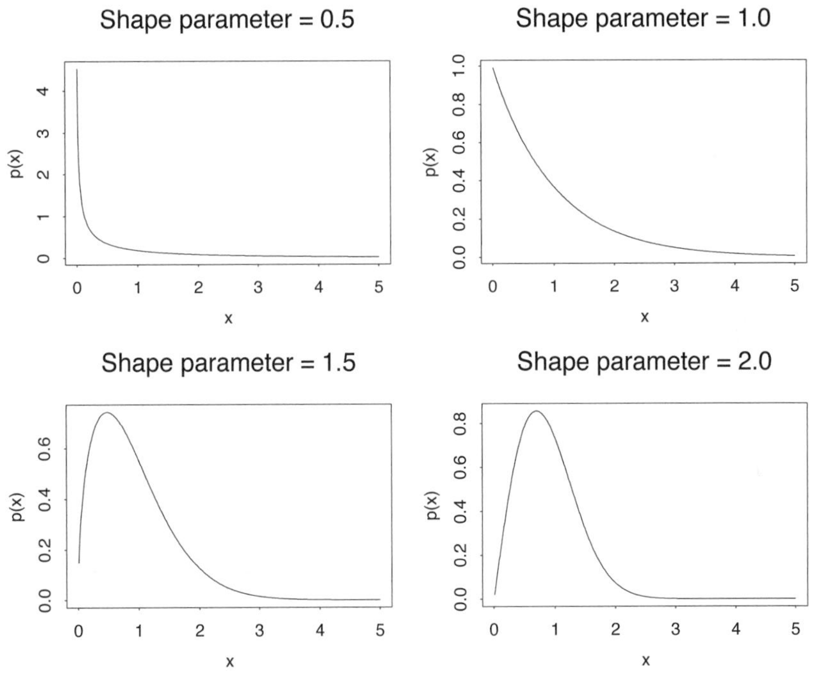

than the gamma density. In contrast, the Weibull density exhibits an extremely rapid asymptotic decay, reflecting the fact that the shape parameter appears as a power in the exponent of the density function.

A particularly appealing feature of the Weibull distribution is that its cumulative distribution function is also simple:

$$F(x) = 1 - \exp\left[-\left(\frac{x - \xi}{\alpha}\right)^c\right]. \qquad (4.105)$$

Consequently, simple expressions for the quantiles of this distribution may be derived:

$$F(x_q) = q \Rightarrow x_q = \xi + \alpha\left[\ln\left(\frac{1}{1 - q}\right)\right]^{1/c}. \qquad (4.106)$$

As in the case of the logistic distribution, however, one of the prices for this simplicity of the CDF is increased complexity of the moment expressions. Specifically, the rth moment of the standard Weibull distribution ($\xi = 0$, $\alpha = 1$) is given by [162, p. 632]:

$$m_r = \Gamma\left(\frac{r}{c} + 1\right). \qquad (4.107)$$

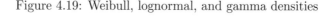

Figure 4.19: Weibull, lognormal, and gamma densities

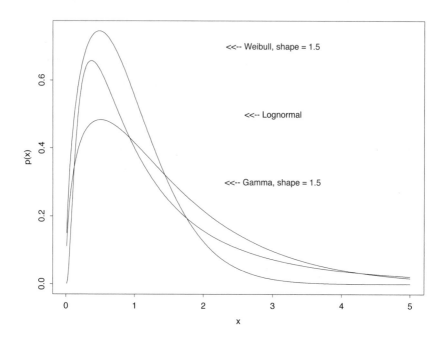

The first four standardized moments may be computed from these central moments, but the resulting expressions do not offer much insight into the behavior of the Weibull distribution. Johnson *et al.* [162, ch. 21] present mean and variance approximations for large values of the shape parameter c, and they give numerical results for the first four standardized moments for c between 1.2 and 4.0. As $c \to \infty$, the mean approaches the limiting value $\xi + \alpha$ and the variance approaches zero as:

$$\mu_2 \simeq \frac{\alpha^2 \pi^2}{6c^2} \simeq \frac{1.64493\alpha^2}{c^2}. \tag{4.108}$$

The skewness is negative for c smaller than $\sim$3.6, and positive for larger values of c. The kurtosis is positive for c smaller than $\sim$2.2, and negative for larger values of c, approaching a minimum of approximately -0.29 for $c \simeq 3.35$.

4.6   Exercises

1. Prove the relationship between the central moments μ_n and the non-central moments m_r given in Eq. (4.15).

2. Suppose the density $p(x)$ is symmetric about x_0. Prove that the mean satisfies $\bar{x} = x_0$.

3. Suppose the density $p(x)$ is symmetric about x_0. Prove that the median satisfies $x^\dagger = x_0$.

4. Suppose the density $p(x)$ is unimodal and symmetric about x_0. Prove that the mode satisfies $x^* = x_0$.

5. Consider the Pareto distribution:

$$p(x) = \frac{ak^a}{x^{a+1}}, \quad x \geq k > 0, a > 0.$$

For this distribution, compute:

    a. the mean $\bar{x}$,
    b. the median $x^\dagger$,
    c. the mode x^*.

6. Consider the Laplace distribution:

$$p(x) = \frac{1}{2\phi} e^{-|x-\mu|/\phi}, \quad \phi > 0.$$

    a. For this distribution, compute the exact probability $\mathcal{P}\{|x-\mu| > c\phi\}$;
    b. Construct either a plot or a small table, comparing these exact values with those given by Chebyshev's inequality.

7. (Computer exercise) Using any software package with a Gaussian random number generator, generate 100 sequences of zero-mean, unit-variance Gaussian random numbers, each of length $N = 100$. For each sequence, estimate the first four standardized moments: the mean μ, the standard deviation σ, the skewness γ, and the kurtosis κ. Summarize these results with a boxplot. What can you say about the general behavior of these moment estimates?

Chapter 5

Fitting Straight Lines

Sections and topics:

This chapter presents a detailed discussion of a simple but extremely important data analysis problem: that of fitting straight lines to points in the plane. This problem arises frequently in practice, either directly or indirectly, and one of the primary points of this chapter is that this problem—like most data analysis problems—may be formulated in several different ways that generally lead to different solutions. To understand which formulation is appropriate for a particular application, it is necessary to consider the working assumptions on which the different formulations are based. Consequently, this chapter places considerable emphasis on specifying and understanding these working assumptions. Many of the results presented here extend directly to a wide variety of other data analysis problems, and most of the central ideas presented in this chapter have some counterpart in almost all data analysis problems.

Key points of this chapter:

1. The problem of fitting the linear equation $y = ax + b$ to a set of observed (x, y) data pairs occurs very often in practice, either directly or indirectly through some simple transformation of the original data variables.

2. Many different techniques exist for fitting lines to a given dataset, and they are generally *not* equivalent: applying different methods to the same dataset will yield different estimates of the *best fit* parameters $\hat{a}$ and $\hat{b}$.

3. As in almost all other data analysis problems, differences in the results obtained by different methods reflect differences in the underlying working assumptions on which these methods are based, particularly with respect to the inherent variability, uncertainty, or irregularity in the dataset.

4. As a practical matter, the utility of the results obtained by different methods will depend on both how well the available data conforms to their respective working assumptions and how sensitive the ultimate application is to differences caused by violations of these assumptions.

5. Two broad classes of problem formulations are the *errors-in-model* (EM) and *errors-in-variables* (EV) formulations. For line-fitting problems, the EM formulation corresponds to the assumption that the *true* relationship between variables is not quite linear. *Mathematically*, this assumption is equivalent to assuming measurement errors in the y variable alone, although the *physical significance* of these two interpretations may be quite different. The EV formulation assumes errors in both variables.

6. The most popular line-fitting procedure is the method of *ordinary least squares* (OLS), an EM formulation that is in some respects equivalent to assuming Gaussian observation errors on y. In particular, the OLS procedure represents the *maximum likelihood estimator* for the EM problem with Gaussian observation errors.

7. An important limitation of the OLS approach is its severe outlier sensitivity. The method of *least absolute deviations* (LAD) is a more outlier-resistant alternative to OLS, although it does suffer from nonuniqueness. This procedure corresponds to the maximum likelihood estimator for the EM problem assuming errors on y that have a *Laplace* distribution.

8. Another limitation of the OLS method is that it is an EM formulation; in cases where an EV formulation is more appropriate, the method of *total least squares* (TLS) is a reasonable alternative, although it, too, suffers from outlier sensitivity.

9. Other uncertainty descriptions are possible and this chapter briefly considers one that was introduced in Chapter 1: the *unknown-but-bounded* or *set-theoretic* model. This model is appealing in its conceptual simplicity, but it generally leads to more complicated computational problems than methods like OLS, LAD, and TLS do.

5.1 Why do we fit straight lines?

The following four examples provide four general motivations for fitting straight lines. First, Sec. 5.1.1 briefly discusses *linear constitutive relations*, widely used to describe phenomena like electrical conduction in resistive materials (e.g., Ohm's law discussed in Chapter 1) and analogous relations for heat transfer, thermal expansion, friction, and many other phenomena. One argument in favor of linear constitutive relations—and linear relations between variables more generally—is that many important mathematical functions we encounter in applications are *analytic*, meaning that they may be expanded as a *Taylor series*. Often, it is reasonable to assume this series may be truncated to its first term, leading to a linear approximation, an idea discussed in detail in Sec. 5.1.2. Then, Sec. 5.1.3 considers a large class of problems that, although not originally linear, can be transformed to the problem of fitting straight lines, analogous to Ohm's original dataset discussed in Chapter 1. Finally, Sec. 5.1.4 introduces a number of *functional equations* that arise when we impose certain *behavioral restrictions* on the relations between variables. In cases where these functional equations can be solved, they frequently yield simple functional forms involving a few free parameters, and the problem of estimating these parameters from data can often be transformed into a problem of fitting straight lines.

5.1.1 Linear constitutive relations

The Ohm's law example discussed in Chapter 1 is a typical illustration of a *linear constitutive relation* between two variables, x and y, of the form:

$$y = \alpha x, \tag{5.1}$$

where α is some constant. In the case of Ohm's law, x and y represent the current flowing in the conductor and the voltage drop across it, respectively, and α represents its electrical resistance, measured in ohms. In this particular example, α is a property of the *component* defined by the conductor, depending on its length ℓ, its cross-sectional area A, and the material of its construction. Specifically, this resistance is given by:

$$R = \frac{\rho\ell}{A}, \tag{5.2}$$

where ρ is the *resistivity* of the material. Alternatively, Ohm's law may also be viewed as a relation between the *electric field* $\mathcal{E}$ (in volts per meter) and the *current density* $\mathcal{J}$ (in amperes per square meter) inside the material:

$$\mathcal{E} = \rho\mathcal{J}. \tag{5.3}$$

The distinction between these two formulations is that the resistivity ρ (in units of ohm-meters) is a characteristic of the *material* from which the resistive conductor is made, whereas the resistance R depends on both this material and the sample geometry.

Linear constitutive relations arise in many different fields; the following examples are taken from the book by Haliday *et al.* [131]. A capacitor is an arrangement of electrodes separated by one or more dielectric materials that serves to store an electric charge Q in response to an applied voltage V; the relation between these quantities is given by:

$$Q = CV, \tag{5.4}$$

where C is a property of the capacitor depending on both its geometry and the dielectric materials involved, as in the resistor example just discussed. For the case of a parallel-plate geometry, this relationship is given by $C = \epsilon A/d$, where A is the cross-sectional area of the plates, d is their separation, and ϵ is the dielectric constant of the material separating the plates. As a non-electrical example, note that materials respond to heat sources by changing their temperature, and the relationship between the quantity H of heat absorbed and the temperature change ΔT is given by the linear constitutive relation:

$$H = c\Delta T, \tag{5.5}$$

where c is the heat capacity, a property of the material under consideration. Similarly, the linear relation describing sliding friction is:

$$F_f = \mu F_N, \tag{5.6}$$

where F_f is the friction force that opposes sliding motion, F_N is the force acting normal (i.e., perpindicular) to the sliding surface, and μ is the coefficient of kinetic friction. Various linear relations also exist between applied stresses and changes in material dimensions; the general form of these relations is:

$$\text{stress} = \text{modulus} \times \text{strain}, \tag{5.7}$$

where stress is an applied force per unit area (a pressure, in the case of fluids), strain is a normalized change in length or volume, and the resulting modulus has a different name depending on which stresses and strains are involved. In particular, Young's modulus E relates changes in length to tensile or compressive stresses, the shear modulus G relates a measure of shearing displacement to an applied shear stress, and the bulk modulus B relates the change in volume of a compressible fluid to an applied pressure. Finally, the coefficient of linear expansion relates the fractional change in length of a material to a change in temperature. This list of common linear constitutive relations is by no means exhaustive, even of those discussed by Halliday *et al.* [131], and there are many other important examples not discussed in that book.

A less common linear constitutive relation is that arising from the *Costa-Ribeiro effect* or the *thermodielectric effect* [100]. This effect describes the generation of an electrostatic charge at the solid/liquid interface where a phase change occurs. Specifically, in 1944 Costa-Ribeiro measured the current I induced in a pair of capacitor plates immersed in a dielectric system that included both solid and liquid phases. Measuring the rate G at which the solid phase grew (on crystalization) or shrank (on melting), he noted that the current appeared to vary linearly with the growth rate:

$$I = kG. \tag{5.8}$$

The constant k is a material property, termed the *thermodielectric constant*. Interestingly, it has been noted that this effect is *reversible*: if a current is applied to capacitor electrodes immersed in a dielectric solution near its freezing point, crystal growth may be accelarated significantly [99]. The key point of this example is that, although linear constitutive relations are the simplest possible relationships between two variables, their discovery and analysis can lead to unexpected results.

Finally, it is useful to conclude this discussion of linear constitutive relations with two observations. First, if the roles of the independent and dependent variables are reversed, the *structure* of their relationship remains the same, although the constant involved in this relationship changes. That is, Eq. (5.1) may be rewritten as the equivalent constitutive relation between y and x:

$$x = (1/\alpha)y. \tag{5.9}$$

As a specific example, the resistive behavior of metals is often characterized in terms of the *conductivity* $\sigma = \rho^{-1}$, converting Eq. (5.3) to the equivalent form:

$$\mathcal{J} = \sigma\mathcal{E}. \tag{5.10}$$

The second concluding observation is that all of the linear relations considered in this discussion have been of the form:

$$y = ax + b, \tag{5.11}$$

subject to the constraint that $b = 0$. To estimate the constant a that is of interest in these constitutive relations, we may proceed in either of two ways. The first is to explicitly impose this constraint, leading to some particularly simple estimators for a: given a set of N pairs (x_k, y_k), we can immediately compute a sequence of estimates $a_k = y_k/x_k$ and combining these estimates (e.g., by averaging) leads to an estimate of a. The disadvantage of this approach is that it leads to simple estimates of the constant a even in cases where no linear constitutive relation holds. Alternatively, if we apply the methods described in this chapter to estimate the parameters a and b both in Eq. (5.11), we can subsequently check to see whether $b \simeq 0$. If so, this result provides further evidence that the linear constitutive model is reasonable. This idea—of first fitting a model and then examining its reasonableness—is extremely important in practice and is discussed in detail in Chapter 15.

5.1.2 Taylor series expansions

As noted in the introduction to this section, Taylor series expansions represent another common source of linear relations between variables, obtained by approximations that are assumed to hold over a *sufficiently narrow* range of variation in some independent variable x, related to the dependent variable y by a *smooth* function, $y = f(x)$. More specifically, these variables are often— although, it is important to note, not always—related through a function $f(\cdot)$ that is *analytic*, an assumption that has two key consequences. First, these functions are smooth in the sense of being *infinitely differentiable*, commonly arising as solutions of ordinary or partial differential equations. The second key point is that these functions may be represented by the following Taylor series expansion [213, sec. 4.3]:

$$f(x) = f(x_0) + [x - x_0]f'(x_0) + \frac{[x - x_0]^2}{2}f''(x_0) + \cdots . \qquad (5.12)$$

Here, x_0 denotes a specified reference value of x, $f'(x_0)$ represents the first derivative of the function at this reference value, and $f''(x_0)$ represents the second derivative evaluated at x_0. The neglected terms in this series involve higher order derivatives of $f(x)$ evaluated at x_0 and higher powers of the difference $x - x_0$. For any analytic function, then, there will exist some neighborhood of points x where $|x - x_0|$ is small enough that terms of second and higher order may be neglected. Consequently, within this neighborhood, it follows that:

$$f(x) \simeq [f(x_0) - x_0 f'(x_0)] + [f'(x_0)]x \equiv \alpha x + \beta. \qquad (5.13)$$

In other words, any analytic function may be approximated by a straight line sufficiently near the reference point x_0. In practice, this approximation is often reasonable over a physically important range, leading to linear equations like the linear constitutive relations considered in the previous section.

Conversely, it is also important to note that this relationship does fail if the approximation is stretched too far. As a specific example, note that electrical components that have been designed to be resistors (i.e., to satisfy Ohm's law) generally only do so over a specified operating range. A nice illustration of this point is provided by the *varistor*, an electronic component designed specifically to exhibit a nonlinear current-voltage relationship of the form:

$$I = AV^\alpha. \qquad (5.14)$$

Note that when $\alpha = 1$, this equation reduces to Ohm's law with $A = 1/R$, but for certain surge protection applications, much larger values of α are desirable [138, p. 17]. Generally, these devices are fabricated from zinc oxide (ZnO), doped with controlled amounts of specific impurities; the nonlinearity index α for these devices typically varies from $\alpha \sim 4$ to $\alpha \sim 50$ [186], and it depends strongly on the details of the doping and firing of the resulting ceramic [128].

Finally, it is worth emphasizing the point noted earlier that not all functions we encounter are analytic. For example, consider the square root:

$$f(x) = \sqrt{x} \;\Rightarrow\; f'(x) = \frac{1}{2\sqrt{x}}. \qquad (5.15)$$

It follows that $f'(x) \to \infty$ as $x \to 0$, meaning the function is not infinitely differentiable at $x = 0$ and therefore cannot be expanded as a Taylor series about $x_0 = 0$. Similarly, since analyticity implies differentiability of all orders and differentiability implies continuity, it follows that functions exhibiting discontinuities, cusps (i.e., slope discontinuities), or other such irregular behaviors are not analytic and do not have Taylor series expansions.

5.1.3   Allometry

The original dataset of Georg Ohm discussed in Chapter 1 did not exhibit a linear relation between the independent variable (conductor length) and the dependent variable (measured torsion). As is often the case, however, it was possible to transform this relationship into a linear one between related variables computable from the original data. An extremely broad application of this idea is the field of *allometry* [51], roughly defined as the study of how geometry influences biology. In fact, this topic dates at least as far back as Galileo in the early seventeenth century, who studied the relationship between bone geometry and strength, noting that if certain scaling laws did not hold, large animals' bones would be crushed under their own weight [268, p. 234]. More generally, the *allomeric equation*:

$$Y = aM^b, \tag{5.16}$$

has been used to describe observed scaling relations between the body mass M of an animal and many many physiological, morphological, and ecological variables [51, p. 26]. Specific examples include skeletal mass, metabolic rate, territorial range, running speed, incubation period, life span, and population density. A particular advantage of this equation is that it may be linearized, exactly as in the brain-weight, body-weight example considered in Chapter 1. Specifically, logarithmic transformations of both variables M and Y yield the linear equation:

$$\log Y = b \log M + c, \tag{5.17}$$

where $c = \log a$. Consequently, the methods described later in this chapter for fitting straight lines to observed data pairs may be used to estimate b and c from $\log M$ and $\log Y$; the original constant a may then be computed from c as $a = e^c$ if natural logarithms are used or $a = 10^c$ if common logarithms are used.

   In the discussion of linear approximations based on Taylor series arguments, it was noted that these arguments break down if the range of the independent variable x is allowed to vary too widely. Conversely, Calder [51, p. vii] emphasizes the importance of making the range of variation in the independent variable (i.e., the body mass) *large enough* to overcome variability due to secondary effects. In particular, he argues that allometric analysis become less reliable as the focus is restricted to progressively narrower classes of animals, noting that there are neither "moose-sized mice" nor "mouse-sized moose." Roughly, he argues that if variability on the order of $\pm 20\%$ is expected for some particular variable Y, this corresponds to the effects of $\sim 2\%$ variation in the mass across the class Mamalia, versus $\sim 25\%$ to $\sim 50\%$ variation across a single species. This observation illustrates the importance of accounting for uncertainty and inherent variability in observed datasets, one of the central points of this book.

5.1.4   Behavior and functional equations

Galileo's argument concerning the scaling between a bone's length and its diameter necessary to avoid exceeding material strength limits may be viewed as a *behavioral restriction*, a mathematical relationship between variables that is imposed on the basis of simple physical arguments. Often, such descriptions may be translated into *functional equations*, which relate the behavior of one or more unknown mathematical functions. Perhaps surprisingly, many functional equations that have arisen in this context turn out to have fairly simple solutions involving relatively few free parameters. In addition, it is often possible to transform the problem of estimating these parameters from observed data into one of fitting straight lines.

As a specific example, Galileo also considered the motion of falling bodies under the influence of gravity, where he made explicit use of the following characterization of the quadratic function $f(x) = c(x - x_0)^2$ [3, p. 359]:

$$\frac{f([n+1]x) - f(nx)}{f(nx) - f([n-1]x)} = \frac{2n + 1}{2n - 1}. \tag{5.18}$$

In fact, Galileo actually demonstrated empirically that the motion of falling bodies satisfied Eq. (5.18), consequently establishing the quadratic dependence of position on time. This characterization of quadratic functions goes back to Oresme in the fourteenth century, who showed that such functions are the only solution of this functional equation. If we wished to estimate the constants c and x_0 from observed data, one possible approach would be to use a nonlinear regression method like those discussed briefly in Chapter 15, but a much simpler option is to transform this problem to a line-fitting problem, just as in the allometry example discussed in the previous section. Specifically, note that if $y = c(x - x_0)^2$, then either $y \geq 0$ for all x (if $c \geq 0$) or $y < 0$ for all x (if $c < 0$). In the first case, $z = \sqrt{y}$ is well defined and is related to x by the linear equation:

$$z = \alpha x + \beta, \tag{5.19}$$

where $\alpha = \sqrt{c}$ and $\beta = -x_0\sqrt{c}$. Similarly, in the second case, $z = \sqrt{-y}$ is well defined and satisfies Eq. (5.19) with $\alpha = \sqrt{-c}$ and $\beta = -x_0\sqrt{-c}$.

Another simple example of an important functional equation is the *Cauchy functional equation:*

$$f(x_1 + x_2) = f(x_1) + f(x_2), \tag{5.20}$$

for all real x and y [3, p. 14]. It turns out that the linear constitutive relation (5.1) is the only *well-behaved* solution of this functional equation. More specifically, Eq. (5.20) does have other (i.e., nonlinear) solutions, but these are *incredibly* badly behaved: unbounded on every interval (no matter how small), nonmonotonic on every interval (again, no matter how small), and discontinuous at every point [3, p. 15]. Hence, if we can postulate on physical grounds that $y_1 = f(x_1)$ and $y_2 = f(x_2)$ imply that $x_1 + x_2$ should elicit the response $y_1 + y_2$, it follows that x and y must be related by a linear constitutive relation like those discussed in Sec. 5.1.1.

More generally, various authors [4, 191, 193, 245] have considered different aspects of the following problem. Suppose an observable variable y is to be explained in terms of n independent variables $x_1, \ldots, x_n$; in mathematical terms, this statement means we are seeking a function $f(\cdot)$ such that:

$$y = f(x_1, \ldots, x_n). \tag{5.21}$$

Now, suppose we change the units in which the independent variables x_i are measured and ask that this change result in only a unit change in the dependent variable y. Aczel et al. [4] consider 12 special cases of this problem that arise for different notions of changing units. For example, the conversion of mass from grams to kilograms is multiplicative: $x_i \rightarrow r_i x_i$ where $r_i > 0$; Aczel et al. refer to variables that undergo this type of conversion as *ratio scale variables*. Conversely, the conversion of temperature from the Farenheit to the Celsius scale is of the form $x_i \rightarrow r_i x_i + p_i$; Aczel et al. refer to variables subject to such affine scalings as *interval scale variables*. Depending on the type of scaling assumed for the variables x_i and y_i and the relationship (if any) between these different scalings, 12 different special cases may be identified, and the principal result of Aczel et al. is to derive and solve functional equations associated with each of these cases.

As a specific illustration, suppose each independent variable x_i represents a nonnegative *independent ratio scale variable* (i.e., $x_i \rightarrow r_i x_i$ with no relation assumed between the scalings r_i other than $r_i > 0$ for all i) and the dependent variable is also nonnegative and satisfies the ratio scaling condition $y \rightarrow Ry$ as a consequence. This case was originally considered by Luce [192] and represents one of the 12 considered by Aczel et al. [4]. Substituting this scaling requirement into Eq. (5.21) leads to the functional equation:

$$f(r_1 x_1, \ldots, r_n x_n) = R(r_1, \ldots, r_n) f(x_1, \ldots, x_n). \tag{5.22}$$

The general solution of this functional equation is then shown to be of the form:

$$f(x_1, \ldots, x_n) = a M_1(x_1) \cdots M_n(x_n), \tag{5.23}$$

where the functions $M_i(x)$ are each *multiplicative*, satisfying *Cauchy's power equation* [3, p. 29]:

$$M(xy) = M(x)M(y), \tag{5.24}$$

for all $x > 0$, $y > 0$. It turns out that the only continuous, nonzero solution of this functional equation is $M(x) = x^b$ for some constant b. Consequently, the only mathematical models of the general form (5.21) satisfying these scaling requirements are of the form:

$$y = a x_1^{b_1} \cdots x_n^{b_n}, \tag{5.25}$$

where a is an arbitrary positive constant and the constants b_1 through b_n are completely arbitrary. Finally, note that since $x_i > 0$ for all i and $y > 0$, we may take the logarithm of both sides of this equation, to obtain:

$$\log y = \log a + b_1 \log x_1 + \cdots + b_n \log x_n. \tag{5.26}$$

In other words, applying a logarithmic transformation to this general solution results in a linear regression model of the general type discussed in Chapter 13. For the special case, $n = 1$, this result reduces to the allomeric equation discussed in Sec. 5.1.3. This observation may explain the considerable utility of the allomeric model in describing certain biological scaling phenomena: Calder [51] devotes an entire 431-page book to this topic, covering species from hummingbirds to ostriches and from shrews to elephants. In addition, note that this equation also describes the current-voltage relationship of the zinc oxide varistors discussed in Sec. 5.1.2.

5.2 Do we fit y on x or x on y?

Often, datasets of points in the plane consist of one variable that may be regarded as an *independent* or *stimulus* variable and a second *dependent* or *response* variable. In such cases, it may be reasonable to suppose that the independent variable x is known more accurately than the dependent variable y, leading us to consider approximate models of the form:

$$y \simeq ax + b, \tag{5.27}$$

where the symbol "$\simeq$" means "is approximately equal to" and will be interpreted more precisely in subsequent discussions. Here, the point is that, once we adopt this description of a particular dataset, we can use a standard technique like the method of ordinary least squares (OLS) discussed in the next section to obtain estimates for the parameters a and b. Alternatively, if we exchange the roles of these variables in this model, we can also consider the approximate model:

$$x \simeq cy + d. \tag{5.28}$$

Once again, given this formulation, we can apply methods like OLS to estimate the parameters c and d from the available data. Further, once we have a model of the form (5.28), we may convert it into the form (5.27) since:

$$x = cy + d \; \Rightarrow \; y = \left(\frac{1}{c}\right) x + \left(-\frac{d}{c}\right). \tag{5.29}$$

Now, suppose we fit the dataset both ways and convert the second model to the form $y \simeq a'x + b'$ according to Eq. (5.29): how different will this model be from that given by Eq. (5.27)? Also, if these models are significantly different, what—if anything—does this difference tell us?

Figure 5.1: Two least squares fits to the same dataset

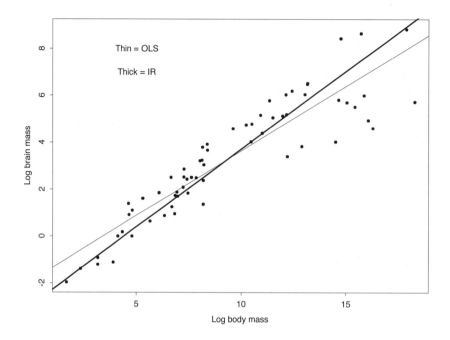

The results of such a comparison are shown in Fig. 5.1. There, the log-transformed brain-weight/body-weight dataset introduced in Chapter 1 is plotted, together with two lines fit to this transformed dataset. The thin line, designated "OLS" in the plot, was obtained by applying the least squares fitting procedure described in Sec. 5.3.2 to the dataset with $x = \log$ (body mass) and $y = \log$ (brain mass). The thick line, designated "IR" for "inverse regression," was obtained by the role-reversal procedure just described: first, a model of the form $x \simeq cy + d$ was fit to the dataset, and then the slope and intercept parameters a' and b' for the model $y = a'x + b'$ were computed from c and d as in Eq. (5.29). It is clear from the plot that the resulting lines have somewhat different slopes and intercepts. In this particular example, this difference is largely due to the cluster of six points lying well below both lines at the extreme right of the plot. The character of these points and their influence are examined in some detail in Chapter 7, where this example is revisited. Here, the key points are, first, that apparently small changes in the way we set up and solve a problem can cause significant differences in the solution, and second, that this particular difference reflects an asymmetry in the way the OLS procedure treats errors in the x and y variables, a topic examined in detail in the next section.

5.3 Three approaches to fitting lines

One of the main points of this chapter is that the problem of fitting lines to a given dataset may be approached in many different ways, each one generally giving a different numerical result and often permitting different interpretations or extensions. Consequently, before discussing any of these methods in detail, it is useful to first take a broad look at the range of approaches and the fundamental ways in which they differ. As a first step in this direction, it is useful to distinguish between the following two common working assumptions:

EM *errors-in-model* assumption: the observations x and y are precise, and all inaccuracy arises from inadequacies of the assumed linear model structure;

EV *errors-in-variables* assumption: the linear model structure is correct, but both x and y observations are uncertain.

Assumption EM is much more popular in practice than the EV formulation because it leads to simpler and better known results, but there are important cases where the EV formulation is inherently more reasonable. Because the EM assumption is more popular, the following sections begin with a discussion of this case, but Sec. 5.3.4 takes up the EV problem and introduces its most popular solution: the method of *total least squares* (TLS).

For the simpler EM formulation, many different solutions are available, and it is useful to partition these solutions into the following four classes:

a. optimization approaches,
b. random-variable approaches,
c. unknown-but-bounded approaches,
d. various "other" approaches.

Optimization-based approaches are generally the most popular in practice, and these methods are based on the idea that the observed data points should lie as close to some *best fit* line as possible. To compute this line, some measure of lack of fit is defined, and the slope and intercept parameters of a generic line are adjusted until this lack of fit measure is minimized. In contrast, the random variable approach assumes that the differences between the best-fit line and the available data points correspond to a sequence of random variables. In the *maximum likelihood approach* described in Sec. 5.4, slope and intercept parameters are chosen for the best-fit line so that the fit errors resulting from this line are most consistent with this probability distribution (i.e., the probability of observing this particular error sequence is maximum under the assumed random variable model). It is possible to solve the maximum likelihood estimation problem analytically for certain popular random variable models, and the resulting solutions turn out to be popular optimization-based approaches. As a specific and historically important example, assuming a Gaussian distribution for the line fit errors leads to the popular method of ordinary least squares discussed in Sec. 5.3.2. A few other random variable models also lead to simple optimization-based fitting procedures, but most lead to maximum likelihood fitting problems that are much more difficult to solve.

One alternative to random variable approaches that is increasing somewhat in popularity is the *unknown-but-bounded* or *set-theoretic* fitting problem. There, it is assumed that the deviations of all observed data points from the best-fit line must all lie within some specified *set*. Given a description of this set, unknown-but-bounded data-fitting methods seek to compute the corresponding *set* of all parameters (here, slopes and intercepts) that are consistent with the available data and the set specification for the data-fit errors. In general, exact solutions of this problem are computationally difficult, so much effort has gone into developing computationally efficient approximation procedures for these sets of feasible parameters. Because the basic idea is simple and the working assumptions are less restrictive than those of the random variable model, a brief discussion of this approach is given in Sec. 5.6, along with some useful references for those interested in exploring these methods further. As noted in Chapter 1, a number of "other" approaches to data-fitting problems have also been studied (e.g., the fuzzy set approach), but because these techniques are somewhat more complicated to explain and use, they have not yet come into widespread use in exploratory data analysis and are not considered in this book.

5.3.1   Optimization-based problem formulations

Optimization-based data-fitting methods are popular in practice because they generally lead to simple computational procedures, and in the most popular cases, these methods have strong connections with random variable models, as noted in the preceeding discussion. The usual optimization-based problem formulation involves fitting a model of the form:

$$\hat{y}(k) = a\hat{x}(k) + b, \tag{5.30}$$

to a dataset $\mathcal{D}$ of N data pairs, $(x(k), y(k))$. The quantities $\hat{x}(k)$ and $\hat{y}(k)$ appearing in this equation are approximations of the data values $x(k)$ and $y(k)$, respectively, that are defined more precisely in the following discussions. To simplify subsequent notation, define the vectors:

$$\mathbf{v}_k = (x(k), y(k)), \qquad \hat{\mathbf{v}}_k = (\hat{x}(k), \hat{y}(k)).$$

All of the approaches described here are specific implementations of the following general idea: the unknown coefficients a and b in Eq. (5.30) are chosen to minimize some measure $J(a, b)$ of the total distance between the predicted (x, y) pairs $\hat{\mathbf{v}}_k$ and the observed pairs $\mathbf{v}_k$. To accomplish this objective, it is common to first choose some measure $\rho : R^2 \to R$ of distance between points in the plane, and define d_k as the distance between the kth predicted and observed data point:

$$d_k = \rho(\hat{\mathbf{v}}_k, \mathbf{v}_k). \tag{5.31}$$

The "degree of mismatch" of the line defined by the parameters a and b in Eq. (5.30) and the dataset $\mathcal{D}$ is then given by:

$$J(a, b) = \sum_{k=1}^{N} d_k, \qquad (5.32)$$

and the *best fit* parameters a and b are determined by minimizing $J(a, b)$.

The parameters obtained by this approach, the detailed computational procedures required to obtain them, and the interpretation of the resulting models depend strongly on our choice of distance measure $\rho(\cdot, \cdot)$, as the following sections will illustrate. First, however, it is useful to note that if $\rho(\cdot, \cdot)$ satisfies the following two conditions:

$$\rho(\mathbf{v}, \mathbf{w}) \geq 0 \qquad \text{for all } \mathbf{v}, \mathbf{w} \text{ in } R^2,$$
$$\rho(\mathbf{v}, \mathbf{w}) = 0 \Rightarrow \mathbf{v} = \mathbf{w}, \qquad (5.33)$$

then $J(a, b)$ satisfies the conditions:

$$J(a, b) \geq 0 \qquad \text{for all } a \text{ and } b,$$
$$J(a, b) = 0 \Rightarrow \hat{\mathbf{v}}_k = \mathbf{v}_k \text{ for all } k. \qquad (5.34)$$

In other words, the minimum possible value of $J(a, b)$ is 0, corresponding to a "perfect fit" between the model in Eq. (5.30) and the dataset $\mathcal{D}$.

To proceed further, it is necessary to specify the function $\rho(\cdot, \cdot)$. The next three sections consider the following choices of the function $\rho(\cdot, \cdot)$, all of which satisfy the conditions (5.33):

1. $\rho(\hat{\mathbf{v}}_k, \mathbf{v}_k) = [\hat{y}(k) - y(k)]^2$,

2. $\rho(\hat{\mathbf{v}}_k, \mathbf{v}_k) = |\hat{y}(k) - y(k)|$,

3. $\rho(\hat{\mathbf{v}}_k, \mathbf{v}_k) = [\hat{y}(k) - y(k)]^2 + [\hat{x}(k) - x(k)]^2$.

The first of these cases corresponds to the method of *ordinary least squares*, without question the most popular data-fitting method in practice. Despite its popularity, however, this method suffers from at least two potential shortcommings, and both are serious. The first is sensitivity to outliers, a point discussed further in Chapters 7 and 13. One alternative that is less sensitive to the presence of outliers is the method of *least absolute deviations*, corresponding to the second choice of $\rho(\cdot, \cdot)$ listed here; this approach is closely related to the median introduced in Chapter 1 and a number of other useful ideas discussed elsewhere in this book. The second difficulty with the ordinary least squares method is that it is an EM formulation, assuming that the independent variable $x(k)$ is measured perfectly (i.e., observed without error); as noted previously, this assumption is not always reasonable in practice. The third choice of $\rho(\cdot, \cdot)$ listed here is a squared error criterion that assumes observation errors in both $x(k)$ and $y(k)$, corresponding to the EV problem formulation introduced at the beginning of this section.

5.3.2   The ordinary least squares (OLS) fit

The method of ordinary least squares, based on the errors in model (EM) formulation, is unquestionably the most popular data-fitting method in practice. The method goes back to Gauss in the early nineteenth century and has been almost the only approach widely employed in practice until fairly recently. Both because of this historical importance and because of its continuing popularity, it is sensible to begin our discussion of data fitting with this case. Substituting the first of the three expressions for $\rho(\cdot, \cdot)$ listed previously and the definition of $\hat{\mathbf{v}}_k$ implied by Eq. (5.30) into Eq. (5.32) leads to the following expression for $J(a, b)$:

$$J_{OLS}(a, b) = \sum_{k=1}^{N} [\hat{y}(k) - y(k)]^2 = \sum_{k=1}^{N} [y(k) - ax(k) - b]^2, \qquad (5.35)$$

where the subscript "OLS" is used to designate the least squares mismatch criterion. To find the best fit parameters a and b with respect to this criterion, note that $J_{OLS}(a, b)$ is a smooth function of both parameters, so the parameters minimizing $J_{OLS}(a, b)$ must satisfy the conditions $\partial J/\partial a = 0$ and $\partial J/\partial b = 0$.

It is easier to first differentiate with respect to b, leading to the result:

$$\sum_{k=1}^{N} [y(k) - a_{OLS}x(k) - b_{OLS}] = 0. \qquad (5.36)$$

Defining the averages $\bar{x}$ and $\bar{y}$:

$$\bar{x} = \frac{1}{N} \sum_{k=1}^{N} x(k), \qquad \bar{y} = \frac{1}{N} \sum_{k=1}^{N} y(k), \qquad (5.37)$$

and rearranging Eq. (5.36) then yields:

$$b_{OLS} = \bar{y} - a_{OLS}\bar{x}. \qquad (5.38)$$

Substituting this relationship into Eq. (5.35) leads to the following expression, depending only on the unknown parameter a:

$$J(a, b_{OLS}) = \sum_{k=1}^{N} \{[y(k) - \bar{y}] - a[x(k) - \bar{x}]\}^2. \qquad (5.39)$$

To simplify subsequent results, define:

$$\hat{\sigma}_x^2 = \frac{1}{N} \sum_{k=1}^{N} [x(k) - \bar{x}]^2, \qquad \hat{\sigma}_y^2 = \frac{1}{N} \sum_{k=1}^{N} [y(k) - \bar{y}]^2,$$

$$\hat{c}_{xy} = \frac{1}{N} \sum_{k=1}^{N} [x(k) - \bar{x}][y(k) - \bar{y}], \qquad (5.40)$$

which represent estimates of the *variance* of x and y, and the *covariance* between x and y, respectively. The expression for $J(a, b_{OLS})$ given above may then be simplified to:

$$J(a, b_{OLS}) \quad = \quad N[\hat{\sigma}_y^2 - 2a\hat{c}_{xy} + a^2\hat{\sigma}_x^2]. \tag{5.41}$$

Differentiating this expression with respect to a and setting the derivative to zero leads immediately to the result:

$$a_{OLS} \quad = \quad \frac{\hat{c}_{xy}}{\hat{\sigma}_x^2}. \tag{5.42}$$

The *correlation coefficient* between $\{x(k)\}$ and $\{y(k)\}$ is discussed further in Chapter 11 and is defined as:

$$\hat{\rho}_{xy} \quad = \quad \frac{\hat{c}_{xy}}{\sigma_x \sigma_y}. \tag{5.43}$$

It follows from the Cauchy-Schwartz inequality [39, p. 42] that $|\hat{\rho}_{xy}| \leq 1$, and this inequality is strict unless $y(k) = cx(k)$ for all k; in particular, $\hat{\rho}_{xy} = +1$ if $c > 0$ and $\hat{\rho}_{xy} = -1$ if $c < 0$. It is instructive to note that the least-squares slope estimate a_{OLS} can be reexpressed in terms of this correlation coefficient as:

$$a_{OLS} \quad = \quad \hat{\rho}_{xy} \frac{\sigma_y}{\sigma_x}. \tag{5.44}$$

Hence, for the EM problem formulation, the slope a_{OLS} of the least squares line fit to the data is equal to the correlation coefficient $\hat{\rho}_{xy}$, scaled by the ratio of the range of variation in y to the range of variation in x. Also, note that a_{OLS} and b_{OLS} define a line through the point $(\bar{x}, \bar{y})$ that may be regarded as the "center" of the data sequence $\{(x(k), y(k))\}$.

5.3.3 The least absolute deviations (LAD) fit

The second data-fitting problem introduced at the beginning of this section—the least absolute deviations problem—appears to be even older than the least squares method, having been used by Galileo in the mid-seventeenth century [91]. Also, it was noted previously that this method is less sensitive to outliers than is the method of least squares, but it is not as popular as least squares in part because the LAD criterion $J_{LAD}(a, b)$ is not smooth, and this fact complicates the optimization problem somewhat. Even worse, the solution of the LAD problem is not always unique, in contrast to the OLS problem; this point is illustrated by the simple case study discussed in Sec. 5.5.1. Proceeding as in the least squares problem, the mismatch $J_{LAD}(a, b)$ can be written as:

$$J_{LAD}(a, b) = \sum_{k=1}^{N} |y(k) - ax(k) - b|. \tag{5.45}$$

Because the absolute value function $f(x) = |x|$ is not differentiable at zero, it is not possible to develop the same type of analytic solution for the LAD problem that we have for the OLS problem. Instead, our best alternative is to reformulate the problem as a linear programming problem; prior to the simplex algorithm of Dantzig [80], these problems were difficult to solve, so the LAD approach was not a practical alternative to the method of least squares.

To convert the LAD problem considered here into a linear programming problem, first define the *nonnegative* variables e_k^- and e_k^+:

$$e_k^- = -\min\{0, e(k)\}, \qquad e_k^+ = \max\{0, e(k)\}, \tag{5.46}$$

where $e(k)$ is the prediction error, defined by:

$$e(k) = y(k) - ax(k) - b. \tag{5.47}$$

The advantage of defining the quantities e_k^- and e_k^+ is that both $e(k)$ and $|e(k)|$ may be expressed in terms of them:

$$e(k) = e_k^+ - e_k^-, \qquad |e(k)| = e_k^+ + e_k^-. \tag{5.48}$$

Hence, the LAD problem reduces to one of minimizing the *linear* objective function:

$$J_{LAD}(a, b) = \sum_{k=1}^{N} [e_k^+ + e_k^-], \tag{5.49}$$

subject to the *linear* constraints:

$$ax(k) + b + e_k^+ - e_k^- = y(k), \quad e_k^+ \geq 0, \quad e_k^- \geq 0, \tag{5.50}$$

for $k = 1, 2, \ldots, N$. A computationally efficient algorithm for solving this linear programming problem has been developed by Barrodale and Roberts [28], and extensions to handle problems with additional inequality constraints on the admissable parameter values are also available [29].

5.3.4   The total least squares (TLS) fit

A fundamental working assumption underlying the OLS method described in Sec. 5.3.2 is that the sequence $\{x(k)\}$ is essentially free of errors. Even when this assumption is not strictly satisfied, such EM problem formulations are often employed—particularly OLS formulations—and it may be argued that these results provide useful reference standards for comparison. Conversely, if the the sequence $\{x(k)\}$ does exhibit significant observation errors, it is known that EM solutions will be *biased*, a point illustrated and discussed in Sec. 5.5.2. Here, the main points are the following: first, the presence of errors in the data sequence $\{x(k)\}$ causes a degradation of the performance of the OLS estimation procedure described in Sec. 5.3.2, and second, an alternative methodology is available for errors-in-variables problems, commonly known as *total least squares* (TLS).

For the case of the simple linear model (5.30) considered here, the TLS problem is formulated as follows. Substituting the third distance measure listed at the end of Sec. 5.3.1 into Eq. (5.32) leads to the following expression for the TLS mismatch criterion:

$$J_{TLS}(a,b) = \sum_{k=1}^{N} \{[\hat{y}(k) - y(k)]^2 + [\hat{x}(k) - x(k)]^2\}$$

$$= \sum_{k=1}^{N} \{[y(k) - a\hat{x}(k) - b]^2 + [\hat{x}(k) - x(k)]^2\}. \qquad (5.51)$$

To simplify this expression further, it is necessary to determine $\hat{x}(k)$: the OLS problem considered in Sec. 5.3.2 corresponds to the choice $\hat{x}(k) = x(k)$, consistent with the idea that the sequence $\{x(k)\}$ is observed without error. In the TLS problem considered here, we determine $\hat{x}(k)$ by solving the following optimization problem: for *fixed* parameters a and b and data values $x(k)$ and $y(k)$, we choose $\hat{x}(k)$ so that the distance:

$$\rho(\hat{\mathbf{v}}_k, \mathbf{v}_k) = [\hat{x}(k) - x(k)]^2 + [y(k) - a\hat{x}(k) - b]^2, \qquad (5.52)$$

is minimized. Because $\rho(\hat{\mathbf{v}}_k, \mathbf{v}_k)$ is a smooth function of $\hat{x}(k)$, we may proceed as in the OLS problem, differentiating it with respect to $\hat{x}(k)$ and setting the derivative to zero. This process leads immediately to the following expression for $\hat{x}(k)$ in terms of a, b, $x(k)$, and $y(k)$:

$$\hat{x}(k) = \frac{x(k) + a[y(k) - b]}{1 + a^2}. \qquad (5.53)$$

Substituting this result into the linear model equation (5.30) gives:

$$\hat{y}(k) = \frac{a^2 y(k) + ax(k) + b}{1 + a^2}. \qquad (5.54)$$

Geometrically, the point $\hat{\mathbf{v}}_k = (\hat{x}(k), \hat{y}(k))$ represents the prediction of $\mathbf{v}_k = (x(k), y(k))$ that lies on the line $y = ax + b$ and minimizes the orthogonal distance to $\mathbf{v}_k$. For this reason, the TLS method is also referred to as *orthogonal distance regression* or ODR. Given these expressions for $\hat{x}(k)$ and $\hat{y}(k)$, $\rho(\hat{\mathbf{v}}_k, \mathbf{v}_k)$ may be simplified to:

$$\rho(\hat{\mathbf{v}}_k, \mathbf{v}_k) = \frac{[y(k) - ax(k) - b]^2}{1 + a^2}. \qquad (5.55)$$

Hence, the TLS mismatch criterion $J_{TLS}(a, b)$ may be expressed as:

$$J_{TLS}(a,b) = \frac{1}{1 + a^2} \sum_{k=1}^{N} [y(k) - ax(k) - b]^2 = \frac{J_{OLS}(a,b)}{1 + a^2}. \qquad (5.56)$$

As in the OLS problem, $J_{TLS}(a, b)$ is a smooth function of a and b, so we may proceed as before to obtain analytic expressions for the best-fit model parameters a_{TLS} and b_{TLS}.

Once again, it is simpler to first solve for b_{TLS} in terms of a_{TLS} by computing $\partial J(a,b)/\partial b$ and setting it to zero. In fact, it follows from Eq. (5.56) that:

$$\frac{\partial J_{TLS}(a,b)}{\partial b} = \frac{1}{1+a^2}\frac{\partial J_{OLS}(a,b)}{\partial b}. \tag{5.57}$$

Consequently, $\partial J_{TLS}/\partial b = 0$ implies $\partial J_{OLS}/\partial b = 0$, so the intercept b_{TLS} bears exactly the same relationship to the slope a_{TLS} that b_{OLS} bears to a_{OLS}; i.e.:

$$b_{TLS} = \bar{y} - a_{TLS}\bar{x}, \tag{5.58}$$

where $\bar{x}$ and $\bar{y}$ are the data means defined in Eq. (5.37). Similarly, substituting Eq. (5.58) into the expression for $J_{TLS}(a,b)$ simplifies the problem to the minimization of a smooth function of the single variable a_{TLS}. Before making that simplification, however, it is useful to first note that:

$$\frac{\partial J_{TLS}(a,b)}{\partial a} = \frac{1}{1+a^2}\left\{\left(\frac{-2a}{1+a^2}\right)J_{OLS}(a,b) + \frac{\partial J_{OLS}(a,b)}{\partial a}\right\}, \tag{5.59}$$

from which it follows that $\partial J_{OLS}(a,b)/\partial a = 0$ implies $\partial J_{TLS}(a,b)/\partial a \neq 0$ unless one of the following two conditions hold:

1. $J_{OLS}(a,b) = 0$,
2. $a = 0$.

Hence, it is only in these two special cases that the OLS and TLS solutions will coincide. The first of these conditions implies that the line $y = ax+b$ is a *perfect fit* to all of the available data, and the second case can occur only when the OLS solution is the constant model $\hat{y}(k) = b$. Geometrically, the OLS solution to the line-fitting problem is the line that minimizes the sum of the squared *vertical* distances between the data points and the line $y = ax + b$; in contrast, the TLS solution minimizes the *orthogonal* distance, as noted. Condition 2 holds when these two distances are the same: the orthogonal distance from the horizontal line $y = b$ is vertical. Also, note that this solution arises in the least squares problem when $\hat{c}_{xy} = 0$ or, equivalently, $\hat{\rho}_{xy} = 0$. Because this case is special, the following discussion first considers the general TLS solution, for which $\hat{c}_{xy} \neq 0$.

Substituting Eqs. (5.41) and (5.58) into Eq. (5.56) yields:

$$J_{TLS}(a,b_{TLS}) = N\left[\frac{\hat{\sigma}_y^2 - 2a\hat{c}_{xy} + a^2\hat{\sigma}_x^2}{1+a^2}\right]. \tag{5.60}$$

Differentiating this expression with respect to a and setting the result to zero yields the following quadratic equation for the best fit model parameter a_{TLS}:

$$a^2 + \left[\frac{\hat{\sigma}_x^2 - \hat{\sigma}_y^2}{\hat{c}_{xy}}\right]a - 1 = 0. \tag{5.61}$$

Defining $\lambda = [\hat{\sigma}_x^2 - \hat{\sigma}_y^2]/\hat{c}_{xy}$ and solving Eq. (5.61) yields:

$$a_{TLS} = \frac{-\lambda \pm \sqrt{\lambda^2 + 4}}{2}. \tag{5.62}$$

This equation has two solutions, one corresponding to the desired best-fit solution, and the other corresponding to a local *maximum* in the degree of mismatch between the data and the linear fit. To select the best fit solution, it is necessary to examine the second derivative of $J_{TLS}(a, b)$ with respect to a:

$$\frac{\partial^2 J_{TLS}(a, b_{TLS})}{\partial a^2} \Big|_{a=a_{TLS}} = \left[\frac{2N}{(1+a^2)^2}\right] (\lambda + 2a)\hat{c}_{xy}. \qquad (5.63)$$

It follows from Eq. (5.62) that:

$$\lambda + 2a_{TLS} = \pm\sqrt{\lambda^2 + 4}, \qquad (5.64)$$

reducing Eq. (5.63) to:

$$\frac{\partial^2 J_{TLS}(a, b_{TLS})}{\partial a^2} \Big|_{a=a_{TLS}} = \left[\frac{2N\sqrt{\lambda^2 + 4}}{(1+a^2)^2}\right] (\pm\hat{c}_{xy}). \qquad (5.65)$$

Note that a will minimize $J_{TLS}(a, b_{TLS})$ if the sign of this second derivative is positive, implying that we must take the sign in Eq. (5.62) to be the same as that of $\hat{c}_{xy}$; note that since σ_x and σ_y are positive quantities, it follows that this is the same as the sign of the correlation coefficient $\hat{\rho}_{xy}$ defined in Eq. (5.44). Further, note that since $\sqrt{\lambda^2 + 4} > |\lambda|$, it follows that a_{TLS} is positive if the positive sign is taken and negative if the negative sign is taken, qualitatively consistent with the OLS result.

Returning to the special case $\hat{c}_{xy} = 0$, proceed by substituting this value into Eq. (5.60) and rearranging to obtain:

$$J_{TLS}(a, b_{TLS}) = N \left[\frac{\hat{\sigma}_y^2 - \hat{\sigma}_x^2}{1 + a^2} + \hat{\sigma}_x^2\right]. \qquad (5.66)$$

Note that if $\hat{\sigma}_x^2 > \hat{\sigma}_y^2$, the first term is negative, and J will be minimized by choosing $a = 0$, corresponding to the OLS solution discussed in Sec. 5.3.2. On the other hand, if $\hat{\sigma}_x^2 < \hat{\sigma}_y^2$, the function J defined in Eq. (5.66) does not have a finite minimizer but approaches its infimum as $|a| \to \infty$, corresponding to a vertical line. Also, note that this case would correspond to a line of zero slope if we reversed the roles of x and y. Finally, if $\hat{c}_{xy} = 0$ and $\hat{\sigma}_x^2 = \hat{\sigma}_y^2$, it follows from Eq. (5.60) that $J_{TLS}(a, b_{TLS}) = N\hat{\sigma}_y^2$, independent of a. In this case, the TLS solution is not uniquely defined.

5.4 The method of maximum likelihood

The optimization-based fitting procedures described in the previous section each represent a different implementation of the general idea that slope and intercept parameters a and b should be chosen so that the resulting fit errors e_k are "small" in some useful sense. This section introduces an alternative philosophy, although one that can lead us to the same data fitting procedures in some cases.

Specifically, the *method of maximum likelihood* assumes that the fit errors $\{e_k\}$ may be viewed as a statistically independent, identically distributed sequence of random variables, all drawn from a common (and known) probability distribution. Under this assumption, given a *candidate* line (i.e., slope and intercept parameters) and the available data, it is possible to compute the resulting fit error sequence $\{e_k\}$ *and assign to this sequence a probability*. Because this probability depends on the parameters of the candidate line, this probability may be associated with these parameters. Adjusting these parameters until this probability is maximized leads to the method of maximum likelihood. Sec. 5.4.1 presents a more detailed discussion of the mechanics of this method (i.e., how do we go from a specific probability distribution to something we can actually compute from data?), and Sec. 5.4.2 specializes these results to three important cases: the Gaussian, Laplace, and uniform probability distributions.

5.4.1   The basic concept

The method of maximum likelihood is an extremely general approach that may be applied to many different problem formulations. Here, to illustrate the basic ideas involved, consider the errors-in-model (EM) formulation of the line-fitting problem. In this problem formulation, given any candidate parameter values a and b defining a line in the plane and a dataset $\mathcal{D}$ of data pairs (x_k, y_k), the fit errors associated with this line are:

$$e_k(a, b) = \hat{y}_k - y_k = ax_k + b - y_k. \tag{5.67}$$

In the method of maximum likelihood, a random variable model is assumed for the fit error sequence, and the parameters a and b are chosen to make the *observed* fit error sequence $\{e_k(a, b)\}$ as consistent as possible with this probability model. The simplest problem formulation assumes that all members of this fit error sequence are both *statistically independent* and *identically distributed*, all described by some common probability density function $p(e)$. Such sequences are called *independent, identically distributed* (i.i.d.) random variable sequences and are extremely important in practice because they appear as a basic working assumption in many different data analysis procedures. For this reason, this notion is discussed at some length in Chapter 10, but for now it is enough to note two points: first, that this assumption is invoked very frequently in practice, and second, that this assumption greatly simplifies the derivation of the basic maximum likelihood method.

Under the assumption that $\{e_k(a, b)\}$ is identically distributed, it follows that each individual member of this sequence may be associated with a common probability density function $p(e)$. The independence assumption then allows us to combine these individual error descriptions to obtain the following *joint probability density* $p(\mathbf{e})$ for the entire sequence of fit errors:

$$
\begin{aligned}
p(\mathbf{e}) &= p(\{e_k = ax_k + b - y_k, k = 1, 2, \ldots, N\}) \\
&= \prod_{k=1}^{N} p(e_k = ax_k + b - y_k). \tag{5.68}
\end{aligned}
$$

In particular, the consequence of the statistical independence assumption here is that we may write the joint density $p(\mathbf{e})$ associated with the entire sequence $\{e_k\}$ as the product of the individual densities associated with each element of the sequence, an idea discussed further in Chapters 7 and 11. Given a probability density $p(e)$ and line parameters a and b, the direct interpretation of Eq. (5.68) would be as a probability of observing the dataset $\mathcal{D}$ of data pairs (x_k, y_k). The idea of maximum likelihood estimation is to reverse this interpretation: assuming the dataset $\mathcal{D}$ is given, treat $p(\mathbf{e})$ as a probability that the parameters a and b are consistent with the assumed common density $p(e)$. Under this assumption, the quantity $p(\mathbf{e})$ defined here is usually designated $L(a, b)$ and called the *likelihood* associated with the parameters a and b. The basic idea of maximum likelihood estimation is to choose a and b to maximize $L(a, b)$.

Often, it is much more convenient to solve the equivalent problem of maximizing the *log likelihood* function $\ell(a, b)$, defined as the logarithm of $L(a, b)$:

$$\ell(a, b) = \ln L(a, b) = \sum_{k=1}^{N} \ln\, p(e_k = ax_k + b - y_k). \qquad (5.69)$$

The advantages of this transformation are, first, that it converts the product defining $L(a, b)$ into a sum, and second, that the logarithm of many popular probability density functions is mathematically simpler than the density function itself. This point is demonstrated in the next section, which presents detailed descriptions of the mechanics of minimizing $\ell(a, b)$ for three particular choices of $p(e)$.

5.4.2   Three specific maximum likelihood solutions

The question of which probability distribution is most reasonable to choose in problems involving random variables is discussed at some length in Chapter 3. The following discussion focuses on three particular choices, for several reasons. First, all three of the random variable models considered here are reasonably popular as approximate error distributions, and each may be regarded as "most reasonable" in the sense of maximum entropy under the appropriate assumptions (see Chapter 3 for a brief discussion of this idea). Second, the three distributions considered here all lead to simple and popular data-fitting procedures, two of which have already been examined in detail: the methods of ordinary least squares (OLS) and least absolute deviations (LAD). Finally, these three choices provide simple illustrations of the range of optimization problems that result when we attempt to maximize likelihood.

Without question, the Gaussian distribution has enjoyed the greatest historical popularity, a point that has been noted in Chapters 2 and 4. For the Gaussian maximum likelihood problem, we assume that $\{e_k\}$ is a zero-mean, Gaussian i.i.d. sequence with variance σ^2, and under this assumption, the density $p(e_k)$ associated with any individual element of this error sequence is:

$$p(e_k) = \frac{1}{\sqrt{2\pi}\sigma} \exp\left[-\frac{e_k^2}{2\sigma^2}\right]. \qquad (5.70)$$

Substituting this expression into Eq. (5.69) then leads to the following expression for the log likelihood:

$$
\begin{aligned}
\ell(a,b) &= \sum_{k=1}^{N}\left[-\frac{1}{2}\ln(2\pi) - \ln\sigma - \frac{(ax_k + b - y_k)^2}{2\sigma^2}\right] \\
&= -\frac{N}{2}\ln(2\pi) - N\ln\sigma - \frac{1}{2\sigma^2}\sum_{k=1}^{N}(y_k - ax_k - b)^2. \quad (5.71)
\end{aligned}
$$

Note that the first two terms in this expression do not depend on the line parameters a and b, so maximization of $\ell(a,b)$ depends only on maximization of the last term in this expression. Further, this last expression involves a sum of nonnegative terms, multiplied by the negative factor $-1/2\sigma^2$, which is independent of the parameters a and b. Hence, maximizing $\ell(a,b)$ is equivalent to minimizing this sum of squared prediction errors. *In other words, the maximum likelihood solution for the Gaussian error distribution is simply the OLS estimator discussed in Sec. 5.3.2.* That is, we have the following correspondence for the Gaussian case:

$$
\max_{a,b}\ \ell(a,b) \iff \min_{a,b}\sum_{k=1}^{N}(y_k - ax_k - b)^2. \quad (5.72)
$$

One important practical observation is that large errors e_k are often observed more frequently in practice than expected under the Gaussian error model, a point discussed at some length in Chapter 7. A useful alternative in such cases is the *Laplace distribution* that was introduced in Chapter 4. As noted there, this distribution is described by the probability density function:

$$
p(e_k) = \frac{1}{2\phi}e^{-|e_k|/\phi}, \quad (5.73)
$$

where ϕ is a positive constant quantifying the spread of the distribution, analogous to the variance parameter σ^2 appearing in the Gaussian distribution. Substituting this error distribution into Eq. (5.69) for the log likelihood function yields:

$$
\begin{aligned}
\ell(a,b) &= \sum_{k=1}^{N}\left[-\ln 2 - \ln\phi - \frac{|y_k - ax_k - b|}{\phi}\right] \\
&= -N\ln 2 - N\ln\phi - \frac{1}{\phi}\sum_{k=1}^{N}|y_k - ax_k - b|. \quad (5.74)
\end{aligned}
$$

As in the Gaussian case just considered, note that the first two terms in this expression are independent of the line parameters a and b, and the last term is a sum of nonnegative quantities, multiplied by the negative factor $-1/\phi$. Hence, it again follows that maximization of $\ell(a,b)$ is equivalent to minimizing the sum appearing in this last term. *In other words, the maximum likelihood estimator for the Laplace distribution is simply the LAD solution described in Sec. 5.3.3.* More specifically, we have the following correspondence for the Laplace case:

$$\max_{a,b} \ell(a,b) \ \Leftrightarrow \ \min_{a,b} \sum_{k=1}^{N} |y_k - ax_k - b|. \tag{5.75}$$

Finally, the last example assumes that the error sequence $\{e_k\}$ is *uniformly distributed* on some interval $[-e^*, e^*]$, and this case leads to a new estimator that has not been discussed previously. Specifically, the error distribution in this case is assumed to be:

$$p(e_k) = \begin{cases} 1/2e^* & -e^* \le e_k \le e^*, \\ 0 & \text{otherwise}. \end{cases} \tag{5.76}$$

Here, the resulting maximum likelihood problem is a little more involved than in the previous two examples. In particular, note that if $|e_k(a,b)| > e^*$ for any k, it follows that $p(e_k) = 0$, implying $L(a,b) = 0$ and, as a consequence, $\ell(a,b) = -\infty$. Hence, to have a well-posed maximization problem, it is necessary to impose the following constraint for all k:

$$-e^* \ \le \ y_k - ax_k - b \ \le \ e^*. \tag{5.77}$$

If this constraint is satisfied for all k, it follows that $p(e_k) = 1/2e^*$ for all k. If it is assumed that e^* is fixed, analogous to the assumptions made in the previous examples that σ^2 and ϕ were fixed, any parameter values a and b satisfying the inequalities (5.77) have the same likelihood, $\ell(a,b) = -N \ln 2 - N \ln e^*$. Conversely, if we regard e^* as a free parameter to be optimized, it follows that $\ell(a,b)$ is maximized by minimizing e^*, subject to the constraints (5.77). This strategy leads to the *min-max* or *Chebyshev* estimator [289, ch. 2]:

$$\max_{a,b} \ell(a,b) \ \Leftrightarrow \ \min_{a,b} \ \max_{k=1,2,\dots,N} |y_k - ax_k - b|. \tag{5.78}$$

In view of this last example, it is useful to revisit the role of the distribution parameters σ^2 and ϕ in the previous two examples. In the derivations given here for the maximum likelihood estimators, these parameters were assumed to be fixed. Instead, if we regard them as optimization variables as in the uniformly distributed example just considered, we obtain maximum likelihood estimates for these parameters as well. Derivations are left as exercises (see Sec. 5.8), but the results we obtain are the following:

$$\hat{\sigma^2} \ = \ \frac{1}{N} \sum_{k=1}^{N} (y_k - \hat{a}x_k - \hat{b})^2,$$

$$\hat{\phi} \ = \ \frac{1}{N} \sum_{k=1}^{N} |y_k - \hat{a}x_k - \hat{b}|, \tag{5.79}$$

where $\hat{a}$ and $\hat{b}$ are the parameter estimates obtained from each of these maximum likelihood solutions.

Finally, it is useful to note that all three of these solutions may be viewed as minimizations of a *norm* of the error sequence $\{e_k\}$ [177, sec. 2.2], commonly given the following designations:

L_1 LAD fit: $\|\mathbf{e}\|_1 = \sum_{k=1}^{N} |e_k|$,

L_2 OLS fit: $\|\mathbf{e}\|_2 = \left[\sum_{k=1}^{N} e_k^2\right]^{1/2}$,

L_∞ Chebyshev fit: $\|\mathbf{e}\|_\infty = \max_{k=1,2,\ldots,N} |e_k|$.

5.5 Two brief case studies

The preceeding sections of this chapter have introduced a number of different approaches to the problem of fitting straight lines to a given dataset, and most of the discussion has focused on differences in the character of each and the underlying working assumptions on which they are based. To help in the task of deciding which method to use in practice, the following subsections present two brief case studies comparing specific methods or formulations to focus on various important aspects of the line fitting problem.

5.5.1 Case study 1: L_1 vs. L_2 vs. L_∞

This first case study compares the three maximum likelihood EM solutions introduced in Sec. 5.4.2, corresponding to the L_1 or LAD fit, the L_2 or OLS fit, and the L_∞ or Chebyshev fit. This case study is based on an extremely small, simulation-based dataset that is not representative of what we are likely to see in practice, but that does clearly illustrate some of the important practical differences between these methods. Specifically, this example is based on perturbations of the following four-point dataset:

$$y_k = x_k, \quad x_k = k - 1, \quad k = 1, 2, 3, 4. \tag{5.80}$$

The perturbations considered here each consist of a single additive outlier corrupting the y_ℓ value at the position $\ell = 1$, 2, 3, or 4:

$$y_k \;\rightarrow\; x_k + o_k, \quad o_k = \begin{cases} 1 & k = \ell, \\ 0 & k \neq \ell. \end{cases}$$

Plots of these datasets are shown in Fig. 5.2, which also show the OLS solutions obtained for each case. Several points are noteworthy, particularly in comparison with the other two methods considered here. First, note that none of the four lines shown in Fig. 5.2 fit any of the data points exactly: they pass near some of the points in each example, but none of the lines pass *through* any of these points. Also, the effect of each outlying point is to draw the OLS line away from the ideal line $y_k = x_k$, in the direction of the outlier. This feature is characteristic of OLS fits and reflects the fact that the OLS distance criterion penalizes large fit errors heavily, forcing the OLS solution to *accommodate* the outliers. One consequence of this accommodation is that the fit errors associated with outliers

Figure 5.2: Four contaminated datasets and their OLS fits

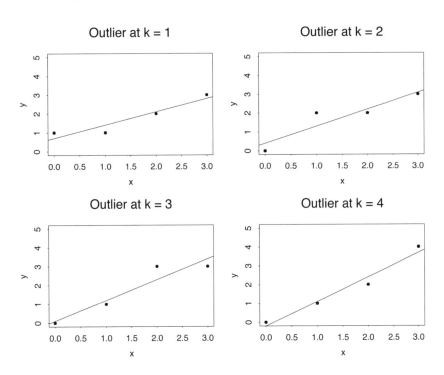

can be *smaller* than those associated with nonoutying data points. This point may be seen clearly in the upper left plot, where the outlier occurs at $k = 1$ but the largest fit error occurs for the regular data point $k = 2$. Finally, note that the outlier sensitivity of the OLS fit depends strongly on where the outlier occurs: here, outliers at $k = 1$ and $k = 4$ have a larger effect than do those at $k = 2$ or $k = 3$. This phenomenon is called *leverage*, is extremely important in practice, and is discussed further in Chapter 15.

Fig. 5.3 shows the best fit lines obtained by minimizing the L_∞ or Chebyshev error criterion: the slope and intercept parameters are chosen to minimize the maximum fit error. No analytic solution exists for the Chebyshev problem, but computational procedures are available [289, ch. 2]. Also, the solution of the Chebyshev fitting problem is known to be unique if the *Haar condition* discussed in Chapter 1 is satisfied [289, p. 32]. For the line-fitting problem, as in the case of the overfitting problem discussed in Chapter 1, the Haar condition reduces to one of distinctness of the x_k values; that is, the condition is satisfied if $j \neq k$ implies $x_j \neq x_j$, as it does in this example. In addition, Fig. 5.3 illustrates the fact that the Chebyshev fitting procedure forces the fit errors for all points to be equal in magnitude; that is, $|e_k| = e^*$ for all k, where e^* is the magnitude of the maximum fit error that is minimized by the Chebyshev fitting procedure.

Figure 5.3: Four contaminated datasets and their Chebyshev fits

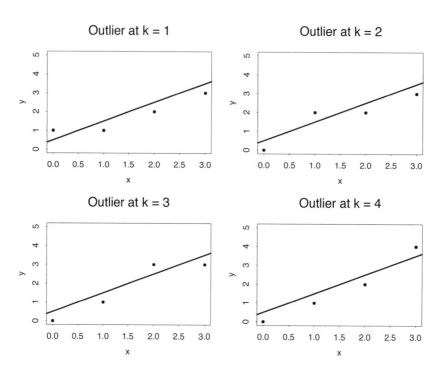

This observation has two immediate consequences: first, the outlier sensitivity of the Chebyshev procedure is severe (worse even than OLS sensitivity), and second, the Chebyshev procedure is *not* sensitive to leverage effects. Both of these points may be seen clearly in Fig. 5.3: in all cases, the slope of the line is equal to the slope of the ideal line, but the intercept is shifted by an amount equal to half of the magnitude of the outlier o_k, independent of where it appears in the dataset.

Fig. 5.4 shows the best fit lines obtained by minizing the LAD criterion: in all cases, the result is identical to the exact line obtained for the outlier-free dataset, independent of where the outlier appears in the dataset. Unfortunately, this result does not tell the whole story because, as noted in Sec. 5.3.3, the LAD fit is not unique. This point is illustrated in Fig. 5.5, which shows two lines, both achieving the same total L_1 error sum of 1: the lighter line corresponds to the lower left plot in Fig. 5.4, while the darker line fits the outlier at $k = 1$ and the good data point at $k = 3$ exactly but yields $|e_k| = 1/2$ for $k = 2$ and $k = 4$. In both cases, note that the best fit lines correspond to an *exact fit* to a subset of the data, a feature characteristic of the LAD procedure [289, ch. 6]. If this subset corresponds to error-free nominal data, the resulting fit can be perfect,

Figure 5.4: Four contaminated datasets and their LAD fits

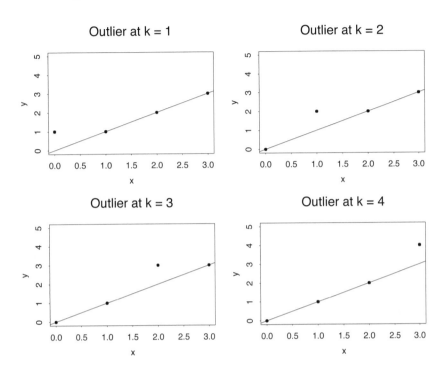

as in Fig. 5.4, but if this subset contains outliers, the resulting fit can be worse than that obtained by either OLS or Chebyshev methods, as in Fig. 5.5.

A comparison of the results obtained by the three different fitting procedures considered here is given in Table 5.1. Overall, the advantages and disadvantages of each method can be summarized for this example as follows. The OLS procedure has the advantage of exhibiting a unique solution, but this fit is sensitive to outliers and exhibits the phenomenon of leverage: the influence of these outliers depends on where they occur in the dataset. The Chebyshev approach also yields a unique solution, and it has the advantage of being insensitive to leverage, but it is also quite sensitive to outliers; in fact, because the best fit is determined entirely by the worst error, Chebyshev solutions generally exhibit even worse outlier sensitivity than OLS solutions. Finally, the LAD procedure can yield the exact fit for this example, unlike either of the other two approaches, but the best-fit solution is not unique, and the slope and intercept estimates obtained by this method can be worse than those obtained by these other methods. This point is illustrated in Table 5.1, where the parameter values in parentheses for $k = 1$ and $k = 4$ achieve the same fit errors as the "true" values of a and b. In addition, this method also suffers from leverage effects like those seen in the OLS solution.

Figure 5.5: Nonuniqueness of the LAD fit

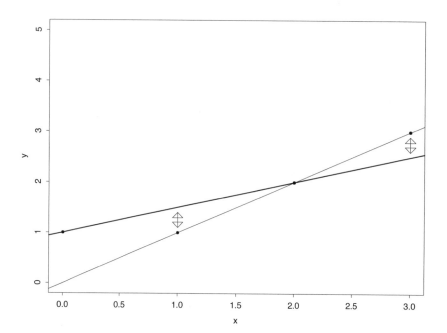

5.5.2   Case study 2: OLS vs. TLS

The following case study compares the EM and EV problem formulations for
the L_2 goodness-of-fit measure. Specifically, this case study compares the EM-
based OLS estimator with the EV-based TLS estimator for the following eight
scenarios, to illustrate the influence of the actual character of the data, the
assumed character of the data, and the size of the available dataset:

1. OLS estimate, EM simulation, $N = 10$;
2. OLS estimate, EV simulation, $N = 10$;
3. TLS estimate, EM simulation, $N = 10$;
4. TLS estimate, EV simulation, $N = 10$;
5. OLS estimate, EM simulation, $N = 100$;
6. OLS estimate, EV simulation, $N = 100$;
7. TLS estimate, EM simulation, $N = 100$;
8. TLS estimate, EV simulation, $N = 100$.

In all cases, the independent variable sequence $\{x_k\}$ is of length N and is generated from the model:

$$x_k = \frac{k - 1}{N - 1} + \nu_k, \tag{5.81}$$

Table 5.1: Comparison of L_1, L_2, and L_∞ parameter estimates

Outlier location	OLS a_2	Chebyshev a_∞	LAD a_1	OLS b_2	Chebyshev b_∞	LAD b_1
$k = 1$	0.7	1.0	1.0 (0.5)	0.7	0.5	0.0 (1.0)
$k = 2$	0.9	1.0	1.0	0.4	0.5	0.0
$k = 3$	1.1	1.0	1.0	0.1	0.5	0.0
$k = 4$	1.3	1.0	1.0 (1.5)	-0.2	0.5	0.0 (-0.5)
Exact:	1.0	1.0	1.0	0.0	0.0	0.0

where $\{\nu_k\}$ is defined in one of two ways: for errors-in-model (EM) simulations, $\nu_k = 0$ for all k, and for errors-in-variables (EV) simulations, $\{\nu_k\}$ is a sequence of N statistically independent, zero-mean Gaussian random variables with standard deviation 0.05. The dependent variable sequence $\{y_k\}$ is also of length N, generated according to the following model:

$$y_k = a \left(\frac{k-1}{N-1} \right) + b + \epsilon_k, \tag{5.82}$$

where $\{\epsilon_k\}$ is a sequence of N statistically independent, zero-mean Gaussian random variables with standard deviation 0.05, exactly like the sequence $\{\nu_k\}$ used in the EM simulations. It is important to note, however, that the sequences $\{\nu_k\}$ and $\{\epsilon_k\}$ are statistically independent from each other (for a more detailed discussion of the statistical independence of two random variable sequences, refer to Chapter 10). The parameter values assumed in all simulations were $a = 0.6$ and $b = 0.4$, yielding response values y_k in the approximate range from 0.4 to 1.0 as x_k varied from approximately 0 to 1 (exactly so in the EM simulations). Two EM simulation datasets and two EV simulation datasets, each of length $N = 100$, are shown in Fig. 5.6: the upper two plots correspond to EM simulations, and the lower two plots correspond to EV simulations. The key point is that the differences between datasets simulated under these two different error models are not visually obvious in these data plots.

The differences in the results obtained from these eight simulation experiments are shown in figs. 5.7 and 5.8, which present boxplot summaries of the estimated slope parameter a and intercept parameter b, respectively, obtained from 100 simulations of each of these eight scenarios. The boxplots shown on the left of the double vertical line in both figures correspond to the results obtained

Figure 5.6: Simulated EM and EV datasets

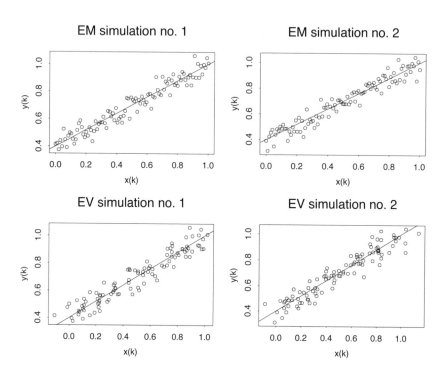

from the small datasets ($N = 10$), and these results exhibit a much wider range of variation than those obtained from the larger datasets ($N = 100$) shown to the right of the double vertical line. These differences in the variability of the parameter estimates simply reflect the differences in dataset size and are the consequence of general results discussed at some length in Chapter 6: estimates computed from larger datasets of size N' are typically less variable than those computed from smaller datasets of size N by a factor of $\sqrt{N'/N}$. This observation suggests a variability reduction by ~ 3 for the larger datasets, consistent with the results observed here.

The differences that are more specific to this example are those due to mismatch between the assumed formulation (EM or EV) and the actual formulation used to generate the data. These differences are most easily seen in the results for the larger datasets, corresponding to experiments 5 through 8. In particular, in both experiments 5 and 8, there is no mismatch between the assumed formulation and the true formulation, and the results are, on average, correct; this behavior is illustrated by the nearly perfect agreement between the median value in both boxplots (the white line in the center of the plot) and the horizontal lines at the true parameter values $a = 0.6$ and $b = 0.4$ in these plots. In contrast, in experiments 6 and 7, there is a mismatch between assumed and

Figure 5.7: Boxplots, estimated a parameters

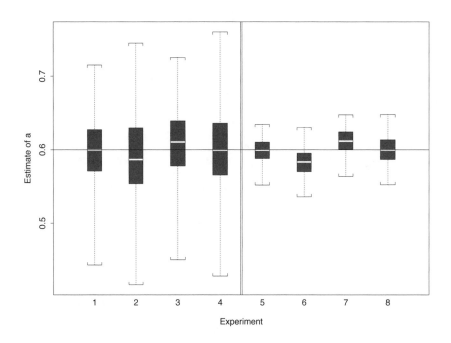

true problem formulations, and this mismatch causes a *bias error* in the results: on average, the estimated parameter value is not equal to the true value, by an amount that is essentially independent of the sample size. This point may be seen by comparing the median values for experiments 1 through 4 with those for experiments 5 through 8.

5.6   The unknown-but-bounded formulation

One practical alternative to probability-based uncertainty models is the *set-theoretic* or *unknown-but-bounded model*, briefly introduced in Chapter 1; for more complete introductions, consult the references [67, 261, 288]. For the case of fitting lines, the general set-theoretic formulation is:

$$
\begin{aligned}
y(k) &= \hat{y}(k) + \epsilon_k, \ \epsilon_k \in \mathcal{S}, \\
x(k) &= \hat{x}(k) + \nu_k, \ \nu_k \in \mathcal{T}, \\
\hat{y}(k) &= a\hat{x}(k) + b.
\end{aligned}
\tag{5.83}
$$

Here, the set $\mathcal{S}$ is assumed to be an interval of the form $\mathcal{S} = [-\epsilon^*, \epsilon^*]$ where ϵ^* is a known, nonnegative constant. For EM problems, $x(k)$ is assumed error-free,

Figure 5.8: Boxplots, estimated b parameters

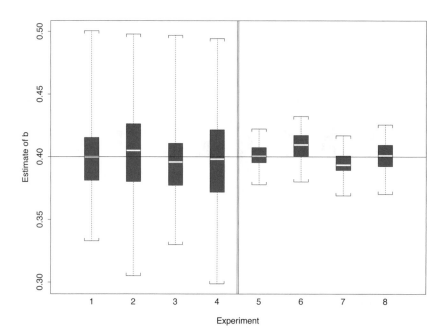

corresponding to $\mathcal{T} = \{0\}$ (i.e., $\nu_k = 0$ for all k), but EV problems may also be treated by assuming $\mathcal{T} = [-\nu^*, \nu^*]$ for some known constant $\nu^* > 0$. The advantage of the unknown-but-bounded formulation is the extreme simplicity of its nonrestrictive working assumptions, but it also suffers from a number of important disadvantages, as the following discussion illustrates. Also, there are circumstances where the set-theoretic formulation is inherently much more natural than random variable formulations; examples include mechanical tolerances (e.g., "all dimensions ± 0.1 millimeter"), impurity specifications (e.g., "less that 0.5% XYZ by weight"), and analog-to-digital converter quantization errors.

Four examples of the set-theoretic problem formulation are shown in Fig. 5.9. The upper left plot shows the most common errors-in-model situation where the data values $x(k)$ are assumed to be known precisely and all uncertainty is absorbed into the y values, assumed to fall in the interval $[y(k) - \epsilon^*, y(k) + \epsilon^*]$. The opposite situation is shown in the upper right plot, where it is assumed that $y(k)$ is precisely known and all uncertainty is absorbed into the x values. The lower left plot shows the standard errors-in-variables situation, where moderate errors of comparable magnitude are allowed in both x and y variables. Finally, the lower right plot shows the same situation but with errors of substantially larger magnitude.

Figure 5.9: Four set-theoretic formulation examples

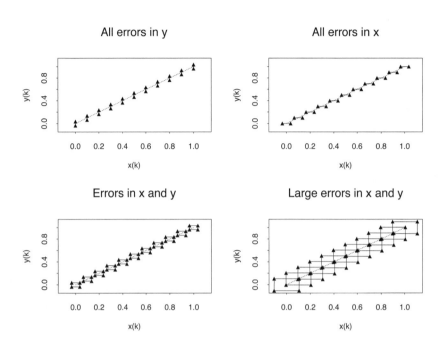

For the EM problem shown in the upper left plot in Fig. 5.9, it follows immediately from Eq. (5.83) and the equation for the line that the slope and intercept parameters a and b are required to satisfy:

$$y(k) - \epsilon^* \leq ax(k) + b \leq y(k) + \epsilon^*,$$

for $k = 1, 2, \ldots, N$. Note that this set of N simultaneous inequalities is precisely the same as the constraint (5.77) imposed on the maximum likelihood estimation problem for the case of uniformly distributed errors on the *fixed* interval $[-\epsilon^*, \epsilon^*]$. In fact, the set-theoretic and uniform maximum likelihood problem formulations are quite different in certain important respects, a point discussed further at the end of this section. Taken together, these N constraints define a convex polygon like the one shown in the center of Fig. 5.10. Specifically, note that each value of k defines the following pair of bounding lines in the (a, b) plane:

$$b \leq b^+ = -x(k)a + y(k) + \epsilon^*,$$
$$b \geq b^- = -x(k)a + y(k) - \epsilon^*.$$

Any intercept value b that is consistent with the data point $(x(k), y(k))$ and the uncertainty bound $|e_k| \leq \epsilon^*$ must necessarily fall between these two limiting values. Consequently, all admissible (a, b) parameter pairs must fall in the strip in the (a, b) plane bounded by these two lines. If all of the $x(k)$ values are distinct

Figure 5.10: A three-point parameter estimate set

(note that this is the Haar condition assumed in the discussion of Chebyshev line fitting in Sec. 5.5.1), it follows that the pair of parallel bounding lines obtained for each k have different slopes; hence, the set of parameter values determined by all of these constraints together is necessarily bounded, as shown in Fig. 5.10 for the case of three data points. In particular, the consistent parameter values for this example fall in the six-sided polygon defined by the pair of lines with positive slope, the pair with negative slope, and the pair of horizontal lines.

Algorithms have been developed to compute the exact bounding sets for fairly general unknown-but-bounded parameter estimation problems [210], but in unfortunate cases, the number of sides of this exact solution polygon can be on the order of N^2, and the situation becomes much worse for more complex problems involving more than two parameters. Consequently, much of the set-theoretic parameter estimation literature is concerned with the construction of geometrically simpler *bounding sets* like ellipsoids or parallelepipeds that contain the exact solution set but that have more efficient mathematical descriptions [67, 288]. In addition, similar considerations carry over to errors-in-variables problems [33, 225] and to other, more complex problem formulations (e.g., situations where the unknown parameters do not appear as simple linear multipliers of known data values [32]); in both the case of EV problems and the case of nonlinear problems, it should be noted that the parameter estimate sets are generally no longer convex and can even become disconnected [32, 225]. In all cases, one of the reasons that set-theoretic approaches have not achieved the popularity of older methods like ordinary least squares is that the computations are somewhat more involved.

Finally, it is important to emphasize the fundamental difference between the set-theoretic data uncertainty model and the uniform random variable model because these two models are sometimes confused. The key difference is that the uniform random variable model asserts that any error $e_k \in [-\epsilon^*, \epsilon^*]$ is *equally likely*, whereas the set-theoretic model asserts only that e_k lies in this set and says *nothing* about the relative likelihood of different values within the set. In particular, the set-theoretic model allows e_k to be uniformly distributed on this set, to be constant, to be a deterministic chaotic sequence that lies in this set, to be a deterministic sinusoid of arbitrary frequency and amplitude no larger than ϵ^*, or infinitely many other possibilities. This difference is important because there are many common data analysis procedures that exploit certain features inherent in most popular random variable models that are not present in the set-theoretic formulation. As a specific example, consider the arithmetic average of N numbers:

$$\bar{x}_N = \frac{1}{N} \sum_{k=1}^{N} x_k,$$

which commonly arises when we wish to estimate an unknown constant c from the data model $x_k = c + e_k$, where e_k is an observation error. If we adopt the set-theoretic model that $|e_k| \leq e^*$ for all k, it follows easily that the average satisfies:

$$\bar{x}_N = c + \bar{e}, \text{ where } \bar{e} = \frac{1}{N} \sum_{k=1}^{N} e_k \in [-e^*, e^*].$$

In other words, under the set-theoretic error model, averaging does not reduce our uncertainty at all, independent of how many data points we include in the average. In contrast, if we assume $\{e_k\}$ has *any symmetric distribution* (including the uniform distribution) on $[-e^*, e^*]$, it is not difficult to show that the variability of the average reduces to zero as $N \to \infty$, a point discussed in detail in Chapter 6. The key point here is that the uniform random variable model is quite different from the set-theoretic model, both because it requires the *additional assumption* that the errors are evenly distributed over their admissible range and because it inherits certain theoretical and practical advantages from these assumptions.

5.7 Which method do we use?

This chapter has examined several different methods of fitting straight lines to point pairs in the plane, demonstrating that these methods generally yield different results and exhibit different sensitivities to various features in the data. These differences raise the practical question of which method we should use in analyzing a given dataset. Unfortunately, it is not possible to say that one method is "always best" or even "usually best"—the best choice will depend on the dataset, and this dependence may be quite strong. Nevertheless, the following general observations may be useful.

First, despite their significant shortcomings (specifically, outlier sensitivities), ordinary least squares approaches have been historically the most popular choice for a long time, and they remain quite popular today. Reasons for this popularity include computational simplicity (OLS problems often reduce to the solution of sets of simultaneous linear equations, for which computationally efficient and reliable algorithms are widely available), uniqueness of the best fit solution (in contrast to the LAD fit and certain other more outlier-resistant methods), and their connections with Gaussian maximum likelihood problems (the Gaussian distribution remains the most popular error model despite *its* significant practical shortcommings, a point discussed at length in Chapter 7). Further, many useful auxillary results have been developed for OLS estimators, like the computationally efficient deletion diagnostics discussed in Chapter 15. *For all of these reasons, OLS fits should probably be computed as a part of most comprehensive analyses unless there are clear reasons to regard them as inadequate.* For example, in situations where some other distribution is more natural (e.g., lognormal or uniform distributions—see Chapter 4 for a further discussion of such cases), it may be more reasonable to adopt some other strategy, although even there, this strategy may involve a linearizing transformation like those discussed in Sec. 5.1, followed ultimately by a least squares fitting procedure. In such cases, however, it may be necessary to consider alternatives like *weighted least squares* discussed in Chapter 13.

Having said that OLS procedures should be included as part of routine data analyses, it is also important to emphasize that OLS results should be regarded primarily as a basis for comparison and not as "the answer." In particular, it is important to remember that outliers do arise in real-world datasets—especially in exploratory data analysis—and that OLS procedures are extremely sensitive to outliers. The question of how to deal with outliers is one that will occupy us repeatedly throughout the rest of this book, but two general ideas are worth emphasizing. *First, wherever possible, perform the analysis in at least two ways and compare the results.* These methods should be chosen so that their expected outcomes are roughly the same; if the results obtained are in reasonable agreement, this comparison provides added support of their plausability. Conversely, if there are large differences, we should investigate to find out what appears to be responsible for these differences; again, as noted in Chapter 2, such surprises can ultimately yield extremely beneficial consequences (e.g., Nobel Prizes or new commercial products).

The second generally useful idea is that, after "anomalous" or "unusual" portions of the dataset have been identified, it is often beneficial to repeat the analysis without them. This idea is advocated by Rousseeuw and Leroy [252], who apply extremely robust methods to obtain a preliminary result that is largely unaffected by outliers, use this preliminary result to detect outliers, and then fit and interpret what remains using OLS procedures. This approach has the advantage of allowing the use of the auxilliary analysis machinery that has been developed for OLS procedures over the last two centuries. Because of their importance, both of these ideas will be revisited frequently in later chapters of this book, specialized where possible to different analysis procedures.

Another inherent OLS limitation, noted repeatedly in discussions here, is that it is an errors-in-model (EM) problem formulation, which is not always appropriate: often, the errors-in-variables (EV) problem formulation better approximates the true distribution of variability in the dataset. This observation means that, even if we restrict consideration to a least squares formulation (gaining simplicity at the expense of outlier-sensitivity), the resulting computational problem is significantly more complex than the OLS problem, for two reasons. First is the need to specify the ratio of relative uncertainties in the variables involved, although this difficulty is in fact a phantom since adopting the OLS solution corresponds to assuming all errors appear in the response variable y, corresponding to a limiting case of this ratio (i.e., zero). The second difficulty is real: the computations are more involved and not as well developed, although this situation is improving significantly. Consequently, as these methods become more widely available, they should probably also be included in the results of a comprehensive analysis and compared with both OLS results and robust alternatives. In particular, it is important to emphasize that the differences in the results obtained between EM and EV problem formulations can be extremely signficant. One example is that described by High and Danner [139], who fit experimental gas adsorption data to a nonlinear model for pure components and then used the resulting model to predict the behavior of binary and ternary mixtures. The authors used both EM and EV formulations, obtaining prediction errors that were between ∼17% and ∼37% smaller with the EV formulation.

Finally, it is important to say something about the other methods discussed in this chapter. First, note that the maximum likelihood approach introduced in Sec. 5.4 has many theoretical advantages (some of which are discussed in Chapter 6), and these become very significant practical advantages when the resulting maximization problem is computationally tractable. This is certainly true for the Gaussian, Laplace, and uniform maximum likelihood problems considered here. For more general cases like the gamma distributions introduced in Chapter 4, however, the computations become much more complex, so the resulting procedures find little use in practice. Similarly, as noted in Chapter 1, the set-theoretic approach discussed in Sec. 5.6 also has some quite significant theoretical advantages—e.g., if the exact parameter uncertainty sets can be computed, they tell us *exactly* which parameter values are possible and which ones are not—but the computations can become difficult and the results tend to be conservative. Further, these procedures are not supported nearly as well by commercial software as are the other approaches described here.

5.8 Exercises

1. Derive the expressions given in Eq. (5.79) for the maximum likelihood estimates $\hat{\sigma}^2$ and $\hat{\phi}$ of the Gaussian and Laplace scale parameters.

2. Derive Eqs. (5.53), (5.54), and (5.55) for the TLS line-fitting solution.

3. Prove that the second derivative of $J_{TLS}(a, b)$ with respect to a at $a = a_{TLS}$ is given by Eq. (5.63).

4. In 1923, Subotin proposed the following family of probability distributions [163, p. 195]:

$$p(x) = \frac{1}{C\phi} \exp \left[-\frac{1}{2} \left| \frac{x - \theta}{\phi} \right|^{2/\delta} \right],$$

$$C = 2^{(\delta/2)+1} \Gamma \left(\frac{\delta}{2} + 1 \right),$$

where θ is an arbitrary real constant and ϕ and δ are positive constants. Suppose the line fit errors $\{e_k\}$ form a statistically independent, identically distributed, zero-mean sequence corresponding to this distribution with $\theta = 0$ and $\delta > 0$ *fixed and known*. Reduce the maximum likelihood estimation problem to an optimization problem. What norm of $\{e_k\}$ is being minimized?

Chapter 6

A Brief Introduction to Estimation Theory

Sections and topics:

This chapter is concerned with the characterization of *estimators*, broadly defined as any mapping T from an observed dataset $\mathcal{D}$ to an *estimate* $\hat{\theta}$ of some parameter θ of interest. Chapters 1 and 2 emphasized the nonnegligible variability or uncertainty inherent in real data sequences, from which it follows that $\hat{\theta}$ is also variable or uncertain. Thus, in the design or selection of an estimator it is desirable to make $\hat{\theta}$ as accurate and precise as possible, and this chapter introduces some useful concepts and techniques for characterizing estimator accuracy and precision. Specific examples considered here include traditional data characterizations like the mean and standard deviation of a single sequence of numbers, along with some important alternatives; subsequent chapters consider estimators for correlation coefficients and regression model parameters that relate two or more variables. The basis for these characterizations is probability theory, so this chapter makes extensive use of some of the ideas introduced in Chapter 4. The primary objective of this chapter is to provide a practically motivated introduction to some of the most important ideas from estimation theory. For more detailed discussions, refer to books like those of Arnold *et al.* [16], Huber [151], or especially Lehmann [180].

Key points of this chapter:

1. An *estimator* $T(\theta)$ is any procedure for computing an *estimate* $\hat{\theta}$ of some unknown parameter θ from an observed dataset $\mathcal{D}$.

2. If we follow common practice and adopt the random variable model described in Chapters 3 and 4 for the uncertainty inherent in the observed dataset $\mathcal{D}$, it follows that $\hat{\theta}$ inherits a related random variable description.

3. The resulting distribution of the estimate $\hat{\theta}$ depends on:

   a. the distribution of the data,
   b. the size of the dataset,
   c. the nature of the estimator.

4. This chapter assumes that the dataset $\mathcal{D}$ consists of a single sequence $\{x_k\}$ of N *independent, identically distributed* (i.i.d.) random variables. This assumption is both extremely standard and extremely restrictive; its nature and consequences are discussed further in Chapter 10, and alternative assumptions form the basis for dynamic data analysis.

5. Many—*but not all*—estimators exhibit *asymptotic normality*, meaning that $\hat{\theta}$ is approximately Gaussian for sufficiently large sample sizes N. A common (but, again, not universal) result is that the distribution of $\hat{\theta}$ is approximately $N(\theta, \sigma_T^2/N)$ for sufficiently large N, where σ_T^2 depends on the estimator T and the distribution of the dataset $\mathcal{D}$.

6. A closely related and extremely important result is the Central Limit Theorem (CLT), which gives conditions under which sums and averages exhibit Gaussian limiting distributions.

7. Conversely, the CLT does not always apply, a point discussed in detail in Sec. 4.5. In particular, the CLT fails to hold for the class of *non-Gaussian stable random variables*, the best-known example of which is the Cauchy distribution, but it can also fail to hold in other cases.

8. The *information inequality* or *Cramer-Rao lower bound* establishes a lower bound on the variance of the practically important class of *unbiased estimators* for finite sample sizes N; this result provides useful insight into how accurately we can hope to estimate θ under the best of circumstances.

9. Many popular estimators are based on the moment characterizations discussed in Chapter 4, but it was noted that moments have certain inherent limitations, a point emphasized further in Chapter 7 in connection with *outliers*. Estimators based on *order statistics* provide one important practical alternative that often exhibit vastly superior resistance to outliers; the basic notions of order statistics are introduced in Sec. 6.6, which introduces the class of *L-estimators* as an important practical application.

6.1 Characterizing estimators

Normally, the uncertainty or variability inherent in a dataset translates into uncertainty or variability in any quantity we compute from it. Thus, the general performance of any estimator (i.e., good, bad, or indifferent) can depend strongly on the characteristics of the dataset. One of the best ways to illustrate this point—which is one of the key points of this chapter—is with a simple example. Consequently, Sec. 6.1.1 begins with a brief discussion of the *location estimation problem*, comparing the performance of four different estimators as a function of sample size and data distribution. Following this example, secs. 6.1.2 and 6.1.3 define the important notions of estimator *bias* and *consistency*, and Sec. 6.1.4 briefly introduces some other estimator characterizations that are discussed further in later sections of this chapter.

6.1.1 Location estimators

The basic estimation problem considered here is the following: given a data sequence $\{x_k\}$, how do we estimate its "center"? In the statistics literature, possible answers to this question are called *location estimates*, and they represent the computable counterpart to the location parameters $\bar{x}$, $x^\dagger$, and x^* discussed in Chapter 4. Without question, the most popular location estimator is is the arithmetic mean:

$$\bar{x}_N = \frac{1}{N} \sum_{k=1}^{N} x_k, \tag{6.1}$$

where the subscript N denotes the sample size and distinguishes this estimator from the location parameter $\bar{x} = E\{x\}$ discussed in Chapter 4 for a *data distribution*. The distinction is important because $\bar{x}_N$ can be *computed* from a given data sequence $\{x_k\}$, whereas the location parameter $\bar{x}$ cannot since it depends on the distribution, rather than the data. If the data distribution is such that $\bar{x}_N$ is a *consistent estimator* of $\bar{x}$, then $\bar{x}_N \to \bar{x}$ as $N \to \infty$, an idea discussed further in Sec. 6.1.3. The popularity of $\bar{x}_N$ as a location estimator is illustrated by the following comment from Andrews *et al.* [11, p. 243]:

> ... as everybody familiar with practical statistics (not to mention a number of theories) knows, the mean is *the* estimator of location, to be applied under almost all circumstances (expecially more complex models).

As the ellipses in this quote suggests, these comments do not convey the complete story of the mean $\bar{x}_N$, which does exhibit significant practical limitations, but these comments do fairly convey its popularity; Chapter 7 examines the rest of this quote, which addresses some of these limitations.

To give a preliminary assessment of the performance of $\bar{x}_N$ as a location estimator, it is instructive to compare it with three other, closely related, alternatives. These comparisons are made for two different sample sizes ($N = 50$

Figure 6.1: Four different data distributions

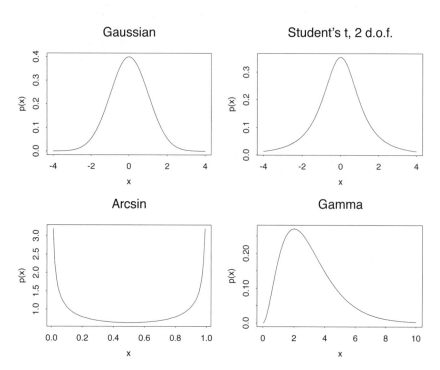

and $N = 500$) and four different data distributions: the standard normal distribution, Student's t-distribution with 2 degrees of freedom, the bimodal arc-sine distribution discussed in Chapter 4 (i.e., the beta distribution with shape parameters $p = q = 1/2$), and the gamma distribution with shape parameter $\alpha = 3$. Plots of these density functions are shown in Fig. 6.1; the distribution means are $\bar{x} = 0$ for the Gaussian and Student's t-distributions, $\bar{x} = 1/2$ for the arc-sine distribution, and $\bar{x} = 3$ for the gamma distribution. The basic question addressed in this case study is how well the four different location estimators considered estimate $\bar{x}$ from the available data; typical data sequences from each of these distributions are shown in Fig. 6.2.

The four estimators compared here all belong to the family of *symmetric trimmed means*, discussed further in Sec. 6.6.5 and defined as:

$$x_N^{[p]} = \frac{1}{N - 2p} \sum_{i=p+1}^{N-p} x_{i:N}, \qquad (6.2)$$

where p is an integer between 0 and $(N-1)/2$, and $x_{i:N}$ denotes the ith element in the *rank-ordered* data sequence:

$$x_{1:N} \leq x_{2:N} \leq \cdots \leq x_{(N-1):N} \leq x_{N:N}. \qquad (6.3)$$

Figure 6.2: Typical data sequences, length $N = 500$

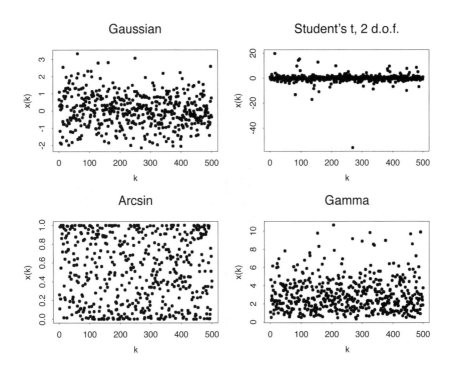

This rank-ordered sequence is the basis for the class of *order statistics* discussed in Sec. 6.6. Note that if we take $p = 0$, we recover the arithmetic mean $\bar{x}_N$, whereas if we take $p = (N - 1)/2$ for N odd, we obtain the *sample median* discussed in Chapter 1. For symmetric, unimodal distributions, this estimator is popular because it achieves the smallest possible maximum bias among all translation-invariant estimators [151, p. 75]; the notion of estimator bias is discussed in detail in Sec. 6.1.2, but the key point here is that, like the mean, the median is a logical choice of location estimator that exhibits certain desirable properties. The purpose of this example is to compare these two estimators, along with two others, under a several different working assumptions.

The other two location estimators considered here correspond to intermediate choices of p in Eq. (6.2), advocated by two different authors. Lehmann [180, p. 362] presents a table comparing the performance of various trimmed means, concluding:

> This table provides impressive support for the idea that a moderate amount of trimming can provide much better protection than (the arithmetic mean $\bar{x}_N$) against fairly heavy tails, as represented for example by t_3 (i.e., Student's t-distribution with 3 degrees of freedom), while at the same time giving up little in the normal case.

Figure 6.3: Comparison of trimmed means, Gaussian data

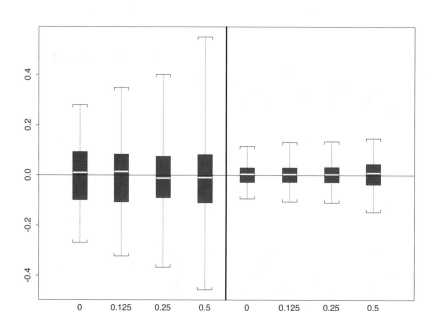

More specifically, Lehmann suggests 12.5% trimming as a reasonable choice, corresponding to $p \simeq N/8$. Conversely, Andrews *et al.* [11, p. 240] argue that:

> trimmed means which don't trim too little on either end, e.g., the 25%-trimmed mean ("midmean"), are also fairly safe and still quite good.

Since all four of these choices ($p = 0$, $p \simeq N/8$, $p \simeq N/4$, and $p \simeq N/2$) appear to have both advantages and advocates, it is interesting to compare them.

Fig. 6.3 illustrates this comparison for the case of Gaussian data, presenting boxplots for each of these four trimmed means, indexed by the trimming fraction p/N, for sample sizes $N = 50$ (leftmost boxplots) and $N = 500$ (rightmost boxplots). In all cases, the boxplots summarize the estimates obtained from 100 simulations, each based on a sequence of N statistically independent samples with the $N(0, 1)$ distribution. The most striking feature of these plots is much smaller variability seen in the results computed from the larger samples ($N = 500$) vs. the smaller samples ($N = 50$); also, note that in all cases the median value over the 100 simulations is fairly accurate. Both of these observations are consequences of the fact that symmetrically trimmed means are *consistent* estimators of $\bar{x}$ for Gaussian data, an important notion discussed in Sec. 6.1.3.

Figure 6.4: Comparison of trimmed means, t_2-distributed data

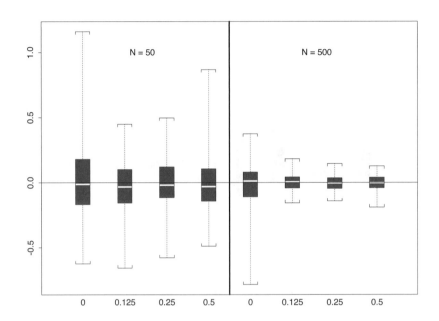

Conversely, note that the variability, for fixed sample size, appears to increase somewhat with the trimming fraction. These observations are consistent with expectations, since $\bar{x}_N$ is the *uniformly minimum variance unbiased* or UMVU estimator [180, p. 84] of the distribution mean for Gaussian data. In fact, it has been suggested that there are only two circumstances under which $\bar{x}_N$ should *not* be used to estimate the mean of a *Gaussian* distribution [162, p. 123]:

1. in the case of *censored data*, where extreme observations have not been recorded, or
2. when outliers may be present.

Further, it is also known that the standard deviation of the median $(p = 0.5)$ for Gaussian data is larger than that of the mean by a factor of $\sqrt{\pi/2} \simeq 1.25$ [162, p. 123], consistent with the results seen in Fig. 6.3.

Fig. 6.4 presents the analogous results for 100 data sequences each having Student's t-distribution with 2 degrees of freedom, denoted t_2 for simplicity. The same reduction in variability with increasing sample size seen in the Gaussian case is also seen in this example, although there is a pronounced difference: 12.5% symmetric trimming appears to reduce the variability of the location estimate substantially here, in contrast to the Gaussian case where the effect

Figure 6.5: Comparison of trimmed means, arc-sine distributed data

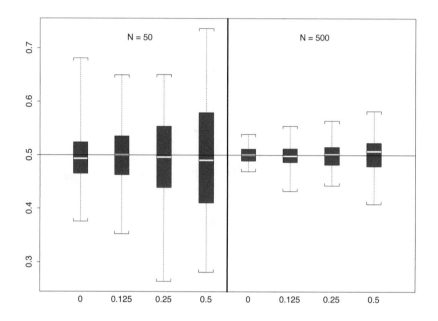

was a slight increase in variability. This result reflects the extremely heavy tails of the t_2 distribution, a feature seen clearly in the upper right data plot in Fig. 6.2. The t_2 distribution is better behaved than the t_1 (Cauchy) distribution, but both of these distributions exhibit infinite variance, and as a consequence, the untrimmed mean $\bar{x}_N$ is an *inconsistent* estimator of $\bar{x}$ for these distributions. This example supports Lehmann's argument in favor of the moderately trimmed mean, a point considered further in Chapter 5. Also, note that in this example, further increasing the trimming percentage p/N appears to have little effect on the variability of the estimator, at least in the case of large samples.

Fig. 6.5 shows the corresponding boxplots for the arc-sine distributed data. In marked contrast to the t_2 data, these results show a clear *increase* in variability with the trimming fraction p/N, more pronounced than that seen in the Gaussian data. This difference reflects the profound difference in tail behavior between the bimodal arc-sine distribution and the heavy-tailed t_2 distribution: for the arc-sine data, the effect of trimming is to remove *representative* values that cluster around the extremes of this bimodal distribution; this removal tends to increase the variability of the estimator since the *nonrepresentative* values from the center of the data range are weighted more heavily. In contrast, the extreme values are much less representative for the Student's t-distribution,

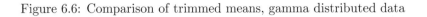

Figure 6.6: Comparison of trimmed means, gamma distributed data

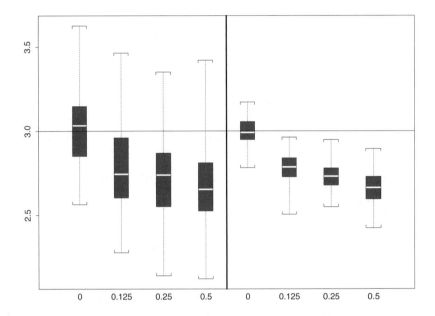

and the effect of removing them is to make the estimator *less* variable. This observation illustrates an important point that is discussed further in Chapter 7: estimators that perform well in the presence of outliers often perform badly (or at least strangely) when the data distribution is light-tailed, as in this example.

Finally, Fig. 6.6 compares the four location estimators considered here for the asymmetric gamma distribution. As in all three previous examples, increasing the sample size from $N = 50$ to $N = 500$ results in a significant reduction in the variability of all four location estimates, but the most striking effect seen in these boxplots is the systematic shift in the median of the 100 location estimates as the trimming percentage is increased from 0 to 50%. This shift illustrates the notion of *estimator bias* discussed in the next section and is a consequence of the difference between the location parameters $\bar{x}$ and $x^\dagger$ for the gamma distribution. That is, recall from the discussion in Chapter 3 that the three logical location parameters $\bar{x}$, $x^\dagger$, and x^* are not equal for asymmetric distributions. In the presence of outliers, the situation is further complicated by the different outlier sensitivity of the "natural" choices of estimator for these different location parameters (e.g., $\bar{x}_N$ for the distribution mean $\bar{x}$ vs. the sample median $x_N^{[(N-1)/2]}$ for the distribution median $x^\dagger$); the problem of outliers in asymmetric data distributions is discussed further in Chapter 7.

6.1.2 Estimator bias

If we adopt a random variable model for the data sequence $\{x_k\}$, it follows that the estimator $T(\theta)$ is also a random quantity, as noted in previous discussions. Consequently, $T(\theta)$ may be at least partially characterized in terms of its moments, as discussed in Chapter 4. The simplest of these characterizations is the first moment, and the estimator $T(\theta)$ of the unknown parameter θ is said to be *unbiased* if the first moments agree, meaning that the *bias* $b(\theta)$ is zero:

$$b(\theta) \equiv E\{T(\theta)\} - \theta = 0. \tag{6.4}$$

Although it is not true that *all* estimators commonly used in practice are unbiased, many of them are, and lack of bias is a highly desirable characteristic. In fact, a less formal term for estimator bias is *systematic error*. A more complete discussion of estimator bias is given by Lehmann [180, ch. 2], where it is noted that, despite their desirability, unbiased estimators do not always exist.

As a simple example of an unbiased estimator, consider the arithmetic mean $\bar{x}_N$ of a sequence $\{x_k\}$. If $\{x_k\}$ is an i.i.d. sequence drawn from a distribution with mean μ, the expected value of $\bar{x}_N$ is:

$$E\{\bar{x}_N\} = E\left\{\frac{1}{N}\sum_{k=1}^{N} x_k\right\} = \frac{1}{N}\sum_{k=1}^{N} E\{x_k\} = \mu. \tag{6.5}$$

In words, this result means that the arithmetic mean $\bar{x}_N$ is an unbiased estimator of the distribution mean μ. Further, note that this result holds for *any* i.i.d. random variable sequence, provided only that the distribution mean μ is finite. In fact, we do not even need the assumption of statistical independence: it is enough that the data samples x_k are all drawn from distributions with a common mean. As noted in Sec. 6.1.1, $\bar{x}_N$ is an extremely popular location estimator, used either as an informal quantitative data description or as an estimate of the first moment $E\{x\}$ of the underlying data distribution. Conversely, one of the key points of Chapter 7 is that moment-based estimators like $\bar{x}_N$ are extremely sensitive to the presence of outliers in the dataset $\{x_k\}$, so alternatives like those discussed in Sec. 6.6.5 are more appropriate in some circumstances.

An important relaxation of the notion of an unbiased estimator is that of an *asymptotically unbiased estimator*: the estimator $T(\theta)$ is asymptotically unbiased if $b(\theta) \to 0$ as $N \to \infty$. Clearly, any unbiased estimator is also asymptotically unbiased, corresponding to $b(\theta) = 0$ for all N, but there are important situations where $b(\theta)$ only approaches zero asymptotically. As a specific example, the expected value of the variance estimator $\hat{\sigma}^2$ considered in Sec. 6.2 differs from the correct variance σ^2 by the factor $(N-1)/N$, which approaches 1 as $N \to \infty$. Conversely, note that this difference is not negligible for small samples: the error is 10% for $N = 10$, for example. Finally, note that if the exact bias of an estimator is known—as in this example—it is always possible to apply a correction factor to obtain an unbiased estimator, a point illustrated in Sec. 6.2 for this variance estimator and revisited in Chapter 7 in connection with the outlier-resistant MAD scale estimator.

6.1.3   Variance and consistency

A qualitative characterization of *estimator consistency* was given at the beginning of this section, and it is defined more precisely here, but it is useful to begin with the following observation. Suppose $\hat{\theta}_N$ is an unbiased estimator of the parameter θ, and consider the variance σ_N^2 of the estimation error $e_N = \hat{\theta}_N - \theta$. It follows from the Chebyshev inequality, discussed in Chapter 4, that:

$$\mathcal{P}\left\{|\hat{\theta} - \theta| \geq \alpha\right\} \leq \frac{\sigma_N^2}{\alpha^2}, \tag{6.6}$$

for *any* distribution with finite variance. Hence, if $\sigma_N^2 \to 0$ as $N \to \infty$, the probability of the estimator $\hat{\theta}_N$ assuming any value other than the exact value θ will go to zero. More generally, we will say that the estimator $\hat{\theta}_N$ is *consistent* if the following two conditions are met [180, p. 332]:

1. $E\{\hat{\theta}_N\} \to \theta$ as $N \to \infty$,
2. $\mathrm{var}\{\hat{\theta}_N - \theta\} \to 0$ as $N \to \infty$.

Note that the first of these conditions is simply that the estimator be asymptotically unbiased, whereas the second condition is a quantitative version of the rough characterization given earlier: the variability of a consistent estimator can be made arbitrarily small by making N large enough. Conversely, an estimator that is not consistent is called *inconsistent*. Most commonly, this terminology is applied to estimators whose variance does not approach zero as $N \to \infty$; specific examples illustrating this point are given in secs. 6.2 and 6.4.

The arithmetic mean $\bar{x}_N$ provides a convenient illustration of the notion of a consistent estimator: its variance describes its variability as an estimator of the distributional mean μ, and the result developed here shows how this variability decreases with increasing sample size N. Derivation of an explicit expression for this variance is simplified somewhat by first noting that:

$$\bar{x}_N - \mu = \frac{1}{N} \sum_{k=1}^{N} [x_k - \mu], \tag{6.7}$$

from which it follows that:

$$\mathrm{var}\{\bar{x}_N\} = E\{(\bar{x}_N - \mu)^2\} = \frac{1}{N^2} E\left\{\left(\sum_{k=1}^{N}[x_k - \mu]\right)^2\right\}$$

$$= \frac{1}{N^2} E\left\{\sum_{k=1}^{N}\sum_{\ell=1}^{N}[x_k - \mu][x_\ell - \mu]\right\}$$

$$= \frac{1}{N^2} \sum_{k=1}^{N}\sum_{\ell=1}^{N} E\{[x_k - \mu][x_\ell - \mu]\}. \tag{6.8}$$

The fact that $\{x_k\}$ is an i.i.d. sequence simplifies this result considerably (for a more general derivation that does not make this independence assumption, see Chapter 10). In particular, since x_k is statistically independent of x_ℓ unless $k = \ell$, we have:

$$E\{[x_k - \mu][x_\ell - \mu]\} = \begin{cases} E\{[x_k - \mu]\}E\{[x_\ell - \mu]\} = 0 & k \neq \ell, \\ E\{[x_k - \mu]^2\} = \sigma^2 & k = \ell. \end{cases} \qquad (6.9)$$

Consequently, all terms in the last double sum in Eq. (6.8) vanish unless $k = \ell$, in which case they are simply equal to σ^2. Since there are N such nonzero terms, this contribution cancels one of the factors of N in the denominator, and the final result is:

$$\text{var}\{\bar{x}_N\} = \frac{\sigma^2}{N}. \qquad (6.10)$$

Since $\bar{x}_N$ is an unbiased estimator of μ and $\text{var}\{\bar{x}_N\} \to 0$ as $N \to \infty$ for any finite-variance data sequence $\{x_k\}$, it follows that $\bar{x}_N$ is a consistent estimator of μ. Further, note that the standard deviation of the arithmetic mean $\bar{x}_N$ is $\sigma/\sqrt{N}$, leading to the terminology $\sqrt{N}$-*consistent* for consistent estimators whose standard deviation decreases asymptotically like $1/\sqrt{N}$. This consistency result for the mean is extremely well known, establishing something of a standard for estimator behavior as a function of sample size. Although not all consistent estimators behave this way, enough of them do that exceptions are noteworthy; see, e.g., the comment about $N^{-1/3}$ vs. $N^{-1/2}$ estimators in the Princeton robustness survey [11, p. 224].

6.1.4 Other characterizations

The concepts of bias and consistency introduced in the previous sections represent useful but incomplete characterizations of an estimator. It is logical to ask when more complete characterizations can be obtained, ideally in the form of an exact expression for the distribution of $\hat{\theta}_N$, given the distribution of the starting data $\{x_k\}$. Unfortunately, as subsequent discussions illustrate, such complete characterizations are generally not practical, forcing us to consider simpler alternatives. Typically, these alternatives take one of the following three forms:

1. determination of *approximate* finite-sample distributions for $\hat{\theta}_N$;
2. determination of *asymptotic* characterizations of $\hat{\theta}_N$ as $N \to \infty$;
3. determination of *lower bounds* on the finite-sample variance of an unbiased, consistent estimator.

Roughly speaking, these characterizations are listed in order of their desirability and in reverse order of their difficulty. In particular, although the exact finite-sample distribution of an estimator $\hat{\theta}_N$ is the most complete characterization possible, such results are only known for a few estimators and a few data distributions. As a specific example, Sec. 6.2.2 presents the exact finite-sample distribution for the variance estimator obtained by the method of moments when the data sequence $\{x_k\}$ is normally distributed. This result follows from those given in Sec. 6.3.1 for sums of gamma distributed random variables. *Approximate* finite-sample distributions can sometimes be obtained, but in practice, these results are almost as rare as exact distributions, so for most estimators we must settle for one of the last two characterizations on this list: asymptotic results or lower bounds. As noted previously, many estimators exhibit approximately Gaussian distributions for sufficiently large sample sizes, a result that is closely related to the Central Limit Theorem discussed in Sec. 6.3, where some of the practical consequences of asymptotic normality are also discussed briefly. The most common lower bound on estimator variance is the *information inequality*, also known as the *Cramer-Rao lower bound* and discussed in Sec. 6.5.

6.2 An example: variance estimation

The estimation of means and either standard devations or variances are problems that arise quite frequently in practice. The statistical characterization of the mean $\bar{x}_N$ of a sequence $\{x_k\}$ of N statistically independent, identically distributed random variables was discussed in the previous section because it nicely illustrates the notions of estimator bias and consistency. The following discussion considers the slightly more complicated problem of variance estimation, both because it also illustrates a number of important points and because its statistical characterization is not as widely appreciated among nonstatisticians as that of the mean. In particular, it is instructive to consider how the behavior of this estimator depends on the distribution of the data sequence from which it is computed. First, however, it is useful to say something about how this estimator is derived.

6.2.1 The standard estimators $\bar{x}_N$ and $\hat{\sigma}^2$

The method of moments was introduced in Chapter 4 as one way of estimating the probability density $p(x)$ that best fits an observed data sequence $\{x_k\}$. Applying this idea to the first two moments leads immediately to the following estimators for the mean μ and the variance σ^2 of the underlying distribution:

$$\bar{x}_N = \frac{1}{N} \sum_{k=1}^{N} x_k,$$

$$\hat{\sigma}^2 = \frac{1}{N} \sum_{k=1}^{N} (x_k - \bar{x}_N)^2. \qquad (6.11)$$

Taking the square root of this last results yields the method of moments estimator for the standard deviation σ:

$$\hat{\sigma} = \left[\frac{1}{N} \sum_{k=1}^{N} (x_k - \bar{x}_N)^2 \right]^{1/2}. \qquad (6.12)$$

It is important to note that these results are obtained by simply matching the estimated first and second moments for a finite data sequence to the theoretical values μ and σ associated with the otherwise unspecified distribution $p(x)$. In particular, nothing beyond finiteness of these moments has been assumed about this underlying density.

It is also possible to derive estimators for μ and σ via the method of maximum likelihood introduced in Chapter 5, *provided the form of the distribution* $p(x)$ *is specified*. In the problem considered here, we assume that $\{x_k\}$ is a sequence of N statistically independent Gaussian random variables with unknown mean μ and variance σ^2. To obtain estimators for μ and σ, we maximize the log likelihood function defined and discussed in Chapter 5:

$$\ell(\hat{\mu}, \hat{\sigma}) = \sum_{k=1}^{N} \ln p(x_k | \hat{\mu}, \hat{\sigma}). \tag{6.13}$$

Substituting the Gaussian density function for $p(x_k | \hat{\mu}, \hat{\sigma})$ leads to the following explicit expression for the log likelihood function:

$$\ell(\hat{\mu}, \hat{\sigma}) = -N \ln \sqrt{2\pi} - N \ln \hat{\sigma} - \frac{1}{2\hat{\sigma}^2} \sum_{k=1}^{N} (x_k - \hat{\mu})^2. \tag{6.14}$$

Differentiating this expression with respect to the parameters $\hat{\mu}$ and $\hat{\sigma}$ and setting the results to zero leads immediately to the moment estimators $\hat{\mu} = \bar{x}_N$ and $\hat{\sigma}$ defined in Eqs. (6.11) and (6.12).

Finally, it is useful to note the existence of two simple upper bounds for the variance estimator $\hat{\sigma}^2$ defined in Eq. (6.11). Specifically, Bhatia and Davis [36] have shown that, if $m \leq x_k \leq M$ for all k, then

$$\hat{\sigma}^2 \leq (M - \bar{x}_N)(\bar{x}_N - m) \leq \frac{(M - m)^2}{4}. \tag{6.15}$$

Further, both of these bounds are achievable with binary data: the tighter bound holds with equality if and only if half of the x_k values are equal to m and half are equal to M, whereas the looser bound holds with equality if and only if every x_k value is equal to either m or M [36, proposition 1].

6.2.2   Exact distribution for the Gaussian case

Since $\hat{\sigma}^2$ is the maximum likelihood estimator for σ^2 under the assumption of a Gaussian i.i.d. data sequence $\{x_k\}$, it is sensible to ask what can be said about this estimator when this assumption holds. The resulting distribution for $\hat{\sigma}^2$ was derived in the middle of the nineteenth century [162, p. 415] and follows from useful results presented later in this chapter. Specifically, the exact finite-sample distribution of $N\hat{\sigma}^2/\sigma^2$ is the *chi-squared distribution with* $\nu = N - 1$ *degrees of freedom*, denoted χ_ν^2 and defined by the following density for all $x \geq 0$:

$$p(x) = \frac{x^{(\nu/2)-1} e^{-x/2}}{2^{\nu/2} \Gamma(\nu/2)}. \tag{6.16}$$

This distribution corresponds to a special case of the gamma family discussed in Chapter 4 and has been characterized extensively. In particular, the first four moments of this distribution are [162, p. 420]:

$$E\{x\} = \nu, \qquad \text{var}\{x\} = 2\nu,$$

$$\gamma\{x\} = \sqrt{\frac{8}{\nu}}, \qquad \kappa\{x\} = \frac{12}{\nu}. \tag{6.17}$$

It follows from these results that the variance estimator $\hat{\sigma}^2$ has the following mean, variance, skewness, and kurtosis:

$$E\{\hat{\sigma}^2\} = \left(\frac{N-1}{N}\right)\sigma^2, \qquad \text{var}\{\hat{\sigma}^2\} = \frac{2(N-1)\sigma^4}{N^2},$$

$$\gamma\{\hat{\sigma}^2\} = \sqrt{\frac{8}{N-1}}, \qquad \kappa\{\hat{\sigma}^2\} = \frac{12}{N-1}. \tag{6.18}$$

Consequently, $\hat{\sigma}^2$ is asymptotically unbiased and consistent, with the approximate asymptotic variance $2\sigma^4/N$. In fact, the small-sample bias may be corrected as discussed in Sec. 6.1.2, leading to the modified variance estimator:

$$\hat{\sigma}^2 = \frac{1}{N-1}\sum_{k=1}^{N}(x_k - \bar{x}_N)^2, \tag{6.19}$$

which can be shown to be unbiased for any i.i.d. data sequence with finite mean and variance (Exercise 1). Here, the key points to note are that the moments of $\hat{\sigma}^2$ for Gaussian data are determined by both the true distributional parameter σ^2 and the sample size N.

6.2.3 What about non-Gaussian cases?

More generally, the behavior of any estimator depends on both the sample size and the distribution of the data from which it is computed. For the unbiased variance estimator defined in Eq. (6.19), it is not difficult to show (Exercise 2) that the *estimator variance* is:

$$\text{var}\{\hat{\sigma}^2\} = \frac{(\kappa+2)\sigma^4}{N} + \frac{2\sigma^4}{N(N-1)}, \tag{6.20}$$

where κ is the kurtosis of the data sequence $\{x_k\}$, defined and discussed in Chapter 4, where it was also noted that $\kappa \geq -2$ for all distributions. Hence, it follows that both of these variance contributions are positive for all distributions, approaching zero as $N \to \infty$. Also, note that the first term is *generally* dominant, approaching zero more slowly than the second term, and it reflects the dependence of the estimator variance on the distribution of the data. The sole exception occurs for binary sequences, which achieve the lower bound $\kappa = -2$, implying $\text{var}\{\hat{\sigma}^2\} = 2\sigma^4/N(N-1)$. This result establishes that the i.i.d. binary

Figure 6.7: A comparison of four variance estimates

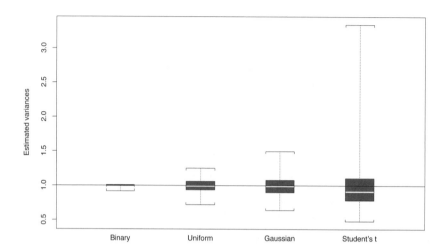

sequence (i.e., the Bernoulli sequence) exhibits the smallest achievable estimator variance, decaying to zero approximately like $1/N^2$ vs. the more usual $1/N$ convergence rate. The key point here is that, in contrast to the results presented in Section 6.1.2 for the mean estimator $\bar{x}_N$, the moments of the variance estimator $\hat{\sigma}^2$ involve *higher* moments of the data distribution.

To illustrate the general distributional dependence of the estimator $\hat{\sigma}^2$, Fig. 6.7 shows a boxplot summarizing 200 variance estimates, each computed from $N = 100$ data points, for four different distributions. The leftmost boxplot summarizes the variance estimates obtained for an i.i.d. binary sequence, assuming values $x_k = \pm 1$ with equal probability. As noted, the variance of $\hat{\sigma}^2$ for this example achieves the smallest possible value; for the sample size $N = 100$ considered here, the corresponding standard deviation of the variance estimate is approximately 1.4%. Next to the boxplot for this binary sequence is that for a data sequence that is uniformly distributed on the interval $[-\sqrt{3}, \sqrt{3}]$; this interval was chosen so the data sequence has zero mean and unit variance like all of the other examples considered here. For this case, the kurtosis is $\kappa = -6/5$, and the resulting standard deviation for $\hat{\sigma}^2$ is approximately 8.9%, consistent with the much larger spread seen in these variance estimates relative to those for the binary data. The next boxplot in this sequence (second to the right in Fig. 6.7) was obtained for a standard normal (i.e., $N(0,1)$) data sequence. Consequently, these estimates exhibit the exact distribution discussed in Sec. 6.2.2 with a standard deviation of $\sim$14.1%, 10 times larger than that for

Figure 6.8: Q-Q plots of the four variance estimates

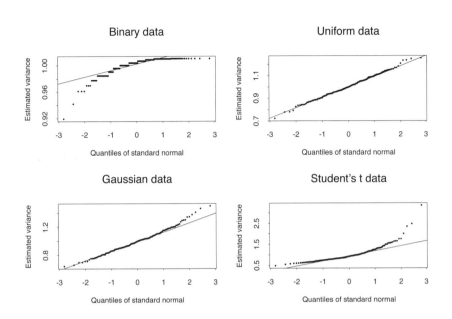

the binary data. Finally, the rightmost boxplot in Fig. 6.7 was obtained from 200 simulations of a Student's t-distribution with 4 degrees of freedom, scaled by $1/\sqrt{2}$ so the resulting sequence also has zero mean and unit variance. Here, however, since the kurtosis of this distribution is infinite, it follows that $\hat{\sigma}^2$ is an *inconsistent estimator* with an infinite variance, a fact reflected in the great variability seen in these 200 estimates.

Another feature seen in the box plots in Fig. 6.7 is the significant skewness of the distribution of $\hat{\sigma}^2$ when the data sequence is strongly non-Gaussian. This point may be seen even more clearly in quantile-quantile plots, like those shown in Fig. 6.8 (refer to Chapter 8 for a detailed discussion of the construction and interpretation of Q-Q plots). In particular, both the upper left plot corresponding to the binary data sequences and the lower right plot corresponding to the Student's t-distributed data sequences show clear evidence of pronounced asymmetry. For the other two cases, there is little or no evidence of asymmetry or other nonnormal behavior. In fact, it is a standard result [162, p. 421] that the exact χ^2_ν distribution obtained for the Gaussian case approaches a Gaussian limit as $\nu \to \infty$.

6.3 The CLT and asymptotic normality

If we add two statistically independent random variables, both drawn from the same distribution, the distribution of the sum is generally not the same as that of the original two variables. In favorable cases, the distribution of the sum has a simple expression, but most cases are not so favorable, and in really unfavorable cases, we may be left with nothing simpler than the general result that the density of the sum may be expressed as the convolution of the two starting densities [37, 222]. Since the distribution of the average of a data sequence can generally be related simply to the distribution of the sum, it follows as a corollary that the exact distribution of averages is also complicated, in general. Fortunately, in most cases, the *Central Limit Theorem (CLT)* holds, which gives a simple approximate result for averages of sufficiently long data sequences: the distribution of the average approaches a Gaussian limit as the number of data samples becomes infinitely large. This result is extremely useful in characterizing estimators since it can often be used to establish their asymptotic normality, the practical advantages of which are discussed briefly in Sec. 6.3.3. Conversely, the CLT may also be one of the most widely *misunderstood* results in engineering and scientific applications, where it is often invoked in a hand-waving manner to argue that *data sequences* should exhibit approximately normal distributions. In particular, if we assume that measurement errors arise as the result of *additively combining* a sufficiently large number of statistically independent, finite variance error sources, the approximate normality of these measurement errors would follow from the CLT. In practice, however, such expectations are not always met, motivating the extensive discussion of non-Gaussian distributions given in Chapter 4. The primary objective of this section is to give a fairly precise statement of the CLT and to illustrate its consequences in cases where it is applicable. The primary objective of Sec. 6.4 is to illustrate that there are important situations in which the CLT *does not* apply.

6.3.1   Distributions of sums and averages

If x_1 and x_2 are two statistically independent random variables with densities $p_1(x)$ and $p_2(x)$, respectively, the density $p_x(x)$ of the sum $x = x_1 + x_2$ is given by the *convolution* of the individual densities, as noted above [37, 222]:

$$p_s(x) = \int_{-\infty}^{\infty} p_1(z)p_2(x - z)dz. \qquad (6.21)$$

The convolution integral (6.21) is extremely important in the theory of linear dynamical systems [41], where it is typically denoted with the symbol $*$, i.e.: $p_s(x) = p_1(x) * p_2(x)$. Without going through the details of the derivations, the following four examples illustrate the utility of this general result.

First, consider the sum of two statistically independent Gaussian random variables, $x \sim N(\mu, \sigma^2)$ and $y \sim N(\nu, \tau^2)$. Applying Eq. (6.21) and simplifying ultimately yields the result that $x + y \sim N(\mu + \nu, \sigma^2 + \tau^2)$. In words, this result means that the sum of two independent Gaussian random variables is another Gaussian random variable whose mean is the sum of the original means and whose variance is the sum of the original variances. In fact, the result that the mean of the sum of independent random variables is simply the sum of the means *always* holds, as does the result that the variance of the sum is simply the sum of the variances. The special features of the Gaussian case—which generally *do not* hold—are, first, that the distribution of the sum has the same form (i.e., Gaussian) as the distribution of the original variables, and second, that the mean and the variance give a complete characterization of the resulting distribution. The fact that these two results do hold for the Gaussian distribution represent part of the reason for its extreme historical popularity.

It follows by repeated application of the result just given that if $x_k \sim N(\mu_k, \sigma_k^2)$ is a sequence of N statistically independent Gaussian random variables, the sum $s_N = x_1 + \cdots + x_N$ has the distribution $s_N \sim N(\mu_N, \sigma_N^2)$ where:

$$\mu_N = \sum_{k=1}^{N} \mu_k \quad \text{and} \quad \sigma_N^2 = \sum_{k=1}^{N} \sigma_k^2. \tag{6.22}$$

As noted above, the distribution of the average often follows simply from the distribution of the sum, and the Gaussian case provides an especially useful illustration. First, note that if $x \sim N(\mu, \sigma^2)$, then $ax \sim N(a\mu, a^2\sigma^2)$ for any real a. Taking $a = 1/N$ leads immediately to the following result:

$$x_k \sim N(\mu, \sigma^2) \quad \Rightarrow \quad \bar{x}_N \sim N(\mu, \sigma^2/N). \tag{6.23}$$

It is worth emphasizing here that this result gives the *exact* finite-sample distribution for the mean of Gaussian data; part of the significance of this result stems from the fact that this distribution holds *approximately* for sufficiently large N for many other distributions of the i.i.d. sequence $\{x_k\}$, a consequence of the CLT discussed in Sec. 6.3.

As another application of the convolution result for the sum of independent random variables, consider the following example. Suppose x_1 and x_2 are both exponentially distributed random variables with common distribution:

$$p(x) = \begin{cases} e^{-x} & x \geq 0, \\ 0 & x < 0. \end{cases} \tag{6.24}$$

Now, consider the distribution of the *difference* $y = x_1 - x_2$: this distribution may be computed from Eq. (6.21) by first rewriting the difference as the sum $x_1 + (-x_2)$ and noting that if x_2 has density $p(x)$, then $-x_2$ has the density $p(-x)$. The density for the difference y is then given by (Exercise 6):

$$\phi(y) = \frac{1}{2} e^{-|y|}. \tag{6.25}$$

Hence, it follows that if x_1 and x_2 have identical exponential distributions, their difference exhibits a Laplace distribution. This observation provides a convenient way of generating Laplace random variables since exponential random variables are easy to generate via transformation from widely available uniform random number generators (see Chapter 12 for details).

As a third example, consider the sum y of two gamma distributed random variables x_1 and x_2 with common scale parameter β, but possibly different shape parameters α_1 and α_2. That is, suppose $x_1 \sim \Gamma(\alpha_1, \beta, \xi_1)$ and $x_2 \sim \Gamma(\alpha_2, \beta, \xi_2)$, where $v \sim \Gamma(\alpha, \beta, \xi)$ means that v has the standard three-parameter gamma density discussed in Chapter 4:

$$p(v) = \frac{(v - \xi)^{\alpha-1} \exp[-(v - \xi)/\beta]}{\beta^\alpha \Gamma(\alpha)}, \tag{6.26}$$

for $v > \xi$, where $\alpha > 0$, $\beta > 0$ and $\Gamma(\alpha)$ is the gamma (factorial) function discussed in Chapter 4. Evaluating the density of the sum from the general convolution result yields, after some algebraic simplification, the result that $y \sim \Gamma(\alpha_1 + \alpha_2, \beta, \xi_1 + \xi_2)$. More generally, it follows from this result that the sum of N statistically independent gamma distributed random variables with common scale parameter β has the distribution:

$$\sum_{k=1}^{N} x_k \sim \Gamma\left(\sum_{k=1}^{N} \alpha_k, \beta, \sum_{k=1}^{N} \xi_k\right). \tag{6.27}$$

One reason this result is important in practice is that it specializes to the case of χ^2 random variables discussed in Sec. 6.2.2 in connection with the standard variance estimator for Gaussian data sequences. In particular, the χ_ν^2 distribution is simply the distribution $\Gamma(\nu/2, 2, 0)$; hence, the sum of N such variables, each with ν_k degrees of freedom, has a $\chi_{\nu*}^2$ distribution with $\nu^* = \sum_{k=1}^{N} \nu_k$ degrees of freedom. This result is important in applications because if x is a standard Gaussian random variable, x^2 has the χ_1^2 distribution, a result discussed in some detail in Chapter 12. Hence, the sum of the squares of ν statistically independent Gaussian random variables exhibits the χ_ν^2 distribution.

To obtain the corresponding result for the average, note that if $x_k \sim \Gamma(\alpha, \beta, \xi)$, then $ax_k \sim \Gamma(\alpha, a\beta, a\xi)$ for any $a > 0$. Taking $a = 1/N$ and combining this result with Eq. (6.27) gives the exact distribution for the mean:

$$x_k \sim \Gamma(\alpha, \beta, \xi) \Rightarrow \bar{x}_N \sim \Gamma(N\alpha, \beta/N, \xi). \tag{6.28}$$

In contrast to the previous result for the Gaussian distribution, note that the effect of averaging gamma distributed random variables is to change the *shape* of the distribution, a fact reflected in the change of shape parameter from α to $N\alpha$. In fact, for large N, it is a standard result [162, p. 340] that $\Gamma(N\alpha, \beta/N, \xi) \simeq N(\xi, \beta/N)$. This result also follows directly from the CLT discussed in Sec. 6.3.2.

The fourth example considered here is the Cauchy density $C(\mu, \beta)$ discussed in Chapter 4, given by:

$$p(x) = \frac{1}{\pi\beta}\left[1 + \left(\frac{x - \mu}{\beta}\right)\right]^{-1}, \tag{6.29}$$

for arbitrary real μ and $\beta > 0$. It is a standard result [162, p. 301] that if $x_k \sim C(\mu_k, \beta_k)$, then:

$$\sum_{k=1}^{N} a_k x_k \sim C\left(\sum_{k=1}^{N} a_k \mu_k, \sum_{k=1}^{N} |a_k| \beta_k\right). \tag{6.30}$$

In contrast to the the gamma distribution just discussed, note that the *shape* of the Cauchy distribution is preserved by the sum. This point is important and is closely related to the fact that the Cauchy distribution *does not* satisfy the conditions of the CLT. Specifically, it follows from Eq. (6.30) that if $x_k \sim C(\mu, \beta)$, then the sum of the sequence exhibits the distribution $C(N\mu, N\beta)$, and it follows directly from the expression for the Cauchy density that $ax_k \sim C(a\mu, a\beta)$. Taking $a = 1/N$ then yields the surprising result that:

$$x_k \sim C(\mu, \beta) \;\Rightarrow\; \bar{x}_N \sim C(\mu, \beta). \tag{6.31}$$

In other words, averaging Cauchy distributed random variables does not reduce their variability, in marked contrast to all of the other distributions considered in this discussion. This observation is closely related to the fact that the Cauchy distribution belongs to the family of *stable random variables*, defined and discussed further in Sec. 6.4.

Finally, it is worth noting that although the previous four examples all ultimately yielded simple expressions for the distribution of the sum of N independent random variables, this behavior is the exception rather than the rule. To see this point, consider the case of the uniform distribution, arguably the simplest possible distribution. The distribution of the sum s_N of N independent random variables x_k, each distributed uniformly on the unit interval $[0, 1]$ is the *Irwin-Hall distribution* [163, p. 296]. The density $p_N(x)$ for this distribution is zero unless x lies between 0 and N. For $x \in [0, N]$, let k be the unique integer for which $k \leq x \leq k + 1$; the density is then given by (Exercise 3):

$$p_N(x) = \frac{1}{(N-1)!} \sum_{j=0}^{k} (-1)^j \binom{N}{j} (x - j)^{N-1}. \tag{6.32}$$

Note that this distribution is a continuous piecewise polynomial of order $N - 1$ that becomes smoother with increasing N. In fact, it follows from the CLT discussed in Sec. 6.3 that $p_N(x)$ approaches a Gaussian limit as $N \to \infty$. Indeed, convergence to this limit is rapid enough that Hoyt [150] has proposed the following simple procedure for generating approximately Gaussian random numbers. First, generate three mutually independent random variables x_i, each uniformly distributed on $[-1, 1]$. It follows by scaling and translating the Irwin-Hall distribution for $N = 3$ that the resulting density is:

$$p(x) = \begin{cases} (3 - x^2)/8, & |x| \leq 1, \\ (3 - |x|)^2/16, & 1 \leq |x| \leq 3, \\ 0, & |x| \geq 3. \end{cases} \tag{6.33}$$

The integral of this density may be computed analytically (Exercise 4) and gives an approximation of the Gaussian cumulative distribution function with an error not exceeding 0.01 [162, p. 112]. Scaling the original sum by $1/N$ leads to the *Bates distribution* [163, p. 297] for the average $\bar{x}_N$:

$$p(x) = \frac{N^N}{(N-1)!} \sum_{j=0}^{[Nx]} (-1)^j \binom{N}{j} \left(x - \frac{j}{N} \right)^{N-1}, \tag{6.34}$$

for $0 \leq x \leq 1$, where $[Nx]$ denotes the largest integer not exceeding Nx. The key point of this example is that, even in cases where the distribution of the original data values is extremely simple, the distribution of sums and averages of these values may not be.

6.3.2   The Central Limit Theorem

To state the CLT concisely, it is useful to introduce the following notation. First, assume that $\{x_k\}$ is a sequence of statistically independent, identically distributed random variables with common mean μ and variance σ^2. Next, define the normalized quantity:

$$\zeta_N = \frac{\sqrt{N}(\bar{x}_N - \mu)}{\sigma}, \qquad (6.35)$$

and note that, since $E\{\bar{x}_N\} = \mu$ for any i.i.d. sequence (see Sec. 6.1.2 for a discussion), it follows that $E\{\zeta_N\} = 0$ for all N. Consequently, it follows that the variance of ζ_N is:

$$\text{var}\{\zeta_N\} = E\{\zeta_N^2\} = \frac{N}{\sigma^2}E\{(\bar{x}_N - \mu)^2\} = \frac{N\,\text{var}\{\bar{x}_N\}}{\sigma^2}. \qquad (6.36)$$

Recall from the discussion in Sec. 6.1.3 that, for any i.i.d. sequence, $\text{var}\{\bar{x}_N\} = \sigma^2/N$, from which it follows that $\text{var}\{\zeta_N\} = 1$ for all N. A standard statement of the CLT is [37, p. 367, theorem 27.1]:

> If $\{x_k\}$ is a sequence of i.i.d. random variables with mean μ and *finite, positive* variance σ^2, then the sequence ζ_N defined in Eq. (6.35) converges in distribution to a Gaussian random variable with zero mean and unit variance.

A discussion of the proof of this theorem is well beyond the scope of this book, but it is important to briefly discuss the assumptions imposed in this statement of the CLT and the implications of those assumptions. First, note that the italicized words *finite* and *positive* appearing in the statement of the theorem are important. The exclusion of sequences with nonpositive (i.e., zero) variance is necessary to make ζ_N well defined and does not rule out anything that is particularly interesting. Conversely, the restriction to sequences with finite variance excludes the class of *stable random variables* defined and briefly discussed in Sec. 6.4.1; this exclusion does have practical implications, as suggested by the unusual behavior of the stable Cauchy distribution under averaging, noted in Sec. 6.3.1 and discussed further in Sec. 6.4.1. The restriction to i.i.d. sequences $\{x_k\}$ in this statement of the CLT also has profound practical implications. Recall that this condition implies two things about $\{x_k\}$: first, that all elements of this sequence are statistically independent, and second, that the probability density associated with each element of this sequence is the same. The first half of this condition—statistical independence—may be weakened to various "asymptotic independence" conditions called *mixing conditions* [37, 250, 304];

Figure 6.9: Exact distributions for $\bar{x}_N$, $N = 1, 2, 4, 9$

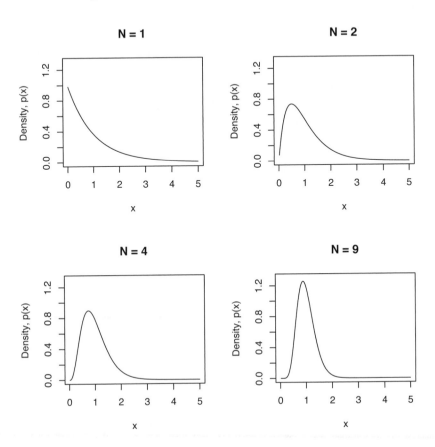

essentially, these conditions specify that correlations between x_k and x_{k-j} decay sufficiently rapidly with increasing distance j between elements of the sequence. The implications of the second half of this condition—that all x_k have the same distribution—are considered further in Sec. 6.4.

The essential character of the CLT is illustrated nicely by the following simple example. Suppose $\{x_k\}$ is a sequence of N i.i.d. samples drawn from the exponential distribution discussed in Chapter 4, with location parameter $\xi = 0$ and arbitrary scale parameter $\beta > 0$. The density for each x_k is therefore:

$$p(x) = \frac{1}{\beta} e^{-x/\beta}. \tag{6.37}$$

Because the exponential distribution belongs to the family of gamma distributions, corresponding to the special case $\alpha = 1$, it follows that the exact distribution for the average of the data sequence $\{x_k\}$ is given by Eq. (6.28) as $\Gamma(N, \beta/N, 0)$. Plots of these density functions for $\beta = 1$ are shown in Fig. 6.9

Figure 6.10: Exact distribution vs. Gaussian approximation, $N = 9$

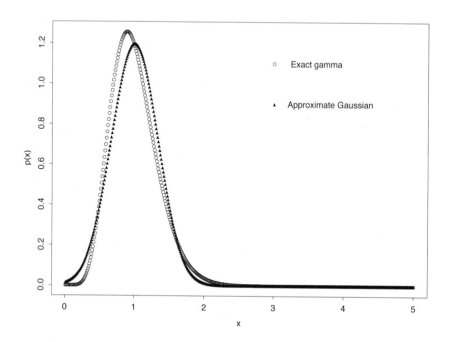

for $N = 1$ (upper left), 2 (upper right), 4 (lower left), and 9 (lower right), from which it is clear that this distribution changes in character from extremely asymmetric and non-Gaussian (e.g., for $N = 1$ or 2) to a reasonable approximation of the Gaussian distribution for $N = 9$.

To see how rapidly the Gaussian limiting distribution is approached, consider this last case, $N = 9$. It follows from the moment results for the gamma distribution given in Chapter 4 that the first four moments of $\bar{x}_N$ are:

$$E\{\bar{x}_N\} = \beta, \quad \text{var}\,\{\bar{x}_N\} = \beta^2/N, \quad \gamma\{\bar{x}_N\} = 2/\sqrt{N}, \quad \kappa\{\bar{x}_N\} = 6/N. \quad (6.38)$$

The first two of these results imply that $\bar{x}_N$ is a consistent, unbiased estimator of the exponential scale parameter β, while the skewness and kurtosis both decay to zero as $N \to \infty$, as they must by the CLT. Further, these results suggest the finite sample Gaussian approximation $\bar{x}_N \sim N(\beta, \beta^2/N)$. Plots of the exact gamma distribution and this Gaussian approximation are shown in Fig. 6.10 for $N = 9$ and $\beta = 1$, showing that this approximation is quite reasonable.

One issue that may seem troubling is that, for the Gaussian approximation, the probability of observing $\bar{x}_N < 0$ is nonzero, even though this event is clearly impossible: $\bar{x}_N$ is the average of N exponentailly distributed random variables, all of which are positive. In fact, the probability of observing negative values from this Gaussian approximation rapidly goes to zero: for $\beta = 1$, we have:

$$\mathcal{P}\{\bar{x}_N < 0\} \simeq 0.07865 \text{ for } N = 2,$$
$$\mathcal{P}\{\bar{x}_N < 0\} \simeq 0.00135 \text{ for } N = 9, \text{ and}$$
$$\mathcal{P}\{\bar{x}_N < 0\} \simeq 7.61 \times 10^{-24} \text{ for } N = 100.$$

This result provides further evidence of how rapidly the Gaussian limit is approached when the CLT does apply.

6.3.3   Asymptotic normality and relative efficiency

As noted, the CLT provides sufficient conditions under which $\bar{x}_N$ approaches a Gaussian limit as $N \to \infty$. This behavior is called *asymptotic normality* and is quite common for practical estimators. In fact, most of the estimators considered in this book exhibit asymptotic normality. Typically, results for an estimator $\hat{\theta}_N$ are obtained by considering the sequence:

$$\zeta_N = \frac{\hat{\theta}_N - \theta}{\sigma/\sqrt{N}}, \qquad (6.39)$$

and showing that, as $N \to \infty$, ζ_N converges either in distribution [37] or in law [180] to a Gaussian random variable with zero mean and unit variance. This result may be expressed in a number of different notations, all interpreteted as limiting behavior for $N \to \infty$:

$$\zeta_N \sim N(0,1), \quad \hat{\theta}_N \sim N(\theta, \sigma^2/N), \quad \text{or} \quad \sqrt{N}(\hat{\theta}_N - \theta) \sim N(0, \sigma^2). \qquad (6.40)$$

One extremely useful application of results of this type is that they provide a means of comparing the effectiveness of different estimators for the same parameter θ. In particular, one useful measure of estimator performance is *estimator efficiency*, which may be defined in a number of different ways. For an unbiased estimator $\hat{\theta}_N$, efficiency is essentially a standardized measure of variance, obtained by comparing it with some reference value. This reference value may be either some absolute bound on achievable variance, leading to the concept of *absolute efficiency*, or the variance of some chosen standard estimator $\tilde{\theta}_N$, leading to the concept of *relative efficiency*. In addition, it is important to distinguish between *finite-sample efficiencies* that characterize the estimator $\hat{\theta}_N$ for fixed N, and *asymptotic efficiencies* that characterize the estimator in the limit as $N \to \infty$.

More specifically, suppose $\hat{\theta}_N$ and $\tilde{\theta}_N$ are both unbiased estimators of the parameter θ. The *finite-sample relative efficiency* of $\hat{\theta}_N$ with respect to $\tilde{\theta}_N$ is defined as:

$$e_N(\hat{\theta}_N, \tilde{\theta}_N) = \text{var}\{\tilde{\theta}_N\}/\text{var}\{\hat{\theta}_N\}, \qquad (6.41)$$

and the *asymptotic relative efficiency* (ARE) of these two estimators is defined as the limit:

$$\text{ARE}(\hat{\theta}_N, \tilde{\theta}_N) = \lim_{N \to \infty} e_N(\hat{\theta}_N, \tilde{\theta}_N). \qquad (6.42)$$

Note that in both of these cases, $\tilde{\theta}_N$ is the reference estimator, so efficiencies less than 1 suggest that $\hat{\theta}_N$ yields poorer results than does this reference estimator and efficiencies greater than 1 suggest that $\hat{\theta}_N$ yields better results. Similarly, if σ_N^2 represents a known lower bound on the variance of $\hat{\theta}_N$, then we may define the *finite-sample absolute efficiency* of $\hat{\theta}_N$ as:

$$e_N(\hat{\theta}_N) = \sigma_N^2/\mathrm{var}\{\hat{\theta}_N\}, \tag{6.43}$$

and the *asymptotic (absolute) efficiency* as the limit:

$$\mathrm{AE}(\hat{\theta}_N) = \lim_{N\to\infty} e_N(\hat{\theta}_N). \tag{6.44}$$

It is important to emphasize that the finite-sample efficiencies defined here depend on the distribution of the data considered, since it is this distribution that determines the distribution of $\hat{\theta}_N$ (and thus its variance). The subject of finite-sample variance bounds for unbiased estimators is considered in some detail in Sec. 6.5, where the *information inequality* or *Cramer-Rao lower bound* is introduced. Both absolute and relative efficiencies are presented for various estimators discussed throughout this book.

Asymptotic relative efficiencies can be particularly useful characterizations for asymptotically normal estimators. In particular, suppose that $\hat{\theta}_N$ and $\tilde{\theta}_N$ are two different unbiased, asymptotically normal, "$\sqrt{N}$-consistent" estimators of θ. This statement means that Eq. (6.40) holds for these two estimators with the asymptotic variances $\hat{\sigma}^2/N$ and $\tilde{\sigma}^2/N$, respectively, and it implies that the asymptotic relative efficiency of these two estimators is:

$$\mathrm{ARE}(\hat{\theta}_N, \tilde{\theta}_N) = \tilde{\sigma}^2/\hat{\sigma}^2. \tag{6.45}$$

The utility of this result is illustrated nicely by the discussion of Lehmann [180, pp. 352–360], who considers the asymptotic relative efficiency of the median with respect to the mean. In particular, suppose $\{x_k\}$ is an i.i.d. sequence with common density $p(x - \theta)$ that is symmetric about θ. Under this symmetry assumption, the mean and the median of the distribution coincide, so both the arithmetic mean of the data and the median of the data are reasonable choices for estimating θ. Lehmann [180, p. 354] proves that the asymptotic variance for the median of a symmetric distribution is $1/4N[p(0)]^2$, provided $p(0) > 0$. Similarly, it follows from the CLT that the asymptotic variance of the arithmetic mean is σ^2/N where σ^2 is the variance of the distribution $p(x)$. Hence, by Eq. (6.45), the ARE of the median with respect to the mean is:

$$\mathrm{ARE}(\tilde{x}_N, \bar{x}_N) = 4[p(0)]^2\sigma^2. \tag{6.46}$$

The utility of this result is that it permits us to decide when—i.e., for which data distributions $p(x)$—the median is a better location estimator than the mean, a question considered further in Chapter 7.

Finally, it is important to note that although asymptotic normality is common among consistent estimators, it is not universal. As a specific example, the

shorth location estimator discussed by Andrews *et al.* [11] as an alternative to the arithmetic average is consistent, but converges to a nonnormal limiting distribution. In addition, it is also interesting to note that the rate at which this limiting distribution is approached is $N^{-1/3}$, as opposed to the more common $N^{-1/2}$ convergence rate noted previously.

6.4 Cases where the CLT does not apply

Despite the relatively weak assumptions on which the CLT rests, it is important to emphasize that this result *does not always apply.* To illustrate this point, the following discussions introduce two important cases where the CLT fails to hold. The first is the case of non-Gaussian *stable* random variables, introduced in Sec. 6.4.1. The best known member of this family is the Cauchy distribution, and this example illustrates a general characteristic of all non-Gaussian stable random variables: infinite variance. Despite this apparently extreme pathology, stable random variable models have been considered in certain signal processing applications [220]; also, some other application areas are mentioned in Samorodnitsky and Taqqu [256, ch. 14], with references to more detailed treatments. The other case considered here where the CLT does not apply is that of *weighted sums* of i.i.d. random variables discussed in Sec. 6.4.2 or, equivalently, sums of statistically independent but *non-identically distributed* random variables. In particular, if the weights in these sums approach zero as $N \rightarrow \infty$, the limiting distribution may be markedly non-Gaussian. This point is illustrated in Sec. 6.4.3 in connection with non-Gaussian ocean noise models.

6.4.1 Stable random variables

It was noted in Sec. 6.3.1 that both the Gaussian distribution and the Cauchy distribution are invariant in form under averaging; in contrast, the other distributions considered in Sec. 6.3.1 approached Gaussian limits as the sample size $N \rightarrow \infty$. This distributional invariance of both the Gaussian and Cauchy distributions reflects the fact that both of these distributions belong to the class of *stable random variables*, which may be defined in terms of this invariance. In contrast, the other distributions considered lie in the domain of attraction of the stable Gaussian distribution, which has the unique characteristic of being the only *finite-variance* stable distribution. Consequently, none of the other members of the family of stable random variables satisfies the conditions of the CLT and indeed, none of them approaches Gaussian limiting distributions on averaging. This class of random variables has been the subject of much recent research because of its highly unusual behavior [220, 256].

Unfortunately, closed-form expressions for densities and cumulative distribution functions are generally unknown for the class of stable random variables. Instead, these variables may be defined either in terms of certain key properties or, more explicitly, in terms of the characteristic function $\Phi(\theta)$ introduced in Chapter 4. Specifically, the characteristic function for the stable random variable X may be expressed as [256, p. 5]:

$$
\begin{aligned}
\Phi(\theta) &\equiv E\{e^{i\theta X}\} &(6.47)\\
&= \begin{cases} \exp\{-\sigma^\alpha|\theta|^\alpha(1 - i\beta(\text{sign }\theta)\tan\frac{\pi\alpha}{2}) + i\mu\theta\} & \text{if } \alpha \neq 1,\\ \exp\{-\sigma|\theta|(1 + i\beta\frac{2}{\pi}(\text{sign }\theta)\ln|\theta|) + i\mu\theta\} & \text{if } \alpha = 1. \end{cases}
\end{aligned}
$$

The four parameters α, β, μ, and σ uniquely define the distribution and have the following names and ranges [256]:

α is the *index of stability* or *characteristic exponent*, with $0 < \alpha \leq 2$,
β is the *skewness parameter*, with $-1 \leq \beta \leq 1$,
μ is the *shift parameter* and may assume any real value,
σ is the *scale parameter* and may assume any non-negative value.

There are three exceptional cases for which densities are known, and we have already encountered two of them. The first is the class of Gaussian densities, corresponding to $\alpha = 2$. In this case, $\Phi(\theta)$ is independent of the skewness parameter β, so it may be taken as $\beta = 0$. The shift and scale parameters μ and σ then reduce to the mean and standard deviation of the Gaussian distribution. It is important to note that the Gaussian distribution is a pathologically special member of the class of stable random variables, since it is the only nondegenerate example for which the variance is finite. More typical of the class in some respects is the Cauchy distribution, obtained by setting $\alpha = 1$ and $\beta = 0$. This distribution has infinite variance, as we have already seen, but it is not entirely representative of the stable distributions because $\alpha = 1$ is the special case singled out in the characteristic function defined in Eq. (6.47).

Possibly a more representative member of the class of stable distributions is the *Levy distribution*, obtained for $\alpha = 1/2$ and $\beta = 1$. Here, the density is defined on the interval (μ, ∞) as:

$$
p(x) = \left(\frac{\sigma}{2\pi}\right)^{1/2}\frac{1}{(x-\mu)^{3/2}}exp\left\{-\frac{\sigma}{2(x-\mu)}\right\}. \tag{6.48}
$$

The Levy distribution may be related to the Gaussian distribution $N(0,1)$ by the following transformation [256, p. 41]:

$$
x = \mu + \frac{\sigma}{z^2}, \tag{6.49}
$$

where $z \sim N(0,1)$. This point is important because, as discussed in Chapter 12, simple transformations of this type arise naturally when we attempt to manipulate problems into more convenient forms. Hence, although it may sound patently unreasonable to consider a random variable with infinite variance like the Levy distribution, such variables are more readily encountered than we might expect. Similarly, recall that some of the special cases of the Zipf distribution discussed in Chapter 2 also exhibit infinite variance.

The principal reason for introducing the class of stable random variables here is to illustrate that the CLT does not apply: sums and averages of non-Gaussian stable random variables do *not* approach Gaussian limits. In fact, the family of stable random variables may be *defined* in terms of the following *generalized Central Limit Theorem* [256, definition 1.1.4]:

> A random variable X has a stable distribution with critical exponent α if $\sum_{i=1}^{N} X_i$ has the same distribution as $N^{1/\alpha} X + D_N$ for some real number D_N, where $X_1, \ldots, X_N$ are N independent copies of the random variable X.

To see the connection between this definition and the CLT, note that this definition implies that if $\{X_i\}$ is an i.i.d. sequence of stable random variables with critical exponent α, the distribution of the arithmetic average will be the same as that of the random variable Z:

$$Z = N^{1/\alpha-1} X + D_N/N. \tag{6.50}$$

For $1 < \alpha \le 2$, note that $N^{1/\alpha-1} \to 0$ as $N \to \infty$. In particular, for $\alpha = 2$, this coefficient is $N^{-1/2}$, and this result correctly implies that the arithmetic mean is a $\sqrt{N}$-consistent estimator of the mean of a Gaussian distribution.

For $1 < \alpha < 2$, this result indicates that the effect of averaging is to narrow the distribution of the individual variables X_i, but not as much as in the Gaussian case. For $\alpha = 1$, it follows that *no* narrowing of the distribution occurs, as discussed in Sec. 6.3.1 for the Cauchy distribution. For $0 < \alpha < 1$, the situation is even worse, since the effect of averaging is to *broaden* the distribution. As a specific example, note that for the Levy distribution, we have $\alpha = 1/2$, and the coefficient in Eq. (6.50) is N, indicating that averaging actually *increases* the variability in this case. This result illustrates a surprising pathology that can occur in certain complex estimation problems where the variability increases rather than decreases with increasing sample size [219, p. 134]. Such behavior may be termed *anticonsistent* to distinguish it from the more common case of an inconsistent estimator whose variance decreases to a nonzero limit as or remains constant as $N \to \infty$.

6.4.2 Weighted averages

Another important special case for which the CLT does not apply is the case of sums of random variables that are dominated by a subset of one or more terms. More specifically, consider the following *weighted average* of an i.i.d. sequence $\{x_k\}$ with finite mean μ and variance σ^2:

$$\omega_N = \sum_{k=1}^{N} w_k x_k, \quad \text{where} \quad 0 \le w_k \le 1 \quad \text{and} \quad \sum_{k=1}^{N} w_k = 1. \tag{6.51}$$

Under these assumptions, it follows from the linearity of the expectation operator that ω_N is an unbiased estimator of the data sequence mean μ:

$$E\{\omega_N\} = \sum_{k=1}^{N} w_k E\{x_k\} = \mu. \tag{6.52}$$

Next, consider the variance of ω_N. Since $\{x_k\}$ is an i.i.d. sequence, it follows that the individual terms $w_k x_k$ are statistically independent, each with mean $w_k \mu$ and variance $w_k^2 \sigma^2$. Hence, the variance of the sum is simply the sum of the variances, implying:

$$\text{var}\{\omega_N\} = \sigma^2 \sum_{k=1}^{N} w_k^2. \tag{6.53}$$

Note that since $0 \le w_k \le 1$, it follows that $0 \le w_k^2 \le w_k$, so the sum satisfies the upper bound:

$$\sum_{k=1}^{N} w_k^2 \le \sum_{k=1}^{N} w_k = 1. \tag{6.54}$$

Similarly, it follows from the Cauchy-Schwartz inequality [39, p. 46] that:

$$1 = \left[\sum_{k=1}^{N} w_k \cdot 1 \right]^2 \le \sum_{k=1}^{N} w_k^2 \sum_{k=1}^{N} 1^2 = N \sum_{k=1}^{N} w_k^2. \tag{6.55}$$

Dividing this equation through by N, we obtain the following lower bound:

$$\sum_{k=1}^{N} w_k^2 \ge 1/N. \tag{6.56}$$

Note that this lower bound is achieved by the unweighted average case $w_k = 1/N$ for all k. Similarly, the upper bound (6.54) is also achievable by taking $w_j = 1$ for some particular j, implying $w_k = 0$ for all $k \ne j$. In the context of the CLT, recall that the $1/\sqrt{N}$ convergence of the mean as $N \to \infty$ follows from the unweighted average result. The fact that this unweighted average variance achieves the smallest possible value suggests that this case is most favorable with respect to rate of convergence. At the other extreme, note that if we keep $w_j = 1$ for some fixed j for all N, then the corresponding weighted average is simply the random variable x_j. Taking the limit as $N \to \infty$ changes nothing so long as j is fixed, so the CLT clearly does not apply here: the distribution of ω_N is the distribution of x_j for all N. As a practical matter, this result suggests—correctly—that if one or a few terms in a weighted sum dominate, the limiting distribution may not be Gaussian. Conversely, note that the sample median of an odd number of data points may be viewed as a weighted average of $\{x_k\}$ where $w_j = 1$ when j indexes the middle value in the sequence, and $w_k = 0$ for all $k \ne j$. In this case, however, there is an important difference: because the index j depends on the data sample, it turns out that the median is an asymptotically normal, consistent estimator like the mean, as discussed in Sec. 6.3.3.

A more realistic case where the Central Limit Theorem does not apply lays the foundation for the application discussed in Sec. 6.4.3. Suppose $w_k > 0$ for all k, but let $N \to \infty$ and assume $w_k \to 0$ as $k \to \infty$. Weighted sums of this type arise commonly in dynamic modeling problems and can lead to situations where the distribution of ω_N approaches a non-Gaussian limit as $N \to \infty$. As a specific example, suppose that $0 < \phi < 1$ and define the weights w_k:

$$w_k = \frac{\phi^{k-1}}{\sum_{j=0}^{N-1} \phi^j} = \frac{(1-\phi)\phi^{k-1}}{1-\phi^N}. \tag{6.57}$$

It follows from Eq. (6.53) that the variance of this weighted average is given by:

$$\begin{aligned}
\text{var}\{\omega_N\} &= \sigma^2 \left[\frac{1-\phi}{1-\phi^N}\right]^2 \sum_{k=1}^{N} \phi^{2(k-1)} \\
&= \sigma^2 \left[\frac{1-\phi}{1-\phi^N}\right]^2 \frac{1-\phi^{2N}}{1-\phi^2} \\
&= \sigma^2 \left(\frac{1-\phi}{1+\phi}\right)\left(\frac{1+\phi^N}{1-\phi^N}\right) \to \sigma^2 \left(\frac{1-\phi}{1+\phi}\right), \tag{6.58}
\end{aligned}$$

as $N \to \infty$. Hence, unlike the result for the CLT, the variance of this weighted average does not approach zero as $N \to \infty$. Consequently, although ω_N is an unbiased estimator of μ, it is not a consistent estimator.

6.4.3   Webster's ambient noise statistics

The ideas presented in the preceeding section provide the basis for analyzing the class of *hyperbolic models* considered by Webster [290] for ambient ocean noise. A key point of Webster's paper is that ocean noise is often seen to exhibit significant deviations from normality (in particular, kurtosis values in the range $-0.7 \le \kappa \le 0.7$), and he notes that such distributions can arise from the additive combination of many *unequally intense* sources. In particular, Webster's results are based on the observation (Exercise 5) that the kurtosis of ω_N is given by the simple expression:

$$\kappa(\omega_N) = \kappa(x)\frac{\sum_{k=1}^{N} w_k^4}{\left[\sum_{k=1}^{N} w_k^2\right]^2}. \tag{6.59}$$

Here, $\kappa(x)$ is the kurtosis of the underlying data sequence, and since kurtosis is scale invariant, this result holds even if the weights $\{w_k\}$ do not sum to 1. Also, note that if $\{x_k\}$ is a Gaussian sequence, then $\kappa(x) = 0$, so $\kappa(\omega_N) = 0$ for all choices of weights $\{w_k\}$. This result is simply a consequence of the fact that the weighted average of Gaussian random variables is just another Gaussian random variable. Further, note that for any non-Gaussian random variable x with a finite kurtosis $\kappa(x)$ in the unweighted average case (i.e., $w_k = 1/N$ for all k), we have $\kappa(\omega_N) = \kappa(x)/N$, which approaches zero as $N \to \infty$. This observation is a consequence of the CLT, which applies in the case of unweighted averages of finite kurtosis random variables (note that the non-Gaussian stable random variables all have infinite moments of order 2 and higher [256, p. 18]).

More unusual results are obtained when $w_k \to 0$ as $k \to \infty$. A particular case considered by Webster [290] is that of *hyperbolic noise models*, for which $w_k = 1/k$. In the limit as $N \to \infty$, these sums may be evaluated in terms of the Riemann zeta function [1, ch. 23], giving:

$$\kappa(\omega_\infty) = \kappa(x)\frac{\zeta(4)}{[\zeta(2)]^2} = \kappa(x)\frac{\pi^4/90}{[\pi^2/6]^2} = \frac{2}{5}\kappa(x). \qquad (6.60)$$

A key observation here is that, since $\kappa(\omega_N)$ does not approach zero as $N \to \infty$ if $\kappa(x) \neq 0$, the distribution of ω_∞ cannot be Gaussian.

Similar arguments hold for other weighted average noise models, such as those based on the exponentially decaying weights w_k defined in Eq. (6.57). In particular, let $\lambda = (1 - \phi)/(1 - \phi^N)$ and note that, from Eq. (6.59), we have:

$$
\begin{aligned}
\kappa(\omega_N) &= \kappa(x)\frac{\sum_{k=1}^{N}\lambda^4\phi^{4(k-1)}}{\left[\sum_{k=1}^{N}\lambda^2\phi^{2(k-1)}\right]^2} \\
&= \kappa(x)\left[\frac{1-\phi^{4N}}{1-\phi^4}\right]\left[\frac{1-\phi^{2N}}{1-\phi^2}\right]^{-2} \\
&= \kappa(x)\left[\frac{1-\phi^2}{1+\phi^2}\right]\left[\frac{1+\phi^{2N}}{1-\phi^{2N}}\right] \to \kappa(x)\left[\frac{1-\phi^2}{1+\phi^2}\right] \qquad (6.61)
\end{aligned}
$$

as $N \to \infty$. Again, for $|\phi| < 1$ and $\kappa(x) \neq 0$, this limiting kurtosis is nonzero, implying the limiting distribution is non-Gaussian.

6.5   The information inequality

This section presents a brief discussion of the *information inequality* [180] or *Cramer-Rao lower bound* [16, 189], which establishes a lower bound on the exact finite-sample variance achievable by any unbiased estimator. In fact, as the following discussion illustrates, the result is somewhat more general than this, but it also has its limitations, a point discussed briefly at the end of this section.

To formulate the information inequality, consider the following situation. Suppose T is an estimator of $g(\theta)$ where $g(\cdot)$ is a known, differentiable function and θ is an unknown parameter. Typically, we will take $g(x) = x$, but the more general result considered here is easily stated and permits consideration of simple transformations that arise frequently (e.g., the square root transformation that maps the variance into the standard deviation). Further, denote the bias in T by $b(\theta)$ and assume this function is also differentiable; i.e., assume:

$$E\{T\} = g(\theta) + b(\theta), \qquad (6.62)$$

where the first derivative $b'(\theta)$ exists. Again, the most common situation considered here is the case of unbiased estimators, for which $b(\theta) = 0$. The information inequality relates the variance of T to the first derivatives of $g(\theta)$ and $b(\theta)$ and to a quantity called the *Fisher information*; more specifically, the information inequality states that [180, p. 122]:

$$\text{var}\{T\} \geq \frac{[b'(\theta) + g'(\theta)]^2}{I(\theta)}, \tag{6.63}$$

where the Fisher information $I(\theta)$ is defined as [16, 180, 189]:

$$I(\theta) = E\left\{\left[\frac{\partial}{\partial\theta}\ln p(\mathcal{D};\theta)\right]^2\right\} = -E\left\{\frac{\partial^2}{\partial\theta^2}\ln p(\mathcal{D};\theta)\right\}. \tag{6.64}$$

Here, $p(\mathcal{D};\theta)$ denotes the joint density of the dataset $\mathcal{D}$ parameterized by the unknown parameter θ (for a discussion of joint densities, refer to Chapter 10). The following results illustrate how the Fisher information may be computed for the situation of interest here, where $\mathcal{D}$ is an i.i.d. sequence of random variables with common density $p(x;\theta)$.

First, suppose that x and y are statistically independent random variables with distributions $p(x;\theta)$ and $q(y;\theta)$, and denote the Fisher information concerning the common parameter θ contained in these random variables by $I_x(\theta)$ and $I_y(\theta)$, respectively. It can be shown [180, p. 121] that the total Fisher information $I(\theta)$ in the combined dataset $\mathcal{D} = \{x, y\}$ is given by:

$$I(\theta) = I_x(\theta) + I_y(\theta). \tag{6.65}$$

It follows as an immediate corollary that if the dataset $\mathcal{D}$ under consideration consists of an i.i.d. sequence of random variables with common density $p(x;\theta)$, the Fisher information appearing in the Cramer-Rao bound (6.63) is given by:

$$I(\theta) = N I_x(\theta), \tag{6.66}$$

where $I_x(\theta)$ is the Fisher information carried by a single observation x_k.

For the important case of an unbiased estimator $T(\theta)$ of the parameter θ from an i.i.d. sequence $\{x_k\}$, Eq. (6.66) may be combined with the information inequality (6.63) to obtain the simple result:

$$\text{var}\{T(\theta)\} \geq \frac{1}{N I_x(\theta)}. \tag{6.67}$$

In practical terms, this result implies that *necessary conditions* for obtaining low-variability estimates of an unknown parameter θ from a dataset $\mathcal{D}$ of N independent observations are, first, to make N large enough, and second, to make the Fisher information $I_x(\theta)$ contained in each individual observation large enough. It is also worth noting that Eq. (6.66) represents an *upper bound* on the total Fisher information $I(\theta)$ contained in the dataset $\mathcal{D}$. In particular, if the observations x_k are not statistically independent, it follows that $I(\theta) < N I_x(\theta)$.

It is instructive to consider the evaluation of $I_x(\theta)$ for the simple but important case of the Gaussian distribution. There, note that:

$$\ln p(x_k;\theta) = -\frac{1}{2}\left[\ln(2\pi) + \ln\sigma^2 + \frac{(x_k - \mu)^2}{\sigma^2}\right]. \tag{6.68}$$

For $\theta = \mu$, it follows from Eq. (6.64) that:

$$I_x(\mu) = -E\left\{\frac{\partial^2}{\partial\theta^2}\ln p(x_k;\theta)\right\} = \frac{1}{2\sigma^2}E\left\{\frac{\partial^2}{\partial\mu^2}(x_k-\mu)^2\right\} = \frac{1}{\sigma^2}. \quad (6.69)$$

Similarly, for $\theta = \sigma^2$, it follows that:

$$\begin{aligned}
I_x(\sigma^2) &= \frac{1}{2}E\left\{\frac{\partial^2}{\partial\theta^2}\left[\ln\theta + \frac{(x_k-\mu)^2}{\theta}\right]\right\} \\
&= \frac{1}{2}\left[-\frac{1}{\theta^2} + \frac{2E\{(x_k-\mu)^2\}}{\theta^3}\right] = \frac{1}{2\sigma^4}.
\end{aligned} \quad (6.70)$$

Combining these results with the Cramer-Rao inequality (6.67), it follows that for any unbiased estimators $\hat{\mu}$ and $\hat{\sigma}^2$:

$$\text{var}\{\hat{\mu}\} \geq \frac{\sigma^2}{N} \quad \text{and} \quad \text{var}\{\hat{\sigma}^2\} \geq \frac{2\sigma^4}{N}. \quad (6.71)$$

Alternatively, taking $\theta = \sigma^2$ and $g(\theta) = \sqrt{\theta}$ in inequality (6.63) yields:

$$\text{var}\{\hat{\sigma}\} \geq \frac{[g'(\theta)]^2}{I(\theta)} = \frac{[-1/2\sqrt{\theta}]^2}{NI_x(\theta)} = \left[\frac{1}{4\sigma^2}\right]\left[\frac{2\sigma^4}{N}\right] = \frac{\sigma^2}{2N}. \quad (6.72)$$

It is interesting to note that all of the moment-based estimators considered in Sec. 6.2.1 approximately achieve these lower bounds for Gaussian data sequences. Conversely, it is important to emphasize that the quality of these bounds depends strongly on the validity of the underlying distributional assumptions, a point discussed further at the end of this section.

Although somewhat specialized, the following simple result is extremely useful. First, define a *location-scale family* as a two-parameter family of probability density functions $p(x;\theta,b)$ of the general form:

$$p(x;\theta,b) = \frac{1}{b}f\left(\frac{x-\theta}{b}\right). \quad (6.73)$$

For such a family, θ is called the *location parameter* and b is called the *scale parameter*. If $f(x) > 0$ for all x and the first derivative $f'(x)$ exists for all x, it follows that the Fisher information $I(\theta)$ is given by [180, p. 120]:

$$I(\theta) = \frac{I_f(\theta)}{b^2}, \quad \text{where} \quad I_f(\theta) = \int_{-\infty}^{\infty}\frac{[f'(x)]^2}{f(x)}dx. \quad (6.74)$$

As an application of this result, note that the Cauchy distribution $C(\mu,\lambda)$ defines a location-scale family with location parameter μ, scale parameter λ, and the function $f(x) = [\pi(1+x^2)]^{-1}$. It is not difficult to show that $I_f = 1/2$ for this case, implying the following lower bound on all unbiased estimators of μ for i.i.d. Cauchy sequences of length N, a result used in Sec. 6.6.1:

$$\text{var}\{\hat{\mu}\} \geq \frac{2\lambda^2}{N}. \quad (6.75)$$

Finally, it is important to say something about the practical limitations of the information inequality. First, note that it may be extended to multiparameter settings [180, 189], and this extension is particularly useful in applications where Gaussian data assumptions are reasonable. Conversely, Huber [151, p. 4] notes that the Cramer-Rao lower bound on estimator variance is extremely sensitive to distributional assumptions. As a specific example, note that the logistic distribution is often regarded as nearly Gaussian (see Chapter 4 for a further discussion of this point), but the normalized Fisher information $I_f(\theta)$ for the location parameter in this distribution is $1/3$, as compared to the Gaussian value of 1 [180, p. 121]. Further, it is important to note that the Cramer-Rao lower bound is generally a conservative one: necessary and sufficient conditions are available for this bound to be sharp, but they are fairly restrictive [180, p. 123].

6.6   Order statistics and L-estimators

Historically, moments have played a dominant role in the characterization of random variables, and as illustrated in later chapters of this book, moment characterizations continue to play a central role in more complex problems like regression analysis. Conversely, moments suffer from a sensitivity to outliers, an extremely important point discussed at some length in Chapter 7. Because of their significantly lower sensitivity to outliers, characterizations based on *order statistics* have become increasingly popular, and the following paragraphs provide a brief introduction to this important topic.

To give some notion of the extent to which order statistics can be useful, Fig. 6.11 illustrates, on the left-hand side, some standard moment-based estimators and, on the right-hand side, their counterparts based on order statistics. For example, the mean, variance, and skewness represent direct applications of the first, second, and third moments of a data sequence as qualitative characterizations of central tendency, spread, and asymmetry. To the right of these moment-based estimators are listed some alternative characterizations based on order statistics: the median, the MADM scale estimate, and Galton's skewness measure, all discussed in some detail in Chapter 7. In addition, it is sometimes advantageous to combine both (low-order) moments and order statistics, as in Hotelling's skewness measure, which is based on the observation that the mean and median are not equal for asymmetric distributions; this skewness measure is also discussed in Chapter 7. Chapter 10 is concerned with relations between variables, and one standard measure introduced there is the product-moment correlation coefficient, closely related to the variance; an extremely useful alternative based on order statistics is the Spearman rank correlation coefficient, also discussed in Chapter 10. Similarly, the ordinary least squares (OLS) regression problem introduced in Chapter 5 is very closely related to the product-moment correlation coefficient, whereas the least absolute deviations (LAD) regression approach may be viewed roughly as an extension of the median from the setting of location estimation to that of fitting regression models.

The essential idea behind order statistics is that of *rank-ordering* a data

Figure 6.11: Moments vs. order statistics

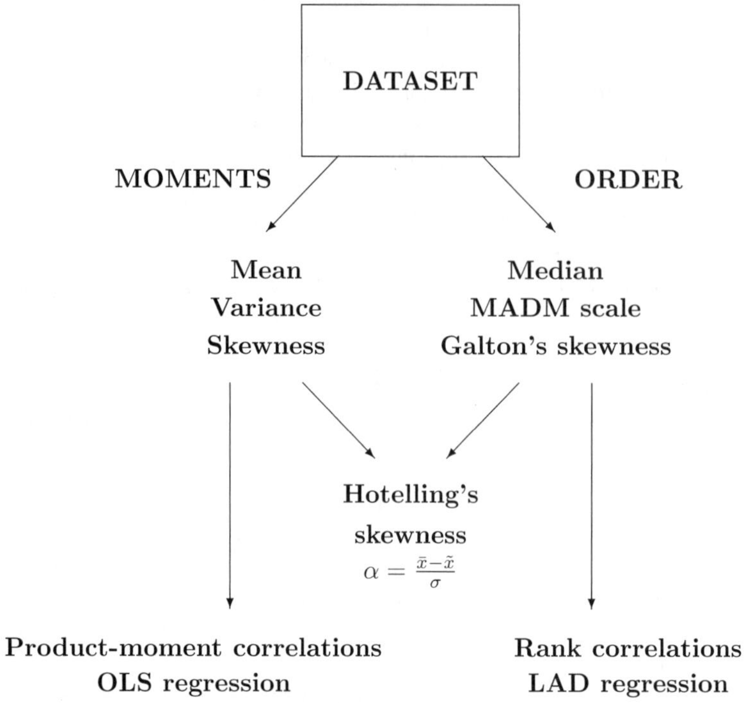

sequence $\{x_k\}$, obtaining the ordered sequence denoted either $\{x_{(i)}\}$ or $\{x_{i:N}\}$:

$$x_{(1)} \leq x_{(2)} \leq \cdots \leq x_{(N)} \quad \text{or} \quad x_{1:N} \leq x_{2:N} \leq \cdots \leq x_{N:N}. \quad (6.76)$$

Particularly important examples of order statistics include:

the *sample minimum* $x_{(1)}$ or $x_{1:N}$,
the *sample median* $x_{([N+1]/2)}$ or $x_{([N+1]/2):N}$ for N odd,
the *sample maximum* $x_{(N)}$ or $x_{N:N}$.

For samples of even length N, the sample median is typically defined as the average of the two central order statistics:

$$\frac{1}{2}[x_{(N/2)} + x_{([N/2]+1)}] \quad \text{or} \quad \frac{1}{2}[x_{[N/2]:N} + x_{([N/2]+1):N}].$$

The following discussions introduce a number of practical applications of order statistics, along with some useful characterizations.

6.6.1 Characterizing the Cauchy distribution

For Gaussian data sequences, the arithmetic mean $\bar{x}_N$ represents *the* estimator of choice (see, e.g., the discussions in Johnson *et al.* [162, p. 123] and Andrews *et al.* [11, p. 243], along with those in Chapter 7 of this book). Conversely, it was shown in Sec. 6.3.1 that if $\{x_k\}$ is an i.i.d. sequence with the Cauchy distribution $x_k \sim C(\mu, \lambda)$, then $\bar{x}_N \sim C(\mu, \lambda)$. This observation implies that, although $\bar{x}_N$ is an unbiased estimator of μ, it also an inconsistent estimator since the Cauchy distribution exhibits infinite variance. Similarly, it follows that second-moment characterizations like the usual variance estimator do not provide a reasonable basis for estimating the Cauchy scale parameter λ. Consequently, it is necessary to consider other alternatives if we wish to estimate distributional parameters for Cauchy-distributed data.

Johnson *et al.* [162, pp. 306–318] consider a number of different estimators for the parameters of the Cauchy distribution, including several based on order statistics. Particularly simple estimators for μ and λ are:

$$\hat{\mu} = \frac{\hat{x}_p + \hat{x}_{1-p}}{2},$$

$$\hat{\lambda} = \left[\frac{\hat{x}_p - \hat{x}_{1-p}}{2} \right] \tan[\pi(1-p)]. \tag{6.77}$$

Here, p is a number lying between $1/2$ and 1, and the $100p\%$ quantile of the distribution is estimated by the order statistic $x_{r:N}$ where r is the largest integer not exceeding $(N+1)p$; this order statistic is denoted $\hat{x}_p$ here for simplicity. These estimators are consistent, with large sample variances approximately given by [162, p. 307]:

$$\text{var}\{\hat{\mu}\} \simeq \frac{\lambda^2 \pi^2 (1-p) \csc^4 \pi p}{2N}$$

$$\text{var}\{\hat{\lambda}\} \simeq \frac{2\pi^2 \lambda^2 (1-p)(2p-1) \csc^2 2\pi p}{N}. \tag{6.78}$$

In these estimators, p is a free parameter which may be chosen to minimize the estimator variance; choosing $p = 0.5565$ leads to an unbiased estimator of μ with an asymptotic variance of approximately $2.33\lambda^2/N$. For comparison, recall from Sec. 6.5 that the Cramer-Rao lower bound for unbiased Cauchy location estimators is $2\lambda^2/N$; hence, the optimal order statistic estimator exhibits an efficiency of approximately 86% relative to this bound. Also, note that the median is another unbiased, consistent estimator of μ, corresponding to $p = 0.5$ and having an asymptotic efficiency of 81% relative to the Cramer-Rao bound.

6.6.2 Distributions of order statistics

One of the useful features of order statistics is that a simple general expression may be derived for their distribution, given the distribution of the original data sequence $\{x_k\}$. Unfortunately, the application of this result is often not as straightforward as it sounds, since it involves both the cumulative distribution function $F(x)$ and the density $p(x)$. Still, there are some extremely important cases that do lead to simple results, as illustrated in secs. 6.6.3 and 6.6.4.

The basis for deriving the distribution of $x_{i:N}$ is the following observation. Denote the cumulative distribution function (CDF) of $x_{i:N}$ by $F_{i:N}(x)$, and recall the definition of the CDF from Chapter 4:

$$F_{i:N}(x) = \mathcal{P}\{x_{i:N} \leq x\}, \qquad (6.79)$$

where $\mathcal{P}\{E\}$ denotes the probability of the event E. Further, note that this defining condition is:

$$\mathcal{P}\{x_{i:N} \leq x\} = \mathcal{P}\{\text{at least } i \text{ values of } x_j \leq x\}$$

$$= \sum_{r=i}^{N} \mathcal{P}\{\text{exactly } r \ x_j \leq x\}. \qquad (6.80)$$

Note that this last condition—that exactly r of the N data values x_j do not exceed x—can occur in $\begin{pmatrix} N \\ r \end{pmatrix}$ ways, all of which are mutually exclusive. Further, each of these events has the probability:

$$P_r = [\mathcal{P}\{x_j \leq x\}]^r [\mathcal{P}\{x_j > x\}]^{N-r} = [F(x)]^r [1 - F(x)]^{N-r}. \qquad (6.81)$$

Combining these results leads to the following expression for the CDF of $x_{i:N}$:

$$F_{i:N}(x) = \sum_{r=i}^{N} \begin{pmatrix} N \\ r \end{pmatrix} [F(x)]^r [1 - F(x)]^{N-r}. \qquad (6.82)$$

By itself, this expression appears fairly formidable, but it may be simplified by invoking the following useful identity. The *incomplete beta function* is defined as [1, p. 263]:

$$I_p(i, N - i + 1) \equiv \int_0^p \frac{N!}{(i-1)!(N-i)!} t^{i-1}(1-t)^{N-i} dt, \qquad (6.83)$$

for $0 \leq p \leq 1$. The utility of this function comes from the following identity [1, p. 263]:

$$\sum_{r=i}^{N} \begin{pmatrix} N \\ r \end{pmatrix} p^r (1-p)^{N-r} = I_p(i, N - i + 1). \qquad (6.84)$$

In particular, combining Eq. (6.82) with Eq. (6.84) leads to the integral representation for $F_{i:N}(x)$:

$$F_{i:N}(x) = \int_0^{F(x)} \frac{N!}{(i-1)!(N-i)!} t^{i-1}(1-t)^{N-i} dt. \qquad (6.85)$$

The advantage of this representation is that it may be differentiated to obtain the following, much simpler, density expression [16, p. 10]:

$$p_{i:N}(x) = \frac{N!}{(i-1)!(N-i)!} [F(x)]^{i-1} [1 - F(x)]^{N-i} p(x). \qquad (6.86)$$

This result is general, holding for all $1 \leq i \leq N$ and all distributions $p(x)$ of the original i.i.d. data sequence $\{x_k\}$. The practical difficulty with this result is that it involves both the density $p(x)$ and the cumulative distribution function $F(x)$: even in the rare cases where both of these functions are explicitly known and have a simple form, the resulting combination is often quite complicated. Still, it is worth noting the following special cases: assuming $N = 2n + 1$ is odd, the densities for the sample minimum, the sample median, and the sample maximum are given by:

$$
\begin{aligned}
p_{1:N}(x) &= N[1 - F(x)]^{N-1}p(x), \\
p_{(n+1):(2n+1)}(x) &= \frac{(2n+1)!}{(n!)^2}[F(x)]^n[1 - F(x)]^np(x), \\
p_{N:N}(x) &= N[F(x)]^{N-1}p(x).
\end{aligned}
\tag{6.87}
$$

The following section considers a special case where all of these results are simple enough that the resulting general expressions are extremely useful.

6.6.3 Uniform order statistics

If $\{x_k\}$ is an i.i.d. sequence, uniformly distributed on the interval $[0, 1]$, the cumulative distribution function $F(x)$ and the density $p(x)$ are given by:

$$
F(x) = \begin{cases} 0 & x < 0 \\ x & 0 \leq x \leq 1 \\ 1 & x > 1, \end{cases} \quad p(x) = \begin{cases} 0 & x < 0 \\ 1 & 0 \leq x \leq 1 \\ 0 & x > 1. \end{cases}
\tag{6.88}
$$

Substituting these expressions into Eq. (6.86) for the density of the general order statistic $x_{i:N}$ then yields:

$$
p_{i:N}(x) = \frac{N!}{(i-1)!(N-i)!}x^{i-1}(1-x)^{N-i}.
\tag{6.89}
$$

In fact, this density corresponds to the beta distribution discussed in Chapter 4, with shape parameters $a = i$ and $b = N - i + 1$.

Because the beta distribution has been studied for over a century, many results are available [163, ch. 25]. In particular, expressions for the mean and variance of $x_{i:N}$ are readily obtained from the results given in Chapter 4:

$$
\begin{aligned}
E\{x_{i:N}\} &= \frac{i}{N+1} \equiv p_i, \\
\text{var}\,\{X_{i:N}\} &= \frac{p_i(1-p_i)}{N+2} = \frac{i(N+1-i)}{(N+1)^2(N+2)}.
\end{aligned}
\tag{6.90}
$$

It is interesting to note that that this variance is maximized for $p_i = 1/2$, corresponding to the sample median, for which $\sigma^2 = 1/[4(N+2)]$. Conversely, the variance is minimum for the extreme order statistics $x_{1:N}$ and $x_{N:N}$, which both exhibit the variance $\sigma^2 = N/[(N+1)^2(N+2)] \simeq 1/N^2$ for large samples.

Figure 6.12: Distribution of the sample minimum vs. N

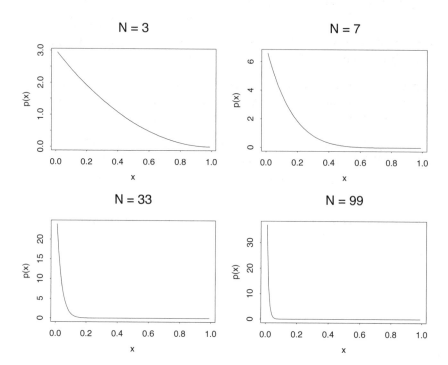

Fig. 6.12 shows plots of the densities $p_{1:N}(x)$ of the sample minimum for sequences of samples of sizes $N = 3$, $N = 7$, $N = 33$, and $N = 99$. These plots illustrate the general behavior: because $a = 1$ and $b = N > 1$ for this beta distribution, $p_{1:N}(x)$ is J-shaped, exhibiting a maximum at $x = 0$ and decaying monotonically as $x \rightarrow 1$. As N increases, the rate of this decay also increases, consistent with the observations made above about the small variance of the sample minimum in large samples. In particular, the probability of observing a value significantly different from $p_1 = 1/(N + 1)$ becomes very small for large N. The densities $p_{N:N}(x)$ for the sample maximum are the mirror images of those for the sample minimum, reflected about $x = 1/2$. The result is a reverse J-shaped distribution not shown here; as with the sample minimum, the probability of observing any value significantly different from $p_N = N/(N + 1)$ becomes very small for large N.

For samples of odd size N, the sample median corresponds to the order statistic $x_{i:N}$ with $i = (N + 1)/2$, which has the expected value $p_i = 1/2$. As noted, this order statistic exhibits the maximum possible variance $\sigma^2 = 1/[4(N + 2)]$, significantly greater than that of the extreme order statistics $x_{i:N}$ for $i \simeq 1$ or $i \simeq N$. This point is reflected clearly in the significantly greater width of the distributions for the median plotted in Fig. 6.13 for the same

Figure 6.13: Distribution of the sample mediam vs. N

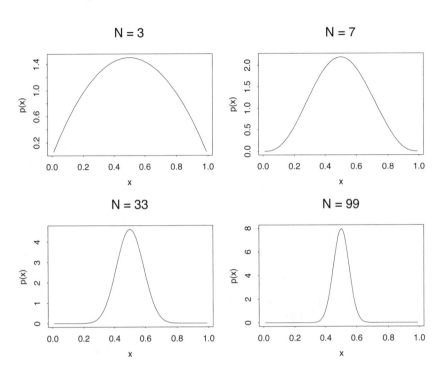

values of N considered for the sample minimum in Fig. 6.12. In addition, the distribution of the medain is seen to be symmetric, corresponding to the fact that the beta distribution shape parameters $a = i$ and $b = N + 1 - i$ are equal in this case; as a consequence, the skewness of the median distribution is zero and the kurtosis is

$$\kappa(x_{i:N}) = -\frac{6}{N+4}. \tag{6.91}$$

Note that this kurtosis value is always negative, increasing toward zero from that of the single uniformly distributed sample ($\kappa = -6/5$) for $N = 1$ to the Gaussian value ($\kappa = 0$) as $N \to \infty$. This result is consistent with the asymptotic normality of the median, discussed in Sec. 6.3.3.

6.6.4 A maximum likelihood estimation problem

Suppose we wish to fit an i.i.d. data sequence $\{x_k\}$ of length N to the two-parameter exponential distribution $E(\theta, \sigma)$, defined by the density:

$$p(x_k | \theta, \sigma) = \begin{cases} \sigma^{-1} \exp\left[-\frac{x_k - \theta}{\sigma}\right] & x_k \geq \theta, \\ 0 & x_k < \theta. \end{cases} \tag{6.92}$$

To accomplish this objective, we proceed as discussed in Chapter 5, first forming the log likelihood function:

$$\ell(\theta, \sigma) = \sum_{k=1}^{N} \ln p(x_k | \theta, \sigma)$$

$$= -N \ln \sigma - (1/\sigma) \sum_{k=1}^{N} x_k + (N/\sigma)\theta, \qquad (6.93)$$

and then choosing θ and σ to maximize $\ell(\theta, \sigma)$. Since ℓ is a smooth function of σ, we may proceed in the customary manner, differentiating ℓ with respect to σ and setting the derivative to zero, yielding the maximum likelihood estimator:

$$\hat{\sigma} = \bar{x}_N - \hat{\theta}, \qquad (6.94)$$

where $\hat{\theta}$ is the maximum likelihood estimator for θ. Conversely, some care is required in determining $\hat{\theta}$ since Eq. (6.93) for $\ell(\theta, \sigma)$ is only valid if $x_k \geq \theta$ for all k. Otherwise, $p(x_k | \theta, \sigma) = 0$ for some k, implying $\ell(\theta, \sigma) = -\infty$. Consequently, maximization of ℓ must be performed subject to the constraint that $x_k \geq \theta$ for all k. Further, note that so long as this constraint is satisfied, $\ell(\theta, \sigma)$ is an increasing function of θ, so it is maximized by making θ as large as possible, subject to the constraint that $\theta \leq x_k$ for all k. Hence, it follows that $\ell(\theta, \sigma)$ may be maximized with respect to θ by choosing $\hat{\theta} = \min\{x_k\} = x_{1:N}$.

One advantage of this result is that the exact distribution of the sample minimum is easily computed for the exponential distribution. In particular, note that the cumulative distribution function may be determined analytically by integrating $p(x)$ to obtain:

$$F(x) = 1 - \exp\left[-\frac{(x - \theta)}{\sigma}\right], \qquad (6.95)$$

for $x \geq \theta$. Substituting Eqs. (6.92) and (6.95) into the general expression (6.87) for $p_{1:N}(x)$ then yields the following simple result:

$$p_{1:N}(x) = \begin{cases} (N/\sigma) \exp\left[-N\left(\frac{x-\theta}{\sigma}\right)\right] & x \geq \theta \\ 0 & x < \theta \end{cases} = p(x|\theta, \sigma/N). \qquad (6.96)$$

Consequently, it follows that the maximum likelihood estimator $\hat{\theta} = x_{1:N}$ for this case is itself exponentially distributed: $\hat{\theta} \sim E(\theta, \sigma/N)$.

Having the exact distribution for $\hat{\theta}$, it is possible to give a complete characterization, which illustrates some interesting and unusual features. First, note that in contrast to most of the other estimators considered in this book, $\hat{\theta}$ is *not* asymptotically normal. Next, it follows from the moment results presented in Chapter 4 that:

$$E\{\hat{\theta}\} = \theta + \frac{\sigma}{N}, \quad \text{and} \quad \text{var}\{\hat{\theta}\} = \frac{\sigma^2}{N^2}, \qquad (6.97)$$

implying $\hat{\theta}$ is asymptotically unbiased and consistent since the variance goes to zero as $N \to \infty$. In fact, the convergence of this estimator is unusually rapid since the variance goes to zero like $1/N^2$ instead of $1/N$.

6.6.5 L-estimators and their properties

One of the best-known applications of order statistics is the use of the sample median as a location estimator. A simple extension of this idea is the use of *L-estimators*, based on *linear combinations of order statistics* [151, 180]:

$$T = \sum_{i=1}^{N} w_i x_{i:N}. \tag{6.98}$$

Generally, the weights are required to satisfy the conditions $0 \le w_i \le 1$ and $\sum_{i=1}^{N} w_i = 1$, and L-estimators satisfying these additional conditions will be called *strict L-estimators*. An important subclass of strict L-estimators are the *trimmed means* introduced in the next paragraph, but other examples of L-estimators include the Cauchy scale estimator described in Sec. 6.6.6, Gastwirth's estimator described in Sec. 6.6.7, Gini's mean difference described in Sec. 6.6.9, and the maximum likelihood estimators for the parameters of the uniform distribution described in Sec. 6.6.10. It is also possible to exploit the structure of L-estimators to derive conditions under which they are asymptotically normal and to choose the weights w_i to minimize the asymptotic variance of these estimators; these topics are discussed briefly in Sec. 6.6.8.

The basic notion of the trimmed mean is quite simple and is in fact fairly widely used in scoring contestants and/or student performances: some fraction of the highest and lowest scores awarded by a panel of judges or obtained on examinations are discarded and the remainder are averaged to obtain a final ranking or grade. More precisely, suppose $0 \le p_1 < p_2 \le 1$ and define $n_1 = [Np_1]$ and $n_2 = [Np_2]$ as the largest integers not exceeding Np_1 and Np_2, respectively. The trimmed mean defined by these integers is [16, p. 231]:

$$T_N = \frac{1}{n_2 - n_1} \sum_{i=n_1+1}^{n_2} x_{i:N}. \tag{6.99}$$

More typically, *symmetric* trimmed means are considered as these are more appropriate location estimators for symmetric distributions, a point illustrated in Sec. 6.1.1. These estimators are obtained by taking $k = [N\alpha]$ for some $\alpha < 1/2$ and considering [180, p. 360]:

$$\bar{x}_\alpha = \frac{1}{N - 2k} \sum_{i=k+1}^{N-k} x_{i:N}. \tag{6.100}$$

So long as $F^{-1}(q)$ is continuous at the quantiles $q = 0.5 \pm \alpha$, it can be shown that the trimmed mean is a consistent, asymptotically normal estimator of the population mean of a symmetric distribution [16, p. 232].

In addition, these asymptotic normality results can be used to derive some useful results concerning the asymptotic relative efficiency (ARE) of trimmed means. For example, Lehmann [180, p. 362] shows that the ARE of the symmetric α-trimmed mean $\bar{x}_\alpha$ relative to the untrimmed arithmetic mean $\bar{x}_N$ is bounded below by $(1 - 2\alpha)^2$ for all symmetric distributions with continuous, positive densities. Unfortunately, this efficiency approaches zero as $\alpha \to 1/2$, but if $p(x)$ is also unimodal, the ARE is bounded below by $1/(1 + 4\alpha)$, which is no worse than $1/3$. In fact, this lower bound is achieved by the median ($\alpha = 1/2$) for the uniform distribution and no other [180, p. 359].

In practice, symmetric trimmed means can be extremely useful because they are intermediate in their outlier sensitivity between the disasterously sensitive mean, corresponding to $\alpha = 0$, and the median, corresponding to $\alpha \simeq 1/2$ and having the best possible outlier resistance, but at the price of somewhat unusual and generally undesirable behavior in other situations. The use of trimmed means permits us to steer between these two extremes and is considered further in Chapter 7, devoted to the problem of outliers and their practical implications.

6.6.6   L-estimators for Cauchy parameters

Johnson *et al.* [162, p. 309] describe simple L-estimators for both the location and scale parameters of the Cauchy distribution. These estimators were proposed by Chernoff *et al.* [61] and are asymptotically optimal (i.e., they minimize the asymptotic variance over the class of L-estimators). To simplify subsequent notation, define:

$$\zeta_i = \pi \left[\frac{i}{N+1} - \frac{1}{2} \right]. \tag{6.101}$$

The weights for the optimal L-estimator for μ are given by:

$$w_i = \frac{\sin 4\zeta_i}{(N+1)\tan\zeta_i}, \tag{6.102}$$

and those for the optimal L-estimator for λ are given by:

$$w_i = \frac{8\tan\zeta_i}{(N+1)\sec^4\zeta_i}. \tag{6.103}$$

(Note: there is an error in the weights presented by Johnson *et al.* [162]; the weights given there must be divided by $N+1$ to obtain asymptotically unbiased estimators. The results given here agree with those given originally by Chernoff *et al.* [61].)

Plots of these optimal weights for $N = 25$ and $N = 100$ are shown in Fig. 6.14. Note that although the the weights for the location L-estimator sum to 1, they include some negative weights, a point discussed further in Sec. 6.6.8. Conversely, the weights for the scale L-estimator sum to zero.

Figure 6.14: Optimal weights, Cauchy L-estimators

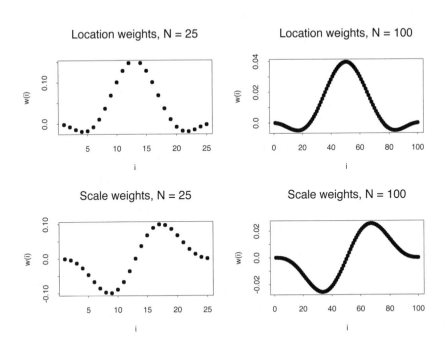

6.6.7 Gastwirth's location estimator

Gastwirth's location estimator is the L-estimator defined by the following weights [112], for N even:

$$T_G = 0.3x_{(a:N)} + 0.4x_{(b:N)} + 0.3x_{(c:N)}, \tag{6.104}$$

where $a = [N/3+1]$ is the largest integer not larger than $N/3+1$, $b = N/2$ and $c = N - [N/3]$. This estimator was one of those considered in the Princeton robustness survey [11] and was one of the group of 14 estimators classified as particularly useful by F.R. Hampel out of the 68 considered in the study [11, p. 236] because they exhibited high breakdown points and relatively narrow confidence limits over all distributions considered (see Chapter 7 for a discussion of breakdown points). Of these 14 estimators, Gastwirth's is the simplest, being the only one that does not involve iterative computations.

6.6.8 Asymptotic normality of L-estimators

Lehmann [180, ch. 5] presents a number of results establishing conditions on the weights $\{w_i\}$ such that the L-estimator defined by Eq. (6.98) exhibits a normal limiting distribution as $N \to \infty$. In particular, suppose $\lambda(x)$ is a bounded function mapping the unit interval $[0,1]$ into the real line that is continuous almost everywhere, symmetric about $1/2$, and integrates to 1. Further, suppose $\{x_k\}$ is an i.i.d. sequence whose common density $p(x)$ is symmetric about θ and satisfies $p(\theta) > 0$. Finally, consider the weights:

$$w_i = \alpha\lambda\left(\frac{i}{N+1}\right), \qquad (6.105)$$

where α is chosen so the weights sum to 1. Under these circumstances, Lehmann [180, p. 371] shows that the L-estimator $T(\theta)$ defined by these weights satisfies the asymptotic normality condition:

$$\sqrt{N}[T(\theta) - \theta] \rightarrow N(0, \sigma^2), \qquad (6.106)$$

where σ^2 depends on both the function $\lambda(\cdot)$ defining the weights and the distribution $F(\cdot)$ of the data sequence $\{x_k\}$.

In fact, it is possible to exploit this dependence to obtain optimal L-estimators for the location parameter θ of specified distributions. In particular, if $p(x)$ is twice differentiable almost everywhere and $p(x) \rightarrow 0$ as $x \rightarrow \pm\infty$, define:

$$\gamma(x) = -\frac{d}{dx}\ln p(x). \qquad (6.107)$$

Lehmann [180, p. 373] shows that the optimal weights are given by Eq. (6.105) where:

$$\lambda(x) = \frac{\gamma'[F^{-1}(x)]}{\int_{-\infty}^{\infty}\gamma^2(z)p(z)dz}. \qquad (6.108)$$

Further, the resulting estimator is shown to achieve the Cramer-Rao lower bound, establishing that no other estimator (in particular, none outside the class of L-estimators) can achieve smaller estimator variance. Application of this result to the Gaussian distribution leads to the unweighted average $\bar{x}_N$ as the optimal L-estimator, and application to the logistic distribution discussed in Chapter 4 leads to the nonuniform optimal weights:

$$w_i = \alpha\left(\frac{i}{N+1}\right)\left(1 - \frac{i}{N+1}\right). \qquad (6.109)$$

Lehmann [180, p. 375] also notes that heavy-tailed distributions like the Student's t-distribution with few degrees of freedom can lead to *negative* optimal weights. This observation is consistent with the behavior of the L-estimator for the Cauchy location parameter considered in Sec. 6.6.6, which corresponds to the most heavy-tailed of the Student's t-distributions (i.e., the Student's t-distribution with 1 degree of freedom). Also, note from Eq. (6.108) that a necessary and sufficient condition for $\lambda(x) \geq 0$ for all x is that $\gamma'(x) \geq 0$ for all x, which in turn implies that $-\ln p(x)$ is convex; distributions satisfying this condition are called *strongly unimodal* [180, p. 375] and are interesting because the convolution of any unimodal distribution with a strongly unimodal distribution remains unimodal [163, p. 677].

6.6.9 Gini's mean difference

Although the variance and the standard deviation are probably the best-known measures of the scale or "spread" of an i.i.d. sequence $\{x_k\}$, many other measures have been proposed. One possible scale measure is the expectation $E\{|x_k - x_j|\}$, which measures how far, on average, distinct samples from the sequence lie from each other. If we simply average over all possible values of $|x_k - x_j|$, we obtain *Gini's mean difference*, introduced in Chapter 3, which may be expressed either as [16, p. 230]:

$$ g_N = \frac{1}{N(N-1)} \sum_{j=1}^{N} \sum_{k \neq j} |x_k - x_j|, \qquad (6.110) $$

or in the slightly more efficient form [162, p. 133]:

$$ g_N = \binom{N}{2}^{-1} \sum_{i=1}^{N} \sum_{j>i} |x_i - x_j|. \qquad (6.111) $$

It is not obvious from either of these expressions that Gini's mean difference belongs to the class of L-estimators, but it is also possible to reexpress g_N in the following form [16, p. 230], from which it is clear that g_N is an L-estimator:

$$ g_N = \frac{2(N+1)}{N(N-1)} \sum_{i=1}^{N} \left(\frac{2i}{N+1} - 1 \right) x_{i:N}. \qquad (6.112) $$

Like the scale estimator for the Cauchy distribution considered in Sec. 6.6.6, it is interesting to note that these weights also sum to zero rather than 1. In fact, these weights increase linearly from approximately $-2/N$ for $i = 1$ to approximately $+2/N$ for $i = N$. Consequently, the conditions presented in Sec. 6.6.8 for asymptotic normality do not apply here, but it can be shown that g_N is asymptotically normal anyway [16, p. 230]. Computationally, the L-estimator formulation of Gini's mean difference is much better than either of the expressions given in Eqs. (6.110) and (6.111) because the L-estimator implementation runs *much* faster for long data sequences.

If $\{x_k\}$ is an i.i.d. Gaussian sequence, it is known that $E\{g_n\} = (2/\sqrt{\pi})\sigma$ [162, p. 133], so $G_N = (\sqrt{\pi}/2)g_N$ is an unbiased L-estimator of the standard deviation for Gaussian data. Further, this estimator is almost as efficient as the optimal L-estimator for σ in this case, with an efficiency approaching 97.8% as $N \to \infty$ [162, p. 133]. In addition, G_N is also known to be reasonably robust against outliers [16, p. 230], in marked contrast to the usual moment-based standard deviation estimator, a point discussed further in Chapter 7.

6.6.10 Uniform maximum likelihood estimators

Suppose $\{x_k\}$ is an i.i.d. data sequence that is uniformly distributed on some interval $[a, b]$ where $a < b$ are unknown constants. It is possible to derive maximum likelihood estimators for this problem that are in some respects quite similar to those obtained in Sec. 6.6.4 for the exponential distribution. To proceed, first reformulate this problem in terms of the location parameter θ and the scale parameter ϕ, defined as:

$$\theta = \frac{a+b}{2} \quad \text{and} \quad \phi = \frac{b-a}{2}. \tag{6.113}$$

In terms of these parameters, x_k is uniformly distributed on $[\theta - \phi, \theta + \phi]$, implying the density:

$$p(x_k|\theta, \phi) = \begin{cases} 1/2\phi & \theta - \phi \le x_k \le \theta + \phi, \\ 0 & \text{otherwise.} \end{cases} \tag{6.114}$$

Proceeding as in Sec. 6.6.4, we form the log likelihood function, which for this problem is:

$$\ell(\theta, \phi) = \ln \prod_{k=1}^{N} p(x_k|\theta, \phi) = -N \ln 2\phi, \tag{6.115}$$

provided $\theta - \phi \le x_k \le \theta + \phi$ *for all* k. Otherwise, $p(x_k|\theta, \phi) = 0$ for some k and $\ell(\theta, \phi) = -\infty$, exactly as in the exponential MLE problem.

From these results, it follows that ℓ is maximized by minimizing ϕ subject to the two constraints:

$$\begin{aligned} \theta - \phi &\le x_{1:N} \le \theta + \phi, \\ \theta - \phi &\le x_{N:N} \le \theta + \phi. \end{aligned} \tag{6.116}$$

Subtracting the first of these inequalities from the second (note that the order of the first inequality must be reversed in doing this), it follows that $x_{N:N} - x_{1:N} \le 2\phi$. Consequently, the minimum admissable value of ϕ is the maximum likelihood L-estimator:

$$\hat{\phi} = \frac{x_{N:N} - x_{1:N}}{2}. \tag{6.117}$$

Substituting this result into the inequalities in (6.116) then yields the following two inequalities for θ:

$$\begin{aligned} \theta &\le x_{1:N} + \hat{\phi} = \frac{x_{1:N} + x_{N:N}}{2}, \\ \theta &\ge x_{N:N} - \hat{\phi} = \frac{x_{1:N} + x_{N:N}}{2}. \end{aligned} \tag{6.118}$$

Hence, it follows that this common bound defines the maximum likelihood L-estimator for θ:

$$\hat{\theta} = \frac{x_{1:N} + x_{N:N}}{2}. \tag{6.119}$$

Like the exponential MLE considered in Sec. 6.6.4, these estimators are consistent but not asymptotically normal. In fact, the variance of these estimators each converges to zero approximately like $2\phi^2/N^2$ [163, p. 286], exhibiting the same unusually rapid convergence as the exponential MLE. Also, these estimators are interesting because they are *uncorrelated but not statistically independent*, an important distinction discussed further in Chapter 10.

6.7 Exercises

1. Suppose $\{x_k\}$ is a sequence of N independent, identically distributed random variables with mean μ and variance σ^2. If μ and σ^2 are finite, show that the following estimator of σ^2 is unbiased:

$$\hat{\sigma}^2 = \frac{1}{N-1} \sum_{k=1}^{N} (x_k - \bar{x}_N)^2.$$

2. For the unbiased estimator considered in Exercise 1, show that the estimator variance is given by:

$$\text{var}\,\{\hat{\sigma}^2\} = \frac{(\kappa+2)\sigma^4}{N} + \frac{2\sigma^4}{N(N-1)},$$

where κ is the kurtosis of the common distribution of $\{x_k\}$.

3. Using the convolution representation for the density of the sum of two independent random variables, determine the distribution of the sum of N independent random variables, each uniformly distributed on the interval $[0, 1]$. (Hint: start with the simplest case $N = 2$ and then proceed by induction to obtain the Irwin-Hall distribution discussed in Sec. 6.3.1.)

4. The rescaled Irwin-Hall density for $N = 3$ was considered in Sec. 6.3.1:

$$p(x) = \begin{cases} (3 - x^2)/8 & |x| \leq 1, \\ (3 - |x|)^2/16 & 1 \leq |x| \leq 3, \\ 0 & |x| \geq 3, \end{cases}$$

where it was noted that this density provides a reasonable basis for approximating the Gaussian cumulative distribution function (CDF). Integrate this density to obtain this approximation to the Gaussian CDF.

5. Consider the following weighted sum of a sequence $\{x_k\}$ of independent, identically distributed random variables:

$$z_N = \sum_{k=1}^{N} \zeta_k x_k.$$

Show that the kurtosis $\kappa(z_N)$ of this sum is related to the kurtosis $\kappa(x)$ of the data sequence via:

$$\kappa(z_N) = \kappa(x) \left[\frac{\sum_{k=1}^{N} \zeta^4}{\left(\sum_{k=1}^{N} \zeta^2\right)^2} \right].$$

6. Derive Eq. (6.25): suppose x_1 and x_2 are independent, exponentially distributed random variables with the common distribution:

$$p(x) = \begin{cases} e^{-x} & x \geq 0, \\ 0 & x < 0, \end{cases}$$

and let $y = x_1 - x_2$. Show that y has the Laplace distribution:

$$\phi(y) = \frac{1}{2}e^{-|y|}.$$

Chapter 7

Outliers: Distributional Monsters (?) That Lurk in Data

Sections and topics:

This chapter provides a brief introduction to the topic of *outliers* or *anomalous data points*, focusing on their nature, their prevalence in real datasets, and the type of damage they can do to our analysis if left undetected. In addition, this chapter also introduces some simple techniques for detecting outliers in a dataset, emphasizing the importance of using techniques that do not *themselves* fail if outliers are present. In fact, this paradoxical situation occurs in the popular *3σ-edit rule* discussed in Sec. 7.1.2 due to an effect called *masking*. Fortunately, effective alternatives do exist, and one of these—the *Hampel identifier*—is introduced in Sec. 7.5.1. This chapter also introduces the notion of *robust estimation*, or the use of "bullet-proof" procedures that are relatively (and sometimes extremely) insensitive to the presence of outliers. Finally, this chapter briefly describes some of the practical difficulties resulting from asymmetry, light-tailed data distributions, and coarse quantization that can arise when using robust estimators. Extensions of many of the ideas presented here to specific data analysis procedures are presented in subsequent chapters.

Key points of this chapter:

1. This chapter adopts the very general working definition of an outlier as an anomalous data point and explores a few simple specializations of this definition that permit the development of quantitative procedures for outlier detection.

2. Real data examples are presented in this chapter to illustrate that outliers do arise commonly in measurement data and to give an idea of what typical outliers look like in a dataset.

3. Simple analysis examples are presented, using both real data and simulation data, to illustrate the profound consequences that outliers can have on even routine analysis (e.g., estimation of the standard deviation of a data sequence).

4. Four basic procedures are described for dealing with outliers: simple omission, replacement, detection and examination, and the use of robust analytical procedures that are not highly sensitive to the presence of outliers. All of these approaches have their place, each with their own sets of advantages and disadvantages, a topic discussed in some detail here.

5. Three of these four approaches to dealing with outliers begin with their detection in the dataset, so this problem is considered at length in this chapter. Ironically, probably the best known outlier detection procedure (the 3σ edit rule discussed in Sec. 7.1.2) often fails completely when applied to datasets containing multiple outliers (this effect is called *masking*).

6. A generally more effective outlier detection procedure is the Hampel identifier, introduced in Sec. 7.5.1, although this procedure is sometimes too aggressive, finding more outliers in a dataset than are present (this effect is called *swamping*).

7. Asymmetric data distributions cause a number of difficulties: the traditional asymmetry measure (the skewness γ discussed in Chapter 3) exhibits *extreme* outlier sensitivity, and although alternative asymmetry measures exist that have better outlier resistance, they also tend to be highly variable even in the absence of outliers. This topic is discussed further in secs. 7.6 and 7.8.2.

8. Light-tailed, bimodal, and discrete data distributions can cause robust estimators to behave badly, as examples presented in Sec. 7.7 illustrate.

9. Overall, an extremely useful strategy in dealing with outliers is to analyze the dataset using both traditional and robust methods and compare the results: reasonable agreement provides some increase in our confidence in the outcome, but large differences often indicate the presence of unusual structure in the data, which should be examined further.

7.1 Outliers and their consequences

The basic notion of an outlier as an anomalous data point was introduced in Chapter 1, and a simple working definition was given in Chapter 2, where outliers were defined as data points inconsistent with the behavior seen in most of the rest of the data. While this definition is conceptually useful and was illustrated with a number of practical examples in Chapter 2, it is too vague to serve as a basis for developing outlier detection procedures or analyzing the outlier sensitivity of different data analysis procedures. One of the primary objectives of this chapter is to introduce some simple mathematical outlier models that are specific enough for these applications. The first of the following three examples illustrates a case of considerable historical importance: the presence of a single anomalous observation in an otherwise Gaussian data sequence. As this example illustrates, the consequences of even a single outlier can be quite pronounced, particularly for characterizations involving higher moments. The second example, presented in Sec. 7.1.2, illustrates the point noted earlier that outlier detection procedures can themselves fail due to the presence of outliers in the dataset. Finally, Sec. 7.1.3 introduces the *contaminated normal outlier model*, a mathematical outlier model that has been extremely useful in both analyzing the outlier sensitivity of standard analysis procedures and in developing outlier-resistant alternatives.

7.1.1 The outlier sensitivity of moments

Moment estimators were introduced in Chapter 4, and their historical importance was discussed briefly. One of the reasons for their continuing popularity is their mathematical simplicity, but as the following example illustrates, their outlier sensitivity is significant and it becomes worse with increasing moment order. To see this point, consider the following pair of *simulated* datasets, one consisting of $1,024$ statistically independent, zero-mean, unit variance Gaussian samples, and the other obtained by replacing one of these points with the value $+8$. The advantage of considering simulated datasets in this example is that we then know precisely what the answers should be: the mean, skewness, and kurtosis should all be zero, and the standard deviation should be 1. The results presented in Table 7.1 illustrate the consequences of this single outlier on the mean, standard deviation, skewness, and kurtosis values estimated from these datasets, along with the corresponding results for smaller versions of these two datasets, each of size $N = 128$. These estimates are based on the definitions given in Chapter 4 and the first four moments of the sequence:

$$\bar{x}_N = \frac{1}{N} \sum_{k=1}^{N} x_k, \qquad \hat{\mu}_n = \frac{1}{N} \sum_{k=1}^{N} [x_k - \bar{x}_N]^n, \quad n = 2, 3, 4. \qquad (7.1)$$

The estimated mean is small (relative to the standard deviation $\sigma = 1$) whether the outlier is present or absent, as is the difference between these two estimates.

Table 7.1: Estimated moments from $N(0, 1)$ data

N	Outlier	Mean	Std. dev.	Skewness	Kurtosis
1024	none	−0.006	1.000	−0.002	0.553
1024	+8σ	+0.002	1.032	+0.454	3.894
128	none	−0.075	0.875	−0.156	0.115
128	+8σ	−0.012	1.137	+2.721	18.820

Hence, we conclude that the estimated mean is approximately correct (i.e., approximately zero) and is not strongly influenced by this single anomalous data point. The standard deviation is estimated to be about 3% larger when the outlier is present than when it is absent, a significant change given that only about 0.1% of the data has been corrupted. Still, this result is not dramatic. The result *is* dramatic for the estimated skewness, however: there, the estimated value changes from −0.002 to +0.454, solely on the basis of the treatment (inclusion or exclusion) of this single data point. The behavior of the kurtosis estimates is even more dramatic, with the addition of the outlier changing the estimated value from 0.553 to 3.894. Although it could be argued that *neither* of these values seems that close to the true value of zero, the real point here is the influence of this single point on the computed estimate. The situation is even worse for smaller datasets, as seen in the results for $N = 128$ presented in Table 7.1. Although we expect greater variability for smaller datasets, here we observe much larger changes in variance due to the outlier and and *enormous* changes in the estimated skewness and kurtosis. The kurtosis, in particular, changes from a somewhat reasonable value of 0.115 to a patently absurd value of 18.820, all on the basis of a single outlier.

In fact, it is worth examining the kurtosis estimates more carefully since even in the absence of contamination, these values appear somewhat larger than we might expect for a large Gaussian dataset. Fig. 7.1 presents a boxplot summary of four sets of simulations that illustrate the general behavior of the estimated kurtosis, under four different scenarios. The leftmost boxplot summarizes 100 kurtosis estimates, each computed from an uncontaminated Gaussian sequence of length $N = 1024$; these estimates span the range from about −0.303 to +0.418, with a median value of −0.042, quite close to the correct value of zero. The second-left boxplot again summarizes 100 kurtosis estimates, this time computed from the same data sequence but with a single value replaced by +8, consistent with the results presented in Table 7.1. Here, the range of estimated kurtosis values lies between 2.423 and 3.928, with a median value of 3.192. The difference between these two boxplots further illustrates the outlier sensitivity of the kurtosis just discussed, emphasizing that this shift is well in

Figure 7.1: Boxplots of estimated kurtosis values

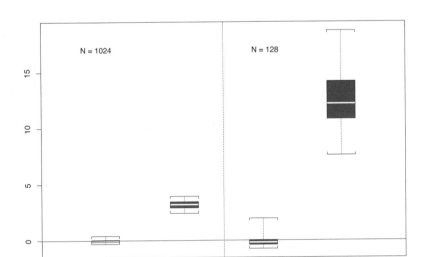

excess of the high natural variability of the estimated kurtosis, large as that variability is. The two boxplots to the right of the vertical dashed line present the corresponding results obtained for samples of size $N = 128$. There, the range of estimated kurtosis values is substantially wider in the uncontaminated case, from about -0.735 to 1.950 with a median value of -0.184, reflecting the influence the smaller sample size. Again, the presence of a single outlier at $+8$ shifts the kurtosis to an anomalously large value that is well outside this nominal range: for the contaminated data sequence, the estimated kurtosis varies from a minimum of 7.594 to a maximum of 18.717.

The real point of this example is that even *one* outlier can *dramatically* influence moment estimates. Further, the effect becomes worse as the size of the dataset decreases and as the order of the moment increases. This example is revisited in connection with a number of more complicated cases in subsequent chapters—dynamic characterization (e.g., time-dependent moment estimators), linear regression analysis (e.g., two-variable moment estimators), and more general problems (e.g., many-variable moment estimators). As a general rule of thumb, we offer a variation of Murphy's law: the more complicated the situation, the more things can (and thus will) go wrong. The real point, however, is not to advocate Dante's advice ("Abandon hope, ye that enter here"), but rather to emphasize that outliers are a part of real life that we cannot afford to ignore and hope they will go away—they won't.

Figure 7.2: Makeup flow rate data with 3σ edit limits

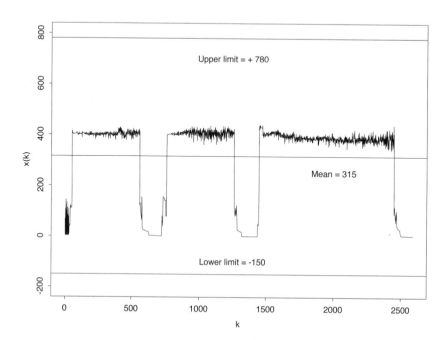

7.1.2 Failure of the 3σ-edit rule

Probably the most common procedure for outlier detection is the *3σ-edit rule*, known more formally as the *extreme Studentized deviation (ESD) identifier* [82]. The basic idea is quite simple: given a single observation x of a standardized Gaussian random variable (i.e., $x \sim N(0,1)$), the probability that $|x| > 3$ is approximately 0.27%. This observation leads to the following, apparently reasonable, procedure for detecting outliers:

1. compute the mean of the data sequence $\bar{x}_N = \frac{1}{N} \sum_{k=1}^{N} x_k$,
2. estimate the standard deviation $\hat{\sigma} = \left[\frac{1}{N-1} \sum_{k=1}^{N} (x_k - \bar{x}_N)^2 \right]^{1/2}$,
3. declare x_k "suspicious" if $|x_k - \bar{x}_N|/\hat{\sigma} > t = 3$.

The threshold parameter t in the last step of this procedure clearly influences its performance, but the primary point of this discussion is that, *regardless of how we choose t, this procedure fails in the presence of enough outliers.*

Fig. 7.2 shows an enlarged view of the makeup flow rate data discussed in Chapter 2, together with lines indicating the outlier detection limits for the ESD identifier. The upper line corresponds to the $+3\sigma$ limit of ~780, nearly twice the maximum makeup flow rate observed in this data sequence (439.59). The middle

Table 7.2: Makeup flow rate data, $\bar{x}_N$ and $\hat{\sigma}$

Sequence	$\bar{x}_N$	$\hat{\sigma}$
Complete dataset	315.46	155.04
Subsequence 1	403.28	7.67
Subsequence 2	406.62	11.50
Subsequence 3	390.38	14.75

line represents the mean value $\bar{x}_N \simeq 315$, which falls below *all* of the nominal makeup flow rate values in this dataset; in fact, in this example, the mean value represents an effective threshold value for separating the nominal values from the shutdown episodes. Finally, the bottom line in Fig. 7.2 represents the -3σ lower detection limit of approximately -190, corresponding to an unphysical reversal of the flow direction in this feed stream. *Consequently, even though it has been argued on physical grounds that this dataset divides cleanly into a nominal subset and a shutdown subset, the popular 3σ-edit rule fails to detect any outliers at all.*

This example provides a real data illustration of the effect of *masking*, defined as the failure of an outlier detection rule in the presence of the outliers themselves. Here, the basic cause of this difficulty is that the mean $\bar{x}_N$ and the standard deviation $\hat{\sigma}$ are highly susceptible to the influence of outliers. In particular, the presence of multiple outliers tends to cause severe *variance inflation*, in which the estimated variance becomes much larger than the variance of the nominal portion of the data sample. This point is seen clearly in Table 7.2, which shows the estimated means and standard deviations for the complete recycle flow rate dataset and three subsequences, each corresponding to a contiguous episode of normal process operation. These subsequences are discussed further in Sec. 7.2.1, but the point here is that the results shown in Table 7.2 clearly illustrate the mean shift and variance inflation caused by the shutdown episodes. In particular, note that the mean of the original data sequence is $\sim$30% smaller than the mean of these subsequences, reflecting the tendency of the outlying shutdown episodes to shift the mean toward zero. Significant as these differences are, they pale in comparison to the *enormous* differences between the estimated standard deviations: $\hat{\sigma}$ for the original sequence is 10–20 times larger than that for the individual subsequences. It is precisely this variance inflation that is responsible for the failure of the 3σ-edit rule to detect *any* of the outliers present in this dataset.

Figs. 7.3 through 7.5 further illustrate the behavior of the 3σ-edit rule, presenting boxplot summaries of the results of a small simulation study. In all, 24 different scenarios are described by these boxplots, corresponding to six differ-

Figure 7.3: Outliers identified vs. present, $N_0 = 1, 2$

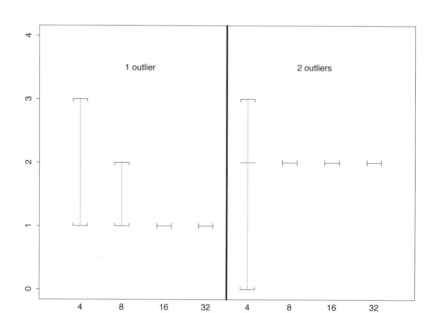

ent contamination levels and four different outlier magnitudes. In each scenario, $M = 100$ data sequences were generated as follows. First, a nominal sequence of length $N = 100$ was generated, having a standard normal distribution (i.e., $x_k \sim N(0,1)$), and then N_0 of these data values were replaced with outliers, each of magnitude x_0. The 3σ-edit rule was then applied to each of these contaminated data sequences, and the number of outliers detected was recorded and summarized in the boxplots shown in figs. 7.3 through 7.5. Results obtained for $N_0 = 1$ and $N_0 = 2$ are shown in Fig. 7.3, those for $N_0 = 4$ and 8 are shown in Fig. 7.4, and those for $N_0 = 9$ and 10 are shown in Fig. 7.5. In all of these boxplots, the results obtained for the two different values of N_0 are separated by a heavy vertical line, and the results are plotted against the outlier intensity, $x_0 = 4$, 8, 16, and 32.

For $N_0 = 1$, the number of outliers detected is always at least 1, but it can be larger, particularly for *marginal* outliers like $x_0 = +4$. This effect is called *swamping* and is the opposite of masking: the presence of outliers cause nominal data points to be misclassified as outliers. This problem is discussed further in Sec. 7.5.2, where it is noted that masking and swamping are somewhat complimentary phenomena; in particular, estimators that are resistant to masking tend to be prone to swamping and vice versa, although ironically it is possible

Figure 7.4: Outliers identified vs. present, $N_0 = 4, 8$

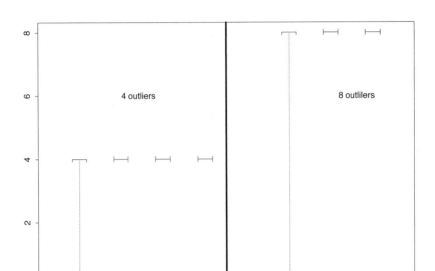

to construct estimators that exhibit poor performance with respect to both criteria, as noted by Davies and Gather [82]. Swamping is also seen for $N_0 = 2$, but only for the marginal case $x_0 = +4$: sometimes, three outliers are detected when only two are present. In addition, the results for $N_0 = 2$ with $x_0 = +4$ also illustrate masking: sometimes, no outliers are detected, even though two are present. Conversely, with $N_0 = 2$ and $x_0 = +8$, $+16$, or $+32$, the ESD identifier always correctly identifies two outliers in the sample.

Fig. 7.4 shows the corresponding results for $N_0 = 4$ and $N_0 = 8$. For $N_0 = 4$, marginal outliers with $x_0 = +4$ may or may not be masked but for $x_0 = +8$, $+16$, or $+32$, four outliers are always identified. The situation is somewhat worse when eight outliers are present: masking is severe enough in this case that no outliers are ever detected in the marginal case $x_0 = +4$, and fewer than eight outliers may be detected even when $x_0 = +8$. As in the previous examples, if the outliers are extreme enough in value (e.g., $x_0 = +16$ or $+32$), the 3σ-edit rule correctly identifies all of them: neither swamping nor masking occur.

Essentially the same behavior is also seen for $N_0 = 9$, as shown in the left-hand boxplots in Fig. 7.5: masking is complete when $x_0 = +4$ (i.e., no outliers are identified) and present for $x_0 = +8$, but the correct results are obtained for $x_0 = +16$ or $+32$. The situation changes abruptly and dramatically for

Figure 7.5: Outliers identified vs. present, $N_0 = 9, 10$

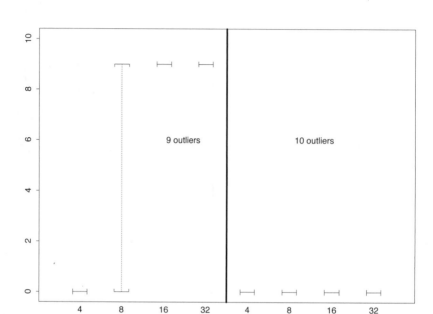

$N_0 = 10$, as seen in the right-hand boxplots in Fig. 7.5. There, regardless of the magnitude of the outliers, *none* is detected, despite the fact that $N_0 = 10$ are present. This is precisely the masking phenomenon seen in the recycle flow rate data discussed at the beginning of this section. The basic cause of this masking effect is illustrated in Fig. 7.6, which shows one of the simulations for $N_0 = 10$, with lines indicating the true nominal mean of zero, the estimated mean $\bar{x}_N \simeq 3.14$, and the estimated standard deviation $\hat{\sigma} \simeq 9.71$. The 10 outliers present in this example are also indicated, each with a value of $x_0 = +32$, which falls below the upper outlier detection limit of $\bar{x}_N + 3\hat{\sigma} \simeq 32.27$.

In fact, it is possible to obtain a simple analytical result that approximately describes the masking behavior of the 3σ-edit rule. Suppose a dataset of N values $\{x_k\}$ is contaminated with a fraction ϵ of outliers, all equal to A where $A > 0$ is large enough that $|x_k| \ll \epsilon A$ for all of the $(1-\epsilon)N$ nominal data values. Under this assumption, the mean of the data sequence is approximately:

$$\bar{x}_N = \frac{1}{N} \sum_{k=1}^{N} x_k \simeq \frac{1}{N}[(1 - \epsilon)N \cdot 0 + \epsilon N A] \simeq \epsilon A.$$

Similarly, note that the deviations from the mean $x_k - \bar{x}_N$ assume one of two approximate values: $x_k - \bar{x}_N \simeq -\epsilon A$ for the nominal data points and $x_k - \bar{x}_N \simeq$

Figure 7.6: Typical masking example, $N_0 = 10$, $x_0 = +32$

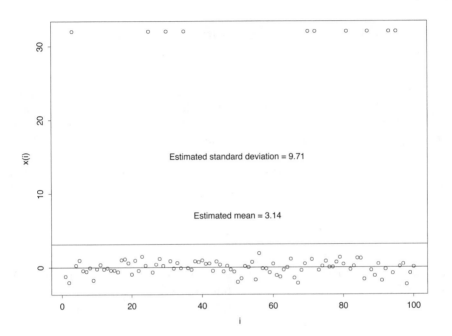

$(1 - \epsilon)A$ for the outliers. Hence, the variance is approximately:

$$\hat{\sigma}^2 \simeq \frac{1}{N}\{(1 - \epsilon)N \cdot (\epsilon A)^2 + \epsilon N[(1 - \epsilon)A]^2\} = \epsilon(1 - \epsilon)A^2.$$

Consequently, the approximate standard deviation is $\hat{\sigma} \simeq [\epsilon(1 - \epsilon)]^{1/2}A$, and the upper limit in the 3σ-edit rule becomes:

$$\bar{x}_N + 3\hat{\sigma} \simeq (\epsilon + 3[\epsilon(1 - \epsilon)]^{1/2})A.$$

The significance of this result lies in the fact that no outliers are detected if this upper detection limit exceeds the outlier value A. This condition holds if:

$$\epsilon + 3[\epsilon(1 - \epsilon)]^{1/2} > 1 \quad \Rightarrow \quad 3[\epsilon(1 - \epsilon)]^{1/2} > 1 - \epsilon$$
$$\Rightarrow \quad 9\epsilon(1 - \epsilon) > (1 - \epsilon)^2$$
$$\Rightarrow \quad 9\epsilon > 1 - \epsilon$$
$$\Rightarrow \quad \epsilon > 0.1.$$

In fact, this approximation gives almost exactly the result seen in the previous simulation example and suggests that the 3σ-edit rule is inherently incapable of handling contamination levels greater than $\sim$10%, a value often cited as either a typical or a conservative number of outliers to expect in real datasets.

Clearly, if we are to reliably detect multiple outliers in real datasets, a more effective procedure than the 3σ-edit rule is required.

7.1.3   The contaminated normal outlier model

In assessing the performance degradation of different estimators in the presence of outliers, it is useful to have a simple model for contaminated data. One of the most popular such models is the *contaminated normal model*, a member of the family of mixture distributions introduced in Chapter 4. Essentially, this model assumes that most of the data sequence $\{x_k\}$ is well represented by an independent, identically distributed sequence of Gaussian random variables with mean μ and variance σ^2, but that some fraction ϵ is drawn from a different distribution. Letting $\phi(\mu, \sigma; x)$ denote the density of the $N(\mu, \sigma^2)$ distribution and $\psi(x)$ denote the density of the contaminating distribution, the overall distribution of the data is given by:

$$p(x) = (1 - \epsilon)\phi(\mu, \sigma; x) + \epsilon\psi(x). \tag{7.2}$$

In general, it is possible to consider any contaminating distribution $\psi(x)$, but the most popular outlier model assumes normal contamination with the same mean and a larger variance $\tau^2 > \sigma^2$, implying $\psi(x) = \phi(\mu, \tau; x)$. Normally contaminated normal data models are often denoted $CN(\mu_1, \mu_2, \sigma_1, \sigma_2, \epsilon)$, corresponding to the normal mixture density:

$$p(x) = (1 - \epsilon)\phi(\mu_1, \sigma_1; x) + \epsilon\phi(\mu_2, \sigma_2; x). \tag{7.3}$$

As a specific example, Huber [151, p. 3] argues in favor of the $CN(\mu, \mu, \sigma, 3\sigma, \epsilon)$ error model:

> In the physical sciences typical "good data" samples appear to be well modeled by an error law of the form (7.3) with ϵ in the range between 0.01 and 0.1.

The standard normal density $N(0, 1)$ and the contaminated normal density $CN(0, 0, 1, 3, 0.1)$ corresponding to Huber's proposed error model are shown in the upper two plots in Fig. 7.7. It is clear from these plots that very careful comparison is required to see the differences between these densities. In fact, the tails of the contaminated normal density do decay somewhat slower than those of the standard normal density, and this point is most easily seen in the bottom two plots in Fig. 7.7, which show 200 sample data sequences drawn from each of these distributions. The standard normal data sequence fall between the limits of -2.63 and 2.39, consistent with the expectations of the 3σ-edit rule, but the contaminated normal samples fall between -5.40 and 6.28, well outside this expected range. Estimated standard deviations for these sequences are 0.998 for the standard normal data and 1.377 for the contaminated normal data, clearly illustrating the variance inflation caused by the outliers in this data sample. In

Figure 7.7: Standard vs. contaminated normal error models

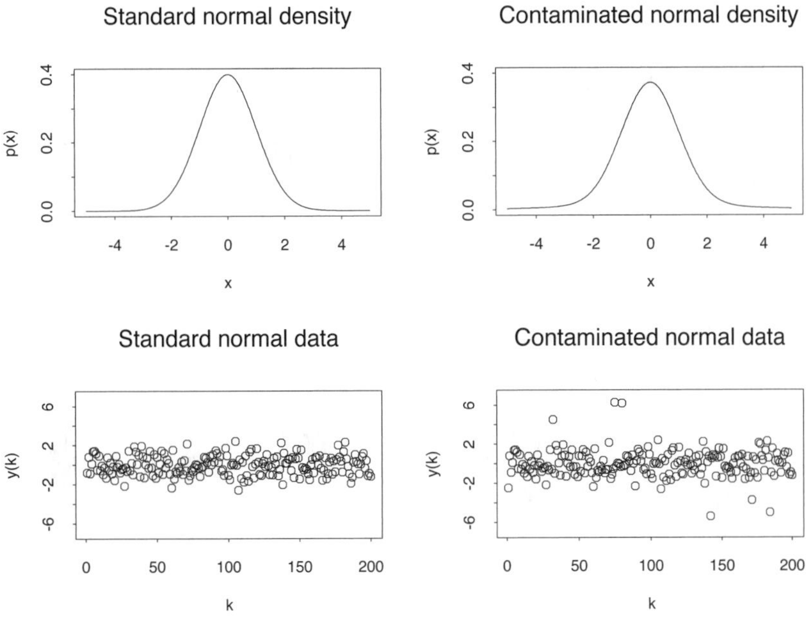

this particular example, the 3σ-edit rule performs reasonably, declaring the five most extreme points in the contaminated sample as outliers.

7.2 Four ways of dealing with outliers

In dealing with outliers, we have four basic options:

1. detect them and omit them from subsequent analysis,
2. detect them and replace them with more reasonable values,
3. detect them and set them aside for special analysis,
4. use outlier-resistant analytical procedures.

The best option to select will depend on the available data, the practicality of each option, and the ultimate objectives of the analysis. The following subsections discuss each of these options in turn, presenting one or more typical examples and briefly discussing where each option might or might not be appropriate. To provide a simple, common example to facilitate comparison, all of these approaches are applied to the makeup flow rate dataset discussed in Sec. 7.1.2 and shown in the upper left plot in Fig. 7.8. The other three plots show the three normal operation subsequences extracted from this dataset using the Hampel identifier described in Sec. 7.5 and these subsequences are consid-

Figure 7.8: Makeup flow rate data and subsequences

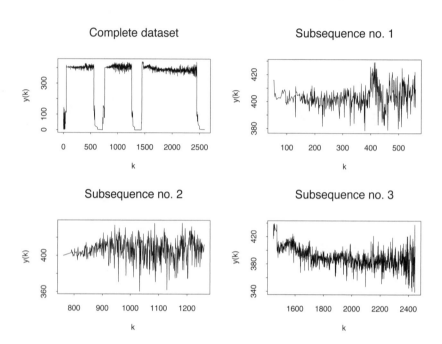

ered here to provide a frame of reference for various approaches to characterizing the overall data sequence. In particular, since this dataset is real and not simulation-based, we cannot know what the "true" characterizations of the data sequence are, making it difficult to evaluate the performance of different analysis procedures directly. The strategy adopted in the discussions that follow is to compare results obtained using the different outlier treatment strategies with those obtained from these three normal operation subsequences.

7.2.1 Detect and omit

Conceptually, the simplest approach to dealing with outliers is to simply omit them from our analysis. In practice, this approach requires that we have a practical tool for identifying outliers in a dataset, a topic examined in some detail in Sec. 7.5; for now, it is enough to note that outlier detection procedures do exist that are at least sometimes fairly reliable. Using such a procedure, the option of omitting outliers is probably reasonable if:

1. we have some basis for believing these data points are of limited interest,
2. subsequent analysis is not adversely affected by missing data,
3. the fraction of omitted data values is not deemed excessive.

Figure 7.9: Makeup flow rate data with outliers omitted

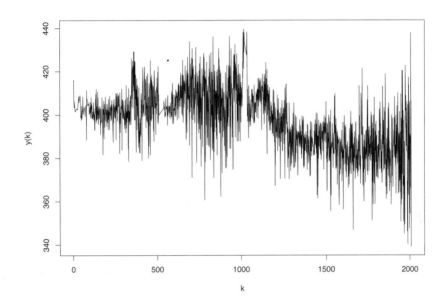

Probably the most common case where outliers are of secondary interest is that of gross measurement errors that can be traced back to a known cause, but it is important to emphasize that highly nonrepresentative data values can be sources of extremely valuable information, a point discussed in detail in Sec. 7.2.3. In the data analysis problems considered in this book, omission of outliers poses no fundamental difficulties other than the loss of precision associated with the reduced size of the dataset. In dynamic analysis, however, the practical consequences of missing data can be much more severe. Finally, the question of what fraction of outliers is excessive is a somewhat controversial one, examined further in Sec. 7.8.1.

The data sequence obtained by simply removing the outliers from the makeup flow rate data is shown in Fig. 7.9. Table 7.3 shows the mean $\bar{x}_N$, the standard deviation $\hat{\sigma}$, the skewness $\hat{\gamma}$, and the kurtosis $\hat{\kappa}$ estimated from the complete data sequence (designated CS), from each of the three subsets (designated S1, S2, and S3), and from the data sequence shown in Fig. 7.9 with the outliers omitted (designated OO). In addition, results are also presented for a sequence designated MR that is discussed in Sec. 7.2.2, obtained by replacing outliers with the median of the original data sequence. Note that the values obtained for the mean and, more especially, the standard deviation for sequence OO are *much* closer to those obtained for each of the three nominal data subsequences S1, S2, and S3 than the values obtained from the unmodified data sequence

Table 7.3: Estimated moments of the makeup flow data

Sequence	N	$\bar{x}_N$	$\hat{\sigma}$	$\hat{\gamma}$	$\hat{\kappa}$
CS	2589	315.5	155.05	−1.36	−0.06
S1	508	403.3	7.67	0.15	1.48
S2	502	406.6	11.50	−0.50	1.03
S3	999	390.4	14.75	0.27	1.04
OO	2009	397.7	14.51	−0.31	0.48
MR	2589	396.7	12.91	−0.13	1.24

CS. Hence, despite the evident differences between these subsequences, the outlier deletion strategy appears to give much more representative results for this example than simply analyzing the original dataset does.

In cases where all data samples are treated equally (i.e., where individual samples are *exchangeable*, a notion discussed in detail in Chapter 10), the effect of omitting outliers is to effectively shorten the available data record. This assumption holds in the examples considered here (i.e., for $\bar{x}_N$, $\hat{\sigma}$, $\hat{\gamma}$, and $\hat{\kappa}$), and the practical consequence is to increase the variability of these estimates. Although this increased variability is undesirable, the gross distortions introduced by outliers are generally far worse, so in cases where it is applicable, omitting outliers is probably a better strategy than including them in subsequent analysis. Again, however, it is important to emphasize that anomalous values *need not be physically unrepresentative* but may simply be unexpected. For this reason, it is always a good idea to examine outliers before simply disregarding them, a point discussed further in Sec. 7.2.3.

7.2.2   Detect and replace

The last row in Table 7.3 shows the mean, standard deviation, skewness, and kurtosis estimated by *replacing* all detected outliers with the median of the original data sequence, designated MR for "median replacement." Although this value is not *correct* (i.e., the "true" makeup flow rate values are those included in the original data sequence, corresponding to the physical shutdowns that occurred at those times), the median is *more representative of the nominal data values than the observed data values are*. This point may be seen clearly in Fig. 7.10, which shows the MR data sequence. Comparing the estmated moments obtained from this sequence with those obtained from the original data sequence CS and the normal operation subsequences S1, S2, and S3, it is clear that median replacement yields estimates that are much more representative than those obtained from the original data sequence. In addition, the standard

Figure 7.10: Makeup flow rate data, modified by median replacement

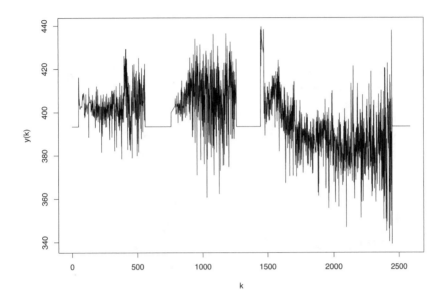

deviation, skewness, and kurtosis estimates all appear to be more consistent with the general behavior of the three subsequences than the estimates obtained from the data sequence OO obtained by simply omitting the shutdown episodes.

In the classical statistics literature, the problem of replacing *missing* data values with estimates based on the available data is called *imputation*, and a variety of strategies have been proposed for it, including *hot deck imputation* in which missing values are replaced by available data values that are deemed representative, *mean imputation* in which missing values are replaced by the mean of the dataset or some subset, and *regression imputation* in which missing data values are replaced by predictions from a regression model based on the available data [187, p. 6]. These ideas are discussed further in Chapter 16, but the key point here is that it is usually better to replace data values that are *clearly nonrepresentative* with values that, although almost certainly *not correct*, are at least *representative* than it is to simply analyze, unmodified, a badly contaminated data set.

7.2.3 Detect and scrutinize

Fig. 7.11 shows plots of the four shutdown episodes seen in the original makeup flow rate dataset. In this particular case, examination of these anomalous data subsequences serves only to confirm what was noted previously: the observed

Figure 7.11: Makeup flow rate data, shutdown episodes

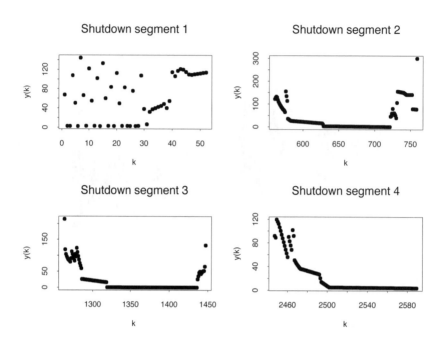

flow rate values are unusually small relative to the normal operation episodes, consisting primarily of values near zero. Conversely, it is generally good practice to carefully examine outliers before discarding them, even in cases where we have reason to believe they are nonrepresentative. This point was noted in Chapter 2 in connection with the discoveries of argon (leading to Lord Rayleigh's Nobel Prize) and Teflon (registered trademark of the DuPont company, which built a large business on the discovery). The key difference between the makeup flow rate dataset and Lord Rayleigh's nitrogen dataset lies in their *physical interpretation*, and this point is important: *anomalous* values may be detected using mathematical procedures like those described in Sec. 7.5, but once found, it is their physical interpretation that determines whether they are interesting.

More generally, it is important to emphasize that the search for anomalies is an integral part of science, and it can lead to extremely significant discoveries. For example, materials with anomalous physical properties often lead to entirely new fields of research, development, and application. Specific illustrations include the phenomenon of superdiffusivity in solid helium, in which a metal wire diffuses through a solid sample [50], and the development of a superhard form of silicon nitride, based on the discovery of a new crystalline phase of the material [303]. The importance of keeping an open mind in examining data is further emphasized by the following two examples, the first from chemistry and the second from physics.

Chandrasekhar [58] discusses carbon-carbon bond lengths in organic materials, beginning with the following observation:

> There is a general feeling amongst chemists that C-C bond lengths
> are entirely predictable and correspondingly uninteresting.

He goes on to note that these bond lengths typically range from $\sim$1.18 angstroms in triple-bonded compounds like acetylene to $\sim$1.53 angstroms in single-bonded hydrocarbons. Conversely, he also makes the following observation:

> The above generalizations hold good for a majority of systems. But
> a critical evaluation of known C-C distances shows a surprisingly
> large variation which deserves to be highlighted.

It is noted that exceptionally long bonds may result from steric hindrance, and examples are cited with bonds as long as 1.67 angstroms. Even longer bonds are found in the class of cyclophane compounds, including the longest bond known at the time the article was written (1.96 angstroms). At the other extreme, the shortest carbon-carbon bond known was 1.145 angstroms in a salt of a triple-bonded compound containing iodine. The key points of this example are, first, that violations of the rules of thumb on which our intuition is built do occur frequently, and second, that these violations often have interesting underlying explanations. For example, the search for unusual bond lengths and their underlying mechanisms may lead to interesting new classes of compounds.

The second example involves a question of larger scope: in 1931, Dirac postulated the existence of the *magnetic monopole*, essentially the magnetic counterpart to the electrically charged electron, to explain the quantization of electric charge. Extensive experimental searches for Dirac monopoles, however, repeatedly failed to find any evidence for their existence [25]. More recent *Grand Unified Theories* (GUT's) that attempt to combine the three fundamental forces of nature (the strong nuclear force, the weak nuclear force, and the electromagnetic force) also predict the existence of magnetic monopoles, although monopoles different in character from those postulated by Dirac [25, 147]. The detection of magnetic monopoles poses extremely formidible experimental challenges, however, and a critically important practical question that arises frequently is whether some detected "event" is in fact evidence for a magnetic monopole or only some sort of instrumental anomaly. At the "Monopole '83" meeting, Barish [25] noted that:

> The only positive evidence remains the original candidate of Cabrera
> from an induction experiment in a superconducting loop.

Barish presents a detailed discussion of the experimental difficulties involved in Cabrera's experiment, which requires the detection of magnetic flux changes six orders of magnitude smaller than Earth's background magnetic field; it is noted that this is possible with commercial superconducting quantum interference devices (SQUID's), but that *great* care is required in shielding the experiment from interference effects. Because of these experimental difficulties, the possibility that any interesting response is an instrumental artifact (i.e., an outlier) is generally the subject of much discussion. One example is the exchange of letters between Guy [127] and Caplin *et al.* [54] concerning a candidate event labeled "event 160." The event was seen experimentally by Caplin *et al.* in the response of a SQUID-based monopole detection system, and offered as a possible monopole sighting. Conversely, Guy disagrees with this assessment, arguing that:

> although the precise cause of event 160 remains unknown the experimental record of the event contains enough evidence to discount any connection with a magnetic monopole.

The key point here is that, as in this example, the clear detection and unambiguous interpretation of even a single, highly unusual event can be a matter of great interest, potentially vindicating or overturning fundamental theories about the nature of the universe.

7.2.4   Use outlier-resistant analytical procedures

The fourth option for dealing with outliers listed at the beginning of this section is that of using outlier-resistant analysis procedures. This idea is illustrated in Table 7.4, which compares the performance of the outlier-sensitive mean $\bar{x}_N$ and standard deviation $\hat{\sigma}$ with that of the outlier-resistant median $x^\dagger$ and the MADM scale estimate defined and discussed in Sec. 7.4. First, recall that the median was discussed in Chapter 4 as a location parameter, equal to the mean for symmetric distributions. Hence, if the data distribution is approximately symmetric, the median represents a reasonable alternative to the mean and the results presented in Table 7.4 show how much less sensitive the median is than the mean to the presence of the shutdown episodes in the recycle flow rate dataset. In particular, note that for the nominal subsequences S1, S2, and S3, the estimated means and medians differ by ~0.25% or less, and the median computed from the complete dataset lies in the range spanned by the medians of these subsequences; in contrast, the mean of the complete dataset lies ~20% below the mean values for the nominal subsequences. It is also interesting to note that the median value obtained from the original data sequence is very close to that obtained from the modified dataset OO, with the outliers omitted. Not surprisingly, if the outliers are replaced by the median computed from the complete dataset, there is no change at all in the median computed from the modified dataset MR.

The last column in Table 7.4 shows the values computed from each of these data sequences for the *MADM scale estimator S*, an outlier-resistant alternative to the standard deviation $\hat{\sigma}$ that is described in Sec. 7.4.2. Although the values of S computed from the six different data sets considered here differ more than the values of the median do, this range of variation is only a factor of ~3.5, compared with a factor of 20 for the standard deviation estimates. Also, although the agreement is not as good as that between the median and the mean, the agreement between S and $\hat{\sigma}$ for the nominal data subsequences S1, S2, and S3 is reasonable. Also, note that as in the case of the moment estimators $\hat{\sigma}$, $\hat{\gamma}$,

Table 7.4: Location and scale estimates for the makeup flow data

Sequence	N	$\bar{x}_N$	$x^\dagger$	$\hat{\sigma}$	S
CS	2589	315.5	393.4	155.05	20.22
S1	508	403.3	402.8	7.67	5.86
S2	502	406.6	407.2	11.50	10.10
S3	999	390.4	389.4	14.75	12.60
OO	2009	397.7	399.5	14.51	13.37
MR	2589	396.7	393.4	12.91	11.71

and $\hat{\kappa}$ considered in Sec. 7.2.2, the results obtained for dataset MR, cleaned by median replacement, are somewhat more consistent with those of the nominal subsequences than the results obtained by simple omission.

More generally, analysis procedures that are insensitive to outliers are called either *robust* or *resistant*, and the development of such estimators for different types of data analysis continues to be an active area of statistical research. The following section introduces some estimator characterizations that are useful in quantifying robustness, and subsequent sections apply some of these ideas to the problems of location and scale estimation, outlier detection, and other important practical problems like the assessment of asymmetry.

7.3   Robust estimators

Mallows [196] suggests that a procedure is robust if it possesses *resistance*, *smoothness*, and *breadth*. As noted in Chapter 1, resistance is one of the "four R's of exploratory data analysis" and is defined as insensitivity to either small changes in most of the data or large changes in a small fraction of the data. *Smoothness* requires that the results of a procedure change *gradually* with changes in the data or the underlying model on which they are based. Hence, smooth procedures will not ignore small changes in the data or the model, responding suddenly and dramatically when these changes exceed some critical threshold. Finally, *breadth* refers to a procedure that is applicable in a wide variety of situations. Although these characterizations of robustness are useful, they are not quantitative; to aid in comparing estimators, the following subsections introduce quantitative measures for each of these three characteristics. Specifically, Sec. 7.3.1 introduces the *breakdown point* of an estimation procedure, which represents a quantitative measure of its resistance to outliers. Similarly, the *influence function* of an estimator with respect to a specified probability distribution provides a quantitative measure of the smoothness of the estimator's behavior under departures from the specified reference distribution; this function is defined and discussed in Sec. 7.3.2. Finally, the notion of *efficiency robustness* provides a quantitative measure of an estimator's breadth and is discussed in Sec. 7.3.3. Other quantitative robustness assessments are possible, but these three taken together give a reasonably useful picture of an estimator's performance under a wide variety of different circumstances.

7.3.1   The breakdown point: a measure of resistance

The *breakdown point* of an estimator may be qualitatively defined as the largest fraction of the dataset that may be changed arbitrarily without changing the resulting estimate. To give a quantitative definition, we proceed along the lines discussed by Rousseeuw and Leroy [252]. Consider a dataset S consisting of N data points $\{x(k)\}$, and let T be any real-valued estimator applied to this dataset (e.g., mean, median, standard deviation, MADM scale estimate, etc.). For comparison, consider a second dataset S' for which $m < N/2$ of the original data points have been changed by arbitrarily large amounts. If we compare the values $\hat{e} = TS$ and $\hat{e}' = TS'$ of the estimator T applied to these two datasets, the change tells us about the sensitivity of T to a fraction m/N of gross outliers in the data.

More specifically, define $\Delta(m; T, S)$ as the maximum change:

$$\Delta(m; T, S) = \sup_{S'} |TS' - TS|, \tag{7.4}$$

where the supremum indicated is taken over all possible corrupted sets S'. The *finite sample breakdown point* $\epsilon_N(T, S)$ for the estimator T is defined as the smallest fraction m/N of contamination in the data that can cause the estimator T to be arbitrarily far from the value TS; more precisely:

$$\epsilon_N(T, S) = \min \{ \frac{m}{N} \mid \Delta(m; T, S) = +\infty \}. \tag{7.5}$$

As a specific example, it is shown at the end of this section that a single extreme data point is sufficient to make the mean μ assume values arbitrarily far from the correct value for the rest of the dataset. It follows from this observation that its finite sample breakdown point is:

$$\epsilon_N(\mu, S) = 1/N. \tag{7.6}$$

Hence, the breakdown point for the mean approaches 0% in the limit of large datasets (i.e., as $N \to \infty$). In contrast, it can be shown that the breakdown point for the median approaches 50% [151, p. 13], implying insensitivity to gross changes in anything less than half of the original dataset. In terms of resistance, these two values represent the range of behavior possible for any estimator: a 0% breakdown point represents the worst possible outlier sensitivity, whereas a 50% breakdown point represents the best possible outlier sensitivity.

The discussion of breakdown point presented by Rousseeuw and Leroy [252] is given in terms of vector-valued datasets, illustrating that this idea carries over directly to more general estimation problems than the means and medians discussed here. In general, "standard" (e.g., least squares) procedures based on underlying Gaussian statistical assumptions have an elegant simplicity, but the price paid for this simplicity seems to be a 0% breakdown point. In contrast, it is often possible to construct high breakdown estimators whose breakdown points approach 50%. Not surprisingly, there are some prices to be paid for this performance, too, a point discussed further in subsequent sections of this chapter. In particular, these alternative estimators can also exhibit highly undesirable— and, to the uninitiated, highly unexpected—behavior in some situations. These difficulties arise from the fact that high-breakdown estimators are *ignoring* just barely less than half of the data. If strong conflicts exist between two halves of a dataset, the behavior of high-breakdown estimators can be a little strange, as illustrated in Sec. 7.7.3.

7.3.2 The influence function: a measure of smoothness

In his book on robust statistics, Huber [151, p. 13] describes the *influence function* as "perhaps the most useful heuristic tool of robust statistics." This notion was introduced by Hampel [132], who describes it qualitatively as, "essentially the first derivative of an estimator, viewed as a functional, at some distribution." Alternatively, the influence function (or *influence curve*, as it was originally introduced) may be viewed as a measure of the influence of a single observation added to an infinitely large dataset. This measure depends on three things: the estimator T under consideration (e.g., the mean, the median, etc.), the assumed distribution F of the data, and the value x of the single added observation. Normally, the estimator T and the distribution F are fixed, and the influence function is considered as a function of x.

More specifically, the influence function $IC(x; F, T)$ is defined as follows. Suppose T is any real-valued estimator, well defined for some set of probability distributions that includes the reference distribution F. Then, let δ_x denote the degenerate distribution concentrated at the point x and consider mixture distributions of the form $(1 - \epsilon)F + \epsilon\delta_x$ for $0 < \epsilon < 1$ (see Chapter 4 for a more detailed discussion of mixture distributions). For every real value x, the influence function $IC(x; F, T)$ is defined as:

$$IC(x; F, T) = \lim_{\epsilon \downarrow 0} \frac{T[(1 - \epsilon)F + \epsilon\delta_x] - T(F)}{\epsilon}, \qquad (7.7)$$

and it tells us about the sensitivity of an estimator to a single outlier added at any point x. If this function is unbounded for some finite x or in the limit as x becomes infinite, the implication is that the estimator T is infinitely sensitive, either to the presence of outliers at some critical finite value or to the presence of *gross outliers*. In practice, finite discontinuities in the influence function can also have undesirable consequences, a point illustrated in Sec. 7.7.3.

To make these ideas more concrete, consider the influence function for the nth central moment $M_n = E\{(x - \mu)^n\}$ for any probability distribution F with mean μ and finite n^{th} moment. The influence function for M_n is given by:

$$
\begin{aligned}
IC(x; F, M_n) &= \lim_{\epsilon \downarrow 0} \frac{E\{(z - \mu)^n | (1 - \epsilon)F + \epsilon \delta_x\} - E\{(z - \mu)^n | F\}}{\epsilon} \\
&= \lim_{\epsilon \downarrow 0} \frac{(1 - \epsilon)E\{(z - \mu)^n | F\} + \epsilon(x - \mu)^n - E\{(z - \mu)^n | F\}}{\epsilon} \\
&= (x - \mu)^n - E\{(z - \mu)^n | F\}.
\end{aligned}
\tag{7.8}
$$

Here, the notation $E\{(z - \mu)^n | F\}$ has been used to indicate the nth central moment of the probability distribution F. The interpretation of this result is that the influence function for the nth central moment is unbounded as $x \to \pm\infty$ for all $n \geq 1$. Further, the rate at which the influence function grows increases with increasing n, implying that moment estimators become more sensitive with increasing n, a point illustrated in Sec. 7.1.1. Finally, note that this result holds for any reference distribution F for which the moment estimators are defined.

7.3.3   Efficiency robustness: a measure of breadth

Useful as they are, it is important to note that the notions of resistance and smoothness are *not* sufficient to specify a "good" estimator. In particular, note that a constant estimator, independent of the data, is both resistant and smooth but not especially useful (e.g., "The answer is 6: what was the question?"). To avoid such pathologies, it is necessary to introduce some other utility criterion, such as the notion of breadth described by Mallows [196]: it would be difficult to argue that the constant estimator just described exhibits "applicability in a wide variety of situations." One useful quantification of breadth is via the notion of *efficiency robustness*: recall from Chapter 6 that a parameter estimator is called *efficient* if it achieves the lowest possible uncertainty in some well-defined sense. Here, as in the case of the influence function, we consider the performance of an estimator T relative to some reference distribution F. Specifically, the estimator T is intended to estimate some characteristic of this reference distribution (e.g., the mean, median, standard deviation, absolute first moment, etc.), and we are interested in characterizing its performance relative to the best achievable. If there exists a minimum-variance, unbiased estimator [180, p. 75] T_F for the characteristic of interest for the distribution F, we may take its variance, σ_F^*, as a reference value.

The *absolute efficiency* $E_T(F)$ of the estimator T for the reference distribution F is defined by the variance ratio:

$$
E_T(F) = \frac{\sigma_F^*}{\sigma_F(T)},
\tag{7.9}
$$

where $\sigma_F(T)$ is the variance of the estimator T for the reference distribution F. As a specific example, suppose we consider a set of distributions like the contaminated normal family discussed in Sec. 7.1.3 and ask how the efficiency of a given estimator varies across this family (e.g., as a function of the contamination level ϵ in the contaminated normal family). An estimator is called *efficiency robust* if the efficiency remains sufficiently high over the entire family. Alternatively, for comparing two different estimators, say T and T^*, over a set of distributions F, it is useful to consider *relative efficiency* measures like:

$$R_T(F) \;=\; \frac{\sigma_F(T^*)}{\sigma_F(T)}. \tag{7.10}$$

Note that relative efficiencies may exceed 1.

7.3.4 A comparison of the mean and the median

To clarify these different robustness measures, it is useful to apply them to a familiar (and extremely important) practical example: a comparison of the mean and the median. As noted in Chapter 4, the mean and the median are generally distinct for asymmetric distributions, so it makes sense to consider only symmetric distributions for which they are equal. Further, there is no loss of generality in assuming their common value occurs at $z = 0$. For such a distribution, Hampel [132] gives the following expression for the influence function for the median $\tilde{x}$:

$$IC(x; F, \tilde{x}) \;=\; \begin{cases} 1/2f(0) & x > 0, \\ -1/2f(0) & x < 0. \end{cases} \tag{7.11}$$

For comparison, the influence function of the mean $\bar{x}_N$ for this same class of distributions is simply:

$$IC(x; F, \bar{x}_N) \;=\; x. \tag{7.12}$$

It is clear from these results that the median is less sensitive to gross outliers than the mean because the median's influence function is bounded, whereas that for the mean is not. Note, however, that the influence function for the median is discontinuous at zero, and this discontinuity does cause certain difficulties, a point illustrated in Sec. 7.7.3.

Various authors [132, 151, 252] have noted that the breakdown point for the mean is 0%, whereas that for the median is 50%. The result for the mean follows from the fact that if one point out of N, say x_i, is replaced by the contaminated value $x_i + c$, the mean changes from $\bar{x}_N$ to $\bar{x}_N + c/N$. By making c sufficiently large, the mean can thus be changed to any arbitrary value by contaminating a single data point. This result is a direct consequence of the mean's unbounded influence function as $|x| \to \infty$, and it implies that the finite-sample breakdown point is $1/N$, as noted in Eq. (7.6). Taking $N \to \infty$, the 0% breakdown point for the mean follows from this result. For the median, consider a sample of size $N = 2m + 1$ for some integer m. For this estimator to exhibit breakdown, it is necessary to introduce sufficient contamination to make the median assume any arbitrary value, c. However, for c to be the median of the data sample, there must be $m + 1$ data points $x(k) \geq c$, implying that half of the sample must be contaminated. The 50% breakdown point of the median follows from this result.

Although it is clear that the median is generally better behaved than the mean with respect to its influence function and its breakdown point, it suffers in efficiency for Gaussian data. Specifically, since the sample mean is the minimum variance unbiased estimator of the mean of the Gaussian distribution, it exhibits 100% absolute efficiency. Unfortunately, expressions for the distribution of the median are rather messy [16, p. 27], so it is not easy to compute the efficiency of the median relative to the mean for the reference distributions of greatest interest (e.g., Gaussian) for finite sample sizes [16, p. 88]. If we consider *asymptotic* relative efficiencies, however, more general results are available: for densities symmetric about 0 with $p(0) > 0$, the median and the mean are both asymptotically normal, unbiased estimators [180, p. 354]. By comparing the limiting variances of these estimators, we may obtain the following asymptotic efficiency for the median, relative to the mean [180, p. 356]:

$$E = 4f^2(0)\sigma^2. \tag{7.13}$$

Lehmann [180] discusses the efficiency of the median relative to the mean for a variety of distributions, including both the Student's t and contaminated normal. Not surprisingly, for heavy-tailed distributions, like t_3 and t_4, the relative efficiency of the median exceeds 100% (e.g., $E \simeq 1.62$ for t_3). In particular, the efficiency decreases monotonically from an infinite value for 1 degree of freedom (the Cauchy limit) to a value of 63.7% in the Gaussian limit. Further, Lehmann [180, p. 359] shows that for any distribution that is symmetric about zero and satisfies the condition $f(z) \leq f(0)$ for all z (e.g., unimodal densities), the asymptotic efficiency of the median relative to the mean is bounded below by $E \geq 1/3$, and he notes that this minimum efficiency is achieved by the uniform distribution and no other.

7.4   Robust alternatives to $\bar{x}_N$ and $\hat{\sigma}$

As noted in Sec. 7.1.2, the basic difficulty with the 3σ-edit rule is that both of the estimators on which it is based—the mean $\bar{x}_N$ and the standard deviation $\hat{\sigma}$—are themselves quite sensitive to outliers. The previous example has illustrated the improved robustness of the median relative to the mean, and the next subsection considers robust location estimates more generally. Following that discussion, the MADM scale estimator is introduced in Sec. 7.4.2, described by Huber [151, p. 107] as, "the single most useful ancillary estimate of scale."

7.4.1   The Princeton robustness study

The Princeton robustness study [11] compared the performance of 68 different location estimators, all required to satisfy the following two constraints:

1. the estimator must result from a computable algorithm,
2. the estimator must be location- and scale-invariant: transforming the data $x_k \rightarrow ax_k + b$ must transform the estimate as $T \rightarrow aT + b$.

These estimators were divided into four groups: the L-estimators introduced in Chapter 6, the M-estimators discussed in Chapter 13, *skipped procedures* involving iterative detection and rejection of outliers based on order statistics, and "other" approaches including **shorth**, defined as the mean of the "shortest half of the sample" (i.e., the mean of the $N/2$ most closely spaced order statistics). Detailed finite sample characterizations are presented for 65 of these 68 estimators, based on 30 different distributional scenarios and sample sizes ranging from $N = 5$ to $N = 40$. A number of these estimators are discussed in Chapter 6, including the arithmetic mean $\bar{x}_N$, the median $x^\dagger$, various trimmed means, and Gastwirth's estimator. FORTRAN subroutine listings are given for all of the 68 estimators considered, together with tables of simulation results, graphical summaries, and detailed discussions of these tables and graphs. The summary given here is offered, first, to call attention to the wealth of detail contained in this study, and second, to briefly discuss a few of its conclusions.

One of these conclusions is the following characterization of the mean, given by one of the authors of the study (F. Hampel) [11, p. 243], part of which was discussed previously in Chapter 4:

> The mean M deserves some special remarks. As everybody familiar with robust estimation knows, the mean is a horrible estimator, except under the strict normal distribution; it gets rapidly worse even under very mild deviations from normality, and at some distance from the Gaussian distribution, it is totally disasterous. However, as everybody familiar with practical statistics (not to mention a number of theories) knows, the mean is *the* estimator of location, to be applied under almost all circumstances (especially in more complex models); and it is even used by some very good statisticians, or so it seems at first glance.

In clarifying this last point, Hampel notes that "a good statistician (or research worker) never uses the mean, even though he may honestly claim to do so." The real point is that *before* estimating the mean of a data sequence, it is imperative to identify and eliminate gross outliers like the shutdown episodes in the makeup flow rate data sequence considered in Sec. 7.2. Once we have done this, the mean may not be a bad estimator in practice, and it certainly permits us to take advantage of much statistical theory that is not generally either as well developed or as convenient for alternative location estimators.

Another contributor to the study was P.J. Huber, who offered the following useful observations on the rate at which the estimator variances approach their asymptotic limits [11, p. 252]. For uncontaminated normal data, convergence was noted to be very rapid for most of the estimators considered in the study, approximating asymptotic behavior reasonably well for samples as small as $N = 5$. The estimator **shorth** was noted explicitly as an exception to this observation, and it was also noted that convergence is generally slower for longer tailed data distributions, although it was suggested that asymptotic approximations may remain reasonable for samples as small as $N = 20$ even there.

Finally, it is worth noting that separate summaries were presented by five of the authors of this study, and not surprisingly, there was not unanimous agreement in these summaries, but a few estimators that repeatedly appear as "good" choices include heavily trimmed means (e.g., the midmean, corresponding to the average of the middle 50% of the data; the median; and Gastwirth's estimator discussed in Chapter 6) and various M-estimators, discussed in Chapter 13.

7.4.2   The MADM scale estimate

The concept of the median is fairly well known (e.g., the term appears in newspapers) and its robustness relative to the mean is also fairly widely appreciated. The analogous robust alternative to the standard deviation appears to be much less well known outside the statistics community, although the situation does seem to be improving. Specifically, an extremely practical alternative to the usual moment estimate for the standard deviation is the MADM scale estimator, an acronym for "median absolute deviation from the median," defined as follows. Given a dataset $\{x(k)\}$, first compute the median $\tilde{x}$, and then compute the *nonnormalized* scale estimate S_0 defined by:

$$S_0 = med\{|x(k) - \tilde{x}|\}. \tag{7.14}$$

Note that this quantity gives a measure of the natural variation in the data: it is the median absolute distance of the data values $x(k)$ from the reference value $\tilde{x}$. Because this median distance S_0 is insensitive to extreme data values, it gives a highly representative measure of how far an arbitrary data value $x(k)$ in the dataset lies from the nominal value $\tilde{x}$. Further, because this nominal value is also insensitive to extreme points in the dataset, the nonnormalized scale estimate S_0 is a very robust measure of the natural spread in the data, a point discussed further in Sec. 7.4.3.

In practice, the MADM scale estimator is almost always normalized in the following manner:

$$S = \frac{S_0}{0.6745} \simeq 1.4826 S_0. \tag{7.15}$$

This normalization constant is chosen to make the expected value of S equal to the standard deviation σ if the data sample actually has a Gaussian distribution. It is useful to briefly consider how this result is obtained, both to understand it better and because the derivation leads to another useful result as a by-product.

To obtain these results, compute the expected value of S_0, assuming that y is a random variable with mean μ and a probability density function $p(\cdot)$ that is symmetric about μ. Because of this symmetry, it follows that $med\{y\} = \mu$, so we have:

$$E\{S_0\} = med\{|y - med\{y\}|\} = med\{|y - \mu|\}. \tag{7.16}$$

Subsequent discussions simplify somewhat by noting that the random variable $z = y - \mu$ is symmetrically distributed about zero. It follows from the definition of the median that $\text{med}\{|y - \mu|\} = \text{med}\{|z|\}$ is the number α such that:

$$\mathcal{P}\{|z| \le \alpha\} = \frac{1}{2}. \tag{7.17}$$

This probability is given by:

$$\mathcal{P}\{|z| \le \alpha\} = \mathcal{P}\{-\alpha \le z \le \alpha\} = \int_{-\alpha}^{\alpha} p(z)dz = 2\int_0^{\alpha} p(z)dz, \tag{7.18}$$

where the last result follows from the symmetry of $p(z)$ about 0. Combining Eqs. (7.17) and (7.18), it follows that:

$$\int_0^{\alpha} p(z)dz = \frac{1}{4}. \tag{7.19}$$

For any symmetric distribution, note that half of the total cumulative probability lies below the mean:

$$\int_{-\infty}^0 p(z)dz = \frac{1}{2}. \tag{7.20}$$

Adding the integrals appearing in Eqs. (7.19) and (7.20) yields:

$$\int_{-\infty}^{\alpha} p(z)dz = \frac{3}{4}. \tag{7.21}$$

This result is extremely useful since the integral appearing on the left-hand side is simply the cumulative probability distribution $F(\alpha)$ for the random variable z. Hence, *for any symmetrically distributed, zero-mean random variable* y, the expected value of S_0 is:

$$E\{S_0\} = F^{-1}\left(\frac{3}{4}\right). \tag{7.22}$$

If the mean $\bar{y}$ of y is nonzero, note that we must modify this expression to $E\{S_0\} = F^{-1}(3/4) - \bar{y}$.

For the Gaussian distribution, $F^{-1}(\frac{3}{4}) \simeq 0.6745\sigma$, yielding the mysterious normalization constant appearing in the denominator of Eq. (7.15). For symmetric non-Gaussian distributions, we may compare the behavior of the scale estimator S_0 with the more usual scale estimator σ (i.e., the standard deviation). For example, these values are given in Table 7.5 for the family of Student's t-distributions with ν degrees of freedom; specifically, values of σ, S_0, and S are given for $\nu = 1, 2, 3, 4, 5, 10, 25, \infty$. As discussed in Chapter 4, this family exhibits a very wide range of tail behavior, including both the Cauchy distribution ($\nu = 1$) and the Gaussian distribution ($\nu \to \infty$) as extreme limits. The key point is that although the MADM scale estimate increases as ν decreases, it remains finite for all ν, unlike the variance, which is infinite for $\nu < 3$. More generally, the MADM scale estimate is much less strongly dependent on the distribution of the data than the variance is.

Table 7.5: Scale estimates for Student's t-distributions

ν	σ	S_0	S
1	∞	1.0000000	1.4825797
2	∞	0.8164966	1.2105213
3	1.7320508	0.7648923	1.1340138
4	1.4142136	0.7406971	1.0981425
5	1.2909944	0.7266868	1.0773711
10	1.1180340	0.6998121	1.0375272
25	1.0425721	0.6844300	1.0147220
∞	1.0000000	0.6744898	0.9999848

7.4.3   Robustness of the MADM scale estimate

Hampel [132] describes the MADM scale estimate as the most robust scale estimator possible, in terms of both breakdown point (50%) and gross error sensitivity γ^*, defined as the supremum of the influence function:

$$\gamma^* \;=\; sup_x \, |IC(x; F, T)|. \tag{7.23}$$

Note that this value represents the sensitivity to an outlier at the worst case location; for 0% breakdown estimators like the mean and standard deviation, this worst case limit is $\gamma^* = \infty$, typically approached as $x \to \pm\infty$. For distributions F that are symmetric about zero, the influence function for the MADM scale estimator S is [151, p. 138]:

$$IC(x; F, S) \;=\; \begin{cases} 1/4f(S(F)) & x < -S(F), \\ -1/4f(S(F)) & -S(F) < x < S(F), \\ 1/4f(S(F)) & x > S(F). \end{cases} \tag{7.24}$$

Here, $S(F)$ denotes the value of the MADM scale at the reference distribution F. Hence, like the median, the MADM scale estimate is bounded for all x, with a gross error sensitivity $\gamma^* = 1/4f(S(F))$ that is finite for all reference distributions for which $S(F) \neq 0$. Note, however, that this influence function is discontinuous, also like that of the median, and this discontinuity may cause difficulties in certain applications, a point illustrated in Sec. 7.7.2.

Another price paid for the superior breakdown point and gross error sensitivity of the MADM scale estimator is a loss of efficiency relative to the standard deviation for the Gaussian reference distribution. In fact, Hampel [132] notes that Gauss himself briefly considered the MADM scale estimator, dismissing it from further consideration because of its low efficiency ($\sim$36.7% [151, p. 112]). Ironically, Hampel [132] notes that if the reference distribution is changed to the slightly heavy-tailed but approximately normal Student's t-distribution with 5 degrees of freedom, the absolute efficiency of the standard deviation is about the same as the MADM scale estimate.

Overall, despite its low Gaussian efficiency, the MADM scale estimator appears to be extremely useful because of its simplicity, its high breakdown, and its excellent gross error sensitivity. In practice, Hampel [132] recommends it over the standard deviation for the following types of applications:

1. as a rough but fast scale estimate where no higher accuracy is needed,
2. as a check for more refined computations,
3. as a basis for rejecting outliers,
4. as a starting point for iterative (or one-step) procedures.

The use of the MADM scale estimate in outlier detection is the principal subject of the next section.

7.5 Outlier detection

It was demonstrated in Sec. 7.1.2 that the 3σ-edit rule suffers badly from the phenomenon of masking in the presence of multiple outliers. As noted, this difficulty is due to the inherent outlier sensitivity of the mean $\bar{x}_N$ and standard deviation $\hat{\sigma}$ estimators on which this rule is based. Replacing these quantities by the more robust median and MADM scale estimators leads to the *Hampel identifier* [82], which is much less sensitive to masking effects, although it does suffer somewhat from the opposite problem, that of *swamping* in which nominal data points are declared outliers. The following discussions describe the Hampel identifier in detail (Sec. 7.5.1), examines the issues of masking and swamping (Sec. 7.5.2), and presents some practical application guidelines (Sec. 7.5.3).

7.5.1 The Hampel identifier

Replacing the mean with the median and the estimated standard deviation with the MADM scale estimator in the 3σ-edit rule yields the *Hampel identifier* [82]:

1. compute the median of the data sequence, $x^\dagger = \text{median } \{x_k\}$,
2. compute the MADM scale estimate, $S = 1.4826 \text{ median}\{|x_k - x^\dagger|\}$,
3. declare x_k suspicious if $|x_k - x^\dagger|/S > t = 3$.

As in the case of the extreme Studentized deviation (ESD) identifier discussed in Sec. 7.1.2, the performance of this procedure depends on the choice of the threshold t, but the key point here is that by replacing the outlier-sensitive mean and standard deviation with more robust alternatives, we obtain an outlier detection rule that is often *far* more effective, as the following example illustrates.

Fig. 7.12 shows the makeup flow rate dataset considered in Sec. 7.1.2, plotted together with lines representing the median and the upper and lower outlier detection limits obtained with the Hampel procedure. In particular, the middle line represents the median value of $x^\dagger \simeq 393.36$, passing through approximately

Figure 7.12: Makeup flow rate data with Hampel edit limits

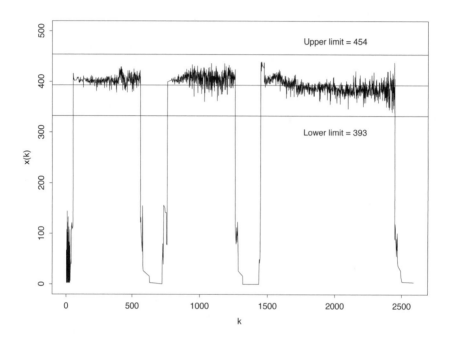

the center of all three of the normal operation data subsequences considered earlier. The lower outlier detection limit (~332.68) partitions the dataset into shutdown values below this limit and nominal values above it, entirely consistent with their physical interpretation. Similarly, all of the data values fall below the upper outlier detection limit of 454.03, ~3% larger than the maximum observed value. All in all, the performance of the Hampel identifier is quite reasonable for this dataset, providing a useful partitioning into nominal and anomalous data values. This behavior stands in marked contrast to that of the more widely known 3σ-edit rule, which failed to find *anything* unusual in this dataset.

Figs. 7.13 through 7.15 summarize the performance of the Hampel identifier for the same case study considered in Sec. 7.1.2 for the 3σ-edit rule. These plots summarize 100 simulations, each based on datasets of size $N = 100$, nominally drawn from the standard $N(0, 1)$ distribution, but contaminated with N_0 outliers, all of the same magnitude x_0. Fig. 7.13 presents boxplot summaries for $N_0 = 1$ and 2, plotted against $x_0 = +4$, $+8$, $+16$, and $+32$. Comparing these results with those obtained for the 3σ-edit rule emphasizes the greater tendency toward swamping for the Hampel identifier: whereas no swamping occurred for the 3σ-edit rule when $x_0 = +16$ or $+32$, the Hampel identifier declares as many

Figure 7.13: Outliers identified vs. present, $N_0 = 1, 2$

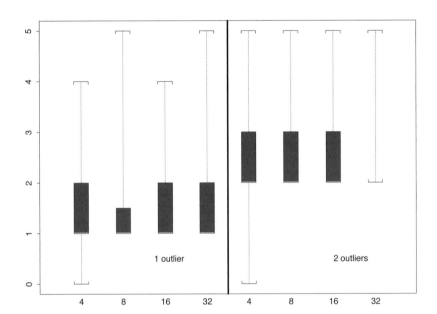

as four nominal data points to be outlying. Also, note that both identifiers exhibit similar masking of marginal outliers ($x_0 = +4$) for $N_0 = 2$.

Fig. 7.14 gives the corresponding results for the Hampel identifier when $N_0 = 4$ and 8, which are mostly similar to those for $N_0 = 1$ and 2. Specifically, note the similar behavior of both the Hampel and ESD identifiers with respect to masking of marginal outliers for $N_0 = 4$ and the much greater tendency of the Hampel identifier to exhibit swamping for nonmarginal outliers, both for $N_0 = 4$ and $N_0 = 8$. Conversely, the ESD identifier shows preliminary signs of its complete breakdown for $N_0 = 8$, failing to find *any* marginal outliers, whereas the Hampel identifier finds anywhere between 0 and 10 outliers in these datasets.

The results for $N_0 = 9$ and 10 are given in Fig. 7.15, where the most dramatic differences between the Hampel and ESD identifiers are observed. In contrast to the results obtained for the 3σ-edit rule, note that masking is *not* observed for $N_0 = 10$, except for the case of marginal outliers, $x_0 = +4$. Even there, masking is not *always* observed, as it was for the 3σ-edit rule for all values of x_0 considered. The Hampel identifier still exhibits tendency toward masking for marginal outliers, and swamping is consistently a problem, but overall, the performance of the Hampel identifier at high contamination levels is much more reasonable than that of the 3σ-edit rule, which breaks down completely.

Figure 7.14: Outliers identified vs. present, $N_0 = 4, 8$

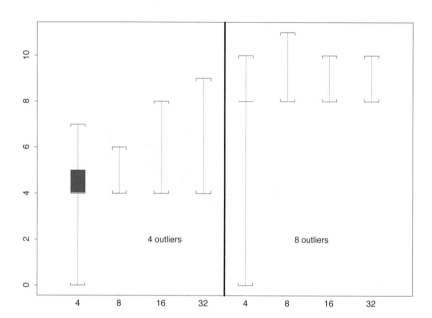

7.5.2   Masking and swamping breakdown points

Davies and Gather [82] give a quantitative definition of the *largest nonidentifiable outlier* ρ in a dataset and say that *masking breakdown* has occurred in an outlier identification procedure when $\rho \to \infty$. Analogously to the breakdown point defined in Sec. 7.3.1, they define the *masking breakdown point* ϵ^M as:

$$\epsilon^M = \frac{k^M}{k^M + N},\tag{7.25}$$

where k^M is the smallest number of outliers that can cause the identifier to exhibit masking breakdown when added to an outlier-free sample of size N. Not surprisingly, they show that the ESD identifier has a masking breakdown point of $\epsilon^M = 1/(N+1)$, asymptotically going to zero in the limit of large datasets. Conversely, they also show that the Hampel identifier exhibits a 50% masking breakdown point, which is the best possible. Similarly, Davies and Gather [82] also define the *swamping breakdown point* ϵ^S as:

$$\epsilon^S = \frac{k^S}{k^S + N},\tag{7.26}$$

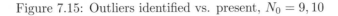

Figure 7.15: Outliers identified vs. present, $N_0 = 9, 10$

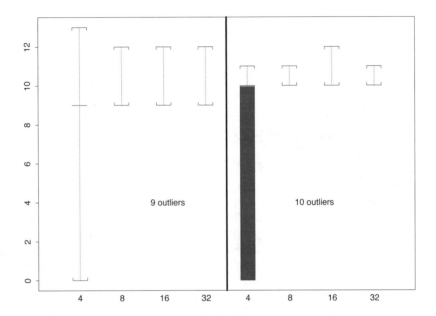

where k^S is the smallest number of outliers that can cause a point that is not an outlier to be declared one. The maximum possible value of ϵ^S is 1, and this value is achieved by the ESD identifier. In fact, Davies and Gather note that masking and swamping are complementary phenomena, so outlier identification procedures that have good masking performance are susceptible to swamping and vice versa. This point was seen clearly in the simulation case studies presented in secs. 7.1.2 and 7.5.1: the masking behavior of the Hampel identifier is *much* less severe than that of the ESD identifier, but the Hampel identifier exhibits some degree of swamping in all of the examples considered. These results also illustrate that swamping can occur in finite samples with the ESD identifier, but this difficulty appears to be most prevalent in the presence of marginal outliers (e.g., $x_0 = +4$). Conversely, the Hampel identifier does achieve a 50% swamping breakdown point, the best possible for any estimator with a 50% masking breakdown point [82]. Finally, although it is not possible to have the best of both worlds—a masking breakdown point of $\epsilon_M = 0.5$ and a swamping breakdown point of $\epsilon_S = 1$—it *is* possible to have the worst: Davies and Gather [82] mention an outlier identification procedure that exhibits both 0% masking breakdown and 0% swamping breakdown.

7.5.3   Practical details in outlier detection

Both the ESD outlier detection procedure described in Sec. 7.1.2 and the Hampel
identifier considered in Sec. 7.5.1 may be represented as:

$$x_k = \begin{cases} \text{outlier} & \text{if } |x_k - \hat{c}|/\hat{S} > t, \\ \text{nonoutlier} & \text{otherwise.} \end{cases} \tag{7.27}$$

The difference between these procedures lies in the choice of the reference value
$\hat{c}$ (i.e., mean vs. median), and the scale parameter $\hat{S}$ (i.e., standard deviation
vs. MADM scale estimator). In either case, we are left with the task of selecting
t. It should be clear that the larger we choose t, the less susceptible we are to
swamping, but the more susceptible we are to masking, and vice versa. Hence,
although this choice is, within reason, arguably less important than the choice
of $\hat{c}$ and $\hat{S}$, it is still important.

Davies and Gather [82] consider the problem of selecting this threshold,
based on an assumed Gaussian nominal data model, and choose t to achieve
specified probabilities of correct classification. A detailed discussion of this approach is beyond the scope of this book, but it is instructive to look at some
of the threshold values they recommend. Note that because the classification
probabilities from which these threshold values were determined depend on the
size N of the dataset, these threshold values also depend on N. Two different
strategies are discussed, designated "Option 1" and "Option 2," and both are
applicable to either the ESD identifier or the Hampel identifier. Typical threshold values obtained by Davies and Gather [82] are shown in Table 7.6 for sample
sizes $N = 20$, 50, and 100; very roughly, these threshold values were chosen to
give a 95% probability of correct classification. The key point here is that these
threshold values range in value from ~2 to ~6. Overall, note that these values
are comparable with both the traditional value $t = 3$ and the somewhat more
liberal recommendation $t = 2.5$ of Rousseeuw and Leroy [252] in their book
on robust regression and outlier detection. Because we are generally looking
for *suspicious* data values, ideally to be subjected to further scrutiny and special treatment, it does not make sense to agonize greatly over the choice of the
threshold value t. Generally speaking, values in the range $2.5 \leq t \leq 5$ are probably reasonable, with the larger values leading to more conservative detection
schemes. The smaller the dataset, the larger we probably want to make t to
protect against excessive zeal in rejecting subsets of a small sample.

The outlier identification procedures discussed so far have been based on
the idea of estimating upper and lower limits for "reasonable data values." A
philosophically different approach to outlier detection is to focus on the extreme
values of the sequence and ask, for each individual value or for a group of m
extreme values, "are these points outliers?" This philosophy leads to the notions
of inward and outward testing, originally developed for identifying a few outliers
(e.g., $m = 1$, 2, or 3) in a small dataset [108]. Various authors discuss this
approach, including Barnett and Lewis [27], Fung and Paul [108], and Davies
and Gather [82], and a number of different algorithms are available. *Inward
procedures* start with the complete dataset and ask, "is the most extreme value

Table 7.6: Threshold parameters for outlier identification

N	ESD		Hampel	
	Option 1	Option 2	Option 1	Option 2
20	2.99	2.71	5.67	5.82
50	2.55	3.13	4.31	5.53
100	2.30	3.38	3.81	5.52

inconsistent with the rest of the dataset?" If so, the most extreme point is rejected from the dataset and the process is repeated until no further outliers are detected. Along similar lines, Davies and Gather [82] propose the following inward version of the Hampel identifier:

1. apply the Hampel identifier to the original dataset S_0 and remove any outliers identified to obtain dataset S_1,
2. apply the Hampel identifier to dataset S_i and remove any outliers identified to obtain dataset S_{i+1},
3. repeat step 2 until no further outliers are found.

They argue that this iterative procedure performs better than the Hampel identifier, at the expense of "a small increase in the swamping effect" [82].

   In contrast, *outward procedures* start by systematically removing the m most extreme points from the original dataset and then testing them in reverse order to see whether they should be reclassified as nominal data points. In general, outward procedures are more complicated to implement, for two reasons. First, data points are removed sequentially from the original dataset, so at each stage new location and scale estimates must be computed to provide a basis for deciding which observations are most extreme. The second complication is that outward procedures require specification of the maximum number m of outliers that can be in the sample. Davies and Gather [82] describe a procedure that starts with the maximum possible number of outliers: $m = [N/2] + 1$ (here $[x]$ indicates the integer part of x). Barnett and Lewis [27, p. 132] advocate significantly smaller values for m, citing $m = \sqrt{N}$ as a specific example. These authors also argue that outward procedures are essentially immune to masking, *provided the number of outliers in the sample does not exceed the prespecified bound m* [27, p. 131]. This last point is important, because in the makeup flow rate data example considered in this chapter, Barnett and Lewis's guidelines would suggest $m = \sqrt{2500} = 50$, smaller than the actual number of outliers in the sample by a factor of 10. Similarly, for the Davies and Gather [83] example, with $N = 2001$, the recommended cutoff would be $m \simeq 45$, compared with the 396 outliers identified with the Hampel identifier.

7.6   The problem of asymmetry

The Princeton robustness study includes a very brief discussion of robustness under asymmetric contamination, noting that [11, p. 109]:

> Except in a few instances there may be no reason to believe the underlying distribution is symmetric.

In the results presented there, asymmetric contamination is modeled by a zero-mean, unit variance Gaussian nominal sequence contaminated by a second unit variance Gaussian sequence with mean either 2 or 4. This contamination model represents a special case of the general *slippage model* [27, p. 49], discussed briefly in Chapter 4. The results indicate a clear bias-variance trade-off, with the mean exhibiting the largest average error of the 68 estimators compared but the smallest variance, and the **shorth** estimator mentioned in Sec. 7.4.1 exhibiting the smallest bias and the largest variance. Beyond a single table of means and variances for the 68 estimators considered under these two scenarios, little more is said about asymmetric contamination, aside from the observation that "much more must be learned." In fact, one of the conclusions given by study participant J.W. Tukey was [11, p. 226]:

> We did a little about unsymmetric situations, but we were not able to agree, either between or within individuals, as to the criteria to be used.

In general, the problem of characterizing asymmetric data sequences is somewhat complicated, as discussed in Chapters 4 and 6, but the problem of characterizing an asymmetric data sequence containing outliers is even more complicated, as the following two subsections illustrate. Specifically, Sec. 7.6.1 considers the problem of robust asymmetry characterization, somewhat analogous to the problem of robust location and scale estimation considered in Sec. 7.4. Following that discussion, Sec. 7.6.2 briefly describes a location-free scale estimator that may be appropriate in strongly asymmetric situations where it is not obvious what location parameter should be used as a center for scale estimation; in particular, note that the standard deviation is a mean-centered scale estimate and the MADM scale estimate is median centered.

7.6.1   Robust asymmetry measures

One of the primary points of the example presented in Sec. 7.1.1 was the increasing sensitivity of moment-based estimators to outliers with increasing moment order. In particular, the outlier sensitivity of the traditional skewness estimate γ is dramatic enough to render it essentially useless in the presence of any contamination. The following discussion introduces two alternative asymmetry measures that are more resistant to outliers and compares their performance for an example where exact values can be determined. Specifically, the following simulation experiment is considered: a total of $M = 100$ results are compared,

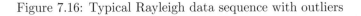

Figure 7.16: Typical Rayleigh data sequence with outliers

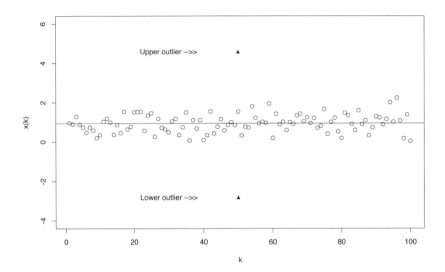

each based on a data sequence of length $N = 100$ generated according to the Rayleigh distribution discussed briefly in Chapter 4 and defined by the density:

$$p(x) = \frac{x}{\beta^2} e^{-x^2/2\beta^2}, \tag{7.28}$$

with $\beta = 1/2$. For each sequence, the three skewness measures considered here are computed under three different contamination scenarios: the uncontaminated reference case, the case of a single *upward outlier*, 8 standard deviations above the mean at $x_+ = 4.59$, and the case of a single *downward outlier*, 8 standard deviations below the mean at $x_- = -2.82$. A typical example is plotted in Fig. 7.16, showing the Rayleigh distributed data sequence and both upper and lower outliers.

The first of the three asymmetry measures considered here is the traditional third moment-based skewness parameter γ introduced in Chapter 4 and examined further in Sec. 7.1.1. The second asymmetry measure is *Galton's skewness measure* [162, p. 3]:

$$\gamma_G = \frac{x_{0.75} + x_{0.25} - 2x_{0.50}}{x_{0.75} - x_{0.25}}$$
$$= \frac{(\text{upper quartile} - \text{median}) - (\text{median} - \text{lower quartile})}{\text{interquartile range}}. \tag{7.29}$$

Because this asymmetry measure is based on order statistics, it is well defined for all distributions, and we expect it to be reasonably outlier resistant. The third asymmetry measure considered here is *Hotelling's skewness measure* [218], mentioned in Chapter 6 and defined as:

$$\gamma_H = \frac{\bar{x}_N - x^\dagger}{\hat{\sigma}}. \tag{7.30}$$

This estimator is based on the observation that the mean $\bar{x}_N$ and the median $x^\dagger$ are not equal for asymmetric distributions. Further, because this estimator uses both moments and order statistics, we can expect its outlier sensitivity to lie between that of the moment-based estimator γ and the order statistic-based estimator γ_G. It is not difficult to show that $|\gamma_G| \leq 1$ (Exercise 2), and it is also known that $|\gamma_H| \leq 1$ [218]. In contrast, note that the traditional skewness parameter γ can assume any real value.

The motivation for considering the Rayleigh distribution here is that both its order statistics and its first three moments are easily computable. Specifically, the mean, standard deviation, and skewness are given by:

$$\bar{x} = \beta\sqrt{\frac{\pi}{2}} \simeq 1.25311\beta,$$

$$\sigma = \beta\sqrt{\frac{4-\pi}{2}} \simeq 0.6551364\beta,$$

$$\gamma = \frac{2(\pi-3)\sqrt{\pi}}{(4-\pi)^{3/2}} \simeq 0.63111. \tag{7.31}$$

Similarly, the median, upper, and lower quartiles are easily computed from the following result [162, p. 458]:

$$x_{[p]} = \left[2\ln\left(\frac{1}{1-p}\right)\right]^{1/2}\beta \quad \Rightarrow \quad x_{0.25} \simeq 0.7585276\beta,$$

$$\Rightarrow \quad x_{0.50} \simeq 1.17741\beta,$$

$$\Rightarrow \quad x_{0.75} \simeq 1.665109\beta. \tag{7.32}$$

From these results, both Galton's and Hotelling's skewness measures may be computed:

$$\gamma_G \simeq 0.075908, \quad \gamma_H \simeq 0.11586. \tag{7.33}$$

Note that all three of these values are positive but relatively small, suggesting a slight upward distributional asymmetry, consistent with both the plot of the density shown in Chapter 4 and the data sequence plotted in Fig. 7.16.

The outlier sensitivity of the traditional skewness estimator γ was illustrated in Sec. 7.1.1, and this point is further emphasized in Fig. 7.17. Specifically, the boxplot in the center of this figure shows the skewness estimated from the un-contaminated data, indicated by the "0" on the horizontal axis. It is clear from this boxplot that the median value is approximately equal to the correct value $\gamma \simeq 0.63$ indicated by the horzontal line across the figure. Further, although the range of variation of these estimates is significant, the estimated skewness is al-ways positive for the uncontaminated data. In marked contrast, the effect of the

Figure 7.17: Boxplots of traditional skewness estimates

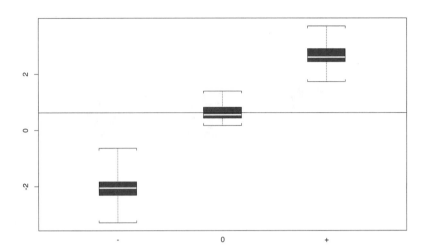

single downward outlier is to shift the estimated skewness dramatically downward, as indicated by the boxplot marked "−" on the horizontal axis. There, the median value is approximately −2 and the estimated skewness never even has the correct sign, reflecting the extreme outlier sensitivity of this moment-based estimator. Exactly analogous behavior is seen for the upward outlier, as indicated by the boxplot marked "+" on the horizontal axis. There, the median estimate of γ is shifted upward to approximately 2.5 and the smallest estimate is larger than the largest estimate obtained from the uncontaminated data sequences.

Fig. 7.18 shows the corresponding boxplots for Galton's skewness measure γ_G. As expected, this skewness measure is *much* less sensitive to outliers; in fact, the outliers have no visible effect here, as these boxplots all appear essentially identical. In particular, almost no variation is seen in the middle 50% of these estimates, and only very slight shifts are seen in the upper and lower extremes. Conversely, the natural variation in this asymmetry measure appears to be distressingly large: although the median value appears approximately correct in all cases, Galton's skewness estimate has a significant probabiltiy of giving the wrong sign. Further, the range of variation of these estimates covers approximately 30% of the total range of admissible values for this skewness estimate, raising serious questions of its utility, at least for this example.

Finally, Fig. 7.19 compares the performance of Hotelling's skewness measure γ_H for these three contamination scenarios. Not surprisingly, because it is based

Figure 7.18: Boxplots of Galton's skewness estimates

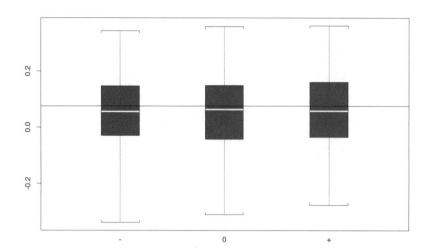

on both moments and order statistics, the general behavior of this asymmetry measure is intermediate between that of the traditional moment-based skewness γ and Galton's order-statistic based skewness γ_G, both with respect to its outlier sensitivity and with respect to its natural variability. In this case, it appears that downward outliers are somewhat more damaging than are upward outliers, but the influence is small compared to that seen for γ. Similarly, the range of variation of the estimated skewness is still large, but smaller than that for Galton's skewness measure: $\sim \pm 20\%$ of the admissible range vs. $\sim \pm 30\%$.

Overall, the primary point of this discussion is to emphasize that the practical assessment of distributional asymmetry is somewhat challenging, particularly if outliers are present.

7.6.2   Location-free scale estimates

It was noted in Chapter 4 that the mean, median, and mode coincide for symmetric, unimodal distributions, and Hotelling's skewness measure discussed in the previous example exploits the fact that the mean and the median differ for asymmetric distributions. Further, the MADM scale estimator S was presented in Sec. 7.4.2 as a robust alternative to the standard deviation $\hat{\sigma}$. It is important to note, however, that these two scale estimates both attempt to measure the typical distance a data point lies from some reference location: for $\hat{\sigma}$, this reference location is the mean, and for S, this reference location is the median. For

Figure 7.19: Boxplots of Hotelling's skewness estimates

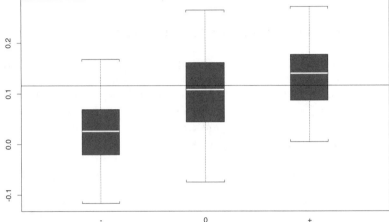

strongly asymmetric distributions, the fact that these reference locations are not the same is troubling. This observation motivates interest in *location-free scale estimates*, which attempt to measure the typical distance between distinct data points in the sample rather than the typical distance to a reference location like the mean or the median.

A classical location-free scale estimator is Gini's mean difference, discussed in Chapters 3 and 6: this estimator is the mean of the distances between x_i and x_j for all $i \neq j$, given by either of the following equivalent expressions:

$$g_N = \frac{1}{N(N-1)} \sum_{i=1}^{N} \sum_{j \neq i} |x_i - x_j| = \frac{2}{N(N-1)} \sum_{i=1}^{N} \sum_{j > i} |x_i - x_j|. \qquad (7.34)$$

Replacing this average by a median leads to a more robust location-free scale estimator discussed briefly by Rousseeuw and Leroy [252], who note that this estimator had been discussed previously by other authors and shown to have a 29% breakdown point but a high Gaussian efficiency ($\sim$86%), provided it was suitably scaled to make it an unbiased estimator of σ. As an alternative, Rousseeuw and Croux [251] propose a more robust estimator, with a 50% breakdown point and a Gaussian efficiency of $\sim$82%, defined by:

$$Q_N = d\{|x_i - x_j| \mid i < j\}_{(k:N)}, \quad k = \binom{[N/2]+1}{2} \simeq \frac{N(N-1)}{8}, \qquad (7.35)$$

where $d = 2.2291$ is a normalization constant that makes Q_N an unbiased estimator of σ for Gaussian data. Brute force evaluation of Eq. (7.35) requires rank-ordering the list of the $N(N-1)/2$ differences $|x_i - x_j|$ for $i < j$, leading to long computations for large N, but a computationally efficient algorithm is available [74]. These authors also describe another location-free scale estimator, based on repeated medians [74, 251].

7.7 Other practical considerations

The previous section emphasized the complications that asymmetric data distributions introduce into the problem of selecting, designing, or understanding robust estimators. The following subsections briefly describe three other important practical complications: the effects of light-tailed or bimodal nominal data distributions (Sec. 7.7.1), the consequences of working with coarsely quantized data (Sec. 7.7.2), and the practical consequences of the discontinuous influence curve of the median, on which many popular robust estimators are based (Sec. 7.7.3). All of these discussions illustrate important but unexpected phenomena that can arise in practice.

7.7.1 Light-tailed and bimodal distributions

Most of the robust statistics literature is concerned with the development and characterization of estimators that tolerate nonrepresentative data points lying unusually far out in the tails of the nominal reference distribution. Conversely, these estimators can behave in unexpected ways if this nominal data distribution is either *platykurtic* (i.e., light-tailed) or *bimodal*. Recall from Chapter 4 that under the traditional interpretation of the kurtosis κ, light-tailed distributions correspond to $\kappa < 0$, and further, that $\kappa > -6/5$ for symmetric, unimodal distributions, so $\kappa < -6/5$ implies the existence of at least two modes. The family of symmetric beta distributions (i.e., the beta distributions discussed in Chapter 4 with both shape parameters equal, $p = q$) exhibits both of these forms of behavior, with kurtosis $\kappa = -6/(2p+3)$ varying between limiting values of -2 as $p \to 0$ and 0 as $p \to \infty$. To see the influence of light-tailed and bimodal distributions in a typical robust estimation algorithm, it is instructive to consider the behavior of the MADM scale estimator for members of the beta distribution family. These results are presented in Table 7.7, which lists exact values for the standard deviation σ, the normalized MADM scale estimate S, and the kurtosis κ for five different symmetric beta distributions. The first two of these distributions, $p = 0.1$ and $p = 0.5$ are bimodal, the next ($p = 1.0$) is uniform, and the last two ($p = 2.0$ and $p = 4.0$) are unimodal. The key observation here is that in all cases, the MADM scale estimate S is *larger* than the standard deviation σ, particularly for the bimodal distributions where S is *much* larger than σ. *This behavior is unusual, and it can provide a practical basis for detecting light-tailed data distributions: estimate both S and σ and take $S > \sigma$ as an indication of light tails or multiple modes.* Applications of this idea to real datasets are presented in Chapter 8 to illustrate its utility in practice.

Table 7.7: Symmetric beta distributions: p, σ, S, and κ

p	σ	S	κ
0.1	0.456	0.740	−1.875
0.5	0.354	0.524	−1.500
1.0	0.289	0.371	−1.200
2.0	0.224	0.257	−0.857
4.0	0.167	0.180	−0.545

In addition, this observation is important because it illustrates a failure mechanism for the Hampel identifier. Specifically, for light-tailed data, the MADM scale estimator tends to overestimate the standard deviation, just as the moment-based estimator $\hat{\sigma}$ does for heavy-tailed or contaminated data. Consequently, the Hampel identifier may exhibit worse performance than the 3σ-edit rule if the nominal data distribution is light-tailed. This point is illustrated in Fig. 7.20, which shows a sequence of $N = 100$ samples from the bimodal arcsine distribution (i.e., the beta distribution with $p = q = 0.5$), contaminated with a single outlier; the open circles in the figure represent the 99 nominal data points and the solid circle in the middle of the figure represents the single outlier, lying slightly more than three standard deviations above the mean. In addition, the outlier detection thresholds for the 3σ-edit rule (ESD identifier) and the Hampel identifier are indicated on the figure, illustrating that in this case, the ESD identifier is *more effective* than the Hampel identifier, which does not detect this outlier. This example illustrates that if the MADM scale estimate S is significantly smaller than the estimated standard deviation, $\hat{\sigma}$, it is probably better to use the 3σ-edit rule, at least if there is a reasonable basis for supposing the nominal data distribution is approximately symmetric.

7.7.2 Discrete distributions (quantization)

Although the Hampel identifier is less prone to masking than the ESD identifier, it is more prone to swamping, reflecting a tendency of the MADM scale estimator to underestimate the standard deviation. An extreme example of this underestimation can occur for coarsely quantized data, corresponding to a *discrete* distribution. The root of this difficulty lies in the fact that for continuous data distributions, the event $x_k = x_j$ for $j \neq k$ has zero probability, but for discrete distributions this event generally has finite probability and can, in extreme cases, cause *implosion* of the MADM scale estimator: in contrast to the variance inflation phenomenon where $\hat{\sigma}$ essentially "explodes," the MADM scale estimator can "implode," yielding a value of *zero* for a nonconstant data

Figure 7.20: Failure of the Hampel identifier

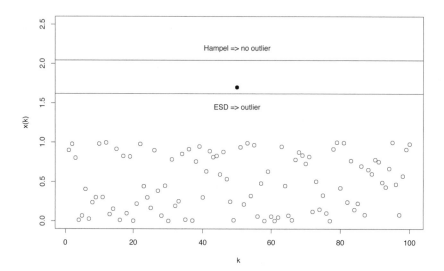

sequence. Since scale estimates are frequently required for normalization (i.e., they appear in denominators), implosion can render an estimator completely useless for some applications.

To illustrate this difficulty, consider a dataset containing an odd number of points and express this number as $N = 2n + 1$. If $m \geq n + 1$ of these data points have the same value, say, x_0, it follows that the median is $x^\dagger = x_0$. Further, the absolute deviations $d_k = |x_k - x^\dagger|$ form a sequence of nonnegative numbers, m of which are zero. Since $m \geq n + 1$, the median of the sequence $\{d_k\}$ is zero, implying the MADM scale estimate S is zero for this data sequence, independent of the values of the $N - m$ data points that are not equal to x_0. In contrast, the standard deviation always gives some indication of data variation when it is present since $\hat{\sigma} = 0$ implies $x_k = \bar{x}_N$ for all k (Exercise 4). Consequently, if the Hampel identifier is applied to coarsely quantized data like the example just described, it will partition the dataset into a nominal part, consiting of the m values x_0, and an outlying part consisting of the rest of the dataset, *regardless of how near or far these data values lie from* x_0.

This point is illustrated in Fig. 7.21, which shows two data sequences, each based on a sequence of $N = 100$ zero-mean, Gaussian random numbers with standard deviation $\sigma = 0.15$. The light line corresponds to the 7-digit quantization of the random number generator, and the solid circles correspond to a 1-digit quantization, obtained by retaining only the leading digit of the original data sequence. Denoting the original data sequence by $\{r_k\}$ and the coarsely

Figure 7.21: A Gaussian data sequence and a coarse quantization

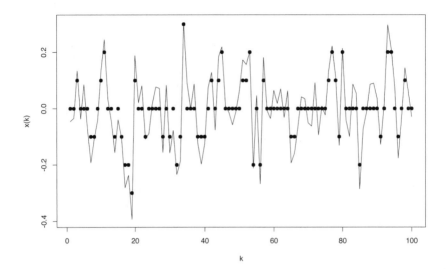

quantized data sequence by $\{q_k\}$, it is instructive to compare the means, medians, standard deviations, MADM scale estimates, and the performance of the ESD and Hampel outlier dectection procedures. For both the mean and the median, quantization has little effect:

$$\bar{r}_N = -0.0029 \quad \text{vs.} \quad \bar{q}_N = -0.0030,$$
$$r^\dagger = -0.0033 \quad \text{vs.} \quad q^\dagger = 0,$$

although the median value $q^\dagger = 0$ does appear somewhat suspicious. Quantization has a significant effect on the estimated standard deviation, reducing it by about 25%, but the effect on the MADM scale estimator here is catastrophic, leading to scale implosion:

$$\hat{\sigma}_r = 0.1366 \quad \text{vs.} \quad \hat{\sigma}_q = 0.0989$$
$$S_r = 0.1317 \quad \text{vs.} \quad S_q = 0.$$

If we apply either the 3σ-edit rule or the Hampel identifier to the data sequence $\{r_k\}$, we obtain the correct result: no outliers are detected in the dataset by either procedure. For the quantized data sequence $\{q_k\}$, because the standard deviation is underestimated, the 3σ-edit rule exhibits a slight swamping effect, erroneously declaring two of the 100 data points as outliers. Conversely, the complete breakdown of the MADM scale estimator in this case causes severe swamping: 42 outliers are detected in this data sequence, corresponding to the 42% of the data values whose leading digit is not 0.

Figure 7.22: A segment of the helicopter collective pitch sequence

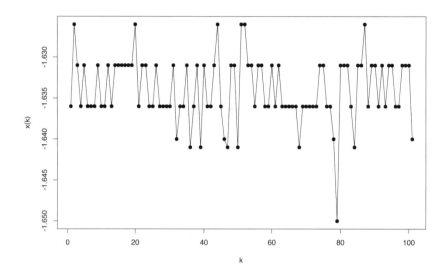

Although this example is simulation based, it is important to emphasize that the problem considered here is real. For example, industrial process data recorded to the nearest 0.1 degree is not uncommon, and sometimes the quantization is even coarser. As another example, Fig. 7.22 shows a 100-point subsequence of the helicopter collective pitch dataset discussed in Chapter 2. It is clear from this figure that the character of this data sequence is exactly analogous to that of the coarsely quantized Gaussian data sequence just considered, where robust estimation procedures like the MADM scale estimator and the Hampel identifier can perform very poorly. Such examples are important because they illustrate the phenomenon of scale implosion in cases where the iterpretation is clear: here, implosion was a direct consequence of using a high-breakdown scale estimator on quantized data. In more complicated data analysis problems, scale implosion and other such unexpected phenomena may be harder to interpret. The key point is that pathological behavior like scale implosion can occur and should be viewed as a warning signal that some significant—and possibly unexpected—structure is lurking in the data.

7.7.3 Discontinuity—a cautionary tale

As noted in Sec. 7.3.2, the influence function for the median is discontinuous, and this discontinuity has some undesirable consequences. The following example illustrates one of these consequences: in highly heterogeneous datasets, the

Figure 7.23: A highly heterogeneous dataset and three medians

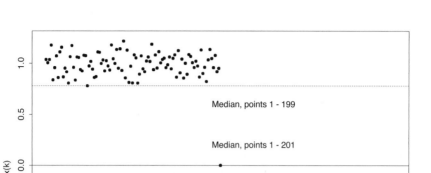

median may be extremely sensitive to small changes in the structure of the dataset. Specifically, Fig. 7.23 shows a sequence of 201 points, obtained as follows:

- points 1 through 100 are distributed as $N(1, 0.1^2)$
- point 101 has the value 0,
- points 102 through 201 are distributed as $N(-1, 0.1^2)$.

The median value for the complete dataset is $x^\dagger = 0$, corresponding to the fact that the first 100 points in the dataset are strictly larger than this value, whereas the last 100 points in the dataset are strictly smaller. The horizontal line passing through data point 101 provides a visual indication of this median value relative to the rest of the dataset. Here, because the dataset consists of two essentially opposing subsets, each comprising almost half of the data, the median value represents an unreasonable compromise that is not representative of either of these subsets.

If we omit either the first two points or the last two points from this dataset, the effect is dramatic, as indicated by the two dashed lines in the figure. Specifically, the upper line represents the median computed from data points 1 through 199, and the lower line represents the median computed from data points 3

Table 7.8: Influence of dataset truncations

Points	Median	Mean
1 - 201	0.0000	0.0002
1 - 199	0.7789	0.0101
3 - 201	−0.7759	−0.0101

through 201. In both cases, reducing the size of the dataset by 1% results in a large change in the median value. This effect may be viewed as a "swing vote" phenomenon: given two dissimilar subsets, each comprising nearly 50% of the data, small changes in the size of the dataset can shift the "balance of power" between these subsets, resulting in very large changes in the median. In contrast, the mean is much better behaved in this example, exhibiting responses to these dataset changes that are in the same direction as those observed for the median, but almost 80 times smaller. These changes are summarized in Table 7.8. The key point of this example is that the discontinuity of the median and other related estimators can result in unexpected—and often highly undesirable—forms of behavior.

7.8   General recommendations

The main points of this chapter have been, first, that outliers arise frequently in practice, and second, that if we do not treat these points carefully, they can invalidate our analytical results completely. The examples presented here have attempted to illustrate both of these points and to provide some general ideas of how we can avoid the undesirable consequences of outliers in the data. The following paragraphs provide a brief summary of some of the important practical conclusions that may be drawn from these examples.

7.8.1   How robust is enough?

Before the advent of highly robust exploratory analysis methods, outlier detection was generally approached from the confirmatory perspective, assuming a single nominal distribution for the data (often Gaussian, though not always) and testing for the presence of a single discrepant data point. Tests for multiple outliers in this setting were also proposed, and this was recognized as a more difficult problem since it required deciding how many outliers to test for. In the conclusion of their paper on outlier detection, Fung and Paul [108] refer to this difficulty, noting that the decision "depends on the investigator's experience of the practical situation." They argue that for "small to moderately large samples" (specifically, datasets of size $N = 5$ to $N = 20$), "testing for up to three outliers should be adequate." One difficulty with this viewpoint is that techniques like the 3σ-edit rule that are extremely prone to masking in the presence of multiple outliers may be fairly effective in detecting a single outlier. Recognition of the practical importance of masking in datasets with more than a few outliers has led to the development of much more robust procedures like those described by Rousseeuw and Leroy [252] that can essentially ignore up to $\sim$50% outliers in a dataset.

Advocates of confirmatory procedures are often extremely uncomfortable with these high-breakdown procedures. For example, Fung and Paul [108] argue that "if there are more than a few outliers present, the underlying model is suspect and the problem ceases to be one of outlier detection." Consequently, there is some debate concerning how robust we should make our data analysis procedures. Huber [152] suggests that "contaminations between 1% and 10% are sufficiently common that a blind application of procedures with a breakdown point of less than 10% must be considered outright dangerous." Conversely, he also argues that the disadvantages (e.g., computational complexity, efficiency loss at the nominal distribution, etc.) associated with very high breakdown (e.g., 50%) estimators make them difficult to justify. Overall, Huber seems to advocate a breakdown point of about 25%.

These apparently opposing views (i.e., at most 3 outliers vs. 25% breakdown point) converge somewhat if we consider sample size. Specifically, note that 3 outliers amount to 15%–60% contamination in the datasets considered by Fung and Paul [108], so their recommendations are roughly consistent with Huber's. As the size of the dataset increases, however, absolute limits like Fung and Paul's "at most 3 outliers" stray into the "less than 10% breakdown" region that Huber declares to be dangerous. For example, recall that the makeup flow rate dataset considered in this chapter was a *real* example with about 500 outliers out of about 2500 data points. Here, Huber's 25% breakdown point would suffice, but Fung and Paul's upper limit of 3 outliers is far too conservative.

In general, the practical cases requiring high-breakdown estimators are those in which we are dealing with large datasets collected under conditions that *may be heterogeneous*, as in the case of the makeup flow rate data just noted. There, the heterogeneity was obvious, and it could be argued that the spurious shutdown data points should have been eliminated from the outset. Blind application of such arguments is dangerous, however, for two reasons. First, although such prescreening often can and should be done, equally often it is *not* done, for a variety of reasons. In the face of such reality, it is blatantly irresponsible for data analysts not to protect themselves against the consequences of such oversights. The second reason such arguments are dangerous is that often the heterogeneous nature of the data is simply not known in advance. Finding it when it is present can then lead to significant advances in our understanding.

7.8.2 Overall recommendations

Generally, one of the important first steps in exploratory data analysis is screening for outliers. A useful part of this screening step is to examine each variable

Figure 7.24: Second pressure dataset and outlier limits

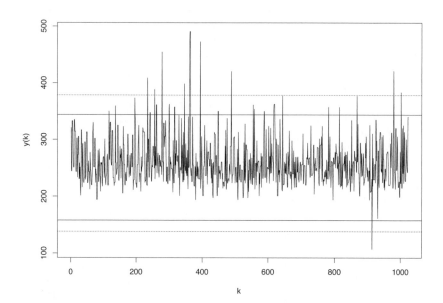

individually, looking for suspicious data points. As noted in Sec. 7.2.3, if we find any such data values, we should, if at all possible, classify them as *anomalous*, *questionable*, or *unremarkable* based on *physical understanding* of their source and acquisition. Still, to make this classification, we must start with a reliable approach to detecting suspicious points in the original dataset. Although they are susceptible to swamping (i.e., excessive zeal in finding suspicious points), the two techniques that appear the most promising are the following:

1. the Hampel identifier, with threshold $t \sim 3$ (e.g., 2.5 to 4.0),
2. the inward Hampel identifier described in Sec. 7.5.3, possibly with a more conservative threshold t (e.g., 3.0 to 5.0).

These recommendations are for datasets of moderate size (e.g., $N \geq 100$); for smaller datasets, it may be reasonable to increase these threshold values slightly (e.g., 5.0 to 6.0).

As noted in Sec. 7.6, asymmetry poses particular difficulties, for which there is no clear remedy. This point is illustrated in Fig. 7.24, which shows the second of the two pressure data sequences introduced in Chapter 1, together with upper and lower outlier detection limits for the ESD and Hampel identifiers. The ESD identifier limits are shown as dotted lines and are wider than those of the Hampel identifier (solid lines), reflecting the influence of the large upward deviations from the mean on the estimated standard deviation. The key point here, however, is that although both of these detection procedures find the larger of the two downward outliers, neither procedure finds the smaller of these points. Further, both procedures find many *upward outliers*, despite the fact that these points appear *visually* consistent with the general asymmetry of the data sequence. These results are not surprising since both the ESD and Hampel identifiers are inherently based on assumptions of a symmetric nominal distribution. One practical alternative is the use of the quantile-quantile plots described in Chapter 8 as a *graphical* means of detecting both asymmetry and outliers in a dataset.

Overall, given the general lack of panaceas in dealing with outliers in a dataset, how do we proceed? Probably the best advice along these lines is the following suggestion from Mallows [196]:

> A simple and useful strategy is to perform one's analysis both robustly and by standard methods and to compare the results. If the differences are minor, either set can be presented. If the differences are not, one must perforce consider why not, and the robust analysis is already at hand to guide the next steps.
>
> The importance of these considerations is enhanced when we are dealing with large amounts of data, since then examining all the data in detail is impractical, and we are forced to contemplate working with data that is, almost certainly, partially bad and with models that are almost certainly inadequate.

7.9 Exercises

1. For the Laplace distribution defined by:

$$p(x) = \frac{1}{2\phi} e^{-|x-\mu|/\phi},$$

   compute the normalization factor γ required so that the MADM scale estimate $S = \gamma S_0$ is an unbiased estimator of ϕ.

2. For Galton's skewness measure γ_G discussed in Sec. 7.6.1, prove that $|\gamma_G| \leq 1$ provided $x_{0.75} \neq x_{0.25}$. (Hint: define $\alpha = x_{0.75} - x_{0.25}$ and $\beta = x_{0.50} - x_{0.25}$. What do we know about α and β?)

3. Consider the following data sequence:

$$x_k = \begin{cases} \mu & k = 1, 2, \dots, N-p, \\ \lambda\mu & k = N-p+1, \dots, N. \end{cases}$$

   This sequence may be viewed as a collection of $N-p$ nominal data points with value μ and p outlying data points with value $\lambda\mu$.

   a. Suppose $\lambda > 1$ and define $\epsilon = p/N$ as the contamination level. What is the maximum contamination permitted if the error in estimating the nominal mean μ is less than 10%?

b. What is the limiting value of this contamination as $\lambda \to \infty$?

c. Under what conditions will the sample median fail to give the desired result (i.e., μ)? Do these conditions depend on λ?

4. Prove that $\hat{\sigma} = 0$ implies $x_k = \bar{x}_N$ for all k, where $\hat{\sigma}$ is the traditional standard deviation estimate and $\bar{x}_N$ is the arithmetic mean of the sample.

5. (Computer exercise) Using any computational platform, implement the MADM scale estimator (using the standard Gaussian normalization, $\gamma = 1.4826$), the Hampel identifier with $t = 3$, and the 3σ-edit rule.

a. For $\epsilon = 0$, 0.05, and 0.10, generate 100 sequences of length $N = 100$, each drawn from the contaminated normal model $CN(0, 0, 1, 3, \epsilon)$.

b. Generate boxplots comparing, for all three cases:

    i. means and medians,

    ii. standard deviations and MADM scale estimates,

    iii. the number of outliers detected with each identifier.

Chapter 8

Characterizing a Dataset

Sections and topics:

This chapter introduces the following four techniques, developed to give useful insights into the distributional character of an observed data sequence:

1. quantile-quantile or Q-Q plots,
2. diagnostic plots for discrete distributions,
3. histograms,
4. kernel density estimates.

The first two of these techniques may be regarded as informal graphical diagnostics for examining specific distributional hypotheses: if the hypothesis holds, the resulting plot should approximate a straight line, and systematic deviations from linearity provide useful information about *how* the data sequence violates distributional assumptions (e.g., heavy or light tails, asymmetry, outliers, etc.). Histograms are crude density estimates, introduced here because they are well-known and simple and provide important background for kernel density estimates, which may be regarded as smoothed histograms. Closely related to kernel density estimates are *scatterplot smoothers*, which can be useful in characterizing and/or removing long-term variations in a data sequence. All of these techniques are applied to real datasets to illustrate how they work and the nature of the insights that they can provide.

Key points of this chapter:

1. Quantile-quantile or Q-Q plots provide informal graphical assessments of specific distributional assumptions: if these assumptions are approximately satisfied, the plot should approximate a straight line. If not, the character of the deviations from linearity can be quite informative, providing evidence for outliers, asymmetry, heavy- or light-tailed behavior, or bimodaltiy.

2. Although normal Q-Q plots are the most popular in practice, Q-Q plots can be constructed for any distribution of fixed shape, but any shape parameters (e.g., for the gamma or beta distributions) must be specified.

3. Analogous diagnostic plots can be constructed for discrete distributions; specific examples described here include the *Poissonness plot* and the *negative binomialness plot* to assess the reasonableness of these two distributions for count data.

4. Histograms represent extremely simple nonparametric density estimates that, despite their historical importance and continuing popularity, are such crude approximations that they are of limited practical utility. Kernel density estimators represent a more practical alternative, and they are closely related, motivating the space devoted to histograms here.

5. In their simplest form, kernel density estimators are specified by two components: a kernel function $K(x)$ and a positive bandwidth parameter h. While both of these components significantly influence the resulting density estimates, the influence of the bandwidth parameter is both greater and easier to vary; hence, comparing the results obtained for different values of h can be very useful, particularly in the early stages of a long analysis of many related datasets.

6. Closely related to kernel density estimates are kernel smoothers, which attempt to estimate the general form of a relationship between two variables. Like kernel density estimators, kernel smoothers are *nonparametric* procedures that, in contrast to the regression analysis methods discussed elsewhere in this book, do not assume an explicit mathematical form for this relationship. As examples presented here illustrate, both the method chosen and the bandwidth parameter specified can strongly influence the final results obtained.

7. Many of the techniques presented in this chapter are extremely useful in making a preliminary assessment of a new dataset: what kinds of variables are included, and what are they like? The chapter concludes with an illustrative example, characterizing the chronic fatigue syndrome dataset introduced in Chapter 1.

8.1 Surveying and appreciating a dataset

There was a time when data collection and analysis were pretty much the domain of subject matter experts: the Ohm's law example presented in Chapter 1 provides a representative illustration. While this situation had the advantage that the analyst could rely on a great deal of subject matter expertise, it also had two glaring disadvantages relative to the situation today. First, since data observations were collected manually, typically as the result of painstaking experimental work, there were usually only a few of them available to analyze. Second, since modern computational hardware and software were not available then, data analysis had to rely on techniques simple enough to permit hand calculation.

The typical situation today is very different. Now, large datasets are common and widely available, as are substantial computational resources for analyzing them. A natural consequence is that data analysis is now frequently *not* done by subject matter experts, but rather by data analysis experts of one type or another (e.g., statisticians of various types, data mining specialists, machine learning specialists, and a continually growing list of others). As a corollary, this means it is now quite common for a data analyst to be given (or to seek out) a relatively large dataset that was generated entirely by other people, often a relatively large group of other people, and increasingly a group of people with whom the data analyst has little or no direct contact. In fact, this same situation often holds even when the analyst *is* a subject matter expert, since the availability of large datasets collected by others can provide an extremely useful frame of reference in evaluating results obtained from datasets that the analyst was directly involved in collecting. In all of these cases, a critically important first step in analyzing a dataset obtained from elsewhere is a preliminary survey of the dataset to see exactly what it does—and does not—contain.

Chapter 2 introduced the notion of *metadata* that describes what is—or what should be—in a dataset, but it was also noted that metadata is seldom as complete as we would like and is sometimes blatantly incorrect. A representative example discussed in Chapter 2 was the Pima Indians diabetes dataset in the UCI Machine Learning Archive, where the metadata stated that there were no missing observations but subsequent analysis revealed the presence of physiologically impossible zero values (e.g., for blood pressure) that were being used as codes for missing data. As this example illustrates, part of the task of initially characterizing a dataset is to carefully examine and, where possible, augment the metadata with information from other sources. Similarly, it is also important to examine the data in light of the metadata to assess the degree of agreement between the two and identify any apparent descrepancies for further examination.

Specific tasks in this initial examination of a dataset include answering certain key questions. The focus of this chapter is on simple, effective techniques for addressing these and other similar questions to gain a useful initial understanding of a new dataset. Representative examples include the following questions:

1. How many records N does the dataset contain, and how many variables M? Are these numbers consistent with the available metadata, if any?

2. What types are each of the M records: binary, nominal, ordinal, integer, real-valued?

3. Do any of these values exhibit range restrictions, based either on explicit statements in the metadata or on limits inferred from variable definitions (e.g., the metadata states that variable X is a fractional composition, from which we infer that it must lie between 0 and 1; further, there may be other fractional compositions and it may be necessary that these sum to a total equal to 1 or less than 1)?

4. What are the *observed* ranges of these variables? How do these compare with anything stated or implied in the metadata?

5. What fraction of each variable exhibits missing values? Are any of these cases where missing values are expected (e.g., characterizations that are not applicable to a particular data record)?

6. Are any records "unusual" in any way (e.g., exhibiting outlying values for some real variable, or all exhibiting exactly the same value for a real variable, etc.)?

7. Do either missing values or anomalous values exhibit unusual patterns? (e.g., in a multisource dataset, do all of the outliers appear to come from one source?)

8.2   Three useful visualization tools

One of the best ways of initially examining a dataset is to examine the variables graphically, a point noted in Chapter 1 and illustrated in a number of examples in subsequent chapters. For this reason, the primary focus of this chapter is on simple graphical procedures that can be used to understand the character of a new dataset. The following three discussions introduce three specific techniques, all of which are described in detail in later sections of this chapter.

8.2.1   The normal Q-Q plot

As noted in Chapter 4, although the Gaussian distribution is widely adopted as the "standard" distribution for real-valued data samples this distributional assumption is not always warranted. This point was further emphasized in Chapter 7 in connection with the issue of *outliers* in real-valued data sequences, even a few of which can cause Gaussian-based analysis procedures to give strongly biased results. For this reason, *quantile-quantile plots* or *Q-Q plots* have evolved as a very useful informal graphical test of this distributional assumption. The idea, discussed in much more detail in Sec. 8.3, is to construct a graphical representation of the data that should approximately conform to a straight line if the

Figure 8.1: Data plots and their corresponding normal Q-Q plots for two variables from the Pima Indians diabetes dataset

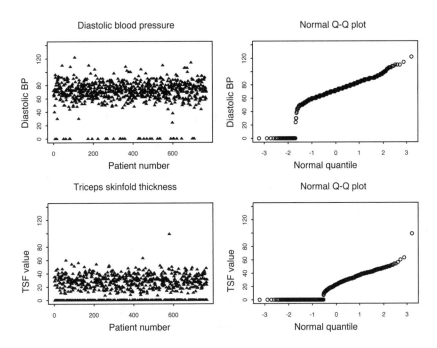

Gaussian assumption is reasonable. As the discussion in Sec. 8.3 demonstrates, this plot can be used to detect a variety of systematic deviations from normality, including both "heavy tails" and significant asymmetry. In addition, Q-Q plots can also highlight data anomalies like outliers and *inliers*, discussed in Chapter 2, that often appear as unusually frequent ocurrences of a single data value that falls well within the nominal data range. The following example illustrates the ability of Q-Q plots to detect both of these types of data anomalies.

Fig. 8.1 shows plots of two variables from the UCI Pima Indians diabetes database introduced in Chapter 2, together with their normal Q-Q plots. Specifically, the left-hand plots show the data values plotted against a numerical patient index for the recorded diastolic blood pressure values (upper left) and for the recorded triceps skinfold thickness values (lower left). Next to each of these plots are the corresponding normal Q-Q plots: the fact that neither of these plots are well approximated by straight lines gives an immediate indication that these data sequences are not well described by the Gaussian distribution. In particular, note the pronounced flat portions in the lower left of both of these plots, which correspond to the unusually large number of zero values seen in both data sequences, relative to all other possible values. As discussed in Chapter 2, these zeros represent a special code for missing data in these variables, since a value of "zero" for either diastolic blood pressure or triceps skinfold thickness is not

Table 8.1: Summary of total Prussian army horse-kick deaths by year, for the 20 years from 1875 to 1894

Year	Count	Year	Count	Year	Count	Year	Count
1875	3	1880	18	1885	5	1890	17
1876	5	1881	6	1886	11	1891	12
1877	7	1882	14	1887	15	1892	15
1878	9	1883	11	1888	6	1893	8
1879	10	1884	9	1889	11	1894	4

realistic. The point of this example is that the Q-Q plots draw our attention immediately to this feature of the data sequences. In addition, the Q-Q plot in the lower right also draws our attention to an outlying triceps skinfold thickness value at the upper right end of the plot. While it may be argued that both of these data characteristics—the large number of zero values in both sequences and the outlying point in the triceps skinfold thickness sequence—may be seen clearly in the original data sequences, the point here is that the Q-Q plots make these features more pronounced than they are in the data sequences.

8.2.2   The Poissonness plot

One of the points noted in Sec. 8.3 is that Q-Q plots can be constructed for essentially any continuous reference distribution. The following example illustrates that this idea can be extended to count distributions as well, an idea discussed in more detail in Sec. 8.4. Since the Poisson distribution plays a role for count data that is, in many respects, analogous to that of the Gaussian distribution for continuous data, it is appropriate to demonstrate this idea for the Poisson distribution. Further, since the Prussian army horse-kick data introduced in Chapter 2 played such a significant role in the history of the Poisson distribution, it seems only reasonable to demonstrate this informal graphical test of the Poisson hypothesis for this dataset.

A summary of the total number of horse-kick deaths in the 14 Prussian army corps considered by von Bortkewitsch is given in Table 8.1. By themselves, these numbers do not tell us much, other than to confirm first, that these events ocurred and second, that the number of such occurrances was not large. The Poissonness plots defined and described in Sec. 8.4.1 are much more informative in this regard. Like the Q-Q plots discussed in the previous example, the Poissonness plot represents an informal test of the hypothesis that a collection of counts like that listed in Table 8.1 is reasonably well-described by the Poisson distribution.

Figure 8.2: Poissonness plot for the number of horse-kick deaths in 14 Prussian army corps from 1875 to 1894

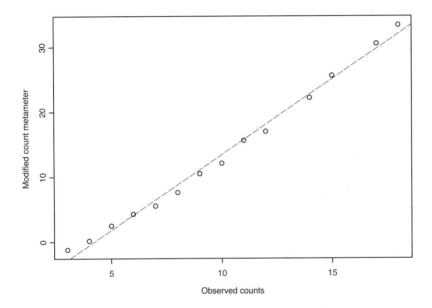

A Poissonness plot for the total number of horse-kick deaths observed during the 20 years from 1875 to 1894 is shown in Fig. 8.2. This plot also includes a reference line to help judge the linearity of the points, and since all of the points in Fig. 8.2 lie quite near this reference line, it is clear that these observations conform reasonably well to the Poisson distribution. In fact, the slope of the line corresponds to the natural logarithm of the distribution parameter λ, and the intercept is equal to $-\lambda$, providing both a possible basis for estimating λ and a basis for assessing the reasonableness of the Poisson distribution. More important, however, significant departures from linearity in this plot would suggest the need for an alternative distribution to describe the data, a point illustrated in Sec. 8.4.1.

8.2.3   Nonparametric density estimators

Most data analysis problems make an implicit assumption of *homogeneity*, i.e., that the entire dataset may be described by *one* underlying relationship. Although this assumption is quite common in practice and is frequently adequate, there are at least two circumstances in which it is not. The first is the problem of outliers discussed in Chapter 7, which typically corresponds to a highly asymmetric two-cluster data model: most of the data values exhibit the nomi-

Figure 8.3: Q-Q plots and density estimates for the *Old Faithful* dataset

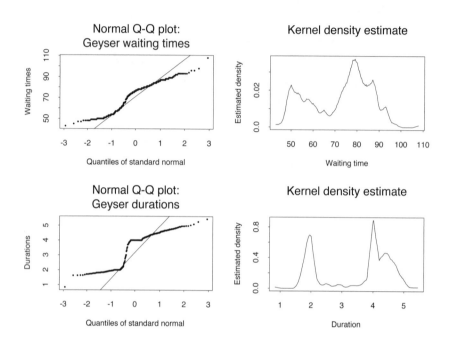

nal relationship implied by the assumption of homogeneity, but a small fraction is anomalous and violates this nominal relationship. This case motivates the use of robust analysis procedures, a point discussed throughout this book in connection with different types of data analysis. In these problems, the issue of primary interest is generally that of characterizing the nominal relationship that holds for most of the data without undue influence from the anomalous elements in the dataset. The second circumstance under which the assumption of homogeneity fails is that of datasets generated by multiple mechanisms, all of which may be of equal interest. In these cases, we are faced with two important tasks: that of detecting the heterogeneity present in the dataset, and that of modeling it. A very useful class of models for heterogeneous data is the class of *mixture distributions* introduced in Chapter 10, and methods that are sometimes useful in detecting significant data heterogeneity are the nonparametric density estimation procedures introduced in Sec. 8.6.

The following example illustrates this point. One of the standard datasets provided with either the *R* or *S-Plus* implementations of the *S* programming language is the Old Faithful geyser eruption data. Two characterizations of this dataset are shown in the four plots in Fig. 8.3. Specifically, this figure shows Q-Q plots (left-hand plots) and nonparametric density estimates (right-hand plots) for sequences of observed waiting times between eruptions (upper plots) and the durations of those eruptions (lower plots) for the Old Faithful geyser in

Yellowstone National Park in the western United States. A characteristic feature of both Q-Q plots is the apparent discontinuity near the center of the plot, particularly pronounced in the Q-Q plot for the durations shown in the lower left. The source of these discontinuities is the bimodal behavior of these two distributions, seen clearly in the kernel density estimates shown in the right-hand two plots. These plots, which may be regarded as smoothed histograms, are estimates of the probability densities of the two data sequences. The presence of two clear, well-separated peaks in these plots provides strong evidence of bimodality, again especially for the geyser duration data shown in the lower right plot. A detailed discussion of the kernel density estimation methods used to generate the right-hand pair of plots is given in Sec. 8.6.

8.3 Quantile-quantile plots

This section describes the construction of the Q-Q plots introduced in Chapter 1 and discussed further in secs. 8.2.1 and 8.2.3. As noted in these discussions, the purpose of a Q-Q plot is to give a graphical comparison of an observed data sequence $\{x_k\}$ with some standard reference distribution. While *normal Q-Q plots* based on the ubiquitous Gaussian distribution are the most commonly seen in practice, these plots can be constructed for any reference distribution for which quantiles can be constructed. In fact, they are most simply constructed for the uniform reference distribution, and this case is considered first in the following discussion.

8.3.1 The basic idea

Given a sequence $\{x_k\}$ of N real-valued data observations, the basic construction of a Q-Q plot involves the following three steps:

1. rank-order the observed data sequence $\{x_k\}$ to obtain the sequence $\{x_{(i)}\}$,
2. under the assumption that $\{x_k\}$ was drawn from a specified reference distribution, compute the *approximate quantiles* q_i, corresponding to the values $\{x_{(i)}\}$ *should* have,
3. plot $\{x_{(i)}\}$ vs. $\{q_i\}$ and see whether the result is a straight line.

The first step in this sequence is always the same and is independent of the reference distribution considered, whereas the second step depends on the reference distribution but is independent of the data. As noted, the most common choice of reference is the Gaussian distribution, but the basic construction is much more general, and Q-Q plots may be used to assess the reasonableness of many of the other distributions described in Chapter 4. In fact, the simplest case is the uniform Q-Q plot, described next.

For the uniform distribution, a reasonable choice of q_i for large N is simply the expected value of the uniform order statistic $u_{(i)}$, given in Chapter 6 as:

$$E\{u_{(i)}\} = \frac{i}{N+1} \equiv p_i. \tag{8.1}$$

Hence, a *uniform Q-Q plot* is simply a plot of $x_{(i)}$ vs. p_i, and significant departures from linearity in this plot may be taken as evidence *against* the hypothesis that $\{x_k\}$ is uniformly distributed. This emphasis is important because it is in general not possible to *accept* a statistical hypothesis, but only to *reject* one by arguing that the observed data sequence is clearly inconsistent with it. Conversely, in practice we routinely adopt *provisional acceptance* of a hypothesis if tests like the one described here do not provide us a basis for rejection. These ideas form the basis for the theory of formal hypothesis testing and are discussed in more detail in Chapter 9. There, specific numerical criteria are developed and used to either formally reject a specified *null hypothesis* or to declare the results "nonsignificant," meaning we do not have sufficient evidence to reject the null hypothesis at a stated significance level. In contrast, graphical techniques like the Q-Q plots described here may be regarded as *informal hypothesis tests* since they provide evidence in support of or against a given working hypothesis but are not generally used as the basis for formal hypothesis tests of the kind just described.

To illustrate the utility of these graphical characterizations, Fig. 8.4 compares four uniform Q-Q plots. The upper left plot was obtained from 500 samples, uniformly distributed on the interval $[0, 1]$. Specifically, this plot shows the rank-ordered data $\{u_{(i)}\}$ plotted against the quantiles $p_i = i/(N + 1)$. As expected, this plot conforms quite well to a straight line with slope $+1$ and y-axis intercept 0. For comparison, the upper right plot shows the corresponding uniform Q-Q plot obtained from a sequence of 500 samples uniformly distributed on the interval $[-3, 3]$. Again, the plot is well approximated by a straight line, but here the slope is $+6$ and the y-axis intercept is -3. This difference corresponds to the fact that if x is uniformly distributed on $[-3, 3]$, it may be written as $x = 6u - 3$, where u is uniformly distributed on $[0, 1]$; this point is important and is discussed further in the next section in connection with general Q-Q plots. The lower left plot shows the uniform Q-Q plot obtained from the first 50 samples of the 500-point sequence of uniform random variables shown in the upper left plot. The difference in the appearance of these plots reflects the influence of sample size, illustrating that—in common with many other data analysis procedures—Q-Q plots yield better results with larger samples than with smaller ones, another point that is discussed further (see Sec. 8.3.3). Finally, the lower right plot shows the Q-Q plot obtained for 500 normally distributed samples: the systematic, symmetric deviation from linearity at both ends of this plot reflects the nonuniformity of the data distribution in this case.

In practice, it is common to modify the reference quantiles $p_i = i/(N + 1)$ to obtain plots that are somewhat better behaved for small sample sizes N and to obtain more reliable behavior near the sample extremes. Specifically, the following modification is quite common [76]:

$$q_i = \frac{i - c}{N - 2c + 1}, \tag{8.2}$$

for some constant c between 0 and 1. Particularly popular choices are $c = 0$ (for which $q_i = p_i = i/(N + 1)$), $c = 1/3$ [141, p. 435], and $c = 0.5$ [76, 133, 272].

Figure 8.4: Four examples of uniform Q-Q plots

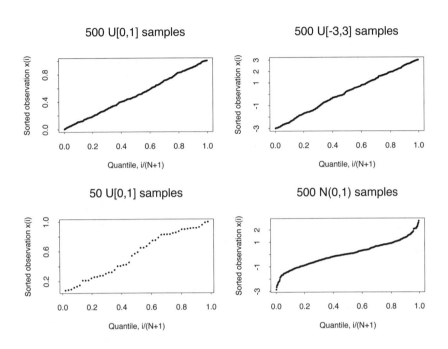

Note that these approximations only differ significantly either for small sample sizes (i.e., N small) or near the ends of the distribution (i.e., i near 1 or N).

8.3.2 The general construction

Many—although not all—cumulative distribution functions $F(\cdot)$ are continuous, strictly increasing functions from some set $\mathcal{S}$ onto the interval $[0, 1]$. Consequently, it follows [174, p. 181] that $F^{-1}(\cdot)$ exists, and from the general transformation results presented in Chapter 12, the transformed uniform random variable $z = F^{-1}(u)$ has the cumulative distribution function $F(\cdot)$. *Conversely, it follows that the transformed variable $u = F(z)$ is uniformly distributed on* $[0, 1]$. Consequently, we could—in principle—apply the uniform Q-Q plots just described to this transformed data sequence, plotting $u_{(i)}$ against reference quantiles q_i and checking to see whether the result conforms reasonably to a straight line. Alternatively, since the function $F(\cdot)$ and its inverse are both strictly increasing in the case considered here, note that the transformations $u = F(z)$ and $z = F^{-1}(u)$ are both order preserving. Hence, it follows that:

$$z_{(i)} = [F^{-1}(u)]_{(i)} = F^{-1}(u_{(i)}) \simeq F^{-1}(q_i). \tag{8.3}$$

This observation provides the basis for constructing Q-Q plots for any reference distribution with a continuous, strictly increasing CDF: simply plot the sorted data values $z_{(i)}$ against the reference quantiles $F^{-1}(q_i)$ with q_i given by Eq. (8.2). If the chosen reference distribution is reasonable for the data sequence $\{z_k\}$, the resulting plot will be approximately linear with unit slope and a vertical axis intercept of 0.

Many distributions define *location-scale* families whose *shape* is specified by some reference function $F(z)$ and whose individual members are obtained by linear transformations $x = x_0 + \sigma z$, yielding the cumulative distribution function:

$$G(x) = F(z) = F\left(\frac{x - x_0}{\sigma}\right). \qquad (8.4)$$

Here, x_0 describes the translation of the center of the distribution and σ is a positive scale factor that varies the spread. In such cases, the rank-ordered data $\{x_{(i)}\}$ will be described by $x_{(i)} = x_0 + \sigma z_{(i)}$ where $z_{(i)}$ is approximated by Eq. (8.3). Hence, if the shape of the reference distribution is correct but the center and width of the data distribution are not the same as the reference distribution, the plot just described will still yield approximately a straight line, but the slope and intercept of this line will now reflect the center and width corrections. Recall that this behavior is precisely that seen in the upper right plot in Fig. 8.4 with $x_0 = -3$ and $\sigma = 6$.

The procedure just described permits the construction of Q-Q plots for any reference distribution $G(\cdot)$ defined by a location-scale family via Eq. (8.4) for some known parent distribution $F(\cdot)$. These plots are extremely useful exploratory data analysis tools, offering the following kinds of insights:

- visual assessment of the appropriateness of the reference distribution,
- visual evidence of the number and severity of outliers in the data,
- qualitative assessment of *systematic* differences between the data distribution and the reference distribution (e.g., asymmetry or elongation).

As discussed in Chapter 4, the normal distribution is often regarded as a natural choice for data sequences that can assume any real value, so the most popular Q-Q plots are those constructed for the normal distribution to assess this working hypothesis. In cases where data values cannot be negative, inherently nonnegative densities are sometimes required. One candidate is the lognormal distribution and construction of lognormal Q-Q plots is extremely straightforward: recall that if x is lognormally distributed, then $y = \ln x$ is normally distributed. Hence, a lognormal Q-Q plot may be obtained by simply plotting $y_{(i)} = \ln x_{(i)}$ vs. the normal quantiles used to construct the normal Q-Q plot. Other inherently nonnegative distributions include the gamma distribution, the Pareto distribution, and the Weibull distribution. Both the gamma and Weibull distributions involve shape parameters that must be specified in order to compute the quantiles required in the construction of the Q-Q plot. Similar observations apply to the beta distribution, a logical choice for data sequences that are necessarily restricted to a bounded interval (e.g., concentrations or other variables that must lie between 0 and 1). For a more detailed discussion of Q-Q plots for different distributions, see the book by Fowlkes [105].

Essentially, Q-Q plots are a graphical tool used to examine distributional hypotheses through the use of an expected relationship of the form $x_{(i)} \simeq \mu + \sigma \rho_i$ where $x_{(i)}$ is computed simply from available data and ρ_i is computed from known characteristics of the reference distribution. *Reference lines* are an extremely useful addition to Q-Q plots, providing visual assistance in determining the degree to which these expectations are or are not met. In addition, the construction of reference lines for Q-Q plots provides a nice illustration of the practical problem discussed in Chapter 5 of selecting a line-fitting method. Specifically, the following two observations are relavent here:

1. any outliers present in the original data sequence $\{x_k\}$ are more likely to influence the order statistics $x_{(i)}$ for $i \sim 1$ and $i \sim N$ than for $i \sim N/2$;
2. the reference values ρ_i are *computed* and thus free of measurement errors.

The consequence of the second of these observations is that we need not consider an errors-in-variables formulation of this problem since the reference values may be assumed to be known with substantially greater accuracy than the rank-ordered data values, which are subject to observation errors. Similarly, it follows from the first observation that we are unlikely to require the use of highly robust fitting methods if we restrict consideration to the middle portion of the data where outliers are unlikely.

Combining these observations leads to the following general procedure for constructing reference lines for Q-Q plots, for any reference distribution:

1. restrict consideration to the middle 50% of the data, $x_{(m)}$ through $x_{(n)}$ where $m \simeq N/4$ and $n \simeq 3N/4$;
2. using ordinary least squares, fit this data to the quantiles q_m through q_n.

This procedure is simple to implement and tends to be highly resistant to outliers in the data since it excludes both the upper 25% and the lower 25% of the data values. This procedure is essentially that implemented in the *S-Plus* statistical software package for normal Q-Q plots and is the basis for the reference lines shown in the Q-Q plots presented in the remainder of this chapter.

8.3.3   Normal Q-Q plots

As noted in the preceeding discussion, the most popular Q-Q plot is undoubtedly the normal Q-Q plot, obtained by specializing the general prescription described above to the standard normal reference distribution. Construction of this distribution follows the general procedure outlined previously, specializing the second step to the Gaussian distribution:

1. rank-order the observed data sequence $\{x_k\}$ to obtain the sequence $\{x_{(i)}\}$,
2. compute the approximate normal quantiles q_i,
3. plot $\{x_{(i)}\}$ vs. $\{q_i\}$ and see whether the result is a straight line.

Since the Gaussian distribution defines a location-scale family with location parameter μ and scale parameter σ, it follows from the results presented in

Figure 8.5: Normal Q-Q plots vs. sequence length N

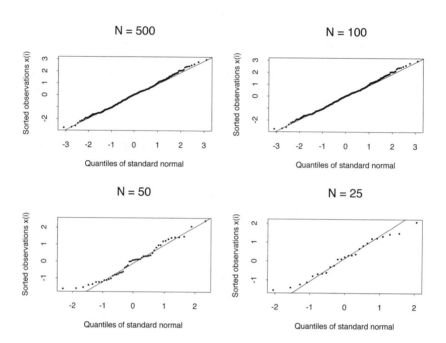

Sec. 8.3.2 that if $\{x_k\}$ is normally distributed, the normal Q-Q plot defined here will have positive slope σ and a y-axis intercept at μ. The approximate quantiles required for this plot are computed as:

$$q_i \;=\; \Phi^{-1}\left(\frac{i-c}{i-2c+1}\right), \tag{8.5}$$

typically with $c = 0.5$. Here, $\Phi(x)$ represents the cumulative distribution function for the Gaussian distribution with mean $\mu = 0$ and standard deviation $\sigma = 1$. Both this function and its inverse are readily available both in tables [1, p. 976] and in various software packages [272, 297].

Four examples of normal Q-Q plots are shown in Fig. 8.5, constructed from four Gaussian random sequences of differing lengths but all with mean $\mu = 0$ and standard deviation $\sigma = 1$. The upper left plot represents a moderately long sequence ($N = 500$), and here, the Q-Q plot conforms quite well to our expectations: the plot is quite linear, with unit slope and zero intercept. As with the uniform Q-Q plots considered in Sec. 8.3.1, normal Q-Q plots work better for larger samples than for smaller ones, which exhibit greater deviations from linearity, particularly in the tails of the distribution (i.e., near both extremes of the plot). This point is illustrated by the differences between the four plots shown here, corresponding to $N = 500$ (upper left), $N = 100$ (upper right), $N = 50$ (lower left), and $N = 25$ (lower right). In fact, D'Agostino and Stephens

Figure 8.6: Normal Q-Q plots for t_2-distributed data

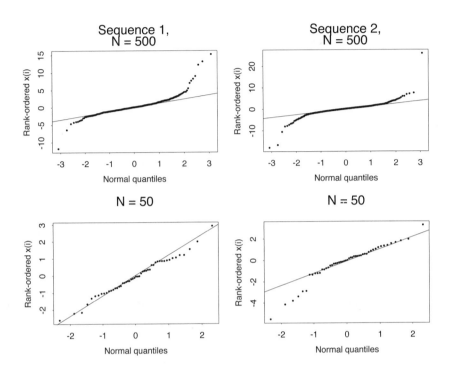

[76] note that normal Q-Q plots for samples smaller than $N = 25$ can appear to exhibit significant departures from linearity even when the data sequence is normally distributed, citing Daniel and Wood [78] and Hahn and Shapiro [133] for specific examples and concluding with the advice:

> We caution the reader against placing too much reliance upon a plot in these situations.

To illustrate the general behavior of normal Q-Q plots, it is instructive to consider the results obtained for various other data distributions. An extensive collection of such plots has been published [105], and the following examples make no attempt to duplicate that collection, but it is instructive to have a few examples available. Fig. 8.6 shows normal Q-Q plots for four data sequences having the Student's t-distribution discussed in Chapter 4, with 2 degrees of freedom. This choice results in an extremely heavy-tailed distribution that, like the Cauchy distribution (corresponding to the Student's t-distribution with 1 degree of freedom), exhibits infinite variance. The upper two plots characterize two independently generated sequences, each of length $N = 500$. The symmetric, heavy-tailed nature of these distributions is evident from the pronounced and roughly equal departure above the reference line at the right end of the plot

Figure 8.7: Normal Q-Q plots for gamma distributed data

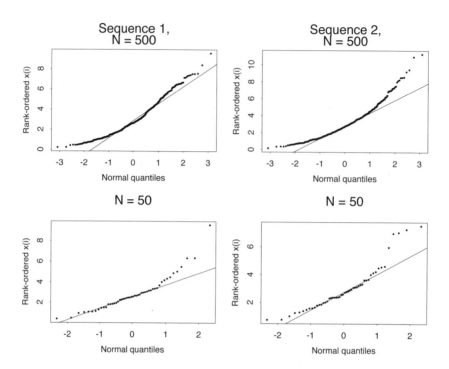

and below the reference line at the left end of the plot. Conversely, this distributional character is much less apparent in the lower two plots, corresponding to smaller samples of size $N = 50$. In particular, the lower left plot appears quite linear, offering no evidence at all of the pronounced non-Gaussian nature of this data sequence. Similarly, the lower right plot exhibits significant deviations from linearity only in the lower tail (i.e., the left end of the plot). As the next example illustrates, this behavior is typical of asymmetric data distributions. Taken together, both of these bottom two plots re-emphasize the caution quoted above [76] about the difficulty of interpreting Q-Q plots for small samples.

To illustrate the behavior of the normal Q-Q plot for asymmetrically distributed data, consider the gamma distribution with shape parameter $\alpha = 3$. Four Q-Q plots constructed from sequences drawn from this distribution are shown in Fig. 8.7, again for $N = 500$ (the upper two plots) and $N = 50$ (the lower two plots). The asymmetry of this distribution is reflected by the fact that both the upper tail (i.e., the right end of the plot) and the lower tail (the left end of the plot) lie *above* the Gaussian reference line. In fact, here, this asymmetry is apparent even in the small sample plots, in marked contrast to the previous example. Conversely, since outliers appear as isolated points in the tails of the distribution, the points that provide evidence of distributional

Figure 8.8: Normal Q-Q plots for arcsin distributed data

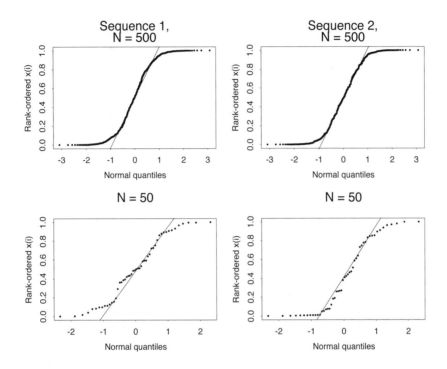

asymmetry in the smaller samples might instead be regarded as outliers. This is particularly the case in the lower right plot, where the five points in the upper tail (representing 10% of the data) might well be regarded as a group of outliers.

Finally, Fig. 8.8 shows the normal Q-Q plots obtained from four data samples drawn from the symmetric beta distribution with common shape parameter $p = q = 1/2$, known as the *arcsin distribution*. Again, the format of these plots is the same as before: the upper plots correspond to two independently generated samples of length $N = 500$, and the lower plots correspond to two independently generated samples of length $N = 50$. Recall from the discussion of beta distributions given in Chapter 4 that this density function is U-shaped, exhibiting modes at both limiting values $x = 0$ and $x = 1$. More generally, any symmetric beta distribution with $p = q < 1$ may be regarded as significantly light-tailed, in contrast to the heavy-tailed Student's t-distributions considered earlier. As a consequence, these light-tailed distributions tend to exhibit the behavior seen clearly in all four of the Q-Q plots shown in Fig. 8.8. In particular, note the steep slope in the center of the plot, separating two relatively flat tails. In contrast to the heavy-tailed sample considered previously, note that here the upper tail lies below the reference line and the lower tail lies above it. Also, the symmetry of these plots reflects the symmetry of the underlyng distribution.

8.3.4   Data comparison plots

The Q-Q plots described in the previous sections have all been constructed by first, rank-ordering an observed data sequence $\{x_k\}$ to obtain $\{x_{(i)}\}$, and then, plotting this rank-ordered sequence against computed reference quantiles. This basic idea extends immediately to an even simpler procedure for comparing two data sequences, $\{x_k\}$ and $\{y_k\}$. In particular, it is often insightful to ask whether two sequences that in some sense "should have the same distribution" really do. Note that if this assumption is true, then the plot of $\{y_{(i)}\}$ vs. $\{x_{(i)}\}$ should also be linear, regardless of what the underlying common distribution is. That is, suppose the following approximations hold:

$$x_{(i)} \simeq x_0 + \sigma_x z_{(i)}, \text{ and } y_{(i)} \simeq y_0 + \sigma_y z_{(i)}, \tag{8.6}$$

relating the rank-ordered samples of x and y to those of an unknown but common reference distribution z. Eliminating $z_{(i)}$ from these equations yields the following approximate relationship between the rank-ordered observations of x and y:

$$y_{(i)} \simeq \left[y_0 - \left(\frac{\sigma_y}{\sigma_x} \right) x_0 \right] + \left(\frac{\sigma_y}{\sigma_x} \right) x_{(i)}. \tag{8.7}$$

This idea is illustrated with four examples in Fig. 8.9, all based on the Pima Indians dataset introduced in Chapter 2. Recall that the serum insulin value (INS) was recorded as zero for almost half of the patients in the dataset, representing missing data observations (specifically, INS was missing for 374 of 768 patients). An important question is whether those patients with missing INS values differ systematically in other respects from those with nonmissing INS values, and this example illustrates that data comparison plots can be useful in exploring this question. Specifically, each plot in Fig. 8.9 compares the distribution of values for one of the other variables included in the Pima Indians dataset between the subset of patients with missing INS values (vertical axis) and those with nonmissing INS values (horizontal axis). As a specific example, the upper left plot compares the age distributions for the missing INS and nonmissing INS data subsets, together with a reference line corresponding to equality of the distributions. It is clear from this plot that the patients with missing INS values were generally somewhat older than those with nonmissing values since the rank-ordered AGE values for the missing INS subset fall consistently above the reference line. In contrast, the upper right plot shows that most of the BMI values exhibit essentially the same distribution, except for those in the upper and lower tails of the distribution. The upper tail here corresponds to BMI outliers, which are more prevalent among the patients with nonmissing INS values than among those with missing INS values, while the lower tail corresponds to missing BMI values, also coded as zero, which are more prevalent among patients with missing INS values. This same lower tail difference is seen in the plasma glucose data (PGL), shown in the lower left plot. Finally, the lower right plot in Fig. 8.9 shows the corresponding data comparison plot for the number of times pregnant (NPG). The fact that the points in this plot all lie fairly close to the reference

Figure 8.9: Comparisons of data sequences defined by four different variables from the Pima Indians dataset, comparing cases where INS is recorded as zero vs. those where INS is nonzero.

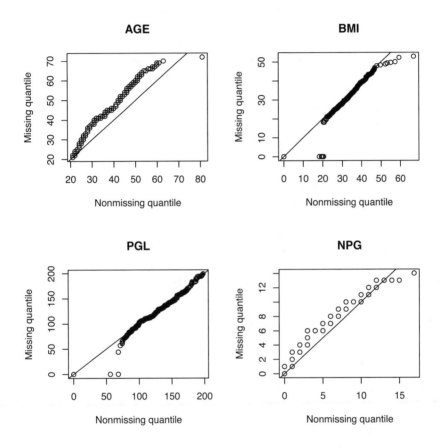

line suggest that there are no marked distributional differences for this variable between the patients with missing INS values and those with nonmissing INS values. The much coarser appearance of this plot relative to the other three in Fig. 8.9 reflects the fact that NPG is a discrete variable (i.e., a count variable taking values between 0 and 14). As a practical matter, data comparison plots are of somewhat limited utility for discretely distributed variables like this one; better alternatives are the Poissonness plot and its extensions described in the next section.

An important aspect of the four examples just described is that, in all cases, the two data sequences being compared were of different lengths. As a practical matter, this means that the data comparison plot construction procedure described at the beginning of this section must be modified slightly to account for this difference. One possible approach is to simply truncate the longer data sequence to the length of the shorter sequence, but a better alternative—and the one adopted in constructing the plots shown in Fig. 8.9—is to form a random subsample of the longer data sequence, of the same length as the shorter sequence. This approach reduces the influence of any unsuspected serial correlations or other order-related effects that may be present in the data.

8.4   Plots for discrete distributions

The statistical analysis of the *Federalist Papers* was discussed briefly in Chapter 2 as a representative example to illustrate how count data analysis problems frequently arise in practice. It was also noted in Chapter 2 both that the Poisson distribution is often adopted at least as an initial working assumption in dealing with counts of "rare events" and that it is not always a completely adequate description of these events. The following discussions introduce two useful tools in informally assessing these assumptions. The first, previously introduced in Sec. 8.2.2 but described in greater detail in Sec. 8.4.1, is the *Poissonness plot*, a graphical procedure for examining the reasonableness of the Poisson distribution as an approximate description of a collection of count data. The second, introduced in Sec. 8.4.2, is the analogous plot for what is probably the most common alternative to the Poisson distribution for count data, the negative binomial distribution introduced in Chapter 2. Since Sec. 8.2.2 demonstrated the Poissonness plot for the Prussian horse-kick example from Chapter 2, often used to introduce the Poisson distribution, both of the following discussions focus primarily on the *Federalist Papers* example, where the Poisson distribution was found to be inadequate.

8.4.1   Poissonness plots

As with the Q-Q plots discussed previously for continuous distributions, the fundamental idea behind the Poissonness plot is to construct a sequence of number pairs from the available data that *should*, if the Poisson distributional assumption is reasonable, correspond approximately to a straight line. In the case of count data, assume that n_k represents the total number of times the count k occurs (e.g., the number of years for which k Prussian calvery soldiers are killed by horse kicks), and let N denote the total of these counts:

$$N = \sum_{k=0}^{K} n_k, \tag{8.8}$$

where K is the maximum count observed in the sequence. The basis for what follows is the observation that the simplest estimator of the probability p_k of observing the count k in this sequence is the relative frequency:

$$\hat{p}_k = \frac{n_k}{N}. \tag{8.9}$$

The basic idea behind the Poissonness plot is to assess the degree of agreement between the sequence $\{\hat{p}_k\}$ of relative frequencies defined in Eq. (8.9) and their values under the Poisson distribution, which are:

$$p_k = \frac{exp(-\lambda k)\lambda^k}{k!}. \tag{8.10}$$

As discussed in detail in Chapter 12, one way of fitting a sequence of numbers like $\{\hat{p}_k\}$ to a mathematical model like that defined in Eq. (8.10) is to apply a linearizing transformation. In this case, taking the logarithm of both probabilities yields the approximation:

$$\ln \hat{p}_k \simeq k \ln \lambda - \lambda - \ln k! \tag{8.11}$$

where λ is the (unknown) Poisson distribution parameter. Next, define the *Poissonness count metameter* $\phi(k)$ as:

$$\phi(k) = \ln \hat{p}_k + \ln k! = \ln n_k - \ln N + \ln \Gamma(k+1), \tag{8.12}$$

where $\Gamma(\cdot)$ is the gamma function [1, 162, 164]. The advantage of writing $\phi(k)$ in terms of the gamma function is that convenient computational approximations are available for the log of the gamma function [1]. In particular, note that the S language has a built-in function **lgamma** that evaluates $\ln \Gamma(\cdot)$. Combining Eqs. (8.10) and (8.11) yields the basis for the Poissonness plot, which is a plot of the count metameter $\phi(k)$ against the count value k:

$$\phi(n_k) \simeq (\ln \lambda)k - \lambda. \tag{8.13}$$

Thus, if the counts $\{n_k\}$ are consistent with a Poisson distribution, a plot of $\phi(k)$ vs. k should yield a straight line with slope $\ln \lambda$ and an intercept $-\lambda$. The Prussian horse-kick data discussed in Sec. 8.2.2 provided a useful illustration of this idea: there, the count metameter $\phi(k)$ exhibited a very linear variation with k over the range of observed counts.

The discussion of the *Federalist Papers* presented in Chapter 2 noted that, despite initial expectations that the occurrences of specific words like "from" or "may" in these documents would exhibit a Poisson distribution, they did not. The Poissonness plot just described illustrates this point clearly for counts of the word "from" in documents known to have been written by either Hamilton or Madison. Specifically, Fig. 8.10 shows the Poissonness plots constructed from these two count sequences, representing the results for Hamilton as solid triangles and those for Madison as open circles. While these two plots exhibit visually obvious differences that may imply significant differences in distribution between the two authors, neither of these plots conform well to a straight line.

In their discussion of Poissonness plots, Hoaglin *et al.* [141, ch. 9] note that $\phi(k)$ tends to be highly variable when n_k is small. For this reason, counts k that only occur once (i.e., values of k for which $n_k = 1$) are maximally unreliable, and they recommend plotting these points with a different symbol to emphasize their special nature. Fig. 8.11 shows a Poissonness plot constructed this way

Figure 8.10: Poissonness plots for counts of the word "from" in documents written by Hamilton (solid triangles) and those written by Madison (open circles)

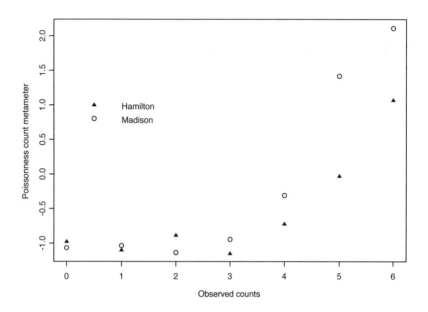

for counts of the word "may" in documents known to have been written by Madison. There, "frequent counts" corresponding to cases with $n_k > 1$ are plotted as solid circles, while those for which $n_k = 1$ are represented as open circles; in this particular example, $n_k = 1$ only for $k = 5$ and 6. To illustrate the influence of these point, two reference lines are shown in Fig. 8.11. The dotted line with positive slope was obtained from a least squares fit to all of the $\phi(k)$ values, while the dashed line with negative slope was obtained when the case $k = 5$ and $k = 6$ were omitted. The problem of assessing the quality of Poissonness plots will be revisited in Chapter 9, where approximate confidence intervals are constructed for $\phi(k)$. Here, the main point is that Fig. 8.11 provides reasonably clear evidence against the Poisson working hypothesis for this data sequence. The following discussion describes some techniques that can be useful in seeking alternative distributional assumptions for count data.

8.4.2   Negative binomialness plots

The negative binomial distribution was introduced in Chapter 2, where it was noted that it is often used as an alternative to the Poisson distribution for count data that exhibits *overdispersion*. That is, since the Poisson distribution is fully specified by the single parameter λ in Eq. (8.10), this parameter determines

Figure 8.11: Poissonness plot for counts of the word "may" in documents written by Madison. Counts k for which $n_k = 1$ are shown as open circles, while counts for which $n_k > 1$ ("frequent counts") are shown as solid circles

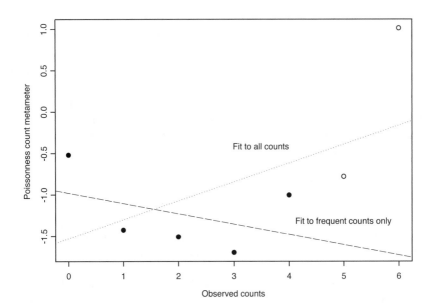

both the mean and the variance of Poisson counts; in fact, these two values are equal. In cases of overdispersion, the variance is too large relative to the mean to be compatible with the Poisson model, and an alternative distribution is required. The negative binomial distribution has two parameters, and the variance is always larger than the mean for this distribution, so it is a natural alternative for overdispersed count data.

The negative binomial distribution can be parameterized many different ways, and here, it will be convenient to parameterize it as:

$$p_k = \binom{r+k-1}{r-1}(1-p)^k p^r. \qquad (8.14)$$

Here, $p = (1+P)^{-1}$ in terms of the notation introduced in Chapter 2, and this parameter is restricted to the range $0 < p < 1$; the second parameter can be any positive (real) number: $r > 0$. In terms of the parameters p and r, the mean (m_1) and variance (μ_2) are given by:

$$m_1 = \frac{(1-p)r}{p}, \quad \text{and} \quad \mu_2 = \frac{(1-p)r}{p^2} = \frac{m_1}{p}. \qquad (8.15)$$

Since $p < 1$, it follows from this last expression that $\mu_2 > m_1$.

Probably the simplest graphical test for negative binomialness is Ord's plot, constructed as follows [164, p. 214]. It is easily shown (Exercise 1) that the negative binomial distribution satisfies:

$$\frac{kp_k}{p_{k-1}} = (1-p)k + (1-p)(r-1). \tag{8.16}$$

Replacing p_k with the frequency estimates $\hat{p}_k = n_k/N$ in this expression yields the basis for the *Ord plot*, which plots the *Ord metameter*, u_k, against k, where:

$$u_k = \frac{k\hat{p}_k}{\hat{p}_{k-1}} = \frac{kn_k}{n_{k-1}}, \tag{8.17}$$

for $k = 1, 2, \ldots, K$, where K is the largest count observed. If the negative binomial distribution is appropriate, this plot should approximate a straight line with a slope determined by the distributional parameter p and an intercept determined by both p and r. In fact, closely related results hold for other distributions, including the Poisson, where the resulting line has *zero slope:*

$$\frac{kp_k}{p_{k-1}} = \lambda. \tag{8.18}$$

Also, in contrast to the negative binomialness plot discussed below, the Ord plot does not require specification of the distribution parameter r. Unfortunately, despite these two very attractive features—i.e., applicability to multiple distributions and independence of the r parameter in the negative binomial case—the Ord plot tends to be extremely variable. This point is illustrated in the following example, and it represents the main reason that Hoaglin *et al.* generally do not recommend it, other than as a crude assessment tool [141, pp. 393–396].

Fig. 8.12 shows four Ord plots to illustrate their general behavior. The upper left plot is based on the Prussian horse-kick data discussed in Sec. 8.2.2, where it was used to introduce the Poissonness plot in view of its historically important connection with the Poisson distribution, discussed in Chapter 2. From the preceeding discussion, we expect the best-fit straight line to the corresponding Ord plot to have zero slope. The least squares fit to the Ord plot data is also shown in the upper left plot in Fig. 8.12, where it appears to have a slight positive slope but the scatter of the data around this line is large enough to call the quality of fit into question. This example is revisited in Chapter 9 in connection with the issue of significance characterization: when is the slope of a line large enough to be considered nonzero? The point here is that, in contrast to the Poissonness plot constructed in Sec. 8.2.2 for this example, the Ord plot for this example in Fig. 8.12 is *not* "strongly suggestive of a Poisson model."

The other three plots in Fig. 8.12 are Ord plots for three *Federalist Papers* count sequences. The upper right plot was constructed from counts of the word "may" in documents known to have been written by Madison, and the lower two plots were constructed from counts of the word "from" in documents known to have been written by Hamilton (lower left) and Madison (lower right). The least squares lines fit to these three Ord plots are also shown in each case, and

Figure 8.12: Four examples of Ord's plot, designed to assess either Poissonness or negative binomialness of count data sequences

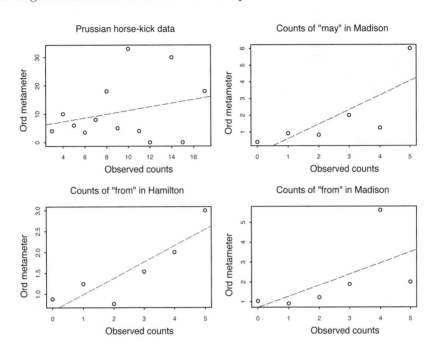

all of these lines are seen to have a positive slope—consistent with a negative binomial distribution—but the scatter around these lines is again large enough to be troubling.

As with the Poisson distribution, fitting the observed frequencies $\{\hat{p}_k\}$ to the probabilities $\{p_k\}$ defined by the negative binomial distribution provides a basis for constructing a useful graphical diagnostic. Again, the basic idea is to construct a plot of $\ln \hat{p}_k$ vs. $\ln p_k$ and show that the result should be a straight line. Taking the logarithm of Eq. (8.14) gives (Exercise 2):

$$\ln p_k = r \ln p + k \ln(1 - p) + \ln \Gamma(r + k) - \ln \Gamma(r - 1) - \ln \Gamma(k + 1). \quad (8.19)$$

Substituting $\hat{p}_k$ for p_k in this expression and rearranging motivates the following definition for the negative binomial count metameter:

$$\phi(k) = \ln \hat{p}_k + \ln \Gamma(r + k) - \ln \Gamma(r) - \ln \Gamma(k + 1). \quad (8.20)$$

Note that *for given* r, we can compute $\phi(k)$ from the count data n_k, and it follows from Eqs. (8.19) and (8.20) that it should approximately satisfy:

$$\phi(k) \simeq [\ln(1 - p)]k + r \ln p. \quad (8.21)$$

The fact that we must specify r complicates the use of the negative binomialness plot, but as the following example illustrates, it can still be extremely useful.

Figure 8.13: Four negative binomialness plots for the counts of the word "from" in documents known to have been written by Hamilton. Values for the distributional parameter r are indicated above each plot

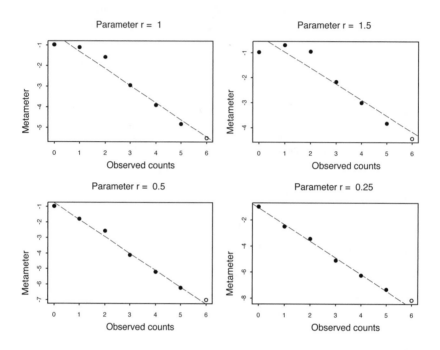

Fig. 8.13 shows four negative binomialness plots constructed from counts of the word "from" in documents known to have been written by Hamilton, for different values of the parameter r. In all cases, the value of r is indicated above the plot, and reference lines obtained from a least squares fit of $\phi(k)$ to k are shown in each plot. Comparing these plots with their reference lines, it appears that the upper two plots represent poorer fits than do the lower two plots, and these results are borne out by the standard deviations σ_r of the corresponding residuals sequences. These values provide a useful measure of the quality of the fit of these lines, a point discussed in detail in Chapter 15. The point here is that the smaller the standard deviations σ_r the better the fit, and here these values are:

$\sigma_r = 0.361$ for $r = 1$;

$\sigma_r = 0.455$ for $r = 1.5$;

$\sigma_r = 0.203$ for $r = 0.5$;

$\sigma_r = 0.156$ for $r = 0.25$.

Informally, then, these plots support the idea that, at least for the word "from" in documents written by Hamilton, the negative binomial distribution is a more reasonable model for the data than the Poisson model. Further, these plots suggest a distribution parameter of $r \sim 0.25$; in particular, the negative binomialness plot for $r = 0.10$ (not shown) gives a residual standard deviation of $\sigma_r = 0.362$. Conversely, note that a variety of methods exist for directly estimating the distribution parameters r and p from the count data [164, pp. 215–220], and these methods represent a better alternative for obtaining a detailed distribution model for the count data, if such a model is desired. The key point here is that the Poissonness plots and negative binomialness plots support the claim noted in Chapter 2 that the *Federalist Papers* counts are better described by the negative binomial distribution than by the Poisson distribution.

8.5 Histograms: crude density estimates

Perhaps the best-known graphical technique for characterizing distributions is the *histogram*, which may be viewed as a piecewise-constant approximation of the density $p(x)$. The principal advantages of the histogram are its familiarity and its ease of construction, but in practical terms it is the least useful of the three techniques considered here—the Q-Q plot, the histogram, and the kernel density estimate. In particular, kernel density estimates may be regarded as smoothed histograms and yield *much* better approximations of $p(x)$ in almost all cases. Further, kernel density estimates are only slightly more complex to implement than histograms, making them a much better alternative in practice. Because the histogram is more familiar and is closely related to kernel density estimators, however, Sec. 8.5.1 gives a detailed introduction to the construction and general behavior of the histogram. In fact, it is possible to obtain a reasonably complete and extremely useful characterization of the histogram, in terms of its bias (discussed in Sec. 8.5.2) and its variance (discussed in Sec. 8.5.3). Following these discussions, Sec. 8.6 introduces the basic ideas of kernel density estimation and builds on the histogram results to establish general characteristics and practical trade-offs for kernel density estimates.

8.5.1 The basic histogram

As noted, the histogram is essenitally a piecewise-constant estimate of a probability density $p(x)$, constructed from a finite set $\{x_k\}$ of N data samples as follows. First, some value $x_0 \leq x_k$ for all k is taken as the *origin* of the histogram, and *bins* B_j are defined as the semiopen intervals:

$$B_j = \{x | x_0 + (j-1)h \leq x_k < x_0 + jh\}, \text{ for } j = 1, 2, \ldots, J. \qquad (8.22)$$

Defining $\#\{x_k \in B_j\}$ as the number of points from the data sequence $\{x_k\}$ in the bin B_j, the histogram density estimate is defined as

$$\hat{p}(x) = \frac{\#\{x_k \in B_j\}}{Nh} = \frac{\#\{x_k \in B_j\}/N}{h}, \qquad (8.23)$$

Figure 8.14: Histogram approximation to the $N(0, 1)$ density

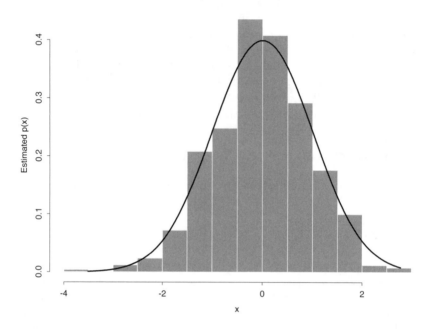

for all $x \in B_j$. In other words, the density estimate is constant in each bin B_j, where it is equal to the fraction of the total number of data points that belong to the bin, divided by its width, h.

This basic idea is illustrated in Fig. 8.14, which shows a histogram constructed from $N = 500$ data samples, drawn from the standard normal distribution, with origin $x_0 = -4$ and bin width $h = 0.5$. For comparison, the solid curve shows the exact density function for the data distribution. It is clear from this figure that the histogram gives a general indication of the shape of the density function, but a very crude one. To construct these density estimates, we must specify two parameters: the origin x_0 and the bin width h. Of these, the bin width h is much more influential, and the next two subsections are devoted to an examination of the influence of h on the bias and variance of the histogram, viewed as an estimator of the density function $p(x)$. Both because it is less influential and because it disappears when we consider the higher quality kernel density estimates discussed in Sec. 8.6, the influence of the origin of the histogram is not considered here. For further discussion and an example that illustrates this influence, refer to Silverman [264, p. 9]; more generally, for more detailed discussions of the histogram than that given here, refer to the books by Silverman [264], Simonoff [265], or Thompson and Tapia [277].

8.5.2   Histogram bias

Because the histogram is computed from what is assumed to be a random data sequence, it is a random quantity that may be characterized in terms of its mean and variance. Further, for a given data distribution $p(x)$, it is reasonable to ask whether the expected value $E\{\hat{p}_h(x)\}$ is equal to $p(x)$ and, if not, how much it differs from this value. To compute this expectation, it is convenient to reparameterize the problem in terms of the *bin centers*:

$$\tilde{x}_j = x_0 + \left(j - \frac{1}{2} \right) h, \tag{8.24}$$

and the "box function":

$$B(x) = \begin{cases} 0 & x < -1/2, \\ 1 & -1/2 \le x \le 1/2, \\ 0 & x > 1/2. \end{cases} \tag{8.25}$$

In terms of these quantities, the condition $x \in B_j$ is equivalent to either of the following conditions:

$$\tilde{x}_j - \frac{h}{2} \le x \le \tilde{x}_j + \frac{h}{2} \Leftrightarrow B\left(\frac{x - \tilde{x}_j}{h} \right) = 1. \tag{8.26}$$

The advantage of this observation is that it allows us to rewrite Eq. (8.23) as:

$$\hat{p}_h(x) = \frac{1}{Nh} \sum_{k=1}^{N} B\left(\frac{x_k - \tilde{x}_j}{h} \right). \tag{8.27}$$

In both this expression and the original histogram expression, note that the dependence of the density estimate on x enters through the condition $x \in B_j$, which may be regarded as a discrete random event with some probability p_j that is determined by the density $p(x)$ we wish to estimate. The key to evaluating both the mean and the variance of $\hat{p}_h(x)$ lies in computing this probability.

More specifically, note that the expected value of $\hat{p}_h(x)$ may be obtained by applying the expectation operator to Eq. (8.27) to obtain:

$$E\{\hat{p}_h(x)\} = \frac{1}{Nh} \sum_{k=1}^{N} E\left\{ B\left(\frac{x_k - \tilde{x}_j}{h} \right) \right\}. \tag{8.28}$$

Here, it is necessary to invoke the assumption that $\{x_k\}$ is an independent, identically distributed (i.i.d.) sequence, a notion discussed further in Chapter 10. Under this assumption, it follows that the expectations appearing inside the sum all have the same value, yielding:

$$E\{\hat{p}_h(x)\} = \frac{1}{h} E\left\{ B\left(\frac{x - \tilde{x}_j}{h} \right) \right\}. \tag{8.29}$$

Further, note that the box function assumes precisely two values: zero if $x \notin B_j$ and 1 if $x \in B_j$, from which we have:

$$E\left\{B\left(\frac{x-\tilde{x}_j}{h}\right)\right\} = 1 \cdot \mathcal{P}\left\{B\left(\frac{x-\tilde{x}_j}{h}\right) = 1\right\} = \mathcal{P}\{x \in B_j\} = p_j. \qquad (8.30)$$

To evaluate this probability, note that:

$$
\begin{aligned}
p_j &= \mathcal{P}\{\tilde{x}_j - h/2 \le x \le \tilde{x}_j + h/2\} \\
&= \int_{\tilde{x}_j-h/2}^{\tilde{x}_j+h/2} p(x)dx = h\int_{-1/2}^{1/2} p(\tilde{x}_j + hz)dz,
\end{aligned}
\qquad (8.31)
$$

where we have made the variable substitution $x = \tilde{x}_j + hz$ to simplify the integral. If, as is usually the case, the density is smooth (i.e., infinitely differentiable), we can form the Taylor series expansion about the bin center $\tilde{x}_j$ to obtain:

$$p(\tilde{x}_j + hz) \simeq p(\tilde{x}_j) + p'(\tilde{x}_j)hz + \frac{p''(\tilde{x}_j)}{2}(hz)^2, \qquad (8.32)$$

where terms of third and higher order have been neglected. In practice, we generally take h to be "small" (or at least "not too large"), so this approximation is often a reasonable one. Substituting this result into the last integral in Eq. (8.31) then yields an expression for p_j. In particular, note that the first term is constant (i.e., independent of the integration variable z), whereas the second term exibits odd symmetry over the integration interval $[-1/2, 1/2]$, so its integral vanishes. Thus, we obtain the following expression for the probability p_j that $x \in B_j$:

$$p_j \simeq hp(\tilde{x}_j) + \frac{h^3 p''(\tilde{x}_j)}{24}. \qquad (8.33)$$

Combining this result with Eqs. (8.29) and (8.30) then yields the following approximation for the expected value of the histogram:

$$E\{\hat{p}_h(x)\} \simeq p(\tilde{x}_j) + \frac{h^2 p''(\tilde{x}_j)}{24}, \qquad (8.34)$$

for all $x \in B_j$. Note that at the center of the bin, $x = \tilde{x}_j$ and the bias $b_h(x) = E\{\hat{p}_h(x)\} - p(x)$ assumes its smallest value, which is approximately quadratic in h. The largest bias generally occurs at the edges of the bin where $|x - \tilde{x}_j|$ is largest; again considering the Taylor series expansion about $\tilde{x}_j$, we have:

$$p(x) \simeq p(\tilde{x}_j) + (x - \tilde{x}_j)p'(\tilde{x}_j). \qquad (8.35)$$

Further, since $|x - \tilde{x}_j| \le h/2$ for all $x \in B_j$, it follows that the bias of the histogram is approximately bounded, for any $x \in B_j$, by the worst-case value of this first-order term in h:

$$|b_h(x)| \lesssim \frac{h|p'(\tilde{x}_j)|}{2}. \qquad (8.36)$$

Overall, the key results here are the following. First, the histogram is a biased estimator, and the bias $b_h(x)$ depends on both x and the true density $p(x)$; in particular, note that the bias is more severe in regions where the density is changing rapidly (i.e., where $|p'(x)|$ is large). Second and most important, note that the bias of the histogram goes to zero as $h \to 0$. Hence, *to minimize the bias of the histogram, it is desirable to make the bin width small.*

8.5.3 Histogram variance

To obtain an expression for the variance of the histogram, consider a fixed bin B_j and define the sequence $\{\beta_i\}$:

$$\beta_i = B\left(\frac{x_i - \tilde{x}_j}{h}\right) = \begin{cases} 1 & \text{with probability } p_j, \\ 0 & \text{with probability } 1 - p_j. \end{cases} \tag{8.37}$$

An alternative way of obtaining Eq. (8.30) would be to note that the random variable β_i represents a Bernoulli trial, discussed in Chapter 3. There, the mean and variance results were given as:

$$E\{\beta_i\} = p_j, \quad \text{and} \quad \text{var}\{\beta_i\} = p_j(1 - p_j). \tag{8.38}$$

It follows from this result and Eq. (8.27) that the variance of the histogram is given by:

$$\begin{aligned} \text{var}\{\hat{p}_h(x)\} &= \text{var}\left\{\frac{1}{Nh}\sum_{k=1}^{N}B\left(\frac{x_k - \tilde{x}_j}{h}\right)\right\} \\ &= \frac{1}{N^2h^2}\sum_{i=1}^{N}\text{var}\{\beta_i\} = \frac{p_j(1 - p_j)}{Nh^2}. \end{aligned} \tag{8.39}$$

In obtaining this result, use has again been made of the assumption that $\{x_k\}$ is an i.i.d. sequence, as in the bias result derived in the preceeding section.

As before, if we assume the unknown density $p(x)$ is smooth, we may invoke the Taylor series argument to obtain Eq. (8.33) for the probability p_j. Retaining the lowest order term in h, we have:

$$p_j \simeq p(\tilde{x}_j)h \quad \Rightarrow \quad \text{var}\{\hat{p}_h(x)\} \simeq \frac{p(\tilde{x}_j)}{Nh}. \tag{8.40}$$

This result has two practically important consequences. First, note that *for fixed h*, the variance of the histogram approaches zero as $N \to \infty$, meaning that the histogram is a consistent estimator of the unknown density $p(x)$. The second important point is that, *for fixed N*, $\text{var}\{\hat{p}_h(x)\} \to \infty$ as $h \to 0$. This last observation is particularly important because it means that histograms exhibit a *bias/variance trade-off*: the larger we make h, the smaller the variance but the larger the bias, and vice versa. To illustrate the nature of this bias-variance trade-off, Fig. 8.15 shows four histograms, each estimated from normally distributed data sequences of length $N = 500$. The upper left plot shows the

Figure 8.15: Four histograms computed from normally distributed data

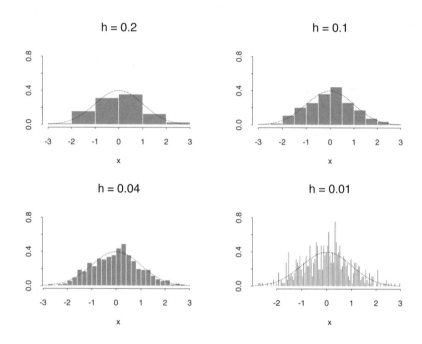

results obtained for the widest bins considered here ($h = 0.2$), and the lower right plot shows the results obtained for the narrowest bins ($h = 0.01$), with the other two plots showing the results for the intermediate values $h = 0.1$ (upper right) and $h = 0.04$ (lower left). Comparing the two extremes, it is clear that the variability seen for $h = 0.01$ is much larger than that seen for $h = 0.2$, but careful examination of the upper left plot shows that this histogram underestimates the true density most of the time. Generally, the intermediate values $h = 0.1$ and $h = 0.04$ seem to give better results, that are closer on average to the true density than those obtained for $h = 0.2$ but much less variable than those obtained for $h = 0.01$. Also, although there are too few distinct values in these esimates to draw definite conclusions, note that the variability of these estimates appears to generally be larger near the peak than in the tails of the distribution, consistent with the approximation (8.40). Analogous results (not shown) are obtained for histogram estimates of other densities.

8.6   Kernel density estimators

It was noted at the beginning of Sec. 8.5 that, although it is well known and remains somewhat popular in practice, the histogram is a rather crude density estimate, a point illustrated nicely in Fig. 8.15. Fortunately, it is not difficult to construct alternative density estimators that generally give much better results: these are the *kernel density estimators*, which may be viewed essentially as *smoothed* histograms. The basic structure of kernel density estimators is introduced in Sec. 8.6.1, building on the results presented in Sec. 8.5. As with the histogram, because kernel density estimates are computed from random data, they may be usefully regarded as random variables and characterized in terms of their bias and variance. Detailed discussions of kernel density estimator bias and variance are given in secs. 8.6.2 and 8.6.3, respectively, analogous to the results presented in Sec. 8.5 for histograms. Sec. 8.6.4 then briefly compares histograms and three different kernel density estimates for four beta distributions to illustrate how these estimators perform in practice. In addition, the kernel density estimator is also closely related to the *scatterplot smoother*, an idea introduced in Sec. 8.7.

8.6.1 The basic kernel estimator

Kernel density estimators are closely related to the histograms discussed in the previous section, and their general structure is best appreciated by comparing it with that of the histogram. Recall that one general expression for the histogram was Eq. (8.27), repeated here for convenience:

$$\hat{p}_h(x) = \frac{1}{Nh} \sum_{k=1}^{N} B\left(\frac{x_k - \tilde{x}_j}{h}\right). \tag{8.41}$$

Note that the dependence of the right-hand side of this expression on x is implicit, arising from the fact that $\tilde{x}_j$ depends on x:

$$\tilde{x}_j = x_0 + \left(j(x) - \frac{1}{2}\right)h, \tag{8.42}$$

where $j(x)$ is the unique integer satisfying:

$$j(x) - 1 \leq \frac{x - x_0}{h} < j(x). \tag{8.43}$$

Kernel density extimates make the following two modifications of the histogram: $\tilde{x}_j$ is replaced with x, and the box function $B(\cdot)$ is replaced with a more general *kernel function* $K(\cdot)$. More explicitly, the general form of the kernel density estimator is:

$$\hat{p}_h(x) = \frac{1}{Nh} \sum_{i=1}^{N} K\left(\frac{x_i - x}{h}\right). \tag{8.44}$$

Since the origin parameter x_0 appearing in the definition of the histogram appeared only in $\tilde{x}_j$, it follows that this parameter does not appear in the kernel density estimator, as noted earlier. Conversely, although the parameter h no longer has the interpretation as a bin width, it still strongly influences both the bias and the variance of the kernel density estimate and is called the *bandwidth* parameter.

Figure 8.16: Silverman's "naive estimator" for Gaussian data

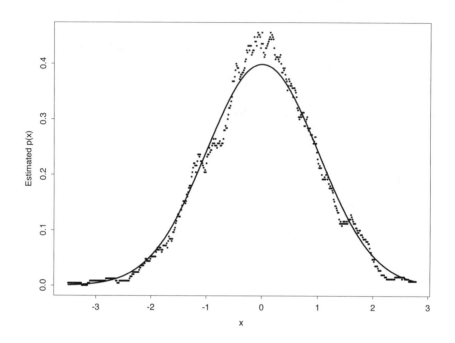

To see the inflence of replacing $\tilde{x}_j$ with x, it is useful to consider what Silverman [264] calls the "naive estimator," obtained by retaining the box kernel:

$$K(x) = B(x) = \begin{cases} 1 & |x| \leq 1/2, \\ 0 & |x| > 1/2. \end{cases} \tag{8.45}$$

The results obtained with $h = 1$ from a 500 point standard normal i.i.d. data sequence are shown in Fig. 8.16, overlaid with the exact distribution. It is clear from this result that the simple replacement of $\tilde{x}_j$ with x brings about a dramatic improvement in the quality of the estimated density, relative to the histogram. More detailed comparisons in terms of bias and variance are given in secs. 8.6.2 and 8.6.3, but first it is useful to make the following observation.

Suppose $K(\cdot)$ is chosen as any proper density function. It then follows that $K(x) \geq 0$ for all x, implying by Eq. (8.44) that $\hat{p}_h(x) \geq 0$ for any choice of the bandwidth h. Similarly, $K(x)$ satisfies the normalization condition discussed in Chapter 4, implying that $\hat{p}_h(x)$ also satisfies this condition for all $h > 0$, i.e.:

$$\int_{-\infty}^{\infty} \hat{p}_h(x)dx = \frac{1}{N}\sum_{i=1}^{N}\int_{-\infty}^{\infty} K\left(\frac{x_i - x}{h}\right)\frac{dx}{h} = \frac{1}{N}\sum_{i=1}^{N}\int_{-\infty}^{\infty} K(z)dz = 1. \tag{8.46}$$

Because of these useful constraints on the behavior of the resulting density estimate, kernel functions are generally chosen to be proper density functions, a restriction imposed on all of the kernel functions considered in this book.

8.6.2 Bias in kernel estimates

The defining equation for the kernel density estimate, Eq. (8.44), provides the basis for deriving bias expressions analogous to those obtained in Sec. 8.5.2 for the histogram. Specifically, taking the expectation of this equation yields:

$$
\begin{aligned}
E\{\hat{p}_h(x)\} &= \frac{1}{Nh} \sum_{i=1}^{N} E\left\{K\left(\frac{x_i - x}{h}\right)\right\} \\
&= \frac{1}{h} E\left\{K\left(\frac{x_i - x}{h}\right)\right\},
\end{aligned} \tag{8.47}
$$

where it has been assumed that the data sequence $\{x_i\}$ is an i.i.d. sequence, as in the histogram bias and variance results derived earlier. Specifically, this assumption implies that the expectation appearing in the sum in this expression is independent of the summation index i. Evaluating this expectation then leads to the following result:

$$
\begin{aligned}
E\{\hat{p}_h(x)\} &= \frac{1}{h} \int_{-\infty}^{\infty} K\left(\frac{y - x}{h}\right) p(y) dy \\
&= \int_{-\infty}^{\infty} K(z) p(x + hz) dz.
\end{aligned} \tag{8.48}
$$

Again as in the histogram results derived earlier, this expression may be simplified if we assume that $p(x)$ is smooth, permitting us to make a Taylor series expansion of $p(x + hz)$:

$$
p(x + hz) \simeq p(x) + hzp'(x) + \frac{h^2 z^2}{2} p''(x), \tag{8.49}
$$

where terms of order h^3 and higher have been neglected, based on the observation that h is typically a small parameter. Substituting this result into Eq. (8.48) and exploiting the normalization result, Eq. (8.46), then yields the approximation:

$$
E\{\hat{p}_h(x)\} \simeq p(x) + hp'(x) \int_{-\infty}^{\infty} zK(z)dz + \frac{h^2 p''(x)}{2} \int_{-\infty}^{\infty} z^2 K(z)dz. \tag{8.50}
$$

Note that if we choose the kernel function $K(z)$ to be symmetric about zero (i.e., $K(-z) = K(z)$), the integral involving $zK(z)$ vanishes, leaving us with the following result for the dominant bias contribution:

$$
b_h(x) = E\{\hat{p}_h(x)\} - p(x) \simeq \frac{h^2 p''(x)}{2} \int_{-\infty}^{\infty} z^2 K(z)dz \equiv \frac{h^2 p''(x)}{2} \mu_2(K), \tag{8.51}
$$

where $\mu_2(K)$ represents the second moment of the kernel function. Note also that if we choose a symmetric kernel function, the third-order contribution to this bias expression also vanishes by symmetry, so the first neglected term in Eq. (8.51) is of order h^4.

Two key observations here are, first, that the bias of the kernel density estimator is generally lower than that of the histogram, going to zero faster as $h \to 0$, and second, that the bias of the kernel density estimate is worse in regions of high curvature than in regions of low curvature. This point may be seen clearly in the box kernel results shown in Fig. 8.16, where the estimated density lies farthest from the true density around the peak in the Gaussian density function where the curvature is the highest.

8.6.3   Variance of kernel estimates

If we assume, as in the derivation of the bias results just presented, that $K(x)$ is a symmetric kernel function and that $\{x_i\}$ is an i.i.d. sequence, we obtain the following variance results for the kernel density estimate defined by Eq. (8.44):

$$\text{var}\{\hat{p}_h(x)\} = \text{var}\left\{\frac{1}{Nh}\sum_{i=1}^{N} K\left(\frac{x_i - x}{h}\right)\right\} \tag{8.52}$$

$$= \frac{1}{N^2}\sum_{i=1}^{N} \text{var}\left\{\frac{1}{h}K\left(\frac{x_i - x}{h}\right)\right\} \tag{8.53}$$

$$= \frac{1}{N}\text{var}\left\{\frac{1}{h}K\left(\frac{x_i - x}{h}\right)\right\}. \tag{8.54}$$

To evaluate this result, it is useful to introduce the following random variable:

$$\xi = \frac{1}{h} K\left(\frac{x_i - x}{h}\right), \tag{8.55}$$

noting that its distribution is independent of i under the assumption that $\{x_i\}$ is an i.i.d. sequence. Specifically, we have:

$$\text{var}\{\hat{p}_h(x)\} = \frac{\text{var}\{\xi\}}{N} = \frac{E\{\xi^2\} - [E\{\xi\}]^2}{N}. \tag{8.56}$$

For a symmetric kernel, neglecting terms of order h^2 and higher, it follows from Eq. (8.50) that $E\{\xi\} \simeq p(x)$. To obtain an expression for $E\{\xi^2\}$, first note that:

$$E\{\xi^2\} = \frac{1}{h^2}\int_{-\infty}^{\infty} K^2\left(\frac{z - x}{h}\right) p(z)dz = \frac{1}{h}\int_{-\infty}^{\infty} K^2(s)p(x + hs)ds, \tag{8.57}$$

where $s = (z - x)/h$. As before, under the assumption that the unknown density $p(x)$ is smooth, $p(x + hs)$ may be expanded as a Taylor series; retaining terms of order up to h^2 in this expansion ultimately yields:

$$E\{\xi^2\} \simeq \frac{p(x)}{h}\int_{-\infty}^{\infty} K^2(s)ds + p'(x)\int_{-\infty}^{\infty} sK^2(s)ds + \frac{hp''(x)}{2}\int_{-\infty}^{\infty} s^2K^2(s)ds. \tag{8.58}$$

Note that for a symmetric kernel function, $K^2(s)$ is also symmetric, so the integral involving $sK^2(s)$ in this expression vanishes for symmetric kernels. Combining this result with Eq. (8.56) then yields the following approximate variance expression:

$$
\begin{aligned}
\text{var}\{\hat{p}_h(x)\} &\simeq \frac{p(x)}{Nh}\int_{-\infty}^{\infty}K^2(s)ds - \frac{p^2(x)}{N} + \frac{hp''(x)}{2N}\int_{-\infty}^{\infty}s^2K^2(s)ds \\
&\simeq \frac{p(x)}{Nh}\int_{-\infty}^{\infty}K^2(s)ds, \qquad\qquad (8.59)
\end{aligned}
$$

in the usual case where N is large and h is small. The key point here is, as in the case of the histogram, there is an inherent bias/variance trade-off: as $h \to 0$, the bias improves, but the variance diverges, and vice versa. An important practical difference is that we have an additional degree of freedom in kernel density estimation, since the choice of kernel function influences both bias and variance.

8.6.4 A comparison of four examples

To illustrate both the differences between kernel density estimators and histograms and the influence of different choices of kernel functions, the following discussion briefly compares four different nonparametric density estimates for four special cases of the beta distribution discussed in Chapter 4. Specifically, recall that the beta distribution is defined by the density function:

$$
p(x) = \left[\frac{\Gamma(p+q)}{\Gamma(p)\Gamma(q)}\right]x^{p-1}(1-x)^{q-1}, \qquad\qquad (8.60)
$$

on the interval $0 \le x \le 1$. The examples considered here correspond to the following four special cases of this distribution:

1. the arcsin distribution, $p = q = 1/2$,
2. the uniform distribution, $p = q = 1$,
3. a "pseudo-Gaussian" distribution, $p = q = 4$,
4. an asymmetric distribution, $p = 2, q = 4$.

For each of these distributions, an i.i.d. sequence of 500 samples was generated and the following four density estimates were computed:

1. the histogram discussed in Sec. 8.5,
2. the box kernel introduced in Sec. 8.6.1,
3. the following triangular kernel:

$$
K(x) = \begin{cases} 1 - |x| & |x| < 1, \\ 0 & |x| \ge 1, \end{cases}
$$

4. the Gaussian kernel:

$$
K(x) = \frac{1}{\sqrt{2\pi}}e^{-x^2/2}.
$$

Figure 8.17: Estimated densities for the arcsin distribution

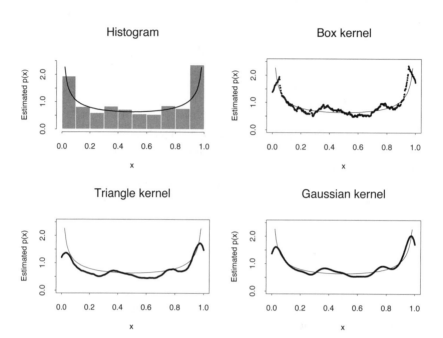

In all cases, the bandwidth parameter h was taken as 0.1; also, note that the three kernel functions considered are symmetric.

Fig. 8.17 compares the four density estimates obtained for the arcsine distributed data sequence. The upper left plot shows the histogram, overlaid with a plot of the true density function; it is clear from this plot that, despite the coarseness of the approximation, the histogram does exhibit the correct general shape of the density. Much better results are obtained with the box kernel ("naive estimator") shown in the upper right plot; in fact, it can be argued that, in this case, the box kernel gives more reasonable estimates than the triangular kernel, shown in the lower left. In particular, note that the triangular kernel tends to underestimate the density throughout the unit interval, and it exhibits markedly poorer estimates than the box kernel near the ends of the interval where the true density exhibits its maxima. Probably the best results are obtained with the Gaussian kernel, shown in the lower right plot, but this estimate requires much longer to compute than the others because for each value of x, the estimate $\hat{p}_h(x)$ involves contributions from all of the data points in the data sequence, whereas the box and triangular kernels vanish identically outside a finite set of data observations; with the triangular kernel, for example, the only contributions to $\hat{p}_h(x)$ come from those points x_i satisfying $|x_i - x| < h$.

Fig. 8.18 shows the corresponding set of four density estimates obtained from a uniformly distributed data sequence, again with the true constant den-

Figure 8.18: Estimated densities for the uniform distribution

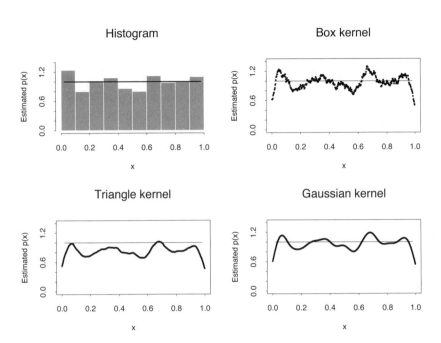

sity shown as a solid reference line in all four plots. In general, all four of these density estimates appear pretty comparable for this example: the kernel density estimates are substantially smoother, but the overall deviations from the true constant density look quite similar for the four estimates. As in the previous example, the triangular kernel generally underestimates the true density. The primary differences between the Gaussian kernel and the box kernel in this example are, first, that the Gaussian kernel estimate is significantly smoother than the box kernel estimate, and second, that the Gaussian kernel estimate again takes much longer to compute than the box kernel estimate. Overall, however, the general deviations from the true constant density are very comparable in this example for the box and Gaussian kernels.

Fig. 8.19 presents the analogous results for the pseudo-Gaussian distribution. Note here that both the histogram presented in the upper left plot and the box kernel estimate shown in the upper right plot are quite similar to the results presented earlier for Gaussian data sequences (specifically, figs. 8.14 and 8.16). Here, in addition to being markedly smoother, the Gaussian kernel estimate appears to also give results that are generally closer to the true density than the box kernel estimate. Once again, the triangle kernel gives results that are similar to the Gaussian kernel in terms of smoothness, but the density is consistently underestimated, particularly near the peak where the curvature is the highest and the bias can therefore be expected to be the worst.

Figure 8.19: Estimated densities for the pseudo-Gaussian distribution

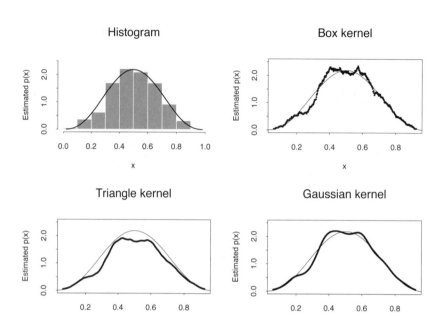

The results for the asymmetric example are shown in Fig. 8.20, in the same format as the previous three examples, and the results are qualitatively quite similar. Specifically, the histogram gives a clear indication of the distributional asymmetry but the kernel density estimates are much smoother and give much better density approximations overall. The deviations seen in the box kernel density estimate from the true density are slightly larger than those obtained with the Gaussian kernel, and the Gaussian kernel estimate is much smoother, but again at the expense of significantly longer computing time. Once again, even though the triangular kernel gives a smoother result than the box kernel, it consistently underestimates the density. In fact, in this example the underestimation is pronounced enough that the box kernel estimate might actually be preferred, despite its lack of smoothness.

Finally, Fig. 8.21 illustrates the influence of the bandwidth parameter h on the density estimates for the asymmetric density example. Specifically, the upper right plot shows the estimated density obtained using the triangular kernel just considered with the same bandwidth, $h = 0.1$. The upper left plot illustrates the increase in estimator variance that results when the bandwidth is decreased to $h = 0.05$; conversely, although this result consistently underestimates the true density as in the case $h = 0.1$, this underestimation is not as severe. Increasing the bandwidth to $h = 0.2$ gives the estimate shown in the lower left plot: this estimated density is significantly smoother than those obtained with smaller

Figure 8.20: Estimated densities for the asymmetric distribution

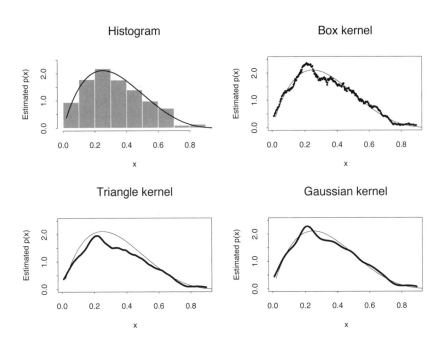

bandwidth parameters and the bias is clearly worse. Both of these changes become even more pronounced for $h = 0.5$, shown in the lower right plot. Taken together, these plots clearly illustrate the bias/variance trade-off discussed in Sec. 8.6.3.

Overall, the point of this discussion has been to provide a few realistic performance comparisons between histograms and kernel density estimators in cases where the correct result is known. These simulation-based examples illustrate, first, that the kernel density estimates both are smoother than histograms and generally give better approximations of the unknown density. Second, these examples illustrate that the quality of the results can depend significantly on the kernel chosen, as can the computational costs. This point is particularly apparent from the comparisons between the triangular kernel and the Gaussian kernel: the Gaussian kernel gave consistently better results, but at the price of substantially greater computational requirements. Conversely, it is important to note that the effects of changes in the bandwidth parameter on the overall quality of the density estimate are even more pronounced, suggesting that this tuning parameter should be examined carefully in practice, possibly by comparing the results obtained for several different values. Finally, these results emphasized the point made earlier in the discussion of kernel density estimator bias, that densities exhibiting high curvature are more difficult to estimate than densities of lower curvature. This point is perhaps best illustrated in the

Figure 8.21: Influence of bandwidth on estimated densities

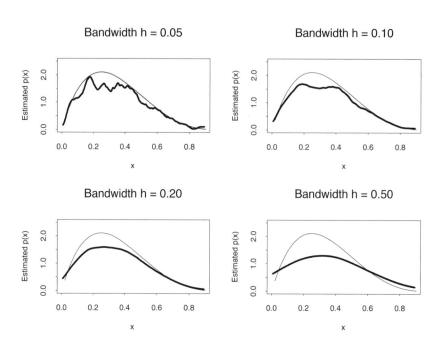

asymmetric example, where all three estimators perform extremely well if we move far enough into either the upper or the lower tail, but these estimators also all exhibit significant biases at the peak of the density where the curvature is greatest.

8.7 Scatterplot smoothers

The kernel density estimator just discussed is closely related to the idea of *scatterplot smoothing:* given a sequence of (x_k, y_k) data pairs, we seek an estimate of the form:

$$y_k = m(x_k) + e_k, \tag{8.61}$$

where $m(x_k)$ is a smooth function whose form is not specified, and e_k is a zero-mean random variable, statistically independent of x_k. The notion of *conditional expectations* is discussed in Chapter 10, but the essential idea is that the conditional expectation of one random variable y given an observation of another random variable x is the expected value of y when x is treated as a constant, equal to the observed value. Under the assumptions of this problem formulation, the conditional expectation of y_k given x_k is equal to:

$$E\{y_k|x_k\} = E\{m(x_k)|x_k\} + E\{e_k|x_k\} = m(x_k) + E\{e_k\} = m(x_k). \tag{8.62}$$

The idea of nonparametric regression is to exploit this relationship, giving:

$$m(x) = E\{y|x\} = \int_{-\infty}^{\infty} yp(x|y)dy, \tag{8.63}$$

where $p(x|y)$ is the conditional density of x given y, a quantity defined and discussed further in Chapter 10. The key point in this discussion is that a simple extension of the nonparametric density estimation procedures discussed previously leads to nonparametric estimators for $p(x|y)$ and, from there, to *kernel smoothers*, which are nonparametric estimators of the function $m(x)$.

The development of the basic form of the kernel smoother, often called the *Nadaraya-Watson kernel estimator* [265, p. 136] is as follows. First, a result discussed in Chapter 10 is that the conditional density $p(y|x)$ of y given x is related to the *joint density* $p(x,y)$ of x and y by:

$$p(y|x) = \frac{p(x,y)}{p_x(x)}, \tag{8.64}$$

where $p_x(x)$ is the (marginal) density of x, obtained by integrating the joint density $p(x,y)$ over all possible y values to average out its effect:

$$p_x(x) = \int_{-\infty}^{\infty} p(x,y)dy. \tag{8.65}$$

Combining Eqs. (8.63), (8.64), and (8.65) leads to the following result:

$$m(x) = \frac{\int_{-\infty}^{\infty} yp(x,y)dy}{\int_{-\infty}^{\infty} p(x,y)dy}. \tag{8.66}$$

The second key observation in developing the kernel smoother considered here is to note that two univariate kernel functions like those discussed previously, say, $K_x(\cdot)$ and $K_y(\cdot)$, may be combined to obtain a two-dimensional kernel density estimator of the following form:

$$\hat{p}(x,y) = \frac{1}{Nh_xh_y} \sum_{k=1}^{N} K_x\left(\frac{x-x_k}{h_x}\right) K_y\left(\frac{y-y_k}{h_y}\right), \tag{8.67}$$

where h_x and h_y are the corresponding bandwidth parameters for each univariate smoother. The Nadaraya-Watson kernel estimator is obtained by substituting this nonparametric estimator for the joint density $p(x,y)$ into Eq. (8.66). Under the assumption that $K_y(\cdot)$ is a symmetric proper density function, it can be shown (Exercise 4) that the resulting nonparametric estimator of $m(x)$ can be written in terms of the x-kernel $K(\cdot) = K_x(\cdot)$ alone:

$$\hat{m}(x) = \frac{\sum_{k=1}^{N} y_k K\left(\frac{x-x_k}{h}\right)}{\sum_{k=1}^{N} K\left(\frac{x-x_k}{h}\right)}, \tag{8.68}$$

where h denotes the corresponding bandwidth h_x in Eq. (8.67).

Figure 8.22: Exact curve and three nonparametric estimates

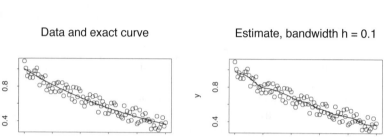

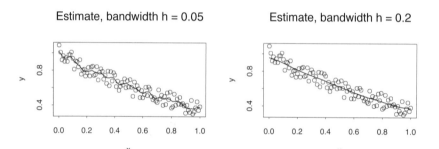

   To illustrate the performance of the Nadaraya-Watson kernel estimator, it is instructive to consider the following simple example. Suppose x and y are related via the simple exponential decay curve $y = e^{-x}$, where x varies over the range $0 \leq x \leq 1$. Further, suppose 101 noisy observations of y are observed, corrupted by random noise that is uniformly distributed on $[-0.1, 0.1]$. This observed data sequence is shown in the upper left plot in Fig. 8.22, along with the exact relationship between x and y, indicated by the solid line in the plot. The other three plots in Fig. 8.22 show nonparametric regression estimates computed from these observed data points, using a triangular kernel $K(x)$ and three different smoothing bandwidths: $h = 0.1$ (upper right plot), $h = 0.05$ (lower left plot), and $h = 0.2$ (lower right plot). The general character of these estimates and their behavior as a function of h is quite analogous to that of the closely related nonparametric density estimates introduced in Sec. 8.6.1.
   Fig. 8.23 shows a plot of the estimation errors for these three nonparametric regression models $\hat{m}(x)$, evaluated at each of the data points (x_k, y_k). It is clear from this plot that, as in the case of kernel density estimates, the variability of these errors decreases as h increases. This observation is confirmed by the standard deviations of these error sequences, which decrease from 0.021 for $h = 0.05$, to 0.014 for $h = 0.1$, to 0.010 for $h = 0.2$. The best bandwidth for any given application will depend on the underlying curve being estimated (obviously unknown in typical applications, since that motivates the estimation problem to

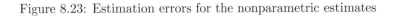

Figure 8.23: Estimation errors for the nonparametric estimates

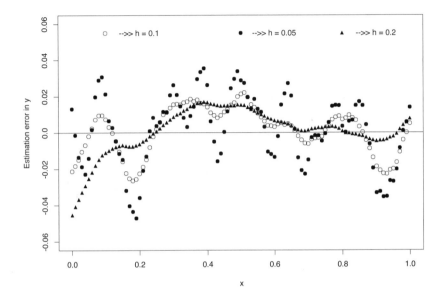

begin with), the kernel function used, and the variance of the noise present. For a useful discussion of bandwidth selection that addresses these and other issues in detail, refer to the treatment by Simonoff [265, ch. 5]. The key point here is that nonparametric regression can be an extremely useful technique in practice that is fairly simple to implement; as in the case of nonparametric density estimation, the bandwidth parameter can have an extremely significant influence on the estimated curve we obtain from any given dataset. Hence, it is often useful to compute several different nonparametric estimates and compare them, particularly if qualitative insights are available about the real-world source of the data that give some idea of how smooth the underlying function should be.

8.7.1 Supsmu: an adaptive smoother

An approach that can give substantially better results than any fixed-bandwidth kernel smoother is the use of an adaptive procedure like the *supersmoother*, available in both *S-Plus* [272] and *R* [240] as the procedure **supsmu**. Essentially, this procedure estimates the local curvature of the data, uses a small smoothing bandwidth in regions of high curvature to avoid excessive bias, and uses a large bandwidth in regions of low curvature to minimize variance. The result is a fairly complex procedure, and a detailed description is beyond the scope of this book, but it is introduced here to illustrate the potential advantages of adaptive

Figure 8.24: Exact curve and observed data values

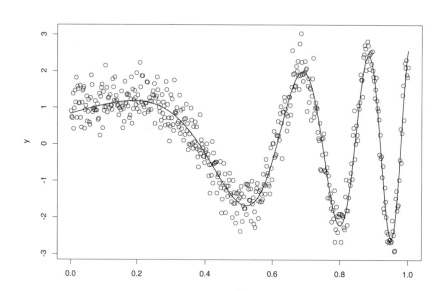

smoothing procedures that adjust the bandwidth based on the estimated curve. A much more complete discussion is given by Härdle [134, p. 181].

The following example provides a useful illustration of how the supersmoother works in practice. Fig. 8.24 shows a plot of 501 data points, generated as follows. The smooth curve to be estimated is given by:

$$y = e^x \sin e^{3x}, \qquad (8.69)$$

on the interval $0 \le x \le 1$. The open circles in this plot correspond to the observed data points, which are generated by adding zero-mean, Gaussian random errors with standard deviation 0.4 to these exact data values.

The results obtained using the fixed bandwidth smoother discussed in Sec. 8.7 are shown in Fig. 8.25. The heavy line in this figure represents the kernel smoother estimate, and the light line is the exact, noise-free curve defined by Eq. (8.69). It is apparent from this figure that, as in the nonadaptive kernel density estimates considered previously, the bias of this estimate is most severe in the highly oscillatory parts of this figure where the curvature is the greatest. In addition, note the oscillatory ripples in the estimate between $x = 0$ and $x \sim 0.2$, which appear in a region of low curvature. This behavior illustrates the point noted earlier: the bandwidth parameter h is too large in this case to give good estimates in regions of high curvature and too small to give good

Figure 8.25: Exact curve and kernel smoother estimate

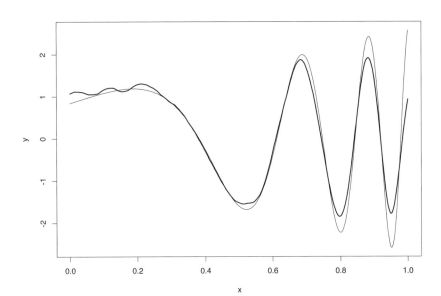

estimates in regions of low curvature. Because the curvature of this function varies significantly, it follows that no fixed bandwidth smoothing procedure can perform uniformly well over the entire range of the data.

The results obtained with the supersmoother are shown in Fig. 8.26, in the same format as before: the heavy line represents the supersmoother estimate, and the light line is the exact curve, included for comparison. As with the fixed-bandwidth procedure, the supersmoother exhibits a large bias in the highly oscillatory regions where the curvature is greatest; in fact, detailed comparisons of figs. 8.24 and 8.26 reveal that the bias is actually somewhat worse for the supersmoother in this case. Conversely, the notable advantage of the supersmoother is the absence of the ripples seen in the low-curvature portion of the function, between $x = 0$ and $x \sim 0.2$. Overall, while it cannot be argued that either method is *uniformly better*, comparing these two figures does suggest that the supersmoother results are *generally better* for this example.

8.7.2 The lowess smoother

It was emphasized in Chapter 7 that the presence of outliers in a dataset often causes classical statistical methods to perform poorly. As the following example illustrates, this is also true of the kernel smoother, but as with characterizations like location and scale estimators, it is possible to develop alternative procedures

Figure 8.26: Exact curve and supersmoother estimate

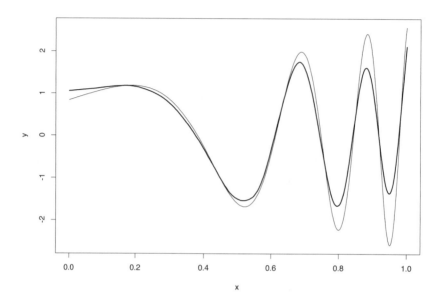

with better outlier resistance. A particularly popular alternative is the *lowess smoother*, also frequently called the *loess smoother*. (Note, however, that the *S-Plus* procedures **lowess** and **loess** are quite different: **loess** is a model-based procedure, while **lowess** is not.) As with the supersmoother introduced in Sec. 8.7.1, a detailed discussion of the lowess smoother is beyond the scope of this book, but since it is both readily available and extremely popular, it is useful to give a brief introduction here. For more detailed discussions, refer to the books by Härdle [134] or Simonoff [265].

The lowess procedure is based on local polynomial fits to the data, typically in the form of a locally linear model that minimizes the following misfit criterion, for every real value x:

$$J(x) = \frac{1}{N} \sum_{i=1}^{N} W_{ki}(x)(y_i - \beta_0 - \beta_1 x)^2, \qquad (8.70)$$

were $\{W_{ki}\}$ is a sequence of *k-nearest neighbor weights*. These weights are equal to N/k if x_i is one of the k nearest neighbors of x from the data sequence $\{x_k\}$, and equal to zero otherwise. Here, k is a tuning parameter that must be specified by the user, but which is typically provided with a default value. For example, in *S-Plus*, this default value is based on the fraction $f = 2/3$ of the total number of data points N. As was noted in the example presented in

Figure 8.27: Plots of a portion of the product storage tank dataset (upper left), two kernel smoothers with different bandwidths (upper right and lower left), and the lowess smoother

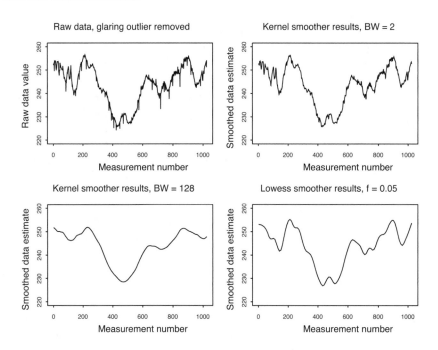

Chapter 1, however, this default does not always give reasonable results, so it is important to explore the effects of different choices of this tuning parameter in practice. In the lowess procedure, a fit minimizing the criterion $J(x)$ defined in Eq. (8.70) is first obtained and then refined using essentially the *iteratively reweighted least squares* procedure discussed in Chapter 11. The basic idea is that, once a preliminary fit $\hat{y}_i$ is obtained for the response data sequence $\{y_i\}$, the residuals $\epsilon_i = \hat{y}_i - y_i$ are computed and used to modify the weights. The specific modification described by Härdle is to replace the weights W_{ki} with $\delta_i W_{ki}$ where δ_i is computed for each data point (x_i, y_i) as:

$$\delta_i = K(\epsilon_i/6S), \qquad (8.71)$$

where $S = \text{median}\{|\epsilon_i|\}$ is similar in spirit to the MAD scale estimate discussed in Chapter 7. The advantage of this iterative refinement is that it downweights the effects of "large" residuals ϵ_i since the smoothing kernel $K(\cdot)$ causes δ_i to be small for these data observations. Härdle [134, p. 192] gives a particularly succinct description, in the form of pseudocode, for the lowess procedure.

The general behavior of the lowess procedure was introduced in Chapter 1, where it was applied to the product storage tank dataset. Fig. 8.27 gives a slightly different perspective on this procedure, comparing its performance with two different kernel smoother examples. Specifically, the upper left plot shows a line plot of the original storage tank data sequence, with the glaring missing data outlier removed (a value of zero, $\sim$30 standard deviations below the mean, creating an enormous spike in the data sequence). Even so, it is clear from this plot that the remaining data sequence contains a number of fairly pronounced local outliers, appearing as spikes in the data sequence. Applying the *S-Plus* kernel smoother with the default box kernel (i.e., $K(x) = 1$ for $-1 \le x \le 1$ and $K(x) = 0$ otherwise) and the default bandwidth ($h = 0.5$) causes no smoothing at all in this case. Increasing the bandwidth parameter to $h = 2.0$ gives the result shown in the upper right plot in Fig. 8.27, while the lower left plot shows the results obtained with the much bandwidth $h = 128$. Comparing both of these kernel smoothing results with the original data sequence, it is clear that the smaller bandwidth result follows the local variations well but does not give a smooth result, while the larger bandwidth gives a result that is much smoother but only captures the large-scale variations seen in the original data sequence. In contrast, the lowess smoothing result with $f = 0.05$ shown in the lower right plot gives the best of both worlds for this example: the result follows the local variations in the data as well as the smaller bandwidth kernel smoother does, but gives a result that is as smooth as the larger bandwidth kernel smoother.

8.8   The preliminary data survey

The chronic fatigue syndrome (CFS) dataset was introduced in Chapter 1 as a representative example of a moderate-sized medical dataset. As noted, this dataset characterizes 227 patients, each of whom either had been diagnosed with CFS in some degree of severity, or who represented non-CFS patients who nevertheless exhibited some symptoms associated with the disease. A detailed analysis of this dataset is beyond the scope of this book, but it is presented here to illustrate, first, what a representative, moderate-sized medical dataset looks like, and second, the kind of preliminary data survey that should be undertaken whenever we are faced with a new dataset.

This dataset contains 67 categorical variables (summarized in Tables 8.2 through 8.5), 18 numerical variables, and one date variable. Besides month, day, and year, this date variable also includes a time field that can be used to specify the time in hours, minutes, and seconds, down to the level of milliseconds. Such fields often appear in medical datasets, although the time fields are almost always blank, as they are in this dataset. It is also worth noting that date variables, while not real values in and of themselves, can be converted into real values, typically expressed in days from some generally accepted time origin. This conversion permits the easy evaluation of time intervals (e.g., the interval between the date patients were prescribed a specific drug and when they first exhibited a particular medical response, either desirable or undesirable). Also, it is important to note that this conversion is sometimes made internally, as in systems that represent dates as a separate data type. This can be a significant source of confusion, since the dates can be displayed in a

Table 8.2: Summary of the first 16 of the 67 CFS categorical variables

No.	Variable	Levels	H_B	H_G	H_{Sh}	H_{Si}
1	Intake.Classific	5	0.197	0.207	0.035	0.026
2	Empiric	15	0.577	0.692	0.289	0.138
3	CLUSTER	3	0.226	0.226	0.051	0.051
4	Onset	2	0.784	0.784	0.507	0.615
5	sex	2	0.639	0.639	0.319	0.408
6	race	6	0.931	0.958	0.840	0.868
7	ethnic	2	0.974	0.974	0.899	0.948
8	Exclusion	2	0.712	0.712	0.405	0.507
9	MDDM.Current	3	0.791	0.847	0.592	0.634
10	MDD.Current	3	0.932	0.941	0.808	0.870
11	Sore.Throat	2	0.515	0.515	0.201	0.266
12	Freq.Sore.Throat	4	0.618	0.733	0.433	0.366
13	Severity.Sore.Throat	3	0.400	0.527	0.215	0.209
14	Tender.Nodes	2	0.674	0.674	0.359	0.454
15	Freq.Nodes	4	0.450	0.514	0.195	0.166
16	Severity.Nodes	3	0.419	0.514	0.229	0.202

form that makes them *look* like a character variable (e.g., "01-27-2009"), but they are treated numerically by arithmetic operations. In other cases, dates may actually be represented as character variables and must be converted to numbers before these computations become meaningful. Unfortunately, both internal and external representations can be highly system-dependent and even inconsistent within the same system. For these reasons, it is important to check any calculations involving dates carefully to be certain the results are correct.

The 67 categorical variables exhibit between 2 and 15 levels, as noted in Tables 8.2 through 8.5, which also lists the name of each variable and the corresponding values for the four heterogeneity measures defined in Chapter 2. Recall that, ideally, these values range from 0, for a uniformly distributed variable, to 1, for a variable that is entirely concentrated in one of its levels. While the relative heterogeneity assessments are not in complete agreement between the four measures listed in the tables, all four measures agree on the two extremes. The most heterogeneous categorical variable in this dataset is "Ethnic" (no. 7 in Table 8.2), which assumes two possible values: *Hispanic* appears 3 times while *non-Hispanic* appears 224 times. This tremendous degree of imbalance is responsible for the very high heterogeneity values assigned by all four measures. While the absolute rankings differ between the four measures for the variables "MDD.Current" and "Race," all four measures do rank them

Table 8.3: Summary of CFS categorical variables 17 through 34

No.	Variable	Levels	H_B	H_G	H_{Sh}	H_{Si}
17	Diarrhea	2	0.198	0.198	0.029	0.039
18	Freq.Diarrhea	4	0.579	0.707	0.382	0.330
19	Severity.Diarrhea	3	0.484	0.626	0.359	0.295
20	Post.Exertion.Fatigue	2	0.057	0.057	0.002	0.003
21	Freq.Post.Exertion.Fatigue	4	0.355	0.427	0.131	0.110
22	Severity.Post.Exertion.Fatigue	3	0.355	0.411	0.120	0.136
23	Muscle.Pain	2	0.489	0.489	0.180	0.239
24	Freq.Muscle.Pain	4	0.209	0.215	0.036	0.033
25	Severity.Muscle.Pain	3	0.343	0.402	0.115	0.128
26	Joint.Pain	2	0.137	0.137	0.013	0.019
27	Freq.Joint.Pain	5	0.421	0.496	0.194	0.131
28	Severity.Joint.Pain	4	0.566	0.695	0.368	0.345
29	Fever	2	0.736	0.736	0.437	0.541
30	Freq.Fever	2	0.333	0.333	0.082	0.111
31	Severity.Fever	3	0.800	0.833	0.574	0.643
32	Chills	2	0.410	0.410	0.125	0.168
33	Freq.Chills	3	0.254	0.269	0.067	0.065
34	Severity.Chills	3	0.410	0.522	0.225	0.206

either second or third, well ahead of the fourth-place variables. The variable "MDD.Current" exhibits the three values *yes* (7 cases), *no* (217 cases), or missing (3 cases), illustrating that "missing" can be regarded as a valid level for a categorical variable. Similarly, "Race" exhibits six possible values, including *AmerInd/Alaska* (1 case), *Asian* (1 case), *Black* (8 cases), *Multiple Race* (2 cases), *Refused* (1 case), and *White* (214 cases). Again, some of the levels for this variable are either ambiguous (i.e., *Multiple Race*) or code missing data (i.e., *Refused*). The least heterogeneous variable, again by all four measures, is "Post.Exertion.Fatigue," which exhibits 120 values of *no* and 107 values of *yes*.

It is also worth noting that many of these categorical variables occur in closely related groupings. For example, variables 35 through 37 are "Unrefreshing.Sleep," "Freq.Unrefreshing.Sleep," and "Severity.Unrefreshing.Sleep," conforming to a general pattern that is repeated 19 times (i.e., "X," "Freq.X," and "Severity.X," where "X" represents a medical condition). In these groupings, the first variable is a binary indicator of the presence or absence of a medical condition, the second is a four-level ordinal variable indicating how frequently the patient experiences the condition (e.g., "rarely," "occasionally," "regularly," or "frequently"), and the third is a three-level ordinal variable indicating the

Table 8.4: Summary of CFS categorical variables 35 through 49

No.	Variable	Levels	H_B	H_G	H_{Sh}	H_{Si}
35	Unrefreshing.Sleep	2	0.454	0.454	0.154	0.206
36	Freq.Unrefreshing.Sleep	4	0.281	0.382	0.106	0.097
37	Severity.Unrefreshing.Sleep	3	0.218	0.236	0.042	0.049
38	Sleep.Problems	2	0.542	0.542	0.224	0.294
39	Freq.Sleep.Problems	4	0.225	0.295	0.060	0.053
40	Severity.Sleep.Problems	3	0.263	0.349	0.087	0.091
41	Headache	2	0.119	0.119	0.010	0.014
42	Freq.Headache	4	0.394	0.454	0.184	0.133
43	Severity.Head.Ache	3	0.252	0.276	0.066	0.065
44	Memory	2	0.551	0.551	0.231	0.303
45	Freq.Memory	4	0.327	0.438	0.137	0.118
46	Severity.Memory	3	0.176	0.216	0.034	0.036
47	Concentration	2	0.427	0.427	0.136	0.183
48	Freq.Concentration	4	0.359	0.395	0.115	0.102
49	Severity.Concentration	3	0.223	0.292	0.061	0.064

severity of these episodes (e.g., "mild," "moderate," or "severe"). This observation suggests two things. First, these variables should probably be regarded as "triplets" for purposes of analysis: i.e., consideration should be given to ways of analyzing "X," "Freq.X," and "Severity.X" together. In particular, note that these variables are strongly associated: if "X" has the binary value corresponding to "condition not present," then "Freq.X" and "Severity.X" values are either missing or meaningless. Second, note that once we have developed a reasonable strategy for combining these related variables for one medical condition "X," it should be possible to adapt this strategy to other medical conditions.

Table 8.6 summarizes the 18 numerical variables included in the CFS dataset, giving their names, their minimum, maximum, and median values, and the number of records for which each value is missing. Note that in all but two cases, this number is zero, implying that this dataset is extremely clean. One of the two cases where missing values occur is "Yrs.Ill," which specifies the number of years the patient had been diagnosed with CFS when the dataset was constructed. Since the nonmissing values range from a minimum of 3.6 to a maximum of 52.8 but the dataset contains patients who had not been diagnosed with CFS, it is likely that "Yrs.Ill" values were not recorded for these patients, explaining much of this missing data. In contrast, the variable "X0" is *completely missing*, suggesting either that a data field was included in the original dataset but never filled in, or that a data conversion error created an extra, unpopulated

Table 8.5: Summary of CFS categorical variables 50 through 67

No.	Variable	Levels	H_B	H_G	H_{Sh}	H_{Si}
50	Nausea	2	0.595	0.595	0.273	0.354
51	Freq.Nausea	3	0.402	0.522	0.219	0.205
52	Severity.Nausea	3	0.543	0.630	0.297	0.318
53	Abdominal.Pain	2	0.366	0.366	0.099	0.134
54	Freq.Abdominal.Pain	4	0.519	0.602	0.285	0.225
55	Severity.Adominal.Pain	3	0.292	0.347	0.096	0.094
56	Sinus.Nasal	2	0.251	0.251	0.046	0.063
57	Freq.Sinus.Nasal	4	0.178	0.202	0.028	0.026
58	Severity.Sinus.Nasal	3	0.352	0.352	0.138	0.124
59	Shortness.of.breath	2	0.498	0.498	0.187	0.248
60	Freq.Shortness.of.breath	3	0.316	0.386	0.118	0.114
61	Severity.Shortness.of.breath	3	0.500	0.632	0.319	0.302
62	Photophobia	2	0.242	0.242	0.043	0.059
63	Freq.Photophobia	4	0.202	0.256	0.054	0.044
64	Severity.Photophobia	3	0.326	0.372	0.120	0.112
65	Depression	2	0.198	0.198	0.029	0.039
66	Freq.Depression	4	0.300	0.348	0.104	0.081
67	Severity.Depression	3	0.456	0.604	0.303	0.274

field somewhere in the process of building the dataset and turning it into an *S-Plus* analysis data frame. The noninformative field name "X0" further supports this conjecture. In any case, since this field contains no data, it can be omitted from subsequent analysis. Another variable that can probably be omitted from subsequent analysis is the patient identifier field "ABTID" which is simply a recoding of the row numbers in the data table. Conversely, such record identifiers sometimes encode additional information (e.g., identity of the data source) that can turn out to be highly influential, although it is often difficult to determine exactly what this additional information is, greatly complicating the interpretation of the results.

Fig. 8.28 shows normal Q-Q plots for four of these numerical variables. The upper left plot shows the results for "Yrs.Ill," from which it is immediately apparent that this variable exhibits a strongly asymmetric distribution. In contrast, the distribution of "Age" values shown in the upper right plot appears much more nearly normally distributed, although close examination of the Q-Q plot suggests significant quantization. Indeed, examination of the data values shows that all age observations are integer-valued. The lower two Q-Q plots are for the variables "Phys.Funct" (lower left) and "Motiv.Reduc" (lower right), and

Table 8.6: Summary of the 18 real variables in the CFS data frame

No.	Variable	Min	Median	Max	Missing
1	ABTID	10043905.0	24245703.0	29499005.0	0
2	Yrs.Ill	3.6	12.1	52.8	56
3	Age	25.0	51.0	69.0	0
4	BMI	16.0	29.0	40.0	0
5	Phys.Funct	5.0	75.0	100.0	0
6	Role.Physic	0.0	75.0	100.0	0
7	Bodily.Pain	0.0	61.0	100.0	0
8	Gnrl.Hlth	10.0	72.0	100.0	0
9	Vitality	0.0	40.0	100.0	0
10	Social.Funct	0.0	75.0	100.0	0
11	Role.Emotional	0.0	100.0	100.0	0
12	Mental.Hlth	12.0	76.0	100.0	0
13	Gen.Fat	4.0	15.0	20.0	0
14	Phys.Fat	4.0	11.0	20.0	0
15	Activ.Reduc	4.0	10.0	20.0	0
16	Motiv.Reduc	4.0	9.0	20.0	0
17	Mental.Fat	4.0	11.0	20.0	0
18	X0	—	—	—	227

both of these variables show pronounced quantization effects, as indicated by the coarse "stepwise" appearence of these plots. Also, note that both of these lower Q-Q plots show strong evidence of saturation effects: the variable "Phys.Funct" in the lower left plot exhibits a large number of observations with the maximum value 100, and the variable "Motiv.Reduc" in the lower right plot exhibits a large number of observations with the minimum value 4. These observations, together with the general staircase-like appearance of the plots suggests that rank-based methods or other analysis approaches that are sensitive to coarse discretization artifacts should probably not be used in characterizing these variables.

8.9 Exercises

1. Show that for the negative binomial distribution, the following relation holds:

$$\frac{kp_k}{p_{k-1}} = (1-p)k + (1-p)(r-1).$$

This result forms the basis for the Ord plots discussed in Sec. 8.4.2.

Figure 8.28: Normal Q-Q plots for four numeric variables from the CFS dataset

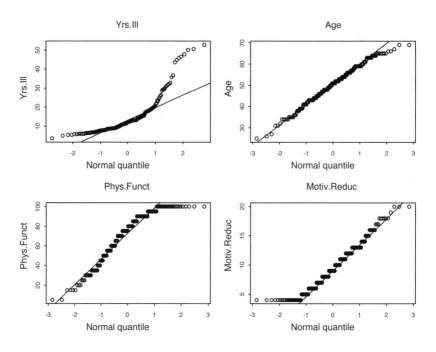

2. The negative binomialness plot is based on the following result for the negative binomial distribution:

$$\ln p_k = r \ln p + k \ln(1 - p) + \ln \Gamma(r + k) - \ln \Gamma(r - 1) - \ln \Gamma(k + 1).$$

Prove this result.

3. Consider the box and triangle kernel density estimates discussed in Sec. 8.6. For a fixed data distribution $p(x)$, a fixed sample size N, and a fixed bandwidth h:

   a. which kernel has smaller bias?
   b. which kernel has smaller variance?

4. Show that the Nadaraya-Watson kernel estimator does not depend on the kernel $K_y(\cdot)$, provided it is a symmetric, proper density function; that is, derive Eq. (8.68).

5. (Computer exercise) Write a program to generate logistic Q-Q plots. Generate four sequences of $N = 200$ Gaussian random numbers and examine their logistic Q-Q plots. Are the differences in these distributions apparent?

6. (Computer exercise) Implement a kernel density estimator based on the triangular kernel. Using the results of Exercise 2, generate a sequence of 500 logistic random variables and obtain kernel density estimates for the following bandwidth parameters: $h = 0.01, 0.05, 0.1,$ and 0.2. Plot these estimates, overlaid with the exact density, given in Chapter 4.

7. (Computer exercise) Implement the Nadaraya-Watson kernel estimator discussed in Sec. 8.7. Generate $N = 500$ points using the following data model:

$$y_k = f(x_k) + e_k, \ 0 \le x_k \le 1, \ e_k \sim N(0, \sigma^2),$$

$$f(x) = \begin{cases} 3x & 0 \le x < 1/3, \\ 1 - 3x/4 & 1/3 \le x < 2/3, \\ 1/2 & 2/3 \le x \le 1. \end{cases}$$

Compare the results obtained for $\sigma = 0.1$ and $\sigma = 0.5$ and for the bandwidths $h = 0.01, 0.05, 0.1,$ and 0.2.

Chapter 9

Confidence Intervals and Hypothesis Testing

Sections and topics:

As noted in Chapter 1, confirmatory data analysis is very much concerned with the design of experiments and the evaluation of their results using the methods of formal hypothesis testing. Hypothesis testing also arises regularly in exploratory data analysis, most commonly in connection with the assessment of data models. As a specific example, if we fit a regression model to a given dataset, as discussed in Chapters 5, 11, and 13, the result is a collection of estimated parameters that define a specific member of a model class that best fits the available data. As discussed in Chapter 15, once we have build this data model, the next step is to assess it: how well does it fit the data? Can it be simplified? Should we consider alternative models? The tools of formal hypothesis testing and the closely related notion of confidence intervals provide a very useful basis for addressing all of these questions. This chapter gives a brief introduction to these topics to serve as the foundation for detailed treatments of these questions given in subsequent chapters.

Key points of this chapter:

1. Results computed from uncertain data are necessarily somewhat imprecise, and an extremely useful way of handling this imprecision is to construct *confidence intervals* for these results, defined as sets of *possible results* constructed to contain the *true result* with a specified probability.

3. The construction of confidence intervals requires an explicit model of the uncertainty in the data from which the result is computed, and it depends on the nature of the characterization under consideration.

4. One of the most important applications of this idea, both historically and in current practice, is the construction of confidence intervals for the mean of a data sequence, and this problem is considered in detail here.

5. Closely related to the notion of a confidence interval is that of a *formal hypothesis test,* which typically has the following form: given a set of working assumptions, is the probability that a certain *null hypothesis* is true less than a specified threshold? An important specific case considered here is the test of whether two computed mean values are equal. There, the specific question posed is whether two *estimated* means differ *enough* in value that the probability the *exact* means are equal is less than a specified value α, typically taken as 5% or 1%.

6. Other important hypothesis testing applications considered here include the χ^2 test of the hypothesis that a categorical variable is described by a specific vector of probabilities, the F-test of the hypothesis that two variances are equal, and tests of the hypothesis that two estimated binary probabilities are equal.

7. As noted, both confidence intervals and hypothesis tests are based on specific distributional assumptions. Classically, these assumptions are imposed, but it is also possible to develop methods based on empirical distributions derived from the data values themselves. This chapter is concerned with classical confidence intervals and hypothesis tests based on fixed distributional assumptions; confidence intervals and hypothesis tests based on empirical distributions (e.g., bootstrap methods) are discussed in Chapter 14.

8. The level of a hypothesis test specifies the probability of falsely rejecting the null hypothesis when it is true. To obtain useful results, it is necessary to make this probability nonzero, and it follows from this fact that if we make a large enough sequence of related hypothesis tests, the probability of at least one of these null hypotheses being falsely rejected can become unacceptably high. This difficulty is commonly called the *multiple comparison problem* in the statistics literature, and many techniques have been developed to address it. The end of this chapter describes three of these that are simple to implement and reasonably popular in practice.

One of the points of Chapter 6 was that if we adopt the random variable model to describe the unavoidable uncertainty inherent in a dataset $\mathcal{D}$, it follows that any characterization $\theta(\mathcal{D})$ that we compute from this dataset can also be characterized as a random variable. This observation provided the foundation for estimator characterizations like bias and consistency discussed in Chapter 6, and it forms the basis for the two topics introduced in this chapter: confidence intervals and hypothesis tests. A detailed discussion of confidence intervals is given in Sec. 9.1, but the basic idea is that a confidence interval for $\theta(\mathcal{D})$ is a set S_α such that the true value for θ lies in set S_α with probability at least $1 - \alpha$, where α is a small number (typically 5% or less). A closely related idea, discussed in detail in Sec. 9.3, is that of a *formal hypothesis test*, which asks whether the observed value for some characterization $\theta(\mathcal{D})$ is consistent with the available data and a specified set of working assumptions, again with probability at least $1 - \alpha$. Specific examples considered in this chapter are tests of the hypotheses that two means are equal, that two variances are equal, that two binomial probabilities are equal, or that a categorical variable has a specified (e.g., uniform) distribution. Both confidence intervals and hypothesis tests are discussed together in this chapter because they are closely related, as the following discussions demonstrate.

9.1   Confidence intervals

The primary characterizations for an estimator $\theta(\mathcal{D})$ introduced in Chapter 6 were based on its first and second moments. As noted above, the *confidence interval* for $\theta(\mathcal{D})$ is a set S_α of values such that the true value θ^* for $\theta(\mathcal{D})$ lies in S_α with probability at least $1 - \alpha$. That is, a confidence interval at level α for θ is *any* set S_α such that:

$$\mathcal{P}\{\theta^* \in S_\alpha\} \geq 1 - \alpha. \tag{9.1}$$

Several points are worth noting here. First, confidence intervals are not unique. In particular, suppose S_α and S_β are confidence intervals at levels α and β, respectively, and that $\beta < \alpha$ (i.e., the defining requirements for S_β are more stringent than those for S_α). It follows immediately from Eq. (9.1) that S_β also qualifies as a confidence interval at level α since:

$$\mathcal{P}\{\theta^* \in S_\beta\} \geq 1 - \beta > 1 - \alpha. \tag{9.2}$$

In fact, as a logical extension, note that if $\theta(\mathcal{D})$ is a real-valued characterization of $\mathcal{D}$, then the entire real line represents a confidence interval at level $\alpha = 0$, implying it is also a confidence interval at level α for *any* $\alpha > 0$. As a practical corollary, it follows that the most useful confidence intervals at level α are the narrowest.

The remainder of this discussion restricts consideration to real-valued characterizations $\theta(\mathcal{D})$, arguably the situation in which confidence intervals are most useful. In addition, we also restrict consideration to the case where S_α is an *interval*, again the most useful and most commonly encountered case in practice, and it simplifies subsequent results to represent this interval in the form $S_\alpha = [\theta(\mathcal{D}) - \phi_-, \theta(\mathcal{D}) + \phi_+]$. (Note that since ϕ_- and ϕ_+ are arbitrary numbers, there is no loss of generality in representing the result this way.) For this case, it is also advantageous to introduce the following representation:

$$\theta(\mathcal{D}) = \theta^* + \delta, \tag{9.3}$$

where θ^* represents the (unknown) true value for $\theta(\mathcal{D})$ in the absence of uncertainty, and δ represents the uncertainty in $\theta(\mathcal{D})$ induced by the uncertainty in the dataset $\mathcal{D}$. Combining these ideas leads to the following result:

$$\begin{aligned}
\mathcal{P}\{\theta^* \in S_\alpha\} &= \mathcal{P}\{\theta(\mathcal{D}) - \phi_- \le \theta^* \le \theta(\mathcal{D}) + \phi_+\} \\
&= \mathcal{P}\{\theta(\mathcal{D}) - \phi_- \le \theta(\mathcal{D}) - \delta \le \theta(\mathcal{D}) + \phi_+\} \\
&= \mathcal{P}\{-\phi_+ \le \delta \le \phi_-\} \\
&= \int_{-\phi_+}^{\phi_-} p(\delta) d\delta.
\end{aligned} \tag{9.4}$$

Note that $\theta(\mathcal{D})$ is a known quantity, computed from the dataset $\mathcal{D}$, and we are seeking constants ϕ_- and ϕ_+ such that, first, condition (9.1) holds and, second, the width $w = \phi_+ + \phi_-$ of the resulting confidence interval is as small as possible. To determine these values, it is necessary to know the distribution $P(\delta)$ of δ, and in classical statistics, this distribution is based on explicit working assumptions. It is also possible to obtain confidence intervals based on empirical data distributions, leading to ideas like the bootstrap discussed in Chapter 14.

Before considering specific examples, the following points are worth noting. First, if $\theta(\mathcal{D})$ is an unbiased estimator, so that $E\{\theta(\mathcal{D})\} = \theta^*$, it follows from Eq. (9.3) that $E\{\delta\} = 0$. If we further suppose that the error distribution is symmetric—another extremely common assumption in practice—it is reasonable to seek symmetric confidence intervals, obtained by setting $\phi_- = \phi_+ = \phi$, leaving us one constant to determine. In fact, these assumptions simplify the problem greatly, leading immediately to an exact solution. Specifically, note that these assumptions imply that we are seeking the smallest positive number ϕ such that the following condition holds:

$$\int_{-\phi}^{\phi} p(\delta) d\delta = 2 \int_0^{\phi} p(\delta) d\delta \ge 1 - \alpha. \tag{9.5}$$

From the definition of the cumulative distribution function $F(\cdot)$ associated with the density $p(\delta)$ and the symmetry of this density, it follows that:

$$F(\phi) = \int_{-\infty}^{\phi} p(\delta) d\delta = \frac{1}{2} + \int_0^{\phi} p(\delta) d\delta. \tag{9.6}$$

Combining Eqs. (9.5) and (9.6) gives:

$$2F(\phi) - 1 \ge 1 - \alpha \Rightarrow F(\phi) \ge 1 - \alpha/2. \tag{9.7}$$

Thus, any ϕ satisfying condition (9.7) will define a symmetric confidence interval at level α. Since $F(\cdot)$ is a monotonically increasing function, it follows that the narrowest such confidence interval will be that defined by the value of ϕ that satisfies this condition with equality, implying:

$$\phi = F^{-1}(1 - \alpha/2). \tag{9.8}$$

Combining all of these results, it follows that the narrowest symmetric confidence interval at level α for an unbiased characterization $\theta(\mathbf{D})$ with a symmetric error distribution $F(\cdot)$ is:

$$S_\alpha = [\theta(\mathcal{D}) - F^{-1}(1 - \alpha/2), \theta(\mathcal{D}) + F^{-1}(1 - \alpha/2)]. \tag{9.9}$$

To obtain explicit numerical values, it is necessary to specify a distribution $F(\cdot)$, as the next two examples illustrate.

9.1.1 Application: systematic errors

It was noted in Chapter 2 that the Gaussian white noise model is a very popular working assumption for measurement noise in real-valued data like temperature, pressure, or viscosity. That is, for a sequence $\{x_k\}$ of repeated measurements of the variable x in a setting where its true underlying value is unknown but constant (i.e., $x = c$), this model assumes:

$$x_k = c + e_k, \tag{9.10}$$

where $\{e_k\}$ is an independent, identically distributed (i.i.d.) sequence of Gaussian random variables with mean zero and constant variance σ^2. Given the extreme popularity of this model, it is reasonable to ask to what extent it can be validated experimentally. To do so, we could perform the experiment defined by the description of the error model just given. That is, we could make a sequence $\{x_k\}$ of repeated measurements in a setting where the true value of the measurand x was both constant and *known*. For example, in measuring temperature, these measurements could be made at the boiling point of water at known atmospheric pressure. Alternatively, for electrical measurements like voltage or resistance, we could use a standard voltage reference or a standard resistor, whose value is traceble to the National Institute of Standards and Technology. In any case, if we assume c is both constant and known, we can compute the measurement error values as $e_k = x_k - c$.

Given the sequence $\{e_k\}$, an informal assessment of the normal distribution working assumption can be made using the Q-Q plots introduced in Chapter 8, and the utility of this characterization can be enhanced using the bootstrap confidence intervals described in Chapter 14. In addition, there exist a number of formal tests for the Gaussian hypothesis [195, ch. 1], closely related to the ideas discussed in this chapter, but the Q-Q plots and their extensions described in Chapter 8 are simpler, are generally adequate for exploratory applications, and provide additional information beyond whether the normality assumption seems reasonable or not. Similarly, formal tests of the statistical independence hypothesis (i.e., that $\{x_k\}$ is an i.i.d. sequence) are described by Priestley [238, pp. 478–485] and Brockwell and Davis [44, pp. 312–313] in connection with dynamic data analysis (e.g., time-series modeling and spectrum estimation). Here, we consider the case where we are willing to at least provisionally accept the Gaussian i.i.d. working hypotheses and ask the following question: given the sequence $\{e_k\}$ of measurement errors just described, is there evidence of a *systematic* error? That is, is there evidence that the error distribution is $N(\mu, \sigma^2)$ for some nonzero mean μ?

First, consider the case where the variance σ^2 is known. While this assumption is usually unrealistic in practice, the result we obtain is simple and illuminating, providing the foundation and a frame of reference for results discussed in Sec. 9.1.2. Under this assumption, it follows from the results presented in Chapter 6 that the mean of the observation errors also exhibits a Gaussian distribution, with mean μ and variance σ^2/N where N is the number of data observations:

$$\bar{e} = \frac{1}{N} \sum_{k=1}^{N} e_k \sim N(\mu, \sigma^2/N). \tag{9.11}$$

Thus, $\bar{e}$ is an unbiased estimator of the mean, so it follows from Eq. (9.9) that the level α confidence interval for μ is:

$$S_\alpha = [\bar{e} - F^{-1}(1 - \alpha/2), \ \bar{e} + F^{-1}(1 - \alpha/2)]. \tag{9.12}$$

Here, $F^{-1}(1 - \alpha/2)$ is the upper tail probability for the $N(0, \sigma^2/N)$ distribution associated with the variations δ in $\bar{e}$ about the mean μ. To determine this value, first note that if $u \sim N(0, 1)$ is a standard normal random variable and $x = (\sigma/\sqrt{N})u$, then the distribution of x is $x \sim N(0, \sigma^2/N)$. Thus, we can replace $F^{-1}(1 - \alpha/2)$ in Eq. (9.12) with $z_{1-\alpha/2}\sigma/\sqrt{N}$, where $z_{1-\alpha/2}$ is the corresponding quantile for the standard normal distribution, defined by:

$$\frac{1}{\sqrt{2\pi}} \int_{-\infty}^{z_{1-\alpha/2}} e^{-x^2/2} dx = 1 - \alpha/2. \tag{9.13}$$

While there is no simple analytical expression for $z_{1-\alpha/2}$, it is widely tabulated for common threshold values α [162, p. 84] and it is available as a built-in function in many statistical software packages, including both *S-Plus* and *R*. Probably the two most popular choices are the 5% significance level, $\alpha = 0.05$, for which $z_{1-\alpha/2} \simeq 1.96$, and the 1% level, $\alpha = 0.01$, for which $z_{1-\alpha/2} \simeq 2.58$. Combining these results, the symmetric α level confidence interval for the unknown mean μ in this example is:

$$S_\alpha = [\bar{e} - z_{1-\alpha/2}\sigma^2/\sqrt{N}, \ \bar{e} + z_{1-\alpha/2}\sigma^2/\sqrt{N}]. \tag{9.14}$$

Note that the width of this interval is $w = 2z_{1-\alpha/2}\sigma/\sqrt{N}$, which represents a measure of the uncertainty in our estimate of μ. Further, note that this width depends on the variance σ^2 (i.e., the less variable the data, the more certain our results), the confidence limit α (i.e., the more certain we insist on being that the true value μ falls in the interval, the wider we must make that interval), and the sample size N (i.e., the more data we have, the more certain we can be of our results).

Returning to the question of whether our hypothesis of no systematic error (i.e., that $\mu = 0$) is consistent with the available data, note that if it is consistent, the confidence interval given in Eq. (9.14) should include the value $\mu = 0$. Conversely, if this interval does *not* include $\mu = 0$, it follows that the probability that there is no systematic measurement error is less than α. As a practical matter, then, if we choose α small (e.g., 5%, 1%, or 0.1%) and the confidence interval defined in Eq. (9.14) does not include the value $\mu = 0$, this observation provides evidence of systematic measurement error, *assuming the working hypotheses of an i.i.d. sequence of Gaussian random errors with known variance σ^2 is reasonable*. This point is discussed further in Sec. 9.3, where the notion of formal hypothesis testing is discussed.

9.1.2 The case of unknown variance

Arguably, the weakest assumption in the example just presented was that the variance σ^2 was known. Historically, the case of unknown variance was treated as just described, with σ^2 simply replaced by its estimate, obtained as described in Chapter 6:

$$\hat{\sigma}^2 = \frac{1}{N-1} \sum_{k=1}^{N} (e_k - \bar{e})^2. \tag{9.15}$$

In 1908, W.S. Gossett (publishing under the pseudonym "Student") derived the explicit distribution for the following quantity:

$$t = \frac{\sqrt{N}(\bar{e} - \mu)}{\hat{\sigma}}, \tag{9.16}$$

showing that this distribution can be markedly non-Gaussian. Specifically, he showed that t defined in Eq. (9.16) exhibits what has come to be known as *Student's t-distribution with $N - 1$ degrees of freedom*, discussed in Chapter 4. Recall that for fewer than 3 degrees of freedom, this distribution has infinite variance, and for fewer than 5 degrees of freedom, it has infinite kurtosis. The practical consequence is that while the Gaussian confidence intervals discussed previously are asymptotically correct, becoming accurate in the limit of large sample sizes, they are too narrow for finite sample sizes. That is, the level α confidence interval presented in Eq. (9.14) for the Gaussian case must be replaced with the following interval:

$$S_\alpha = [\bar{e} - t_{1-\alpha/2,N-1}\sigma/\sqrt{N},\ \bar{e} + t_{1-\alpha/2,N-1}\sigma/\sqrt{N}], \tag{9.17}$$

where $t_{1-\alpha/2,N-1}$ is the upper $1 - \alpha/2$ quantile for the t distribution with $N-1$ degrees of freedom. To see the difference between these distributions, note that for $\alpha = 0.05$, the Gaussian multiplier is $z_{1-\alpha/2} \simeq 1.96$ while the Student's t

result is $t_{1-\alpha/2,N-1} \simeq 2.26$ for $N = 10$, about 15% larger than the Gaussian value; for $N = 30$, $t_{1-\alpha/2,N-1} \simeq 2.05$, only about 5% larger, demonstrating the point noted above that the difference becomes less important with increasing sample size. More important, however, the differences become more pronounced as we move farther out into the tails: for $\alpha = 0.01$, $z_{1-\alpha/2} \simeq 2.58$, vs. $t_{1-\alpha/2,N-1} \simeq 3.25$ for $N = 10$ (~26% larger) and $t_{1-\alpha/2,N-1} \simeq 2.76$ for $N = 30$ (~7% larger).

The two key points of this example are, first, that using estimated variances instead of known variances increases the width of the resulting confidence intervals. Intuitively, this is what we should expect, since we are replacing a precisely known value with one that must be estimated from imprecise data. The second point is that even in a very simple case like that considered here—determination of the confidence interval for the mean of a sequence $\{e_k\}$ of observed, Gaussian data values—the exact distributions on which classical confidence intervals are based may not be simple, well behaved, or easily derived. This observation motivates the computational approaches for constructing confidence intervals (e.g., bootstrap methods) introduced in Chapter 14, which are extremely general.

9.2 Extensions of the Poissonness plot

An extremely useful application of confidence intervals is to provide a frame of reference for interpreting graphical displays. This idea is illustrated further in Chapter 14, where the bootstrap confidence intervals are introduced, derived from the empirical data distribution rather than distributional assumptions. The following example illustrates how approximate confidence intervals can be used to enhance the Poissonness plot introduced in Chapter 8 as an informal graphical method for assessing the reasonableness of the Poisson model for a sequence $\{n_k\}$ of count data. Recall that this display consists of a plot of the Poissonness count metameter:

$$\phi(n_k) = \ln n_k - \ln N + \ln \Gamma(k+1), \qquad (9.18)$$

where N is the sum of the individual counts $\{n_k\}$. The following discussion introduces two extensions of this plot: the first is the addition of confidence intervals for the metameter values $\phi(n_k)$, and the second is a modification of the count sequence $\{n_k\}$ that makes $\phi(n_k)$ fall more nearly in the center of these confidence intervals. These results are approximate, based on arguments presented by Hoaglin *et al.* [141, ch. 9], which should be consulted for detailed derivations and more detailed discussions.

The modification of the count sequence proposed by Hoaglin *et al.* involves replacing n_k with n_k^* defined by:

$$n_k^* = \begin{cases} n_k - 0.67 - 0.8n_k/N & n_k \geq 2, \\ 1/e & n_k = 1, \\ \text{undefined} & n_k = 0. \end{cases} \qquad (9.19)$$

Note that the only uncertain term in the count metameter $\phi(n_k)$ is the logarithm $\ln n_k$, which is replaced by $\ln n_k^*$ in Hoaglin *et al.*'s modified count metameter definition:

$$\phi(n_k^*) = \ln n_k^* - \ln N + \ln \Gamma(k+1). \qquad (9.20)$$

For $n_k \geq 2$, the approximate confidence intervals for $\ln n_k^*$ are defined by the limits:

$$\ln n_k^* \pm \frac{1.96\sqrt{1 - n_k/N}}{n_k - (0.47 + 0.25 n_k/N)\sqrt{n_k}}, \qquad (9.21)$$

while for $n_k = 1$, the approximate confidence interval is:

$$[\ln n_k^* - 2.67, \ \ln n_k^* + 2.717 - 2.3/N] = [-3.67, 1.717 - 2.3/N]. \qquad (9.22)$$

Combining these results yields the basis for a modified Poissonness plot with three sets of points: the central curve is obtained by plotting $\phi(n_k^*)$ against k, while upper and lower curves are defined by plotting the upper and lower confidence limits on $\phi(n_k^*)$ derived from Eqs. (9.21) and (9.22) against k.

This construction is illustrated in Fig. 9.1, based on the *Federalist Papers* example discussed in Chapter 8. Specifically, this figure shows the modified Poissonness plot just described based on counts of the word "may" appearing in documents known to have been written by Madison. Recall that this example was used in Chapter 8 to illustrate the influence of points based on a single count (i.e., k for which $n_k = 1$) on least squares lines fit to the points defining the original plot. The advantage of the modified Poissonness plot is that it shows clearly the large uncertainty associated with these points, as well as other points based on other small counts. In particular, note that the least uncertainty is associated with $k = 0$, corresponding to the relatively large number of text blocks authored by Maidson that did not contain the word "may" at all. In contrast, the widest confidence intervals are those associated with single counts, where the points are represented as open circles.

9.3 Formal hypothesis tests

Hypothesis testing is central to confirmatory data analysis, where the objective is to establish the statistical significance of a result. A classic example is that of a clinical trial: two treatments (or one treatment and a placebo) are compared to determine whether there is evidence that one is more effective than the other in a given population. In exploratory data analysis, hypothesis tests are more commonly applied to assess the statistical significance of parameters in a data model. For example, we fit a straight line to a collection of points in the plane, and we compute the slope to be 0.032. Is this number large enough to be significant, or could the true slope equally well be zero? The answer to this question depends on the scatter in the data, and this particular example is discussed in detail in Chapter 14. The purpose of the discussion presented here is to introduce some of the basic ideas and terminology and to illustrate these with some common examples of hypothesis tests that arise frequently.

Figure 9.1: Modified Poissonness plot for counts of the word "may" in documents written by James Madison. The central points correspond to the modified count metameter $\phi(n_k^*)$, while the upper and lower points define approximate 95% confidence limits

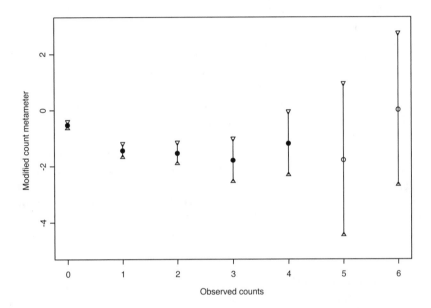

The basic notion of a formal hypothesis test is the following. By imposing certain working assumptions, it is often possible to define a *test statistic* that satisfies two criteria: first, its distribution is known in the case where the working assumptions hold, and second, it can be computed from observed data. When these criteria are met, we can compute the value of the test statistic and assess its probability relative to the working assumptions. If this probability is sufficiently low, we conclude that the working assumptions do not hold.

The following example illustrates the basic idea. Consider the problem formulation given in Sec. 9.1.1, where we have a sequence $\{x_k\}$ of repeated measurements of some reference standard that should give a known value c, allowing us to compute a sequence of N observation errors, $e_k = x_k - c$. If we now *assume* that $\{e_k\}$ is a sequence of zero-mean, Gaussian i.i.d. random variables with known variance σ, it follows that the distribution of the mean observation error $\bar{e}$ is $N(0, \sigma^2/N)$, as noted in the preceeding discussion. Given a nonzero observed value $\bar{e}$, we can then compute the probability of observing a value this large or larger:

$$p_+ = \mathcal{P}\{x \geq \bar{e} | x \sim N(0, \sigma^2/N)\} = 1 - F(\bar{e}), \qquad (9.23)$$

where $F(\cdot)$ is the cumulative distribution function for the $N(0, \sigma^2/N)$ distribution. If this probability is smaller than some specified threshold value α (e.g., 5%, 1%, or 0.1%), we reject the working hypothesis (here, that there are no systematic measurement errors) at this level of confidence.

Several points are worth noting here. First, the set of working assumptions tested—i.e., that the observation errors are an i.i.d. Gaussian sequence with zero mean and known variance σ in this case—is called the *null hypothesis*, and a formal hypothesis test is fundamentally concerned with *rejecting* the null hypothesis. That is, the null hypothesis represents a set of working assumptions that allow us to determine the distribution of our test statistic. If the computed value of the test statistic has a sufficiently small probability under the null hypothesis, this means that the observed data values are inconsistent with these working assumptions, so we must reject the null hypothesis, at the stated level of significance. Conversely, if the test statistic has a large probability under the null hypothesis, we do not "accept" this hypothesis, we only "fail to reject it." That is, in cases where the evidence against the null hypothesis is not significant at the stated level, it is still possible that some other mechanism could be responsible for the observed outcome (e.g., in the example considered here, $\bar{e}$ could be the mean of a Gaussian distribution with some nonzero mean and a sufficiently large variance). Conversely, if $\bar{e}$ lies too far from zero, relative to σ^2/N, we can firmly conclude that the $N(0, \sigma^2/N)$ model postulated under the null hypothesis is untenable.

Another point to note about the previous example is that it represents what is called a *one-sided hypothesis test*. That is, we are asking whether a value as large as $\bar{e}$ or larger is consistent with the null hypothesis. Alternatively, we could ask whether a value as small as $\bar{e}$ or smaller is consistent with the null hypothesis, and this would be a different one-sided hypothesis test. In the example just considered, however, it is probably more reasonable to ask whether the observed *magnitude* of $\bar{e}$ is too large to be consistent with the lack of a systematic error, allowing for the possibility that the systematic error might have either sign (i.e., it might be either a consistent overestimate, or a consistent underestimate). This case corresponds to a *two-sided hypothesis test* where we are asking whether the probability of observing an average error with magnitude as large as $|\bar{e}|$ or larger is less than some specified value. For a zero-mean, symmetric distribution like the standard normal, it is easily demonstrated that (Exercise 1):

$$\mathcal{P}\{|x| \geq |\bar{e}|\} = 2[1 - F(|\bar{e}|)]. \tag{9.24}$$

9.4 Comparing means

One of the best-known hypothesis tests asks whether the means of two sequences of numbers are equal. As in the measurement noise examples discussed earlier, the classical setting for this problem assumes that both sequences are i.i.d. Gaussian sequences, and the question of whether the variances are known or not plays an important role. In fact, as the following discussion emphasizes, the problem considered here is very closly related to the examples discussed already.

9.4.1   The classical t-test

The classical t-test considers the following basic problem. Given a sequence $\{x_k\}$ of n_x observations and $\{y_k\}$ of n_y observations, each corresponding to i.i.d. samples from Gaussian distributions, independent from each other, is it feasible that the means of these distributions are the same? That is, in the general case, we assume $x_k \sim N(\mu_x, \sigma_x^2)$ and $y_k \sim N(\mu_y, \sigma_y^2)$ and take as the null hypothesis that $\mu_x = \mu_y$. As in the examples considered in secs. 9.1.1 and 9.1.2, what we assume about the variances plays an important role here.

The simplest case assumes that both variances are known, allowing the problem to be reduced exactly to that considered in Sec. 9.1.1. Specifically, let $\bar{x}$ and $\bar{y}$ denote the means of the sequences $\{x_k\}$ and $\{y_k\}$, respectively, and note that, under the hypotheses stated, $\bar{x} \sim N(\mu_x, \sigma_x^2/n_x)$ and $\bar{y} \sim N(\mu_y, \sigma_y^2/n_y)$. Further, since it is assumed that $\{x_k\}$ and $\{y_k\}$ are statistically independent of one another, it follows that the distribution of the difference is:

$$\bar{x} - \bar{y} \sim N(\mu_x - \mu_y, \sigma_x^2/n_x + \sigma_y^2/n_y). \tag{9.25}$$

Thus, under the null hypothesis that $\mu_x = \mu_y$, the difference in means should have an expected value of zero and a variance of $\sigma_x^2/n_x + \sigma_y^2/n_y$. This observation implies the following result:

$$t = \frac{\bar{x} - \bar{y}}{\sqrt{\frac{\sigma_x^2}{n_x} + \frac{\sigma_y^2}{n_y}}} \sim N(0, 1). \tag{9.26}$$

In words, Eq. (9.26) means that the t-statistic defined here is computable from the available data (subject to the assumption of known variances) and has a known distribution. Thus, if we compute t and its value is too large in magnitude to be consistent with the $N(0, 1)$ distribution, we reject the hypothesis that the means are equal.

While this result provides a nice, simple illustration of the ideas behind the t-test of the hypothesis that two means are equal, it is seldom useful because the known variance assumption is usually unrealistic. As discussed in Sec. 9.1.2, an apparently reasonable alternative is to simply replace the variances σ_x^2 and σ_y^2 in Eq. (9.26) with their estimates. As in the case considered in Sec. 9.1.2, this replacement changes the distribution of the test statistic, but here the problem is more subtle, requiring us to distinguish between two subcases.

First, if σ_x and σ_y are unknown *but assumed to be equal*, the t-statistic defined in Eq. (9.26) may be rewritten as:

$$t = \frac{\bar{x} - \bar{y}}{\sigma\sqrt{\frac{1}{n_x} + \frac{1}{n_y}}}, \tag{9.27}$$

where σ is the common value of σ_x and σ_y. In this case, replacing σ by its estimate leads to the Student's t-distribution just as in Sec. 9.1.2. That is, if $\sigma_x = \sigma_y = \sigma$, we estimate this unknown variance using:

$$\hat{\sigma} = \frac{1}{n_x + n_y - 2} \left[\sum_{k=1}^{n_x} (x_k - \bar{x})^2 + \sum_{k=1}^{n_y} (y_k - \bar{y})^2 \right]. \tag{9.28}$$

Substituting this result into Eq. (9.26) then yields the result:

$$t = \frac{\bar{x} - \bar{y}}{\hat{\sigma}\sqrt{\frac{1}{n_x} + \frac{1}{n_y}}} \sim t_{n_x+n_y-2}. \tag{9.29}$$

More explicitly, if we adopt these working assumptions—that $\{x_k\}$ and $\{y_k\}$ are statistically independent, Gaussian i.i.d. sequences with means μ_x and μ_y and a common unknown variance σ—we will reject the hypothesis that $\mu_x = \mu_y$ with confidence $1 - \alpha$ (e.g., 95% if $\alpha = 0.05$) if the following condition is met:

$$|t| > t_{1-\alpha/2,n_x+n_y-2}. \tag{9.30}$$

The general case where the variances are both unknown and unequal is known as the *Behrens-Fisher problem* and has been discussed fairly extensively in the statistics literature (see, e.g., Lehmann and Romano's discussions of this problem [181]). Both these authors [181, pp. 447–448] and Madansky [195, pp. 216-217] describe *Welch's approximate t-test* that assumes t as defined in Eq. (9.26) with the variances σ_x^2 and σ_y^2 replaced by their estimates exhibit a Student's t-distribution with a modified number of degrees of freedom, computed from n_x, n_y, $\hat{\sigma}_x$, and $\hat{\sigma}_y$. A detailed expression is not given here because it is a special case of the Yuen-Welch test discussed in Sec. 9.4.4, where the general result is given.

9.4.2   Limitations of the t-test

Despite its historical importance and continuing popularity, the t-test can perform very badly in the face of non-normality. Wilcox [295, pp. 157–158] discusses this point in detail, noting that the problem is particularly acute when the two distributions differ in symmetry. To illustrate this point, the following discussion presents a brief simulation-based case study that compares the number of times in 1,000 statistically independent simulations that the t-test as defined by Eqs. (9.28) through (9.30) gives the correct result. In all cases, samples $\{x_k\}$ and $\{y_k\}$ of length $n_x = n_y = 100$ are compared, and the tests are evaluated at the 5% significance level ($\alpha = 0.05$). Overall, 19 different cases are considered: in each one, the mean of the $\{x_k\}$ sequence is zero, while the mean of the $\{y_k\}$ sequence is zero in 6 cases and nonzero in the other 13.

The results of these comparisons are summarized in Table 9.1. The $\{x_k\}$ sequence exhibits a standard normal (i.e., $N(0,1)$) distribution in cases 1 through 12, and an exponential distribution with scale parameter $\alpha = 1$ in cases 13

Table 9.1: Summary of 1,000 t-test results for 19 cases

Case	$\{x_k\}$	$\{y_k\}$	$\mu_x = \mu_y$?	Rejected
1	$N(0,1)$	$N(0,1)$	Yes	3.9%
2	$N(0,1)$	$N(0.5,1)$	No	94.7%
3	$N(0,1)$	$N(1,1)$	No	100.0%
4	$N(0,1)$	$N(0,2)$	Yes	4.6%
5	$N(0,1)$	$N(0.5,2)$	No	59.9%
6	$N(0,1)$	$N(1,2)$	No	99.4%
7	$N(0,1)$	$C(0,1)$	Yes	2.0%
8	$N(0,1)$	$C(0.5,1)$	No	6.8%
9	$N(0,1)$	$C(1,1)$	No	20.2%
10	$N(0,1)$	$E(1) - 1$	Yes	5.3%
11	$N(0,1)$	$E(1) - 0.5$	No	94.8%
12	$N(0,1)$	$E(1)$	No	100.0%
13	$E(1) - 1$	$E(1) - 1$	Yes	5.5%
14	$E(1) - 1$	$E(1) - 0.5$	No	92.7%
15	$E(1) - 1$	$E(1)$	No	100.0%
16	$E(1) - 1$	$E(2) - 2$	Yes	5.7%
17	$E(1) - 1$	$E(2) - 1.5$	No	59.8%
18	$E(1) - 1$	$E(2) - 2.5$	No	62.4%
19	$E(1) - 1$	$E(2) - 1$	No	99.8%

through 19, but shifted to have a zero mean. Recall from Chapter 4 that the exponential distribution with parameter α has both mean and variance equal to α, so the distribution of $\{y_k\}$ exhibits a mean of 1 in cases 12, 15, and 19, a mean of 0.5 in cases 11, 14, and 17, and a mean of -0.5 in case 18. Also, note that while the mean of the Cauchy distribution is not strictly defined, the center of symmetry of the Cauchy distribution $C(\mu, \beta)$ is μ, which provides a basis for defining a mean value; thus the mean value for the $\{y_k\}$ sequence is 0.5 in case 8 and 1 in case 9. Table 9.1 gives the correct response for the t-test (i.e., is $\mu_x = \mu_y$?) and the percentage of the 1,000 simulations for which the t-test gave the correct result.

It is clear from these results that the t-test performs very much as expected in the setting for which it was developed (e.g., comparison of Gaussian sequences as in cases 1 through 6), rejecting the null hypothesis of equality of means when it is false at approximately its 5% target rate (cases 1 and 4), and consistently rejecting the null hypothesis when it is false, at least for sufficiently large differences. In particular, note that the rejection rate is lower for a mean difference of 0.5 than for a mean difference of 1.0 (e.g., cases 2 vs. 3 and 5 vs. 6). Further, the effect of unequal variances becomes more pronounced when the mean difference is small and detection is more difficult: compare the rejection rates for cases 3 (equal variances) and 6 (unequal variances). In contrast, when the second data sequence exhibits the heavy-tailed Cauchy distribution, the t-test essentially fails altogether, very rarely detecting a difference in means even when it is present (cases 8 and 9). This failure is a consequence of the inflation of the estimated variance caused by the extreme data values in the Cauchy sequence, and is closely related to the outlier sensitivity of standard estimators discussed in Chapter 7. Finally, the results for cases 10 through 19 involving the asymmetric exponential distribution are quite comparable to those for the corresponding Gaussian cases. In particular, the probability of rejecting the null hypothesis when it is true is approximately 5% in all of the cases considered here, and the probability of rejecting it when it is false is quite high in all cases except cases 17 and 18, where the differences in scale between the two exponential distributions involved appear to play essentially the same role as in case 5.

9.4.3   The Wilcoxon rank-sum test

One of the most popular alternatives to the t-test for cases where the normality assumption underlying the t-test is questionable is the *Wilcoxon rank-sum test* [129, p. 96]. The null hypothesis for this test is that the two samples $\{x_k\}$ and $\{y_k\}$ have the same (arbitrary) distribution. The starting point for this test is to form the *pooled sample:*

$$z_k = \begin{cases} x_k & k = 1, 2, \ldots, n_x, \\ y_{k-n_x} & k = n_x + 1, n_x + 2, \ldots, n_x + n_y. \end{cases} \tag{9.31}$$

This pooled sample is then rank-ordered to obtain:

$$z_{(1)} < z_{(2)} < \cdots < z_{(n_x + n_y)}, \tag{9.32}$$

and the ranks $R(x_k)$ of the $\{x_k\}$ sequence are defined as:

$$R(x_k) = j \text{ such that } z_{(j)} = x_k. \tag{9.33}$$

Wilcoxon's test statistic is:

$$S = \sum_{k=1}^{n_x} R(x_k). \tag{9.34}$$

The idea behind this test is that if the null hypothesis is true (i.e., if $\{x_k\}$ and $\{y_k\}$ are independent samples drawn from the same distribution), the rank-sum S should not be either "too large" or "too small." In particular, this result would imply either that the x_k values were consistently larger or consistently smaller than the y_k values, providing evidence against the null hypothesis.

If n_x and n_y are sufficiently small, the possible values for S can be enumerated exactly, providing the basis for exact calculations of the tail probabilities of the distribution of S. A simple illustration of this point is presented in Chapter 14 in connection with the closely related idea of *permutation tests*. In the R and *S-Plus* statistics packages, the Wilcoxon rank-sum procedure, **wilcox.test**, provides the option of computing this exact distribution. For sufficiently large values of $n_x + n_y$, these computations become prohibitively complex and even if the exact distribution is requested, the result returned is based on the asymptotic approximation that is most widely used in practice.

The basis for this approximation is the fact that the test statistic S exhibits asymptotic normality. Specifically, the mean and variance of S are [129, p. 97]:

$$
\begin{aligned}
E\{S\} &= \frac{n_x(n_x + n_y + 1)}{2}, \\
\text{var}\{S\} &= \frac{n_x n_y (n_x + n_y + 1)}{12},
\end{aligned}
\tag{9.35}
$$

and the following normalized test statistic approaches a standard normal distribution in the limit of infinitely large samples [129, p. 98]:

$$
S_0 = \frac{S - E\{S\}}{\sqrt{\text{var}\{S\}}} \to N(0, 1).
\tag{9.36}
$$

This result provides the basis for the approximate test that rejects the hypothesis that the distributions of $\{x_k\}$ and $\{y_k\}$ are the same if:

$$
|S_0| > z_{1-\alpha/2} \Leftrightarrow \left| S - \frac{n_x(n_x + n_y + 1)}{2} \right| > z_{1-\alpha/2} \sqrt{\frac{n_x n_y (n_x + n_y + 1)}{12}}.
\tag{9.37}
$$

Table 9.2 summarizes the performance of the Wilcoxon rank-sum test for the same 19 examples considered in Sec. 9.4.2. Comparing these results with those given in Table 9.1 shows that the rank-sum test performs quite comparably for the Gaussian comparisons (cases 1 through 6), but the performance is *dramatically* better for the Gaussian/Cauchy comparisons: the rejection probability is closer to the target 5% level when there is no difference in means (5.9% vs. 2.0% for case 7), but the power to reject the null hypothesis when it is false is much improved for the rank-sum test (69.0% vs. 6.8% for the more challenging $C(0.5, 1)$ comparison in case 8, and 99.5% vs. 20.2% for the easier $C(1, 1)$ comparison in case 9). Conversely, a price paid for this performance improvement for the heavy-tailed Cauchy distribution is a significant degradation in performance with respect to the asymmetric exponential distribution. In particular, note that the rate of rejection of the null hypothesis when it is true in the Gaussian/exponential comparison in case 10 is 14.4%, nearly three times the 5% target level for the test. Far worse, the rank-sum test exhibits a false rejection rate of 65% in case 16 where the two exponential distributions compared differ in scale. Also, note that the true mean difference in case 17 is $+0.5$ while that in case 18 is -0.5: in the t-test results presented in Sec. 9.4.2, the test rejected both cases about equally often (59.8% vs. 62.4%), but the rank-sum test exhibits radically different results for these two cases (8.4% vs. 99.7%). It is clear from these results that the Wilcoxon rank-sum test suffers as much in the face of pronounced asymmetry as the t-test does in the face of heavy-tailed distributions.

Table 9.2: Summary of 1,000 Wilcoxon rank-sum test results for 19 cases

Case	$\{x_k\}$	$\{y_k\}$	$\mu_x = \mu_y$?	Rejected
1	$N(0,1)$	$N(0,1)$	Yes	4.4%
2	$N(0,1)$	$N(0.5,1)$	No	93.4%
3	$N(0,1)$	$N(1,1)$	No	100.0%
4	$N(0,1)$	$N(0,2)$	Yes	5.3%
5	$N(0,1)$	$N(0.5,2)$	No	56.8%
6	$N(0,1)$	$N(1,2)$	No	99.0%
7	$N(0,1)$	$C(0,1)$	Yes	5.9%
8	$N(0,1)$	$C(0.5,1)$	No	69.0%
9	$N(0,1)$	$C(1,1)$	No	99.5%
10	$N(0,1)$	$E(1)-1$	Yes	14.4%
11	$N(0,1)$	$E(1)-0.5$	No	82.4%
12	$N(0,1)$	$E(1)$	No	100.0%
13	$E(1)-1$	$E(1)-1$	Yes	4.3%
14	$E(1)-1$	$E(1)-0.5$	No	99.8%
15	$E(1)-1$	$E(1)$	No	100.0%
16	$E(1)-1$	$E(2)-2$	Yes	65.0%
17	$E(1)-1$	$E(2)-1.5$	No	8.4%
18	$E(1)-1$	$E(2)-2.5$	No	99.7%
19	$E(1)-1$	$E(2)-1$	No	98.1%

The Wilcoxon rank-sum test rests on two fundamental working assumptions. The first is that the distributions are identical in form, which implies that the variances of the two samples $\{x_k\}$ and $\{y_k\}$ are the same, as are the skewness values, kurtosis values, etc. In cases where this condition does not hold, it is known that the variance estimate given in Eq. (9.35) is incorrect, and this problem manifests itself in effects like the complicated dependence on asymmetry seen in the results just discussed. Attempts to address these problems have led to extensions of the test like the method of Mee described by Wilcox [295, sec. 5.7.1]. The second key working assumption is that the two sequences $\{x_k\}$ and $\{y_k\}$ are samples drawn from continuous distributions so there are no ties (i.e., $x_i \neq x_j$ and $y_i \neq y_j$ for all $i \neq j$). In cases where tied values are possible (e.g., if $\{x_k\}$ and $\{y_k\}$ are integer-valued counts), a modification of the method is required, like the method of Cliff described by Wilcox [295, sec. 5.7.2].

9.4.4 The Yuen-Welch and Welch rank-based tests

Although not as well known as the Wilcoxon rank-sum test, another alternative to the classical t-test that addresses the problem of non-Gaussian data distributions is the Yuen-Welch test [300], an extension of the Welch test mentioned at the end of Sec. 9.4.1 to address the unequal variance problem (i.e., the Behrens-Fisher problem) that arises frequently in practice. The essential idea of the Welch test is to modify the degrees of freedom used in evaluating the test statistic to correct for the effects of unequal, unknown sample variances. The essential idea behind the Yuen-Welch test is is to replace the means appearing in the Welch test with trimmed means to reduce the dependence of the results on the tails of non-Gaussian data distributions. A reasonably detailed discussion of the Yuen-Welch test is given by Wilcox [295, sec. 5.3], who recommends 20% symmetric trimmed means as a practical basis for implementing this test.

Recall from the discussion in Chapter 6 that the symmetric γ-trimmed mean of a data sequence $\{x_k\}$ of length N is obtained by first rank-ordering it to obtain the sequence $\{x_{(i)}\}$ satisfying:

$$x_{(1)} \leq x_{(2)} \leq \cdots \leq x_{(N)}. \tag{9.38}$$

Next, let $g = [\gamma N]$ where $[x]$ denotes the largest integer that does not exceed x and let $\tilde{x}$ denote the trimmed mean, given by:

$$\tilde{x} = \frac{x_{(g+1)} + \cdots + x_{(N-g)}}{N - 2g}. \tag{9.39}$$

The null hypothesis for the Yuen-Welch test is that the symmetric γ trimmed means μ_{tx} and μ_{ty} for the samples $\{x_k\}$ and $\{y_k\}$ are equal, and the test statistic is:

$$t = \frac{\tilde{x} - \tilde{y}}{\sqrt{d_x + d_y}}, \tag{9.40}$$

where d_x and d_y are defined in the next paragraph. This test statistic has an approximate t-distribution with ν degrees of freedom, where:

$$\nu = \frac{(d_x + d_y)^2}{\frac{d_x^2}{n_x - 1} + \frac{d_y^2}{n_y - 1}}, \tag{9.41}$$

so the null hypothesis that the trimmed means are equal is rejected if:

$$|t| > t_{1-\alpha/2}, \tag{9.42}$$

where $t_{1-\alpha/2}$ is the upper $\alpha/2$ quantile of the t-distribution with ν degrees of freedom. The $1 - \alpha$ confidence interval for the difference in trimmed means is given by:

$$\mu_{tx} - \mu_{ty} \in [(\tilde{x} - \tilde{y}) - t_{1-\alpha/2}\sqrt{d_x + d_y}, (\tilde{x} - \tilde{y}) + t_{1-\alpha/2}\sqrt{d_x + d_y}]. \tag{9.43}$$

Taking $\gamma = 0$ in this test replaces the trimmed means with ordinary means and yields Welch's extension of the t-test to handle the case of unknown, unequal variances under the assumption of Gaussian data distributions.

The quantities d_x and d_y on which the Yuen-Welch test are based are based on the *Windsorized sample variance*, defined as follows. First, from the rank-ordered data sample defined in Eq. (9.38), compute the corresponding *Windsorized data sample:*

$$W_k = \begin{cases} x_{(g+1)} & \text{if } x_k \leq x_{(g+1)}, \\ x_k & \text{if } x_{(g+1)} < x_k < x_{(N-g)}, \\ x_{(N-g)} & \text{if } x_k \geq x_{(N-g)}. \end{cases} \qquad (9.44)$$

The *Windsorized sample mean* $\bar{W}$ is the mean of the Windsorized sample:

$$\bar{W} = \frac{1}{N} \sum_{k=1}^{N} W_k, \qquad (9.45)$$

and the *Windsorized sample variance* s_W^2 is obtained by applying the usual variance estimator to the Windsorized sample:

$$s_W^2 = \frac{1}{N-1} \sum_{k=1}^{N} (W_k - \bar{W})^2. \qquad (9.46)$$

The quantities d_x and d_y on which the Yuen-Welch test is based are defined as:

$$d_x = \frac{(n_x - 1)s_{Wx}^2}{(n_x - 2g_x)(n_x - 2g_x - 1)}, \qquad (9.47)$$

with d_y defined analogously. Note that for the untrimmed case $g_x = 0$ and d_x reduces to $d_x = s_x^2/(n_x - 1)$.

Table 9.3 summarizes the results obtained with two cases of the Yuen-Welch test for the 19 examples considered in secs. 9.4.2 and 9.4.3. The specific cases of the Yuen-Welch test considered here are that with no trimming ($\gamma = 0$), corresponding to Welch's modification of the t-test to handle the unequal variance problem, and the case recommended by Wilcox ($\gamma = 0.2$). For the six Gaussian comparisons (cases 1 through 6), the results are quite comparable to those obtained with the standard t-test presented earlier, and are largely independent of the choice of trimming fraction γ. The effects of trimming are clear for the Cauchy comparisons (cases 7, 8, and 9), where the performance for $\gamma = 0.2$ is much better than that for $\gamma = 0$. In fact, the results for $\gamma = 0$ are approximately as bad as those for the classical t-test for this case, while the results for $\gamma = 0.2$ are essentially as good as those for the Wilcoxon rank-sum test for these cases.

In the face of asymmetry, the differences between the results for $\gamma = 0$ and for $\gamma = 0.2$ are sometimes equally pronounced, but in these cases the effects of increased trimming are generally quite undesirable. In particular, the results for $\gamma = 0$ are virtually identical to those for the classical t-test for cases 10 through 19. In 3 of these 10 cases, the means of the two sequences are equal and the test for $\gamma = 0$ yields rejections (i.e., false positives) $\sim$5% of the time, as it should. This is also true for $\gamma = 0.2$ in case 13, where the two distributions are identical,

Table 9.3: Summary of 1,000 Yuen-Welch test results for the 19 cases considered in Table 9.1, for two different trimming fractions γ

Case	$\{x_k\}$	$\{y_k\}$	$\mu_x = \mu_y$?	% Reject $(\gamma = 0)$	% Reject $(\gamma = 0.2)$
1	$N(0,1)$	$N(0,1)$	Yes	3.9%	4.5%
2	$N(0,1)$	$N(0.5,1)$	No	94.7%	90.4%
3	$N(0,1)$	$N(1,1)$	No	100.0%	100.0%
4	$N(0,1)$	$N(0,2)$	Yes	4.5%	5.3%
5	$N(0,1)$	$N(0.5,2)$	No	59.6%	54.7%
6	$N(0,1)$	$N(1,2)$	No	99.4%	98.7%
7	$N(0,1)$	$C(0,1)$	Yes	1.8%	5.4%
8	$N(0,1)$	$C(0.5,1)$	No	6.6%	66.7%
9	$N(0,1)$	$C(1,1)$	No	20.1%	99.4%
10	$N(0,1)$	$E(1) - 1$	Yes	5.3%	39.2%
11	$N(0,1)$	$E(1) - 0.5$	No	94.8%	47.9%
12	$N(0,1)$	$E(1)$	No	100.0%	100.0%
13	$E(1) - 1$	$E(1) - 1$	Yes	5.5%	4.3%
14	$E(1) - 1$	$E(1) - 0.5$	No	92.7%	96.6%
15	$E(1) - 1$	$E(1)$	No	100.0%	100.0%
16	$E(1) - 1$	$E(2) - 2$	Yes	5.7%	25.3%
17	$E(1) - 1$	$E(2) - 1.5$	No	59.8%	23.0%
18	$E(1) - 1$	$E(2) - 2.5$	No	62.2%	93.7%
19	$E(1) - 1$	$E(2) - 1$	No	99.8%	98.0%

but the false-positive rate is much larger than it should be in cases 10 and 16, where the distributions differ in either asymmetry or scale. In particular, note that the performance for case 10 (Gaussian vs. shifted exponential) is even worse than that for the Wilcoxon rank-sum test (39.2% false positives vs. 14.4%), although it is better for case 16 (exponential with different scales, 25.3% false positives vs. 65.0%). Also, comparing cases 18 and 19 shows the same general behavior seen for the Wilcoxon rank-sum test: the direction of the asymmetry matters, with a mean shift in one direction giving a correct rejection rate of only 23.0% against a rate of 93.7% for the same shift in the opposite direction.

The examples presented here are far from complete and therefore cannot provide the basis for clear recommendations of one method over another. In particular, these examples only considered a few different data distributions, and they did not examine the influence of differences in sizes n_x and n_y between the two samples. Still, these examples do suggest a few general conclusions. First, it is clear that the classical t-test is extremely sensitive to the effects of heavy-tailed distributions, caused by the variance inflation effects discussed in Chapter 7. Both the Wilcoxon rank-sum test and the Yuen-Welch test are much better suited to these cases, and both appear to exhibit comparable performance, at least for the examples considered here. Conversely, both of these outlier-resistant tests suffer badly in the face of pronounced asymmetry, where the classical t-test appears to perform much better. Cribbie *et al.* [73] present a more detailed comparison of extensions of the t-test developed to address problems of non-equal variances and non-Gaussian distributions (especially asymmetry), and they conclude that the best of the methods they compared is generally the *Welch rank-based test*, corresponding to the Welch test (i.e., the Yuen-Welch test with zero trimming percentage) applied to the *ranks* of the individual sample values relative to the pooled data sample (i.e., the same ranking scheme used in the Wilcoxon rank-sum test).

9.5 The χ^2 distribution and χ^2 tests

The χ^2 distribution described in Sec. 9.5.1 arises frequently in classical statistic because many historically popular characterizations have been shown to exhibit this distribution in the limit of large sample sizes. This observation provides the basis for a number of different hypothesis tests that are suitable for "sufficiently large sample sizes." One of the most important of these is what is traditionally known as "the χ^2 test," proposed in 1900 by Karl Pearson, described in Sec. 9.5.2. A simple application of this test is discussed in Sec. 9.5.3.

9.5.1 The χ^2 distribution

As noted in Chapter 4, the χ^2 distribution with ν degrees of freedom, commonly denoted χ^2_ν, is a special case of the gamma distribution, having the density:

$$p(x) = \frac{x^{(\nu/2)-1}e^{-x/2}}{2^{\nu/2}\Gamma(\nu/2)}, \quad \text{for all } x \geq 0. \tag{9.48}$$

It follows from the results for the gamma distribution that the mean and variance of the χ^2_ν distribution are:

$$\begin{aligned} E\{\chi^2_\nu\} &= \nu, \\ \text{var}\{\chi^2_\nu\} &= 2\nu. \end{aligned} \tag{9.49}$$

This distribution arises naturally in two contexts. The first is from the fact that if $\{U_i\}$ are a collection of ν standard normal random variables, their sum of squares has a χ^2_ν distribution:

$$\sum_{i=1}^{\nu} U_i^2 \sim \chi^2_\nu. \tag{9.50}$$

A closely related result is that if $\{X_i\}$ is a sequence of N i.i.d. Gaussian random variables with mean μ and variance σ^2, then the following normalized ratio has a χ^2_{N-1} distribution, where $\bar{X}$ is the mean of the sequence $\{X_i\}$:

$$Y = \frac{\sum_{i=1}^{N}(X_i - \bar{X})^2}{\sigma^2} \sim \chi^2_{N-1}. \tag{9.51}$$

As discussed in Sec. 9.6, this result forms the basis for the classical F-test that two variances are equal.

The second situation where the χ^2 distribution arises naturally is the following asymptotic result. As Johnson *et al.* note [162, p. 415], it was shown in 1838 by Bienaymé that, if the k counts $\{N_i\}$ have a joint multinomial distribution with parameters n and p_1 through p_k, then the following random variable asymptotically approaches a χ^2_{k-1} distribution in the limit as $n \to \infty$:

$$\sum_{i=1}^{k} \frac{(N_i - np_i)^2}{np_i} \sim \chi^2_{k-1}. \tag{9.52}$$

This result forms the basis for the χ^2 goodness-of-fit test discussed next.

9.5.2 The χ^2 test

The multinomial distribution discussed in the preceeding section arises frequently in connection with categorical data. Specifically, consider a sequence of n statistically independent trials, each of which yields one of k possible outcomes. Denote the probability of observing the ith outcome on any given trial as p_i, and note that these probabilities must satisfy the normalization condition:

$$\sum_{i=1}^{k} p_i = 1. \tag{9.53}$$

The multinomial distribution describes the probability of observing the joint outcome that the ith possible value arises N_i times where:

$$\sum_{i=1}^{k} N_i = n. \tag{9.54}$$

More specifically, the multinomial probability is given by:

$$P\{n_1 = N_1, n_2 = N_2, \ldots, N_k = n_k\} = \frac{n! p_1^{N_1} p_2^{N_2} \cdots p_k^{N_k}}{N_1! N_2! \cdots N_k!}. \tag{9.55}$$

Based on the asymptotic result given in Eq. (9.52), Karl Pearson proposed the following goodness-of-fit test for the multinomial model. Consider the null hypothesis that $p_i = \pi_i$ for $i = 1, 2, \ldots, k$ where $\{\pi_i\}$ is a specified sequence of nonzero probabilities summing to 1. The χ^2-test rejects this null hypothesis in favor of the alternative that $p_j \neq \pi_j$ for at least one index j if the following quantity is too large to be consistent with the χ^2_{k-1} distribution:

$$Q_n = \sum_{i=1}^{k} \frac{(N_i - n\pi_i)^2}{n\pi_i} > c_{k-1,1-\alpha}, \tag{9.56}$$

where $c_{k-1,1-\alpha}$ is the upper $1 - \alpha$ significance limit of the χ^2_{k-1} distribution. The following section considers the application of this idea to uniformity testing.

Before proceeding to this example, however, it is worth noting that the χ^2 test, despite its historical importance and continuing popularity in practice, can perform poorly in some circumstances. A specific example is discussed in Sec. 9.7, where it is noted that the classical test for the comparison of two independent binomal samples represents a special case of the χ^2 test, one that has been shown to perform poorly in some cases even when the sample size is what has been historically classified as "moderate." Another case where χ^2 approximations are known to be inadequate is in the analysis of sparse $I \times J$ contingency tables (i.e., tables where some cells have small counts, or even zero counts) [8, sec. 9.8.4].

9.5.3 An application: uniformity testing

If the k possible values of the categorical variable considered in Sec. 9.5.2 are all equally likely, it follows that $p_i = 1/k$ for $i = 1, 2, \ldots, k$ and the χ^2 test defined in Eq. (9.56) reduces to:

$$Q_n = \frac{1}{nk} \sum_{i=1}^{k} (kN_i - n)^2 > c_{k-1,1-\alpha}. \tag{9.57}$$

While this result is simple, it suffers from two limitations. First, it provides only a binary assessment of uniformity: either the categorical variable is homogeneously distributed over its possible values (if the test result is not significant so the null hypothesis is provisionally accepted), or it is not (if the inequality in (9.57) holds and the null hypothesis is rejected). In particular, the test result tells us nothing about *how heterogeneously* the variable is distributed if we reject the null hypothesis of homogeneity. The second limitation of this result is its asymptotic character. That is, the distribution of Q_n is not *exactly* χ^2_{k-1} for finite sample size n, it only approaches this limiting distribution as $n \to \infty$. The question of "how large must n be for the asymptotic result to be a good approximation?" is one that depends on a number of things and is one where there appears to be little consensus. For example, Madansky [195, p. 43] presents a range of recommendations for the minimum acceptable expected values $n\pi_i$, ranging from "a single category with $n\pi$ as small as $1/2$ is acceptable" to "every category should have $n\pi \geq 20$." There does seem to be a consensus that the χ^2_{k-1} approximation for the distribution suffers in cases where case counts are too small (in particular, unobserved categories) [8, sec. 9.8.4].

As a specific application, consider the UCI mushroom dataset introduced in Chapter 2. Table 9.4 summarizes the values of the χ^2 statistic Q_n computed from the 8,124 mushrooms in the dataset for each of the 22 categorical variables included there. Comparing these computed Q_n values with the corresponding threshold values $c_{k-1,1-\alpha}$ for $\alpha = 0.05$ shown in the last column of the table, it is clear that none of these variables are uniformly distributed over their ranges of possible values.

Table 9.4: Summary of the Shannon heterogeneity measure, its z-score relative to 1,000 independent uniform reference samples, and the χ^2 test statistic Q_n

No.	Variable	Levels	I_S	z-Score	Q_n	$c_{k-1,1-\alpha}$
1	Cap shape	6	0.361	3337.0	9743.3	11.1
2	Cap surface	4	0.212	1927.6	2924.3	7.8
3	Cap color	10	0.244	2184.0	8043.5	16.9
4	Bruises	2	0.021	168.1	231.7	3.8
5	Odor	9	0.268	2298.0	12038.1	15.5
6	Gill attachment	2	0.827	6752.7	7305.7	3.8
7	Gill spacing	2	0.362	2956.9	3723.5	3.8
8	Gill size	2	0.108	879.3	1182.9	3.8
9	Gill color	12	0.155	1353.9	5592.8	19.7
10	Stalk shape	2	0.013	106.0	146.8	3.8
11	Stalk root	5	0.215	2050.3	5421.7	9.5
12	Stalk surface above ring	4	0.389	3536.7	7987.6	7.8
13	Stalk surface below ring	4	0.300	2728.9	6702.8	7.8
14	Stalk color above ring	9	0.389	3332.6	18683.5	15.5
15	Stalk color below ring	9	0.376	3220.8	17967.0	15.5
16	Veil type	1	—	—	—	—
17	Veil color	4	0.902	8194.3	22800.8	7.8
18	Ring number	3	0.735	6838.0	12714.8	6.0
19	Ring type	5	0.339	3233.7	7345.2	9.5
20	Spore print color	9	0.305	2612.2	9332.9	15.5
21	Population	6	0.225	2081.5	7557.8	11.1
22	Habitat	7	0.190	1688.5	6336.4	12.6

It is interesting to compare these results with those obtained using the Shannon heterogeneity measure introduced in Chapter 2. Recall that this is a normalized measure taking the value 0 for a perfectly uniform distribution of values and the value 1 when all values are concentrated in one of the possible values. It was noted in Chapter 2 that, if $\{x_k\}$ corresponds to a random sample drawn from a uniform distribution over k possible values, the heterogeneity measure will not be *exactly* zero, but can be expected to take on a "small" nonzero value. Thus, to provide a useful frame of reference, we can generate a large number of synthetic random samples that are uniformly distributed over k possible values, compute the corresponding heterogeneity measure, and compare our data-based result with this synthetic reference. A visual indication of the significance of these results was given in Chapter 2 using boxplots, but the hypothesis testing ideas presented here provide the basis for quantitative significance assessments.

Specifically, given a computed heterogeneity value I_x for a data sequence $\{x_k\}$ like any of those considered here, one way of assessing significance is to compute the z-score for I_x with respect to B reference results I_j computed from independent, uniformly distributed reference samples for $j = 1, 2, \ldots, B$. This z-score is given by:

$$z = \frac{I_x - \bar{I}}{\sigma_I}, \tag{9.58}$$

where $\bar{I}$ is the average of the B reference results $\{I_j\}$ and σ_I is the corresponding standard deviation. If this z-score is large and positive, we reject the hypothesis that $\{x_k\}$ is uniformly distributed. This idea is used extensively in Chapter 15, where specific guidelines are given for deciding when z is large enough to declare the result significant. The key point here is that the z-scores shown in Table 9.4 are extremely large and, like the extremely large Q_n values, lead us to firmly reject the hypothesis of uniformity for these variables.

9.6 The F-test

One of the classic hypothesis tests that was once quite popular is the *F-test*, designed to test the hypothesis of equal variances. Unfortunately, this test is known to be extremely sensitive to deviations from normality [181, p. 446]. Indeed, Wilcox goes so far as to say that, while many methods have been proposed for testing the equality of variances, "virtually all have been found to be unsatisfactory with small to moderate sample sizes" [295, p. 170]. Despite its considerable limitations, the F-test is introduced here because it is widely used in analysis of variance (ANOVA), discussed in Chapter 13.

The basic idea behind the F-test is the following. Suppose $\hat{\sigma}_1$ is the estimated variance for a Gaussian data sample of size $N_1 = \nu_1 + 1$ and $\hat{\sigma}_2$ is the estimated variance for a second Gaussian data sample of size $N_2 = \nu_2 + 1$. It then follows from the results discussed in Sec. 9.5.1 for the distribution of estimated variances that $X_1 = \nu_1 \hat{\sigma}^2_1$ has a χ^2 distribution with ν_1 degrees of freedom and $X_2 = \nu_2 \hat{\sigma}^2_2$ has a χ^2 distribution with ν_2 degrees of freedom. If these estimates are statistically independent, it is known that the ratio of these variance estimates has the F-distribution [163, ch. 27]:

$$\frac{\hat{\sigma}^2_1}{\hat{\sigma}^2_2} = \frac{X_1/\nu_1}{X_2/\nu_2} \sim F_{\nu_1, \nu_2}. \tag{9.59}$$

More formally known as the *F-distribution with ν_1, ν_2 degrees of freedom*, this distribution is defined by the density:

$$p(x) = \left[\frac{(\nu_1/\nu_2)^{\nu_1/2}}{B(\nu_1/2, \nu_2/2)} \right] \frac{x^{(\nu_1/2)-1}}{(1 + \nu_1 \nu_2^{-1} x)^{(\nu_1+\nu_2)/2}}, \tag{9.60}$$

for $x > 0$. Here, ν_1 and ν_2 are positive-valued shape parameters and $B(x, y)$ is the beta function introduced in Chapter 4. *Note that the order in which these*

Figure 9.2: $F(\nu_1, \nu_2)$ densities, degrees of freedom indicated on plot

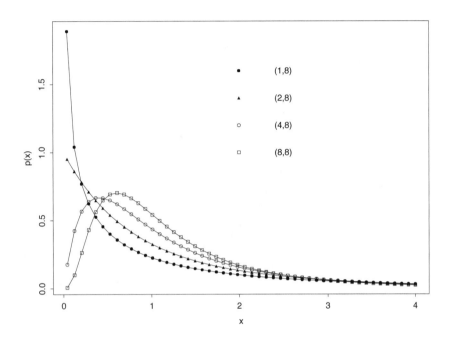

parameters appear is important; in particular, if f has the F_{ν_1,ν_2} distribution,
then $1/f$ has the F_{ν_2,ν_1} distribution [163, p. 322]. If $\nu_1 > 2$, the density is
unimodal with the mode:

$$x^* = \frac{\nu_2(\nu_1 - 2)}{\nu_1(\nu_2 + 2)}. \tag{9.61}$$

For $\nu_1 \le 2$, the density is J-shaped, decreasing monotonically with increasing
x. These points are illustrated in Fig. 9.2, which shows plots of the density
function for $\nu_1 = 1$, 2, 4, and 8, all with $\nu_2 = 8$. Similarly, Fig. 9.3 shows plots
of the density function for $\nu_1 = 8$ and $\nu_2 = 4$, 8, 16, and 64.

The F-distribution is also known as a *beta distribution of the second kind*,
related to the beta distributions discussed in Chapter 4 by the following result
[163, p. 327]: if f has the F_{ν_1,ν_2} distribution, then the transformed variable:

$$x = \frac{\nu_1 f}{\nu_2 + \nu_1 f}, \tag{9.62}$$

exhibits the standard beta distribution with shape parameters $p = \nu_1/2$ and
$q = \nu_2/2$. Unfortunately, general moment expressions are extremely complex,
even for first and second moments. For example, if ν_1 and ν_2 are both even

Figure 9.3: $F(\nu_1, \nu_2)$ distributions, degrees of freedom indicated on plot

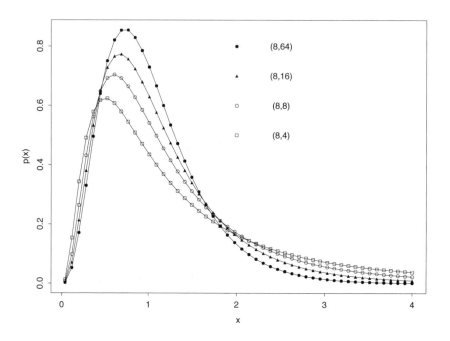

integers, the mean of the distribution is given by [162, p. 330]:

$$m_1 \;=\; \frac{1}{2}\left[\log\left(\frac{\nu_2}{\nu_1}\right) - \sum_{j=j_-}^{j_+} j^{-1}\right], \qquad (9.63)$$

where the limits of the sum are given by:

$$j_- = \frac{\min(\nu_1, \nu_2)}{2} \quad\text{and}\quad j_+ = \frac{\max(\nu_1, \nu_2)}{2} - 1. \qquad (9.64)$$

Despite this complexity of characterization, densities, cumulative distribution functions, and quantiles (i.e., inverse cumulative distribution functions) are widely available, either in the form of tables or as built-in functions in statistics packages like *R* and *S-Plus*.

9.7 Binomial random variables

A problem that commonly arises is the following. We consider a situation with two possible outcomes, with an unknown probability p of observing one of these outcomes. It follows essentially from the frequentist definition of probability that a reasonable estimate of p is the number n of times in N independent trials that we observe this outcome:

$$\hat{p} = \frac{n}{N}. \tag{9.65}$$

Further, it is a standard result that this estimator is unbiased, consistent, and asymptotically normal, and for a long time the following confidence intervals were generally accepted as valid [181, p. 435]:

$$|\hat{p} - p| \leq z_{1-\alpha/2} \left[\frac{\hat{p}(1-\hat{p})}{N} \right]^{1/2} \tag{9.66}$$

$$\Rightarrow p \in \left[\hat{p} - z_{1-\alpha/2} \left\{ \frac{\hat{p}(1-\hat{p})}{N} \right\}^{1/2}, \hat{p} + z_{1-\alpha/2} \left\{ \frac{\hat{p}(1-\hat{p})}{N} \right\}^{1/2} \right].$$

Despite the fairly widespread acceptance of this result as standard, it has been shown to be inadequate [181, p. 435]:

> Unfortunately, an accumulating literature has shown that the coverage of the interval (defined by Eq. (9.66)) is quite unreliable even for large values of N or $Np(1-p)$, and varies quite erratically as the sample size increases.

As a better alternative, Lehmann and Romano [181, p. 435] recommend the following method suggested by Brown *et al.* [47]:

$$p \in \left[\tilde{p} - z_{1-\alpha/2} \frac{\sqrt{N}}{\tilde{N}} \left\{ \hat{p}\hat{q} + \frac{z_{1-\alpha/2}^2}{4N} \right\}, \tilde{p} + z_{1-\alpha/2} \frac{\sqrt{N}}{\tilde{N}} \left\{ \hat{p}\hat{q} + \frac{z_{1-\alpha/2}^2}{4N} \right\} \right], \tag{9.67}$$

where the additional terms appearing in this expression are defined as:

$$\tilde{p} = \frac{n + (1/2)z_{1-\alpha/2}^2}{\tilde{N}},$$

$$\tilde{N} = N + z_{1-\alpha/2}^2,$$

$$\hat{q} = 1 - \hat{p}. \tag{9.68}$$

A second, very closely related problem that also arises frequently is that of comparing two proportions. There, the natural model is the following. We consider two similar but statistically independent binary random sequences, of lengths N_1 and N_2, where the outcome of primary interest occurs n_1 and n_2 times, respectively. Denoting the probability of observing this outcome as p_1 for the first sequence and p_2 for the second, it follows that the logical estimators for these probabilities are:

$$\hat{p}_1 = \frac{n_1}{N_1} \quad \text{and} \quad \hat{p}_2 = \frac{n_2}{N_2}. \tag{9.69}$$

The hypothesis that $p_1 = p_2 = p$ is equivalent to the hypothesis that $p_1 - p_2 = 0$, and reasoning analogous to that used in deriving Eq. (9.66) above leads to the following test statistic [66, p. 33]:

$$z = \frac{\hat{p}_1 - \hat{p}_2}{\sqrt{\hat{p}(1 - \hat{p})\left[\frac{1}{N_1} + \frac{1}{N_2}\right]}}, \tag{9.70}$$

where $\hat{p}$ is the pooled estimate of the assumed common probability p, given by:

$$\hat{p} = \frac{n_1 + n_2}{N_1 + N_2}. \tag{9.71}$$

Asymptotically, z defined in Eq. (9.70) has a standard normal distribution, so for sufficiently large samples, the classical test rejects the null nypothesis that $p_1 = p_2$ if $|z| > z_{1-\alpha/2}$.

As in the case of confidence intervals for individual binary probabilities $\hat{p}$ discussed above, the hypothesis test just described often performs badly in practice. As more reliable alternatives, Wilcox [295, sec. 5.8] describes two methods, noting that it is not obvious which one is to be preferred in any given situation. The following discussion briefly describes one of these—Beal's method [31]—which has the practical advantage that it provides both a test of the hypothesis that $p_1 = p_2$ and confidence intervals for the difference $p_1 - p_2$, while the other method—proposed by Storer and Kim [274]—only provides a formal criterion for rejecting the null hypothesis. Beal's method proceeds by first defining the following constants:

$$\begin{aligned}
a &= \hat{p}_1 + \hat{p}_2, \\
b &= \hat{p}_1 - \hat{p}_2, \\
c &= z^2_{1-\alpha/2}, \\
u &= \frac{1}{4}\left(\frac{1}{N_1} + \frac{1}{N_2}\right), \\
v &= \frac{1}{4}\left(\frac{1}{N_1} - \frac{1}{N_2}\right), \\
V &= u[(2 - a)a - b^2] + 2v(1 - a)b, \\
A &= \sqrt{c[V + cu^2(2 - a)a + cv^2(1 - a^2)]}, \\
B &= \frac{b + cv(1 - a)}{1 + cu}.
\end{aligned} \tag{9.72}$$

The level α confidence interval for the difference $p_1 - p_2$ is given by:

$$p_1 - p_2 \in \left[B - \frac{A}{1 + cu}, B + \frac{A}{1 + cu}\right]. \tag{9.73}$$

Thus, we reject the null hypothesis that $p_1 = p_2$ at the level α if the value 0 does not lie in this confidence interval.

Before leaving this discussion, it is worth noting the intimate connection between the classical hypothesis test result based on the assumed normality of z as defined in Eq. (9.70) and the χ^2 test discussed in Sec. 9.5.2. Specifically, the fact that z is asymptotically distributed as $N(0,1)$ implies that z^2 is asymptotically distributed as χ_1^2. As Collett demonstrates [66, sec. 2.3.3], z^2 is equal to the classical χ^2 test statistic Q_n defined in Eq. (9.56) for $k = 2$. This observation is significant for two reasons: first, it illustrates how frequently χ^2 distributions and χ^2 tests arise in statistical data characterizations, and second, it illustrates the point made at the end of Sec. 9.5.2 that χ^2 tests can perform poorly.

9.8 Testing multiple hypotheses

The discussion of hypothesis testing presented so far has considered a single hypothesis to be tested. Often, however, situations arise involving multiple hypothesis tests, and in these situations a new issue arises: even if a specific hypothesis works exactly as advertised for a single hypothesis, applying the same procedure to multiple hypotheses and combining the results does not give the expected results. The following paragraphs give a more detailed introduction to this problem, commonly called the *multiple comparison problem*, and offer two specific solutions: the simple but conservative Bonferroni correction, and the slightly more complex but less conservative Holm stepdown procedure. For more detailed treatments of the multiple comparison problem, refer to Lehmann and Romano [181, ch. 9] or Westfall and Young [293].

9.8.1 The multiple comparison problem

Suppose we wish to test not one, but $m > 1$ hypotheses, all identical in form and thus amenable to the same type of test. For example, we might be interested in comparing the means of p different data sequences. If we approach this problem by making all possible pairwise comparisons between $\bar{x}_i$ and $\bar{x}_j$ for all $i \neq j$, this amounts to $m = p(p-1)/2$ comparisons. Now, suppose that the null hypothesis holds in every case, and denote these null hypotheses as H_0^i for $i = 1, 2, \ldots, m$. If the test we are using to test each individual hypothesis is made at significance level α, it follows that the probability of rejecting the null hypothesis H_0^i when it is false (e.g., declaring two means significantly different when they are) is at least $1 - \alpha$. Further, if these hypotheses are statistically independent, it follows that the probability of rejecting all m of these null hypotheses is the product of the probabilities of rejecting each one:

$$\mathcal{P}\{\text{reject } H_0^i \text{ for } i = 1, 2, \ldots, m\} \geq (1 - \alpha)^m. \qquad (9.74)$$

Unfortunately, the probability $(1 - \alpha)^m$ is less than $1 - \alpha$ for all $m > 1$, and for large m, this probability can be *much* smaller than $1 - \alpha$. For example, Lehmann and Romano note [181, p. 349] that if the probability of false rejection for a single hypothesis is 5%, it is approximately 10% for two hypotheses, 23% for five hypotheses, and 92% for 50 hypotheses.

The difficulty just described is well recognized and has led to the development of many different approaches to the multiple comparison problem: Venables and Ripley characterize the options available in one specific software package available for R and *S-Plus* as "a bewildering variety of methods" [283, p. 181]. Consequently, the following discussion is limited to three methods, all fairly simple and reasonably popular in practice. The first two are intended to control the *familywise error rate (FWER)*, which is the probability of rejecting one or more null hypotheses in a family of m hypotheses when they are all true. Specifically, both the Bonferroni procedure described in Sec. 9.8.2 and the Holm stepdown procedure described in Sec. 9.8.3 are designed to satisfy the condition:

$$\text{FWER} \leq \alpha. \tag{9.75}$$

A limitation of these procedures for very large m is that they protect against false positives (i.e., rejections of individual null hypotheses H_0^i when they are true) at the cost of a reduced ability to detect one or a few violations of these null hypotheses. As a specific example, gene expression microarray experiments look for evidence of gene expression changes (e.g., between a tumor sample and a normal tissue sample) in tens of thousands of genes simultaneously. Viewing each of these individual comparisons as a test of the null hypothesis of no change in gene expression, we have a situation with very large m and an expectation that most of the null hypotheses are true (i.e., most genes are not differentially expressed between samples), but some of these null hypotheses are false (i.e., the genes of primary interest are those that *are* differentially expressed between the tumor sample and the normal tissue sample). In these cases, m is large enough that multiple comparison issues are overwhelmingly important, but so is the ability to reject the small subset of null hypotheses that are false. Problems of this type have led to the development of methods that control not FWER, but the *false discovery rate (FDR)*, defined as the expected number of violations of null hypotheses H_0^i that are true. The idea is that we accept the possibility of some "false positives" (i.e., rejections of null hypotheses that are true) to improve our ability to detect "true positives" (i.e., to identify those null hypotheses that really are violated). It turns out that FDR is bounded above by FWER, but typically much smaller than FWER, so FDR-based multiple hypothesis testing procedures are generally much more liberal than FWER-based procedures in allowing false positives, but at a substantial improvement in the ability to detect individual violations of null hypotheses that are false. A brief description of the Benjamani-Hochberg procedure based on FDR control instead of FWER control is given in Sec. 9.8.4.

9.8.2 The Bonferroni correction

Consider the case where we have m hypotheses we are interested in testing simultaneously, and suppose that all null hypotheses H_0^i are true for $i = 1, 2, \ldots, m$. Further, let A_i denote the event that our test is consistent (i.e., the test does not reject H_0^i), for $i = 1, 2, \ldots, m$. The *Bonferroni inequality* can be used to relate the probability that all of our tests are consistent to the probabilities that each individual test is inconsistent (i.e., that the ith hypothesis test rejects H_0^i even though it is true). Specifically, the Bonferroni inequality is [198, p. 4]:

$$\mathcal{P}\{A_1 \cap A_2 \cap \cdots \cap A_m\} \geq 1 - \sum_{i=1}^{m} \mathcal{P}\{A_i^c\}, \qquad (9.76)$$

where A_i^c denotes the logical complement of A_i, corresponding to the event that the ith hypothesis test erroneously rejects the null hypothesis H_0^i. A simple rearrangement of Eq. (9.76) leads to the following result:

$$\text{FWER} = 1 - \mathcal{P}\{A_1 \cap A_2 \cap \cdots \cap A_m\} \leq \sum_{i=1}^{m} \mathcal{P}(A_i^c). \qquad (9.77)$$

Note that if all m of our hypothesis tests are at level α, then $\mathcal{P}(A_i^c) = \alpha$ for all i, implying that the familywise error rate is bounded above by:

$$\text{FWER} \leq m\alpha. \qquad (9.78)$$

Thus, if we want FWER $\leq \alpha$, it follows from the Bonferroni inequality that we can achieve this objective by requiring each individual hypothesis test to be of level α/m. The implementation of this strategy is often called the *Bonferroni correction*, and it provides a simple way of correcting for the effects of multiple comparisons.

Two characteristics of this procedure should be noted. First, the Bonferroni inequality is a direct consequence (Exercise 1) of the finite subadditivity of probability measures, commonly known as *Boole's inequality* [37, p. 22]:

$$\mathcal{P}\left(\bigcup_{i=1}^{m} X_i\right) \leq \sum_{i=1}^{m} \mathcal{P}(X_i). \qquad (9.79)$$

Since Boole's inequality applies to all random events, without regard to statistical independence, it follows that the Bonferroni correction is applicable whether the individual hypothesis tests are statistically independent or not. This point is important in applications where some or all of the tests involved are characterizations of the same dataset, since then the tests are likely to exhibit some (usually unknown) degree of statistical dependence. Conversely, the second key observation about the Bonferroni correction is that it tends to be quite conservative in practice, particularly for large m. This observation provides the primary motivation for the slightly more complicated but less conservative tests described in the next two sections.

9.8.3 The Holm stepdown procedure

The conservatism of the Bonferroni correction arises from the fact that all of the m hypotheses are tested at the same level, α/m, which becomes extremely stringent for large m. The *Holm stepdown procedure* provides a way of testing some hypotheses at less stringent levels while still meeting the constraint that FWER $\leq \alpha$ [181, p. 350]. This procedure first computes the p-values $\{p_i\}$ for each of the m individual hypothesis tests and rank-orders them to obtain the sequence:

$$p_{(1)} \leq p_{(2)} \leq \cdots \leq p_{(m)}. \tag{9.80}$$

Each of these ordered p-values is associated with its corresponding null hypothesis, denoted $H_{(i)}$ for $i = 1, 2, \ldots, m$. The Holm procedure corresponds to the following sequence, which continues until all m hypotheses have been tested:

1. If $p_{(1)} \geq \alpha/m$, accept all null hypotheses $\{H_0^i\}$ and stop. If $p_{(1)} < \alpha/m$, reject the null hypothesis $H_{(1)}$ and go to step 2,

2. If $p_{(1)} < \alpha/m$ but $p_{(2)} \geq \alpha/(m-1)$, accept the $m-1$ null hypotheses $\{H_{(i)}\}$ for $i = 2, 3, \ldots, m$ and stop. If $p_{(2)} < \alpha/(m-1)$, reject the null hypothesis $H_{(2)}$ and go to step 3,

$\vdots$

m. If $p_{(m-1)} < \alpha/2$ but $p_{(m)} \geq \alpha$, accept the null hypothesis $H_{(m)}$. Otherwise, if $p_{(m)} < \alpha$, reject $H_{(m)}$. Stop.

Note that this procedure examines the m hypotheses in increasing order of p-values, and the rejection criteria become less stringent as the sequence proceeds. In particular, rejection of the most significant hypothesis $H_{(1)}$, corresponding to the smallest p-value $p_{(1)}$, occurs at the same level as the Bonferroni correction. In contrast, if the $m-1$ most significant hypotheses are all rejected, the significance level for the final test $H_{(m)}$ is the unmodified significance level α. For the general case, if the j most significant hypotheses $H_{(1)}$ through $H_{(j)}$ have been rejected, the next most significant one is tested at the level $\alpha/(m-j)$.

Other stepdown procedures have been developed to correct for multiple comparison, but a more complete discussion of these procedures lies beyond the scope of this book. A useful reference is the book by Lehmann and Romano [181, ch. 9], who note that "it is possible to improve upon the Holm method by incorporating the dependence structure of the individual tests." The general structure of these tests is essentially the same as that of the Holm stepdown procedure, and the key to their improved performance lies in the appropriate choice of the significance levels for rejecting each individual hypothesis $H_{(i)}$ corresponding to the ordered p-values $\{p_{(i)}\}$.

9.8.4 The Benjamani-Hochberg procedure

As noted in the beginning of this discussion of multiple comparisons, the Holm stepdown procedure just described is designed to control the familywise error rate in a way that is less conservative than the Bonferroni correction. In contrast, the Benjamani-Hochberg procedure is a different stepdown procedure that was developed to control the false discovery rate (FDR) to a specified level α.

The structure of this procedure is quite similar to that of the Holm stepdown procedure just described, but the criteria for rejecting hypotheses at each step are less stringent, resulting in a higher likelihood of rejecting null hypotheses that are false at the cost of rejecting more when they are true [208]. Specifically, like the Holm stepdown procedure, the Benjamani-Hochberg procedure first rank-orders the individual p-values as in Eq. (9.80) and then proceeds as follows:

1. If $p_{(1)} \geq \alpha/m$, accept all null hypotheses $\{H_0^i\}$ and stop. If $p_{(1)} < \alpha/m$, reject the null hypothesis $H_{(1)}$ and go to step 2,

2. If $p_{(1)} < \alpha/m$ but $p_{(2)} \geq 2\alpha/m$, accept the $m-1$ null hypotheses $\{H_{(i)}\}$ for $i = 2, 3, \ldots, m$ and stop. If $p_{(2)} < 2\alpha/m$, reject the null hypothesis $H_{(2)}$ and go to step 3,

$$\vdots$$

m. If $p_{(m-1)} < (m-1)\alpha/m$ but $p_{(m)} \geq \alpha$, accept the null hypothesis $H_{(m)}$. Otherwise, if $p_{(m)} < \alpha$, reject $H_{(m)}$. Stop.

It is known that if the m tests are statistically independent and all of the null hypotheses are true, then FDR and FWER are equal, so this procedure controls FWER [208]. Alternatively, note that the individual rank-ordered p-values $\{p_{(i)}\}$ can be modified to the following values $\{\tilde{p}_{(i)}\}$, which are all compared with the desired significance level α [208] (Exercise 2):

$$\tilde{p}_{(m)} = p_{(m)},$$
$$\tilde{p}_{(i)} = \min\left\{\tilde{p}_{(i+1)}, \frac{m}{i}p_{(i)}\right\} \quad \text{for } i = m-1, m-2, \ldots, 1. \quad (9.81)$$

9.9 Exercises

1. It was noted in Sec. 9.4.4 that the Yuen-Welch test is a generalization of the Welch test of the hypothesis that two means are equal in the case where the variances are both unknown and unequal, obtained by replacing ordinary means with trimmed means. The original Welch method computes the same test statistic as the standard t-test, but evaluates it relative to a t-distribution with a modified number of degrees of freedom ν. From the results given for the Yuen-Welch test, derive an expression for ν for the classical Welch test.

2. Prove that the Benjamani-Hochberg FDR control procedure can be expressed as a sequence of tests of hypotheses at a constant target level α using the modified individual rank-ordered p-values defined in Eq. (9.81):

$$\tilde{p}_{(m)} = p_{(m)},$$
$$\tilde{p}_{(i)} = \min\left\{\tilde{p}_{(i+1)}, \frac{m}{i}p_{(i)}\right\} \quad \text{for } i = m-1, m-2, \ldots, 1.$$

3. The Holm stepdown procedure for controlling FWER in multiple comparison problems can also be written in terms of modified ranked p-values, like
   those given in Eq. (9.81) for the Benjamani-Hochberg procedure. Derive
   these modified p-values.

Chapter 10

Relations among Variables

Sections and topics

Previous chapters have been principally concerned with the characterization of a single, real-valued data sequence $\{x_k\}$, or possibly the comparison of such characterizations across a finite collection of these sequences. Often, however, we are interested in the more complex question of how two or more variables are related. The simplest version of this question is that of whether two variables are related at all, a question naturally addressed using measures like the product-moment correlation coefficient, Spearman's rank correlation coefficient, or Kendall's τ. All three of these measures are discussed in this chapter, along with their extensions to cases where the variables involved are not necessarily either real-valued or even of the same type. As the discussions presented here demonstrate, these association measures are intimately related to the concepts of joint and conditional distributions, concepts examined in some detail. As in the univariate case, the Gaussian distribution plays a central role in these discussions. The multivariate non-Gaussian case is much more complicated, and while this chapter does not attempt a thorough treatment, some of the issues that arise in the non-Gaussian case are introduced here, with references to more complete discussions. Another extremely useful notion that is closely related to joint and conditional distributions is that of *mixture distributions*, useful in describing forms of nonideal behavior ranging from outliers to overdispersion.

Key points of this chapter:

1. *Association measures* attempt to provide a quantitative answer to the important practical question of whether two (or more) variables are related.

2. Historically, the most popular association measure is the *product-moment correlation coefficient*, introduced at the end of the nineteenth century and still quite popular as a measure of *linear association* between variables.

3. The product-moment correlation coefficient has strong connections with the *bivariate Gaussian distribution*, and part of its popularity stems from this fact. To understand these connections, this chapter introduces both the bivariate case and the more general *multivariate Gaussian distribution*.

4. The mathematical framework on which these distributions are based includes the ideas of joint and conditional distribution functions and the notion of statistical independence, all of which are discussed in detail here.

5. The product-moment correlation coefficient is sensitive to the presence of both outliers and of nonlinear data transformations. Two useful alternative association measures described here that are less sensitive to these phenomena are *Spearman's rank correlation coefficient* and *Kendall's* τ.

6. An extremely useful concept, intimately related with the notion of conditional distributions, is that of a *mixture distribution*, which can be used to describe many types of data heterogeneity that commonly arise in practice, including both a variety of outlier models and the zero-inflated count models discussed briefly in Chapter 3.

7. In many respects, the multivariate Gaussian distribution represents a simple, "natural" extension of the univariate case. For non-Gaussian distributions, this is typically not the case; for example, "the bivariate exponential distribution" can be defined in several different ways, each exhibiting a different subset of the desirable characteristics of the univariate distribution. A brief introduction to some useful classes of multivariate non-Gaussian distributions is given here, along with some of the mathematical machinery needed to treat more general cases (e.g., the notion of *couplas* that allow for independent specification of marginals and dependence structure).

8. It was emphasized in Chapter 2 that data variables come in different types, including continuous, discrete, ordinal, nominal, and binary. If we want to characterize either the association between two non-continuous variables of the same type or the association between two variables of different types, specialized association measures may be required. Measures appropriate to these cases are discussed in this chapter, with cross-references to methods described in other chapters for those cases where the problem may be usefully recast into another form.

10.1 What is the relationship between popular and electoral votes?

The U.S. Constitution defines the process by which the the president and vice president of the United States are elected. Not simply based on the fraction of the population that prefers one candidate over any others, this procedure is based on *electoral votes* determined by which candidate achieves a majority of the popular vote in each state. Consequently, it may happen—as it did in the presidential elections of 1824, 1876, 1888, and 2000—that the candidate with the greatest number of popular votes is not elected president. Even in cases where both popular and electoral votes agree on the winner, the winning margin—arguably a measure of the popularity or "mandate" of the incoming president—can differ markedly between these two votes. A case in point is the 1936 presidential election in which Franklin Roosevelt defeated Alf Landon by one of the greatest electoral landslides in U.S. history: 523 votes to 8, representing 98.5% of the electoral total, giving Roosevelt more than 65 times as many electoral votes as Landon. In contrast, Roosevelt's popular vote margin was, while substantial, far less dramatic (62.5% to 37.5%, less than 2 to 1).

The following discussion examines the question posed in the title of the section: "what *is* the relationship between popular and electoral votes?" While this discussion cannot pretend to be complete, it does provide a simple illustration of the type of problem with which this chapter is concerned: given two sequences of numbers that may be or should be related, how do we quantify that relationship?

10.1.1 Analysis of the *World Almanac* data

The 2005 *World Almanac* [205, p. 594] summarizes the results of U.S. presidential elections from the first one in 1789 when George Washington was elected with no opposing candidate (at least according to this data source) to the controversial election of 2000 when George W. Bush defeated Al Gore in electoral votes but received fewer popular votes. From 1789 to 1820, this table lists the popular vote as "Unknown," and a footnote to the table indicates that the 1824 election was decided by the House of Representatives because none of the four major candidates achieved a majority of the electoral votes. Consequently, the analysis presented here is based on the popular and electoral votes given for the 44 elections between 1828 and 2000. Also, note that while most of these elections were primarily contests between two leading candidates, additional candidates did feature prominently enough in 12 elections to also be listed in the *World Almanac* table. The preliminary analysis presented here omits these additional candidates, focusing entirely on the top two candidates, but the presence of third-party candidates can have a significant influence on our results, as subsequent discussions illustrate.

Fig. 10.1 plots the total electoral votes listed in the *World Almanac* table for each U.S. presidential election from 1828 to 2000. The point of this plot is

Figure 10.1: Total electoral votes cast in the 44 U.S. presidential elections between 1828 and 2000

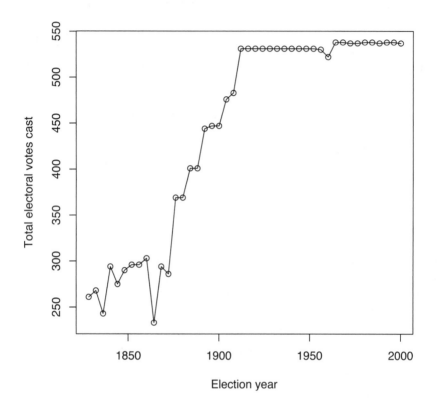

to emphasize the changes in total electoral votes as the country grew in population and as states were added: in 1828, the United States had approximately 12 million people in 24 states, while in 2000, the estimated population was approximately 281 million in 50 states. As a consequence, the total number of electoral votes increased from 261 in 1828 to 538 in 2000, and their allocation among states shifted as new states were added and as the relative populations of the states changed. The nonmonotonic behavior seen at several points in the plot has several distinct causes: the large drop seen in 1864 is due to the Civil War, while most of the other, smaller dips seen in the plot are due to data anomalies in the *World Almanac* table, a topic discussed further in Sec. 10.1.4.

The results examined here were derived from the numbers given in the *World Almanac* table as follows. First, in the 12 elections between 1828 and 2000 where more than two candidates are listed, the top two in terms of popular vote were selected. In 11 of these 12 elections, these same two candidates would also have been selected on the basis of electoral vote (in six of these elections, the leading third-party candidate received no electoral votes at all). The sole exception is the election of 1860 involving Abraham Lincoln, Stephen A. Douglas, John C. Breckenridge, and John Bell: there, the two candidates with the most popular votes were Lincoln and Douglas, but both Breckenridge and Bell won more electoral votes than Douglas did (72 and 39, respectively, vs. 12 for Douglas). Popular vote percentages were computed as the number of popular votes listed for each of the top two candidates, divided by the sum of the popular votes given for all candidates listed. Note that while the popular votes listed for each candidate should be reasonably accurate, the total popular vote estimated in this way neglects votes cast for any additional candidates not listed in the table. The resulting error should not be large, but no attempt has been made here to estimate it. Electoral vote percentages were calculated analogously and are, again, subject to limitations of the *World Almanac* data. Specifically, the results presented here were based on the numbers appearing in the table entries themselves [205, p. 594]; corrections to these numbers based on footnotes to this table and on a second data source are discussed in Sec. 10.1.4.

Fig. 10.2 shows the fraction of the total electoral votes obtained by each candidate plotted against their corresponding fraction of the total popular votes cast. The winning candidates (i.e., those achieving the larger number of electoral votes) correspond to points represented as solid triangles pointing upward, while the losing candidates correspond to points represented as open triangles pointing downward. The two dashed lines in the plot correspond to the lines fit by the method of ordinary least squares, as described in Chapter 5, to the data points for the winning candidate (the upper dashed line) and for the losing candidate (the lower dashed line). In addition, two solid lines are also shown in Fig. 10.2: a horizontal line representing 50% of the electoral vote, separating the winning candidates (above the line) from the losing candidates (below the line), and a vertical line representing 50% of the popular vote. While this second line has no bearing on the outcome of the election, it is interesting because points in the lower right quadrant (less than 50% of the electoral vote, more than 50% of the popular vote) correspond to the elections in 1876 (Rutherford B. Hayes vs. Samuel J. Tilden) and 1888 (Benjamin Harrison vs. Grover Cleveland) when the winning candidates (Hayes and Harrison) received fewer popular votes than the losing candidates did. The first of these cases has been called "the stolen election" and has been the subject of a detailed popular treatment [211]. The more recent case where the winning candidate received fewer popular votes than the losing candidate was the 2000 election in which George W. Bush defeated Al Gore, but this data point does not appear in the lower right quadrant because *neither candidate* received more than 50% of the popular vote, due to the presence of the third-party candidate Ralph Nader.

Data points in Fig. 10.2 falling into the upper left quadrant (less than 50% popular vote, more than 50% electoral vote) represent 11 elections where the winning candidate failed to achieve a majority of the popular vote. These elections include all three of the controversial ones discussed above (1876, 1888, and 2000), along with eight others (those of 1848, 1856, 1860, 1892, 1912, 1948, 1968,

Figure 10.2: Fraction of electoral votes received versus the fraction of popular votes received by U.S. presidential candidates from 1828 to 2000. Results for winning candidates are marked with upward-pointing solid triangles, while results for losing candidates are marked with downward-pointing open triangles. The upper dashed line is a least squares fit to the winning candidate's data, and the lower line is a least squares fit to the losing candidate's data

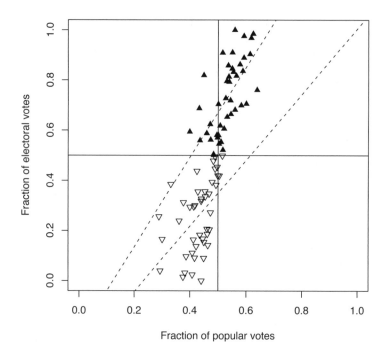

and 1992). In all eight of these cases—as in the election of 2000—third-party candidates received enough of the popular vote to prevent either the winning candidate or the losing candidate with the largest share of the popular vote from achieving a simple majority of the popular vote.

10.1.2   Electoral versus popular vote margins

As noted in the introduction, the margin of victory in a presidential election is often taken as an indication of the strength of the incumbent's "mandate," but here again, the numbers we obtain can depend strongly on the way we compute them: electoral mandate or popular mandate? Fig. 10.3 shows a plot of the electoral margin (i.e., fraction of electoral votes received by the winner

Figure 10.3: Electoral vs. popular vote margins. The vertical dashed line corresponds to a popular margin of zero; the three points (solid circles) to the left of this line represent the elections of 1876, 1888, and 2000. The solid diagonal line corresponds to perfect agreement between the two margins, and the dashed line through the points is a least squares fit to the data

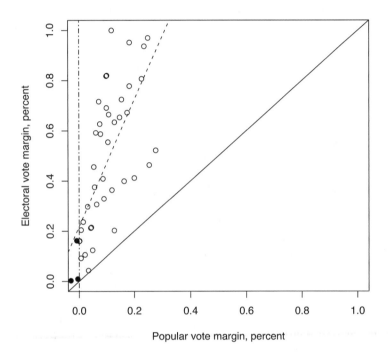

minus fraction received by the loser) against the popular margin, computed analogously from the popular votes. The dashed vertical line in this plot corresponds to a popular vote margin of zero, so the three points to the left of this line, represented by solid circles, correspond to the elections of 1876, 1888, and 2000, where the winning candidate received fewer popular votes than the losing candidate. The dashed line through the points represents the ordinary least squares fit to the data, and the 45° solid line in the plot represents the "perfect agreement line," on which points would fall if the electoral and popular margins were exactly equal.

One of the most striking features of this plot is that *all* of the points lie above the perfect agreement line, indicating that the electoral vote margin always exceeds the popular vote margin. This consistent difference is a direct consequence of the "winner take all" nature of electoral vote assignments: in

Table 10.1: Association measures computed between three pairs of data sequences derived from the *World Almanac* U.S. presidential election data

Dataset	Product-moment correlation, $\hat{\rho}$	Spearman rank correlation, $\hat{\rho}_S$	Kendall's τ measure, $\hat{\tau}$
Winning vote fractions	0.658	0.700	0.502
Losing vote fractions	0.506	0.606	0.444
Victory margins	0.699	0.753	0.577

all states but two (Maine and Nebraska), the candidate receiving the most popular votes within the state is awarded all of that state's electoral votes. As a consequence, while no winning candidate has ever achieved a popular vote margin larger than 30% (the maximum was just under 28%, received by Warren G. Harding over James M. Cox in the 1920 election), the largest electoral vote margin is 100%, for the 1872 election when Ulysses S. Grant defeated Horace Greeley, who died after the election but before the Electoral College met. As a result, Greeley was awarded no electoral votes, and the *World Almanac* lists no other contenders in this election; this case is discussed further in Sec. 10.1.4.

The least squares regression line shown in Fig. 10.3 has a slope of 2.53 and an intercept of 0.22, suggesting that electoral vote margins are typically a bit more than 2.5 times the corresponding popular vote margins. The general scatter of the points about this line, however, makes it clear that this relationship is not perfectly predictive. In fact, the slope of this line is closely related to the product-moment correlation coefficient between the electoral and popular vote results, one of three association measures discussed next.

10.1.3   Association measures

One of the key points of this chapter is that it is possible to compute numerical measures of association between variables. The following paragraphs summarize the results obtained using three such measures: the product-moment correlation coefficient described in Sec. 10.4, the Spearman rank correlation coefficient described in Sec. 10.5, and Kendall's τ described in Sec. 10.7.5. The results illustrate both the utility of these measures and some of their limitations.

Table 10.1 gives values for all three of these association measures, computed for three different pairs of data values: the electoral and popular vote fractions received by the winning candidate, the vote fractions received by the losing candidate, and the winning margins computed from electoral and popular vote fractions. The product-moment correlation coefficient listed in the first column is the best-known association measure, having a value of +1 if the two variables are related by a positive linear rescaling and a value of 0 if they are statistically independent random variables. In contrast, both the Spearman rank correlation coefficient, listed in the second column of Table 10.1, and Kendall's τ, listed in the third column, are measures of *monotone association*, exhibiting the value +1 if there exists a monotonically increasing relationship between variables. Recall that the large difference between the product-moment correlation coefficient and the Spearman rank correlation for the brain weight vs. body weight dataset examined in Chapter 1 was used to suggest the use of logarithmic transformations as a more natural way of looking at that dataset. The difference between these two monotone association measures lies in the way they are computed: as its name implies, the Spearman rank correlation coefficient is computed on the basis of ranks, essentially measuring the tendency for large values of one variable to occur together with large values for the other. In contrast, Kendall's τ is a measure of the degree of *concordance* between the variables: if $x_i > x_j$, how likely is it that $y_i > y_j$?

All three of the association measures listed in Table 10.1 are moderately large, positive numbers, suggesting some tendency for larger electoral vote measures to be associated with larger popular vote measures, for winning vote fractions, losing vote fractions, and winning margins. Further, all three of these association measures give the same ordering for the results: the apparent association is strongest for the winning margins, intermediate for the winning vote fractions, and weakest for the losing vote fractions. Similarly, the three measures exhibit the same ordering of values for all three of these data sequences: the Spearman rank correlations are the largest, the product-moment correlation values are intermediate, and Kendall's τ gives the smallest values.

Overall, these results suggest the following interpretation. First, all three measures give evidence that, as expected, electoral and popular vote measures are related, whether we are considering winning vote fractions, losing vote fractions, or winning margins. Second, the fact that the Spearman rank correlation coefficient is consistently larger than the product-moment correlation coefficient suggests the possibility of either a nonlinear relationship between the variables or the influence of outliers in suppressing the product-moment correlation coefficient. In contrast to the brain weight/body weight example discussed in Chapter 1, however, these differences are not dramatic, suggesting that these effects are not large. Finally, the fact that Kendall's τ is consistently the smallest of the three measures considered, for all three datasets, suggests—correctly, in this case—that discordances between popular and electoral vote results are in important factor in all three of the cases examined here. Again as in the case of the rank correlations, these differences relative to the product-moment correlation coefficient are not large, but they are consistent across all three examples, suggesting a systematic effect.

10.1.4   Data limitations and anomalies

As noted in Sec. 10.1.1, a close examination of Fig. 10.1 suggests some anomalies in the behavior of the electoral vote totals over time. In particular, while it was noted that the nonmonotonic behavior seen in 1864 represents a reduction in electoral votes due to the secession of the southern states during the Civil War, other features—like the dip seen in 1836—are initially quite puzzling. In fact, this nonmonotonic behavior is due to the incompleteness of the *World Almanac* data, which lists only two candidates for that year: Martin Van Buren and William H. Harrison [205, p. 594]. It turns out that there were three other candidates in the 1836 election—Hugh L. White, Daniel Webster, and W.P. Magnum—who gained 51 electoral votes between them. Accounting for these missing electoral votes removes the 1836 dip seen in Fig. 10.1.

Identification of the candidates not listed in the *World Almanac* table [205, p. 594] was based on an alternative presentation of U.S. presidential election results, obtained from the following website:

http://www.infoplease.com/ipa/A0781450.html

Other examples of nonmonotonic behavior in Fig. 10.1 include the dip in total electoral vote seen in 1844—an anomaly not explained by either data source—and another dip in 1872, again a consequence of third-party candidates not listed in the *World Almanac* data table. This particular case was further complicated by the fact that the losing candidate (Horace Greeley) died after the election but before the electoral votes were cast; as a consequence, Greeley's votes were divided among four other candidates, each receiving between 1 and 42 electoral votes, and 17 electoral votes were not counted. Note that these electoral votes were not listed in the *World Almanac* data for this election, leading to the earlier (erroneous) conclusion that Grant received 100% of the electoral vote in 1872. Finally, there is also a dip seen in 1960, which is explained in a footnote to the *World Almanac* table: third-party candidate Harry F. Bird received 15 electoral votes, but was not listed as a candidate in the table.

The possibility of disagreements like these between data sources emphasizes the importance of keeping an open mind regarding the accuracy and completeness of any source. Such open mindedness is especially important in cases like this one where the analysis is based on *secondary* rather than *primary* sources (i.e., republications or summaries obtained from original sources, rather than records from the original sources themselves). In particular, published summaries of larger data collections may be incomplete or inaccurate due to omissions or simplifications necessary to present a succinct account. This point was discussed briefly in Chapter 2 and is treated in greater detail by Jacob [156].

10.2   Joint and conditional distributions

If we model uncertain data variables—of any type—as random quantities, it follows that the relationship between these variables is described by their *joint probability distribution*. These distributions are analogous to the univariate distributions considered up to this point for individual variables, but they describe the probability that *all* of the variables under consideration assume the same value (or range of values) *simultaneously*. A closely related notion is that of the *conditional distribution*, which describes the probability that a subset of one or more of these variables exhibit specified values or ranges of values, given conditions on the other variables. Both of these concepts are extremely useful in describing the relationships between variables, as subsequent discussions will demonstrate.

The following section introduces the mathematical machinery required to describe joint and conditional distributions, starting with the discrete case in Sec. 10.2.1 because it is the simplest. Next, the continuous case is discussed, introducing multivariate distribution and density functions in Sec. 10.2.2 and introducing the key notion of *statistical independence* in Sec. 10.2.3. Finally, conditional distributions and expectations are discussed in Sec. 10.2.4.

10.2.1 Discrete events: the multinomial distribution

Chapter 3 introduced univariate discrete distributions to describe random events that can be described by a single number; e.g., the number of heads observed in a sequence of N coin flips, or the number showing on a single six-sided die. This description extends to more complex settings like the result of rolling two or more fair dice, but only if the result can be characterized by a single number (e.g., the total of the numbers showing on two dice). If we wish to describe the numbers appearing on two or more dice (e.g., "a five and a two" rather than just "a total of seven"), we need a multivariate description that characterizes two or more outcomes simultaneously.

By far the most popular multivariate discrete probability model is the *multinomial distribution*, a distribution whose popularity and widespread use in the discrete case correspond to that of the multivariate Gaussian distribution in the continuous case. The multinomial distribution applies when we have N independent, identical trials, each of which can exhibit one of c possible outcomes. A case in point is the fair dice example noted above: there, $N = 2$ (i.e., two fair dice), and $c = 6$ (i.e., each die has six sides). To define the multinomial distribution, let y_{ij} be an indicator variable, equal to 1 if trial i had outcome j and zero otherwise, and denote the outcome of trial i by the c-component vector:

$$\mathbf{y}_i = [y_{i1}, y_{i2}, \dots, y_{ic}]^T. \qquad (10.1)$$

As an example, in the case of two dice, the combination of a 2 on the first die and a 5 on the second die would correspond to the outcome vectors $\mathbf{y}_1 = [0, 1, 0, 0, 0, 0]^T$ and $\mathbf{y}_2 = [0, 0, 0, 0, 1, 0]^T$. The multinomial distribution describes the numbers of times each outcome occurs in N trials—that is, it is the distribution of the *counts:*

$$n_j = \sum_{i=1}^{N} y_{ij}, \qquad (10.2)$$

for the outcomes $j = 1, 2, \dots, c$. This distribution is based on the individual probabilities π_j that any trial exhibits outcome j, and it is given by:

$$p(n_1, n_2, \dots, n_c) = \left(\frac{n!}{n_1! n_2! \cdots n_c!} \right) \pi_1^{n_1} \pi_2^{n_2} \cdots \pi_c^{n_c}. \qquad (10.3)$$

In the case of two fair six-sided dice, $\pi_j = 1/6$ for $j = 1, 2, \ldots, 6$, and each individual n_i count can take the values 0, 1 or 2. Substituting the values $n_2 = 1$ and $n_5 = 1$ into this expression yields the probability for the (2,5) combination as $1/18$, while the probability for the (1,1) combination ("snake eyes") corresponds to $n_1 = 2$ and has probability $1/36$.

Several important characteristics of the multinomial distribution are worth noting. First, this distribution is redundant in two important respects: neither the c individual counts n_j nor the c individual probabilities π_j are independent. Instead, they satisfy the following constraints:

$$\sum_{j=1}^{c} n_j = n, \quad \text{and} \quad \sum_{j=1}^{c} \pi_j = 1. \tag{10.4}$$

A second key observation is that taking $c = 2$ (binary outcomes) reduces the multinomial distribution to the binomial distribution introduced in Chapter 3; note that since only one of the two probabilities π_j and one of the counts n_j for this case can be independently specified, this distribution is univariate in character and does not require the vector formulation presented here.

A third key point is that the individual probabilities π_j defining the multinomial distribution represent the *marginal probabilities* for this distribution. Specifically, π_j represents the probability of observing outcome j, irrespective of the other outcomes. In fact, these marginal distributions correspond to the binomial distribution defined by the probability π_j. Thus, it follows from the results presented in Chapter 3 that the mean and variance of the count n_j are given by:

$$E\{n_j\} = N\pi_j, \quad \text{and} \quad \text{var}\{n_j\} = N\pi_j(1 - \pi_j). \tag{10.5}$$

The essential new feature that arises in the multinomial distribution is the fact that the individual counts n_j are not independent, due to the sum constraint given in Eq. (10.4). This manifests itself in the following result: the *covariance* between n_j and n_k is given by:

$$\text{cov}\{n_j, n_k\} = E\{(n_j - E\{n_j\})(n_k - E\{n_k\})\} = -N\pi_j\pi_k. \tag{10.6}$$

As discussed further in Sec. 10.2.3, if the counts n_j and n_k were statistically independent, the expectation of the product in Eq. (10.6) could be rewritten as the product of the expectations of the individual terms. Since each of these individual expectations is zero, it follows that the covariance between statistically independent variables is zero. The fact that the covariance here is negative implies that large values of n_j are more likely to arise with smaller values of n_k and vice versa, a direct consequence of the sum constraint.

This last result is closely related to the product-moment correlation coefficient introduced in Sec. 10.4. Commonly called the *correlation coefficient*, this quantity is a data-based estimator of the correlation between two variables, defined as:

$$\rho_{xy} = \frac{E\{(x - E\{x\})(y - E\{y\})\}}{\sqrt{E\{(x - E\{x\})^2 E\{(y - E\{y\})^2\}}} = \frac{\text{cov}\{x, y\}}{\sqrt{\text{var}\{x\}\text{var}\{y\}}}. \tag{10.7}$$

In the case of the binomial distribution, it follows from the results presented above that the correlation between n_j and n_k is:

$$\rho_{n_j,n_k} = -\sqrt{\frac{\pi_j \pi_k}{(1-\pi_j)(1-\pi_k)}}. \tag{10.8}$$

10.2.2 Multivariate distributions and densities

As discussed in Chapter 4, real-valued continuous random variables are defined in terms of a cumulative distribution function (CDF) $F_x(q)$ that—in almost all of the cases considered in this book—may be represented as the integral of a probability density function (pdf) $p(x)$. For multivariate distributions, these ideas extend directly to the *joint CDF* $F_{\mathbf{x}}(\mathbf{q})$ and the *joint pdf* $p(\mathbf{x})$. More specifically, consider the case of a *random vector* $\mathbf{x}$ with N components (denoted $x_1, \ldots, x_N$), and define the corresponding *quantile vector* $\mathbf{q}$ with the N components $q_1, \ldots, q_N$. The joint cumulative distribution function $F_{\mathbf{x}} : R^N \to [0,1]$ is defined as the probability:

$$F_{\mathbf{x}}(\mathbf{q}) = \mathcal{P}\{x_1 \leq q_1, \ldots, x_N \leq q_N\}. \tag{10.9}$$

Note that this definition reduces to the univariate definition for $N = 1$ and that, for $N > 1$, it describes the joint probability that all of the components of $\mathbf{x}$ satisfy the inequalities $x_i \leq q_i$ for $i = 1, \ldots, N$. Also, note that the multivariate function $F_{\mathbf{x}}(\cdot)$ is increasing with respect to each of its arguments separately:

$$r_i > q_i \;\Rightarrow\; F_{\mathbf{x}}(q_1, \ldots, r_i, \ldots, q_N) \geq F_{\mathbf{x}}(q_1, \ldots, q_i, \ldots, q_N), \tag{10.10}$$

for all $i = 1, 2, \ldots, N$. Again as in the univariate case, if $F_{\mathbf{x}}(\cdot)$ is sufficiently smooth, it may be represented in terms of a *joint density* $p(\mathbf{x})$:

$$F_{\mathbf{x}}(\mathbf{q}) = \int_{-\infty}^{q_1} \cdots \int_{-\infty}^{q_N} p(x_1, \ldots, x_N) dx_1 \cdots dx_N. \tag{10.11}$$

In particular, almost all of the continuous multivariate random variable models considered in this book may be described in terms of joint densities.

One of the new features that appears in the multivariate random variable model is the concept of a *marginal distribution*, $F_{x_i}(q_i)$, defined as the probability that $x_i \leq q_i$, regardless of the values of the other components of the random vector $\mathbf{x}$. This distribution may be obtained by integrating the joint density over the complete range of all of the *other* components x_j for $j \neq i$ to average out their influence, resulting in a cumulative distribution function involving only x_i:

$$\begin{aligned} F_{x_i}(q_i) &= \mathcal{P}\{x_i \leq q_i\} \\ &= \int_{-\infty}^{\infty} \cdots \int_{-\infty}^{q_i} \cdots \int_{-\infty}^{\infty} p(x_1, \ldots, x_{i-1}, x_i, x_{i+1}, \ldots, x_N) \\ &\qquad dx_1 \cdots dx_{i-1} dx_i dx_{i+1} \cdots dx_N. \end{aligned} \tag{10.12}$$

Note that this expressions involves integrals over all of the components of the distribution, but only the integral involving the component x_i has a finite upper limit. Differentiating this marginal distribution with respect to this finite upper limit yields the corresponding density $p_{x_i}(x)$ associated with x_i alone, called the ith *marginal density*. Alternatively, this density may be obtained directly by dropping the integral over x_i in Eq. (10.12):

$$
\begin{aligned}
p_{x_i}(x) &= \left. \frac{dF_{x_i}(q_i)}{dq_i} \right|_{q_i=x} \\
&= \int_{-\infty}^{\infty} \cdots \int_{-\infty}^{\infty} p(x_1, \ldots, x_{i-1}, x, x_{i+1}, \ldots, x_N) \\
&\qquad dx_1 \cdots dx_{i-1} dx_{i+1} \cdots dx_N.
\end{aligned}
\tag{10.13}
$$

One of the key points of this chapter is to illustrate the range of relations that can exist between the parent density $p(\mathbf{x})$ and the marginals $p_{x_i}(x)$: in favorable cases (in particular, for the multivariate Gaussian distribution considered in Sec. 10.3), this relation is quite intuitive, but in the general case, the situation is much more complicated, as illustrated in Sec. 10.7.

An important notion explored further in Chapter 14 is that of *exchangeability*, which may be defined in various ways [93, 180]. The basic idea is that observations are exchangeable if the probability of observing any particular joint outcome is invariant under arbitrary reorderings of the components x_i [119, p. 18]. Consequently, the components of $\mathbf{x}$ are exchangeable if the joint CDF $F_{\mathbf{x}}(\cdot)$ and the joint pdf $p(\cdot)$ are invariant under arbitrary reorderings of their N arguments. As a further consequence, note that if the components of $\mathbf{x}$ are exchangeable, all of the marginal densities $p_{x_i}(x)$ are necessarily equal.

Finally, the notion of expectation generalizes the univariate case: given the multivariate density $p(\mathbf{x})$, the expected value of the function $f(\mathbf{x})$ is:

$$
\begin{aligned}
E\{f(\mathbf{x})\} &= \int_{R^N} f(\mathbf{x})p(\mathbf{x})d\mathbf{x} \\
&= \int_{-\infty}^{\infty} \cdots \int_{-\infty}^{\infty} f(x_1, \ldots, x_N)p(x_1, \ldots, x_N)dx_1 \cdots dx_N.
\end{aligned}
\tag{10.14}
$$

Particularly important expectations are the mean vector $\bar{\mathbf{x}}$ and the covariance matrix Σ, defined as:

$$
\bar{\mathbf{x}} = E\{\mathbf{x}\}, \quad \Sigma = E\{(\mathbf{x} - \bar{\mathbf{x}})(\mathbf{x} - \bar{\mathbf{x}})^T\}.
\tag{10.15}
$$

The mean vector $\bar{\mathbf{x}}$ can assume any value in R^N, but the $N \times N$ covariance matrix Σ is a symmetric, positive semidefinite matrix for any distribution. Further, this matrix is usually (although not always) positive definite and thus invertible. The mean vector and covariance matrix completely characterize the multivariate Gaussian distribution and are discussed further in subsequent sections of this chapter.

10.2.3 Statistical independence

An important concept that has already made its appearance in several discussions is that of *statistical independence*, often invoked as a working assumption in analyzing a collection of data observations. The basic idea is that a collection $\{x_i\}$ of N random variables is statistically independent if x_i conveys no information about x_j for $i \neq j$. Its practical importance lies in the fact that it greatly simplifies computations involving random vectors, essentially reducing multivariate characterizations to simple combinations of univariate characterizations.

In the discrete case, this condition corresponds to the following result. Suppose x and y are two discrete random variables, taking one of I and J possible values, respectively, and let $p_x(i)$ denote the probability that x assumes its ith possible value and $p_y(j)$ denote the probability that y assumes its jth possible value. Now, consider the joint probability $p_{xy}(i,j)$ that x assumes its ith possible value and y assumes its jth possible value *simultaneously*. The variables x and y are statistically independent if, for all admissible values of i and j:

$$p_{xy}(i,j) = p_x(i)p_y(j). \qquad (10.16)$$

More generally, in the case of N discrete random variables $\{x_i\}_{i=1}^N$ where x_i can assume any of I_i possible values, let $p_{\mathbf{x}}(i_1, i_2, \ldots, i_N)$ denote the joint distribution. These random variables are statistically independent if this joint distribution can be factored as:

$$p_{\mathbf{x}}(i_1, i_2, \ldots, i_N) = p_{x_1}(i_1)p_{x_2}(i_2) \cdots p_{x_N}(i_N). \qquad (10.17)$$

In the continuous case, the N continuous random variables $\{x_i\}_{i=1}^N$ are statistically independent if the joint CDF $F_{\mathbf{x}}(\mathbf{q})$ may be factored as the product of N univariate CDFs:

$$F_{\mathbf{x}}(q_1, \ldots, q_N) = F_1(q_1) \cdots F_N(q_N). \qquad (10.18)$$

Further, if these CDFs are sufficiently well behaved for the corresponding densities to exist, independence also implies that these joint densities also factor:

$$p(x_1, \ldots, x_N) = p_1(x_1) \cdots p_N(x_N). \qquad (10.19)$$

In fact, substituting Eq. (10.19) into Eq. (10.13) shows that the densities $p_i(x_i)$ are simply the marginal densities $p_{x_i}(x_i)$.

An extremely important specialization of this idea is the following. Suppose $\mathbf{x}$ is a multivariate distribution and that all of the components x_i of this random vector are statistically independent. Further, assume that the marginal densities $p_i(x_i)$ of these components are all identical. These conditions imply that the N components of $\mathbf{x}$ constitute an *independent, identically distributed* (i.i.d.) collection of random variables. In fact, this condition is usually invoked implicitly when applying univariate characterizations like those discussed in Chapter 8 to a finite collection $\{x_i\}$ of data observations. Also, it follows from the commutativity of multiplication that if $\mathbf{x}$ is a sequence of N i.i.d. random variables, these random variables are also *exchangeable*: the joint distribution is invariant under arbitrary reorderings of the components. Conversely, it is important to note that exchangeability is a strictly weaker condition than the i.i.d. requirement, as an example presented in Sec. 10.3.3 illustrates.

10.2.4 Conditional probabilities

When statistical independence *does not* hold, we are dealing with related random variables, and in such cases we are often interested in using this relationship to infer something about one variable from knowledge of other variables. The basic machinery for dealing with these ideas is that of *conditional probabilities*, which describe the probability of observing some particular outcome *given that we have observed some possibly related event*. The following discussions provide an abbreviated introduction to this topic, giving basic definitions and some results that will be useful in subsequent discussions. For a more detailed introduction aimed at students of science and engineering, refer to the book by Papoulis [222]; for a more thorough (and necessarily much more mathematical) discussion, refer to the book by Billingsley [37].

The basic notion of interest here is the following one: suppose $\mathcal{E}$ is some event (i.e., random outcome) with which we can associate a nonzero probability, $\mathcal{P}\{\mathcal{E}\} > 0$. Further, suppose $\mathcal{A}$ is some other event that *may* be statistically related to $\mathcal{E}$. The *conditional probability of $\mathcal{A}$ given $\mathcal{E}$* is defined as:

$$\mathcal{P}\left\{\mathcal{A} \mid \mathcal{E}\right\} = \frac{\mathcal{P}\left\{\mathcal{A} \cap \mathcal{E}\right\}}{\mathcal{P}\left\{\mathcal{E}\right\}}. \tag{10.20}$$

Here, $\mathcal{P}\left\{\mathcal{A} \cap \mathcal{E}\right\}$ represents the probability of observing both events $\mathcal{A}$ and $\mathcal{E}$ together, so this expression defines the conditional probability $\mathcal{P}\left\{\mathcal{A} \mid \mathcal{E}\right\}$ as the probability of observing both events together, normalized by the probability of observing the conditioning event $\mathcal{E}$.

Two useful observations follow immediately from this definition. First, note that two events $\mathcal{A}$ and $\mathcal{E}$ are statistically independent if and only if their joint probability factors; i.e.:

$$\mathcal{P}\{\mathcal{A} \cap \mathcal{E}\} = \mathcal{P}\{\mathcal{A}\} \cdot \mathcal{P}\{\mathcal{E}\}. \tag{10.21}$$

Hence, it follows that these events are statistically independent if and only if the conditional probability of $\mathcal{A}$ given $\mathcal{E}$ is equal to the unconditional probability:

$$\mathcal{A}, \mathcal{E} \text{ independent } \Leftrightarrow \mathcal{P}\left\{\mathcal{A} \mid \mathcal{E}\right\} = \mathcal{P}\left\{\mathcal{A}\right\}. \tag{10.22}$$

The second useful observation is that *if the events $\mathcal{A}$ and $\mathcal{E}$ are mutually exclusive, then:*

$$\mathcal{P}\{\mathcal{A} \cap \mathcal{E}\} = 0 \Rightarrow \mathcal{P}\{\mathcal{A} \mid \mathcal{E}\} = 0. \tag{10.23}$$

In words, this result simply means that if the events $\mathcal{A}$ and $\mathcal{E}$ cannot both occur and we observe $\mathcal{E}$, then the event $\mathcal{A}$ necessarily has probability zero.

Also, the following rearranged form of Eq. (10.20) is frequently useful:

$$\mathcal{P}\{\mathcal{A} \cap \mathcal{E}\} = \mathcal{P}\{\mathcal{A} \mid \mathcal{E}\} \cdot \mathcal{P}\{\mathcal{E}\}. \qquad (10.24)$$

Further, since $\mathcal{A} \cap \mathcal{E} = \mathcal{E} \cap \mathcal{A}$, it follows from Eq. (10.24) that, for any two events $\mathcal{A}$ and $\mathcal{E}$:

$$\mathcal{P}\{\mathcal{A} \mid \mathcal{E}\} \cdot \mathcal{P}\{\mathcal{E}\} = \mathcal{P}\{\mathcal{E} \mid \mathcal{A}\} \cdot \mathcal{P}\{\mathcal{A}\}. \qquad (10.25)$$

In some cases (e.g., maximum likelihood estimation), we have sufficient information to compute the conditional probability $\mathcal{P}\{\mathcal{A} \mid \mathcal{E}\}$, but we are interested in determining the conditional probability $\mathcal{P}\{\mathcal{E} \mid \mathcal{A}\}$. Rearranging Eq. (10.24) gives us a way of doing this, provided we can compute the unconditional probabilities:

$$\mathcal{P}\{\mathcal{E} \mid \mathcal{A}\} = \frac{\mathcal{P}\{\mathcal{A} \mid \mathcal{E}\} \cdot \mathcal{P}\{\mathcal{E}\}}{\mathcal{P}\{\mathcal{A}\}}. \qquad (10.26)$$

This result is known as *Bayes theorem* [40, p. 10], and it forms the basis for *Bayesian statistics*. An introduction to Bayesian statistics lies beyond the scope of this book, but the book by Box and Tiao [40] provides an excellent starting point for those interested in learning more.

10.2.5 Conditional distributions and expectations

Applying the notions of conditional probability just discussed to events defined by continuous random variables leads to the definitions of conditional distributions and expectations. For example, suppose x is a continuous random variable, and consider the event $x \le q$ where q is a fixed real number. This construction leads to the definition of the *conditional distribution*. Specifically, if $\mathcal{E}$ is some event, the *conditional distribution of x given $\mathcal{E}$* is defined as:

$$F_x(q|\mathcal{E}) = \frac{\mathcal{P}\{(x \le q) \cap \mathcal{E}\}}{\mathcal{P}\{\mathcal{E}\}}. \qquad (10.27)$$

To see the utility of this idea, the following paragraphs consider the results obtained for four different choices of the conditioning event $\mathcal{E}$.

First, consider the case where $\mathcal{E}$ is the event $a \le x \le b$. Since this event and the event $x \le q$ are mutually exclusive if $q < a$, it follows from the results presented above that the conditional probability is zero for this case. Similarly, if $q \ge b$, the conditioning event $\mathcal{E}$ (i.e., $a \le x \le b$) implies $x \le q$, which occurs with probability 1. The only case we need to consider, then, is $a \le q < b$, for which the probabilities in Eq. (10.27) are easily evaluated. Specifically, the denominator is given by:

$$\mathcal{P}\{\mathcal{E}\} = \mathcal{P}\{a \le x \le b\} = \int_a^b p(x)dx = F(b) - F(a), \qquad (10.28)$$

where $F(q)$ is the (unconditional) cumulative distribution function for the random variable x and $p(x)$ is its associated density. The numerator probability in

Eq. (10.27) can also be expressed simply in terms of the unconditional CDF:

$$\mathcal{P}\{(x \le q) \cap (a \le x \le b)\} = \mathcal{P}\{a \le x \le q\} = \int_a^q p(x)dx = F(q) - F(a). \quad (10.29)$$

Combining these results yields the following expression for the cumulative distribution function for $x \le q$, given the event $a \le x \le b$:

$$F_x(q|a \le x \le b) = \begin{cases} 0 & q < a, \\ \frac{F(q)-F(a)}{F(b)-F(a)} & a \le q < b, \\ 1 & q \ge b. \end{cases} \quad (10.30)$$

As discussed in Chapter 4, if the function $F(\cdot)$ is sufficiently well behaved, it may be expressed in terms of a density function $p(x)$, equal to $dF(x)/dx$. Differentiating Eq. (10.30) with respect to x in this case leads to the following *conditional density*:

$$p(x|a \le x \le b) = \frac{p(x)}{F(b) - F(a)}. \quad (10.31)$$

This result is particularly useful because it provides the basis for constructing *truncated* densities; specifically, note that this conditional density satisfies the normalization condition (i.e., it integrates to 1) on the bounded interval $a \le x \le b$. As an example, the *truncated Laplace distribution* results if we take $p(x)$ to be the univariate Laplace density function; some of the properties of this density function are explored in Exercise 4.

As a second example of a conditional distribution, suppose x and y together define a bivariate random vector with joint CDF $F_{xy}(q, r)$, and consider the conditional probability that $x \le q$, conditioned on the event $\mathcal{E}$ defined by $y \le r$. It follows from Eq. (10.27) that:

$$F_x(q|y \le r) = \frac{\mathcal{P}\{x \le q, y \le r\}}{\mathcal{P}\{y \le r\}} = \frac{F_{xy}(q, r)}{F_y(r)}, \quad (10.32)$$

where $F_y(r)$ denotes the marginal CDF of y. Rearranging this expression leads to the following result, which is a special case of Bayes' Theorem (Eq. (10.26)):

$$F_{xy}(q, r) = F_x(q|y \le r)F_y(r) = F_y(r|x \le q)F_x(q). \quad (10.33)$$

The third example is a generalization of this result, obtained by conditioning on the event $\mathcal{E}$ defined as $r \le y \le s$. Assuming $f(x, y)$ is the joint density of x and y, it follows from Eq. (10.27) that this conditional distribution is given by:

$$\begin{aligned} F_x(q|r \le y \le s) &= \frac{\mathcal{P}\{(x \le q) \cap (r \le y \le s)\}}{\mathcal{P}\{r \le y \le s\}} \\ &= \frac{\int_{-\infty}^q \int_r^s p(x, y)dxdy}{\int_{-\infty}^\infty \int_r^s p(x, y)dxdy}. \end{aligned} \quad (10.34)$$

To simplify this result, note that:

$$\int_r^s p(x, y)dy = \int_{-\infty}^s p(x, y)dy - \int_{-\infty}^r p(x, y)dy, \quad (10.35)$$

from which it follows that the conditional distribution of x given $r \leq y \leq s$ is:

$$F_x(q|r \leq y \leq s) = \frac{F_{xy}(q, s) - F_{xy}(q, r)}{F_y(s) - F_y(r)}. \tag{10.36}$$

The fourth example is obtained from the third by a limiting argument. Specifically, to obtain the *conditional density* $p(x|y = r)$, first consider the conditional distribution defined by Eq. (10.36) for $s = r + \Delta r$, and then take the limit as $\Delta r \to 0$. That is, note that:

$$F_x(q|r \leq y \leq r + \Delta r) = \frac{[F_{xy}(q, r + \Delta r) - F_{xy}(q, r)]/\Delta r}{[F_y(r + \Delta r) - F_y(r)]/\Delta r}. \tag{10.37}$$

Taking the limit as $\Delta r \to 0$ then yields:

$$\mathcal{P}\{x \leq q|y = r\} = F_x(q|y = r) = \frac{\partial F_{xy}(q, r)/\partial r}{\partial F_y(r)/\partial r} = \frac{\int_{-\infty}^{q} p(x, r)dx}{p_y(r)}. \tag{10.38}$$

If we differentiate this function with respect to x, we obtain the following expression for the conditional density $p(x|y = r)$:

$$p(x|y = r) = \frac{p(x, r)}{p_y(r)}. \tag{10.39}$$

As in the second example, this result may be reformulated to give the following extremely useful relationship between the joint density, the conditional densities, and the marginal densities:

$$p(q, r) = p(q|y = r)p_y(r) = p(r|x = q)p_x(q). \tag{10.40}$$

Conditional expectations are defined as expectations with respect to the conditional densities just defined. That is, if $p(x|\mathcal{E})$ is the density of x conditioned on some event $\mathcal{E}$, the conditional expectation of $f(x)$ is given by:

$$E\{f(x)|\mathcal{E}\} = \int_{-\infty}^{\infty} p(x|\mathcal{E})f(x)dx. \tag{10.41}$$

We can obtain useful results by specializing this expression in either of two ways: specifying the function $f(\cdot)$ or the event $\mathcal{E}$. Two particularly important choices of $f(\cdot)$ are $f(x) = x$, giving us the *conditional mean:*

$$E\{x|\mathcal{E}\} = \int_{-\infty}^{\infty} p(x|\mathcal{E})xdx, \tag{10.42}$$

and $f(x) = [x - E\{x|\mathcal{E}\}]^2$, giving us the *conditional variance:*

$$\text{var}\{x|\mathcal{E}\} = \int_{-\infty}^{\infty} [x - E\{x|\mathcal{E}\}]^2 p(x|\mathcal{E})dx. \tag{10.43}$$

One of the most useful choices of $\mathcal{E}$ is the event $y = r$, for which $p(x|\mathcal{E})$ is given explicitly by Eq. (10.39). Substituting this result into Eq. (10.41) then yields:

$$E\{f(x)|y = r\} = \frac{\int_{-\infty}^{\infty} f(x)p(x,r)dx}{\int_{-\infty}^{\infty} p(x,r)dx}. \tag{10.44}$$

It is important to note here that this conditional expectation is a function of r, but the expectation has averaged over x, so the result does not depend on x.

Finally, the following application of this result is often useful in computing the *unconditional expectation* $E\{f(x)\}$, given the conditional expectation $E\{f(x)|y = r\}$. Specifically, note that the unconditional expectation is given by the following integral:

$$E\{f(x)\} = \int_{-\infty}^{\infty} \int_{-\infty}^{\infty} f(x)p(x,y)dxdy. \tag{10.45}$$

Substituting Eq. (10.40) into this expression then leads to the following rearrangement:

$$
\begin{aligned}
E\{f(x)\} &= \int_{-\infty}^{\infty} \int_{-\infty}^{\infty} f(x)p(x|y = r)p_y(r)dxdr \\
&= \int_{-\infty}^{\infty} p_y(r) \left[\int_{-\infty}^{\infty} f(x)p(x|y = r)dx \right] dr \\
&= \int_{-\infty}^{\infty} E\{f(x)|y = r\}p_y(r)dr \\
&= E_r\{E\{f(x)|y = r\}\}, \tag{10.46}
\end{aligned}
$$

where the subscript on the outer expectation, E_r, has been added to note explicitly that this expectation is taken with respect to r. The interpretation of this result is as follows: the inner expectation regards y as a random variable that has been fixed at the value $y = r$, and the outer expectation simply averages over all possible values r that this random variable can assume, weighted by its relative probability. For convenience, this result is often written less explicitly in one of the following two forms:

$$E\{x\} = E_y\{E\{x|y\}\}, \quad \text{or} \quad E\{x\} = E\{E\{x|y\}\}, \tag{10.47}$$

where the fact that the outer expectation is with respect to the conditioning variable y is assumed to be clear from context in the second expression.

10.3   The multivariate Gaussian distribution

Without question, the most popular multivariate distribution is the multivariate normal or Gaussian distribution. Like its univariate special case, this distribution exhibits many desirable properties, and it serves as a basis for much of our intuition about "related random variables." For example, the product-moment correlation coefficient—used in several previous examples and discussed at some length in Sec. 10.4—is certainly the best-known measure of association between two random variables, and it is intimately related to the bivariate Gaussian distribution. Unfortunately, non-Gaussian distributions do not extend nearly as conveniently, in general, to the multivariate setting, and this fact serves both as further motivation for considering multivariate normal distributions first and as a warning of the additional complications to be faced in the non-Gaussian case. Hence, Sec. 10.3.1 begins with a brief discussion of the definition and general characteristics of the multivariate Gaussian distribution, expressed in terms of vectors and matrices. This representation lends itself nicely to some linear transformation results that are also useful in connection with the linear regression problems discussed in Chapter 11.

10.3.1 Vector formulation

For $N > 1$, the N-variable Gaussian distribution is defined by the joint density function:

$$p(\mathbf{x}) = \frac{1}{(2\pi)^{N/2}|\Sigma|^{1/2}} \exp\left[-\frac{1}{2}(\mathbf{x} - \mu)^T \Sigma^{-1} (\mathbf{x} - \mu)\right]. \qquad (10.48)$$

In this expression, $\mathbf{x}$ denotes the random vector with N components x_1 through x_N, μ is the *mean vector*, a vector with N components μ_1 through μ_N, and Σ is the $N \times N$ *covariance matrix*. Further, for the density given in Eq. (10.48) to be well defined, it is necessary that Σ be positive-definite so that Σ^{-1} exists, generalizing the univariate requirement of positive variance and implying that the determinant $|\Sigma|$ is nonzero. In analogy with the notation used for univariate Gaussian random variables, the multivariate Gaussian random variable defined by the joint density function $p(\mathbf{x})$ in Eq. (10.48) will be denoted $\mathbf{x} \sim N(\mu, \Sigma)$.

An extremely useful alternative expression for the joint density follows from the *singular value decomposition* (SVD) of the matrix Σ, given by:

$$\Sigma = \mathbf{U}\mathbf{D}\mathbf{U}^T, \qquad (10.49)$$

where $\mathbf{U}$ is an $N \times N$ *orthogonal matrix*, which satisfies the defining condition that its inverse exists and is equal to its transpose (i.e., $\mathbf{U}^{-1} = \mathbf{U}^T$), and $\mathbf{D}$ is an $N \times N$ *diagonal matrix* (i.e., $D_{ij} = 0$ unless $i = j$) with strictly positive diagonal values (a consequence of the fact that Σ is positive-definite). It is therefore convenient to write these diagonal matrix elements as σ_i^2:

$$\mathbf{D} = \text{diag}\{\sigma_1^2, \sigma_2^2, \ldots, \sigma_N^2\} = \begin{bmatrix} \sigma_1^2 & 0 & \cdots & 0 \\ 0 & \sigma_2^2 & \cdots & 0 \\ \vdots & \vdots & \vdots & \vdots \\ 0 & 0 & \cdots & \sigma_N^2 \end{bmatrix}. \qquad (10.50)$$

As a consequence of this decomposition, we have the following two useful results.

First, the determinant of Σ may be written explicitly as:

$$|\Sigma| = \prod_{i=1}^{N} \sigma_i^2, \tag{10.51}$$

and second, the inverse matrix Σ^{-1} is given by:

$$\Sigma^{-1} = \mathbf{U}\mathbf{D}^{-1}\mathbf{U}^T, \quad \text{where} \quad \mathbf{D}^{-1} = \begin{bmatrix} \sigma_1^{-2} & 0 & \cdots & 0 \\ 0 & \sigma_2^{-2} & \cdots & 0 \\ \vdots & \vdots & \vdots & \vdots \\ 0 & 0 & \cdots & \sigma_N^{-2} \end{bmatrix}. \tag{10.52}$$

We may exploit these results to obtain a simpler expression for the joint density in terms of a new vector $\mathbf{v}$ defined as:

$$\mathbf{v} = \mathbf{U}^T(\mathbf{x} - \mu) \quad \Leftrightarrow \quad \mathbf{x} = \mu + \mathbf{U}\mathbf{v}. \tag{10.53}$$

Specifically, substituting this expression for $\mathbf{x}$ in terms of $\mathbf{v}$ into Eq. (10.48) gives us the following result:

$$\begin{aligned} p(\mathbf{x}) &= \left[(2\pi)^{N/2} \prod_{i=1}^{N} \sigma_i \right]^{-1} \exp\left[-\frac{1}{2}\mathbf{v}^T\mathbf{D}^{-1}\mathbf{v} \right] \\ &= \left[(2\pi)^{N/2} \prod_{i=1}^{N} \sigma_i \right]^{-1} \exp\left[-\frac{1}{2}\sum_{i=1}^{N} v_i^2/\sigma_i^2 \right] \\ &= \left[(2\pi)^{N/2} \prod_{i=1}^{N} \sigma_i \right]^{-1} \prod_{i=1}^{N} e^{-v_i^2/2\sigma_i^2} \\ &= \prod_{i=1}^{N} \frac{e^{-v_i^2/2\sigma_i^2}}{\sigma_i\sqrt{2\pi}}. \end{aligned} \tag{10.54}$$

In other words, the joint density of $\mathbf{x}$ *has been expressed, through the linear transformation defined in Eq. (10.53), in terms of a collection* $\{v_i\}$ *of N statistically independent, univariate Gaussian components, each with zero mean and variance* σ_i^2.

This result has a number of useful consequences because it essentially reduces many multivariate integrals that arise in computing expectations to collections of scalar integrals. Specific examples include the proof that μ is the mean of the multivariate Gaussian distribution (i.e., that $E\{\mathbf{x}\} = \mu$—see Exercise 1) and that Σ is the covariance matrix (Exercise 2):

$$\text{cov } \mathbf{x} \equiv E\{(\mathbf{x} - \mu)(\mathbf{x} - \mu)^T\} = \Sigma. \tag{10.55}$$

It follows from these results (Exercise 3) that:

$$\mathbf{x} \sim N(\mu, \Sigma) \quad \Rightarrow \quad \mathbf{A}\mathbf{x} + \mathbf{b} \sim N(\mathbf{A}\mu + \mathbf{b}, \mathbf{A}^T\Sigma\mathbf{A}). \tag{10.56}$$

Also, for any symmetric, positive definite $N \times N$ matrix Σ, there exists a unique symmetric, positive definite *square root*, denoted $\Sigma^{1/2}$ and given by:

$$\Sigma^{1/2} = \mathbf{U}\mathbf{D}^{1/2}\mathbf{U}^T, \quad \mathbf{D}^{1/2} = \text{diag}\{\sigma_1, \sigma_2, \ldots, \sigma_N\}. \tag{10.57}$$

Now consider the linear transformation:

$$\mathbf{z} = \Sigma^{-1/2}\mathbf{x} - \Sigma^{-1/2}\mu \iff \mathbf{x} = \Sigma^{1/2}\mathbf{z} + \mu, \tag{10.58}$$

where $\Sigma^{-1/2}$ is the inverse of the positive definite matrix $\Sigma^{1/2}$. It follows from Eq. (10.56) that:

$$\mathbf{z} \sim N(\mathbf{0}, \mathbf{I}_{N \times N}) \implies z_i \sim N(0, 1), \ i = 1, 2, \ldots, N. \tag{10.59}$$

As a practical consequence, Eq. (10.58) means that multivariate Gaussian vectors with any symmetric, positive definite covariance matrix Σ may be generated by applying this linear transformation to sequences $\{z_i\}$ of N statistically independent univariate standard normal random variables. Among other applications, this observation provides a practical way of generating multivariate Gaussian random vectors using readily available univariate random number generators.

10.3.2 Mahalanobis distances

In univariate Gaussian distributions, the following normalized distance plays an essential role in characterizing samples x having the distribution $N(\mu, \sigma^2)$:

$$\delta = \frac{|x - \mu|}{\sigma}. \tag{10.60}$$

In particular, δ determines the probability of observing the sample x, laying the foundation for the 3σ-edit rule discussed in Chapter 7: values $\delta > 3$ have probabilities less than 0.3%. For multivariate Gaussian distributions, the analogous measure of extremeness is the *Mahalanobis distance* [159, p. 100], defined as:

$$d(\mathbf{x}, \mu) = \sqrt{(\mathbf{x} - \mu)^T \Sigma^{-1} (\mathbf{x} - \mu)}. \tag{10.61}$$

In the case of the univariate distribution, the dimension of the vector $\mathbf{x}$ is $N = 1$ and Eq. (10.61) reduces to Eq. (10.60).

More generally, note that the multivariate Gaussian density function $p(\mathbf{x})$ may be written in terms of the Mahalanobis distance as:

$$p(\mathbf{x}) = \frac{1}{(2\pi)^{N/2}|\Sigma|^{1/2}} \exp\left[-\frac{d^2(\mathbf{x}, \mu)}{2}\right]. \tag{10.62}$$

Hence, points $\mathbf{x}$ having the same Mahalanobis distance from the multivariate mean μ have the same probability density. Geometrically, the constraint $d(\mathbf{x}, \mu) = c$ defines an ellipsoid, whose equation is given more explicitly by:

$$(\mathbf{x} - \mu)^T \Sigma^{-1} (\mathbf{x} - \mu) = c^2. \tag{10.63}$$

Figure 10.4: Contours of constant Mahalanobis distance

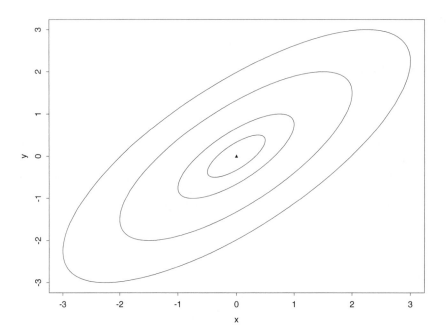

The center of this ellipsoid lies at the multivariate mean μ, its shape is determined by the covariance matrix Σ, and its volume is determined by the constant c. As a specific example, Fig. 10.4 shows four contours of constant Mahalanobis distance for a bivariate Gaussian distribution with correlation coeficient $\rho = 0.75$ and common variances $\sigma_x = \sigma_y = \sigma$.

10.3.3   The bivariate case

The following discussion specializes the multivariate Gaussian distribution to the bivariate case ($N = 2$) for four reasons. First, the bivariate case has strong connections with the product-moment correlation coefficient discussed in detail in Sec. 10.4. Second, the bivariate Gaussian distribution is simple enough that an examination of the joint distribution in terms of individual components yields some useful insights, in contrast to the general case where a componentwise view quickly becomes confusing. Third, the bivariate Gaussian distribution arises frequently in applications, so it is useful to have explicit results that characterize this distribution. Finally, the fourth reason for considering the bivariate case is that samples from the distribution may be plotted easily, in contrast to more general cases where plots are either more complex ($N = 3$) or not directly visualisable ($N > 3$).

The joint density for the bivariate Gaussian distribution corresponds to Eq. (10.48) with $N = 2$; for convenience, the following discussion denotes the two components of the random vector $\mathbf{x}$ as x and y. The mean vector μ is then written as $\mu = [\bar{x}, \bar{y}]^T$ and the covariance matrix is written as:

$$\Sigma = \begin{bmatrix} \sigma_x^2 & \sigma_{xy} \\ \sigma_{xy} & \sigma_y^2 \end{bmatrix} = \begin{bmatrix} \sigma_x^2 & \rho\sigma_x\sigma_y \\ \rho\sigma_x\sigma_y & \sigma_y^2 \end{bmatrix}. \tag{10.64}$$

In this case, $\sigma_{xy} = E\{(x-\bar{x})(y-\bar{y})\}$ and ρ is the correlation coefficient between x and y defined in Sec. 10.2.1 (see also Exercise 5); the matrix Σ is nonsingular so long as $|\rho| < 1$. To illustrate what bivariate Gaussian data samples look like, four examples are plotted in Fig. 10.5, each of size $N = 100$. In all four cases, $\bar{x} = \bar{y} = 0$ and $\sigma_x = \sigma_y = 1$, and the correlation coefficient ρ is as specified above each plot in the figure. Specifically, the upper left plot shows the results for $\rho = 0$, corresponding to a bivariate Gaussian distribution with two statistically independent components. The contours of constant Mahalanobis distance for this case are circles, reflecting the independence of the components. The results shown in the upper right plot in Fig. 10.5 show samples drawn from the bivariate Gaussian distribution with the positive correlation $\rho = 0.5$. It is clear from this plot that the effect of this positive correlation is for the observed data points to cluster somewhat along the $45°$ line $y = x$. The lower right plot shows the corresponding results for the larger positive correlation $\rho = 0.9$, where this effect is much more pronounced. Similarly, the effect of the negative correlation $\rho = -0.5$ is shown in the lower left plot in Fig. 10.5, which causes the data points to cluster around the line $y = -x$.

The joint density for the bivariate Gaussian distribution is given more explicitly as:

$$p(x, y) = \frac{1}{2\pi\sigma_x\sigma_y(1-\rho^2)^{1/2}} \exp\left\{-\frac{1}{2(1-\rho^2)} \cdot \left[\frac{(x-\bar{x})^2}{\sigma_x^2} - \frac{2\rho(x-\bar{x})(y-\bar{y})}{\sigma_x\sigma_y} + \frac{(y-\bar{y})^2}{\sigma_y^2}\right]\right\}. \tag{10.65}$$

One useful application of this result is the following demonstration of the difference between statistical independence and exchangeability. For simplicity, consider the case where both components of the mean μ are equal to $\bar{x}$ and where $\sigma_x = \sigma_y = \sigma$. For a bivariate distribution, there are two possible permutations: the identity, which results in no change, or the exchange of the two components, $x_1 \leftrightarrow x_2$. The effects of this second permutation on the distribution of $\mathbf{x}$ follow from the linear transformation results presented in Sec. 10.3.1:

$$\mathbf{P} = \begin{bmatrix} 0 & 1 \\ 1 & 0 \end{bmatrix} \Rightarrow \mathbf{Px} \sim N(\mathbf{P}\mu, \mathbf{P}\Sigma\mathbf{P}^T), \tag{10.66}$$

Figure 10.5: Four bivariate normal data samples

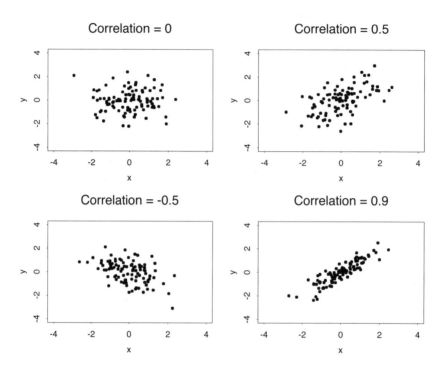

where the transformed mean vector and covariance matrix are easily computed:

$$
\mathbf{P}\mu = \begin{bmatrix} 0 & 1 \\ 1 & 0 \end{bmatrix} \begin{bmatrix} \bar{x} \\ \bar{x} \end{bmatrix} = \begin{bmatrix} \bar{x} \\ \bar{x} \end{bmatrix} = \mu,
$$

$$
\mathbf{P}\Sigma\mathbf{P}^T = \begin{bmatrix} 0 & 1 \\ 1 & 0 \end{bmatrix} \begin{bmatrix} \sigma^2 & \rho\sigma^2 \\ \rho\sigma^2 & \sigma^2 \end{bmatrix} \begin{bmatrix} 0 & 1 \\ 1 & 0 \end{bmatrix}
$$

$$
= \begin{bmatrix} 0 & 1 \\ 1 & 0 \end{bmatrix} \begin{bmatrix} \rho\sigma^2 & \sigma^2 \\ \sigma^2 & \rho\sigma^2 \end{bmatrix} = \begin{bmatrix} \sigma^2 & \rho\sigma^2 \\ \rho\sigma^2 & \sigma^2 \end{bmatrix} = \Sigma.
$$

In other words, the distribution of $\mathbf{x}$ is invariant under this permutation, implying that the components of $\mathbf{x}$ are exchangeable, as discussed in Sec. 10.2.2. Further, as discussed in Sec. 10.2.3, this example illustrates also that exchangeability is a strictly weaker condition than the i.i.d. assumption, which holds here only for $\rho = 0$. An extension of this example is considered in Exercise 6.

Although it is a little messy algebraically, the following derivation leads to convenient expressions for both the marginal and conditional densities for the bivariate Gaussian distribution. The starting point for this derivation is the following rearrangement of the exponent in this joint density [222, p. 182]:

$$\frac{(x-\bar{x})^2}{\sigma_x^2} - \frac{2\rho(x-\bar{x})(y-\bar{y})}{\sigma_x\sigma_y} + \frac{(y-\bar{y})^2}{\sigma_y^2} = \left[\frac{x-\bar{x}}{\sigma_x} - \frac{\rho(y-\bar{y})}{\sigma_y}\right]^2 + \frac{(1-\rho^2)(y-\bar{y})^2}{\sigma_y^2}.$$

(10.67)

Based on this result, define the variable z and rewrite the joint density in terms of z and y:

$$z = \frac{1}{\sqrt{1-\rho^2}}\left[\frac{x-\bar{x}}{\sigma_x} - \frac{\rho(y-\bar{y})}{\sigma_y}\right]$$

$$\Rightarrow p(x,y) = \frac{1}{2\pi\sigma_x\sigma_y(1-\rho^2)^{1/2}} \exp\left\{-z^2/2 - \frac{1}{2}\left(\frac{y-\bar{y}}{\sigma_y}\right)^2\right\}$$

$$= \frac{e^{-z^2/2}}{\sqrt{2\pi}\sigma_x(1-\rho^2)^{1/2}} \cdot \frac{\exp\left\{-\frac{1}{2}\left(\frac{y-\bar{y}}{\sigma_y}\right)^2\right\}}{\sqrt{2\pi}\sigma_y}.$$

(10.68)

The marginal density $p_y(y)$ is obtained by integrating the joint density over the variable x, which has now been subsumed into the new variable z. Noting that $dx = \sigma_x(1-\rho^2)^{1/2}dz$, it follows that:

$$p_y(y) = \sigma_x(1-\rho^2)^{1/2} \int_{-\infty}^{\infty} p(x,y)dz$$

$$= \left[\frac{1}{\sqrt{2\pi}} \int_{-\infty}^{\infty} e^{-z^2/2}dz\right] \cdot \frac{\exp\left\{-\frac{1}{2}\left(\frac{y-\bar{y}}{\sigma_y}\right)^2\right\}}{\sqrt{2\pi}\sigma_y}$$

$$= \frac{\exp\left\{-\frac{1}{2}\left(\frac{y-\bar{y}}{\sigma_y}\right)^2\right\}}{\sqrt{2\pi}\sigma_y},$$

(10.69)

since the term in brackets is simply the normalization integral for the standard $N(0,1)$ random variable and is therefore equal to 1. In words, then, the marginal density $p_y(y)$ just corresponds to the univariate distribution $N(\bar{y}, \sigma_y^2)$.

This result establishes that the marginal density $p_y(y)$ for the second component of the bivariate Gaussian random vector (x,y) is also Gaussian, with mean equal to $\bar{y}$, the second component of the bivariate mean vector μ, and variance equal to σ_y^2, the 2, 2-element of the bivariate covariance matrix Σ. By exactly the same reasoning, the marginal density $p_x(x)$ for the first component is also univariate normal, with mean $\bar{x}$ equal to the first component of the mean vector μ and variance σ_x^2 equal to the 1, 1-element of the covariance matrix Σ. More generally, the marginal densities of a multivariate Gaussian random vector are all univariate normal, with mean equal to the corresponding component of the multivariate mean vector μ and variance equal to the corresponding diagonal element of the covariance matrix Σ. *Conversely, it is important to note that Gaussian marginals are not sufficient to imply that the joint distribution is multivariate Gaussian*, an important point illustrated in Sec. 10.7.

It follows from the relationship between conditional and marginal densities presented in Sec. 10.2.4 that the conditional density $p(x|y = r)$ is given by:

$$p(x|y = r) = \frac{p(x, r)}{p_y(r)} = \frac{e^{-z^2/2}}{\sqrt{2\pi}\sigma_x(1 - \rho^2)^{1/2}}, \tag{10.70}$$

where the second line follows from Eq. (10.68), with the variable z given by:

$$z = \frac{1}{\sqrt{1 - \rho^2}}\left[\frac{x - \bar{x}}{\sigma_x} - \frac{\rho(r - \bar{y})}{\sigma_y}\right]$$

$$= \frac{1}{\sigma_x\sqrt{1 - \rho^2}}\left[x - \left(\bar{x} + \frac{\rho\sigma_x(r - \bar{y})}{\sigma_y}\right)\right]. \tag{10.71}$$

Combining this result with Eq. (10.70) then leads to the conclusion that the *conditional density for x, given $y = r$, is $N(\tilde{x}, \tilde{\sigma}_x)$*, where the mean and standard deviation defining this density are:

$$\tilde{x} = \bar{x} + \frac{\rho\sigma_x(r - \bar{y})}{\sigma_y},$$

$$\tilde{\sigma}_x = \sigma_x\sqrt{1 - \rho^2}. \tag{10.72}$$

Analogously, substituting $\bar{y}$ and σ_y into this result gives the conditional density for y, given $x = r$. Note that these conditional densities reduce to the corresponding marginal densities when $\rho = 0$. Also, note that if the observed value r of y is the mean value $\bar{y}$, then the conditional mean of x is simply the unconditional mean $\bar{x}$, but the conditional variance $\tilde{\sigma}_x^2$ is strictly smaller than the unconditional variance σ_x^2 unless $\rho = 0$. This result reflects the fact that, if x and y are correlated Gaussian random variables, observations of y can be used to improve our predictions of x and vice versa.

Because the conditional distribution of x given $y = r$ is Gaussian, the conditional moments of the distribution are easily obtained. In particular, note that the conditional mean is simply $\tilde{x}$ and the conditional variance is $\tilde{\sigma}_x^2$. Because these results are useful, it is worth noting them explicitly here for reference:

$$E\{x|y = r\} = \bar{x} + \frac{\rho\sigma_x(r - \bar{y})}{\sigma_y} = \left(\frac{\rho\sigma_x}{\sigma_y}\right)r + \left(\bar{x} - \frac{\rho\sigma_x\bar{y}}{\sigma_y}\right),$$

$$\text{var}\{x|y = r\} = \sigma_x^2(1 - \rho^2). \tag{10.73}$$

The conditional mean result has been rewritten here to emphasize that it is linear in the conditioning value r, a characteristic of bivariate Gaussian distributions that can sometimes be used to our advantage. Similarly, one of the key points concerning the conditional variance is that it is independent of the conditioning variable r. It is important to note that these results do *not* extend to non-Gaussian distributions, in general.

10.3.4   Quadrant probabilities

The following result turns out to be useful in a number of problems involving bivariate Gaussian distributions, including the *arcsine law* discussed in Chapter 16, describing the behavior of binary sequences obtained from correlated Gaussian sequences. Here, we assume that $\mathbf{x}$ is a zero-mean bivariate Gaussian random vector with components x_1 and x_2, variances σ_1^2 and σ_2^2, and correlation ρ. The *quadrant probabilities* are simply the probabilities of observing the random vector (x_1, x_2) in each of the four quadrants of the plane:

$$
\begin{aligned}
P_{++} &= \mathcal{P}\left\{x_1 \geq 0, x_2 \geq 0\right\}, \\
P_{+-} &= \mathcal{P}\left\{x_1 \geq 0, x_2 < 0\right\}, \\
P_{-+} &= \mathcal{P}\left\{x_1 < 0, x_2 \geq 0\right\}, \\
P_{--} &= \mathcal{P}\left\{x_1 < 0, x_2 < 0\right\}.
\end{aligned}
\tag{10.74}
$$

It follows from the symmetry of the distribution about zero that $\mathbf{x}$ and $-\mathbf{x}$ have the same distribution, implying $P_{--} = P_{++}$, and $P_{-+} = P_{+-}$. Further, since the four probabilities in Eq. (10.74) are associated with a complete set of mutually exclusive events, one and only one of these events must occur so their associated probabilities sum to 1:

$$
P_{++} + P_{+-} + P_{-+} + P_{--} = 2P_{++} + 2P_{+-} = 1 \Rightarrow P_{+-} = \frac{1}{2} - P_{++}. \tag{10.75}
$$

Therefore, once we have determined the probability P_{++} that the random vector $\mathbf{x}$ lies in the first quadrant of the plane, the other three quadrant probabilities may be determined from P_{++}. To simply the problem further, consider the vector $\mathbf{z} = \mathbf{A}\mathbf{x}$, where:

$$
\mathbf{A} = \begin{bmatrix} \sigma_1^{-1} & 0 \\ 0 & \sigma_2^{-1} \end{bmatrix} \Rightarrow \mathbf{A}^T \Sigma \mathbf{A} = \begin{bmatrix} 1 & \rho \\ \rho & 1 \end{bmatrix}. \tag{10.76}
$$

Since $z_1 = x_1/\sigma_1$ and $z_2 = x_2/\sigma_2$, the condition $x_1 \geq 0$, $x_2 \geq 0$ is equivalent to $z_1 \geq 0$, $z_2 \geq 0$ and P_{++} is equal to the probability that this transformed condition holds. This reformulation has the advantage of simplicity since it eliminates the parameters σ_1 and σ_2 from the problem:

$$
\begin{aligned}
P_{++} &= \mathcal{P}\left\{z_1 \geq 0, z_2 \geq 0\right\} \tag{10.77} \\
&= \int_0^\infty \int_0^\infty \frac{1}{2\pi(1-\rho^2)^{1/2}} \exp\left[-\frac{z_1^2 - 2\rho z_1 z_2 + z_2^2}{2(1-\rho^2)}\right] dz_1 dz_2.
\end{aligned}
$$

This integral may be evaluated by converting it to polar coordinates, defining:

$$
z_1 = r\sin\theta, z_2 = r\cos\theta \Rightarrow dz_1 dz_2 = r\,dr\,d\theta. \tag{10.78}
$$

In this transformed representation, the condition $z_1 \geq 0$, $z_2 \geq 0$ corresponds to $0 \leq r < \infty$ and $0 \leq \theta \leq \pi/2$ and the first quadrant probability becomes:

$$
P_{++} = \frac{1}{2\pi(1-\rho^2)^{1/2}} \int_0^{\pi/2} d\theta \int_0^\infty r\,dr \exp\left[-\frac{r^2[1 - \rho\sin 2\theta]}{2(1-\rho^2)}\right], \tag{10.79}
$$

where use has been made of the trigonometric identities $\sin^2 x + \cos^2 x = 1$ and $\sin 2x = 2 \sin x \cos x$. Defining $v = r^2$, this integral may be simplified to:

$$
\begin{aligned}
P_{++} &= \frac{1}{4\pi(1-\rho^2)^{1/2}} \int_0^{\pi/2} d\theta \int_0^\infty \exp\left[-\left(\frac{1-\rho\sin 2\theta}{2(1-\rho^2)}\right)v\right] dv \\
&= \frac{1}{4\pi(1-\rho^2)^{1/2}} \int_0^{\pi/2} d\theta \left[\frac{2(1-\rho^2)}{1-\rho\sin 2\theta}\right] \\
&= \frac{(1-\rho^2)^{1/2}}{2\pi} \int_0^{\pi/2} \frac{d\theta}{1-\rho\sin 2\theta}.
\end{aligned}
\tag{10.80}
$$

This last integral may be evaluated using a result from Abramowitz and Stegun [1, p. 78, no. 4.3.131] to obtain:

$$
\int_0^{\pi/2} \frac{d\theta}{1-\rho\sin 2\theta} = \frac{2}{(1-\rho^2)^{1/2}}\left[\frac{\pi}{2} + \arctan\frac{\rho}{(1-\rho^2)^{1/2}}\right].
\tag{10.81}
$$

This result may be simplified with the aid of the following inverse trigonometric identity [1, p. 73, table 4.3.45]:

$$
\arctan\frac{\rho}{(1-\rho^2)^{1/2}} = \arcsin\rho.
\tag{10.82}
$$

Combining Eqs. (10.80), (10.81), and (10.82) and using the relationships among the four quadrant probabilities derived earlier ultimately leads to the following results:

$$
\begin{aligned}
P_{++} &= \frac{1}{4} + \frac{1}{2\pi}\arcsin\rho, \\
P_{+-} &= \frac{1}{4} - \frac{1}{2\pi}\arcsin\rho, \\
P_{-+} &= \frac{1}{4} - \frac{1}{2\pi}\arcsin\rho, \\
P_{--} &= \frac{1}{4} + \frac{1}{2\pi}\arcsin\rho.
\end{aligned}
\tag{10.83}
$$

Note that when $\rho = 0$, all four of these probabilities are equal to $1/4$, reflecting the fact that in this case, the distribution is circularly symmetric, so all four quadrants are equally likely. For $\rho > 0$, x and y are more likely to both have the same sign than to have opposite signs, so P_{++} and P_{--} are greater than $1/4$ while P_{+-} and P_{-+} are less than $1/4$. As $\rho \to +1$, P_{++} and P_{--} approach limiting values of $1/2$ and P_{+-} and P_{-+} approach limiting values of zero. For $\rho < 0$, exactly the opposite behavior is observed: the random vector $\mathbf{x}$ is more likely to lie in the second and fourth quadrants, so P_{+-} and P_{-+} are greater than P_{++} and P_{--}, approaching the limits $P_{+-} = P_{-+} = 1/2$ as $\rho \to -1$.

10.4   The product-moment correlation coefficient

The product-moment correlation coefficient $\hat{\rho}_{xy}$ between the data sequences $\{x_k\}$ and $\{y_k\}$ was introduced by Karl Pearson in 1896 [163, p. 557] and continues to be used extensively in practice today. In fact, without further qualification, the term "correlation coefficient" is almost always taken to mean the product-moment correlation coefficient, although other association measures have been defined, such as the Spearman rank correlation coefficient discussed in Sec. 10.5 and Kendall's τ, discussed in Sec. 10.7.5. Because the product-moment correlation coefficient is so important in practice, the following subsections examine it in some detail.

10.4.1 Definition and estimation

The product-moment correlation coefficient is an estimator of the population correlation coefficient ρ_{xy} defined in Sec. 10.2.1, repeated here for convenience:

$$\rho_{xy} = \frac{E\{(x - E\{x\})(y - E\{y\})\}}{\sigma_x \sigma_y}, \tag{10.84}$$

where σ_x and σ_y are the standard deviations of x and y respectively. As noted in Sec. 10.3.3, in the case of a bivariate Gaussian distribution, σ_{xy} corresponds to the correlation parameter ρ defining the distribution, but Eq. (10.84) defines ρ_{xy} for any joint distribution of x and y. It follows from the results presented in Sec. 10.2.3 that $\rho_{xy} = 0$ if $\{x_k\}$ and $\{y_k\}$ are statistically independent, regardless of their distribution. If the joint distribution of x and y is Gaussian, the converse implication also holds: $\rho_{xy} = 0$ implies x and y are statistically independent. The same conclusion also follows if x and y are *binary random variables* [163, p. 557]. *It is important to emphasize, however, that $\rho_{xy} = 0$ does not imply statistical independence in general: unless x and y have either Gaussian or binary joint distributions, they can exhibit strong statistical dependence even when $\rho_{xy} = 0$.* A simple illustrative example is given in Sec. 10.4.6.

The product-moment correlation coeficient estimates ρ_{xy} from data samples $\{(x_k, y_k)\}$ as:

$$\hat{\rho}_{xy} = \frac{\sum_{k=1}^{N}(x_k - \bar{x})(y_k - \bar{y})}{\left[\sum_{k=1}^{N}(x_k - \bar{x})^2 \sum_{k=1}^{N}(y_k - \bar{y})^2\right]^{1/2}}, \tag{10.85}$$

where $\bar{x}$ and $\bar{y}$ are the arithmetic averages of the sequences $\{x_k\}$ and $\{y_k\}$:

$$\bar{x} = \frac{1}{N}\sum_{k=1}^{N} x_k, \quad \bar{y} = \frac{1}{N}\sum_{k=1}^{N} y_k. \tag{10.86}$$

As noted by Johnson *et al.* [163, ch. 32], the product-moment correlation coefficient may be represented in many other equivalent forms, one of which is:

$$\hat{\rho}_{xy} = \cos\theta, \tag{10.87}$$

where θ is the angle between the N-dimensional vectors with components $x_k - \bar{x}$ and $y_k - \bar{y}$. As a consequence, the condition $\hat{\rho}_{xy} = 0$ corresponds to the orthogonality of these data vectors in the plane. It also follows from this result (or, equivalently, from the Cauchy-Schwartz inequality) that $|\hat{\rho}_{xy}| \leq 1$ for any two data sequences of length N. Further, if $\hat{\rho}_{xy} = +1$, it follows that $y_k = \alpha x_k + \beta$ for some $\alpha > 0$; similarly, if $\hat{\rho}_{xy} = -1$, it follows that $y_k = \alpha x_k + \beta$ for some $\alpha < 0$. As a consequence of these observations, $\hat{\rho}_{xy}$ may be viewed as a *measure of linear association* between the variables x and y. A closely related result is that the product-moment correlation is invariant under linear transformations of both x and y so long as algebraic signs are preserved; in particular, $\hat{\rho}_{xy}$ is independent of both origin shifts and unit changes in these data sequences.

The key point of this discussion is that the product-moment correlation $\hat{\rho}_{xy}$ represents an easily computed measure of linear association between two data sequences $\{x_k\}$ and $\{y_k\}$. Further, in the favorable case that these data sequences are jointly Gaussian, $\hat{\rho}_{xy}$ also provides a reliable measure of their statistical association. For these reasons, $\hat{\rho}_{xy}$ remains the most popular measure of association considered today. In particular, if $\hat{\rho}_{xy}$ is sufficiently small in magnitude, this result is often taken as evidence for the independence of the sequences $\{x_k\}$ and $\{y_k\}$. Conversely, as subsequent discussions illustrate, $\hat{\rho}_{xy}$ is quite sensitive to anomalies in the data sequences, and it becomes more difficult to interpret when the joint distribution of x and y is strongly non-Gaussian. In particular, it is possible for x and y to exhibit a strong *nonlinear* dependence even when $\hat{\rho}_{xy} = 0$, a point illustrated in Sec. 10.4.6.

10.4.2   Exact distribution for the Gaussian case

As in the case of the mean, variance, and standard deviation, it is possible to derive an expression for the exact distribution of $\hat{\rho}_{xy}$ when x and y are jointly Gaussian. Unfortunately, the result is rather complex, being representable either in terms of the hypergeometric function $_2F_1(a, b; c; x)$ [1] or the corresponding infinite series. Consequently, various simpler approximations and transformations have been developed. The following discussions present both the exact results and some of these simplifications, along with specialization to the case $\rho = 0$ that arises in statistical independence testing.

If x and y are jointly Gaussian random variables with the population correlation coefficient ρ, the exact distribution for the estimator $\hat{\rho}_{xy}$ computed from N data points is [163, p. 549]:

$$p(r) = \frac{(N-2)(1-\rho^2)^{(N-1)/2}(1-r^2)^{(N-4)/2}}{\sqrt{2}(N-1)B(1/2, N-1/2)(1-\rho r)^{(2N-3)/2}} \, _2F_1\left(\frac{1}{2}, \frac{1}{2}; N - \frac{1}{2}; \frac{1+\rho r}{2}\right).$$
(10.88)

Here, $B(a, b)$ is the beta function and $_2F_1(a, b; c; x)$ is the hypergeometric function. Alternatively, this distribution may be represented as the infinite series [163, p. 548]:

$$p(r) = \frac{(1-\rho^2)^{(N-1)/2}(1-r^2)^{(N-4)/2}}{\sqrt{\pi}\Gamma([N-1]/2)\Gamma([N-2]/2)} \sum_{j=0}^{\infty} \frac{[\Gamma([N-1+j]/2)]^2}{j!}(2\rho r)^j, \quad (10.89)$$

Figure 10.6: Estimated correlation coefficients, Gaussian data

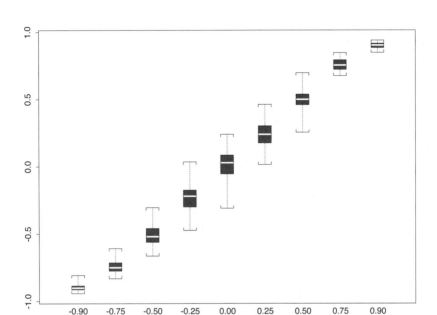

where $\Gamma(x)$ is the gamma (factorial) function. Because the hypergeometric function has been extensively characterized, it is possible to obtain exact expressions for the first four moments of this distribution [163, p. 553]. Like the distribution itself, these results are rather complicated, so they are not given here; instead, the following approximate results appear to be more useful [163, p. 554]:

$$E\{\hat{\rho}_{xy}\} \simeq \rho - \frac{\rho(1 - \rho^2)}{2(N + 6)},$$

$$\mathrm{var}\{\hat{\rho}_{xy}\} \simeq \frac{(1 - \rho^2)^2}{N + 6},$$

$$\gamma(\hat{\rho}_{xy}) \simeq \frac{-6|\rho|}{\sqrt{N + 6}}\mathrm{sign}\,\rho,$$

$$\kappa(\hat{\rho}_{xy}) \simeq \frac{-6(1 - 12\rho^2)}{N + 6}. \tag{10.90}$$

Overall, these results reflect the fact that $\hat{\rho}_{xy}$ is an asymptotically unbiased, consistent estimator of ρ for jointly Gaussian random variables x and y.

The general character of the distribution of $\hat{\rho}_{xy}$ may be seen from Fig. 10.6, which presents boxplot summaries of 100 simulation-based estimates. In each case, a bivariate normal sample of size $N = 100$ was generated, with one of the nine values indicated on the horizontal axis for ρ_{xy}. The moment-based estimator $\hat{\rho}_{xy}$ was then computed from this sample from Eqs. (10.85) and (10.86). It is clear from these results that the variability in these estimates depends strongly on the true correlation ρ_{xy}, being largest for $\rho_{xy} = 0$ and decreasing to zero as $\rho_{xy} \rightarrow \pm 1$. This result is consistent with the approximate moment characterization given in Eqs. (10.90), as is the small bias in these estimates (i.e., the median values of $\hat{\rho}_{xy}$ are in excellent agreement with the population values in all cases). In particular, note that the worst-case bias should be less than 0.5% in all cases.

10.4.3   Fisher's transformation to normality

It is clear from the results presented in the previous section that, although the exact distribution for $\hat{\rho}_{xy}$ is known for jointly Gaussian sequences $\{x_k\}$ and $\{y_k\}$, the resulting distribution is not simple. This fact has motivated various authors to seek transformations of $\hat{\rho}_{xy}$ that yield a simpler distribution, with particular attention to transformations that yield an approximately Gaussian result. One of the most popular of these transformations is that proposed by Fisher in 1915 [163, p. 571]:

$$Z = \tanh^{-1} \hat{\rho}_{xy} = \frac{1}{2} \ln \left(\frac{1 + \hat{\rho}_{xy}}{1 - \hat{\rho}_{xy}} \right). \qquad (10.91)$$

More refined approximations are available, but Z is typically regarded as approximately Gaussian with mean μ and variance σ^2 given by:

$$\mu \simeq \tanh^{-1} \rho_{xy}, \qquad \sigma^2 \simeq \frac{1}{N-3}. \qquad (10.92)$$

One of the primary advantages of this result is that it provides a basis for constructing approximate confidence intervals for $\hat{\rho}_{xy}$ *under the assumption that* $\{x_k\}$ *and* $\{y_k\}$ *are bivariate normal.* In particular, under these working assumptions, the transformed population correlation coefficient ρ_{xy} lies in the interval:

$$\tanh^{-1} \hat{\rho}_{xy} - \frac{1.96}{\sqrt{N-3}} \leq \tanh^{-1} \rho_{xy} \leq \tanh^{-1} \hat{\rho}_{xy} + \frac{1.96}{\sqrt{N-3}}, \qquad (10.93)$$

with 95% probability. A particular application of this result is in testing the hypothesis that $\rho_{xy} = 0$, discussed in the next section.

To illustrate the influence of Fisher's transformation, Fig. 10.7 shows the boxplots obtained when this transformation is applied to the results summarized in Fig. 10.6. The effect of the transformation on the expected value of Z is seen clearly in both the curvature defined by the median values in the box plots and by the range of values, which is expanded beyond the interval $[-1, 1]$ in which $\hat{\rho}_{xy}$ must lie. Conversely, the variance stabilizing effect of the transformation is also clear: the range of variation seen in all of these boxplots is approximately the same, in marked contrast to the strong dependence on ρ_{xy} seen in Fig. 10.6.

Figure 10.7: Fisher's transformed correlations, Gaussian data

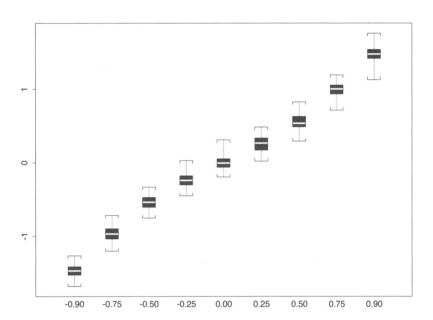

10.4.4   Testing for independence

Under the popular hypothesis that two variables, x and y, are jointly Gaussian, it follows that they are statistically independent if and only if their product-moment correlation coefficient is zero, so there is considerable interest in testing this hypothesis against the alternative hypothesis that $\rho_{xy} \neq 0$. Given the Gaussian joint distributional assumption, it is possible to use the distributional results described in the previous section to construct a formal test of this hypothesis. First, note that the exact distribution simplifies enormously in this case. Setting $\rho = 0$ in Eq. (10.88) reduces the infinite series to the single term obtained for $j = 0$, yielding the following result:

$$p(r) = \frac{\Gamma\left(\frac{N-1}{2}\right)}{\Gamma\left(\frac{1}{2}\right)\Gamma\left(\frac{N-2}{2}\right)}\,(1 - r^2)^{(N-4)/2}. \tag{10.94}$$

Even here, however, the most popular approach in practice is to use the large-sample approximation results, which imply that if x and y are statistically independent then $\hat{\rho}_{xy}$ has the approximate distribution $N(0, 1/N)$. Under this approximation, it follows that there is a 95% probability that the observed estimate $\hat{\rho}_{xy}$ of the correlation coefficient ρ will lie within ± 1.96 standard deviations of the mean. More explicitly, these assumptions imply that $|\hat{\rho}_{xy}|$ should

Figure 10.8: Four bivariate datasets, three with outliers

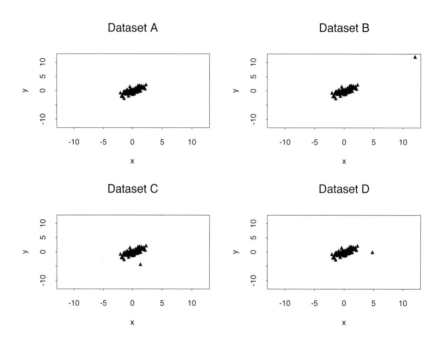

be smaller than $1.96/\sqrt{N}$. Larger values than this may be taken as evidence against the hypothesis of statistical independence. In summary, the most popular independence test has the form:

- if $|\hat{\rho}_{xy}| \geq 1.96/\sqrt{N}$, reject the hypothesis of independence since a value this large has a probability of less than 5% under the working assumptions;

- if $|\hat{\rho}_{xy}| < 1.96/\sqrt{N}$, provisionally accept the hypothesis of independence, since the estimated correlation coeficient provides no evidence against independence at this level of confidence.

It is important to emphasize again that $\hat{\rho}_{xy} = 0$ is only a necessary condition for statistical independence, and not a sufficient one: as examples discussed in secs. 10.4.6 and 10.7 illustrate, strongly dependent non-Gaussian variables can exhibit zero correlations.

10.4.5   The influence of outliers

Fig. 10.8 shows four plots of nominally Gaussian bivariate data sequences, with mean vector $\bar{\mathbf{x}} = [0, 0]^T$, common variances $\sigma_1^2 = \sigma_2^2 = 1$, and correlation coefficient $\rho = 0.8$. In all cases, these datasets are of size $N = 100$, and the uncontaminated dataset is shown in the upper left plot, designated as "Dataset

Table 10.2: Product-moment correlations

Dataset	λ	$\hat{\rho}_{xy}$, $d = 8$	$\hat{\rho}_{xy}$, $d = 128$
A	—	0.738	0.738
B	+1	0.898	0.990
C	−1	0.600	0.311
D	0	0.657	0.327

A." The other three datasets, designated "B," "C," and "D," are each contaminated by replacing one of these points with a single outlier. As discussed
in Sec. 10.3.2, the *Mahalanobis distance* is a scalar distance measure that is
analogous to the normalized distance $|x_i - \bar{x}|/\sigma$ for univariate Gaussian ran-
dom variables. In the three contaminated samples considered here, the outlier
has been placed at a Mahalanobis distance of +8, analogous to the univariate
outliers 8 standard deviations from the mean considered in Chapter 7. Here,
however, because the uncontaminated data cloud has both a *size* and an *orientation*, it is also necessary to specify the orientation of these outliers. In all
cases, these outliers are of the form $(x, \lambda x)$, with $\lambda = 1$ for Dataset B, $\lambda = -1$
for Dataset C, and $\lambda = 0$ for Dataset D, and x chosen so the point has the
desired Mahalanobis distance from the center of the distribution.

Product-moment correlation estimates are given in Table 10.2 for all four
of these datasets. For the uncontaminated dataset (Dataset A), the estimated
correlation is about 8% smaller than the true value of 0.8, within the normal
range of estimation error. For Dataset B, the contamination is placed so that
it increases the effective correlation, in this case by about 22%. In contrast,
note that for Dataset C, contamination of the same magnitude but different
orientation results in a *decrease* in the correlation estimate by about 19%. Similarly, Dataset D also exhibits a decrease in the estimated correlation, here by
about 11%. The key points of this example are, first, to illustrate the significant
outlier sensitivity of the product-moment correlation estimate, and second, to
illustrate the strong *orientation dependence* of these results. In fact, note that
taking $\lambda = 1$ results in the outlier lying along the "perfect correlation line"
$y = x$, which tends to increase the estimated correlation. As we decrease λ from
1, the correlation estimate decreases until $\lambda \simeq 0.38$, at which point the outlier
has essentially no effect. Decreasing λ further to negative values results in outliers like the one in Dataset C that are perpindicular to the perfect correlation
line and are therefore in significant conflict with the positive correlation seen
in the bulk of the data. Not surprisingly, making these outliers more severe in-

Figure 10.9: Influence of one outlier pair on $\hat{\rho}_{xy}$ vs. ρ

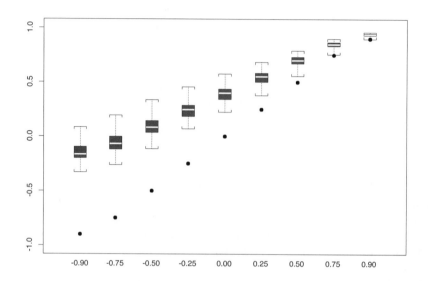

creases their effect correspondingly, as may be seen in the last column of Table 10.2, where the Mahalanobis distance has been increased from 8 to 128.

Fig. 10.9 presents a different view of the outlier sensitivity of $\hat{\rho}_{xy}$, summarizing the results of a small simulation case study. Here, 100 statistically independent bivariate Gaussian data samples were generated for each of the nine correlation values indicated on the horizontal axis of Fig. 10.9, ranging from -0.90 to $+0.90$. Each sample was of size $N = 100$, with each component having zero mean and unit standard deviation, and one data point in each sample was contaminated by adding 8 to both its x and y values, corresponding to a *common mode* contamination of both components of the bivariate data sequence. The boxplots shown in Fig. 10.9 illustrate the range of variation of the estimated correlations $\hat{\rho}_{xy}$, and the solid circles indicate the true correlation value ρ for each case. As in Dataset B in the previous example, note that the effect of this particular form of contamination is to increase the estimated correlation, corresponding to the fact that the outlying point lies near the perfect correlation line $y = x$. Clearly, the effect of this outlying point is much greater for negative correlations than for positive correlations.

Severe as this influence is, it becomes even worse when the contamination level is increased. This point is illustrated in Fig. 10.10, which shows the corresponding results when the contamination consists of a cluster of *four* additive outliers, again having the values $(+8, +8)$. Specifically, 8 was added to both x_k

Figure 10.10: Influence of four outlier pairs on $\hat{\rho}_{xy}$ vs. ρ

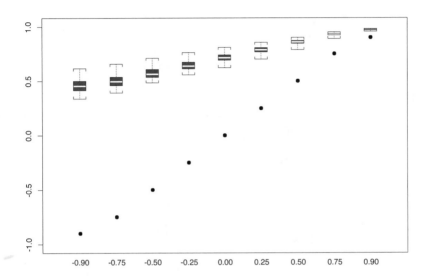

and y_k for $k = 97$, 98, 99, and 100 to obtain the contaminated dataset. In this case, the estimated correlations $\hat{\rho}_{xy}$ are never smaller than approximately $+0.3$ even when the true correlations ρ are as negative as -0.90. Even worse, a careful comparison with Fig. 10.9 reveals that the variability of these badly biased estimates is *reduced* relative to the previous example; intuitively, this observation means the estimated computed from the dataset with 4% contamination yields a value that is both worse *and more insistent* than the dataset with only 1% contamination (i.e., less variable: "yes, I'm certain, the correlation is 0.45 ± 0.15 even though the true correlation is $\rho = -0.9$").

10.4.6 The influence of transformations

As noted earlier in this section, one of the key features of the product-moment correlation coefficient is that it is a measure of *linear* association. As the following three examples illustrate, the value of this association measure can be strongly influenced by the presence of a *nonlinear* relationship between variables.

For the first example, suppose $\{x_k\}$ and $\{v_k\}$ are statistically independent univariate Gaussian data sequences, each of length N, with means zero and common variance σ^2. Define the variable y_k as the following linear combination of these two components:

$$y_k = ax_k + bv_k, \qquad (10.95)$$

where a and b are arbitrary real numbers, not both zero. It follows from the expectation results presented in Chapter 4 that $\{y_k\}$ is a zero-mean sequence with variance $\text{var}\{y_k\} = (a^2 + b^2)\sigma^2$. The correlation between $\{x_k\}$ and $\{y_k\}$ is given by:

$$\rho_{xy} = \frac{E\{x_k y_k\}}{[\text{var}\{x_k\}\text{var}\{y_k\}]^{1/2}} = \frac{aE\{x_k^2\} + bE\{x_k v_k\}}{\sigma^2\sqrt{a^2 + b^2}} = \frac{a}{\sqrt{a^2 + b^2}}. \quad (10.96)$$

Next, suppose we apply the monotone transformation $f(x) = x^3$ to the $\{x_k\}$ sequence and examine its impact on the correlation between $z_k = f(x_k)$ and y_k. Since the distribution of $\{x_k\}$ is Gaussian—and thus symmetric—it follows that $E\{z_k\} = E\{x_k^3\} = 0$, and it follows from the Gaussian moment results presented in Chapter 4 that:

$$\text{var}\{z_k\} = E\{x_k^6\} = 15\sigma^6. \quad (10.97)$$

Hence, the correlation between $\{z_k\}$ and $\{y_k\}$ is given by:

$$\rho_{zy} = \frac{E\{z_k y_k\}}{[\text{var}\{z_k\}\text{var}\{y_k\}]^{1/2}} = \frac{aE\{x_k^4\} + bE\{x_k^3 v_k\}}{[15(a^2 + b^2)\sigma^8]^{1/2}}$$

$$= \frac{3a}{\sqrt{15(a^2 + b^2)}} = \left(\frac{3}{\sqrt{15}}\right)\rho_{xy}, \quad (10.98)$$

where we have used the Gaussian fourth moment result that $E\{x_k^4\} = 3\sigma^4$ from Chapter 4, and the fact that since x and v are statistically independent, so are x^3 and v. Note that $3/\sqrt{15} \simeq 0.775$, so the effect of the cubic transformation applied to x_k is about a 22% reduction in the correlation with y_k.

It is worth emphasizing that the cubic transformation just considered is monotone and continuous, implying that it is invertible, a point discussed further in Chapter 12. The key point here is that this invertibility means that no information is lost: the original variable x_k can be obtained from z_k by the inverse transformation $x_k = z_k^{1/3}$. While this example demonstrates that such transformations can suppress linear correlation, there are other correlation measures (specifically, the Spearman rank correlation and Kendall's τ, both discussed later in this chapter) that are invariant to such information-preserving transformations.

As the next two examples illustrate, the consequences of nonmonotone transformations can be even more severe, completely obscuring any evidence of an association between variables. As a particularly simple example, consider the following quadratic transformation:

$$z_k = x_k^2 - \sigma^2 \quad \Rightarrow \quad E\{z_k\} = 0. \quad (10.99)$$

For this transformation, the expected value of the product $x_k z_k$ is given by:

$$E\{x_k z_k\} = E\{x_k^3\} - \sigma^2 E\{x_k\} = 0, \quad (10.100)$$

Figure 10.11: Two sulfur yield vs. temperature plots

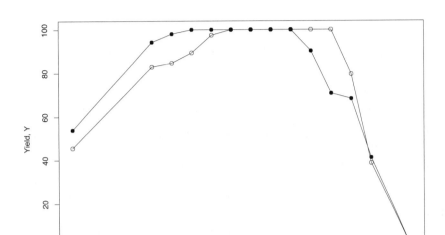

since x_k is both zero-mean and Gaussian so its third moment vanishes; hence, it follows that $\rho_{xz} = 0$ despite the fact that z_k is completely determined by x_k. *This result provides one of several illustrations of an important point presented in this chapter: for bivariate random variables like (x_k, z_k) that are not jointly Gaussian, $\rho_{xz} = 0$ does not imply statistical independence.* Specifically, it is important to note that $\rho_{xz} = 0$ is a *necessary condition* for statistical independence between x and z, but *not a sufficient condition*. The next example illustrates the way in which situations like this one arise in practice.

Fig. 10.11 shows plots of two of the five sulfur yield values obtained in selective oxidation experiments performed to evaluate different catalysts formulated from vanadium, tin, and antimony [183]. The horizontal axis in this plot corresponds to temparature, the vertical axis shows the sulfur yield obtained, in percent, and the two curves correspond to two different catalyst compositions (specifically, vanadium, tin, and antimony ratios of 1 : 0.5 : 1 and 1 : 1 : 1). In fact, the paper reports results for ratios of the form 1 : r : 1 where r assumes the values 0.50, 0.70, 0.85, 1.00, and 1.50, and Table 10.3 shows both the product-moment correlations and the Spearman rank correlations (discussed in the next section) between yield and temperature obtained from these results. In all cases, note that both correlations are consistently negative, becomming more so with increasing tin content r. Also, the numerical agreement between these two association measures improves as r increases, suggesting the possibil-

Table 10.3: Correlations for the sulfur yield data

ρ	$r = 0.5$	$r = 0.7$	$r = 0.85$	$r = 1$	$r = 1.5$
$\hat{\rho}_{xy}$	-0.266	-0.445	-0.419	-0.490	-0.527
ρ_S	-0.064	-0.270	-0.467	-0.435	-0.484

ity that some unusual phenomenon may be present at low tin concentrations (i.e., for $r = 0.5$ and 0.7), but otherwise the results *seem* consistent with the following interpretation: the sulfur yield exhibits some tendency to decrease with increasing temperature, becomming more pronounced with increasing tin content. In fact, these results are *partially* correct, since for sufficiently high temperatures the sulfur yield does decrease, but the temperature at which this occurs does *not* shift monotonically with increasing r, and the low-temperature yield data exhibits precisely the *opposite* behavior. This difference is accurately reflected in the fact that the correlations computed from the low temperature data (specifically, between 180 and 270 degrees Celsius) all lie between $+0.852$ and $+0.966$; similarly, the Spearman rank correlations for the low temperature data fall between 0.802 and 0.991. The key point here is that monotone correlation measures like ρ_{xy} and ρ_S are not adequate to detect nonmonotonic dependences like those seen in this dataset.

10.5   The Spearman rank correlation coefficient

It was noted that the product-moment correlation coefficient is the most popular association measure between two real variables and that it may be viewed as a measure of their linear association. The Spearman rank correlation coefficient was introduced by Spearman in 1904 [269] and is one of two measures of *monotone association* introduced in this chapter (the other is Kendall's τ, introduced in Sec. 10.7.5). In particular, while the product-moment correlation coefficient is invariant under positive linear transformations, the Spearman rank correlation coefficient is invariant under monotonically increasing transformations. In addition, the outlier sensitivity of the Spearman rank correlation coefficient is much less severe than that of the product-moment correlation coefficient, as subsequent examples illustrate.

10.5.1   Definition, estimation, and properties

The *Spearman rank-correlation* $\hat{\rho}_S$ replaces the data values x_k and y_k appearing in the definition of the product-moment correlation coefficient $\hat{\rho}_{xy}$ with their *ranks* $R_x(i)$ and $R_y(i)$. Specifically, define $R_x(i)$ as the index, from 1 to N, of observation x_i in the following rank-ordered list:

$$x_{(1)} \leq x_{(2)} \leq \cdots \leq x_{(N-1)} \leq x_{(N)}. \qquad (10.101)$$

That is, $R_x(i) = 1$ when x_i is the smallest element in the sequence, $R_x(i) = 2$ when x_i is the second-smallest element, and so forth. Because these ranks assume values between 1 and N inclusively, the average of the sequence $\{R_x(i)\}$ is always $\bar{R} = (N+1)/2$, independent of the original data sequence $\{x_k\}$. Similarly, it follows (Exercise 7) that the variance of the sequence $\{R_x(i)\}$ is:

$$\text{var}\,\{R_x(i)\} = \frac{N^2 - 1}{12}. \qquad (10.102)$$

Combining these results yields the estimator [129, p. 61]:

$$
\begin{aligned}
\hat{\rho}_S &= \frac{(1/N)\sum_{i=1}^{N}[R_x(i) - \bar{R}][R_y(i) - \bar{R}]}{[\text{var}\{R_x(i)\}\text{var}\{R_y(i)\}]^{1/2}} \\
&= \frac{12}{N(N^2 - 1)} \sum_{i=1}^{N} \left[R_x(i) - \frac{N+1}{2}\right]\left[R_y(i) - \frac{N+1}{2}\right]. \quad (10.103)
\end{aligned}
$$

Intuitively, Spearman's rank-correlation coefficient measures the extent to which small values of x_k correspond to small values of y_k and large values of x_k correspond to large values of y_k. Further, because $\hat{\rho}_S$ is defined as a the product-moment correlation coefficient between the sequences $\{R_x(i)\}$ and $\{R_y(i)\}$, it follows that $|\hat{\rho}_S| \leq 1$ for all data sequences. In addition, it is not difficult to show (Exercise 8) that the ranks $R_x(i)$ and $R_y(i)$ are invariant under increasing transformations of the original data sequences. Hence, although logarithmic and other increasing transformations can induce radical changes in the product-moment correlation $\hat{\rho}_{xy}$, as seen in Sec. 10.4.6, the Spearman rank correlation $\hat{\rho}_S$ is immune to their effects. Similarly, it can also be shown (Exercise 9) that $\hat{\rho}_S = +1$ if and only if $y_k = f(x_k)$ for some monotone increasing function $f(\cdot)$ and that $\hat{\rho}_S = -1$ if and only if $y_k = g(x_k)$ for some monotone decreasing function $g(\cdot)$. For these reasons, $\hat{\rho}_S$ represents a measure of *monotone association*, in contrast to the product-moment correlation coefficient $\hat{\rho}_{xy}$, which was previously noted to be a measure of linear association. Finally, since the ranks $R_x(i)$ and $R_y(i)$ are bounded between 1 and N for all data sequences, it follows that the Spearman rank correlation coefficient remains well defined even for infinite-variance distributions where product-moment correlations cease to be useful.

Analogous to the approximation presented in Sec. 10.4.4 that the product-moment correlation coefficient exhibits an asymptotic $N(0, 1/N)$ distribution under independence, it is known that the Spearman rank correlation coefficient exhibits an approximate $N(0, 1/(N-1))$ asymptotic distribution under independence [129, p. 124]. For finite samples, it has been shown that the following transformed statistic has an approximate Student's t-distribution with $N - 2$ degrees of freedom [129, p. 179]:

$$T = \frac{\rho_S \sqrt{N - 2}}{\sqrt{1 - \rho^2}}. \qquad (10.104)$$

Table 10.4: Spearman vs. product-moment correlations

Dataset	λ	ρ_S, $d = 8$	$\hat{\rho}_{xy}$, $d = 8$	ρ_S, $d = 128$	$\hat{\rho}_{xy}$, $d = 128$
A	—	0.721	0.738	0.721	0.738
B	+1	0.724	0.898	0.724	0.990
C	−1	0.678	0.600	0.664	0.311
D	0	0.707	0.657	0.707	0.327

10.5.2   The influence of outliers

The Spearman rank correlation coefficient ρ_S exhibits *much* better outlier resistance than its product-moment counterpart $\hat{\rho}_{xy}$. Table 10.4 shows results obtained from the same datasets considered in Sec. 10.4.5, for both $\hat{\rho}_{xy}$ and ρ_S. Recall that for $\hat{\rho}_{xy}$, one outlier in a dataset of size $N = 100$ at a Mahalanobis distance of 8 induced changes as large as $\pm20\%$, depending on the position of the outlier relative to the natural orientation of the point cloud (here defined by a nominal correlation of $\rho = 0.8$). In contrast, the changes in ρ_S range from $\sim -6\%$ for Dataset C to $\sim +0.4\%$ for Dataset B. Even more impressively, increasing the Mahalanobis distance to 128 had a catastrophic effect on $\hat{\rho}_{xy}$, increasing the correlation for Dataset B almost to its upper limit of $+1$, and decreasing the correlation for Dataset C to less than half of its original value. For ρ_S, there is no change to three significant figures for Datasets B and D, and for Dataset C, the estimated correlation decreases by about 2%.

It is also instructive to compare the outlier sensitivity of ρ_S with that of $\hat{\rho}_{xy}$ for the simulation case study discussed in Sec. 10.4.5. Recall that 100 bivariate data sequences, each of length $N = 100$ were generated with specified correlation values ρ and then contaminated with additive outliers having the value $(+8, +8)$. The results for a single outlier, corresponding to 1% contamination, are shown in Fig. 10.12 in the form of boxplots, corresponding to those shown in Fig. 10.9 for $\hat{\rho}_{xy}$. In marked contrast to those results, however, ρ_S shows *much* lower outlier sensitivity. In particular, the nominal range of variation for the product-moment correlations (i.e., the range between the upper and lower quartiles) *never* included the correct correlation value in Fig. 10.9, whereas the nominal range of the Spearman rank correlations includes the correct value whenever $\rho > 0$ in this example. Further, the true correlation value only lies outside the range of computed rank correlations for the most extreme example, $\rho = -0.9$. Conversely, the correct correlation value fell outside the range of computed product-moment correlations for all ρ less than $+0.75$, and the computed correlation had a significant probability of exhibiting the wrong sign whenever $\rho < 0$ for this example.

Figure 10.12: Influence of one outlier pair on ρ_S vs. ρ

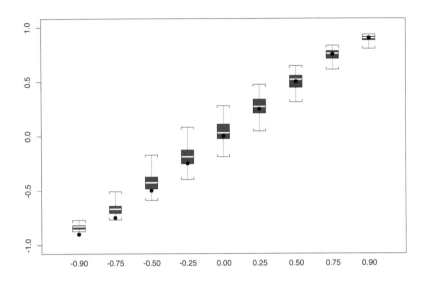

As with the case of the product-moment correlations, the influence of con-tamination becomes more severe when its level increases from 1% to 4%. This point is illustrated in Fig. 10.13, which presents the analogous results for ρ_S to those presented in Fig. 10.10 for $\hat{\rho}_{xy}$. Here, significant biases are observed for ρ smaller than about 0.50, and these biases are large enough to lie well outside the normal range of estimator variation for $\rho \lesssim -0.50$. In the worst case, when $\rho = -0.9$, the mean value of ρ_S is approximately -0.669, roughly 25% too small in magnitude. Conversely, recall from Fig. 10.10 that for this case, the true correlation value *never* fell within the range of computed product-moment correlations, and $\hat{\rho}_{xy}$ never exhibited the correct sign when $\rho < 0$.

10.5.3 An application of rank correlations

The following example illustrates the use of both the Spearman rank correlation ρ_S and the product-moment correlation $\hat{\rho}_{xy}$ in exploratory data analysis. *Since these association measures should be approximately equal under ideal circum-stances, one of the key points here is that detection of significant differences can result in some interesting insights about unexpected structure in a dataset.*

The basis for this example is a small experimental dataset described by Feng and Aldrich [102], who examine the behavior of the foams generated by 21 different fluids in a laboratory floatation cell. The objective of the experiment

Figure 10.13: Influence of four outlier pairs on ρ_S vs. ρ

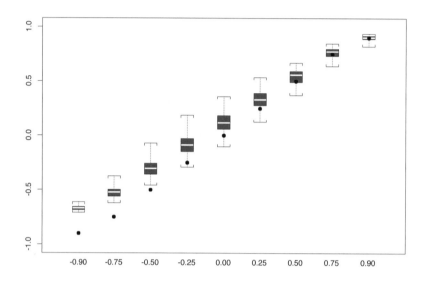

was to learn something about the relationship between the physical properties of the fluids considered and measurable characteristics of the foams generated in the flotation cell. Each of these fluids is characterized by the following four physical properties, shown in the four plots in Fig. 10.14:

1. surface tension (upper left plot),
2. viscosity (upper right plot),
3. density (lower left plot),
4. ionicity (lower right plot).

Note that the viscosity plot in the upper right shows what appears to be a single nonzero value accompanied by 20 zero values; in fact, this apparent nonzero value corresponds to glycerol, the most viscous fluid considered: its reported viscosity is 1223, compared with values of 0.88 to 7.16 for all of the other fluids considered. Similarly, most of the ionicity values shown in the lower right plot appear to be zero; in this case, the appearance is correct. Because of the somewhat anomalous behavior of both of these variables, the results presented here focus on the relationship between measured foam characteristics and the other two variables shown in Fig. 10.14: surface tension and density; for a somewhat more detailed analysis of this dataset, see the paper by Pearson [227].

The measured foam characteristics are plotted in Fig. 10.15 for these 21 different fluids. Specifically, the upper plot shows measured values of a bubble

Figure 10.14: Plots of the fluid property variables

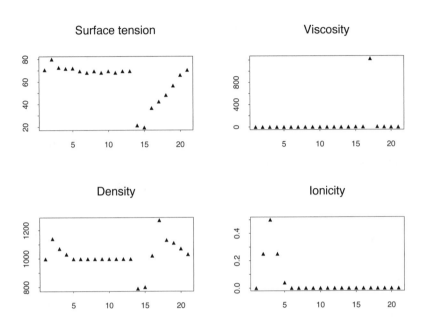

size characterization called *small number emphasis* (SNE), and the lower plot shows the values of a foam instability measure called INSTAB. Both of these characteristics were computed by analyzing sequences of digitized images of foams taken during the course of the experiment. The SNE measure is inversely proportional to the average bubble size in the foam, and the INSTAB measure essentially characterizes how rapidly the foam decays with time. Although they are less glaring than the viscosity outlier (glycerol) noted earlier, careful examination of the INSTAB data plot in Fig. 10.15 does suggest the possibility of a binary clustering of the dataset into 6 "high stability foams" with unusually small INSTAB values (i.e., slowly decaying foams) and 15 "normal" foams with significantly larger INSTAB values. In fact, applying the Hampel identifier to this dataset gives precisely this result: samples 2, 6, 7, 17, 18, and 19 are flagged as outliers.

Table 10.5 presents both the product-moment correlations $\hat{\rho}_{xy}$ and the Spearman rank correlations ρ_S for SNE or INSTAB vs. surface tension or density. For the SNE data, both correlations are comparable, but the INSTAB correlations appear significantly different. In particular, note that the computed correlations with surface tension differ by a factor of 2, suggesting the possibility of outliers, curvature, or other underlying structure in the dataset. Visual support for this hypothesis is evident in Fig. 10.16, which shows a scatterplot of the INSTAB values against the surface tension values for the 21 samples.

Figure 10.15: Plots of the measured foam characteristics

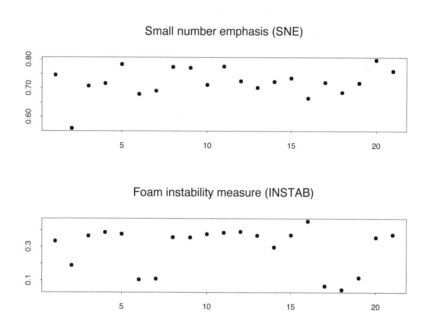

More specifically, the six high stability foams detected by the Hampel identifier applied to the INSTAB data values are shown as triangles in the lower right
of Fig. 10.16, and it appears that INSTAB tends to increase with surface tension
for these six samples, an observation supported by the high correlation values
computed for this subsample: $\hat{\rho}_{xy} = 0.87$ and $\rho_S = 0.71$. The three points
marked with solid circles in the upper left correspond to foams generated by organic liquids quite distinct in character from the other 18 samples in the dataset
(specifically, ethanol, propanol, and aniline). These three samples appear as a
clearly separated cluster, distinct from the others and they also exhibit unusual
behavior with respect to the relationship between INSTAB and log viscosity
data [227]. Finally, the remaining 12 points are shown as the cluster of open
circles in the upper right of the plot. Computed correlations for these observations are much smaller than those for the other two clusters, consistent with the
roughly circular shape of the cluster: $\hat{\rho}_{xy} = 0.14$ and $\rho_S = 0.18$. Also, note that
the range of variation seen in both the INSTAB and the surface tension values
is quite small for this cluster, so one possible explanation is that the scatter
seen in both of these variables is simply due to statistically independent errors
in each of these measurements.

The key point here is that the significant difference seen between the product-
moment and rank correlations computed from this dataset ultimately led to the
cluster decomposition proposed here. The available dataset is not large enough

Table 10.5: Foam characteristic vs. fluid property correlations

(x, y)-Pair	$\hat{\rho}_{xy}$	ρ_S
SNE vs. surface tension	0.00	−0.02
SNE vs. density	−0.25	−0.29
INSTAB vs. surface tension	0.09	0.22
INSTAB vs. density	−0.49	−0.30

to permit more detailed analysis, but the results presented here do illustrate the general nature of exploratory data analysis, summarized in the following quote from Samuel Karlin, cited by MacKay [194, p. 138]:

> The purpose of models is not to fit the data but to sharpen the questions.

In this example, the initial question posed was, "does there appear to exist a relationship between the four reported fluid properties and the two measured foam characteristics?" The sharper question that emerges from the analysis presented here is, "how do the high-stability foam samples differ from the others, and is the apparently much stronger correlation between INSTAB and surface tension for these samples real or merely a consequence of the low variability in surface tension seen in the majority of the cases considered here?" Questions of this sort suggest further experiments, approached from a more focused perspective.

10.6 Mixture distributions

The class of *mixture distributions* is defined on the basis of conditional probabilities, and as the following discussions illustrate, it is extremely useful in describing many different phenomena, including outliers, heterogeneity, and a variety of other departures from ideal modeling assumptions. Because they are defined on the basis of conditional probabilities, mixture distributions provide a number of practical illustrations of ideas introduced earlier in this chapter.

10.6.1 Discrete mixtures

Frequently, the term "mixture distribution" is taken to mean a *discrete mixture distribution*, defined by a density of the general form:

$$p(x) = \sum_{j=1}^{p} \pi_j f_j(x). \tag{10.105}$$

Figure 10.16: Foam instability measure plotted vs. surface tension

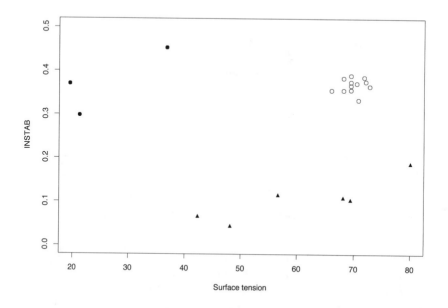

Here, the p constants $\{\pi_j\}$ are usually called the *mixing weights*, and the p functions $\{f_j(x)\}$ are probability density functions, usually called the *component densities*. The underlying idea is that the mixture density $p(x)$ describes a *heterogeneous population*, consisting of p individual component populations. Important restrictions on this model are that the mixing weights be nonnegative, and that they sum to unity; i.e.:

$$\pi_j \geq 0 \quad \text{and} \quad \sum_{j=1}^{p} \pi_j = 1. \tag{10.106}$$

If the functions $f_j(x)$ are proper density functions, it then follows that $p(x)$ also defines a proper density function, called the *mixture density*.

In terms of the urn models discussed in Chapter 3, mixture distributions may be viewed as describing random variables obtained by drawing *independently, from p different urns*; within each urn, the balls drawn are distributed according to the density $f_j(x)$, but the overall distribution of the balls drawn depends on the number of urns p, the probability π_j of selecting the jth urn on each drawing, and the distribution $f_j(x)$ of balls within the selected urn.

The situation just described is what Titterington *et al.* [278] refer to as a *direct application* of a mixture density; specifically, the overall density $p(x)$ arises because p distinct populations are combined in our observed data sample, and we are attempting to describe this heterogeneity explicitly. These applications are quite common, and the authors present a tabular summary of approximately 130 examples, along with more complete discussions of several others, including the modeling of ash particles from the Mount St. Helens volcanic eruption in 1980, characterization of fish sizes, analysis of electrophoresis data, analysis of housing market data, medical diagnosis, multiple target tracking (e.g., for military fighter aircraft), crop analysis via satelite remote sensing data, analysis of the shape (i.e., crust height distribution) of Earth, and characterization of astronomical data (i.e., star clusters). Alternatively, Titterington *et al.* [278] define *indirect applications* of mixture distributions as the use of Eq. (10.106) to approximate some other, more complex distribution. This idea is also extremely important in practice and is discussed further in Sec. 10.6.2 in connection with the class of Gaussian mixture densities.

In either case, it follows from the urn model interpretation that if the selection process described by the probabilities π_j is *independent of the selection process within each urn*, then we may obtain a simple general expression for expectations of the mixture distribution. Specifically, the expected value of any function $\phi(x)$ of the random variable x may be computed from the conditional expectation results given in Sec. 10.2.5 (specifically, Eq. (10.47)) as:

$$E\{\phi(x)\} = E_z\{E\{\phi(x)|z\}\},$$

where the outer expectation is taken with respect to the conditioning variable z. Applying this result to the mixture distribution defined by Eq. (10.105) gives:

$$E\{\phi(x)\} = E_j\{E\{\phi(x)|j\}\} = \sum_{j=1}^{p} \pi_j \int_{-\infty}^{\infty} \phi(x) f_j(x) dx. \tag{10.107}$$

Specializing this result to the case of moments, $\phi(x) = x^n$ gives the following general expression:

$$E\{x^n\} = \sum_{j=1}^{p} \pi_j \, E_j\{x^n\}, \tag{10.108}$$

where $E_j\{x^n\}$ is the nth moment of the jth component density in the mixture. Unfortunately, the results that emerge from this simple general expression tend to become complicated with increasing n, limiting its utility somewhat.

To illustrate this last point, consider the following special cases. For $n = 1$, we have the simple result:

$$E\{x\} = \sum_{j=1}^{p} \pi_j \bar{x}_j, \tag{10.109}$$

where $\bar{x}_j$ is the mean of the jth component distribution. The variance expression is somewhat more complicated, given by (Exercise 10):

$$\text{var}\{x\} = \sum_{j=1}^{p} \pi_j \sigma_j^2 + \sum_{j=1}^{p} \pi_j \bar{x}_j \left[\bar{x}_j - \sum_{k=1}^{p} \pi_k \bar{x}_k \right]. \tag{10.110}$$

A particularly useful observation is the following (Exercise 11):

$$\sum_{j=1}^{p} \pi_j \bar{x}_j \left[\bar{x}_j - \sum_{k=1}^{p} \pi_k \bar{x}_k \right] \geq 0, \tag{10.111}$$

with equality if and only if $\bar{x}_j = \bar{x}_k$ for all j and k. Hence, it follows that the variance of any mixture density is bounded below by the variance of the common-mean case ($\bar{x}_j = \bar{x}^0$ for all j), but if the means are distinct, the variance is strictly greater than this minimum value.

Higher moment expressions rapidly become much more complicated, but one important simplification is noteworthy: if all of the component densities are zero-mean and symmetric, then all of the individual odd-order moments $E_j\{x^{2m+1}\}$ vanish, from which it follows that the odd-order moments of the mixture density also vanish. In particular, the mean and skewness of the mixture density are zero and the variance is given by the weighted sum of component variances defined by Eq. (10.110) when $\bar{x}_j = 0$ for all j. The expression for the kurtosis is more complicated, but the derivation is not difficult (Exercise 12):

$$\kappa(x) = \left[\sum_{j=1}^{p} \pi_j \sigma_j^2 \right]^{-2} \left\{ \sum_{j=1}^{p} \pi_j \kappa_j \sigma_j^4 + 3 \sum_{j=1}^{p} \pi_j \sigma_j^2 \left[\sigma_j^2 - \sum_{k=1}^{p} \pi_k \sigma_k^2 \right] \right\},$$
$$\tag{10.112}$$

where κ_j is the kurtosis of the jth component density. In the general case, it is important to note that both the skewness and the kurtosis of the mixture distribution depend on the means $\bar{x}_j$ of the individual components, except in the case where they are all equal; there, if $\bar{x}_j = \bar{x}^0$ for all j, it follows from Eq. (10.109) that $E\{x\} = \bar{x}^0$ and, as a consequence, that the central moments of x for the mixture density depend only on the central moments of the component distributions μ_j^n (Exercise 13).

10.6.2   Example: Gaussian mixtures

The general class of discrete mixture densities is enormous, so there is considerable interest in special cases, especially in mixtures whose component densities $f_j(x)$ all have the same form. Not surprisingly, the special case of *Gaussian mixture densities*, in which all of the components are Gaussian, is of particular interest because so many results are available for the Gaussian component densities. Despite the analytical advantages of this restriction, it is important to emphasize that the class of Gaussian mixture densities is *extremely* flexible. As an impressive example, Marron and Wand [202] describe a set of 15 Gaussian mixtures used to evaluate the performance of kernel density estimators; the result includes the Gaussian reference distribution, unimodal distributions with varying degrees of asymmetry and elongation, symmetric and asymmetric bimodal distributions, and more general distributions with multiple modes of widely varying widths. Because of this tremendous flexibility, analysis of the general case rapidly becomes complicated even for Gaussian mixture densities;

Figure 10.17: Four Gaussian mixture densities

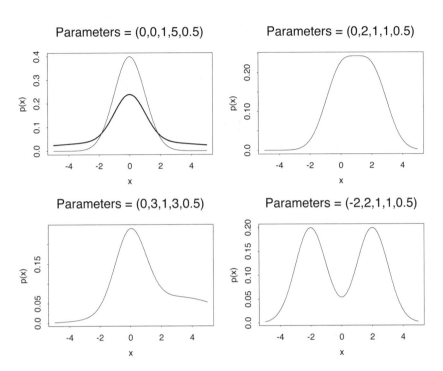

for this reason, the remainder of this section restricts consideration to the case of two-component Gaussian mixtures, which is still flexible enough to be extremely useful, as the following examples illustrate.

Fig. 10.17 shows four members of this family, each representing a special case of the following general expression:

$$p(x) = \alpha\phi(\mu_1, \sigma_1; x) + (1 - \alpha)\phi(\mu_2, \sigma_2; x). \tag{10.113}$$

Here, α is the concentration of the first component in the mixture, corresponding to the parameter π_1 in the general model; note that $\pi_2 = 1 - \pi_1$ for the two-component model, by the normalization requirement on the mixing parameters. The function $\phi(\mu, \sigma; x)$ is the probability density function for a Gaussian random variable with mean μ and standard deviation σ. In Fig. 10.17, the five parameters appearing in Eq. (10.113) are listed at the top of each plot in the following order: μ_1, μ_2, σ_1, σ_2, and α.

In all four of the cases considered here, the mixing parameter is $\alpha = 1/2$. The upper left plot shows two densities: the uncontaminated Gaussian density, corresponding to $\mu_1 = \mu_2 = 0$ and $\sigma_1 = \sigma_2 = 1$, included for reference, and a heavy-tailed distribution obtained for $\mu_1 = \mu_2 = 0$, $\sigma_1 = 1$, and $\sigma_2 = 5$. The upper right plot shows a mildly asymmetric distribution, obtained for $\mu_1 = 0$,

$\mu_2 = 2$, and $\sigma_1 = \sigma_2 = 1$. Separating the means of the component distributions further leads to greater asymmetry, as shown in the lower left plot: there, $\mu_1 = 0$, $\mu_2 = 3$, $\sigma_1 = 1$, and $\sigma_2 = 3$, and the resulting distribution exhibits a broad *shoulder* on the upper side of the main peak. Separating the means still further but keeping the standard deviations equal results in the bimodal distribution shown in the lower right plot: there, $\mu_1 = -2$, $\mu_2 = 2$, and $\sigma_1 = \sigma_2 = 1$.

This last example illustrates that two-component Gaussian mixtures can exhibit bimodal densities; in fact, the following necessary and sufficient conditions are known [162, p. 164]. First, the mixture density cannot be bimodal if the component means are too close together, relative to the harmonic mean of the variances; specifically, bimodality is not possible if:

$$(\mu_1 - \mu_2)^2 < \frac{27\sigma_1^2\sigma_2^2}{4(\sigma_1^2 + \sigma_2^2)} = \frac{27}{4}\left[\frac{1}{\sigma_1^2} + \frac{1}{\sigma_2^2}\right]^{-1}. \tag{10.114}$$

Conversely, bimodality is always possible if the means are far enough apart; specifically, there exist values of α for which the mixture density is bimodal if:

$$(\mu_1 - \mu_2)^2 > \frac{8\sigma_1^2\sigma_2^2}{\sigma_1^2 + \sigma_2^2} = 8\left[\frac{1}{\sigma_1^2} + \frac{1}{\sigma_2^2}\right]^{-1}. \tag{10.115}$$

In this case, α determines the relative heights of the two peaks in the distribution. For $\alpha_1 = 1/2$, $\mu_1 = -\mu_2 = L$, and $\sigma_1 = \sigma_2 = \sigma < L$, we obtain a symmetric distribution with equally intense peaks at $x = \pm L$.

A few useful conclusions regarding the behavior of general mixture densities may be drawn by considering the two-component Gaussian case. Specifically, the first three moments for this case are given by:

$$
\begin{aligned}
E\{x\} &= \alpha\mu_1 + (1-\alpha)\mu_2 = \mu_2 + \alpha(\mu_1 - \mu_2), \\
E\{x^2\} &= \alpha\sigma_1^2 + (1-\alpha)\sigma_2^2 + \mu_2^2 + \alpha(\mu_1^2 - \mu_2^2), \\
E\{x^3\} &= \mu_2(\mu_2^2 + 3\sigma_2^2) + 3\alpha(\mu_1\sigma_1^2 - \mu_2\sigma_2^2) + \alpha(\mu_1^3 - \mu_2^3).
\end{aligned}
\tag{10.116}
$$

It follows from the first moment result that the mean of the mixture distribution can assume any value between the means μ_1 and μ_2 of the two component densities. Further, the first and second moment results may be combined to yield the variance expression:

$$
\begin{aligned}
\text{var}\{x\} &= E\{x^2\} - [E\{x\}]^2 \\
&= \alpha\sigma_1^2 + (1-\alpha)\sigma_2^2 + \alpha(1-\alpha)(\mu_1 - \mu_2)^2.
\end{aligned}
\tag{10.117}
$$

It is clear from this result that the variance of the mixture distribution increases as the distance between the means of the two component distributions increases. Finally, note that although the third moment result is rather complicated, it does illustrate that the symmetry of the mixture distribution will depend on both the means and the variances of the components.

The second two-component Gaussian mixture density example considered here is that of the contanimated normal outlier model, discussed previously in

Chapter 7. This model is defined by the mixture density:

$$p(x) = (1 - \epsilon)p_0(x) + \epsilon p_1(x), \tag{10.118}$$

where $p_0(x)$ represents the normal distribution describing the nominal part of the data, $p_1(x)$ represents a second contaminating distribution, and ϵ represents the fraction of contamination present in the data. In the most general formulation, $p_1(x)$ need not be Gaussian, but in the most popular cases, the following assumptions are made: both $p_0(x)$ and $p_1(x)$ are taken to be zero-mean Gaussian distributions, with variances $\sigma_1^2 > \sigma_0^2$. Further, the contamination level is usually restricted to $0 < \epsilon < 1/2$, with $\epsilon \sim 0.1$ to 0.2 most common; also, the choice $\sigma_1 \simeq 3\sigma_0$ appears quite frequently [151, p. 2].

Assuming that $p_0(x)$ corresponds to the distribution $N(0, \sigma^2)$ and $p_1(x)$ corresponds to $N(0, r^2\sigma^2)$ for some $r > 1$, the resulting mixture density is symmetric and zero-mean, implying that all of its odd-order moments vanish. Similarly, it follows from Eq. (10.117) that:

$$\text{var}\{x\} = [1 - \epsilon + \epsilon r^2]\sigma^2 > \sigma^2. \tag{10.119}$$

The kurtosis of the contaminated normal distribution may be computed from the general expression given in Eq. (10.112) by taking $p = 2$, $\alpha_1 = 1 - \epsilon$, and noting that $\kappa_1 = \kappa_2 = 0$ for this case; simplifying the result leads to the expression:

$$\kappa = \frac{3\epsilon(1 - \epsilon)(r^2 - 1)^2}{[1 + \epsilon(r^2 - 1)]^2}. \tag{10.120}$$

Note that when $r = 1$ or $\epsilon = 0$, we recover the standard normal distribution. Conversely, when r is very large, provided ϵ is not too small (specifically, if $\epsilon(r^2 - 1) >> 1$), the variance of x approaches the contamination percentage times the contaminating variance; i.e., $\sigma_x^2 \to \epsilon r^2\sigma^2$ and the kurtosis approaches the upper limit $\kappa^* = 3(1 - \epsilon)/\epsilon$.

More generally, it is interesting to note that the kurtosis of any Gaussian mixture distribution composed of zero-mean component densities is nonnegative. In particular, since $\kappa_j = 0$ for all j for the case of a Gaussian mixture density, it follows from Eq. (10.112) that the kurtosis of the mixture density is given by (Exercise 14):

$$\kappa(x) = \frac{3}{2}\left(\frac{\sum_{j=1}^{p}\sum_{k=1}^{p}\pi_j\pi_k[\sigma_j^2 - \sigma_k^2]^2}{\sum_{j=1}^{p}\sum_{k=1}^{p}\pi_j\pi_k\sigma_j^2\sigma_k^2}\right) \geq 0. \tag{10.121}$$

Further, this lower bound is attained if and only if all of the variances are equal, implying that the mixture distribution degenerates to the $N(0, \sigma^2)$ density.

Finally, it is worth noting that another popular outlier model that may be expressed as a contaminated normal distribution is the *slippage model* [27, p. 49], discussed briefly in Chapter 7. There, the nominal $N(0, \sigma^2)$ distribution is contaminated by a second normal distribution, but with nonzero mean, typically $N(\mu, \sigma^2)$ for some $\mu > 0$. An important feature of the slippage model is that, in contrast to the symmetric contaminated normal model just described, the resulting distribution is asymmetric, with a nonzero mean that is determined by both the mean of the contaminating distribution and the contamination level. In particular, the mean, variance, and skewness of this slippage model are given by (Exercise 15):

$$
\begin{aligned}
E\{x\} &= \epsilon\mu, \\
\text{var}\{x\} &= \sigma^2 + \epsilon(1-\epsilon)\mu^2, \\
\gamma(x) &= \frac{\epsilon\mu[\mu^2 + 3\sigma^2]}{[\sigma^2 + \epsilon(1-\epsilon)\mu^2]^{3/2}}.
\end{aligned}
\tag{10.122}
$$

In fact, this distribution was used in the brief discussion of asymmetric contamination presented in the Princeton Robustness Survey [11], which considered the slippage models for $\mu = 2\sigma$ and $\mu = 4\sigma$.

10.6.3   Continuous mixtures

A particularly important special case of the discrete mixture density just described is that where the *form* of the component densities $f_j(x)$ is the same and these distributions differ only in their parameters. A specific illustration of this idea is the class of Gaussian mixture densities just described; there, all of the component densities are assumed Gaussian, differing in their means and/or variances. *Continuous mixture densities* extend this idea to the case where one or more parameters of the component distributions vary *continuously*. In particular, suppose $f(x,\theta)$ is a density depending on some r-component parameter vector θ; the continuous mixture density $p(x)$ is then defined by the r-dimensional integral:

$$
p(x) = \int \alpha(\theta)\, f(x,\theta)\, d\theta.
\tag{10.123}
$$

As in the case of discrete mixture densities, the effective *selection probability* $\alpha(\theta)$ must satisfy the normalization condition:

$$
\int \alpha(\theta)\, d\theta = 1,
\tag{10.124}
$$

For example, in a continuous mixture of Gaussian densities with a fixed mean μ but a distribution of standard deviations σ, it follows that $\theta = \sigma$ and this integral is defined over the positive real line R^+; the function $\alpha(\theta)$ then represents a probability density for σ. Similarly, for a continuous mixture of Gaussian densities with fixed standard deviation σ and a distribution of means μ, we have $\theta = \mu$ and the integral is defined over the entire real line R, with $\alpha(\theta)$ representing a probability density for μ. In the most general Gaussian mixture case, both μ and σ will be distributed and the integral indicated in Eq. (10.123) will be a double integral over the appropriate ranges of both μ and σ.

As in the case of discrete mixtures, it is fairly straightforward to write down a general expression for the moments of a continuous mixture density:

$$E\{x^n\} = \int \alpha(\theta)\, E_\theta\{x^n\}\, d\theta, \qquad (10.125)$$

where $E_\theta\{z\}$ is the expected value of the variable z with respect to the component density $f(x, \theta)$. As with discrete mixture densities, determination of moments from Eq. (10.125) may be too difficult to be practical, but one intuitively reasonable result follows immediately: if all components in the mixture have the same mean μ, it follows from Eqs. (10.124) and (10.125) that $E\{x\} = \mu$ for the mixture, regardless of the distribution $\alpha(\theta)$.

The next three sections illustrate specific applications of continuous mixture distributions. The first of these is the phenomenon of *overdispersion*, where the distributional model we wish to assume for the data exhibits a larger variance than it should. Next, it is shown that the ratio of two random variables has a distribution that may be viewed as a continuous mixture density. Following a brief derivation of this general result, it is specialized to obtain two simple models of heavy-tailed (leptokurtic) distributions that may be viewed as *randomly scaled Gaussian random variables*. That is, if $x \sim N(0,1)$ and $y = 1/\sigma$ is a second, independent random variable with mean $1/\sigma_0$, the ratio x/y may be viewed as nominally Gaussian, with mean 0 and a standard deviation that is uncertain, but approximately σ_0. The distribution of y will determine the tail behavior of the ratio, which may range from mildly non-Gaussian (e.g., Student's t with many degrees of freedom) to the classically horrible Cauchy distribution. This idea was used quite successfully in the Princeton Robustness Study [11] to provide a useful range of leptokurtic comparison models.

10.6.4 Example 1: overdispersion

The problem of overdispersion was introduced in Chapter 3, most simply explained for a one-parameter distribution like the binomial or Poisson where the parameter necessarily determines both the mean and the variance. It frequently happens that, if we fit this parameter to the mean of the observed data, the observed variance is larger than that for the distribution given the estimated model parameter. For example, the single parameter λ for the Poisson distribution specifies both the mean and the variance; thus, for a sequence of nominally Poisson-distributed count data, overdispersion corresponds to an observed variance that is substantially larger than the observed mean. As noted in Chapter 3, one common source of overdispersion in simple discrete distributions is *zero inflation:* the number of observed zeros is substantially larger than that predicted by the candidate distribution.

Zero-inflated discrete distributions correspond to two-component discrete mixture models where one component is the degenerate distribution centered at zero (i.e., $x = 0$ with probability 1, implying that all moments of x are zero), and the other is a standard distribution like the binomial or Poisson. As a specific example, the zero-inflated Poisson distribution corresponds to the assumption that x is exactly zero with probability π and is drawn from a Poisson distribution with parameter λ with probability $1 - \pi$. It follows from the general moment results for mixture distributions given in Sec. 10.6.1 that the first two moments of the zero-inflated Poisson distribution are:

$$
\begin{aligned}
E\{x\} &= \pi \times 0 + (1 - \pi) \times \lambda = (1 - \pi)\lambda, \\
E\{x^2\} &= \pi \times 0 + (1 - \pi) \times (\lambda + \lambda^2) = (1 - \pi)\lambda(1 + \lambda). \quad (10.126)
\end{aligned}
$$

Combining this result leads to the following variance expression for the zero-inflated Poisson distribution:

$$
\text{var}\{x\} = E\{x^2\} - [E\{x\}]^2 = (1 - \pi)\lambda(1 + \pi\lambda). \quad (10.127)
$$

Since this variance is larger than the mean of the distribution by a factor of $1 + \pi\lambda$, it follows that the zero-inflated Poisson distribution represents one possible model for overdispersion in situations where the Poisson distribution might be expected.

Another, more general class of overdispersion models is based on continuous mixture distributions. Again for simplicity, consider the case of a one-parameter distribution whose parameter θ necessarily determines both its mean and variance. Introducing uncertainty into this parameter by modeling it as a random variable yields a continuous mixture distribution that exhibits overdispersion relative to the case where θ is fixed (i.e., deterministic). To see this point, first note that the moment results presented in Sec. 10.6.1 imply (Exercise 17):

$$
\text{var}\{x\} = E\{\text{var}\{x|\theta\}\} + \text{var}\{E\{x|\theta\}\}. \quad (10.128)
$$

In the deterministic case where the parameter θ is fixed, the first term in this expression is simply the variance of x defined by this fixed value of θ and the second term is zero since $E\{x|\theta\} = \mu(\theta)$ is a constant and thus has zero variance. Conversely, if θ is a random variable with distribution $p(\theta)$, this second term is positive, leading to overdispersion relative to the fixed parameter model.

To make these results more concrete, again consider the case of the Poisson distribution. Note that by making the parameter λ a positive random variable, we are constructing a continuous mixture distribution that reduces to the Poisson distribution as the variance of λ goes to zero. When the variance is nonzero—whatever the distribution of λ—it follows that:

$$
E\{x|\lambda\} = \lambda \quad \text{and} \quad \text{var}\{x|\lambda\} = \lambda. \quad (10.129)
$$

Denoting the mean of λ by $\bar{\lambda}$, the unconditional mean of x is given by:

$$
E\{x\} = E\{E\{x|\lambda\}\} = E\{\lambda\} = \bar{\lambda}. \quad (10.130)
$$

Combining Eqs. (10.128) and (10.129) gives the unconditional variance for x:

$$
\text{var}\{x\} = E\{\lambda\} + \text{var}\{\lambda\} = \bar{\lambda} + \sigma_\lambda^2, \quad (10.131)
$$

where σ_λ^2 is the variance of λ. Since this variance is greater than the mean $\bar{\lambda}$, the uncertain parameter model exhibits overdispersion relative to the fixed Poisson model. The same line of reasoning can be applied to model overdispersion for binomial data [66, sec. 6.2].

10.6.5 Example 2: ratios of random variables

The ratio $z = x/y$ of two random variables x and y with a joint density $p_{xy}(x, y)$ is given by [222, p. 197]:

$$p_z(z) = \int_{-\infty}^{\infty} |y| p_{xy}(zy, y) dy. \tag{10.132}$$

Derivation of this result is not difficult. First, note that the cumulative distribution $F_{x/y}(z)$ for the ratio $z = x/y$ is

$$F_{x/y}(z) = P\{x/y \leq z\} = P\{x \leq zy|y > 0\} + P\{x \geq zy|y < 0\}. \tag{10.133}$$

Here, use has been made of the facts that first, that multiplication by a negative number reverses the direction of an inequality and second, for a continuous random variable y, the troublesome event $y = 0$ has zero probability. The conditional probability $P\{x \leq zy|y > 0\}$ is easily obtained by integrating the joint density:

$$P\{x \leq zy|y > 0\} = \int_0^{\infty} dy \int_{-\infty}^{zy} dx \; p_{xy}(x, y), \tag{10.134}$$

and the other required conditional probability is given by:

$$P\{x \geq zy|y < 0\} = 1 - P\{x \leq zy|y < 0\} = 1 - \int_{-\infty}^0 dy \int_{-\infty}^{zy} dx \; p_{xy}(x, y). \tag{10.135}$$

To obtain the density $p_z(z)$ for the ratio, Eq. (10.133) is differentiated with respect to z, yielding the result given in Eq. (10.132):

$$p(z) = \int_0^{\infty} y p_{xy}(zy, y) dy - \int_{-\infty}^0 y p_{xy}(zy, y) dy \tag{10.136}$$

$$= \int_0^{\infty} |y| p_{xy}(zy, y) dy + \int_{-\infty}^0 |y| p_{xy}(zy, y) dy = \int_{-\infty}^{\infty} |y| p_{xy}(zy, y) dy.$$

If the random variables x and y are statistically independent—implying that $p_{xy}(x, y) = p_x(x) p_y(y)$—Eq. (10.132) reduces to:

$$p_z(z) = \int_{-\infty}^{\infty} |y| p_x(zy) p_y(y) dy. \tag{10.137}$$

In the examples considered in Sec. 10.6.6, x is assumed to be a standard normal random variable—implying $p_x(x) = (1/\sqrt{2\pi}) e^{-x^2/2}$—and y is assumed to be a positive random variable. Together these observations further simplify Eq. (10.132) to:

$$p_z(z) = \frac{1}{\sqrt{2\pi}} \int_0^{\infty} y e^{-z^2 y^2/2} p_y(y) dy. \tag{10.138}$$

To see the connection between this result and continuous mixture densities, introduce the following alternative representation for Eq. (10.138): define $\sigma = 1/y$ so $z = \sigma x$, and use the result given in Chapter 12 for reciprocal transformations: $p_\sigma(\sigma) = p_y(1/\sigma)/\sigma^2$. Substituting $y = 1/\sigma$ into Eq. (10.138) and simplifying then yields the following continuous mixture density representation:

$$p_z(z) \quad = \quad \frac{1}{\sqrt{2\pi}} \int_0^\infty e^{-z^2/2\sigma^2} p_\sigma(\sigma) d\sigma. \qquad (10.139)$$

The following discussion shows how this result can be used to derive some useful heavy-tailed univariate data distributions.

10.6.6 Example 3: heavy-tailed distributions

Probably the best known application of the result just presented was the derivation by "Student" in 1908 of the *Student's t-distribution with ν degrees of freedom*, discussed in Chapter 9. This distribution describes the ratio of a variable x with the $N(0,1)$ distribution to a second independent random variable with a *chi distribution with ν degrees of freedom*, defined as the square root of the *chi square distribution with ν degrees of freedom* [162, p. 417], also discussed in Chapter 9. Specifically, the random variable:

$$z \quad = \quad \frac{x}{\sqrt{\chi_\nu^2/\nu}}, \qquad (10.140)$$

has the Student's t distribution with ν degrees of freedom. For $\nu = 1$, the denominator becomes the *half-normal distribution*, defined as the distribution of $|\xi|$ where $\xi \sim N(0,1)$ [162, p. 156] and having the CDF:

$$F(y) = 2\Phi(y) - 1, \qquad (10.141)$$

for $y \geq 0$, where $\Phi(y)$ is the standard normal CDF. Because this density is peaked at $y = 0$, the most likely value for y corresponds to an infinite variance $\sigma = 1/y$ for the scaled variable z. This result is essentially the same as the observation that the ratio of two independent standard normal random variables exhibits a Cauchy distribution [162, p. 318]. In any case, it should be clear that the scheme described at the end of the last section—randomly scaling the unit normal distribution—can result in *extremely* non-Gaussian behavior. Conversely, note that as $\nu \to \infty$, the the distribution of the denominator $\chi_\nu/\sqrt{\nu}$ converges to the degenerate limit $y = 1$, implying that z approaches a normal limit and corresponding to the fact that the Student's t distribution approaches the $N(0,1)$ distribution as $\nu \to \infty$ [163, p. 363].

Another useful randomly scaled Gaussian distribution is the *slash distribution* [162, p. 63], obtained by letting y be uniformly distributed on $[0,1]$ to obtain the density:

$$p(x) = \begin{cases} [\phi(0) - \phi(x)]/x^2 & x \neq 0, \\ \phi(0)/2 & x = 0, \end{cases} \qquad (10.142)$$

Figure 10.18: Comparison of Cauchy and slash distributed data

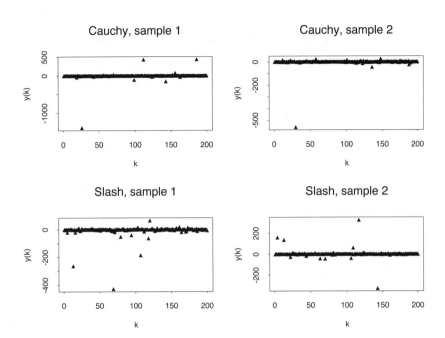

where $\phi(x) = (\sqrt{2\pi})^{-1} \exp[-x^2/2]$ is the density function for the standard normal distribution. Fig. 10.18 shows four data samples of length $N = 200$, two drawn from the slash distribution and two from the Cauchy distribution, illustrating that the slash distribution appears somewhat better behaved (note the difference in scales between these plots). The slash distribution does exhibit the same slow tail decay, but it provides a somewhat less radical interpretation: $1/\sigma$ is uniformly distributed on $[0, 1]$, suggesting the variance σ^2 of the scaled normal random variable is essentially unknown. Again, this should be contrasted with the Cauchy distribution, corresponding to the standard normal distribution scaled by $1/y$ where y is a half-normal random variable whose most probable value is zero, corresponding to infinite variance. Based on this argument, the slash distribution has been suggested as a more reasonable outlier model than the Cauchy distribution [11, p. 226].

Alternatively, if we assume a distribution for the denominator variable y that is more sharply peaked around a nonzero value, the result is a distribution with lighter tails. As a specific example, the Laplace distribution results if y has the following density [11, p. 57]:

$$p(y) = y^{-3} e^{-y^{-2}/2}. \qquad (10.143)$$

Here, the reciprocal transformation $g(y) = 1/y$ yields the Rayleigh density discussed in Chapter 4 for $\sigma = 1/y$; this density is peaked at 1, so the ratio $z = x/y$

exhibits significantly heavier tails than the Gaussian distribution but much less radical behavior than either the Cauchy or the slash distributions.

10.7 Non-Gaussian multivariate distributions

The Gaussian working assumption is even more popular in the multivariate case than in the univariate case because the specification of non-Gaussian alternatives is actually much more difficult in the multivariate case. One of the primary difficulties is that well-defined univariate distributions may be extended to the multivariate setting in more than one way, and these different extensions are not equivalent. As a specific example, note that the multivariate normal distribution exhibits the following desirable properties:

1. the mathematical structure of the multivariate distribution is a simple extension of that for the univariate distribution,
2. analogous to the univariate case, the multivariate distribution is completely specified by the mean vector μ and the covariance matrix Σ,
3. all univariate marginals and conditionals of the multivariate distribution are univariate normal densities,
4. statistical independence of the components of $\mathbf{x}$ is equivalent to diagonality of the covariance matrix Σ.

Unfortunately, such desirable characteristics generally do not hold for non-Gaussian multivariate distributions, so we are forced to decide which of these properties are more important and settle for a subset of them. Different choices can lead to very different multivariate distributions, as the following discussions illustrate.

These discussions are organized as follows. First, Sec. 10.7.1 illustrates the nature of the difficulties just noted by considering the question of how to define a bivariate extension of the univariate exponential distribution. Next, Sec. 10.7.2 briefly considers two surprising "near Gaussian" bivariate distributions that preserve some characteristics of the Gaussian case but exhibit profoundly non-Gaussian behavior in other respects. Sec. 10.7.3 then introduces the class of *elliptical distributions*, a useful class of multivariate non-Gaussian random variables that preserves many of the structural features of the Gaussian case.

Essentially, what elliptical distributions preserve is much of the dependence structure (i.e., the correlation behavior) between components of a multivariate Gaussian random vector. An alternative approach is that based on *copulas*, defined and discussed in Sec. 10.7.4, which uses one multivariate function to specify the dependence between components while independently specifying the univariate marginal distributions of these components. The most useful association measures between these components are measures of monotone association, such as the Spearman rank correlation coefficient introduced in Sec. 10.5 and *Kendall's* τ, introduced in Sec. 10.7.5.

Finally, it is worth noting that these topics are not treated in as much detail as many of the other topics covered in this book (e.g., linear regression, as described in Chapters 5 and 11), but they are introduced here for two reasons. First, they help to explain the continued widespread adoption of multivariate Gaussian distributions in data analysis despite their inherent limitations, and they give a clear indication of some of the complications that must be faced when dealing with problems where multivariate Gaussian working assumptions truly are inadequate. Second, these ideas are introduced here because, although they are not yet as well known as Gaussian-based techniques like product-moment correlations and linear regression models, they are becomming increasingly important in applications. As a case in point, couplas are receiving increasing attention as a basis for building non-Gaussian multivariate models in areas as diverse as financial risk management, predicting the outcomes of soccer matches, and modeling natural catastrophes like tsunamis, earthquakes, and avalanches.

10.7.1 Bivariate exponential distributions

In an article on bivariate exponential distributions, Basu [30] lists a number of desirable characteristics exhibited by the univariate exponential distribution and considers them as desirable criteria for bivariate extensions. Three of these criteria are:

1. the joint distribution exhibits univariate exponential marginals;

2. the joint distribution is absolutely continuous so the distribution exhibits a joint density;

3. the survival function $\bar{F}(x,y) = \mathcal{P}\{X > x, Y > y\}$ exhibits the following *lack of memory property* for all $x, y, t > 0$:

$$\bar{F}(x + t, y + t) = \bar{F}(x,y)\bar{F}(t,t).$$

Unfortunately, it has been shown by Block and Basu [38] that no bivariate distribution exists that satisfies all three of these properties: the best we can have is two out of three.

The Marshall-Olkin model [30, 203] is defined by the following expression for the survival function:

$$\bar{F}(x,y) = \exp\{-\lambda_1 x - \lambda_2 y - \lambda_3 \max(x,y)\}, \qquad (10.144)$$

defined for positive x, y, λ_1, and λ_2 and for $\lambda_3 \geq 0$. This is the only bivariate distribution that exhibits exponential marginals and the lack of memory property defined in criterion 3 above, but the joint distribution function is not absolutely continuous. Finally, note that when $\lambda_3 = 0$, the univariate components x and y are independent, exponentially distributed random variables.

The Block-Basu distribution is also defined in terms of the survival function, which is given by:

$$\bar{F}(x,y) = \left[1 + \frac{\lambda_3}{\lambda_1 + \lambda_2}\right] \exp\{-\lambda_1 x - \lambda_2 y - \lambda_3 \max(x,y)\}$$
$$- \left[\frac{\lambda_3}{\lambda_1 + \lambda_2}\right] \exp\{-(\lambda_1 + \lambda_2 + \lambda_3) \max(x,y)\}, \qquad (10.145)$$

again for positive x, y, λ_1, and λ_2, and $\lambda_3 \geq 0$. This example exhibits an absolutely continuous joint distribution and it satisfies the lack of memory property defined in criterion 3 above, but it does not exhibit exponential marginals. However, as in the case of the Marshall-Olkin distribution, the univariate components x and y do reduce to independent exponentially distributed random variables when $\lambda_3 = 0$.

The Gumbel type I distribution has the following simple joint distribution function [30, p. 328]:

$$F(x,y) = 1 - e^{-x} - e^{-y} + e^{-x-y-\delta xy}, \qquad (10.146)$$

for positive x and y and $0 \leq \delta \leq 1$. This distribution satisfies the first two of the criteria listed above, but not the third: it exhibits exponential marginals and is absolutely continuous, but it does not satisfy the lack of memory property. Like the other two binary exponential distributions defined here, the univariate components x and y are statistically independent when $\delta = 0$.

Taken together, these three examples illustrate the point noted at the beginning of this discussion: while the three desirable criteria for a bivariate exponential distribution cannot all be met simultaneously, any two of the three can be. Nor are these three examples exhaustive: Gumbel proposed two other bivariate exponential distributions (types II and III), and other authors have proposed further alternatives [30]. Analogous problems arise when we attempt to define multivariate analogs of other popular univariate distributions. In fact, the two examples presented in the following section illustrate that ambiguous cases can even arise when we attempt to generalize the univariate Gaussian model: very non-Gaussian distributions can exhibit Gaussian marginal or conditional distributions.

10.7.2   Two surprising "near-Gaussian" examples

It was noted at the beginning of this section that one of the characteristics of the multivariate Gaussian distribution is that both marginal and conditional univariate distributions derived from the multivariate parent are also Gaussian. The following examples illustrate that the converse does not hold: multivariate distributions with univariate marginal or conditional distributions need not be Gaussian. Further, the first of these examples also illustrates the point made earlier that for non-Gaussian bivariate distributions, lack of correlation does *not* imply independence.

The first example is a simple bivariate distribution discussed by Broffit [45]. Consider the joint distribution of the variables X and Y where X is a univariate Gaussian random variable with mean 0 and standard deviation 1, and $Y = XZ$ where Z is a binary random variable that is statistically independent of X with $\mathcal{P}\{Z = 1\} = \mathcal{P}\{Z = -1\} = 1/2$. Because X is symmetrically distributed about zero, it follows that X and $-X$ have the same distribution, further implying the following chain of relationships:

$$\mathcal{P}\{Y \le y \mid Z = 1\} = \mathcal{P}\{XZ \le y \mid Z = 1\} = \mathcal{P}\{X \le y\} \quad (10.147)$$
$$= \mathcal{P}\{-X \le y\}$$
$$= \mathcal{P}\{XZ \le y \mid Z = -1\}$$
$$= \mathcal{P}\{Y \le y \mid Z = -1\}.$$

It was noted in Sec. 10.2.4 that two variables Y and Z are independent if and only if the conditional distribution function $\mathcal{P}\{Y \le y \mid Z\}$ does not depend on Z; hence, it follows in this case that Y and Z are *statistically independent*, even though there is a *functional* relationship between them. Further, since the conditional distribution of Y given Z is simply the unconditional distribution of Y when Y and Z are independent, it also follows that:

$$\mathcal{P}\{Y \le y\} = \mathcal{P}\{Y \le y \mid Z = 1\} = \mathcal{P}\{XZ \le y \mid Z = 1\}$$
$$= \mathcal{P}\{X \le y\}. \quad (10.148)$$

Since $X \sim N(0,1)$, this observation means that $Y \sim N(0,1)$ as well. *In other words, the joint distribution of X and Y defines a bivariate non-Gaussian random variable with Gaussian marginals.*

In addition to establishing that multivariate distributions with Gaussian marginals need not be jointly Gaussian, this example also illustrates the point noted several times before that *uncorrelated random variables need not be statistically independent, unless they are jointly Gaussian or both binary.* Specifically, note that since Z assumes the values ± 1 with equal probability, it follows that $E\{Z\} = 0$. Then, since X and Z are statistically independent, we also have:

$$E\{XY\} = E\{X^2 Z\} = E\{X^2\}E\{Z\} = 0, \quad (10.149)$$

showing that X and Y are uncorrelated. Next, note that if X and Y were *independent*, it would follow that the expectation $E\{X^2 Y^2\}$ could be factored as $E\{X^2\}E\{Y^2\} = 1$; instead, it follows from the definition of Y and the fact that $Z^2 = (\pm 1)^2 = 1$ that:

$$E\{X^2 Y^2\} = E\{X^4 Z^2\} = E\{X^4\} = 3, \quad (10.150)$$

since X is distributed as $N(0,1)$. This example illustrates that it is not enough that the two variables exhibit Gaussian univariate (i.e., marginal) distributions: they must be *jointly* Gaussian for lack of correlation to imply statistical independence.

The second example considered here is a bivariate distribution described by Arnold *et al.* [17] illustrating that multivariate distributions with Gaussian univariate conditional densities can be extremely non-Gaussian. In fact, the authors show that these bivariate distributions can have up to three modes. To establish this result, they use an earlier result of Castillo [56] to show that any bivariate random variable with Gaussian conditionals has a joint density given by:

$$p(x,y) = C \exp[-(\alpha x^2 y^2 + x^2 + y^2 + \beta xy + \gamma x + \delta y)]. \quad (10.151)$$

Here, α is a nonnegative parameter that determines the non-Gaussian character of the joint density; for $\alpha = 0$, this joint distribution is Gaussian with correlation $\rho = \beta/2$ so long as $|\beta| < 2$. In the general case, the conditional densities are:

$$x|y \;\sim\; N\left(-\frac{\beta y + \gamma}{2(\alpha y^2 + 1)}, \frac{1}{2(\alpha y^2 + 1)}\right),$$

$$y|x \;\sim\; N\left(-\frac{\beta x + \delta}{2(\alpha y^2 + 1)}, \frac{1}{2(\alpha y^2 + 1)}\right). \tag{10.152}$$

Arnold *et al.* [17] show that the bivariate distribution defined in Eq. (10.151) can have one, two, or three modes. Results for the trimodal case are complicated, but the following simple condition may be obtained for the symmetric, bimodal case: if $\delta = \gamma$, the joint density is bimodal if and only if:

$$\alpha \delta^2 > 8(2 - \beta). \tag{10.153}$$

Provided $\delta \neq 0$, note that this condition gives a lower bound on the non-Gaussianity parameter α for the joint density to be bimodal. Finally, it is important to note that, although the conditional densities $p(x|y)$ and $p(y|x)$ are Gaussian in these examples, the *marginal densities* $p(x)$ and $p(y)$ are not.

10.7.3   Elliptically distributed random variables

The class of *elliptically distributed random variables* is defined by the following prescription. Let μ be a specified constant N-vector and Σ be an $N \times N$ symmetric, positive-definite matrix. Denoting the determinant of Σ by $|\Sigma|$ and noting that this quantity is a positive number, define the class $E_N(\mu, \Sigma)$ by multivariate densities of the form:

$$p(\mathbf{x}) \;=\; \frac{K_N}{|\Sigma|^{1/2}} \, g\left([\mathbf{x} - \mu]^T \Sigma^{-1} [\mathbf{x} - \mu]\right). \tag{10.154}$$

Here, $g(\cdot)$ is any positive integrable function and K_N is a normalization constant defined by the requirement that the density $p(\mathbf{x})$ integrate to 1 over R^N. Taking $g(z) = \exp[-z/2]$ yields the multivariate Gaussian density discussed in detail in Sec. 10.3, but other choices of $g(\cdot)$ lead to non-Gaussian distributions that share some of the important characteristics of the Gaussian distribution, as the following discussion illustrates.

First, for any elliptically distributed random variable $\mathbf{x}$, the probability density at any point $\mathbf{x}$ in R^N depends only on its Mahalanobis distance from μ, relative to the matrix Σ. It follows as a consequence that the mean and covariance of $\mathbf{x}$ are given by [101, p. 44]:

$$E\{\mathbf{x}\} = \mu, \quad \text{Cov}\{\mathbf{x}\} = \alpha \Sigma, \tag{10.155}$$

for some positive constant α that is determined by the function $g(\cdot)$; note that for $g(x) = e^{-x/2}$, we have $\alpha = 1$. In addition, elliptically distributed random vectors also exhibit analogous behavior to multivariate Gaussian random vectors with respect to linear transformations; specifically, if $\mathbf{x} \sim E_N(\mu, \Sigma)$, then [101, p. 43]:

$$\mathbf{Ax} + \mathbf{b} \quad \sim \quad E_N\left(\mathbf{A}\mu + \mathbf{b}, \mathbf{A}^T \Sigma \mathbf{A}\right). \tag{10.156}$$

In addition, $\mathbf{x}$ is symmetrically distributed, meaning that $\mathbf{x}$ and $-\mathbf{x}$ have the same distribution and that all central moments of odd order vanish. In particular, the skewness of any component x_i is zero, independent of the function $g(\cdot)$. In fact, taking $N = 1$ for an arbitrary positive, integrable function leads to the family of *symmetric univariate* distributions.

Another important characteristic of the class $E_N(\mu, \Sigma)$ is that the components x_i of $\mathbf{x}$ exhibit a common kurtosis value κ that is determined by the function $g(\cdot)$ [101, p. 44]. Further, the following useful fourth moment factorization result generalizes that available for the Gaussian case ($\kappa = 0$), where $R_{xx}(i, j) = E\{x_i x_j\}$ [101, p. 45]:

$$\begin{aligned} E\{x_i x_j x_k x_\ell\} &= (\kappa + 1)[R_{xx}(i, j)R_{xx}(k, \ell) + R_{xx}(i, k)R_{xx}(j, \ell) \\ &\quad + R_{xx}(i, \ell)R_{xx}(j, k)]. \end{aligned} \tag{10.157}$$

Finally, another characteristic that elliptically distributed random vectors share with the multivariate Gaussian case is linearity of the conditional mean. Specifically, suppose $\mathbf{x} \sim E_N(\mu, \Sigma)$ and partition this vector into two subvectors, $\mathbf{x}_1 \in R^r$ and $\mathbf{x}_2 \in R^{N-r}$. The conditional mean of $\mathbf{x}_1$ given $\mathbf{x}_2 = \mathbf{q}_2$ is then given by [170, 52]:

$$E\{\mathbf{x}_1 | \mathbf{x}_2 = \mathbf{q}_2\} = \mu_1 + \Sigma_{12}\Sigma_{22}^{-1}(\mathbf{q}_2 - \mu_2), \tag{10.158}$$

where Σ_{ij} are the four partitions of the matrix Σ induced by the division of $\mathbf{x}$ into subvectors and μ_1 and μ_2 are the corresponding partitions of the mean vector μ. Unfortunately, independence of the conditional covariance from $\mathbf{q}_2$ occurs if and only if the distribution is multivariate Gaussian [170, theorem 7].

10.7.4 Copulas: building from marginals

A *copula* represents a multivariate distribution $F(x_1, x_2, \ldots, x_n)$ in terms of its n univariate marginals $F_i(x_i)$, permitting these marginals to be specified separately from the dependence structure that links them. More specifically, it follows from Sklar's theorem [215] that the cumulative distribution function $F(\cdots)$ may be represented as:

$$F(x_1, x_2, \ldots, x_n) = C(F_1(x_1), F_2(x_2), \ldots, F_n(x_n)), \tag{10.159}$$

for some function $C : [0, 1]^n \to [0, 1]$. Further, the function $C(\cdots)$ is unique if the marginals $F_i(\cdot)$ are continuous. As noted in the introduction to this section, copulas are seeing increasingly widespread use in many applications, motivating the brief introduction given here. For a more complete treatment of copulas, refer to the book by Nelson [215]; for discussions of their applications to finance, refer to the book by Cherubini *et al.* [62] or the recent survey on the growing popularity of copula models in finance by Genest *et al.* [114]; for a discussion of the use of copulas in modeling natural disasters, see the book by Salvadori *et al.* [255]; and for a treatment of international soccor scores using copulas, refer to the report by McHale and Scarf [206]. In addition, note that a number of copula-based analysis tools are supported in the R package **copula** [299], including the evaluation of distribution functions and densities, random number generation, graphical display, and maximum likelihood based data fitting.

To see the connection between copulas and dependence structures, first recall that if the components x_i are statistically independent, the joint distribution is simply the product of the marginals $F_i(x_i)$. Thus, one important copula is the *independence copula:*

$$C(u_1, u_2, \ldots, u_n) = \prod_{i=1}^{n} u_i. \qquad (10.160)$$

More generally, to qualify as a copula, the function $C(\cdots)$ must satisfy a number of conditions to guarantee that the function $F(\cdots)$ defined in Eq. (10.159) satisfies the monotonicity conditions required for a multivariate probability distribution [215]. Further, any copula satisfies the following inequalities:

$$\max\left\{0, \sum_{i=1}^{n} u_i - (n-1)\right\} \le C(u_1, u_2, \ldots, u_n) \le \min\{u_1, u_2, \ldots, u_n\},$$

$$(10.161)$$

for all $\mathbf{u} \in [0, 1]^n$. The lower bound in this expression is called the *Fréchet lower bound*; in the case $n = 2$, this bound corresponds to a copula that represents the strongest possible negative association between components, but for $n > 2$ it no longer defines a copula. The upper bound always defines a copula, called the *Fréchet upper bound* and representing the strongest possible positive association between components.

Another important point to note is that applying continuous, monotone (and thus invertible) univariate transformations to each component of a joint distribution changes the marginals but does not change the copula. That is, suppose $F : R^n \to [0, 1]$ is given by Eq. (10.159) and consider the distribution $G : R^n \to [0, 1]$ obtained by applying univariate transformations $\phi_i : R \to R$ to the marginals:

$$
\begin{aligned}
G(x_1, x_2, \ldots, x_n) &= F(\phi_1(x_1), \phi_2(x_2), \ldots, \phi_n(x_n)) \\
&= C(F_1(\phi_1(x_1)), F_2(\phi_2(x_2)), \ldots, F_n(\phi_n(x_n))) \\
&= C(G_1(x_1), G_2(x_2), \ldots, G_n(x_n)). \qquad (10.162)
\end{aligned}
$$

Since the copula $C(\cdots)$ characterizes the dependence between components, it follows from this result that any numerical dependence measure determined only by the copula is necessarily invariant to monotone data transformations. Thus, while the product-moment correlation coefficient is natural for the multivariate Gaussian case, it is determined by both the copula and the marginals. In contrast, monotone association measures like the Spearman rank correlation coefficient introduced in Sec. 10.5 and Kendall's τ introduced in Sec. 10.7.5 are determined by the copula alone [215].

The range of possible copulas is enormous, but an extremely popular class is that of *Archimedian copulas*, defined by:

$$C(u_1, u_2, \ldots, u_n) = \phi^{-1}(\phi(u_1) + \phi(u_2) + \cdots + \phi(u_n)), \tag{10.163}$$

for some function $\phi : [0,1] \to R^+$. In the binary case ($n = 2$), this function must be convex, decreasing, and satisfies $\phi(1) = 0$ [115], and it can be shown that these conditions are both necessary and sufficient for $C_\phi(x, y)$ to be a valid CDF [260, Thm. 5.4.8]. Further, they are equivalent to the requirement that $1 - \phi^{-1}(x)$ be a unimodal density function on $[0, \infty)$ with mode at $x = 0$ [115]. Important examples of Archimedian bivariate copulas include the *independent copula:*

$$\phi(x) = -\ln x \;\Rightarrow\; C(x, y) = xy, \tag{10.164}$$

the *Gumbel copula* [115]:

$$\phi(x) = [-\ln x]^{\alpha+1} \;\Rightarrow\; C(x, y) = \exp\left[-\left\{[-\ln x]^{\alpha+1} + [-\ln y]^{\alpha+1}\right\}^{1/(\alpha+1)}\right], \tag{10.165}$$

and the *Clayton copula* [65, 68]:

$$\phi(x) = \frac{x^{-\alpha} - 1}{\alpha} \;\Rightarrow\; C(x, y) = (x^{-\alpha} + y^{-\alpha} - 1)^{-1/\alpha}. \tag{10.166}$$

In both of these last two examples, the parameter α may be viewed as a correlation measure, with $\alpha \to 0$ yielding the independent copula, and $\alpha \to \infty$ yielding the Frechet upper bound. In fact, Kendall's τ is determined completely by this parameter, given by:

$$\begin{aligned} \text{Clayton: } \tau &= \frac{\alpha}{\alpha + 2}, \\ \text{Gumbel: } \tau &= \frac{\alpha}{\alpha + 1}. \end{aligned} \tag{10.167}$$

The multivariate Clayton copula exhibits a number of particularly desirable characteristics. For example, the N-dimensional cumulative distribution function for this family is given by:

$$F_{\mathbf{x}}(q_1, \ldots, q_N) = \left[\sum_{i=1}^{N} q_i^{-1/\alpha} - (N - 1)\right]^{-\alpha}. \tag{10.168}$$

Further, associated with this CDF is the following simple density expression:

$$p(x_1, \ldots, x_N) = \frac{\Gamma(\alpha + N)}{\Gamma(\alpha)\alpha^N} \prod_{i=1}^{N} x_i^{(-1/\alpha)-1} \left[\sum_{i=1}^{N} x_i^{-1/\alpha} - (N - 1)\right]^{-(\alpha+N)}. \tag{10.169}$$

Note that since both the CDF and the density are permutation-invariant, this distribution is exchangeable; also, various results are available for correlations and conditional means [68], although they are generally somewhat complicated. In contrast, generation of random samples from this distribution is quite easy: if X has a gamma distribution with shape parameter α and unit scale parameter, and the variables Y_i are statistically independent gamma-distributed random variables with unit shape and scale parameters, then the N-dimensional random vector with the following components exhibits the distribution defined by Eqs. (10.168) and (10.169):

$$U_i = [1 + Y_i/X]^{-\alpha}, \text{ for } i = 1, 2, \ldots, N. \tag{10.170}$$

10.7.5 Kendall's τ

The Spearman rank correlation was introduced in Sec. 10.5, where it was noted that it is a measure of monotone association, in contrast to the product-moment correlation coefficient, which measures linear association. *Kendall's τ* is another monotone association measure, introduced by Kendall in 1938 [172]. For two univariate data sequences $\{x_k\}$ and $\{y_k\}$, Kendall's τ is an estimator of $E\{\text{sign}(x_i - x_j)\text{sign}(y_i - y_j)\}$, given by:

$$\tau = \frac{2}{N(N-1)} \sum_{i=1}^{N} \sum_{j<i} \text{sign}(x_i - x_j)\text{sign}(y_i - y_j). \tag{10.171}$$

Alternatively, τ can be represented in terms of the number C of *concordances* and the number D of *discordances* between pairs (x_i, y_i) and (x_j, y_j) of points in the plane. Here, if $x_i < x_j$ the points are said to be concordant if $y_i < y_j$, and discordant if $y_i > y_j$, and τ is given by:

$$\tau = \frac{2(C-D)}{N(N-1)} = \frac{4C}{N(N-1)}, \tag{10.172}$$

since $C + D = N(N-1)/2$ represents the total number of distinct point pairs to be compared. Note that this second definition assumes that there are no ties, corresponding to points where $x_i = x_j$ or $y_i = y_j$, or both. If there are ties, one recommended modification is Kendall's τ_b, given by [8, p. 68]:

$$\tau = \frac{C-D}{\sqrt{\left(\frac{N(N-1)}{2} - n_x\right)\left(\frac{N(N-1)}{2} - n_y\right)}}, \tag{10.173}$$

where n_x is the number of x-ties (i.e., the number of cases where $x_i = x_j$) and n_y is the number of y-ties. Note that in the absence of ties, $n_x = n_y = 0$ and Eq. (10.173) reduces to Eq. (10.172). Another possible modification of Kendall's *tau* in the case of ties is Goodman and Kruskal's γ, introduced in Sec. 10.8.1.

If $\{x_k\}$ and $\{y_k\}$ are statistically independent, the expected value of τ is zero and the variance is [129, p. 126]:

$$\text{var}\{\tau\} = \frac{2(2N+5)}{9N(N-1)}. \tag{10.174}$$

It has been suggested that the distribution of $\tau/\sqrt{\text{var}\{\tau\}}$ approaches the standard normal distribution faster than the corresponding normalized Spearman rank correlation coefficient, which is approximately $\rho_S\sqrt{N-1}$ [159, p. 114].

10.8 Relations among other variable types

All of the association measures and most of the joint distributions considered to this point have been defined for continuous-valued random variables. As emphasized in Chapter 2, however, many other data types arise in practice, so it is reasonable to ask how we can characterize associations between these other data types, or between variables of different data types. The following subsections consider these questions further.

10.8.1 Discrete, ordinal, and nominal variables

For discrete, integer-valued variables (e.g., count data) where distance-related quantities like averages, differences, and products are well defined, association measures like the product-moment correlation coefficient discussed in Sec. 10.4 for continuous variables remain useful. The same is true of the Spearman rank correlation coefficient introduced in Sec. 10.5 and Kendall's τ introduced in Sec. 10.7.5. An important practical complication with these latter two association measures, however, is that ties—which are a zero-probability event for continuous random variables—are quite likely with count data, and can cause these alternative association measures to behave poorly.

For categorical (i.e., ordinal or nominal) variables, quantities like averages, differences, and products are no longer well defined, so the product-moment correlation coefficient is not applicable. For ordinal variables, however, both ranks and concordances remain well defined, so Spearman's rank correlation coefficient described in Sec. 10.5 and Kendall's τ described in Sec. 10.7.5 can be used with ordinal sequences. Again, as in the case of count data, ties are to be expected with ordinal variables, and these need to be taken into account explicitly. A possible alternative, developed specifically for ordinal variables is Goodman and Kruskal's γ [8, p. 59], which reduces to Kendall's τ in the absence of ties. Define n_{xy} as the number of *double ties* in both x and y (i.e., cases where $x_i = x_j$ and $y_i = y_j$) and note that the total number of distinct data pairs (x_i, y_i) is $N(N-1)/2$, which may be partitioned into concordances, discordances, and ties as:

$$\frac{N(N-1)}{2} = C + D + n_x + n_y - n_{xy}, \tag{10.175}$$

where n_{xy} has been subtracted to correct for the double counting of the double ties in the sum $n_x + n_y$. Goodman and Kruskal's γ is defined as:

$$\gamma = \frac{C - D}{C + D} = \frac{C - D}{\frac{N(N-1)}{2} - n_x - n_y + n_{xy}}. \tag{10.176}$$

Like Kendall's τ_b introduced in Sec. 10.7.5, note that Goodman and Kruskal's γ reduces to Kendall's τ in the absence of ties, but it corrects for them differently than Kendall's τ_b.

The case of nominal variables is the most difficult and, as Agresti notes [8, p. 57], no entirely satisfactory solution exists even though a number of different nominal association measures have been proposed. To illustrate, suppose x is a nominal variable with I possible values and y is a nominal variable with J possible values. Now, consider sequences $\{x_k\}$ and $\{y_k\}$ of length N and let N_{ij} denote the number of times that x_k exhibits the ith possible x-value and y_k exhibits the jth possible y-value. Association measures between these sequences are typically defined in terms of the following three quantities:

$$\pi_{ij} = \frac{N_{ij}}{N}, \quad \pi_{i+} = \sum_{j=1}^{J} \pi_{ij}, \quad \pi_{+j} = \sum_{i=1}^{I} \pi_{ij}. \tag{10.177}$$

As a specific example, Goodman and Kruskal's τ [8, pp. 68–69] can be written as:

$$\tau = \frac{\sum_{j=1}^{J} \pi_{+j} \left(\sum_{i=1}^{I} \frac{\pi_{ij}^2}{\pi_{i+}\pi_{+j}} - \pi_{+j} \right)}{\sum_{j=1}^{J} \pi_{+j}(1 - \pi_{+j})}. \tag{10.178}$$

An important characteristic of this association measure is its asymmetry: reversing the roles of x and y in Eq. (10.178) generally gives a different numerical measure of association, a characteristic shared by a number of other nominal association measures [8, sec. 2.4.2]. In practice, the extent of this asymmetry can be pronounced, and it depends on the numbers of levels I and J of the variables involved.

10.8.2   Mixed data types

In measuring the association between two different variable types, note that one of the two variables will generally exhibit more exploitable mathematical structure than the other one. In particular, real variables are amenable to the widest range of mathematical characterizations since sums, differences, and products of two real variables are well defined, while nominal variables are amenable to the narrowest range of characterizations since none of these quantitative characterizations are meaningful. Ordinal variables represent an intermediate case since at least ranks and concordances remain well defined, as in the real-valued case and in contrast to the nominal case. In light of these observations, mixed variable associations are generally characterized using the full range of techniques available for the less flexible of the two data types.

As a specific example, consider the problem of measuring the association between a real variable and an ordinal variable. Since ranks and concordances are well defined for both variable types, these associations may be characterized using either rank-based methods like the Spearman rank correlation measure or concordance-based methods like Kendall's τ, or—in cases where there are ties in the ordinal variable—Kendall's τ_b or Goodman and Kruskal's γ. Similarly, associations between ordinal and nominal variables may be assessed by ignoring the order-derived characteristics of the ordinal variable and treating it as though it were nominal. This approach leads to the use of nominal association measures like Goodman and Kruskal's τ defined in Sec. 10.8.1.

The two approaches just described could be combined in the case of one real and one nominal variable, applying Goodman and Kruskal's τ to measure the association between the nominal variable and the ranks of the real variable, ignoring the order information. Much is lost in this process, however, and a far better alternative is to use the classical *analysis of variance (ANOVA)* methods described in Chapter 13 (Sec. 13.3). This approach provides both the basis for a formal test of the hypothesis that the two variables are independent and a quantitative model of the dependence between the variables in the case where the independence hypothesis is rejected.

10.8.3 The special case of binary variables

Binary data types may be viewed as the intersection of the classes of real, ordinal and nominal variables. Consequently, binary variables may be treated as members of whichever of these classes is convenient for purposes of analysis, recognizing that different representations have different strengths and weaknesses.

Treating binary variables as real may be useful in applications like variable selection for building regression models where the set of possible explanatory variables includes both continuous and binary data types. For example, if the response variable is continuous, ranking candidate predictor variables in terms of their product-moment correlations with the response variable can be a useful preliminary screening technique for deciding which variables to consider putting into the model, an idea discussed in detail in Chapter 15. Since product-moment correlations between pairs of binary variables are well defined—as are product-moment correlations between binary and real variables—this strategy provides a common basis for comparing the potential utility of all explanatory variables.

The treatment of binary variables as ordinal has no obvious advantage. In particular, note that binary variable ranks correspond to a simple linear rescaling of a $\{0, 1\}$-coded binary representation, so there is no advantage in considering rank-based association measures between binary variables relative to product-moment correlations. In the mixed case—i.e., association between real and binary variables—rank correlations may be useful in dealing with the effects of outliers in or transformations of the real variable. Conversely, care is required since rank-based methods don't always handle ties gracefully and binary variables exhibit *massive* ties (i.e., for N observations, *at least* $N/2$ observations must exhibit the same value).

Table 10.6: Association measures between a binary variable x and a second variable y of arbitrary type

y Type	Measure/method	Discussion
Real	One-way tests:	
	- t-test	Chapter 9 (Sec. 9.4.1)
	- Welch test	Chapter 9 (Sec. 9.4.4)
	- Yuen-Welch test	Chapter 9 (Sec. 9.4.4)
	- Wilcoxon rank-sum test	Chapter 9 (Sec. 9.4.3)
	- Welch rank test	Chapter 9 (Sec. 9.4.4)
Ordinal	Rank-based tests:	
	- Wilcoxon rank-sum test	Chapter 9 (Sec. 9.4.4)
	- Welch rank test	Chapter 9 (Sec. 9.4.4)
Nominal	χ^2 test	Chapter 9 (Sec. 9.5.2)
Binary	Odds ratio	Chapter 13 (Sec. 13.2.1)

Treating binary variables as nominal typically leads to a special case of a well-known characterization, especially in the mixed case where the other variable is not binary. In particular, the problem of measuring the association between a binary variable and a nonbinary variable can usually be converted to a binary hypothesis test. For example, if there is no association between a binary variable x taking the values 0 and 1 and a real variable y, it follows that the conditional distributions $p(y|x = 0)$ and $p(y|x = 1)$ should be the same. Thus, the problem of measuring the association between x and y may be recast as any of the real variable-based hypothesis tests listed in Table 10.6. Similarly, since ranks remain well-defined for ordinal variables, associations between binary and ordinal variables may be assessed using rank-based tests like the Wilcoxon rank-sum test or the Welch rank-based test listed in Table 10.6 for ordinal variables. In the case of nominal y, we can characterize $y|x$ as a multinomial random variable for each possible value of x and ask whether these two distributions are the same. As indicated in Table 10.6, the classical test for this hypothesis is the χ^2 test described in Chapter 9. Finally, note that specialized association measures have been developed for the case where x and y are both binary; see, e.g., the discussion in Agresti's book [8, sec. 2.2]. The binary association measure discussed in this book is the *odds ratio*, described in Chapter 13.

10.9   Exercises

1. Using the transformation defined by Eq. (10.53), show that $E\{\mathbf{x}\} = \mu$ for the multivariate normal random vector with the distribution $N(\mu, \Sigma)$.

2. Again using the transformation defined by Eq. (10.53), show that the covariance of **x** is given by Eq. (10.55):

$$\text{cov } \mathbf{x} \equiv E\{(\mathbf{x} - \mu)(\mathbf{x} - \mu)^T\} = \Sigma.$$

3. Once more using this transformation, show that if $\mathbf{x} \sim N(\mu, \Sigma)$, then $\mathbf{Ax} + \mathbf{b}$ is normally distributed with mean $\mathbf{A}\mu + \mathbf{b}$ and covariance $\mathbf{A}^T\Sigma\mathbf{A}$.

4. Consider the Laplace distribution with mean 0 and scale parameter ϕ, defined by the density:

$$p(x) = \frac{e^{-|x|/\phi}}{2\phi}.$$

Further, consider the truncated version of this distribution, restricted to the interval $-L \leq x \leq L$.

   a. Derive an expression for the truncated density $p(x| - L \leq x \leq L)$.

   b. Show that all odd moments of this distribution vanish.

   c. Derive an expression for the variance of this distribution, as a function of the truncation limit L.

   d. Derive an expression for the kurtosis $\kappa(L)$. Is this distribution platykurtic, leptokurtic, or mesokurtic?

5. For the bivariate Gaussian distribution defined by Eq. (10.65), show that $E\{(x - \bar{x})(y - \bar{y})\} = \rho$.

6. In what follows, suppose **P** is any $N \times N$ permutation matrix.

   a. Define $\mathbf{e} = [1, 1, \ldots, 1]^T \in R^N$ and show that $\mathbf{Pe} = \mathbf{e}$.

   b. Define **E** as the $N \times N$ matrix, all of whose elements are 1. Show that $\mathbf{PE} = \mathbf{E}$ and $\mathbf{EP}^T = \mathbf{E}$.

   c. Consider the N-dimensional random vector $\mathbf{x} \sim N(\mu, \Sigma)$ where:

$$\mu = [\bar{x}, \bar{x}, \ldots, \bar{x}]^T$$

$$\Sigma = \begin{bmatrix} \sigma^2 & \rho\sigma^2 & \cdots & \rho\sigma^2 \\ \rho\sigma^2 & \sigma^2 & \cdots & \rho\sigma^2 \\ \vdots & \vdots & \vdots & \vdots \\ \rho\sigma^2 & \rho\sigma^2 & \cdots & \sigma^2 \end{bmatrix}.$$

   Show that $\mathbf{Px} \sim N(\mu, \Sigma)$, implying that the components of **x** are *exchangeable*, even though they are not *independent* unless $\rho = 0$.

7. Let $R_x(i)$ denote the ranks for the N elements of a data sequence $\{x_k\}$.

   a. Show that the average of the sequence $\{R_x(i)\}$ is $(N + 1)/2$.

   b. Show that the variance of this sequence is $(N^2 - 1)/12$.

8. Show that the ranks $R_x(i)$ for a data sequence are invariant under arbitrary increasing transformations. That is, if $y_k = f(x_k)$ for any increasing function $f(\cdot)$, then $R_y(i) = R_x(i)$.

9. Consider the Spearman rank correlation coefficient ρ_S between two sequences $\{x_k\}$ and $\{y_k\}$ introduced in Sec. 10.5.

   – Show that $\rho_S = +1$ if and only if $y_k = f(x_k)$ for some increasing function $f(\cdot)$.

   – Show that $\rho_S = -1$ if and only if $y_k = f(x_k)$ for some decreasing function $f(\cdot)$.

   (Hint: remember that $R_x(i)$ and $R_y(i)$ must assume every value between 1 and N once and only once. Also, note Exercise 8.)

10. Show that the variance of the the mixture density defined by Eq. (10.105) is given by Eq. (10.110). *Hint: recall that* $var\{x\} = E\{x^2\} - [E\{x\}]^2$.

11. Show that the following expressions are equivalent, provided the coefficients π_j are nonnegative and sum to 1:

$$\sum_{j=1}^{p} \pi_j z_j \left[z_j - \sum_{k=1}^{p} \pi_k z_k \right] = \frac{1}{2} \sum_{j=1}^{p} \sum_{k=1}^{p} \pi_j \pi_k (z_j - z_k)^2 \geq 0.$$

Further, show that this lower bound is achievable if and only if $z_j = z_k$ for all $1 \leq j, k \leq p$.

12. Show that, if all component densities $f_j(x)$ in the mixture density defined by Eq. (10.109) are zero-mean and symmetric, the kurtosis $\kappa(x)$ for the mixture density is given by Eq. (10.112).

13. Show that, if all component densities $f_j(x)$ in the mixture density defined by Eq. (10.109) have the same mean, $\bar{x}_j = \bar{x}^0$, the central moments μ_n of the mixture density depend only on the central moments μ_j^n of the component densities.

14. Show that the kurtosis $\kappa(x)$ for the p-component zero-mean, Gaussian mixture density is given by the result presented in Eq. (10.121). Also, show that $\kappa(x) \geq 0$ with $\kappa(x) = 0$ if and only if $\sigma_1 = \sigma_2 = \cdots = \sigma_p$.

15. Derive Eq. (10.122) for the first three moments of the slippage model. Next, suppose $\mu = \lambda\sigma$ and derive a simplified expression for the skewness that does not depend on σ. Finally, derive an approximate expression for the skewness in the limit as $\lambda \to \infty$.

16. Suppose $r = x/y$ where the distribution of x is arbitrary and y is normally distributed. If x and y are independent, show that r does not have a finite second moment. (Hint: note that $E\{r^2\} = E\{x^2 y^{-2}\}$.)

17. Derive Eq. (10.128) for the variance of x in terms of its conditional mean and conditional variance:

$$\text{var}\{x\} = E\{\text{var}\{x|\theta\}\} + \text{var}\{E\{x|\theta\}\}.$$

Chapter 11

Regression Models I: Real Data

Sections and topics:

A data analysis problem that arises frequently is the following: we have a dataset with values for a response variable of interest and one or more predictor variables that may—or may not—be related to this response variable. To characterize this relationship, we can construct a *regression model* that predicts the values of the response variable from the corresponding values of the predictor variables. One of the simplest regression modeling problems is the line-fitting problem discussed at length in Chapter 5. This chapter extends the ideas presented there to a much wider class of *linear regression problems* that can involve transformations of a single predictor variable (e.g., polynomial models), multiple predictor variables, or both transformations and multiple predictors. The simplest solution approach described here is the method of *ordinary least squares (OLS)*, which is very popular but is not always suitable (e.g., it suffers from severe outlier sensitivity). Therefore, alternatives are also introduced, including *M-estimators* and *least trimmed squares (LTS)*. Questions of how to select specific variables for inclusion in these models, how to interpret them, and how to assess their quality are deferred until Chapter 15. Also, this chapter restricts consideration to real-valued variables; regression models involving categorical variables or counts are considered in Chapter 14.

Key points of this chapter:

1. Regression models arise whenever we wish to predict the value of one variable (the *dependent variable* or *response variable*) from one or more other variables (commonly called *independent variables, predictor variables,* or *covariates*).

2. This chapter is exclusively concerned with *linear regression models* that arise when the response variable depends *linearly* on the unknown model parameters to be estimated from the data. As examples presented in this chapter illustrate, this problem formulation is extremely general. In particular, these models can involve a wide variety of nonlinear transformations of the variables involved, so long as all of the unknown parameters enter the equations linearly.

3. Without question, the most popular approach to fitting linear regression models is the method of *ordinary least squares (OLS)*, introduced in Chapter 5 for the line fitting problem and described here in more general terms.

4. *Nonlinear regression* problems are those where the predicted response variable depends in a nonlinear fashion on the unknown model parameters, and these are generally more difficult to solve than linear regression problems are.

5. In some cases, nonlinear regression problems can be converted into linear regression problems using transformations like those discussed in Chapter 12. This idea forms the basis for the class of *generalized linear models* introduced in Chapter 14.

6. A simple extension of the OLS method that is sometimes extremely useful is the method of *weighted least squares (WLS)*, which forms the basis for the method of *iteratively reweighted least squares (IRWLS)* described here in connection with *M-estimators* and in Chapter 14 in connection with logistic regression.

7. Another simple extension of the OLS method is *restricted least squares*, where model parameters are required to satisfy known equality constraints.

8. A key limitation of the OLS method is its sensitivity to outliers, and this chapter introduces two alternatives with much better outlier resistance. One is the class of *M-estimators* that replace the squared error criterion on which the OLS method is based with another criterion that imposes a smaller penalty on large errors. The other alternative is the method of *trimmed least squares*, which replaces the *sum* of squared errors with a *trimmed mean* of squared errors, resulting in a method with very high outlier resistance, but also much higher computational complexity.

11.1   Building regression models

This chapter considers the regression modeling problem: given a sequence $\{y_k\}$ of *response variable measurements* and one or more sequences $\{x_k^i\}$ of *predictor variable measurements*, how do we construct mathematical models that relate the sequences $\{x_k^i\}$ to the sequence $\{y_k\}$? This general problem can be formulated and solved in many different ways, based on different working assumptions, but a particularly popular approach is that based on *linear regression methods*, defined in Sec. 11.1.1. The practical distinction between linear and nonlinear regression problems is also briefly discussed, and Sec. 11.1.2 considers a specific example that illustrates this difference. Sec. 11.1.3 then considers a second example where physical insights are incorporated into the formulation of the empirical modeling problem, an idea that is often applicable but seems to be seldom used in practice (indeed, too seldom used). A general treatment of the problem of deciding what predictor variables to include in a regression model is deferred until Chapter 15, where the problems of interpreting and refining regression models are discussed in detail.

11.1.1   Linear versus nonlinear regression

The most general problem considered in this chapter is the following extension of the line fitting problem considered in Chapter 5. Given a dataset $\mathcal{D}$ consisting of N observations of the response variable y and $r \geq 1$ explanatory variables x^i, construct a mathematical model of the following form:

$$g(y_k) \simeq \hat{g}(y_k) = F(x_k^1, x_k^2, \ldots, x_k^r), \qquad (11.1)$$

for $k = 1, 2, \ldots, N$. As in the line fitting problem, the approximate equality sign "$\simeq$" may be interpreted in different ways, leading to different solution methods. This chapter is primarily concerned with the most widely used interpretation of this symbol, which leads to the additive error model:

$$g(y_k) \simeq \hat{g}(y_k) \;\Rightarrow\; g(y_k) = \hat{g}(y_k) + e_k, \qquad (11.2)$$

where $\{e_k\}$ is a sequence of *prediction errors* or *modeling errors*. In this view, $\hat{g}(y_k)$ represents a *prediction model* for $g(y_k)$, computed from the predictor variable measurements x_k^i as indicated in Eq. (11.1). One of the simplest special cases of this model is the line-fitting problem, which may be represented in the form of Eq. (11.2) by taking $g(y) = y$, $r = 2$, x_k^1 equal to the single explanatory variable x_k in the original problem formulation, $x_k^2 = 1$ for all k, and $F(x_k^1, x_k^2) = ax_k^1 + bx_k^2$ (the reason for introducing the apparently spurious variable x_k^2 into this problem is explained in the next paragraph). In this problem, the objective is to determine values for the constants a and b that minimize some measure of the prediction error sequence $\{e_k\}$, as discussed at some length in Chapter 5. The problems considered in this chapter are more general, and the purpose of this introductory discussion is to clearly outline the range of specific model-building problems that fall within this framework, making them amenable to the solution procedures described here.

The class of *linear regression problems* considered in this chapter consists of those problems that may be expressed in the following form:

$$g(y_k) = \sum_{j=1}^{p} \alpha_j \phi_j(x_k^1, x_k^2, \ldots, x_k^r) + e_k, \qquad (11.3)$$

for $k = 1, 2, \ldots, N$. In this problem formulation, the functions $g(\cdot)$ and $\phi_j(\cdot)$ are *known* and the model parameters to be determined from the available dataset $\mathcal{D}$ are the p real numbers α_j. The practical advantages of this formulation are great enough to motivate the introduction of the apparently spurious second variable, $x_k^2 = 1$, into the line-fitting problem described in the previous paragraph, bringing it into the linear regression class. As discussed in Chapter 5, the errors-in-model (EM) working assumption is most commonly invoked, meaning that the predictor variables x_k^i are measured with negligible error and that the prediction errors e_k account for measurement errors in y_k, errors in the choices of the functions $\phi_j(\cdot)$ and other defects of the model defined in Eq. (11.3) like omitted variables. It is also possible to consider more general problem formulations but as in the case of the line-fitting problem, these formulations do lead to more complicated solution procedures; still, some of these alternatives (e.g., total least squares for errors-in-variables problems) are important and are discussed briefly in Sec. 11.5.

Most commonly, as in the line-fitting problem, the function $g(\cdot)$ is simply the identity $(g(y_k) = y_k)$, and the functions $\phi_j(\cdot)$ each involve a single argument. One important special case is that for which $p = r$ and $\phi_j(x_k^1, \ldots, x_k^r) = x_k^j$, leading to the following explicit model form:

$$y_k = \alpha_1 x_k^1 + \alpha_2 x_k^2 + \cdots + \alpha_r x_k^r + e_k. \qquad (11.4)$$

Another important special case involves a single explanatory variable x_k (i.e., $r = 1$) but different functions of this explanatory variable, leading to the linear regression model:

$$y_k = \sum_{j=1}^{p} \alpha_j \phi_j(x_k) + e_k. \qquad (11.5)$$

Again, it is important to emphasize that the functions $\phi_j(\cdot)$ are *known*; the monomial functions $\phi_j(x) = x^{j-1}$ represent a historically popular choice but the Reidel equation considered in Sec. 11.1.2 illustrates that other functions do often arise in practice.

Two other important points should be noted here. First, taking $g(y) = y$ is both the simplest and most popular choice in practice, but some problems lend themselves quite naturally to other choices. Examples illustrating this point are considered further in Chapter 14 in the context of *generalized linear models* that are particularly useful in connection with categorical or count response variables. Also, note that some regression problems can be *linearized* or transformed into an expression where the unknown parameters enter linearly as in Eq. (11.3) by the appropriate choice of nonlinear function $g(\cdot)$. This idea was discussed in detail in Chapter 12, along with some difficulties that can arise when applying this procedure. The second important point here is that the models considered in this chapter are *instantaneous*, meaning that there are no time delays or other dynamic effects appearing in these regression models. That is, if k represents a time index at which measurements are made, all of the models considered in this chapter relate the response variable y_k at time instant k to one or more explanatory variables x_k^i *at the same time instant k.* Dynamic effects are important in practice, but they introduce new complications and are not considered in this book.

Finally, it is useful to say something about the more complex problem of *nonlinear regression*, which deals with problems that cannot be cast in to the form of Eq. (11.3). A specific and important example is that considered by Varah [281], of fitting multiple exponential decays to a sequence of observed responses. This problem arises frequently in connection with the decay of chemical species, electrostatic charge, or other physical quantities as a function of time. Specifically, if $\{y_k\}$ is the sequence of responses, observed at times $t_k = t_1 + (k-1)T$, the multiple exponential decay model is of the general form:

$$ y_k = \sum_{j=1}^{p} \alpha_j \exp[-\beta_j t_k] + e_k, \tag{11.6} $$

and the objective of the analysis is to estimate the parameters α_j and β_j by fitting a model of this form to observed (t_k, y_k) data. If the exponents β_j are known, this problem reduces to a special case of Eq. (11.3), but in the more general case where these model parameters are not known, the linear regression procedures discussed in this chapter do not apply. In such cases, it is necessary to use more general—and more complicated—procedures like those discussed by Seber and Wild [262]. Even for simple cases involving only two or three exponentials, this problem can be extremely badly behaved, as noted by Varah [281]: data uncertainties as small as 0.1% can result in changes of the parameter estimates on the order of 30% or more, and in certain cases the least squares formulation exhibits *local minima,* with parameters far from the best-fit values giving results almost as good as the best-fit parameters do. Varah [281] concludes with the observation that, if the exponential parameters β_j are well separated, these difficulties may not arise, but it is difficult to determine whether this condition is satisfied or not from the observed data sequence.

11.1.2   An example: the Riedel equation

The following example provides a nice illustration of the kinds of nonlinear two-variable models that can be fit using the techniques described in this chapter. Specifically, the *Riedel equation* relates the reduced pressure of a gas $p_r = p/p_c$ to the reduced temperature $T_r = T/T_c$ via the expression [284]:

$$ \ln p_r = A - B/T_r + C \ln T_r + D T_r^6. \tag{11.7} $$

The critical temperature T_c and critical pressure p_c are characteristic, measurable properties of a given compound or mixture, and this equation is a generalization of the simpler *Clapeyron equation* [242, p. 182], to which it reduces when $C = D = 0$. The Clapeyron equation arises as the simplest solution of the Clausius-Clapeyron equation describing vapor/liquid equilibrium in pure fluids, but if this equation is solved under different working assumptions, the result is the Rankine or Kirchhoff vapor-pressure equation [242, p. 186], corresponding to Eq. (11.7) with $D = 0$ but $C \neq 0$. The last term in the Riedel equation is included to correct for deviations of these other equations at high temperatures, and it has been suggested that the exponent 6 is not essential and that other values may be reasonable alternatives [242, p. 186]. For example, one such alternative is the Riedel-Planck-Miller vapor-pressure equation, which replaces the exponent 6 in the Riedel equation with 3 [242, p. 190].

In practice, the constants A, B, C, and D appearing in the Riedel equation are generally *not* fit by regression analysis, but instead are first reduced to a smaller set of constants, which are then fit essentially error-free to carefully measured experimental values (e.g., the boiling point of the liquid at a specified pressure). The basis for this reduction is a constraint imposed on the function defined by Eq. (11.7) by either fundamental considerations or observations of a large set of empirical circumstances. As a specific example, Reid [242, p. 187] describes a reduction of the four constants appearing in the original Riedel equation to a single unknown parameter, which is then determined from boiling point data at a pressure of 1 atmosphere.

The primary point of this example is that mathematical models like the Reidel equation arise naturally in a variety of circumstances. A key feature of these equations is their linearity with respect to the unknown parameters to be estimated, despite the presence of significant nonlinear functions of the original data variables (e.g., the logarithm, reciprocal, and sixth power in the Riedel equation). So long as this linearity with respect to the unknown parameters holds, the regression procedures described in this chapter are applicable. Conversely, if we wish to estimate the exponent appearing in the Riedel equation, we must consider more complex nonlinear regression procedures like those discussed by Seber and Wild [262], as in the exponential decay example discussed at the end of Sec. 11.1.1.

11.1.3 A second example: dimensional analysis

One of the important practical problems in analyzing datasets, whether large or small, is that we often have certain physical insights about the data source and we would like to analyze the data in ways that incorporate this insight. Often, this idea is referred to as *gray box modeling*, to distinguish it from *black box modeling*, which treats the dataset as the responses of a "black box" whose internal details are completely unknown to us, and *white box modeling* (also called *fundamental modeling* or sometimes *glass box modeling*), which treats the system as a collection of interconnected components about which we have significant fundamental understanding. The objective of gray box modeling is to deal with intermediate cases, combining fundamental internal knowledge with observed response data. The key practical question is how to accomplish this objective, and a number of different methods have been proposed, including so-called *semiphysical modeling* that uses physical insights to select nonlinear combinations of variables to include in a model [267]. The following discussion gives a detailed illustration of how this idea can be used in the context of regression analysis.

Specifically, this example revisits the foam characterization dataset of Feng and Aldrich [102] considered in Chapter 11; this discussion is a condensed version of one presented elsewhere [227, sec. 4.5], based on a suggestion by J. Ulrich at the Institut für Automatik at ETH, Zürich. It was noted by Feng and Aldrich [102] that the small number emphasis (SNE) characterization varies approximately inversely with bubble size and the instability measure (INSTAB) varies approximately inversely with the foam decay time. One way of incorporating these interpretations into our analysis is to consider relations between these variables and the measured fluid properties that are *dimensionally consistent*, assuming SNE to have units of m^{-1} and INSTAB to have units of s^{-1} and selecting combinations of variables in the regression model so the unknown constants are *dimensionless*. In this example, Feng and Aldrich [102] considered four fluid properties: surface tension σ (in units of N/m = kg s^{-2}), viscosity μ (in units of Pa s = kg m^{-1} s^{-1}), density ρ (in uits of kg m^{-3}), and ionicity η (a dimensionless quantity). Because only 4 of the 21 ionicity values were nonzero, it is not included in the model considered here, which has the form [227, Eq. (7)]:

$$\text{INSTAB} = a_0 f_0(\sigma, \mu, \rho) + a_1 f_1(\sigma, \mu, \rho)\text{SNE} + a_2 f_2(\sigma, \mu, \rho)\text{SNE}^2. \quad (11.8)$$

The functions $f_0(\cdot)$, $f_1(\cdot)$, and $f_2(\cdot)$ appearing in this model are the simplest combinations of the three fluid property variables that make Eq. (11.8) dimensionally consistent. That is, the function $f_0(\sigma, \mu, \rho)$ is chosen to have the same units as INSTAB (s^{-1}), $f_1(\sigma, \mu, \rho)$ has units of m s^{-1} to match the units of INSTAB and SNE, and $f_2(\sigma, \mu, \rho)$ has units of m^2 s^{-1} to match the units of IN-STAB and the quadratic term in SNE. These combinations are given explicitly by the following expressions:

$$f_0(\sigma, \mu, \rho) = \frac{\sigma^2 \rho}{\mu^3}, \quad f_1(\sigma, \mu, \rho) = \frac{\sigma}{\mu}, \quad f_2(\sigma, \mu, \rho) = \frac{\mu}{\rho}. \quad (11.9)$$

The point of this example is that Eq. (11.8) represents a model whose parameters a_0, a_1, and a_2 can be fit to observed data using the methods described in this chapter, but the terms included in this model have been chosen on the basis of physical understanding of the variables involved, an idea that is very broadly applicable.

11.2   Ordinary least squares (OLS)

The basic idea behind the ordinary least squares (OLS) solution is exactly the same in the general regression problem as in the straight line-fitting problem discussed in Chapter 5. That is, given a set $\{y_k\}$ of N observed responses and a corresponding set $\{x_k^1, \ldots, x_k^r\}$ of experimental condition variables, we seek the parameters α_j that minimize the following measure of mismatch between the assumed model defined by Eq. (11.3) and the available data:

$$J_{OLS} = \sum_{k=1}^{N} e_k^2 = \sum_{k=1}^{N} \left[y_k - \sum_{j=1}^{p} \alpha_j \phi_j(x_k^1, \ldots, x_k^r) \right]^2. \tag{11.10}$$

As in the line-fitting problem, one of the overwhelming advantages of the general OLS problem formulation is that J_{OLS} may be minimized by setting $\partial J_{OLS}/\partial \alpha_j$ to zero for $j = 1, 2, \ldots, p$, yielding a set of simultaneous linear equations. Although some care is required in computing the solution of these equations, the procedure is extremely standard and extensively supported by commercial software. Consequently, this chapter is not concerned with the numerical aspects of solving least squares problems, but rather with the formulation of least squares problems that can then be solved using these widely available procedures. Outlier-resistant extensions of the OLS solution are considered in later sections of this chapter, and Chapter 15 introduces some deletion diagnostics based on the structure of the OLS solution that can be useful in identifying unusually influential observations (e.g., possible outliers). The foundation for all of these results is the vector formulation of the OLS problem, described next.

To develop this formulation, first define the following vectors and matrices:

$$\mathbf{y} = [y_1, y_2, \ldots, y_N]^T \in R^N,$$

$$\theta = [\alpha_1, \alpha_2, \ldots, \alpha_p] \in R^p,$$

$$\mathbf{X} = \begin{bmatrix} \phi_1(\mathbf{x}_1) & \phi_2(\mathbf{x}_1) & \cdots & \phi_p(\mathbf{x}_1) \\ \phi_1(\mathbf{x}_2) & \phi_2(\mathbf{x}_2) & \cdots & \phi_p(\mathbf{x}_2) \\ \vdots & \vdots & \vdots & \vdots \\ \phi_1(\mathbf{x}_N) & \phi_2(\mathbf{x}_N) & \cdots & \phi_p(\mathbf{x}_N) \end{bmatrix} \in R^{N \times p},$$

$$\mathbf{e} = [e_1, e_2, \ldots, e_N]^T \in R^N, \tag{11.11}$$

where $\mathbf{x}_k = [x_k^1, x_k^2, \ldots, x_k^r]$. In the traditional interpretation, $\mathbf{y}$ represents the vector of observed responses, the matrix $\mathbf{X}$ defines the conditions under which they are observed, θ represents the vector of unknown parameters to be estimated from the data, and $\mathbf{e}$ is the vector of prediction errors to be minimized by choosing θ appropriately. Note that for any choice of parameters, these quantities are all related via the equation:

$$\mathbf{y} = \mathbf{X}\theta + \mathbf{e}. \tag{11.12}$$

The OLS mismatch criterion may now be expressed in terms of these vectors and matrices as:

$$\begin{aligned} J_{OLS} = \mathbf{e}^T \mathbf{e} &= [\mathbf{y} - \mathbf{X}\theta]^T [\mathbf{y} - \mathbf{X}\theta] \\ &= \mathbf{y}^T \mathbf{y} - \theta^T \mathbf{X}^T \mathbf{y} - \mathbf{y}^T \mathbf{X}\theta + \theta^T \mathbf{X}^T \mathbf{X}\theta. \end{aligned} \tag{11.13}$$

This result expresses J_{OLS} as a smooth function of the unknown parameter vector θ so the optimal value of θ may be obtained by computing the gradient $\nabla_\theta J_{OLS}$ and setting it to zero:

$$\nabla_\theta J_{OLS} = -2\mathbf{X}^T\mathbf{y} + 2\mathbf{X}^T\mathbf{X}\theta = \mathbf{0}. \tag{11.14}$$

Provided the $p \times p$ matrix $\mathbf{X}^T\mathbf{X}$ is invertible, this equation may be solved for the following explicit representation of the OLS solution:

$$\hat{\theta} = (\mathbf{X}^T\mathbf{X})^{-1}\mathbf{X}^T\mathbf{y}. \tag{11.15}$$

It is worth emphasizing that, although this expression gives an algebraically correct representation for the least squares estimate $\hat{\theta}$, it should *not* be taken directly as an algorithmic prescription for computing $\hat{\theta}$. In practice, factorizations of the matrix $\mathbf{X}$ are used to improve the numerical conditioning of this problem, and the matrix $\mathbf{X}^T\mathbf{X}$ is not actually formed. For details, refer to Golub and van Loan [118, ch. 5].

If we adopt the traditional random variable model for the data on which the OLS estimates are based, it follows that the parameter estimate $\hat{\theta}$ is a random quantity whose distribution is determined by that of the data. Most commonly, the errors-in-model formulation introduced in Chapter 5 is adopted, where it is assumed that the prediction error sequence $\{e_k\}$ is *zero-mean*, corresponding to an assumption that there are no *systematic errors* in the observed y_k values. Under these assumptions, we can easily show that $\hat{\theta}$ is an unbiased estimator of the true parameter vector θ, as follows. First, substitute Eq. (11.12) into Eq. (11.15) to obtain:

$$\hat{\theta} = \theta + \mathbf{X}(\mathbf{X}^T\mathbf{X})^{-1}\mathbf{X}^T\mathbf{e}. \tag{11.16}$$

Since $\{e_k\}$ is assumed zero mean for all k, it follows that $E\{\mathbf{e}\} = \{\mathbf{0}\}$, and taking the expectation of Eq. (11.16), we obtain $E\{\hat{\theta}\} = \theta$.

If Σ is the $N \times N$ covariance matrix for the error vector $\mathbf{e}$, it is straightforward to compute the $p \times p$ covariance matrix for the estimate $\hat{\theta}$:

$$\begin{aligned}
\text{Cov}\{\hat{\theta}\} &= E\{(\hat{\theta} - E\{\hat{\theta}\})(\hat{\theta} - E\{\hat{\theta}\})^T\} \\
&= E\{(\mathbf{X}^T\mathbf{X})^{-1}\mathbf{X}^T\mathbf{e}\mathbf{e}^T\mathbf{X}(\mathbf{X}^T\mathbf{X})^{-1}\} \\
&= (\mathbf{X}^T\mathbf{X})^{-1}\mathbf{X}^T\Sigma\mathbf{X}(\mathbf{X}^T\mathbf{X})^{-1}.
\end{aligned} \tag{11.17}$$

Quite commonly, $\{e_k\}$ is assumed to be an independent, identically distributed sequence, implying that $\Sigma = \sigma^2\mathbf{I}$ where σ^2 is the error variance and $\mathbf{I}$ denotes the $N \times N$ identity matrix. This assumption simplifies Eq. (11.17) to:

$$\text{Cov}\{\hat{\theta}\} = \sigma^2(\mathbf{X}^T\mathbf{X})^{-1} \tag{11.18}$$

This result is extremely useful because it says that, under the simplest working assumptions, the variability of the OLS estimate $\hat{\theta}$ depends on two basic things: the magnitude of the inherent variability in the y_k observations, characterized by the error variance σ^2, and the tendency of the matrix $(\mathbf{X}^T\mathbf{X})^{-1}$ to magnify or reduce these errors. Traditionally, the variables that define the $\mathbf{X}$ matrix are experimental conditions that are at least partially under our control, and this observation leads to the idea of *designed experiments*, one of the fundamental notions of confirmatory data analysis: the basic idea is to choose the experimental conditions $\mathbf{X}$ in a way that minimizes the tendency of $(\mathbf{X}^T\mathbf{X})^{-1}$ to amplify data errors. For an introduction to these ideas, refer to the book by Jobson [159, ch. 5]. The key point is that if $\mathbf{X}^T\mathbf{X}$ is *nearly singular*, the required inverse of this matrix can be *extremely* sensitive to small changes in the observed data.

A useful measure of nearness to singularity for a matrix $\mathbf{M}$ is the *condition number* $\kappa(\mathbf{M})$, defined as [118, p. 80]:

$$\kappa(\mathbf{M}) = ||\mathbf{M}|| \cdot ||\mathbf{M}^{-1}||, \tag{11.19}$$

where $|| \cdot ||$ denotes any matrix norm. Most commonly, the 2-norm is used, in which case the condition number reduces to the ratio of the largest to the smallest singular values of $\mathbf{M}$ [118, p. 80]. This ratio is therefore always at least 1, and it approaches $+\infty$ as the matrix becomes singular, since the smallest singular value of the matrix approaches zero in that case. It has been shown that the sensitivity of the OLS solution to parameter variations is roughly proportional to the following quantity [118, p. 230]:

$$\lambda = \kappa(\mathbf{X}) + \mathbf{e}^T\mathbf{e}\kappa(\mathbf{X}^T\mathbf{X}). \tag{11.20}$$

This result means that if the condition number of the $\mathbf{X}$ matrix is large, the parameter estimates will be highly sensitive to small changes in the observed data. As a rough rule of thumb, Jobson [159, p. 281] suggests that condition numbers of 100 or less for $\mathbf{X}^T\mathbf{X}$ do not lead to sensitivity problems, but condition numbers larger than $1,000$ can lead to severe problems.

The following two examples illustrate this point. The first is based on the experimental investigation of physical properties of two-phase foams described by Feng and Aldrich [102] and discussed further in Sec. 11.1.3. As noted in those discussions, the variables involved include two foam characteristics (INSTAB and SNE) and four fluid properties (surface tension, viscosity, density, and ionicity). In their study, Feng and Aldrich [102] consider linear regression models relating these foam characteristics to each of these four explanatory variables. For $N = 21$ samples, the $\mathbf{X}$ matrix has dimensions 21×4 and the condition number of the $\mathbf{X}^T\mathbf{X}$ matrix is approximately 7.5×10^7. This condition number is *huge*, indicating severe numerical ill-conditioning and illustrating why the $\mathbf{X}^T\mathbf{X}$ matrix should *not* be directly constructed and inverted from the available data; instead, a numerically well-conditioned decomposition procedure like the QR decomposition should be employed [118, sec. 5.3], as noted previously. Even using the QR decomposition, this example is extremely sensitive to small changes in the data. Specifically, the OLS regression model obtained from the unperturbed data of Feng and Aldrich [102] relating the INSTAB measurements to the four experimental condition variables is:

$$INSTAB \simeq 0.000174x_1 - 0.000249x_2 + 0.000286x_3 + 0.0238x_4. \tag{11.21}$$

For comparison, a 1.5% change in one of the 21 observed INSTAB values results in the following perturbed model:

$$INSTAB \simeq 0.000180x_1 - 0.000249x_2 + 0.000285x_3 + 0.0166x_4. \qquad (11.22)$$

In other words, a 1.5% change in one of 21 data points results in a $\sim$30% change in the magnitude of the largest regression coefficient. The extreme sensitivities of regression results in problems like this have led to the development of *ridge regression* procedures that deliberately introduce estimator bias to improve the conditioning of the $\mathbf{X}^T\mathbf{X}$ matrix, trading increased bias for reduced variability [159, p. 285].

While this example has illustrated that numerical ill-conditioning can be a problem, the next example illustrates that these difficulties can occur *regularly* in what appear to be simple problem settings. Specifically, consider the important problem of polynomial regression:

$$\phi_j(x) = x^{j-1} \quad \Rightarrow \quad y_k = \sum_{j=1}^{p} \alpha_j x_k^{j-1} + e_k. \qquad (11.23)$$

Also, assume for simplicity that the observations x_k are uniformly spaced on the unit interval $[0, 1]$ as $x_k = k/N$ for $k = 1, 2, \ldots, N$, although it is important to emphasize that the general conclusions drawn from this example remain valid for points x_k spaced uniformly over *any* finite range. The elements of the resulting $\mathbf{X}^T\mathbf{X}$ matrix are easily computed for this example:

$$X_{ki} = \left(\frac{k}{N}\right)^{i-1} \quad \Rightarrow \quad [\mathbf{X}^T\mathbf{X}]_{ij} = \sum_{k=1}^{N} \left(\frac{k}{N}\right)^{i+j-2}. \qquad (11.24)$$

For large N, the sum appearing in this expression can be approximated by an integral, giving:

$$\sum_{k=1}^{N} \left(\frac{k}{N}\right)^{i+j-2} \simeq \int_0^1 x^{i+j-2}dx = \frac{1}{i+j-1}. \qquad (11.25)$$

The $p \times p$ matrix whose i, j entries are $1/(i+j-1)$ is called the *Hilbert matrix* and is denoted H_p. A characteristic feature of this matrix is its horrible ill-conditioning as $p \to \infty$: $\kappa(H_3) \sim 5 \times 10^2$, $\kappa(H_6) \sim 1.5 \times 10^7$, and $\kappa(H_8) \sim 1.5 \times 10^{10}$ [146, p. 341]. Again, the key point of this example is to emphasize that ill-conditioning of the $\mathbf{X}^T\mathbf{X}$ matrix can arise even in fairly simple cases where it comes as an unpleasant surprise. Conversely, this particular problem can be overcome through the use of *orthogonal polynomials*, which leads to a diagonal $\mathbf{X}^T\mathbf{X}$ matrix [95, ch. 22].

11.3   Two simple OLS extensions

As many discussions in this chapter emphasize, the OLS solution has several important practical advantages and a comparable number of important disadvantages. Hence, various extensions have been developed that attempt to address or compensate for the limitations while retaining as many of the advantages as possible. The following discussions briefly describe two of these extensions: weighted least squares (Sec. 11.3.1) and restricted least squares (Sec. 11.3.2).

11.3.1   Weighted least squares

The *weighted least squares* (WLS) problem represents a rather minor but extremely useful modification of the OLS formulation just described. The underlying motivation is that the OLS mismatch criterion J_{OLS} weights all of the individual prediction errors e_k equally, and there are circumstances under which we want to weight some errors more heavily than others. One important application of this idea is the iteratively reweighted least squares algorithm discussed in Sec. 11.4.2. The basic idea behind the WLS formulation is to replace the OLS mismatch criterion J_{OLS} defined in Eq. (11.10) with the modified criterion:

$$J_{WLS} = \sum_{k=1}^{N} w_k \left[y_k - \sum_{j=1}^{p} \alpha_j \phi_j(\mathbf{x}_k) \right]^2 \equiv \sum_{k=1}^{N} w_k e_k^2, \qquad (11.26)$$

where $\{w_k\}$ is a sequence of N nonnegative weights. Like the OLS mismatch criterion, it is also possible to express J_{WLS} in vector representation by defining the $N \times N$ diagonal matrix:

$$\mathbf{W} = \mathrm{diag}\{w_1, w_2, \ldots, w_N\} = \begin{bmatrix} w_1 & 0 & \cdots & 0 \\ 0 & w_2 & \cdots & 0 \\ \vdots & \vdots & \vdots & \vdots \\ 0 & 0 & \cdots & w_N \end{bmatrix}. \qquad (11.27)$$

In terms of this matrix, the WLS mismatch criterion becomes:

$$\begin{aligned} J_{WLS} &= \mathbf{e}^T \mathbf{W} \mathbf{e} \\ &= \mathbf{y}^T \mathbf{W} \mathbf{y} - \theta^T \mathbf{X}^T \mathbf{W} \mathbf{y} - \mathbf{y}^T \mathbf{W} \mathbf{X} \theta + \theta^T \mathbf{X}^T \mathbf{W} \mathbf{X} \mathbf{y}. \end{aligned} \qquad (11.28)$$

As in the OLS problem, note that J_{WLS} is a smooth function of the unknown parameter vector θ, so it may be minimized by setting its gradient with respect to θ equal to zero:

$$\nabla_\theta J_{WLS} = -2\mathbf{X}^T \mathbf{W} \mathbf{y} + 2\mathbf{X}^T \mathbf{W} \mathbf{X} \theta = \mathbf{0}. \qquad (11.29)$$

Solving this equation then yields the analogous expression for the WLS solution:

$$\tilde{\theta} = (\mathbf{X}^T \mathbf{W} \mathbf{X})^{-1} \mathbf{X}^T \mathbf{W} \mathbf{y}. \qquad (11.30)$$

This solution has a number of interesting properties, two of which are particularly worth noting here. First, observe that if $\mathbf{W}$ is scaled by any positive constant λ, the solution $\tilde{\theta}$ is unchanged, since this scaling multiplies $\mathbf{X}^T \mathbf{W} \mathbf{y}$ by λ but multiplies $(\mathbf{X}^T \mathbf{W} \mathbf{X})^{-1}$ by $1/\lambda$, leaving $\tilde{\theta}$ invariant. Consequently, we may choose any convenient normalization or range for these weights; in particular, there is no loss of generality in taking any of the following conventions:

$$\sum_{k=1}^{N} w_k = 1 \quad \text{or} \quad \sum_{k=1}^{N} w_k = N \quad \text{or} \quad 0 \le w_k \le 1 \text{ for all } k. \tag{11.31}$$

The second point is that if $\mathbf{W} = \mathbf{I}$, we recover the OLS solution; hence, by the arguments just presented, this same situation holds if we take $w_k = c$ for all i where c is any positive constant.

11.3.2   Restricted least squares

The *restricted least squares problem* is a variation of the OLS problem that results when we impose a collection of *equality constraints* on the parameter vector θ [95, p. 229]. Specifically, this problem seeks to minimize the mismatch measure J_{OLS} defined in Eq. (11.10), subject to a set of q constraints, which may be written as:

$$\mathbf{C}\theta = \mathbf{h}, \tag{11.32}$$

where $\mathbf{h}$ is a vector of dimension q and $\mathbf{C}$ is a matrix of dimension $q \times p$ where, as before, p is the dimension of the parameter vector θ. The basic solution of this problem is obtained by the method of Lagrange multipliers [104, p. 51], minimizing the modified mismatch criterion:

$$J_{RLS} = \mathbf{e}^T \mathbf{e} + \mathbf{g}^T [\mathbf{h} - \mathbf{C}\theta]. \tag{11.33}$$

The solution is given by the following slightly messy expression:

$$\theta_R = \hat{\theta} + (\mathbf{X}^T\mathbf{X})^{-1}\mathbf{C}^T[\mathbf{C}(\mathbf{X}^T\mathbf{X})^{-1}\mathbf{C}^T]^{-1}(\mathbf{h} - \mathbf{C}\hat{\theta}), \tag{11.34}$$

where $\hat{\theta}$ is the OLS solution defined by Eq. (11.15). Several points are worth noting about this solution. First, the existence of this solution requires the nonsingularity of the matrix $\mathbf{C}(\mathbf{X}^T\mathbf{X})^{-1}\mathbf{C}^T$, which imposes a compatability requirement on the data matrix $\mathbf{X}$ and the constraint matrix $\mathbf{C}$. Next, note that if this condition is satisfied, then it follows from Eq. (11.34) that $\mathbf{C}\theta_R = \mathbf{h}$, which is exactly what we want: the solution of the constrained problem indeed satisfies the constraint. Further, the explicit representation given here illustrates that the restricted solution is unique (again, assuming the required inverses exist). Finally, note that the complicated second term in this expression represents the perturbation that must be applied to the unrestricted least squares solution $\hat{\theta}$ to obtain the restricted least squares solution. In particular, note that this perturbation is a linear function of the difference $\mathbf{h} - \mathbf{C}\hat{\theta}$.

The restricted WLS solution may be obtained through a similar optimization procedure. There, the end result may be obtained by replacing the OLS solution $\hat{\theta}$ with the WLS solution $\tilde{\theta}$ and replacing $\mathbf{X}^T\mathbf{X}$ with $\mathbf{X}^T\mathbf{W}\mathbf{X}$ in Eq. (11.34). Specifically, the restricted WLS result becomes:

$$\theta_{RW} = \tilde{\theta} + (\mathbf{X}^T\mathbf{W}\mathbf{X})^{-1}\mathbf{C}^T[\mathbf{C}(\mathbf{X}^T\mathbf{W}\mathbf{X})^{-1}\mathbf{C}^T]^{-1}(\mathbf{h} - \mathbf{C}\tilde{\theta}). \tag{11.35}$$

It is not difficult to show (see Exercise 1) that θ_{RW} also satisfies the constraint $\mathbf{C}\theta_{RW} = \mathbf{h}$ and is invariant under arbitrary positive scaling of all weights, again provided the required inverses exist.

11.4   M-estimators and robust regression

One of the important disadvantages of OLS methods is their great sensitivity to outliers, and a number of less sensitive alternatives have been developed to address this problem. One general class of methods that is both historically important and practically useful in some circumstances is the class of *M-estimators*, introduced in Sec. 11.4.1 below. In addition to providing improved robustness, these estimators have the advantage that they can often be implemented using *iteratively reweighted least squares* (IRWLS), which builds nicely on standard routines for solving weighted least squares problems; Sec. 11.4.2 briefly describes this basic approach, which is also useful in the logistic regression problem discussed in Chapter 14. For a more complete discussion of the details of iteratively reweighted least squares, refer to the book by Huber [151], which also discusses the link between M-estimators and maximum likelihood estimation, from which the class gets its name.

11.4.1   Basic notions of M-estimators

The severe outlier sensitivity of least squares procedures stems from the fact that the goodness-of-fit criterion J_{OLS} defined in Eq. (11.13) penalizes large errors more than small errors, typically leading to biased parameter estimates when outliers are present in the dataset. One way of overcoming this problem is to consider alternative fitting criteria like least absolute deviations, which penalize the sum of the absolute values of the errors rather than their squares. This change leads to improved outlier resistance, but as shown in Chapter 5, it also leads to a more difficult computational problem whose solution is not unique, in general, and that does not have an analytically convenient closed form solution like OLS does. The class of M-estimators is obtained by replacing the sum of squared errors defining J_{OLS} with an alternative error measure of the form:

$$J_M(\theta) = \sum_{k=1}^{N} \rho(e_k) = \sum_{k=1}^{N} \rho\left(y_k - \sum_{j=1}^{p} \phi_j(\mathbf{x}_k)\right), \qquad (11.36)$$

where $\rho(x)$ is a nonnegative function of x. A particularly important choice is Huber's function:

$$\rho(x) = \begin{cases} x^2/2 & |x| \le c, \\ c|x| - c^2/2 & |x| > c, \end{cases} \qquad (11.37)$$

where c is a positive tuning parameter that determines the outlier sensitivity of the resulting estimation procedure. In particular, note that as $c \to \infty$, we recover the OLS estimator, and as $c \to 0$, we recover the LAD estimator. Plots of this function for $c = 0.5$, 1.0, and 2.0 are shown in Fig. 11.1, illustrating

Figure 11.1: Huber's $\rho(x)$ function

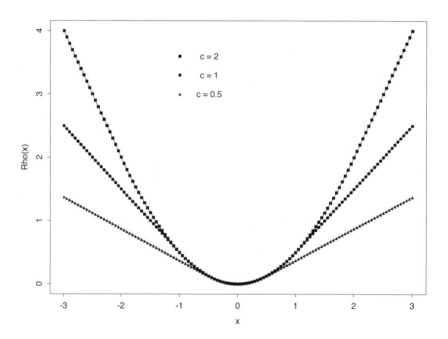

that for $|x| < c$, this function is quadratic like the OLS mismatch criterion, becoming linear like the LAD mismatch criterion for $|x| > c$. Since c determines what size errors are penalized quadratically (i.e., treated nominally) and what size errors are penalized linearly (i.e., treated as outliers), c is typically made proportional to some measure of the scale of the nominal data. Because of its general insensitivity to outliers, the MADM scale estimate S discussed in Chapter 7 is a popular choice. For example, the default value in the *S-Plus* statistics package is $c = 1.345S$ [272]. It is also important to note that many other functions $\rho(x)$ can be used in constructing M-estimators, but a detailed treatment of this topic is beyond the scope of this book; for a much more complete treatment, refer to Huber [151]. The key point here is that if $\rho(x)$ is differentiable, it is easy to derive necessary conditions for θ to minimize $J_M(\theta)$ that may be recast into a form that looks very much like a weighted least squares problem. In fact, the weights in this problem are data-dependent, meaning that the algorithm for computing the optimal parameter vector $\hat{\theta}$ is an iterative one. This procedure is called *iteratively reweighted least squares*, and it is described in the next section.

11.4.2   Mechanics of M-estimators

Defining $\psi(x)$ as the first derivative of $\rho(x)$ and differentiating Eq. (11.36) with respect to α_i leads to the following equation for the M-estimator $\hat{\theta}$ for the vector of model parameters α_j:

$$\frac{\partial J_M(\theta)}{\partial \alpha_i} = \sum_{k=1}^{N} \psi(y_k - \mathbf{z}_k^T \theta)\phi_i(\mathbf{x}_k) = 0, \qquad (11.38)$$

where $\mathbf{z}_k^T$ is the kth row of the $N \times p$ matrix $\mathbf{X}$ defined in Eq. (11.11); that is:

$$\mathbf{X} = \begin{bmatrix} \mathbf{z}_1^T \\ \mathbf{z}_2^T \\ \vdots \\ \mathbf{z}_N^T \end{bmatrix} \Rightarrow \mathbf{z}_k^T = [\phi_1(\mathbf{x}_k), \phi_2(\mathbf{x}_k), \ldots, \phi_p(\mathbf{x}_k)]. \qquad (11.39)$$

To solve for $\hat{\theta}$, first define the residuals r_k and the weights w_k as:

$$r_k = y_k - \mathbf{z}_k^T \theta, \quad w_k = \frac{\psi(r_k)}{r_k}. \qquad (11.40)$$

Eq. (11.38) can be rewritten in terms of these quantities as:

$$\sum_{k=1}^{N} w_k r_k \phi_i(\mathbf{x}_k) = 0, \ i = 1, 2, \ldots, p. \qquad (11.41)$$

Defining $\mathbf{W}$ as the $N \times N$ diagonal matrices whose k, k element is the weight w_k, these equations may be expressed in matrix notation as (Exercise 2):

$$\mathbf{X}^T \mathbf{W} \mathbf{r} = 0, \qquad (11.42)$$

where $\mathbf{r} = [r_1, r_2, \ldots, r_N]^T = \mathbf{y} - \mathbf{X}\theta$. Hence, we have the following result:

$$\mathbf{X}^T \mathbf{W} \mathbf{y} - \mathbf{X}^T \mathbf{W} \mathbf{X}\theta = 0 \Rightarrow \hat{\theta} = (\mathbf{X}^T \mathbf{W} \mathbf{X})^{-1} \mathbf{X}^T \mathbf{W} \mathbf{y}. \qquad (11.43)$$

As written, this result appears to give an explicit solution for the M-estimation problem, identical to the WLS solution; the critical difference in this case is that the weights are data-dependent. Specifically, the elements of the weighting matrix $\mathbf{W}$ are defined by Eq. (11.40), depending on the data through the residuals r_k. The advantage of this formulation is that it leads to the *iteratively reweighted least squares* (IRWLS) procedure, defined as follows. Given an initial estimate θ^j at step j, we first compute the residuals and weights at step j as:

$$\begin{aligned} \mathbf{r}^j &= \mathbf{y} - \mathbf{X}\theta^j = [r_1^j, r_2^j, \ldots, r_N^j]^T \\ \mathbf{W}^j &= \text{diag}\left[\frac{\psi(r_1^j)}{r_1^j}, \frac{\psi(r_2^j)}{r_2^j}, \ldots, \frac{\psi(r_N^j)}{r_N^j}\right]. \end{aligned} \qquad (11.44)$$

As an important example, for Huber's $\rho(x)$ function, the derivative is:

$$\psi(x) = \begin{cases} -c & x < -c, \\ x & |x| \leq c, \\ c & x > c, \end{cases} \tag{11.45}$$

from which it follows that the weights w_k are given by:

$$w_k = \begin{cases} -c/r_k & r_k < -c \\ 1 & |r_k| \leq c \\ c/r_k & x > c \end{cases} = \begin{cases} |c/r_k| & |r_k| > c, \\ 1 & |r_k| \leq c. \end{cases} \tag{11.46}$$

Given $\mathbf{r}^j$ and $\mathbf{W}^j$, the updated solution θ^{j+1} is given by:

$$\begin{aligned} \theta^{j+1} &= (\mathbf{X}^T\mathbf{W}^j\mathbf{X})^{-1}\mathbf{X}^T\mathbf{W}^j\mathbf{y} \\ &= (\mathbf{X}^T\mathbf{W}^j\mathbf{X})^{-1}\mathbf{X}^T\mathbf{W}^j[\mathbf{X}\theta^j + \mathbf{r}^j] \\ &= \theta^j + (\mathbf{X}^T\mathbf{W}^j\mathbf{X})^{-1}\mathbf{X}^T\mathbf{W}^j\mathbf{r}^j. \end{aligned} \tag{11.47}$$

For the Huber $\rho(x)$ function, points whose residuals r_k at the previous step in the iteration are smaller than c in magnitude are fully weighted in the next parameter estimation step, whereas those points whose residuals exceed this threshold value are downweighted by an amount that depends on the extent to which they exceed this threshold. Other choices of weighting functions $\rho(x)$ tend to downweight large values even further, although they can exhibit convergence problems [151, p. 103]. More general convergence results for this procedure are discussed by Huber [151, ch. 7], and it often performs well in practice, although a poor choice of the initial estimate θ^0 can cause the algorithm to get stuck in a local minimum that is far from the optimal parameter estimate [252, p. 149].

11.4.3 Two illustrative examples

To illustrate the behavior of M-estimators, first consider the line fitting case study from Chapter 5, based on the four very small datasets shown in Fig. 11.2. Recall that the nominal dataset consists of the points $\{(0,0),(1,1),(2,2),(3,3)\}$, but the observed datasets are contaminated with a single additive outlier of magnitude $+1$. Hence, the first contaminated dataset consists of the four points $\{(0,1),(1,1),(2,2),(3,3)\}$, and the other three datasets have the same level of contamination but at data points 2, 3, and 4. Results presented in Chapter 5 demonstrated both the outlier sensitivty of the OLS procedure and its dependence on where the outliers appear in the dataset. In contrast, the LAD procedure *could* exhibit much better fits, effectively ignoring these outliers anywhere in the dataset, but for some outlier locations, the LAD procedure exhibits *two* solutions and the second solution is substantially worse than the contaminated OLS solution. For comparison, Fig. 11.2 shows the optimum lines obtained for Huber's M-estimator with $c = 1.345S$ where S is the MADM scale estimate computed from the residuals. As we would expect, the performance is intermediate between that of the OLS and LAD methods: when the outlier occurs

Figure 11.2: Huber's M-estimator applied to the four-point datasets

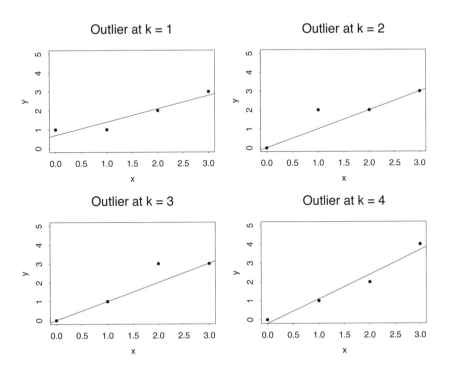

at points $k = 2$ or 3, the M-estimator completely ignores it, just as the LAD estimator does but in marked contrast to the OLS estimator, which always gives biased results. Conversely, when the outlier occurs at $k = 1$ or $k = 4$, Huber's M-estimator gives exactly the same results as OLS. This point may be seen in Table 11.1, which presents both the OLS and M-estimator results for this example. Table 11.1 also gives the final weights w_k determined for each point by the IRWLS procedure, with the weights for the outliers represented in bold face. Ideally, these outlier weights should be zero, and when the outliers occur at $k = 2$ or 3, they are indeed quite small, but when the outliers occur at $k = 1$ or $k = 4$, all data points are weighted equally, reducing the M-estimator to the OLS estimator. Conversely, a significant advantage of the M-estimator over the LAD estimator here is that the M-estimator converges to a single well-defined solution; hence, although this solution is not as good as the better of the two LAD solutions, it is *much* better than the poorer of these two solutions.

The second example is based on the brain weight/body weight dataset. Fig. 11.3 shows a log-log plot of this dataset, overlaid with two lines: the lighter one is the OLS line fit by the methods discussed in Chapter 5 and in Sec. 11.2, and the dark line is the result obtained using Huber's M-estimator, computed exactly as in the previous example. In comparison with other estimators consid-

Table 11.1: Weights and parameters for Huber's M-estimator

Outlier at:	wt. w_1	wt. w_2	wt. w_3	wt. w_4	Huber $\hat{a}$	OLS $\hat{a}$	Huber $\hat{b}$	OLS $\hat{b}$
$k=1$	**1.0**	1.0	1.0	1.0	0.7	0.7	0.7	0.7
$k=2$	1.0	**0.02**	1.0	1.0	0.997	0.9	0.013	0.4
$k=3$	1.0	1.0	**0.02**	1.0	1.003	1.1	0.003	0.1
$k=4$	1.0	1.0	1.0	**1.0**	1.3	1.3	−0.2	−0.2
Exact:	—	—	—	—	1.0	1.0	0.0	0.0

ered both in Chapter 5 (e.g., the inverse regression model) and those considered later in this chapter (e.g., Sec. 11.5), the difference between these estimated lines is quite small. This observation reflects the fact that the outliers in this dataset are anomalous with respect to both their x and y values, illustrating the point that M-estimators provide no protection against outlying x-values.

11.5 Other robust alternatives to OLS

The term *high-breakdown estimators* refers to procedures that, like the median or the MADM scale estimate, exhibit the highest possible breakdown point, meaning that they are resistant to contamination levels up to 50%. As a practical matter, three points are important: first, that such estimators do exist; second, that they can sometimes behave strangely (i.e., badly); and third, that they often lead to difficult computational problems. These last two points led Huber [152] to pose the question, "How high a breakdown point do we really need: 5%, 10%, 25%, or 50%?" In answer, he argues that contamination levels between 1% and 10% are sufficiently common in practice that the use of estimators with breakdown points of less than 10%, "must be considered outright dangerous," as noted in Chapter 7. Conversely, he also argues that breakdown points greater than 25% are difficult to justify if they come at the expense of serious efficiency loss (i.e., much greater estimator variability) or excessive computational complexity. Nevertheless, as the example at the end of this discussion illustrates, high-breakdown (50%) estimators sometimes yield extremely impressive results.

Rousseeuw and Leroy [252, p. 14] note that many approaches to outlier-resistant estimation procedures are based on replacing the sum of squared errors on which the OLS method is based with sums of other error measures. Specific examples include both the LAD method, based on the sum of absolute values

Figure 11.3: Huber and OLS fits to the brain/body dataset

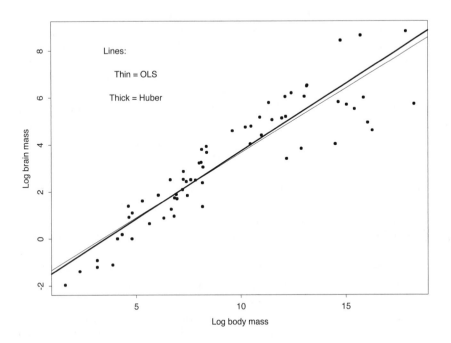

of the errors, and the M-estimators discussed in Sec. 11.4, based on a sum of terms of the form $\rho(e_k)$ for some function $\rho(\cdot)$. As an alternative, Rousseeuw and Leroy [252] note that another possibility is to minimize something other than the *sum* of squared errors. One approach based on this idea is the *least median of squares* (LMS) method, which seeks the parameter θ minimizing:

$$J_{LTS}(\theta) = \text{median}\,\{e_k^2(\theta)\}. \tag{11.48}$$

The primary advantages of this estimator are, first, that it exhibits a 50% break-down point [252, ch. 3], and second, that it is also quite tolerant of outliers in the x-variables, in contrast to the M-estimators described earlier. The primary disadvantages are its poor asymptotic behavior (specifically, the LMS estimator converges at the anomalously slow rate $1/N^{1/3}$ to a nonnormal limit), and the combinatorial complexity associated with computing the exact minimizer of $J_{LTS}(\theta)$. The first of these difficulties may be overcome by considering instead the *least trimmed squares* (LTS) estimator, which replaces the criterion defined in Eq. (11.48) with the alternative measure:

$$J_{LTS}(\theta) = \sum_{i=1}^{h} (e^2)_{i:N}, \tag{11.49}$$

where $(e^2)_{i:N}$ represents the ith order statistic of the sequence of squared prediction errors associated with the regression model parameter θ. It can be shown that this estimator also exhibits a 50% breakdown point [252, ch. 3] and similar insensitivity to x-outliers, a point illustrated in the example considered at the end of this section. In addition, the LTS estimator has been shown to be asymptotically normal with the usual $1/N^{1/2}$ convergence rate, but the problem of combinatorial complexity remains. This problem is extremely significant and is a reflection of the combinatorial complexity of the general subset selection problem discussed in Chapter 10. Still, practical computational algorithms do exist that yield solutions for this problem, although they are not guaranteed to be global minimizers; that is, the result returned by these algorithms may not be the best possible with respect to the LTS criterion. Finally, it is worth noting that high breakdown estimators like LMS and LTS can exhibit unexpected sensitivities to *inliers*, defined as certain critical data points that lie well within the nominal range of variation of the dataset. This point was demonstrated in Chapter 7 in connection with the discontinuity of the median, the best-known (and simplest) member of the family of high-breakdown estimators.

To illustrate the performance possible with the LTS estimator, it is useful to revisit the two examples considered in Sec. 11.4 for Huber's M-estimator. In both cases, the LTS estimate was computed using the built-in procedure **ltsreg** in the *S-Plus* statistics package [272]. Fig. 11.4 shows the results obtained for the four contaminated four-point datasets: in each case, the line is that computed by the LTS procedure, and it corresponds to the exact solution sought: $y_k = x_k$. In contrast to both the OLS estimator and Huber's M-estimator, the results obtained here are not sensitive to the location of the outliers in the dataset (i.e., there are no leverage effects).

The second example is the brain weight/body weight dataset shown in Fig. 11.5: the light line is the OLS solution and the heavy line is the LTS solution, which lies farther from the OLS line than those obtained by any of the other regression procedures considered in this book. Applying the Hampel identifier with the default threshold $t = 3$ to the residuals from this fit leads to the identification of the following nine outliers, all of which are dinosaurs:

1. $k = 48$, protoceratops,
2. $k = 50$, camptosaurus,
3. $k = 54$, stegasaurus,
4. $k = 57$, antosaurus,
5. $k = 58$, iguanodon,
6. $k = 60$, tyrannosaurus,
7. $k = 61$, triceratops,
8. $k = 62$, diplodocus,
9. $k = 64$, brachiosaurus.

The only dinosaur that was *not* detected in this dataset was the allosaurus ($k = 55$), and this case could have been detected with a slightly smaller threshold value for the Hampel identifier, still without declaring any nondinosaurs as

Figure 11.4: LTS lines for the four-point datasets

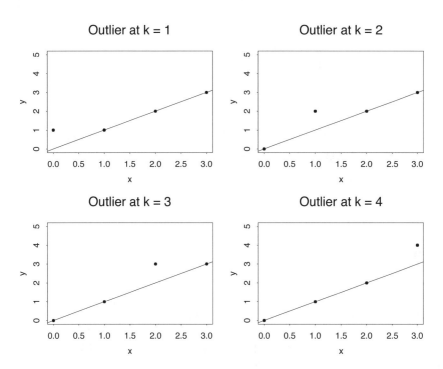

outliers. Overall, the LTS procedure gave the best results for this dataset of any of those considered in this book.

11.6 Exercises

1. Show that the restricted solution θ_{RW} of the weighted least squares problem discussed in Sec. 11.3.2 satisfies the constraint $\mathbf{C}\theta_{RW} = \mathbf{h}$.

2. Show that the condition (11.39) for the M-estimator $\hat{\theta}$ discussed in Sec. 11.4.2 is equivalent to the condition (11.45).

Figure 11.5: LTS and OLS fits for the brain/body dataset

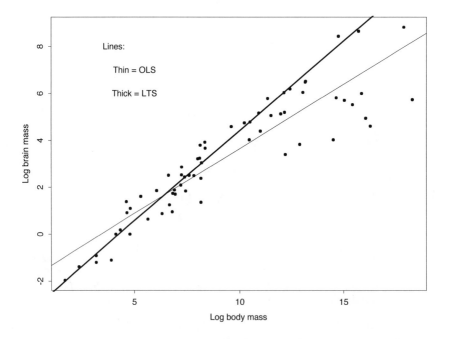

Chapter 12

Reexpression: Data Transformations

Sections and topics:

This chapter is concerned with mathematically well-behaved transformations and their use in data analysis. Such transformations are extremely popular, and as subsequent examples illustrate, they can dramatically improve the results of various types of data analysis. Conversely, other examples also illustrate that seemingly innocuous transformations can also yield results that are *extremely* misleading. More generally, transformations have three primary applications in data analysis: changing the visual emphasis of different details in a data plot, linearizing nonlinear data models, and changing the distribution of random variables. All of these ideas are illustrated with specific examples in this chapter, with particular emphasis on random variable transformations, both to illustrate how these effects can be useful (as in the case of nonuniform random number generation) and to illustrate how they can be surprising (as in the case of reciprocal transformations).

Key points of this chapter:

1. The variables we have available for analysis are not always presented in their most informative representation; further, the notion of a "good representation" is application dependent: an excellent representation for one application may be extremely poor for another application and vice versa.

2. Transformations have three common consequences that may be exploited in some applications and should be kept in mind in other applications:

   a. they change the visual emphasis of data plots;
   b. they can linearize nonlinear relations between variables;
   c. they change the distribution of data sequences.

   Examples presented here illustrate that all of these effects can be dramatic, which may be very good in some applications and very bad in others.

3. Most popular data transformations, including almost all of those considered in this chapter, are *invertible* and *continuous*. Invertibility implies that no information is lost in the transformation, since the inverse transformation may be applied to recover the original data sequence. Continuity implies that arbitrarily small changes in the original data sequence do not lead to macroscopic changes in the transformed sequence.

4. Some of the specific transformations considered in this chapter include simple ones like the square, square root, logarithm, and exponential, more complicated ones like the angular transformation, and families of transformations like the Box-Cox and Aranda-Ordaz transformations.

5. Transformations applied to uniformly distributed random variables can be a practically useful way to generate samples from a number of popular distributions, including the Cauchy, exponential, logistic, Pareto, and Weibull distributions.

6. A particularly interesting transformation is the reciprocal $f(x) = 1/x$, which can cause fundamental difficulties (e.g., nonexistence of moments) when applied to popular distributions like the Gaussian; consequently, specification of distributions for uncertain denominator variables requires particular care, a point examined in detail here.

7. More generally, one of the points illustrated in this chapter is that transformations can have quite unexpected consequences, effectively introducing spurious relations between unrelated variables or suppressing data features that are important and regarded as "inherent."

12.1   Three uses for transformations

As the discussion of the brain-weight/body-weight dataset presented in Chapter 1 illustrates, one important practical factor in the analysis of real-world data is the *view* we adopt of the data: we may either analyze it "as is," retaining its original representation, or we may *transform* it to a different representation. In particular, recall that *reexpression* was introduced in Chapter 1 as one of the "four R's of exploratory data analysis." This chapter presents a more detailed introduction to the topic of reexpression, illustrating some of the options we have available to us in transforming data. In practice, we may consider data transformations for at least three reasons:

1. to change the visual emphasis of data plots,
2. to linearize nonlinear data models,
3. to change the distribution of data sequences.

The first of these motivations was illustrated dramatically by the brain-weight vs. body-weight dataset discussed in Chapter 1, but another amusing illustration is provided by the example discussed in Sec. 12.1.1; the second of these reasons for transforming data is discussed in detail in Sec. 12.1.2, and the third is discussed in Sec. 12.1.3.

12.1.1   Changing visual emphasis

Fig. 12.1 is a drawing by an unknown artist, published in *Chemtech* [13], that shows a collection of 25 different views of a pig. All of these figures were generated from the central one (labeled "linear/linear") by transforming the x and y axes according to the following five transformations:

1. square: $f(x) = x^2$,
2. square root: $f(x) = \sqrt{x}$,
3. linear (no transformation): $f(x) = x$,
4. logarithmic: $f(x) = ln\, x$,
5. exponential: $f(x) = e^x$.

The implied domains in all of these transformations is the positive real line $R^+ = [0, \infty)$ and, on this domain, all five of these transformations satisfy the criteria for useful data transformations discussed in Sec. 12.4. In particular, these transformations are all *invertible*, meaning that all 25 of these pig drawings are *informationally equivalent*: given any one plot (e.g., the "logarithm/squared" entry in the bottom row, second from the left), it is possible to generate any other plot in Fig. 12.1 by the appropriate combination of forward and inverse transformations. Conversely, the key point of this example is that, although these different representations of the pig may be informationally equivalent, they are *not visually equivalent* because each one clearly emphasizes different aspects of the pig's anatomy. For example, square root and logarithmic transformations of the horizontal axis greatly emphasize the pig's snout, whereas the square and exponential transformations eliminate this feature entirely, instead

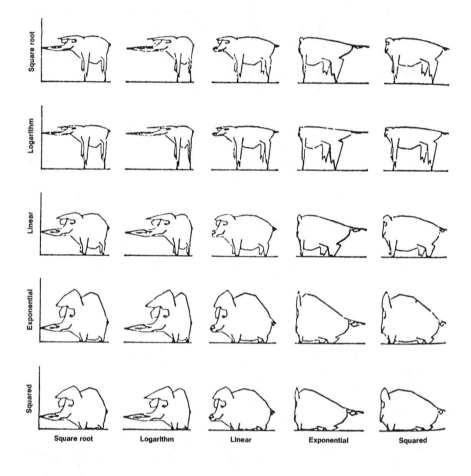

Figure 12.1: The pig in transformed coordinates. From *Chemtech*, October, 1986 [13]; used with permission.

emphasizing the pig's tail. In particular, note that if we apply the exponential transformation to the horizontal axis and the square transformation to the vertical axis, the curl in the pig's tail becomes extremely pronounced, even though it was not visible at all in the original (linear/linear) plot.

12.1.2 Linearizing nonlinear models

It was noted in Chapter 5 that allometry—the study of how various characteristics of animals scale with their size, particularly their body mass—makes extensive use of models that are not linear in the unknown parameters, but which may be simply transformed to linearity. The following discussion examines this idea somewhat more generally, considering Hoerl's family of regression models that can be linearized by applying the logarithmic transformation to the response variable. Specifically, Hoerl [143] noted that the logarithmic transformation will linearize all nonlinear models of the form:

$$y = ax^b e^{cx}, \tag{12.1}$$

reducing them to:

$$\ln y = \alpha + b \ln x + cx, \tag{12.2}$$

where $\alpha = \ln a$. This family is disucssed by Daniel and Wood [79], and it includes a number of important special cases. For example, note that all but one of the "basic pig transformations" discussed in the previous section (specifically, the logarithmic transformation used to linearize these models) are members of this family. Also, note that the logarithmic transformation is only applicable to the model defined in Eq. (12.1) if $x > 0$.

To illustrate the range of qualitative behavior exhibited by these models, Fig. 12.2 shows plots of several members of this family, for different values of b and c. Because it does not affect the shape of these curves, the parameter a is fixed at 1 in all of these examples. The upper left plot in Fig. 12.2 shows three curves corresponding to $b = 0$ and $c = -0.2$, 0, and $+0.2$. Note that taking $c = 0$ yields the constant model $y = a$, whereas positive values of c yield exponential growth models and negative values of c yield exponential decay models. The models shown in the upper right plot all have $b = -0.5$ and exhibit divergence as $x \to 0$, an inherent feature of these models when $b < 0$. For $c = 0$, these models exhibit monotonically decaying responses, whereas for $c > 0$, the exponential term ulitmately dominates and $y \to \infty$ as $x \to \infty$. Hence, for $c > 0$, the response exhibits a unique global minimum for some finite value of x. For $c < 0$, the model response exhibits a monotonic decay similar to that obtained for $c = 0$, but faster (i.e., exponential vs. reciprocal power law). The bottom two plots show model responses for $b = 0.5$ and $b = 2$; when $c = 0$, these responses correspond to the square root and square transformations considered in the transformed pig example. Again, for $c > 0$, these models ultimately exhibit exponential growth, whereas for $c < 0$, they ultimately exhibit exponential decay. In fact, note that the case $b > 0$, $c < 0$ corresponds to the reciprocal of a model of the same form with the signs of b and c reversed. Hence, when $b > 0$ and $c < 0$, the model

Figure 12.2: Members of Hoerl's linearizable family

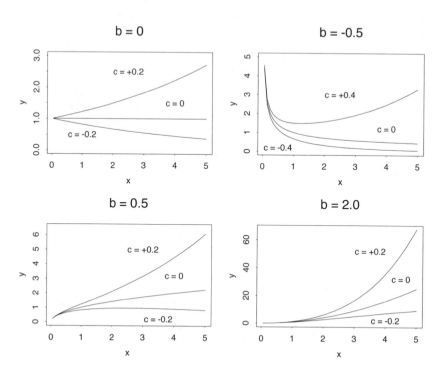

exhibits a unique maximum at some finite value of x, decaying to zero both as $x \to 0$ and as $x \to \infty$. Overall, the key point of this example is to illustrate the considerable range of qualitative behavior exhibited by Hoerl's simple family of linearizable models.

It is important to note, however, that if the least squares error criterion is a reasonable one for the original, untransformed problem, it may not be reasonable for the linearized problem. This situation arises in the Michaelis-Menten model, for example, and it is demonstrated in Sec. 12.2.1 that the resulting complications can be severe. A less dramatic but possibly more representative example is that of Gander *et al.* [110], who consider the problem of fitting circles and ellipses to a set of points in the (x, y) plane, using two different procedures. The first method involves representing the equation for the circle or the ellipse as a quadratic function of the observed data, which may be reduced to a constrained linear regression problem and solved via the methods discussed in Chapter 13. Unfortunately, one of the main points of the paper is that the solution of this transformed problem is *not* equivalent to that obtained by minimizing the geometric distance between the circle and the available data points. The disadvantage of this second problem formulation is that it is a *nonlinear* least squares problem, which is inherently more difficult to solve and requiring methods like those discussed by Seber and Wild [262] that are beyond the scope of this book. Conversely, it is important to emphasize that the simpler transformed problem does not lead to a useful solution in this example. The key point here is one reemphasized by the results presented in Sec. 12.2: if the original problem is transformed to make it easier to solve, it is critically important to examine the solution in the original problem domain to make sure it is a *reasonable* solution and not just "a solution."

12.1.3 Changing data distributions

One reason that transformations can strongly influence the utility of standard fitting procedures—especially least squares procedures—is that they can profoundly change data distributions. This point was demonstrated in Chapter 10 in connection with Fisher's transformation to normality for estimated correlations. In that example, the changes in distribution were beneficial, but as the example discussed in Sec. 12.2.3 demonstrates, these effects can be both unexpected and highly undesirable. To illustrate the general way in which transformations can change data distributions, the following discussion reviews and extends the uniform random variable transformation results presented in Chapter 8 as the basis for Q-Q plots. Sec. 12.5 applies these results to the problem of generating random numbers from several useful distributions by transforming uniformly distributed random samples, and Sec. 12.6 then considers the more general problem of transforming random variables with arbitrary densities.

As discussed in Chapter 8, if x is uniformly distributed on the interval $[0, 1]$, the probability density function $p(x)$ is equal to 1 for all x in this interval and 0 for all x lying outside this interval. The cumulative distribution function is the integral of this density function, given by:

$$U(x) = \begin{cases} 0 & x < 0, \\ x & 0 \le x \le 1, \\ 1 & x > 1. \end{cases} \qquad (12.3)$$

Now, suppose $y = g(x)$ is a transformation of x that is continuous on $[0, 1]$ and strictly monotonic, implying it is invertible [174, p. 181]. Recall from Chapter 4 that that the cumulative distribution function $F_y(\zeta)$ is the probability $\mathcal{P}\{y \le \zeta\}$ that the random variable y assumes a value no larger than ζ. Substituting $y = g(x)$ into this expression yields:

$$F_y(\zeta) = \mathcal{P}\{g(x) \le \zeta\}. \qquad (12.4)$$

Since $g(\cdot)$ is strictly monotonic, it is either order-preserving (if strictly increasing) or order-reversing (if strictly decreasing). Further, its inverse $g^{-1}(\cdot)$ exists and is also either order-preserving or order-reversing the same as $g(\cdot)$. Hence, the condition $g(x) \le \zeta$ is equivalent to one of the following two conditions:

1. $x \le g^{-1}(\zeta)$ if $g(\cdot)$ is order-preserving (i.e., strictly increasing);
2. $x \ge g^{-1}(\zeta)$ if $g(\cdot)$ is order-reversing (i.e., strictly decreasing).

This observation leads to one of two explicit expressions for $F_y(\zeta)$, depending on which of the two conditions above applies. Specifically, if Condition 1 applies, the transformed cumulative distribution function is given by:

$$F_y(\zeta) = \mathcal{P}\{x \leq g^{-1}(\zeta)\} = U(g^{-1}(\zeta)). \tag{12.5}$$

This result simplifies considerably due to the form of $U(x)$: from Eq. (12.3), it follows that:

$$F_y(\zeta) = \begin{cases} 0 & g^{-1}(\zeta) < 0 \\ g^{-1}(\zeta) & 0 \leq g^{-1}(\zeta) \leq 1 \\ 1 & g^{-1}(\zeta) > 1 \end{cases}$$

$$= \begin{cases} 0 & \zeta < g(0) \\ g^{-1}(\zeta) & g(0) \leq \zeta \leq g(1) \\ 1 & \zeta > g(1). \end{cases} \tag{12.6}$$

If Condition 2 applies (i.e., if the transformation $g(\cdot)$ is order-reversing), similar reasoning leads to the following result:

$$F_y(\zeta) = \mathcal{P}\{x \geq g^{-1}(\zeta)\} = 1 - \mathcal{P}\{x \leq g^{-1}(\zeta)\} = 1 - U(g^{-1}(\zeta)). \tag{12.7}$$

Again, this result simplifies due to the form of the uniform CDF $U(x)$:

$$F_y(\zeta) = \begin{cases} 1 & g^{-1}(\zeta) < 0 \\ 1 - g^{-1}(\zeta) & 0 \leq g^{-1}(\zeta) \leq 1 \\ 0 & g^{-1}(\zeta) > 1 \end{cases}$$

$$= \begin{cases} 0 & \zeta < g(1) \\ 1 - g^{-1}(\zeta) & g(1) \leq \zeta \leq g(0) \\ 1 & \zeta > g(0). \end{cases} \tag{12.8}$$

To see the importance of this result, suppose $p(\cdot)$ is any proper density function that is strictly positive on its domain $\mathcal{D}$, a condition satisfied by most of the densities considered in Chapter 4. Because the cumulative distribution function $F(\cdot)$ associated with this density function is simply its integral, it follows that $F(\cdot)$ is a strictly increasing function. Now, suppose we consider the transformation $g(x) = F^{-1}(x)$ applied to a uniformly distributed random variable x. That is, consider the random variable y defined by:

$$y = F^{-1}(x). \tag{12.9}$$

It follows immediately from Eq. (12.5) that the transformed variable y has the cumulative distribution function $F(y)$ and the density $p(y)$.

In principle, this result defines a method for generating samples of any random variable y with a strictly positive density function $p(y)$. In practice, this result is less useful than it might appear, since most popular distributions do not have simple closed-form expressions for their cumulative distribution functions. Still, as examples discussed in Sec. 12.5 illustrate, this result can sometimes be used as the basis for generating nonuniformly distributed random variables from uniform random samples, for which generators are widely available. Similarly, Eq. (12.9) also establishes that the random variable y may be transformed to a *uniformly distributed random variable x on* $[0,1]$ via the transformation:

$$x = F(y). \tag{12.10}$$

In addition, this result also establishes that the random variable y may be transformed to normality. In particular, since the Gaussian cumulative distribution function is invertible it follows that if x is uniformly distributed on $[0,1]$, $z = \Phi^{-1}(x)$ is normally distributed. Consequently, the random variable

$$z = \Phi^{-1}(F(y)), \tag{12.11}$$

is normally distributed. In data analysis applications, we generally do not know the cumulative distribution function $F(\cdot)$, but this result establishes that transformations to normality usually exist, motivating the use of transformations like those introduced in Sec. 12.3 as approximate transformations to normality.

Finally, it is useful to note that these results may be extended to cumulative distribution functions $F(\cdot)$ are not strictly invertible. To accomplish this extension, introduce the *quantile function* $Q(\cdot)$ [16]:

$$Q(x) = \sup\{\zeta|F(\zeta) \le x\}. \tag{12.12}$$

Note that if $F(\cdot)$ is sufficiently well behaved for $F^{-1}(\cdot)$ to exist, then $Q(x) = F^{-1}(x)$. It is a standard result [37, p. 190] that if x is uniformly distributed on $[0,1]$, then $y = Q(x)$ has cumulative distribution function $F(x)$. *The real point of this result is that essentially any continuous random variable may be viewed as a transformation of the uniform distribution, with the transformation given by the quantile function, $y = Q(x)$.* In particular, note that this result relaxes both the strict monotonicity of $F(\cdot)$ assumed above, and the existence of an associated density $p(\cdot)$.

12.2 Four transformation horror stories

One translation of the magnificently horrible little novella *The Metamorphasis* by Franz Kafka [168], opens with the following sentence:

> When Gregor Samsa woke up one morning from unsettling dreams,
> he found himself changed in his bed into a monstrous vermin.

In fact, if they are applied without due care, certain popular data transformations can have comparably unpleasant effects, as the following four examples illustrate. *The key point of these examples is that although "good" data transformations preserve certain qualitative features in our data, they cannot preserve* **all** *qualitative features.*

Figure 12.3: Similation data and exact model response

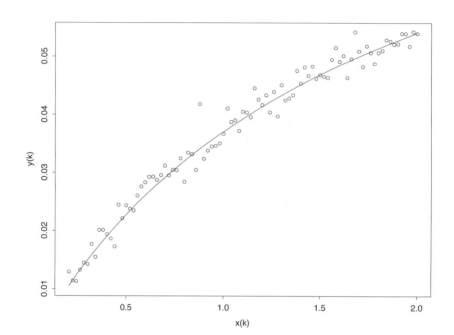

12.2.1   Making unrelated variables appear related

Fig. 12.3 shows a plot of simulated data values for the following nonlinear model:

$$y_k = \frac{0.1x_k}{1.7 + x_k} + e_k. \tag{12.13}$$

Here, $\{e_k\}$ is a sequence of i.i.d. Gaussian random variables with mean zero and standard deviation 0.002 and the plot shows $\{y_k\}$ vs. $\{x_k\}$. A total of 91 values are shown, corresponding to x_k in the range 0.20 to 2.00 in steps of size 0.02. The corresponding range of y_k values is from approximately 0.01 to 0.05, so the fluctuations represented by e_k are significant—in fact, rather large at small x_k—but roughly consistent with the example discussed by Seber and Wild [262, p. 92], on which this simulation is based.

Eq. (12.13) corresponds to a special case of the *Michaelis-Menten model*:

$$y_k = \frac{\alpha x_k}{\beta + x_k}, \tag{12.14}$$

which may be linearized via the transformation:

$$z_k = x_k/y_k \ \Rightarrow \ z_k = \gamma + \lambda x_k. \tag{12.15}$$

Figure 12.4: Transformed data and least squares line

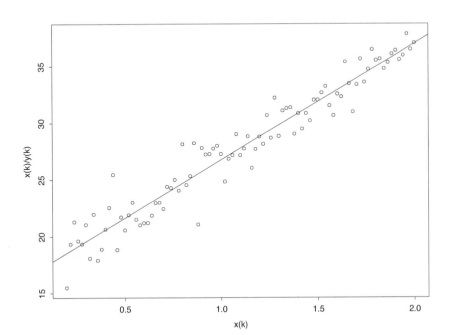

Here, $\gamma = \beta/\alpha$ and $\lambda = 1/\alpha$ are the two parameters of the transformed model, which may be estimated using the least squares procedure discussed in Chapter 5. Unfortunately, it has been noted by various authors [75, 262] that this particular transformation generally leads to unreliable parameter estimates. The example considered here illustrates one aspect of this difficulty.

Fig. 12.4 shows a plot of the transformed variable $z_k = x_k/y_k$ vs. x_k, and the expected linear trend is clearly evident in this plot. In addition, this plot shows the best-fit straight line obtained by the ordinary least squares procedure described in Chapter 5, and it appears that this line fits the transformed dataset reasonably well, although significant scatter about this best fit is clearly evident.

For comparison, Fig. 12.5 shows plots of v_k and $w_k = x_k/v_k$ vs. x_k for the same x_k values. Here, however, the new variable v_k is a sequence of 91 i.i.d. Gaussian random variables with mean 0.02 and standard deviation 0.002 that is *completely unrelated* to x_k. In Fig. 12.5, however, there appears to be a significant linear trend in the transformed data, very much like that seen in Fig. 12.4 for the transformed Michaelis-Menten data. *In fact, the trend in the second example appears precisely because the transformation $v_k \rightarrow x_k/v_k$ introduces a strong correlation between the independent variable x_k and the transformed variable w_k. Further, if we reduce the variance of v_k, the transformed*

Figure 12.5: A second transformed example

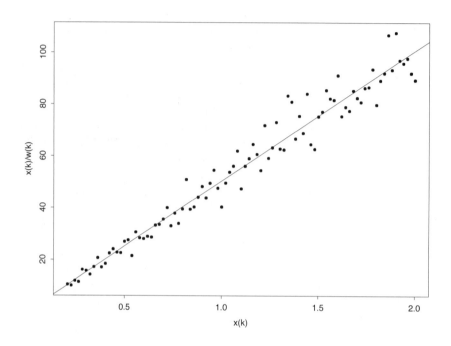

plot looks even better, reflecting the fact that as this variance goes to zero, $x_k/v_k \to x_k/0.02$. In the limit, then, we are simply plotting $50x_k$ against x_k, giving us a perfect fit with a slope of 50 and an intercept of 0.

Note that although the values $y_k \simeq 0.02$ for the second example are approximately consistent with the actual Michaelis-Menten model responses shown in Fig. 12.3, the corresponding slopes and intercepts (51.27 and -0.18) are nothing like the true values of $1/\alpha = 10.0$ and $\beta/\alpha = 17.0$, respectively. In contrast, the simulated Michaelis-Menten data yields estimated slope and intercept values of approximately 9.78 and 17.16, respectively, quite close to these correct values. Consequently, if we have some idea of what constitute "reasonable" parameter values (i.e., from a physical understanding of the problem that generated the data), we are less likely to be mislead by such spurious "fits" to the data. *In particular, note that as $\beta \to 0$ the Michaelis-Menten model reduces to the constant model $y_k = \alpha$.* Since β/α is the intercept of the transformed model, linear fits that appear to go too close to the origin should make us highly suspicious. In fact, the difficulty described here becomes obvious if we compare the two "data" plots shown in Fig. 12.6: the open circles represent the simulated Michaelis-Menten data $y(k)$ plotted against $x(k)$, whereas the solid triangles represent the random sequence $w(k)$ plotted against $x(k)$.

Figure 12.6: The two untransformed data sequences

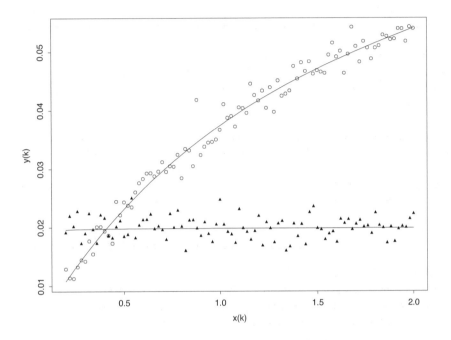

The fundamental basis for the difficulty illustrated in this example lies in the fact that the linearizing transformation for the Michaelis-Menten model involves *both* variables. As a consequence, it is possible for this transformation to induce effective correlations even when none are present in the original data sequences, as this example illustrates. The point is important because linearizing transformations of this type do arise periodically; as a specific example, note the following linearizing transformation:

$$y = \exp\left(-\frac{\alpha x}{\sqrt{x^2 + \beta}}\right) \; \Rightarrow \; \left(\frac{x}{\ln y}\right)^2 = Ax^2 + B, \qquad (12.16)$$

where $A = 1/\alpha^2$ and $B = \beta/\alpha^2$. If α and β represent unknown positive constants to be determined by fitting a sequence of observed (x_k, y_k) pairs, this transformation would lead to a linear regression problem, but it would also be subject to the same difficulties described here for the Michaelis-Menten model.

Figure 12.7: Four quadratic functions with positive curvature

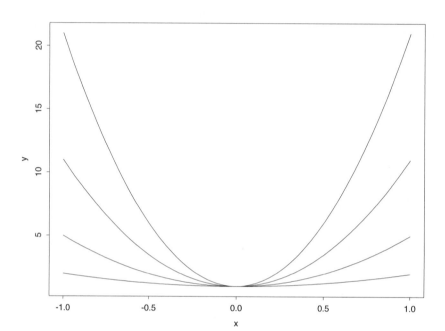

12.2.2   Transformations need not preserve curvature

The curvature γ of a function $f(x)$ at the point x is defined as:

$$\gamma = \frac{f''(x)}{[1 + f'(x)^2]^{3/2}}, \tag{12.17}$$

where $f'(x)$ denotes the first derivative of $f(x)$ and $f''(x)$ denotes the second derivative. In particular, note that the *sign* of the curvature is determined completely by the second derivative. Visually, curvature is a qualitative feature that we can see fairly readily in a plot of $f(x)$ vs. x. Consequently, we are apt to view the curvature suggested by a data sequence $\{(x_k, y_k)\}$ as an "intrinsic property," reflecting the curvature of a function $f(\cdot)$ that relates x_k to y_k. It may come as an unpleasant surprise, then, to note that curvature is in general *not* preserved by standard data transformations. As a specific example, consider the simple quadratic function:

$$f(x) = 1 + \alpha x^2, \tag{12.18}$$

plotted in Fig. 12.7 for $\alpha = 1$, 4, 10, and 20. The curvature of this function is easily computed from Eq. (12.17) as the following value, which is positive for

Figure 12.8: Log transform of previous examples

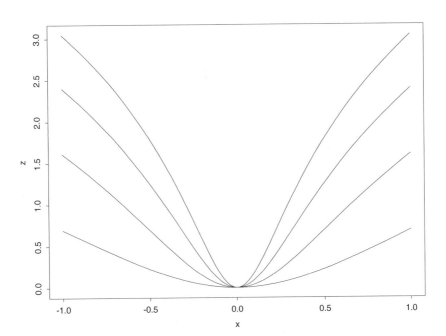

all $\alpha > 0$ and decreases asymptoticaly to zero as $|x| \to \infty$:

$$\gamma_f = \frac{2\alpha}{[1 + 4\alpha^2 x^2]^{3/2}}. \tag{12.19}$$

This behavior of γ_f reflects the fact that the quadratic curve does appear flatter as $x \to \pm\infty$, as may be seen in Fig. 12.7.

Now, suppose we consider the transformed data function $g(x) = \ln f(x)$. Here, the curvature is given by the somewhat more complicated expression:

$$\gamma_g = \frac{2\alpha(1 - \alpha^2 x^4)}{[1 + 4\alpha^2 x^2 + 2\alpha x^2 + \alpha^2 x^4]^{3/2}}. \tag{12.20}$$

As with the original function $f(x)$, the transformed function exhibits a maximum positive curvature of 2α at $x = 0$, but here, for sufficiently large $|x|$, the curvature changes sign. Specifically, note that if $|x| < 1/\sqrt{\alpha}$, the sign of γ_g is the same as that of γ_f, but if $|x| > 1/\sqrt{\alpha}$, the sign of the curvature γ_g reverses. Visually, this means that although a plot of the original function $f(x)$ is always curving upward for $\alpha > 0$, the corresponding plot of the transformed function $g(x)$ shows *downward* curvature for sufficiently large x. This point is illustrated graphically in Fig. 12.8, which shows the effect of the logarithmic transform on the four quadratic examples shown in Fig. 12.7.

Figure 12.9: Transformation $T(\cdot)$ mapping arcsine into Cauchy distributions

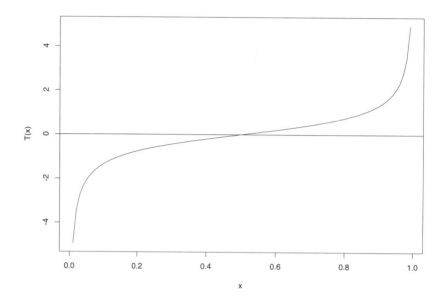

12.2.3   Transformations need not preserve modality

It follows as a direct corollary of the results discussed in Sec. 12.1.3 that essentially any distribution can be transformed into essentially any other distribution. Specifically, suppose x has an invertible CDF $P(x)$ and $Q(\cdot)$ is any other invertible CDF. Since the transformed variable $u = P(x)$ is uniformly distributed on $[0, 1]$, and the transformation $z = Q^{-1}(u)$ takes u into the random variable whose CDF is $Q(\cdot)$, the composite mapping $z = Q^{-1}[P(x)]$ transforms x into a random variable with the desired CDF. Because this procedure is so general, *none* of the characteristics of the distribution that we might regard as "inherent" are preserved by monotone transformations unless we impose further restrictions. The following example illustrates this point explicitly by constructing the monotone transformation shown in Fig. 12.9 that maps the bimodal arcsine distribution, all of whose moments exist:

$$p(x) = \frac{1}{\pi \sqrt{x(1 - x)}}, \ 0 \le x \le 1, \tag{12.21}$$

into the unimodal Cauchy distribution, none of whose moments exist. The reason for the name "arcsine distribution" lies in the fact the density may be

Figure 12.10: Linear and log-log plots of $f(x) = sinh\ x$

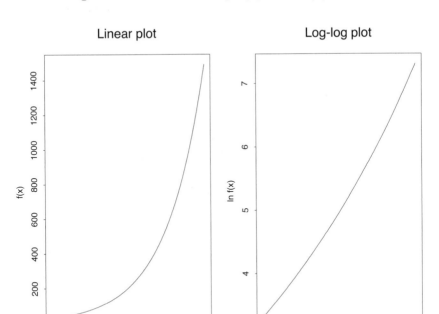

integrated explicitly to obtain the cumulative distribution function:

$$\Phi(x) = \frac{2}{\pi}\ \sin^{-1}\ \sqrt{x}. \tag{12.22}$$

Similarly, as discussed in Chapter 4, the CDF for the Cauchy distribution is:

$$\Psi(x) = \frac{1}{2} + \frac{1}{\pi}\ \tan^{-1}\ x\ \Rightarrow\ \Psi^{-1}(x) = \tan\ \pi\left(x - \frac{1}{2}\right). \tag{12.23}$$

Combining these results leads to the following transformation $T(\cdot)$ mapping the arcsine-distributed random variable x into the Cauchy distributed random variable z:

$$T(x) = \Psi^{-1}[\Phi(x)] = \tan\left(2\sin^{-1}\ \sqrt{x}\ -\ \frac{\pi}{2}\right). \tag{12.24}$$

A plot of this function is shown in Fig. 12.9, from which it may be seen that the function is strictly increasing and continuous.

12.2.4   "Everything looks linear on a log-log plot"

Fig. 12.10 shows two plots of the function $f(x) = \sinh x$. The left-hand plot shows the untransformed function, plotted over the range $4 \leq x \leq 8$, and the

Figure 12.11: Linear and log-log plots of $f(x) = \Gamma(x)$

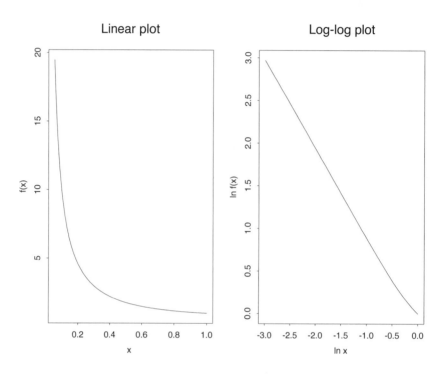

right-hand plot shows the corresponding log-log plot. In fact, $f(x) \simeq (1/2)e^x$ over this range of x values, so a semilog plot (i.e., $\ln y$ vs. x) would be almost linear, but the point here is that the log-log plot suppresses the curvature of this function quite considerably even though $\ln y$ is not really well approximated by a linear function of $\ln x$ over this range. Clearly, the presence of observation noise or other error sources in the data could further obscure what limited evidence of nonlinearity there is in the log-log plot shown here. The importance of this and the next several examples lies in the extreme popularity of the log-log transformation. For example, recall that the allomeric law $y = Ax^\beta$ discussed in Chapter 5 may be linearized by taking logarithms of both variables: $\ln y = \alpha + \beta \ln x$, where $\alpha = \ln A$. The point of this example and the next three is to emphasize, as many others have before, that log-log plots can make a variety of nonlinear functions look quite linear.

Fig. 12.11 shows another pair of plots in the same general format: the curve on the left is the gamma function $f(x) = \Gamma(x)$, plotted over the range $0.05 \leq x \leq 1.00$, and the curve on the right is the corresponding log-log plot. With the exception of the slight upward curvature seen at the right-hand end of this plot, the log-log plot appears extremely linear in this case. As in the previous example, it should be clear that a fit of noisy data to such a log-log plot would

Figure 12.12: Linear and log-log plots of $f(x) = arcsin \sqrt{x}$

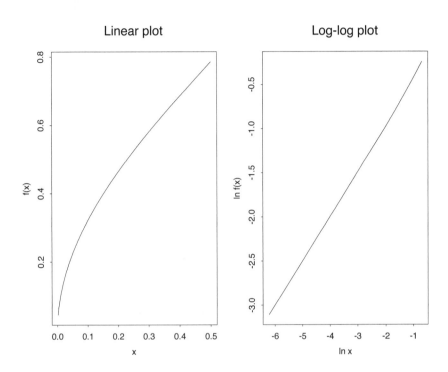

give an excellent fit in this case, corresponding to the following model:

$$\ln y_k \simeq -\ln x_k \Rightarrow y_k \simeq 1/x_k. \tag{12.25}$$

Again, the key point is that the nature of the true nonlinearity—i.e., the gamma function—is well hidden by the log-log transformation.

Fig. 12.12 shows linear and log-log plots of the nonlinearity on which the angular transformation discussed in Sec. 12.3.3 is based:

$$f(x) = \arcsin \sqrt{x}. \tag{12.26}$$

This function is plooted over the range $0.002 \le x \le 0.5$, and the linearity of the log-log plot in this case appears nearly perfect. In fact, the approximation is not bad in this case, since $\sin x \simeq x$ for small x, implying $\arcsin \sqrt{x} \simeq \sqrt{x}$ for sufficiently small x. It is also interesting to note that in the previous two examples, although the log-log transformation reduces the apparent curvature of the plots, the sign of the curvature is not changed. In contrast, what residual nonlinearity does remain in the upper end of this transformed curve exhibits extremely slight positive curvature, opposite to the pronounced negative curvature of the original function. Once again, this example reinforces the point made in Sec. 12.2.2 that transformations need not perserve curvature.

Figure 12.13: Linear and log-log plots of $f(x) = \log\left(\frac{x+1}{x-1/2}\right)$

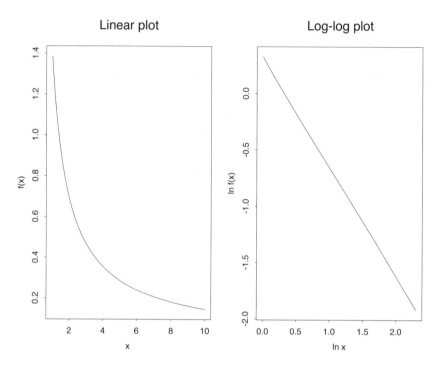

Finally, Fig. 12.13 shows linear and log-log plots of the function:

$$f(x) = \ln\left(\frac{x+1}{x-1/2}\right), \tag{12.27}$$

over the range $1 \le x \le 10$. In this case, the linearity of the transformed curve is *so* good that we would be unlikely to suspect any dependence other than a power-law relationship between x and y even on the basis of very large datasets obtained under extremely good measurement conditions.

In all of these examples, it would be possible to obtain a more accurate assessments of the true nonlinearities if we were to examine log-log plots of these functions over a sufficiently wide range of variation in x. Conversely, it is important to emphasize two practical points here. First, one of the characteristic features of the logarithmic transformation is dynamic range compression, so the range of values required to see significant nonlinearity in a log-log plot is often very large. The second point is that if we are using log-log plots as exploratory analysis tools, the range of available data variation is often beyond our control, anyway. Overall, the main point here is that if this range of variation is not large enough—and "large enough" depends on the true nonlinearity considered—everything *is* likely to look linear on a log-log plot.

Figure 12.14: Examples of the Box-Cox transformation

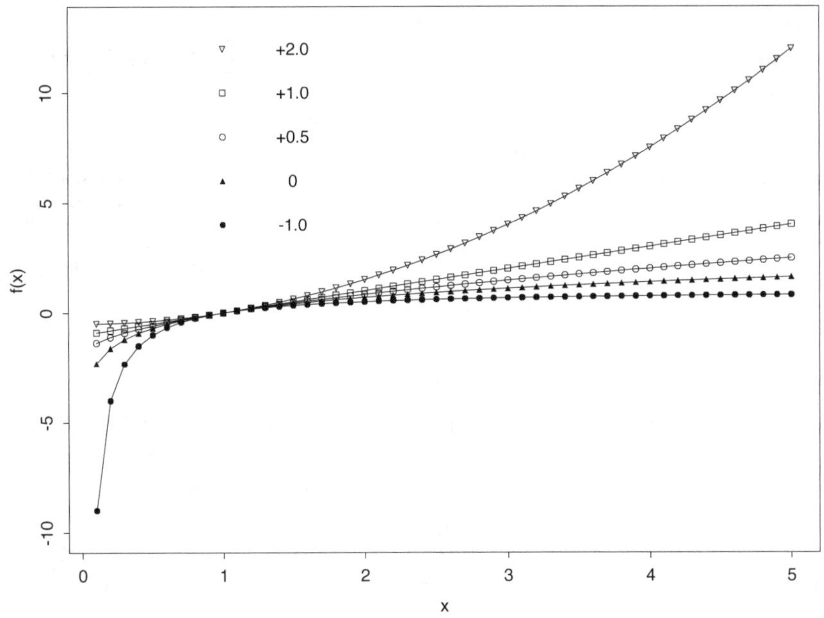

12.3   Three popular transformations

The following subsections introduce three more useful transformations, two of which are actually transformation families. Specifically, Sec. 12.3.1 describes the family of *Box-Cox transformations*, which includes three of the four nonlinear pig transformations discussed in Sec. 12.1.1. Next, Sec. 12.3.2 describes the family of *Aranda-Ordaz* transformations, and Sec. 12.3.3 introduces the *angular transformation*.

12.3.1   Box-Cox transformations

The Box-Cox transformations constitute a family of nonlinear functions $f_\lambda(x)$, dependent on a real-valued parameter λ:

$$f_\lambda(x) = \begin{cases} \frac{x^\lambda - 1}{\lambda} & \lambda \neq 0, \\ \ln x & \lambda = 0. \end{cases} \qquad (12.28)$$

Plots of this function for various values of λ are shown in Fig. 12.14. Note that the transformation for $\lambda = 0$ may be obtained by taking the limit as $\lambda \to 0$ of the transformation for $\lambda \neq 0$ and applying L'Hospital's rule [46, p. 86]. Also, note that $\lambda = 1$ corresponds to essentially no transformaion, simply replacing x by $x - 1$. Similarly, $\lambda = 2$ and $\lambda = 1/2$ correspond approximately to the square and square root transformations discussed in Sec. 12.1.1; hence, four out of the five transformations discussed there are approximately included in the Box-Cox family, excluding only the exponential transformation.

The Box-Cox family has been used fairly extensively in statistical analysis of data, often as a transformation of the response variable y in linear regression problems like those discussed in Chapter 13. In particular, the parameter λ may be estimated by the method of maximum likelihood, often performed approximately as described by Draper and Smith [95, sec. 13.2]. Specifically, these authors advocate first taking a range of λ values (e.g., $\lambda \in [-1, 1]$ or $[-2, 2]$) and partitioning it fairly coarsely into a finite set $\{\lambda_i\}$. Then, the following *modified* Box-Cox transformation is applied to the data sequence:

$$z_k^\lambda = \begin{cases} \frac{y_k^\lambda - 1}{\lambda y_G^{\lambda-1}} & \lambda \neq 0, \\ \\ y_G \ln y_k & \lambda = 0, \end{cases} \qquad (12.29)$$

where y_G is the geometric mean of the data sequence; this modification is included because it often significantly improves the numerical conditioning of subsequent computations. The linear regression problem for the transformed (x_k, z_k^λ) dataset is then solved by the method of ordinary least squares for each λ_i in the selected data range, and the value of λ_i for which the best fit is obtained is taken as a candidate data transformation to be considered further. In particular, Draper and Smith [95] discuss the construction of approximate confidence intervals for the estimated transformation parameter, emphasizing that this transformation parameter is not generally estimated precisely, but is often rounded off to the nearest quarter or even the nearest half.

12.3.2 Aranda-Ordaz transformations

Although the Box-Cox transformation is well defined on the interval $[0, 1]$ for all λ, it was not developed with this domain in mind. More natural if less well known is the Aranda-Ordaz transformation [15, 19], defined by the family of functions:

$$f_\lambda(x) = \begin{cases} \frac{x^\lambda - (1-x)^\lambda}{x^\lambda + (1-x)^\lambda} & \lambda \neq 0, \\ \\ \ln \frac{x}{1-x} & \lambda = 0. \end{cases} \qquad (12.30)$$

Plots of these functions are shown in Fig. 12.15 for $\lambda = 0.5, 1.0$, and 2.0. As in the case of the Box-Cox transformation, the Aranda-Ordaz transformation reduces to a linear rescaling when $\lambda = 1$ (specifically, $f_1(x) = 2x - 1$). It is also true that $f_{-\lambda}(x) = -f_\lambda(x)$, so it is only necessary to consider this transformation family for $\lambda \geq 0$. For all $\lambda \neq 0$, this transformation maps the domain $[0, 1]$ into the range $[-1, 1]$, a point discussed further in Sec. 12.4.

For $\lambda = 0$, the Aranda-Ordaz transformation maps the domain $[0, 1]$ into the entire real line R and is better known as the *logit transformation*, which

Figure 12.15: Aranda-Ordaz transformation, $\lambda = 0.5, 1.0, 2.0$

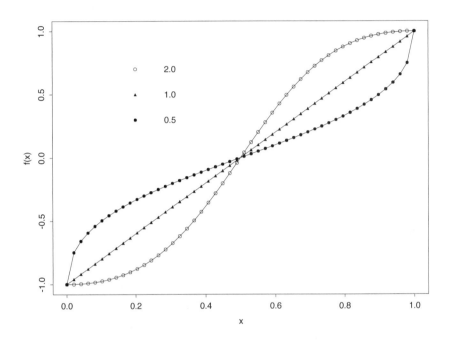

deserves special mention. A plot of this transformation is shown in Fig. 12.16, where it is seen to exhibit similar qualitative behavior to the case $\lambda = 0.5$ shown in Fig. 12.15, but it is not confined to the range $-1 \leq f(x) \leq 1$. This transformation linearizes the logistic model [135, 262]:

$$y = \frac{1}{1 + exp[-(a + bx)]}, \qquad (12.31)$$

a growth model proposed empirically by Verhulst in 1838 [262, p. 330]. Alternatively, the logistic model can be related to a simple mechanistic model involving autocatalysis (i.e., the concentration x increases at a rate proportional to the product $x(1 - \alpha x)$ [262, p. 329]). In any case, this model appears frequently in a variety of applications, particularly in dealing with quantities y like concentrations or probabilities that are required to lie in the range $0 \leq y \leq 1$. In fact, this model forms the basis for *logistic regression*, in which binary responses are related to categorical or continuous explanatory variables, a topic discussed in detail in Chapter 14.

Figure 12.16: Aranda-Ordaz transformation, $\lambda = 0$ (logit transformation)

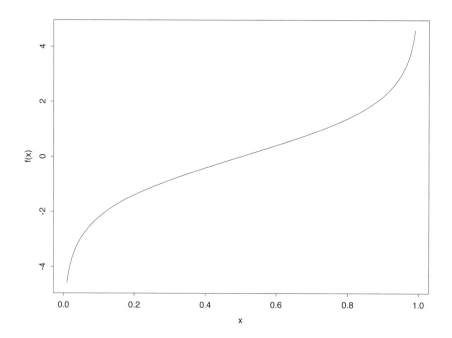

12.3.3   The angular transformation

The *angular transformation* [142] is defined by the function:

$$f(x) \;=\; 2\arcsin(\sqrt{x}) \;-\; \frac{\pi}{2}, \qquad\qquad (12.32)$$

on the domain $\mathcal{D} = [0, 1]$. Although apparently not as well known, the following simpler expression may be obtained via standard trigonometric identities:

$$f(x) \;=\; \arcsin(2x - 1). \qquad\qquad (12.33)$$

Note that this transformation maps $[0, 1]$ into $[-\pi/2, \pi/2]$, taking the center of the domain $x = \frac{1}{2}$ to 0. For $x \simeq \frac{1}{2}$, this transformation is approximately linear, behaving like the Aranda-Ordaz trnasformation for $\lambda = 1$: $f(x) \simeq 2x - 1$. This point is emphasized in Fig. 12.17, which shows both the angular transformation and the line $y = 2x - 1$ for comparison. Conversely, note that the effects of this transformation are most pronounced near the ends of the interval $[0, 1]$.

The angular transformation is commonly applied to percentages and proportions (e.g., concentrations) because it emphasizes the differences between, for example, a change from 1% to 2% and a change from 71% and 72%. In addition,

Figure 12.17: The angular transformation

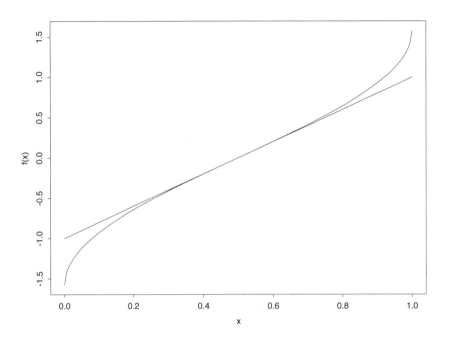

this transformation is symmetric around $x = \frac{1}{2}$, emphasizing the qualitative similarity of changes from 1% to 2% and changes from 99% to 98%.

12.4 Characteristics of good transformations

The five transformations considered in Sec. 12.1.1 and the three transformations considered in Sec. 12.3 all exhibit two properties that are extremely desirable in a data transformation:

1. invertibility,
2. continuity.

Invertibility is desirable because it implies the transformation is *information preserving:* given a data sequence $\{x_k\}$, if we generate the transformed sequence $y_k = f(x_k)$ using a function with a well-defined inverse $f^{-1}(\cdot)$, it is always possible to recover the original data sequence from the transformed one, via the inverse transformation $x_k = f^{-1}(y_k)$. As a practical matter, the ability to transform from our original data representation to another, possibly more convenient one *and back again* is important because we often perform some or all of the quantitative analysis in the transformed domain. As the first of the

Table 12.1: Ranges of some popular transformations, for different domains

Transformation	Domain $[0,1]$	Domain $[0,\infty)$	Domain $(-\infty,\infty)$
log	$(-\infty, 0]$	$(-\infty, \infty)$	
exp	$[1, e]$	$[1, \infty)$	$(0, \infty)$
x^2	$[0, 1]$	$[0, \infty)$	$[0, \infty)$
$\sqrt{x}$	$[0, 1]$	$[0, \infty)$	
$1/x$	$[1, \infty)$	$[0, \infty)$	
Box-Cox, $\lambda > 0$	$[-1/\lambda, 0]$	$[-1/\lambda, \infty)$	
Box-Cox, $\lambda < 0$	$[0, \infty)$	$(-\infty, 1/|\lambda|]$	
Aranda-Ordaz, $\lambda \neq 0$	$[-1, 1]$		
Aranda-Ordaz, $\lambda = 0$	$(-\infty, \infty)$		
Angular	$[-\pi/2, \pi/2]$		

four examples in Sec. 12.2 illustrates dramatically, it is always a good idea to examine the results in the original problem domain because this is usually where we have the best intuition and can therefore best interpret our results.

Similarly, it is desirable that any transformation we consider (and its inverse) be continuous because this means that small changes in the original data sequence will translate into small changes in the transformed data sequence, and vice versa. In particular, note that a *discontinuous* function $f(x)$ exhibits either sudden jumps or singularities (i.e., infinite values) at certain critical values of x [174, p. 185]. Imposing both invertibility and continuity as constraints, it follows that the function $f(x)$ is *strictly monotone* [174, p. 181], meaning that if $x < y$ then either $f(x) < f(y)$ (for an *increasing* or *order-preserving* function) or $f(x) > f(y)$ (for a *decreasing* or *order-reversing* function).

Finally, it is important to note that characterizations like invertibility and continuity are closely related to the *domain* of the transformation we are considering. For example, although the transformation $f(x) = x^2$ is strictly increasing on the interval $[0, \infty)$, it is nonmonotonic on the entire real line. Hence, this transformation is not invertible on the domain R since its inverse, $\sqrt{x}$, is not a real number for $x < 0$. To help avoid such difficulties, Table 12.1 lists the ranges for each of the transformations discussed in Sec. 12.3, for the following typical domains: $[0, 1]$, $[0, \infty)$, and $(-\infty, \infty)$. In cases where the transformation is undefined on the indicated domain, no range is given. Also, it is important to note that the *range of the forward transformation* $f(x)$ is the *domain of the inverse transformation* $f^{-1}(y)$ and vice versa.

In general, note that the Box-Cox transformation can only be defined on positive domains like $[0, 1]$ and $[0, \infty)$, and not on domains like the whole real line R that include negative arguments. Exceptions occur when $\lambda = n$ for some integer n, or when $\lambda = 1/n$ for some *odd* integer n. As a specific example, note that $\lambda = 1/3$ corresponds to a cube root transformation that maps the entire real line R into itself and is reasonably well behaved. Further, the *range* of the general Box-Cox transformation depends both on the domain and the sign of the parameter λ, as indicated in Table 12.1. This point is important since, as noted, the range of the transformation is the domain of the inverse. For the Box-Cox transformation, the inverse transformation is given by:

$$f_\lambda^{-1}(y) = \begin{cases} (1 + \lambda y)^{1/\lambda} & \lambda \neq 0, \\ e^y & \lambda = 0. \end{cases} \qquad (12.34)$$

For $\lambda \neq 0$, note that a domain constraint is necessary to guarantee the quantity $(1 + \lambda y)$ cannot be negative; this constraint will differ depending on whether λ is positive or negative, as indicated in Table 12.1.

As noted in Sec. 12.3.2, the Aranda-Ordaz transformation was developed for quantities like probabilities and concentrations that must lie in the interval $[0, 1]$. As in the case of the Box-Cox transformation, however, this domain can also be extended in special cases (again, $\lambda = 1/3$ represents one specific example), but there seems little motivation to do so. Defined on the domain $[0, 1]$, the range of the Aranda-Ordaz transformation is $[-1, 1]$ for all $\lambda > 0$, as indicated in Table 12.1. On the open interval $(-1, 1)$, the inverse transformation is given by:

$$f^{-1}(y) = \left[1 + \left(\frac{1 - y}{1 + y} \right)^{1/\lambda} \right]^{-1}. \qquad (12.35)$$

At the endpoints of the interval $[-1, 1]$, the inverse transformation is given by $f^{-1}(-1) = 0$ and $f^{-1}(1) = 1$, consistent with the limiting behavior of $f^{-1}(y)$ defined in Eq. (12.35).

For $\lambda = 0$, it was noted that the Aranda-Ordaz transformation reduces to the logit transformation, defined on the entire real line. The inverse of this transformation is the logistic transformation, also discussed in Sec. 12.3.2:

$$f_0^{-1}(y) = \frac{1}{1 + e^{-y}}. \qquad (12.36)$$

The angular transformation is only well defined on the domain $\mathcal{D} = [0, 1]$, mapping this set into the range $\mathcal{R} = [-1, 1]$. The inverse of the angular transformation is well defined, mapping $\mathcal{R}$ into $\mathcal{D}$ via:

$$f^{-1}(y) = \sin^2 \left(\frac{1}{2}[y + \frac{\pi}{2}] \right) = \frac{1 + \sin y}{2}. \qquad (12.37)$$

As in the case of the forward transformation, both of these expressions are equivalent, although the second is much simpler.

12.5 Generating nonuniform random numbers

Probably the two most widely available random number generators are those for the uniform distribution and the Gaussian distribution. In cases where the cumulative distribution function $F(x)$ is known and its inverse exhibits a simple form, the uniform transformation results discussed in Sec. 12.1.3 provide a convenient basis for generating random variables distributed according to $F(x)$ from uniform random variables: specifically, the transformation $x = F^{-1}(u)$ applied to a sequence $\{u_k\}$ of uniformly distributed random samples on $[0, 1]$ yields a sequence $\{x_k\}$ of samples drawn from the distribution defined by $F(x)$. The following sections briefly summarize the use of these transformation results in generating random numbers from a variety of popular distributions; more detailed discussions of each of these distributions are given in Chapter 4.

12.5.1 Exponentially distributed random samples

The exponential distribution is defined by the following density function:

$$p(x) = \frac{1}{\beta} \exp\left[-\left(\frac{x - \xi}{\beta}\right)\right], \tag{12.38}$$

where ξ is a location parameter and β is a scale parameter. As noted in Chapter 4, the cumulative distribution function is given by:

$$F(x) = 1 - \exp\left[-\left(\frac{x - \xi}{\beta}\right)\right], \tag{12.39}$$

from which the inverse CDF is easily obtained as:

$$F^{-1}(x) = \xi + \beta \ln\left(\frac{1}{1 - x}\right), \tag{12.40}$$

for $0 \le x < 1$. Hence, if u is uniformly distributed on $[0, 1]$, the transformed variable $F^{-1}(u)$ will be exponentially distributed on $[0, \infty)$. Note that this transformation is the inverse of the Box-Cox transformation discussed in Sec. 12.3.1 with the transformation parameter $\lambda = -1$.

12.5.2 Cauchy distributed random samples

For the Cauchy distribution with location parameter (median) μ and scale parameter (half the interquartile distance) λ, the distribution function is:

$$F(x) = \frac{1}{2} + \frac{1}{\pi} \arctan\left(\frac{x - \mu}{\lambda}\right). \tag{12.41}$$

The inverse of this function is then given by the simple expression:

$$F^{-1}(x) = \mu + \lambda \tan\left[\pi\left(x - \frac{1}{2}\right)\right]. \tag{12.42}$$

Again, applying this transformation to the standard uniform distribution $U[0, 1]$ yields the corresponding Cauchy distribution. This transformation result is the basis for an important geometric interpretation of the Cauchy distribution. In particular, suppose the signal reflected from some object (e.g., the radar reflection from the metal surface of an aircraft) and is proportional to $\tan \theta$ where θ is the angular orientation of that object. It follows that if this orientation angle is uniformly distributed over the range $-\pi$ to π, the reflected intensity will exhibit a Cauchy distribution. This argument is sometimes invoked to explain the heavy-tailed nature of *glint noise* in radar data.

12.5.3 Logistic distributed random samples

For the logistic distribution discussed in Chapter 4, the defining density is:

$$p(x) = \frac{1}{4b} \operatorname{sech}^2 \left[\frac{1}{2} \left(\frac{x - a}{b} \right) \right], \tag{12.43}$$

which may be integrated to obtain a simple expression for the cumulative distribution function, given in Chapter 4. Further, the inverse cumulative distribution function may be written explicitly as:

$$F^{-1}(x) = a + b \ln \left[\frac{x}{1 - x} \right], \tag{12.44}$$

which is simply a scaled version of the *logit* transformation discussed in Sec. 12.3.2. Hence, applying the logit transformation to a uniformly distributed random variable on $[0, 1]$ yields the standard logistic distribution with $a = 0$ and $b = 1$.

12.5.4 Pareto distributed random samples

The cumulative distribution function for the Pareto type I distribution, denoted $P(I)(k, a)$, is given by:

$$F(x) = 1 - \left(\frac{k}{x} \right)^a, \tag{12.45}$$

on the domain $x \geq k > 0$, where a is a positive constant. As in the previous three examples, this function is easily inverted, yielding:

$$F^{-1}(x) = \frac{k}{(1 - x)^{1/a}}. \tag{12.46}$$

Applying this transformation to a uniformly distributed random variable u on $[0, 1]$ then yields a Pareto type I random variable with parameters k and a.

12.5.5 Weibull distributed random samples

Finally, the cumulative distribution function for the Weibull distribution is:

$$F(x) = 1 - \exp \left[- \left(\frac{x - \xi}{\alpha} \right)^c \right], \tag{12.47}$$

where ξ is a real-valued location parameter defining the lower limit of the distribution (i.e., $F(x) \equiv 0$ for $x < \xi$), α is a positive scale parameter, and c is a positive shape parameter. The inverse of this function is given by:

$$F^{-1}(x) = \xi + \alpha \left[\ln \left(\frac{1}{1-x} \right) \right]^{1/c}. \qquad (12.48)$$

Applying this transformation to a uniform random variable on $[0,1]$ then yields a Weibull distributed random variable with parameters ξ, α, and c.

12.6 More general transformations

If we apply a transformation $f(x)$ to a random variable x, the result is another random variable y, generally with a different distribution. The primary subject of this section is a characterization of this change. Sec. 12.6.1 gives a brief derivation of the general result under the simplifying assumptions that the transformation is both invertible and differentiable. To illustrate the utility of this result, it is applied in Sec. 12.6.2 to the exponential distribution and the pig transformations considered in Sec. 12.1.1. Then, because noninvertible transformations do sometimes arise in practice, Sec. 12.6.3 gives the general result for this case and applies it to obtain the distribution for the χ_1 random variable.

12.6.1 Transforming densities

It is possible to derive explicit results under very weak assumptions on a transformation $T(\cdot)$, and these results are presented in Sec. 12.6.3. First, however, it is useful to present the simpler results that apply under the following two assumptions:

1. differentiability: $T'(x)$ exists;
2. invertibility: $T^{-1}(z)$ exists.

Under these assumptions, the density $\phi(z)$ of the transformed random variable $z = T(x)$ may be determined from standard integration results. Specifically, note that for any function $f(\cdot)$, if $z = T(x)$, the expected value of $y = f(z) = f(T(x))$ must be independent of our representation of z. That is, for any function $f(\cdot)$, the following condition must be satisfied:

$$E_z\{f(z)\} \quad = \quad E_x\{f(T(x))\}. \qquad (12.49)$$

More explicitly, it follows from the definitions of these expectations that:

$$\int_{-\infty}^{\infty} f(z)\phi(z)dz \quad = \quad \int_{-\infty}^{\infty} f(T(x))p(x)dx. \qquad (12.50)$$

If we substitute $z = T(x)$ into the integral on the right-hand side, the following four changes result:

1. $f(T(x)) \rightarrow f(z)$;
2. $p(x) \rightarrow p(T^{-1}(z))$;
3. $dx \rightarrow dz / \left(\frac{dz}{dx} \right)$;
4. the range of integration becomes $S = T^{-1}(R)$.

Combining these results, it follows that:

$$\int_{-\infty}^{\infty} f(z)\phi(z)dz \quad = \quad \int_S f(z) \left[\frac{p(T^{-1}(z))}{\left(\frac{dz}{dx} \right)} \right] dz. \qquad (12.51)$$

To obtain the final expression for the transformed density $\phi(z)$, it is necessary to define the range of integration S explicitly. Since differentiability implies continuity and continuous, invertible functions are strictly monotonic, it follows that $T(\cdot)$ is either order-preserving (if it is strictly increasing) or order-reversing (if it is strictly decreasing). In the first case, if $T(\cdot)$ is strictly increasing, it follows that we can take $S = R$ as our range of integration; in addition, it is also useful to note that:

$$\frac{dz}{dx} \quad = \quad T'(x) \quad = \quad T'(T^{-1}(z)), \qquad (12.52)$$

and since $T(\cdot)$ is strictly increasing, $T'(x) > 0$ for all x. Combining all of these results, it follows that Eq. (12.51) may be rewritten as:

$$\int_{-\infty}^{\infty} f(z)\phi(z)dz \quad = \quad \int_{-\infty}^{\infty} f(z) \left[\frac{p(T^{-1}(z))}{|T'(T^{-1}(z))|} \right] dz. \qquad (12.53)$$

Since $f(\cdot)$ is arbitrary, it then follows that the transformed density $\phi(z)$ must be equal to:

$$\phi(z) \quad = \quad \frac{p(T^{-1}(z))}{|T'(T^{-1}(z))|}. \qquad (12.54)$$

Analogous results are obtained for the strictly decreasing case. Specifically, if $T(\cdot)$ is strictly decreasing, it follows that the range of integration in Eq. (12.51) becomes $S = -R$. Further, note that $T'(x) < 0$ for all x in this case, so $dz/dx = -|T'(T^{-1}(z))|$. Hence, substitutions 3 and 4 listed above each introduce a change of sign, and the net result is Eq. (12.53), just as in the case of order-preserving transformations. By the same logic as before, it follows from the arbitrariness of the function $f(\cdot)$ that the transformed density is given by Eq. (12.54).

12.6.2 Transformed exponential random variables

To illustrate the general transformation results just derived, consider the five pig transformations discussed in Sec. 12.1.1 applied to the exponentially distributed random variable x, with the density:

$$p(x) = e^{-x}, \qquad (12.55)$$

on the domain $[0, \infty)$. Note that all five of the pig transforms are strictly increasing on this domain, so their derivatives are everywhere positive. The first of these transformations is $T(x) = x^2$, implying $T^{-1}(z) = \sqrt{z}$ and $T'(x) = 2x$. Substituting these expressions into the transformation result (12.54) yields:

$$\phi(z) = \frac{e^{-\sqrt{z}}}{2\sqrt{z}}. \tag{12.56}$$

This density function may be recognized as that for the Weibull distribution discussed in Chapter 4, with location and scale parameters $\xi = 0$ and $\alpha = 1$, respectively, and shape parameter $c = 1/2$. The second transformation is the inverse of the first, $T(x) = \sqrt{x}$, implying $T^{-1}(z) = z^2$ and $T'(x) = 1/2\sqrt{x}$. Substituting these results into Eq. (12.54) yields the transformed density:

$$\phi(z) = 2ze^{-z^2}. \tag{12.57}$$

Again, this function may be recognized as a Weibull density, with location and scale parameters $\xi = 0$ and $\alpha = 1$ as before, but shape parameter $c = 2$.

The third pig transformation was the identity transformation $T(x) = x$, from which we recover the original exponential density $p(x)$. The fourth of the five pig transformations is the logarithm $T(x) = \ln x$, which has the inverse $T^{-1}(z) = e^z$ and the derivative $T'(x) = 1/x$. Note that this transformation maps the domain $\mathcal{D} = [0, \infty)$ into the entire real line, so the domain of the transformed distribution is larger than that of the original distribution. Specifically, it follows from Eq. (12.54) that the density of the transformed random variable is:

$$\phi(z) = \frac{e^{-e^z}}{e^{-z}} = e^{z-e^z}. \tag{12.58}$$

This density function is a slight variant of the *extreme value distribution* [163, ch. 22]; specifically, $y = -z$ has the extreme value distribution, defined by the density:

$$p(x) = \theta^{-1}e^{-(x-\xi)/\theta}\exp\left[-e^{-(x-\xi)/\theta}\right]. \tag{12.59}$$

This distribution is asymmetric but unimodal, with mode $x = \xi$, and moments that may be computed from the logarithmic derivatives of the gamma function. The first four standardized moments are given by:

$$\begin{aligned}
\bar{x} &= \xi + \gamma_E\theta \simeq \xi + 0.57722\theta, \\
\mathrm{var}\{x\} &= \frac{\pi^2\theta^2}{6} \simeq 1.64493\theta^2, \\
\gamma &\simeq 1.13955, \\
\kappa &= 2.4,
\end{aligned} \tag{12.60}$$

where $\gamma_E \simeq 0.57722$ is Euler's constant [1, p. 255].

Finally, the last pig transformation is $T(x) = e^x$, for which the inverse is $T^{-1}(z) = \ln z$, and the derivative is $T'(x) = T(x) = e^x$. Hence, the denominator in Eq. (12.54) simplifies to $T'(T^{-1}(z)) = T(T^{-1}(z)) = z$, and the numerator is also simple, yielding $p(T^{-1}(z)) = exp[-lnz] = z^{-1}$. Consequently, the transformed density becomes:

$$\phi(z) = \frac{1}{z^2}, \tag{12.61}$$

which may be recognized as the Pareto density discussed in Chapter 4, with parameter $a = 1$.

12.6.3   The χ_1^2 density

In the most general case of a continuous random variable, recall that a density need not exist; thus, it may be necessary to consider the cumulative distribution function, proceeding along the lines presented above for the transformation of uniformly distributed random variables. The basic notion is the observation [222, p. 118] that if $y = g(x)$, then the cumulative distribution function $F_y(\xi)$ for y is given by:

$$F_y(\xi) = \mathcal{P}\{g(x) \le \xi\} = \mathcal{P}\{x \in S_\xi\}, \tag{12.62}$$

where S_ξ is the set of all x satisfying the condition $g(x) \le \xi$. Knowing the distribution of x and the function $g(\cdot)$, it is possible to compute $F_y(\xi)$, a topic discussed at length by Papoulis [222], who also derives the following useful result, applicable in cases where the original random variable x has a density $p(x)$, the transformation $y = g(x)$ is differentiable, but this transformation is not invertible. In particular, the invertibility condition is weakened to the requirement that the function $f(\cdot)$ is not constant over any finite interval in the domain $\mathcal{D}$ of the density $p(x)$. It follows from this assumption that the equation $z = g(x)$ has at most a countable number of isolated roots $\{x_i\}$ in $\mathcal{D}$ for any z. The transformed density $\phi(y)$ can be shown to be [222, p. 127]:

$$\phi(y) = \sum_i \frac{p_x(x_i)}{|g'(x_i)|}, \quad g'(x) = \frac{dg(x)}{dx}. \tag{12.63}$$

Upper and lower limits are not specified for the sum in Eq. (12.63): it is to be understood that the sum extends over all real roots x_i of the equation $g(x) = y$. Conversely, note that if $g(x) = y$ has no real roots for a specified value of y, the transformed probability density is $\phi(y) = 0$.

As a specific illustration of the application of this result, consider the density obtained by the transformation $g(x) = x^2$ when the original random variable x has the standard normal distribution $N(0, 1)$. In this case, the transformation is differentiable with derivative $g'(x) = 2x$, but it is not invertible when defined on the entire real line, since $g(x) = g(-x)$ for all real x. Hence, for any positive y, the equation $y = g(x)$ has two real roots, one positive and one negative: $y = +\sqrt{x}$ and $y = -\sqrt{x}$. It then follows from Eq. (12.63) that the density $p_y(y)$ for the random variable y is given by:

$$p_y(y) = \frac{\phi(\sqrt{x})}{2\sqrt{x}} + \frac{\phi(-\sqrt{x})}{2\sqrt{x}}, \tag{12.64}$$

where $\phi(x) = e^{-x^2/2}/\sqrt{2\pi}$ is the density for the standard normal distribution. Substituting this explicit density into Eq. (12.64) yields:

$$p_y(y) \quad = \quad \frac{1}{\sqrt{2\pi y}} \, e^{-y/2}, \qquad\qquad (12.65)$$

for $y \geq 0$. It follows from the results presented in Chapter 4 that this density is a member of the family of gamma distributions, with shape parameter $\alpha = 1/2$ and location and scale parameters $\xi = 0$ and $\beta = 2$. This density function was plotted in Chapter 4, where it was seen to be J-shaped, decreasing monotonically from an infinite value at $y = 0$. Given the considerable importance of both the normal distribution and the quadratic transformation, it is not surprising that this particular special case of the gamma distribution arises frequently in practice. In fact, this distribution is a member of the family of *chi-squared distributions*; in particular, this distribution is the χ_1^2 distribution or "chi-squared distribution with one degree of freedom." The general family of chi-squared distributions is important in the characterization of estimators as noted in Chapter 6 and discussed further in Chapter 15.

12.7   Reciprocal transformations

A situation that arises frequently in modeling physical systems is that one variable depends *inversely* on another. For example, the capacitance C of an ideal parallel plate capacitor depends inversely on the separation d of the plates:

$$C \quad = \quad \frac{\epsilon A}{d}, \qquad\qquad (12.66)$$

where ϵ is the dielectric permmitivity of the material between the plates of the capacitor, and A is the area of the plates. More generally, *inverse square laws* arise frequently in physics, both for problems involving gravitational attraction and for problems involving electrostatic interactions between charged bodies. If the denominator variables in these problems are subject to uncertainty, it is reasonable to model them as random variables, leading us to consider transformations of the form $y = x^{-\alpha}$. The key point of the following discussion is that the influence of such a transformation is profound, leading us to distributions that are often rather pathologically behaved. For simplicity, this discussion will focus principally on the reciprocal transformation $y = 1/x$ (i.e., $\alpha = 1$ above), but many of the qualitative results extend to transformations of the more general form $y = x^{-\alpha}$ for arbitrary $\alpha > 0$.

The reciprocal transformation $g(x) = 1/x$ was one component of the linearizing transformation for the Michaelis-Menten model discussed in Sec. 12.2.1, and the Box-Cox transformation reduces essentially to the reciprocal when $\lambda = -1$. Like the square root, this transformation can be well behaved or badly behaved, depending on its domain. In particular, note that, like the square and square root transformations, the reciprocal is strictly monotonic (here, strictly decreasing) on the positive real line $R^+ = (0, \infty)$. Hence, it is order-reversing and invertible—indeed, it is its own inverse—in addition to being infinitely differentiable for all $x > 0$. Consequently, this transformation may be viewed as extremely well behaved on this domain, but it becomes rather badly behaved if we attempt to extend the domain to the entire real line. In particular, note that this transformation is singular at $x = 0$, corresponding to a *discontinuity of the second kind* [294, p. 10].

Given a density $p(x)$ for the random variable x, it is easy to derive the form of the density $\phi(z)$ for the transformed random variable $z = 1/x$. In particular, it follows from Eq. (12.54) that this density is:

$$\phi(z) \quad = \quad \frac{p(1/z)}{z^2}. \tag{12.67}$$

An important feature of this transformed distribution is that, if $p(0) \neq 0$ for the starting density, $\phi(z) \to p(0)/z^2$ for large z, implying heavy-tailed (specifically, Cauchy-like) behavior. Also, note that the tail behavior of the original density determines the qualitative behavior of $\phi(z)$ as $z \to 0$. In particular, if $p(x) \to 0$ rapidly enough as $x \to \pm\infty$, the transformed density will be finite at $z = 0$, but if the original distribution exhibits sufficiently heavy tails, the transformed density will be singular at $z = 0$. Further, the domain on which the original density is defined has a pronounced influence on the character of the transformed density. The following examples illustrate all of these points.

12.7.1   The Gaussian distribution

Because the Gaussian distribution has historically been "the" distribution of choice for describing uncertain quantities, it is useful to begin with a consideration of the reciprocal transformation of a Gaussian random variable. Specifically, assume $x \sim N(\mu, \sigma^2)$ and consider the transformed density $\phi(y)$; it follows from the general result (12.67) that this density is given by:

$$\phi(y) \quad = \quad \frac{e^{-(1-\mu y)^2/2\sigma^2 y^2}}{\sqrt{2\pi\sigma^2 y^2}}. \tag{12.68}$$

Plots of this density are shown in Fig. 12.18 for four different choices of μ and σ, and they illustrate the range of behavior we can expect from this transformation. The plot in the upper right corresponds to $\mu = 1$ and $\sigma = 0.1$, representing a random variable x with a nominal value of 1 and a "worst case uncertainty" of something like 30%. The transformed density exhibits a sharp peak centered at approximately $1/\mu$, and the only strong visual evidence of the non-Gaussian character of this distribution is its pronounced asymmetry. In contrast, the non-Gaussian character of the transformed density for $\mu = 1$ and $\sigma = 1$ shown in the upper right plot in Fig. 12.18 is obvious from the small secondary peak in the density at approximately $x = -1$. In fact, this peak becomes more pronounced with increasing σ, as seen in the lower left plot for $\mu = 1$ and $\sigma = 2$. This bimodality is seen at its most extreme when $\mu = 0$, shown in the lower right plot for $\sigma = 1$. This distribution is symmetric and bimodal, similar in character to the strongly non-Gaussian zero kurtosis example discussed in Chapter 4.

Figure 12.18: Four reciprocal Gaussian densities

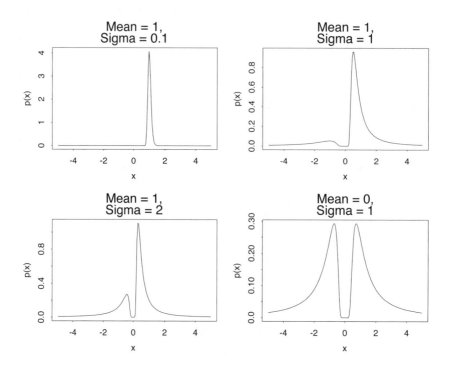

The general behavior just described—i.e., the increasingly non-Gaussian character of $y = 1/x$ as the ratio σ/μ increases—is a reflection of the fact that both "unphysical" values $x \leq 0$ and "troublesome" values $x \gtrsim 0$ occur with increasing probability as $\sigma/\mu \to \infty$. This observation illustrates the importance of selecting a reasonable distribution for the original variable x.

It is also interesting to consider the tail behavior of the transformed density $\phi(y)$ defined in Eq. (12.68). In particular, suppose $\mu > 0$ and consider the behavior of $\phi(y)$ for $y > 1/\mu$. It then follows that:

$$0 < \frac{(1/y - \mu)^2}{2\sigma^2} < \frac{\mu^2}{2\sigma^2} \quad \Rightarrow \quad e^{-\mu^2/2\sigma^2} < \exp\left\{-\frac{(1/y - \mu)^2}{2\sigma^2}\right\} < 1$$

$$\Rightarrow \quad \frac{e^{-\mu^2/2\sigma^2}}{\sqrt{2\pi\sigma^2}y^2} < \phi(y) < \frac{1}{\sqrt{2\pi\sigma^2}y^2}. \qquad (12.69)$$

This last inequality illustrates the point noted above, that the tails of $\phi(y)$ exhibit the same decay rate as the Cauchy distribution. *Consequently, it follows that none of the usual moment characterizations are useful for the transformed Gaussian distribution, just as in the case of the Cauchy distribution.*

Figure 12.19: Reciprocal transform of the Laplace density

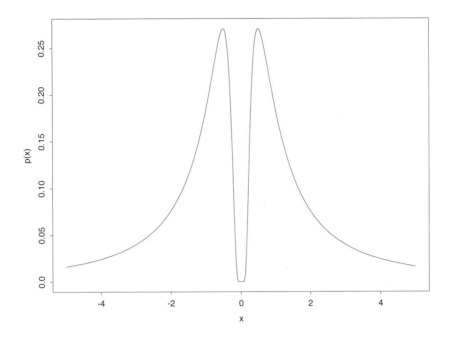

12.7.2 The Laplace distribution

As a second example, consider the zero-mean Laplace distribution:

$$p(x) = \frac{1}{2\alpha} e^{-|x|/\alpha}. \tag{12.70}$$

It follows from Eq. (12.67) that the transformed density is:

$$\phi(y) = \frac{1}{2\alpha y^2} e^{-1/\alpha|y|}, \tag{12.71}$$

which is plotted in Fig. 12.19. As in the previous example, it is interesting to note that the transformed density (12.71) is bimodal. Specifically, note that $\phi(0) = 0$ and $\phi(y) \to 0$ as $y \to \pm\infty$. Differentiating $\phi(y)$ with respect to y, it is not difficult to show that it exhibits maxima at $y = \pm 1/2\alpha$. Also, since $1/\alpha|y| \to 0$ as $y \to \pm\infty$, it follows that $p_y(y) \sim 1/2\alpha y^2$ as $y \to \pm\infty$, which is again the tail behavior of the Cauchy distribution, so the same conclusions apply for reciprocal transformations of Laplace random variables as for the Gaussian case considered above.

12.7.3 The Cauchy distribution

In view of the "Cauchy-like" behavior of the previous two examples, the following result is particularly interesting. If we apply the reciprocal transformation to the standard Cauchy density itself:

$$p_x(x) \quad = \quad \frac{1}{\pi} \frac{1}{1+x^2}, \tag{12.72}$$

we obtain the following result from Eq. (12.67):

$$p_y(y) \quad = \quad \frac{1}{\pi} \frac{1}{1+y^2}. \tag{12.73}$$

In other words, the standard Cauchy density is *invariant* under the transformation $g(x) = 1/x$.

12.7.4 The beta and Pareto distributions

An excellent illustration of the influence of the domain of the random variable x on the transformed density $\phi(y)$ is provided by the example of the beta and Pareto distributions. Specifically, suppose x exhibits the $PI(k,a)$ Pareto density discussed in Chapter 4:

$$p(x) \quad = \quad \frac{ak^a}{x^{a+1}}, \tag{12.74}$$

for $x \geq k > 0$ and $a > 0$. It follows from Eq. (12.67) that the corresponding density for the transformed variable $y = 1/x$ is of the form:

$$\phi(y) \quad = \quad ak^a y^{a-1}, \tag{12.75}$$

for $0 \leq y \leq 1/k$, representing a beta distributions on the interval $[0, 1/k]$ with shape parameters $p = a$ and $q = 1$. In particular, note that the case $k = 1$ corresponds to the standard beta distribution. Also, note that in this case, the domains of the original Pareto distribution and the transformed distribution partition the real line into (essentially) complimentary subsets: x is defined on the interval $[1, \infty)$, while $y = 1/x$ is defined on the interval $[0, 1]$.

A particularly interesting special case is $a = 1$ (and $k = 1$) for which the Pareto density exhibits the same Cauchy-like asymptotic decay rate seen in all three of the previous examples. In this case, the transformed distribution is the standard uniform distribution on $[0, 1]$. For $a > 1$, the Pareto density decays more rapidly than the Cauchy density and the transformed density increases monotonically with increasing y from $\phi(0) = 0$ to a maximum value of $\phi(1) = a$. For $0 < a < 1$, the original Pareto density decays more slowly than the Cauchy density, whereas the transformed density becomes J-shaped, with a singularity at $y = 0$, decreasing monotonically to a minimum value of $\phi(1) = a$.

12.7.5   The lognormal distribution

The lognormal distribution was introduced in Chapter 4, and it is interesting to note that, like the Cauchy distribution, the lognormal distribution is also invariant under the transformation $y = 1/x$. Specifically, recall that if x exhibits a lognormal distribution, then $\ln x \sim N(\mu, \sigma^2)$ for some μ and σ. Thus, it follows that $\ln y = -\ln x \sim N(-\mu, \sigma^2)$, implying that y also exhibits a lognormal distribution, with the only change being the sign of the parameter μ. As a particularly interesting special case, note that if $\mu = 0$, the reciprocal transformation does not change the distribution at all, exactly as in the case of the standard Cauchy distribution.

In fact, this last result extends to any "log symmetric" distribution. That is, suppose $\ln x$ exhibits *any* distribution that is symmetric about zero, including any member of the Pearson VII family of distributions, the logistic distribution, the Laplace distribution, or any of the symmetric beta distributions. It follows that $\ln y = -\ln x$ has the *same* distribution as $\ln x$, and thus that all of these "log symmetric" distributions for x are invariant under the transformation $y = 1/x$.

12.8   Exercises

1. Using the same format as the 25 pig transforms considered in Sec. 12.1.1, generate the corresponding 25 plots for the unit circle centered at the origin:
$$x^2 + y^2 = 1.$$

2. A model that arises frequently is the *thremally activated model:*
$$x(T) = x_0 e^{-E/kT},$$
where x_0 and E are parameters to be detrmined from measurements of the observed response $x(T)$ vs. the absolute temperature T and k is Boltzmann's constant. Derive a linearizing transformation for this model.

3. Consider the triangular density:
$$p(x) = \begin{cases} 0 & x < -1, \\ C(x+1) & -1 \le x < 0, \\ C(1-x) & 0 \le x < 1, \\ 0 & x \ge 1. \end{cases}$$

   a. Determine the normalization constant C.
   b. Derive an expression for the cumulative distribution function $F(x)$ for this density.
   c. Derive an expression for $F^{-1}(x)$.
   d. Use this result to generate triangularly distributed random numbers.

4. The logarithmic transformation is widely used. Suppose x is a random variable with density $p(x)$ that is nonzero only for $x \ge 0$.

    a. Derive a general expression for the density $\phi(z)$ of $z = \ln x$.

    b. Apply this result to the beta distribution with shape parameter $q = 1$ (see Chapter 4).

    c. The resulting distribution is well known: what is it?

    d. Apply this transformation result to the Pareto type I distribution, $P(I)(k, a)$.

    e. How does this distribution differ from that considered in parts [b] and [c]?

5. Determine the density $\phi(z)$ that results when the angular transformation discussed in Sec. 12.3.3 is applied to the uniform distribution on $[0, 1]$. Plot the resulting density.

6. The *Pareto type II distribution* or *Lomax distribution* is defined by the following density [162, p. 575]:

$$p(x) = \frac{a}{(x + 1)^{a+1}}, \quad x > 0,\ a > 0.$$

    a. Consider the transformation $z = \ln x$ and derive an expression for the density of z. Plot this density.

    c. Consider the transformation $r = 1/x$ and derive an expression for the density of r. Plot this density.

Chapter 13

Regression Models II: Mixed Data Types

Sections and topics:

The linear regression models discussed in Chapters 5 and 11 related a real-valued response variable to one or more real-valued explanatory variables (covariates). This chapter considers the problem of developing data models that involve explanatory variables and covariates of mixed data types: real-valued, binary, and/or categorical. To illustrate some of the influences of data type on analysis results, three different representations of the brain-weight/body-weight dataset introduced in Chapter 1 are briefly compared. Next, *analysis of variance (ANOVA) models* are introduced as regression models with categorical covariates and real-valued responses. A general regression framework that can be used to address a number of problems involving mixed data types is that of *generalized linear models*, and two special cases are introduced here: *logistic regression* is useful when the response variable is binary and the covariates are either categorical or real-valued, and *Poisson regression* is appropariate when the response variable is integer-valued (e.g., count data) and the covariates are either real-valued or categorical. Simple application examples are presented to illustrate all of these methods, as are brief discussions of important practical issues that arise in each case.

Key points of this chapter:

1. Quantitative data analysis involves variables that can be represented in different ways. As a specific example, temperature can be represented as a real variable (e.g., in degrees Kelvin, Celsius, or Farenheit), as an ordinal range variable (e.g., degrees expressed in decade ranges), or as a binary variable (above freezing or below freezing).

2. Different analytical methods are appropriate to different data types. For example, associations between real-valued variables can be quantified using the product-moment or rank correlations introduced in Chapter 10, but associations between binary variables are best quantified using other techniques, such as the odds ratio introduced in this chapter.

3. *Analysis of variance (ANOVA) models* relate a real-valued response variable to one or more categorical explanatory variables. While these results are typically represented in tabular form, they can be represented explicitly as more or less standard regression models involving binary covariates derived from the categorical explanatory variables.

4. In their simplest form, *logistic regression models* relate probabilities of binary outcomes to real-valued explanatory variables. The binary representation for categorical explanatory variables used in the regression formulation of ANOVA models can also be used to incorporate categorical explanatory variables into multivariable logistic regression models.

5. *Poisson regression models* relate integer-valued count response variables to one or more real-valued and/or categorical explanatory variables, using techniques very similar to those used in building logistic regression models.

6. The linear regression models discussed in Chapters 5 and 11, logistic regression models, and Poisson regression models all belong to the class of *generalized linear models (GLMs)*. The differences between these special cases lie primarily in the data distributions assumed: Gaussian for the standard linear regression problem, binomial for the logistic regression problem, and Poisson for the Poisson regression problem.

7. An important practical difference between the linear regression models discussed in Chapters 5 and 11 and the GLMs discussed in this chapter is that the latter involve a nonlinear transformation of the response variable. As a consequence, ordinary least squares results are not directly applicable to these GLMs, although specialized iteratively reweighted least squares model-fitting procedures have been developed for them.

8. As will be discussed further in Chapter 15, analysis of variance provides a useful basis for characterizing standard linear regression models; this chapter describes and illustrates the closely related notion of *analysis of deviance* that is appropriate to logistic and Poisson regression models.

13.1 Models with mixed data types

The predictive data models considered so far in this book predict a response variable y from a set of m explanatory variables $\{x^j\}$ by an expression of the general form:

$$y_k \simeq f(x_k^1, \ldots, x_k^m), \quad \text{for } k = 1, 2, \ldots, N. \tag{13.1}$$

When all of these variables are real-valued, this data model reduces to a standard regression model like those discussed in Chapters 5, 11, and 12, but when either the response variable y or the explanatory variables $\{x^j\}$ are not real-valued, alternative model forms are required. This chapter introduces some of the most common of these alternative model forms.

When a real-valued response variable depends on one or more categorical explanatory variables, the most common model form is an *ANOVA model*, whose name is an acronym for "analysis of variance," a standard statistical methodology dating back to the 1920s. While ANOVA models are not typically introduced as regression models, they can be cast into this form, as discussed by both Draper and Smith [94, ch. 9] and Jobson [159, sec. 5.1.3], and this representation provides a useful way of looking at these models in the context of the other models discussed in this chapter. In addition, the ANOVA framework is commonly used in characterizing the results of the generalized linear models described in the next paragraph. When an ANOVA model involves only a single explanatory variable, it is commonly called a *one-way ANOVA model*, while models involving n explanatory variables are called n-way ANOVA models. In the simplest n-way ANOVA models, each explanatory variable has its own, independent influence, but more complex ANOVA models include *interaction terms* involving several variables together. Also, extensions of ANOVA models are possible in settings where some of the explanatory variables are categorical and others are real-valued, leading to what are called *analysis of covariance models*. All of these ideas are discussed further in Sec. 13.3.

When the response variable y is not real-valued, particular members of the family of *generalized linear models (GLMs)* introduced in Sec. 13.4 are often useful. Essentially, these models apply some simple nonlinear transformation $g(\cdot)$ to the response variable to obtain a real-valued quantity $g(y)$ that depends approximately linearly on real-valued explanatory variables. Two particularly important practical applications of the GLM framework are *logistic regression*, discussed further in Sec. 13.5, and *Poisson regression*, discussed further in Sec. 13.6. Logistic regression uses the nonlinear transformation $g(\cdot)$ to construct a real-valued quantity from estimated probabilities, and it is probably the most popular approach in practice to modeling binary responses. Similarly, Poisson regression applies a logarithmic transformation to count data and is probably the most popular approach to modeling count data. Like the analysis of covariance models mentioned above, both of these model classes can accommodate either real-valued or categorical covariates.

13.2   The influences of data type

The main point of the discussion presented in the previous section was that the inclusion of multiple data types in an analysis strongly influences the types of data models we consider. The following discussion illustrates a related point that reemerges several times in subsequent discussions: data type is not unique. That is, it is frequently possible to represent the same variable by a set of real values, as a categorical variable defined by ranges of real values (e.g., quartiles), as an ordinal variable by computing ranks, or as a binary variable by comparing with a specified referece value (e.g., "greater than or equal to the median" vs. "less than the median"). Differences between real and ordinal representations were already seen in Chapter 10 in connection with the differences between product-moment and rank correlation values, and the following two sections extend this discussion to a comparison with binary representations.

13.2.1   Binary association: the odds ratio

As noted in Chapter 2, binary variables arise frequently in practice, particularly in medical applications where classification is important (e.g., patient X has been diagnosed with disease Y or they have not, tissue sample X has been classified as cancerous or not, etc.). Thus, data models involving binary variables, either as explanatory variables or as responses, are important in practice. In cases involving a single binary explanatory variable and a single binary response, this modeling is essentially equivalent to association analysis between binary sequences, and a number of different measures have been developed to characterize these associations [8, 66].

One important measure of association between two binary variables is the *odds ratio*, defined as follows [66, p. 36]. Suppose we have N observations of two binary variables, $\{x_k\}$ and $\{y_k\}$, each taking the values 0 or 1. The following four numbers define the *contingency table* that forms the basis for many different characterizations of the relationship between these variables:

1. n_{00} is the number of observations k for which $x_k = 0$ and $y_k = 0$;

2. n_{01} is the number of observations k for which $x_k = 0$ and $y_k = 1$;

3. n_{10} is the number of observations k for which $x_k = 1$ and $y_k = 0$;

4. n_{11} is the number of observations k for which $x_k = 1$ and $y_k = 1$.

The odds ratio is a measure of the tendency for these two variables to agree, and it is defined by:

$$O_{xy} = \frac{n_{00}n_{11}}{n_{01}n_{10}} = \frac{\hat{p}_{00}\hat{p}_{11}}{\hat{p}_{01}\hat{p}_{10}}, \tag{13.2}$$

where $\hat{p}_{ij} = n_{ij}/N$ is an estimate of the probability that $x_k = i$ and $y_k = j$ for $i, j = 0$ or 1. It is not difficult to show (Exercise 1) that, if $\{x_k\}$ and $\{y_k\}$ are statistically independent, then $O_{xy} = 1$. It follows from this observation that O_{xy} measures association, since $O_{xy} < 1$ implies that the two sequences agree less often than expected under independence, while $O_{xy} > 1$ implies that they agree more often than expected. More generally, note that O_{xy} can vary between 0 when $n_{00} = 0$ or $n_{11} = 0$ and ∞ when $n_{10} = 0$ or $n_{01} = 0$.

Of course, the question of how much larger or smaller than 1 the odds ratio needs to be to represent a "significant departure from independence" is important in practice, as the example presented in the next section demonstrates. A common method of constructing confidence intervals for the odds ratio is based on the observation that a logarithmic transformation gives the odds ratio an approximately normal distribution with standard deviation [66, p. 37]:

$$ \tilde{\sigma} \simeq \sqrt{\frac{1}{n_{00}} + \frac{1}{n_{01}} + \frac{1}{n_{10}} + \frac{1}{n_{11}}}, \tag{13.3} $$

providing the basis for constructing approximate two-sided confidence intervals at level α for $\ln O_{xy}$ as $\ln O_{xy} \pm z_{\alpha/2}\tilde{\sigma}$. Exponentiating these limits then gives the following confidence limits for the true value of the odds ratio:

$$ \text{OR} \in [O_{xy}e^{-z_{\alpha/2}\tilde{\sigma}}, O_{xy}e^{z_{\alpha/2}\tilde{\sigma}}]. \tag{13.4} $$

As Collett notes [66, p. 37], an advantage of this method of constructing confidence intervals is that it guarantees that both limits are positive, an important point since the odds ratio cannot be negative. Conversely, as the following example illustrates, these confidence intervals can be very wide.

13.2.2 Do big animals have big brains?

The following example provides a number of simple but illuminating illustrations of the influence of different data representations on what is nominally the same problem. The brain-weight/body-weight dataset was introduced in Chapter 1 where it was used to illustrate the difference between working with raw data and transformed data. Specifically, while the correlation coefficient estimated between the raw data values was quite small and, together with the extreme appearance of the scatterplot of brain weight vs. body weight, suggestive of little or no association between these two variables. In contrast, either taking logarithms of both variables or converting them to ordinal variables by replacing the values with their corresponding ranks yielded sequences that appeared to have nonnegligible associations. Conversely, it is also important to note that these different types of association analysis are addressing slightly different questions: the product-moment correlation coefficient attempts to quantify the tendency for one variable to vary proportionally to the other, while the rank correlation coefficient attempts to quantify the tendency for larger values of one variable to be associated with larger values of the other. In the body/brain example, then, the product-moment correlation between the original variables asks, effectively, "does brain weight scale linearly with body mass?" In contrast, the Spearman rank correlation coefficient effectively asks, "do heavier animals have heavier brains?" which isn't exactly the same thing. The following example addresses still a third question: do "big" animals have "big" brains?

Figure 13.1: Binarizations of the brain-body dataset: the mean binarizations discussed in the text (left), and the corresponding median binarizations (right)

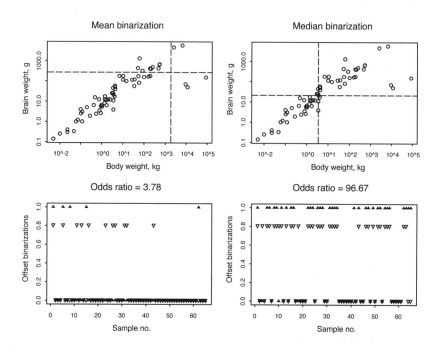

The upper left plot in Fig. 13.1 shows the brain-weight vs. body-weight scatterplot on log-log coordinates, with a vertical dashed line at the mean body weight, and horizontal dashed line at the mean brain weight. The lower left plot shows *binarizations* of these variables, defined as:

$$b(x_k) = \begin{cases} 1 & x_k \geq \bar{x}, \\ 0 & x_k < \bar{x}, \end{cases}$$

$$b(y_k) = \begin{cases} 1 & y_k \geq \bar{y}, \\ 0 & y_k < \bar{y}, \end{cases} \tag{13.5}$$

where x_k represents the kth body weight value in the dataset, y_k represents the corresponding brain weight value, and $\bar{x}$ and $\bar{y}$ are the respective average body and brain weights. Thus, $b(x_k) = 1$ if record k corresponds to a "big animal" and $b(y_k) = 1$ if record k corresponds to a "big brain," and these values are plotted in the lower left in Fig. 13.1. Specifically, the solid triangles represent the binarized body weight values $\{b(x_k)\}$, while the open triangles represent $\{b(y_k)\}$, scaled by a factor of 0.8 so the points in the two sequences are distinctly visible.

In contrast, the reference lines in the upper right plot correspond to the *median* brain and body weights, and points plotted in the lower right represent the corresponding median-based binarizations. That is, the solid triangles in the lower right plot represent the values $b(x_k)$ computed by replacing the mean $\bar{x}$ in Eq. (13.5) with the median body weight, and the open triangles represent the corresponding median-based $b(y_k)$ values, again with the vertical axis scaled to make the points in the two sequences distinctly visible. Comparing these two lower plots, it is clear that the median binarization yields binary sequences that are much more similar than the mean binarization does. Quantitatively, this difference is reflected in the enormous difference in odds ratios computed for the two cases: 96.67 for the median binarizations versus 3.78 for the mean binarizations. The 95% confidence intervals for these estimated odds ratios are $[0.55, 25.88]$ for the mean binarization and $[18.02, 518.53]$ for the median binarization, suggesting that the association is significant when based on median binarizations but not significant when it is based on mean binarizations, since the mean binarization confidence interval includes OR $= 1$. These results also illustrate the point made previously that the standard confidence intervals for the odds ratio can be extremely wide.

The practical importance of this example lies in the fact that it is frequently necessary or desirable to re-represent data values in different forms than the one in which we receive them. Binarizations and more general quantizations (also called *stratifications*) are common, but this example demonstrates that the details of *how* we implement these quantizations are sometimes quite important.

13.3   ANOVA models

The term "analysis of variance" (ANOVA) derives from the idea of decomposing the estimated variance of a real-valued response variable y into components associated with independent sources of variation. This decomposition is based on the fact that the variance of a sum of independent random variables is simply the sum of the individual variances. The resulting analysis is extremely standard in statistics, where it is usually introduced in connection with designed experiments. The following discussion introduces the basic ideas and, as noted in Sec. 13.1, illustrates that ANOVA models can be represented as regression models involving categorical explanatory variables, following the development of Jobson [159, sec. 5.1.3]. For a much more complete treatment of the ideas introduced here, refer to the book by Sahai and Ageel [254]. (A weakness of this book is that the authors argue, based on classical results, that traditional ANOVA methods are robust with respect to departures from standard working assumptions; for an alternative view based on more recent research results, refer to the book by Wilcox [295, ch. 7].)

13.3.1   Analysis of variance (ANOVA)

The simplest formulation of the analysis of variance problem is the one-way problem, where a real-valued response variable y is assumed to depend on the value of a categorical explanatory variable x that can take any one of C possible values. In the standard formulation of the one-way ANOVA problem, it is assumed that n observations of y are made for each value of x, so the overall dataset contains $N = nC$ observations. The one-way ANOVA model is then written as:

$$y_{ij} = \mu + \alpha_i + e_{ij}, \tag{13.6}$$

where y_{ij} represents the jth observation of y when x takes its ith possible value. Note that in this representation, the C distinct values of x have been used in forming the $C \times n$ response matrix $\{y_{ij}\}$ so that x does not appear directly in Eq. (13.6), but is present indirectly through the index i. The constant μ in Eq. (13.6) represents the overall mean of y while the C parameters $\{\alpha_i\}$ describe the influence of each level of x (i.e., each possible value for x) on y. Further, there is no loss of generality in assuming that these parameters satisfy the constraint:

$$\sum_{i=1}^{C} \alpha_i = 0, \tag{13.7}$$

since any nonzero value for this sum can always be incorporated into the constant μ. The term e_{ij} in Eq. (13.6) represents the prediction error associated with this data model.

Once the data matrix $\{y_{ij}\}$ has been formed, estimation of the parameters μ and $\{\alpha_i\}$ in the one-way ANOVA model from the observed data values is extremely easy. Specifically, define the following means:

$$\bar{y}_{i\cdot} = \frac{1}{n} \sum_{j=1}^{n} y_{ij}, \quad i = 1, 2, \ldots, C,$$

$$\bar{y}_{\cdot\cdot} = \frac{1}{C} \sum_{i=1}^{C} \bar{y}_{i\cdot} = \frac{1}{nC} \sum_{i=1}^{C} \sum_{j=1}^{n} y_{ij}. \tag{13.8}$$

It follows directly from Eq. (13.6) for the ANOVA model, the constraint in Eq. (13.7), and the second of the equations in Eq. (13.8) that:

$$\bar{y}_{\cdot\cdot} = \mu + \frac{1}{nC} \sum_{i=1}^{C} \sum_{j=1}^{n} e_{ij}. \tag{13.9}$$

Thus, if $E\{e_{ij}\} = 0$ for all i and j, it follows that $\bar{y}_{\cdot\cdot}$ is an unbiased estimator of the overall mean parameter μ. Similarly, it follows from Eqs. (13.6) and (13.7) and the first of Eqs. (13.8) that:

$$\bar{y}_{i\cdot} = \mu + \alpha_i + \frac{1}{n} \sum_{j=1}^{n} e_{ij}, \tag{13.10}$$

implying that $E\{y_{i\cdot}\} = \mu + \alpha_i$. In fact, under the assumptions that $\{e_{ij}\}$ are zero-mean, uncorrelated random variables with common variance σ_e^2, then $\bar{y}_{\cdot\cdot}$ represents the best linear unbiased estimate (BLUE) for μ, and $\hat{\alpha}_i = \bar{y}_{i\cdot} - \bar{y}_{\cdot\cdot}$.

is the BLUE for α_i. If these assumptions are further strengthened to those of normally distributed errors, it follows that these are the best unbiased estimates possible [254, p. 27].

As noted earlier, the term "analysis of variance" derives from the idea of partitioning estimated variances into contributions from the different terms in the model. Specifically, the basis for this partitioning is the following decomposition of the sums of squares used to estimate these variance components (Exercise 2):

$$\sum_{i=1}^{C}\sum_{j=1}^{n}(y_{ij}-\bar{y}_{..})^2 = n\sum_{i=1}^{C}(\bar{y}_{i.}-\bar{y}_{..})^2 + \sum_{i=1}^{C}\sum_{j=1}^{n}(y_{ij}-\bar{y}_{i.})^2. \qquad (13.11)$$

Commonly, the sum on the left-hand side of this equation is called the *total sum of squares*, SS_T; the first term on the right-hand side is called the *sum of squares between groups*, SS_B; and the second term is called the *sum of squares within groups*, SS_W. It is not difficult to show (Exercise 3) that the expected values for these sums of squares are:

$$E\{SS_B\} = n\sum_{i=1}^{C}\alpha_i^2 + (C-1)\sigma_e^2,$$

$$E\{SS_W\} = C(n-1)\sigma_e^2. \qquad (13.12)$$

The multipliers of σ_e^2 in these expressions represent the degrees of freedom in these estimates (i.e., the effective number of independent random errors contributing to each expression), and the corresponding *mean squares* are defined by dividing each sum of squares by these numbers:

$$MS_B = \frac{n}{C-1}\sum_{i=1}^{C}(\bar{y}_{i.}-\bar{y}_{..})^2,$$

$$MS_W = \frac{1}{C(n-1)}\sum_{i=1}^{C}\sum_{j=1}^{n}(y_{ij}-\bar{y}_{i.})^2. \qquad (13.13)$$

It follows from the expectations for the sums of squares given in Eq. (13.12) that the expectations for the mean squares are:

$$E\{MS_B\} = \sigma_e^2 + \frac{n}{C-1}\sum_{i=1}^{C}\alpha_i^2,$$

$$E\{MS_W\} = \sigma_e^2. \qquad (13.14)$$

All of the results just presented may be derived on the basis of the assumptions that $\{e_{ij}\}$ is a collection of zero-mean, uncorrelated random variables with common variance $E\{e_{ij}^2\} = \sigma_e^2$. Traditionally, ANOVA models are associated with the null hypothesis described next, for which it is necessary to make the stronger distributional assumption that $\{e_{ij}\}$ is a collection of mutually independent, jointly Gaussian random variables.

Table 13.1: Format of the traditional ANOVA table

Source of variation	Degrees of freedom	Sum of squares	Mean square	F-value	p-Value
Between	$C-1$	SS_B	MS_B	$\frac{MS_B}{MS_W}$	$\mathcal{P}\{F \geq \frac{MS_B}{MS_W}\}$
Within	$C(n-1)$	SS_W	MS_W		
Total	$nC-1$	SS_T			

The null hypothesis associated with one-way analysis of variance is that the covariate x in the model has no effect; that is, the null hypothesis assumes that $E\{y_{ij}\} = \mu$, independent of the level i of the categorical explanatory variable x. This assumption is equivalent to the hypothesis:

$$H_0 : \alpha_1 = \alpha_2 = \cdots = \alpha_C = 0. \tag{13.15}$$

It follows from Eq. (13.14) that hypothesis H_0 is equivalent to the condition $E\{MS_B\} = E\{MS_W\}$. Under the assumption that $\{e_{ij}\}$ is an independent collection of zero-mean Gaussian random variables, the appropriate test for this hypothesis is the F-test discussed in Chapter 9. Specifically, under these distributional assumptions, the following ratio has an F-distribution with $C-1$ and $C(n-1)$ degrees of freedom:

$$F = \frac{MS_B}{MS_W} \sim F(C-1, C(n-1)). \tag{13.16}$$

Further, it follows from Eq. (13.14) that if the null hypothesis H_0 does not hold, then $F > 1$, so the appropriate test is one-sided, effectively asking the following question: is the observed value of F large enough that its probability is smaller than a specified significance level γ (e.g., $\gamma = 0.05$) under the $F(C-1, C(n-1))$ distribution? If so, this represents evidence against the null hypothesis, implying that the covariate x does have an effect on the response variable y.

Traditionally, the results of an analysis of variance are represented in the form shown in Table 13.1. A specific example is presented in Sec. 13.3.3, but Table 13.1 illustrates the general layout. The leftmost column lists the sources of variation, which can be more general in two- and higher-way layouts and when the ANOVA framework is applied to summarize regression results under the Gaussian data assumption, a topic discussed in Chapter 15. The next two columns then give the degrees of freedom and the sums of squares associated with these sources of variation. Dividing the between and within sums of squares by their associated degrees of freedom gives the between and within mean square values listed in the fourth column. These values provide the basis for computing the F statistic, which is listed in the fifth column, and the probability of observing a value this large or larger under the appropriate F-distribution is given in the sixth column.

Before turning to the illustrative example given in Sec. 13.3.3, it is useful to note the following points. First, the one-way ANOVA model described here represents what is called a *fixed effects model* because the parameters $\{\alpha_i\}$ are assumed to be unknown but *deterministic* (i.e., nonrandom) quantities. It is also possible to consider models in which these parameters are random, and the simplest of these is the *random effects model* in which they are all assumed to be zero-mean random variables with unknown but constant variance σ_α^2. Combining these ideas leads to *mixed effects models* containing both random and fixed effects. These designations apply to both the one-way ANOVA models considered here and more general n-way models. The discussions presented here are restricted to the case of fixed effects models in order to keep the discussion to a reasonable length; for a discussion of random and mixed effect models—along with a much more extensive discussion of fixed effect models than it is possible to give here—refer to the book by Sahai and Ageel [254] (although see the cautionary notes given in Sec. 13.3.2 below).

13.3.2   Extensions and practical issues

The following sections give brief introductions to a number of important topics in analysis of variance, including its connection with regression analysis, relaxation of the assumption of an equal number of observations at each level of the categorical variable in a one-way analysis, the effects of significantly non-Gaussian data distributions, and the additional issues that arise when multiple explanatory variables are present, possibly of different types.

Regression representation

As both Jobson [159] and Draper and Smith [94] discuss in detail, ANOVA models can be represented in the form of regression models. The following discussion gives a brief introduction to this idea to establish the connection between what may appear to be two very different model classes: the one-way ANOVA model defined by Eq. (13.6) and the linear regression models discussed in Chapters 5, 11, and 12. The basic idea is to replace the categorical variable x in the ANOVA model with a set of $C-1$ binary variables, where C is the number of distinct values that x can assume. These binary variables may be defined in different ways, and the different choices have different practical consequences, but to illustrate the regression formulation of ANOVA models, the following discussion adopts the choice that is conceptually the simplest.

First, assume we have N observations $\{y_k\}$ of the real-valued response y, and N observations $\{x_k\}$ of the explanatory variable x, which assumes one of

C distinct values, $v_1, v_2, \ldots, v_C$. Define the C binary variables $\{b_{i,k}\}$ as:

$$b_{i,k} = \begin{cases} 1 & \text{if } x_k = v_i, \\ 0 & \text{otherwise.} \end{cases} \qquad (13.17)$$

The ANOVA model defined in Eq. (13.6) can now be written as:

$$y_k = \mu + \sum_{i=1}^{C} \alpha_i b_{i,k} + e_k, \qquad (13.18)$$

for $k = 1, 2, \ldots, N$. If we ignore the binary character of the new variables $\{b_{i,k}\}$ and simply write Eq. (13.18) in vector-matrix notation, we obtain what appears to be a standard regression problem. The difficulty is that the constraint on the model parameters $\{\alpha_i\}$ imposed by Eq. (13.8) in the ANOVA model formulation means that the resulting $\mathbf{X}^T\mathbf{X}$ matrix is not invertible. One simple way to overcome this problem is to define the $C - 1$ dummy variables:

$$D_{i,k} = b_{i,k} - b_{C,k}, \qquad (13.19)$$

for $i = 1, 2, \ldots, C - 1$ and $k = 1, 2, \ldots, N$. It follows on combining Eqs. (13.8), (13.18), and (13.19) that:

$$y_k = \mu + \sum_{i=1}^{C-1} \alpha_i D_{i,k} + e_k, \qquad (13.20)$$

for $k = 1, 2, \ldots, N$. In vector-matrix notation, these equations become:

$$\mathbf{y} = \mathbf{X}\theta + \mathbf{e}, \qquad (13.21)$$

where $\mathbf{y}$, θ, and $\mathbf{e}$ are given by:

$$\begin{aligned} \mathbf{y} &= [y_1, y_2, \ldots, y_N]^T, \\ \theta &= [\mu, \alpha_1, \ldots, \alpha_{C-1}]^T, \\ \mathbf{e} &= [e_1, e_2, \ldots, e_N]^T, \end{aligned} \qquad (13.22)$$

and the $N \times C$ matrix $\mathbf{X}$ is given by:

$$\mathbf{X} = \begin{bmatrix} 1 & D_{1,1} & \cdots & D_{C-1,1} \\ 1 & D_{1,2} & \cdots & D_{C-1,2} \\ \vdots & \vdots & \vdots & \vdots \\ 1 & D_{1,N} & \cdots & D_{C-1,N} \end{bmatrix}. \qquad (13.23)$$

The least squares solution of this regression problem is given explicitly by:

$$\hat{\theta} = (\mathbf{X}^T\mathbf{X})^{-1}\mathbf{X}^T\mathbf{y}, \qquad (13.24)$$

exactly as in Sec. 13.3.1 (see Exercise 4).

This connection between ANOVA models and linear regression models is important for a number of reasons. First, it establishes the idea that one-way ANOVA models may be viewed as regression models with a real response variable and one categorical explanatory variable. More important, the approach taken here—introducing binary indicator variables—can be used to include categorical explanatory variables in both logistic and Poisson regression models. Further, the regression formulation extends with minimal modification to cases of unequal numbers of observations at each level of the categorical variable x, to n-way ANOVA models involving several categorical predictors, or to analysis of covariance problems involving both categorical and real-valued predictors.

Unequal numbers of observations

As noted in Sec. 13.3.1, one of the explicit assumptions in the standard formulation of the one-way ANOVA model is that each of the C levels of the explanatory variable x is observed the same number of times, n. As discussed at the end of this section, this restriction can be relaxed by a slight change in the definition of the dummy variables used in the regression formulation of the problem, but if the numbers of observations of each level are not all the same, the explicit solutions to the one-way ANOVA problem change, requiring certain modifications of the standard ANOVA table. This problem is discussed more completely by Sahai and Ageel [254, sec. 2.12], but the following paragraphs give a brief summary of the results for this more general case.

Let n_i denote the number of observations for which the ith value of x is observed and note that the total number of observations is:

$$N = \sum_{i=1}^{C} n_i. \tag{13.25}$$

The one-way ANOVA model corresponds to Eq. (13.6) as before, but the constraint on the coefficients α_i changes to:

$$\sum_{i=1}^{C} n_i \alpha_i = 0. \tag{13.26}$$

Note that if $n_i = n$ for all i as in Sec. 13.3.1, this condition reduces to Eq. (13.8). The means $\bar{y}_{i\cdot}$ and $\bar{y}_{\cdot\cdot}$ are given by:

$$\bar{y}_{i\cdot} = \frac{1}{n_i} \sum_{j=1}^{n_i} y_{ij},$$

$$\bar{y}_{\cdot\cdot} = \frac{1}{N} \sum_{i=1}^{C} \sum_{j=1}^{n_i} y_{ij} = \frac{1}{N} \sum_{i=1}^{C} n_i \bar{y}_{i\cdot}. \tag{13.27}$$

Note that $\hat{\mu} = \bar{y}_{\cdot\cdot}$ is still an unbiased estimator of μ, and $\hat{\alpha}_i = \bar{y}_{i\cdot} - \bar{y}_{\cdot\cdot}$ is still an unbiased estimator of α_i. The between, within, and total sums of squares are given in terms of the means $\bar{y}_{i.}$ and $\bar{y}_{..}$ by the following expressions:

$$SS_B = \sum_{i=1}^{C} n_i(\bar{y}_{i.} - \bar{y}_{..})^2,$$

$$SS_W = \sum_{i=1}^{C} \sum_{j=1}^{n_i} (y_{ij} - \bar{y}_{i.})^2,$$

$$SS_T = \sum_{i=1}^{C} \sum_{j=1}^{n_i} (y_{ij} - \bar{y}_{..})^2. \tag{13.28}$$

These quantities have $C - 1$, $N - C$, and $N - 1$ associated degrees of freedom, respectively, so the between- and within-mean squares are defined by:

$$MS_B = \frac{1}{C-1} \sum_{i=1}^{C} n_i(\bar{y}_{i.} - \bar{y}_{..})^2,$$

$$MS_W = \frac{1}{N-C} \sum_{i=1}^{C} \sum_{j=1}^{n_i} (y_{ij} - \bar{y}_{i.})^2, \tag{13.29}$$

and have the following expected values:

$$E\{MS_B\} = \sigma_e^2 + \frac{1}{C-1} \sum_{i=1}^{C} n_i \alpha_i^2,$$

$$E\{MS_W\} = \sigma_e^2. \tag{13.30}$$

The null hypothesis H_0 that the explanatory variable x has no effect is the same as before and is again equivalent to the assertion that the two expected values given in Eq. (13.29) are equal. Under the Gaussian data assumption, this leads to the same F-test as before, based now on an F-distribution with $C - 1$ and $N - C$ degrees of freedom. (In fact, these numbers are the same as before, since if $n_i = n$ for all i, $C(n-1) = nC - C = N - C$.) Thus, while the terms appearing in the standard ANOVA table given in Table 13.1 are computed differently, they appear in the same places as before and are interpreted the same way.

As in the balanced case where $n_i = n$ for all i, it is possible to cast the unbalanced one-way ANOVA problem in the regression framework, although there is a slight difference. Specifically, since Eq. (13.8) is replaced by Eq. (13.26) in this more general formulation, it follows that Eq. (13.19) must be replaced with (Exercise 5):

$$D_{i,k} = b_{i,k} - \frac{n_i}{n_C} b_{C,k}, \tag{13.31}$$

which reduces to the previous expression when $n_i = n$ for all i. Otherwise, the regression problem formulation remains unchanged.

Robustness questions

As noted in Chapter 9, the classical F-test is extremely sensitive to violations of the Gaussian data working assumptions. While some authors cite older studies claiming robustness for the F-test [254, sec. 2.20], more recent results show that the F-test is quite sensitive to departures from normality [295, sec. 1.5]. One way of overcoming this difficulty is to extend the Yuen-Welch method described in Chapter 9 for comparing two means to the problem of comparing multiple means simultaneously. The following discussion presents a method of this type described by Wilcox [295, sec. 7.1.1].

As in the Yuen-Welch test, the basic idea is to replace means with trimmed means because of their better outlier resistance and test the null hypothesis that the trimmed means are equal for the C different values the categorical predictor x can assume. To do this, we compute $w_i = 1/d_i$ where d_i is the squared error estimate for the trimmed mean that appears in the Yuen-Welch test. It is more convenient to work with w_i directly, which is given by:

$$w_i = \frac{h_i(h_i - 1)}{(n_i - 1)s_{wi}^2}, \tag{13.32}$$

where $h_i = n_i - 2\lfloor \gamma n_i \rfloor$ is the effective size of the ith γ-trimmed sample, and s_{wi}^2 is the Windsorized sample variance for the y_{ij} values, defined in Chapter 9 in connection with the Yuen-Welch test. Analogous to the second equation in Eq. (13.27) for $\bar{y}_{..}$, define $\tilde{y}$ as:

$$\tilde{y} = \sum_{i=1}^{C} \tilde{w}_i \bar{y}_i^{\gamma}, \tag{13.33}$$

where $\bar{y}_i^{\gamma}$ is the γ-trimmed mean of the y_{ij} values and $\tilde{w}_i$ is given by:

$$\tilde{w}_i = \frac{w_i}{\sum_{i=1}^{C} w_i}. \tag{13.34}$$

Next, we compute the following two numbers:

$$A = \frac{1}{C - 1} \sum_{i=1}^{C} w_i (\bar{y}_i^{\gamma} - \tilde{y})^2,$$

$$B = \frac{2(C - 2)}{(C^2 - 1)} \sum_{i=1}^{C} \frac{(1 - \tilde{w}_i)^2}{h_i - 1}. \tag{13.35}$$

The basis for the hypothesis test is that, under the null hypothesis, the following ratio has an approximate F-distribution with ν_1 and ν_2 degrees of freedom:

$$F = \frac{A}{1 + B} \sim F(\nu_1, \nu_2), \tag{13.36}$$

where these numbers are given by the following two expressions:

$$\nu_1 = C - 1,$$

$$\nu_2 = \left[\frac{3}{C^2 - 1} \sum_{i=1}^{C} \frac{(1 - \tilde{w}_i)^2}{h_i - 1} \right]^{-1}. \tag{13.37}$$

Note that in the absence of trimming, the weights w_i appearing in these expressions are inversely proportional to the estimated variance of the y-values observed for the ith value of the explanatory variable x. The essential idea is to reduce the influence of more variable portions of the dataset when these estimated variances are substantially different; in the presence of trimming, the influence of extreme data values is further reduced.

More complex ANOVA models

Higher-way ANOVA models involve more than one categorical predictor variable. The simplest case is the following two-way ANOVA model, where z_{ijk} represents the the kth sample of the response variable z observed when covariate x takes its ith possible value and covariate y takes its jth possible value:

$$y_{ijk} = \mu + \alpha_i + \beta_j + e_{ijk}. \tag{13.38}$$

Note that this model implies that the two explanatory variables, x and y, each act independently: the effect of changing x from level i_1 to level i_2 is the same for all levels j of y, and vice versa. In cases where this is not true, the two factors are said to interact, and in that case, a more general model is required, of the form:

$$y_{ijk} = \mu + \alpha_i + \beta_j + \gamma_{ij} + e_{ijk}, \tag{13.39}$$

where the parameters γ_{ij} account for the interaction. To estimate the parameters $\{\alpha_i\}$, $\{\beta_j\}$ and $\{\gamma_{ij}\}$ in these models, it is necessary to have at least one observation for every combination of x and y values. Datasets satisfying this condition are typically called *two-way crossed classifications*, because the two explanatory variables are "crossed," forming a rectangular array of cells with at least one observation per cell. Further, to estimate the interaction terms γ_{ij} in Eq. (13.39), it is necessary to have more than one observation per cell.

   Unfortunately, space does not permit a detailed discussion of higher-way ANOVA models here. Key points are, first, that these models are useful in practice, and they arise very frequently in statistics, particularly in connection with designed experiments, so it is useful to know what they are. Second, note that like the one-way ANOVA models described above, higher-way ANOVA models can also be represented as regression models involving categorical explanatory variables and real-valued response variables. This representation is particularly useful because it provides a conceptually simple bridge to the class of *analysis of covariance models*. There, multiple explanatory variables are present, some of which are categorical and others are real-valued. For a reasonably detailed discussion of analysis of covariance models in the regression framework, refer to the book by Jobson [159, sec. 5.3]. For an extensive treatment of higher-way ANOVA models, including reasonably detailed discussions of ANOVA models with interactions, strategies for handling missing observations in crossed classifications, and an introduction to experimental design, refer to the book by Sahai and Ageel [254]. As noted previously, however, the F-tests on which classical analysis of variance is based are less robust than these authors suggest, in light of more recent results like those described by Wilcox, whose book should be consulted for outlier-resistant alternatives [295].

13.3.3 Application: bitter pit in apples

The following example illustrates two simple one-way ANOVAs, based on a dataset contained in the book *Data*, by Andrews and Herzberg [12, no. 59]. This dataset was contributed by D.A. Ratkowsky and is discussed in more detail in a paper by Ratkowsky and Martin [241]. A portion of this dataset is summarized in Table 13.2, which gives results obtained from 42 apple trees, including which one of four different treatments were applied, the average weight in grams from apples harvested from the tree, and the percentage of apples exhibiting *bitter pit*, a defect that appears as "small, brown, soft, dried pits of collapsed tissue, 1/16 to 1/4 inch in diameter" [109]. The problem is seldom evident before the apples are harvested and, while apples with bitter pit are edible, they are not attractive to consumers, making bitter pit a significant problem for fruit growers. Indeed, even growers sometimes confuse bitter pit "with damage resulting from insects or pathological diseases, particularly those involving fungus infections or hail injury" [109]. Bitter pit has a number of causes, including "low soil moisture, low leaf calcium, or an imbalance between calcium and potassium within the leaves" [14]. As the following example illustrates, the use of different nitrogen supplements can also influence the prevalence of bitter pit, although it is possible that this is an indirect effect caused by the influence of these treatments on the calcium content, an idea examined further in Sec. 13.5.3.

In the example considered here, nitrogen supplements were given in one of the following four forms: a control group given no supplement (A), a group given nitrogen via urea (B), a group given nitrogen via calcium and potassium nitrate (C), and a group given nitrogen via ammonia and ammonium sulfate (D). The dataset presented by Andrews and Herzberg includes additional variables not considered in this analysis (six chemical composition measurements and a block number specifying how the trees are grouped), although some of these variables are included in subsequent discussions of this example. The analysis presented here considers ANOVA models for both the mean apple weight and the percentage of bitter pit as functions of the supplemental nitrogen treatment applied. For bitter pit percentage, three different ANOVA methods are considered: the classical method based on Gaussian data assumptions, the γ-trimmed mean approach described by Wilcox [295, sec. 7.1.1] with his default value of $\gamma = 0.2$, and the corresponding method with $\gamma = 0$, which does not address non-Gaussian behavior but does account for possible variability differences between treatments, which the classical method does not.

Table 13.2: The apple tree dataset, including sample number, supplemental nitrogen treatment applied (Trt), average apple weight in grams (Wt) for the tree, and percentage of bitter pit (BP) in apples from the tree

No.	Trt	Wt (g)	% BP	No.	Trt	Wt (g)	% BP
1	A	85.3	0.0%	22	C	127.1	9.5%
2	A	113.8	3.2%	23	C	108.5	3.9%
3	A	92.9	0.0%	24	C	99.9	1.6%
4	A	48.9	0.0%	25	C	124.8	27.2%
5	A	99.4	3.6%	26	C	94.5	2.0%
6	A	79.1	0.0%	27	C	99.4	2.7%
7	A	70.0	2.7%	28	C	117.5	13.9%
8	A	86.9	1.8%	29	C	135.0	50.0%
9	A	87.7	6.5%	30	C	85.6	3.6%
10	A	67.3	4.3%	31	C	102.5	14.3%
11	B	117.5	47.0%	32	C	110.8	10.0%
12	B	98.9	39.6%	33	D	77.4	50.0%
13	B	108.5	44.2%	34	D	91.3	54.0%
14	B	104.4	19.0%	35	D	91.3	89.5%
15	B	96.8	10.0%	36	D	81.7	70.5%
16	B	94.5	18.5%	37	D	89.2	37.5%
17	B	90.6	7.3%	38	D	69.6	64.0%
18	B	100.8	23.6%	39	D	69.0	16.0%
19	B	96.0	6.5%	40	D	73.7	39.5%
20	B	99.9	20.4%	41	D	75.1	36.1%
21	B	84.6	0.0%	42	D	87.0	58.6%

The standard ANOVA table for the bitter pit fraction is shown in Table 13.3. Note that this case is slightly imblanced, since the number of trees in the four treatment groups is 11, 10, 10, and 11 for groups A, B, C, and D, respectively. In fact, these results were obtained on the basis of a designed experiment with four plots, each containing 12 trees, and 12 trees were assigned to each treatment group, but six of the trees did not bear fruit, resulting in the slight imbalance seen in the data [12, no. 59]. The large value of F for this example results in an extremely small probability under the F-distribution, so we reject the hypothesis that there is no treatment effect (i.e., that nitrogen supplement has no influence on the prevalence of bitter pit), in agreement with the conclusion we draw examining the numbers in Table 13.2. The overall prevalence of bitter pit estimated from this analysis is $\hat{\mu} = \bar{y}.. = 21.73\%$, and the specific treatment

Table 13.3: Traditional ANOVA table for the prevalence of bitter pit

Source of variation	Degrees of freedom	Sum of squares	Mean square	F-value	p-Value
Between treatments	3	13630.44	4543.480	20.171	5.502×10^{-8}
Within treatments	38	8559.22	225.243		
Total	41	22189.67			

influence coefficients are estimated from the results of the analysis as:

$$\hat{\alpha}_1 = -19.52, \ \hat{\alpha}_2 = -0.26, \ \hat{\alpha}_3 = -9.12, \ \hat{\alpha}_4 = 29.84. \qquad (13.40)$$

These results suggest that no nitrogen supplement gives the best results in terms of minimizing bitter pit, Treatment C gives the second-best results, followed by Treatment B, with Treatment D giving the worst results.

The standard ANOVA table for mean apple weight is given in Table 13.4. As with the prevalence of bitter pit, these results also suggest a significant relationship between supplemental nitrogen treatment and mean apple weight. The estimated overall mean is $\hat{\mu} = 93.68$, corresponding to the average of all of the mean apple weights $y_{...}$. The estimated treatment influence coefficients $\{\alpha_i\}$ are:

$$\hat{\alpha}_1 = -10.55, \ \hat{\alpha}_2 = 5.63, \ \hat{\alpha}_3 = 15.92, \ \hat{\alpha}_4 = -13.15. \qquad (13.41)$$

These coefficients suggest that Treatment D is somewhat worse than no nitrogen supplement, while Treatment B is better and Treatment C is much better in terms of increasing mean apple weight. Combining these results with those of the bitter pit analysis of variance suggests that Treatment D is the least desirable overall since it leads to the smallest average apple weight and the highest incidence of bitter pit. Since Treatment B yields smaller average apple weight and a higher incidence of bitter pit than Treatment C, it appears that Treatment C is the best of the supplemental nitrogen treatments considered. The choice between Treatment C and Treatment A (i.e., no supplemental nitrogen) then reduces to a question of the trade-off between bitter pit and apple weight: if higher average apple weight with somewhat greater incidence of bitter pit ($\sim$10% vs. $\sim$2%) is more acceptable than lower incidence of bitter pit with lower average apple weight ($\sim$80 g vs. $\sim$110 g), Treatment C is to be preferred, while if the opposite is true, Treatment A is preferable.

Table 13.4: Traditional ANOVA table for the mean apple weight

Source of variation	Degrees of freedom	Sum of squares	Mean square	F-value	p-Value
Between treatments	3	5979.84	1993.280	11.277	1.970×10^{-5}
Within treatments	38	6716.94	176.762		
Total	41	12696.78			

Because the bitter pit fractions listed in Table 13.2 appear to contain some outlying values, this example is a reasonable candidate for the robust ANOVA method of Wilcox described in Sec. 13.3.2. Applying the Hampel identifier described in Chapter 7 with a threshold value $t = 3$ to the complete sequence of bitter pit fractions identifies the largest four values as outliers (trees 35, 36, 38, and 42). Alternatively, applying this same procedure to each treatment group individually only identifies a single outlier: tree number 29, which exhibits the largest bitter pit fraction observed for Treatment C.

To examine the influence of these outliers, compare the classical ANOVA results with those obtained using the Wilcox method based on 20% trimmed means. For this example, while the details—specifically, the estimated coefficients and the p-value—are different, the general conclusions are the same. Specifically, the robust ANOVA procedure also rejects the hypothesis of no treatment effect (i.e., the null hypothesis that all means are the same) with $p = 1.965 \times 10^{-4}$, and the estimated coefficients α_i based on the 20% trimmed means are:

$$\hat{\alpha}_1 = -14.41, \ \hat{\alpha}_2 = 3.48, \ \hat{\alpha}_3 = -8.02, \ \hat{\alpha}_4 = 34.31. \qquad (13.42)$$

Note that while the values of these coefficients differ from those based on untrimmed means used in the classical analysis of variance, the order remains the same: the best supplemental nitrogen treatment in terms of prevalence of bitter pit is A (no supplemental nitrogen), followed by C, then B, with D the worst. Setting the trimming fraction γ to zero in the robust procedure reduces it to Welch's method, developed to address the possibility that the variances are not the same between treatments, and here again we obtain results that are different in detail but in qualitative agreement. Specifically, since the parameter estimates α_i are based on standard means, they are exactly the same as in the classical analysis of variance, but the p-value is different, since Welch's procedure modifies both the effective sum of square values, leading to a different computed value for F, and the degrees of freedom used to evaluate this F-value. Specifically, comparing all three results we have:

1. classical ANOVA: $F = 20.171$ with $\nu_1 = \mathrm{df}_B = 3$ and $\nu_2 = \mathrm{df}_W = 38$, implying $p = 5.502 \times 10^{-8}$;

2. Welch's ANOVA: $F = 22.466$ with $\nu_1 = 3$ and $\nu_2 = 16.67$, implying $p = 4.295 \times 10^{-6}$;

3. Wilcox's robust ANOVA: $F = 18.379$ with $\nu_1 = 3$ and $\nu_2 = 10.20$, implying $p = 1.965 \times 10^{-4}$.

The overall implication of these results is that the classical analysis gives the most significant result, but it ignores both the variance differences between the different treatment groups and the presence of outliers in the data. Welch's method accounts for the variance differences but not the outliers, giving a larger (i.e., less significant) p-value, while the robust ANOVA results account for both the variance differences and the outliers, giving the least significant p-value as a consequence. Here, the conclusions are the same in all cases: the nitrogen treatment chosen does influence the prevalence of bitter pit, but in cases where the evidence is less compelling, it is clear that the different methods could lead to different conclusions, with some rejecting the null hypothesis and others not rejecting it. Our choice of method, and thus possibly our conclusion, would then depend on which of the underlying working assumptions we were willing to accept, ranging from most stringent for the classical ANOVA results to least stringent for the robust ANOVA results.

13.4   Generalized linear models

As shown in Sec. 13.3.2, ANOVA models may be viewed as regression models in which the response variable is real-valued and the explanatory variables are categorical. The implementation of this idea involves replacing the original categorical variables each with a collection of binary indicator variables for the different levels of the categorical variable and forming the appropriately weighted linear combinations. This procedure leads to a regression model to which the techniques discussed in Chapters 5 and 11 may be applied. The following discussion introduces the class of *generalized linear models (GLMs)*, which involve transformations of the response variable to obtain something similar to a standard regression problem.

To motivate the standard formulation of the GLM problem, it is useful to first present a slightly modified description of the linear regression problem considered in Chapter 11. There, an observed sequence $\{y_k\}$ of real-valued response variables was related to sequences $\{x_k^i\}$ of p covariates, for $i = 1, 2, \ldots, p$ and $k = 1, 2, \ldots, N$, by a model of the form:

$$y_k = \sum_{i=1}^{p} \beta_i x_k^i + e_k, \tag{13.43}$$

where $\{\beta_i\}$ is a set of p unknown model parameters to be determined, and $\{e_k\}$ is the sequence of prediction errors corresponding to a fixed set of these parameter values. The classical working assumption under which this regression model is analyzed is that $\{e_k\}$ is a sequence of independent, identically distributed (i.i.d.) Gaussian random variables with mean zero and variance σ^2. As a corollary, it follows that if the correct model parameters $\{\beta_i\}$ are inserted into Eq. (13.43), then $\{y_k\}$ is a sequence of i.i.d. Gaussian random variables with the distribution $y_k \sim N(\mu_k, \sigma^2)$ where μ_k is:

$$\mu_k = E\{y_k\} = \sum_{i=1}^{p} \beta_i x_k^i. \tag{13.44}$$

The class of generalized linear models is obtained by replacing Eq. (13.44) with a more general expression and allowing the distribution of $\{y_k\}$ to be something other than Gaussian.

Specifically, generalized linear models replace Eq. (13.44) with:

$$g(\mu_k) = \sum_{i=1}^{p} \beta_i x_k^i = \mathbf{x}_k' \beta, \tag{13.45}$$

where, as before, $\mu_k = E\{y_k\}$ represents the expected value of the observed response under a specified data generation model. The function $g(\cdot)$ appearing in Eq. (13.45) is called the *link*, and it is commonly associated with a "natural" data distribution for the problem at hand. In particular, the identity function $g(\mu) = \mu$ reduces Eq. (13.45) to Eq. (13.44) and is associated with the Gaussian distribution. Usually, this association between the link function and its "natural" distribution is described by saying that $g(\cdot)$ is the *canonical link* for a particular distribution. Thus, the identity link is canonical for the Gaussian distribution, and adopting different canonical links leads to problem formulations that are better suited to other data types, such as binary responses or count data.

The advantage of the generalized linear model formulation is that maximum likelihood estimation equations for GLMs can be simplified significantly if the distribution of y_k is assumed to belong to an exponential family, of the form [204, p. 28]:

$$f(y; \theta, \phi) = \exp\left\{\frac{y\theta - b(\theta)}{a(\phi)} + c(y, \phi)\right\}. \tag{13.46}$$

Here, $a(\cdot)$, $b(\cdot)$, and $c(\cdot)$ are known functions that define the particular distribution under consideration. Specific examples include the Gaussian distribution (Exercise 6), the binomial distribution (Sec. 13.5), and the Poisson distribution (Sec. 13.6). Recall from the discussion in Chapter 5 that the likelihood function is simply the probability density $f(y; \theta, \phi)$ regarded as a function of the data value y that assigns probability (i.e., likelihood) to the parameters θ and ϕ. For an i.i.d. data sequence $\{y_k\}$, the likelihood associated with the entire data sequence is the product of the individual probabilities:

$$L(\{y_k\}; \theta, \phi) = \prod_{k=1}^{N} f(y_k; \theta, \phi). \tag{13.47}$$

Since the distributions considered here belong to an exponential family, it is much easier in practice to work with the log likelihood, given by:

$$
\begin{aligned}
\ell(\{y_k\}; \theta, \phi) = \ln L(\{y_k\}; \theta, \phi) &= \sum_{k=1}^{N} \ln f(y_k; \theta, \phi) \tag{13.48} \\
&= \sum_{k=1}^{N} \left[\frac{y_k \theta - b(\theta)}{a(\phi)} + c(y_k, \phi) \right].
\end{aligned}
$$

Two additional assumptions simplify this result further. First, it is common to take $a(\phi) = \phi/w_k$ where w_k is a data-dependent weight, and second, if the cannonical link is chosen for this model, it follows that $\theta = \mathbf{x}'\beta$. Under these assumptions, the log likelihood may be written as:

$$\ell(\{y_k\}; \theta, \phi) = \frac{1}{\phi} \sum_{k=1}^{N} w_k(y_k \mathbf{x}_k'\beta - b(\mathbf{x}_k'\beta)) + \sum_{k=1}^{N} c(y_k, \phi). \tag{13.49}$$

The advantage of this result is that, for fixed dispersion ϕ, only the first sum in the expression depends on the unknown model parameter vector β, simplifying both the process of estimating the model parameters and the expressions for the *deviance*, the goodness-of-fit measure defined next.

Recall that by increasing the dimension of the parameter vector β so that there is one model parameter per observation, it is possible to achieve a perfect fit to any dataset. For the generalized linear models considered here, this perfect fit is achieved for the parameter vector $\tilde{\beta}$ and augmented covariate vectors $\tilde{\mathbf{x}}_k$ such that, for all k:

$$g(y_k) = \tilde{\mathbf{x}}_k'\tilde{\beta}. \tag{13.50}$$

While this *saturated model* is not useful in data analysis, it does provide an upper limit on the achievable likelihood for a given data sequence $\{y_k\}$ and any other parameter vector β. The *scaled deviance* is defined as twice the difference between the maximum achievable log likelihood and the log likelihood for the parameter vector β associated with the model under consideration:

$$
\begin{aligned}
D^*(\{y_k\}; \beta, \phi) &= 2\ell(\{y_k\}; \tilde{\mathbf{x}}_k'\tilde{\beta}) - 2\ell(\{y_k\}; \mathbf{x}_k'\beta) \tag{13.51} \\
&= \frac{2}{\phi} \sum_{k=1}^{N} w_k(y_k[g(y_k) - \mathbf{x}_k'\beta] - b(g(y_k)) + b(\mathbf{x}_k'\beta)).
\end{aligned}
$$

A key observation is that since the likelihood associated with the saturated model does not depend on the model parameter β under consideration, maximizing the log likelihood with respect to β is equivalent to minimizing the scaled

Table 13.5: Structure of the modified analysis of deviance table used here to summarize GLM results

No.	Covariate	$\hat{\beta}_i$	$\hat{\sigma}_i$	df	Deviance	P_{χ^2}
1	x_1	$\hat{\beta}_1$	$\hat{\sigma}_1$	df_1	D_1	p_1
2	x_2	$\hat{\beta}_2$	$\hat{\sigma}_2$	df_2	D_2	p_2
$\vdots$	$\vdots$	$\vdots$	$\vdots$	$\vdots$	$\vdots$	$\vdots$
m	x_m	$\hat{\beta}_m$	$\hat{\sigma}_m$	df_m	D_m	p_m
—	Null	$\hat{\beta}_0$	$\hat{\sigma}_0$	df_0	D_0	—
—	Total	—	—	df_T	D_T	—

deviance with respect to β. The *deviance* is defined as the scaled deviance multiplied by the dispersion parameter ϕ:

$$D(\{y_k\}; \mathbf{x}'_k\beta) = 2\sum_{k=1}^{N} w_k(y_k[g(y_k) - \mathbf{x}'_k\beta] - b(g(y_k)) + b(\mathbf{x}'_k\beta)). \qquad (13.52)$$

A key observation is that for the linear regression model obtained for the Gaussian distribution with canonical link $g(x) = x$, the deviance is simply equal to the residual sum of squares (Exercise 6). As a consequence, a common characterization of generalized linear models is in terms of an *analysis of deviance table*, analogous to the ANOVA tables introduced in Sec. 13.3.

While it is not quite in the traditional format, the analysis of deviance tables used to summarize GLM model results in this book have the form shown in Table 13.5. This table summarizes results obtained for a model involving m covariates, x_1 through x_m, and the primary modification relative to a traditional analysis of deviance table is the inclusion of the estimated model parameters $\hat{\beta}_i$ and their associated standard errors $\hat{\sigma}_i$. These values are included here because they are important model characteristics, particularly the standard errors, which can be extremely useful in detecting numerical difficulties, as discussed in Sec. 13.5.5. The more traditional elements included in Table 13.5 are the degrees of freedom associated with each covariate, df_i in the column headed "df," the deviance values D_i, and their associated p-values computed from an asymptotic χ^2 approximation. In addition, the last two rows of the table summarize the *null model* defined by the intercept term that is almost always included in the model, and the total degrees of freedom and deviance for the dataset. The utility of these tables are illustrated in the multiple covariate examples discussed in secs. 13.5 and 13.6.

Generalized linear models are typically fit to a dataset using iteratively reweighted least squares (IRWLS) procedures like those discussed in Chapter

11, specialized to the structure of the generalized linear model. Further discussion of these fitting procedures is beyond the scope of this book, but reasonably detailed treatments may be found in the books by McCullagh and Nelder [204], Collett [66, appendix B1], Agresti [8], or Hosmer and Lemeshow [148].

13.5 Logistic regression models

Logistic regression represents one of the most important special cases of the generalized linear modeling formalism just described, and the following discussion gives a more detailed discussion of this special case. These models arise frequently in medical data analysis, which is often concerned with binary response variables, but as the examples presented here demonstrate, logistic regression is applicable to a much wider range of problems. As a particularly interesting application, recall that Chapter 2 discussed the use of beetle fossils in estimating historical temperatures prior to the invention of thermometers. Huppert and Solow have used logistic regression to improve these temperature estimates [155], defining the response variable as the probability that a particular beetle species is present at a given location. Explanatory variables are the mean January temperature T_1 and the mean July temperature T_2 for each location, and a logistic regression model is fit to obtain coefficients relating the probability of ocurrance of each beetle species as a function of these two temperatures. Given these estimated probabilities, based on modern data for which temperature measurements are available, the mean January and July temperatures for an area are then estimated from knowledge of which beetle species are present. This is accomplished by choosing the values of these temperatures that maximize the following χ^2 goodness-of-fit measure:

$$\chi^2(T_1, T_2) = \sum_{k \in \mathcal{S}} \frac{(1 - \hat{p}_k(T_1, T_2))^2}{\hat{p}_k(T_1, T_2)}. \tag{13.53}$$

Here, $\mathcal{S}$ is the set of beetle species observed in the area and $\hat{p}_k(T_1, T_2)$ is the conditional probability of observing species k given January and July temperatures T_1 and T_2. The rationale for this expression is that since the beetles are observed in fossil records for the specified area, the true probabilities of occurrance are $p_k = 1$ for all $k \in \mathcal{S}$, and Eq. (13.53) is simply the χ^2 goodness-of-fit statistic between the estimated and true probabilities discussed in Chapter 9. Numerically determining the values of T_1 and T_2 that maximize this goodness-of-fit measure then gives the most probable values for these temperatures for the specified location, given the beetle fossil record.

13.5.1 Logistic regression

In logistic regression, the response variable is binary and the explanatory covariates may be binary, continuous, or categorical. As Hosmer and Lemeshow demonstrate with a simple example [148, p. 2], plotting a binary data sequence

$\{b_k\}$ against a continuous covariate does not yield an informative graph, but plotting the fraction of $\{b_k\}$ values that are equal to 1 within a well-defined group against the midpoint of a range of values for the covariate can indicate a systematic dependence of the probability that $y_k = 1$ on the covariate. Such a characterization is the fundamental notion underlying logistic regression. That is, rather than fitting the observed responses $\{b_k\}$ to a model that generates binary predictions $\{\hat{b}_k\}$, logistic regression attempts to estimate the conditional probability π_k that $b_k = 1$ given the values of all covariates under consideration. Conceptually, it is therefore easiest to consider the case of *grouped data* where each group or *covariate class* is defined by the values of the covariates and contains some number n_k of binary observations. In this case, there are G groups of n_k observations each, where:

$$\sum_{k=1}^{G} n_k = N, \qquad (13.54)$$

and N is the total number of data observations. In this setting, define y_k to be the fraction of these n_k observations that are 1; in other words, the number of times $b_k = 1$ within this covariate class is $m_k = n_k y_k$. While it is not always the appropriate distribution, the "natural" distribution for this problem is the binomial distribution introduced in Chapter 3, for which:

$$P\{m_k; \pi_k, n_k\} = \binom{n_k}{m_k} \pi_k^{m_k} (1 - \pi_k)^{n_k - m_k}. \qquad (13.55)$$

Logistic regression is typically represented in terms of the fractions $\{y_k\}$ of positive responses (i.e., responses $b_k = 1$), for which the distribution may be expressed as:

$$
\begin{aligned}
P\{y_k; \pi_k, n_k\} = P\{m_k; \pi_k, n_k\} &= \binom{n_k}{n_k y_k} (1 - \pi_k)^{n_k} \left[\frac{\pi_k}{1 - \pi_k} \right]^{n_k y_k}, \\
&= \exp\left[\ln\binom{n_k}{n_k y_k} + n_k \ln(1 - \pi_k) \right. \\
&\qquad \left. + n_k y_k \ln\left(\frac{\pi_k}{1 - \pi_k} \right) \right]. \qquad (13.56)
\end{aligned}
$$

The advantage of this representation is that it is a special case of the exponential family defined in Eq. (13.46) for the following choices of parameters and functions:

$$
\begin{aligned}
\theta &= \ln\left(\frac{\pi_k}{1 - \pi_k} \right), \\
\phi &= 1/n_k, \\
a(\phi) &= \phi, \\
b(\theta) &= \ln(1 + e^{\theta}), \\
c(y_k, \phi) &= \ln\binom{n_k}{n_k y_k} = \ln\binom{1/\phi}{y_k/\phi}. \qquad (13.57)
\end{aligned}
$$

Since the definition of θ defines the cannonical link for a generalized linear model, it follows that the logistic regression model has the form:

$$\theta = \mathbf{x}_k'\beta \;\Rightarrow\; \ln\left(\frac{\hat{\pi}_k}{1 - \hat{\pi}_k}\right) \equiv \text{logit}(\hat{\pi}_k) = \mathbf{x}_k'\beta. \tag{13.58}$$

The practical logistic regression modeling problem then consists of determining the unknown parameter vector β such that Eq. (13.58) best fits the available data. This estimate is obtained using an interative computational procedure like the **glm** procedure in the S programming language, which maximizes the log likelihood or, equivalently, minimizes the deviance, which is given by:

$$D(\{y_k\};\mathbf{x}_k'\beta) = 2\sum_{k=1}^{G} w_k(y_k[\text{logit}(y_k) - \mathbf{x}_k'\beta] + \ln(1 - y_k) - \ln(1 + e^{\mathbf{x}_k'\beta})).$$

$$\tag{13.59}$$

While this expression has the advantage of showing the explicit dependence of the deviance on the model parameter vector β, the deviance can be represented more symmetrically in terms of the estimated probabilities $\hat{\pi}_k$. In particular, substituting Eq. (13.58) into Eq. (13.59) gives the more common expression for the deviance:

$$D(\{y_k\};\hat{\pi}_k) = 2\sum_{k=1}^{G} w_k(y_k[\text{logit}(y_k) - \text{logit}(\hat{\pi}_k) + \ln(1 - y_k) - \ln(1 - \hat{\pi}_k)). \tag{13.60}$$

The advantage of this second deviance expression over the first is that it makes clear what minimizing the deviance is attempting to do: the optimum parameter vector β is the one that makes the logistic regression conditional probability estimates $\hat{\pi}_k$ as close to the simple success frequency estimates $y_k = m_k/n_k$ as possible, subject to the covariate dependence described by the logistic model defined in Eq. (13.58). In particular, note that the deviance is zero if $\hat{\pi}_k = y_k$ for $k = 1, 2, \ldots, G$, but this "perfect fit" value is generally not achievable within the constraints of the logistic regression model.

13.5.2   Ungrouped data

As noted, the basic description of logistic regression given above assumes the availability of *grouped data*, so that the responses y_k are the total number of times a binary variable $\{b_k\}$ takes the value 1 within a given covariate class. It is also possible to fit logistic regression models using the binary data itself, but this does introduce a significant complication. Evidence of this complication may be seen in Eq. (13.60): letting $y_k = b_k$ and $G = N$, we are immediately faced with terms like $\ln y_k$ and $\ln(1 - y_k)$, both of which are infinite if $y_k = 0$ or $y_k = 1$, the only two possible values for this case. In fact, the terms in Eq. (13.60) may be rearranged so that these logarithms appear only in the terms $y_k \ln y_k$ and $(1 - y_k)\ln(1 - y_k)$, both of which may be shown to be zero by L'Hospital's rule [174, p. 245]. As a result, the deviance may be expressed as:

$$D(\{y_k\};\hat{\pi}_k) = -2\sum_{k=1}^{N} w_k\left(y_k \ln\left(\frac{\hat{\pi}_k}{1 - \hat{\pi}_k}\right) - \ln(1 - \pi_k)\right). \tag{13.61}$$

Even though this equation gives a well-defined expression for the deviance, a difficulty with the ungrouped case is that the deviance no longer exhibits an approximate χ^2 distribution, so it cannot be used in assessing the goodness-of-fit between the model and the binary data [8, p. 141]. Similarly, as McCullagh and Nelder note [204, sec. 4.4.5], the deviance is not a reliable absolute measure of goodness-of-fit if many of the counts $\{n_k\}$ in a grouped dataset are small (i.e., if $G \simeq N$). Conversely, they note that the change in deviance remains a useful measure of significance in deciding which terms should be included in a multivariable model [204, p. 122]:

> For instance, we may look for interactions among the covariates or nonlinear effects by adding suitable terms to the model and observing the reduction in deviance. The reduction in deviance thus induced is usually well approximated by a χ^2 distribution.

Agresti makes a similar observation about the utility of deviance in comparing related models [8, p. 176], and he notes that Hosmer and Lemeshow derive an alternative test statistic that gives a better assessment of overall goodness-of-fit for cases where $G \simeq N$. This approach is based on a stratification of the ungrouped data on the basis of estimated probabilities $\hat{\pi}_k$ into g groups. Letting y_{ij} denote the ith observed binary response in the jth group and $\hat{\pi}_{ij}$ represent the corresponding estimated probability from the logistic regression model, Hosmer and Lemeshow develop a goodness-of-fit test based on the χ^2 test discussed in Chapter 9. Specifically, their test statistic is:

$$\tau_{HL} = \sum_{i=1}^{g} \frac{(\sum_j y_{ij} - \sum_j \hat{\pi}_{ij})^2}{(\sum_j \hat{\pi}_{ij})[1 - (\sum_j \hat{\pi}_{ij})/n_i]}, \tag{13.62}$$

which has an approximate χ^2 distribution with $g-2$ degrees of freedom. Agresti emphasizes that this distributional characterization is only approximate and that the computed value of τ_{HL} does depend on the details of the data grouping. For a more complete discussion of this test statistic, refer to Hosmer and Lemeshow's book [148, sec. 5.2.2]. The utility of the deviance in comparing logistic regression models derived from ungrouped data is demonstrated in the example discussed in Sec. 13.5.4.

13.5.3   Application 1: bitter pit revisited

The following example illustrates the application of logistic regression to grouped data; the example presented in Sec. 13.5.4 illustrates the ungrouped case. Table 13.6 presents a different portion of the apple dataset presented by Andrews and Herzberg [12, no. 59] and discussed in Sec. 13.3.3. Specifically, Table 13.6 summarizes both the calcium content of the apple, in parts per million (ppm), and the percentage of apples exhibiting bitter pit, for each of the 42 trees considered in Sec. 13.3.3. Examination of the numbers in this table are consistent with the idea suggested in that discussion that the different nitrogen supplements applied to the apple trees may be affecting the incidence of bitter pit through

Table 13.6: The apple tree dataset, including sample number, supplemental nitrogen treatment applied (Trt), average calcium content in ppm (Ca), and percentage of bitter pit (BP) in apples from the tree

No.	Trt	Ca	% BP	No.	Trt	Ca	% BP
1	A	244	0.0%	22	C	169	9.5%
2	A	142	3.2%	23	C	210	3.9%
3	A	269	0.0%	24	C	172	1.6%
4	A	272	0.0%	25	C	148	27.2%
5	A	202	3.6%	26	C	219	2.0%
6	A	272	0.0%	27	C	224	2.7%
7	A	297	2.7%	28	C	152	13.9%
8	A	225	1.8%	29	C	160	50.0%
9	A	212	6.5%	30	C	211	3.6%
10	A	206	4.3%	31	C	178	14.3%
11	B	138	47.0%	32	C	183	10.0%
12	B	151	39.6%	33	D	148	50.0%
13	B	165	44.2%	34	D	128	54.0%
14	B	151	19.0%	35	D	132	89.5%
15	B	199	10.0%	36	D	120	70.5%
16	B	159	18.5%	37	D	124	37.5%
17	B	158	7.3%	38	D	147	64.0%
18	B	163	23.6%	39	D	255	16.0%
19	B	239	6.5%	40	D	172	39.5%
20	B	180	20.4%	41	D	140	36.1%
21	B	199	0.0%	42	D	181	58.6%

their influence on calcium levels. In particular, these numbers suggest, first, that calcium levels vary systematically with the nitrogen supplement applied, and second, that the incidence of bitter pit is greater in apples with lower calcium content than in apples with higher calcium content. The first of these notions could be examined through a one-way ANOVA of the calcium data like that presented in Sec. 13.3.3 for the average weight and bitter pit incidence data. The purpose of this example is to show how logistic regression can be used to examine the second of these notions—the relationship between calcium content and the prevalence of bitter pit.

First, note that examining a scatterplot of the percentage of bitter pit observed in the apples from each tree against the average calcium content does suggest a relationship between these variables. This may be seen in Fig. 13.2, which shows that the incidence of bitter pit generally declines as calcium con-

Figure 13.2: Scatterplot of the percentage of apples from each tree exhibiting bitter pit vs. the average calcium concentration, in parts per million

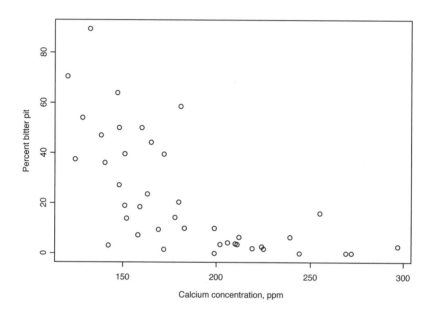

centration increases, an observation supported by the relative large negative correlation coefficient between these two variables ($\hat{\rho} = -0.674$). Since the bitter pit percentage varies between 0 and 100, this relationship cannot be linear over an extended range, again in agreement with the general appearance of Fig. 13.2. Further evidence in support of a tendency for bitter pit percentage to decline with increasing calcium concentration is provided by the Spearman rank correlation coefficient, which is $\hat{\rho}_S = -0.767$ for this case.

Logistic regression provides an alternative way of quantifying this relationship, one that is inherently better suited to the nature of the response variable (i.e., the fact that it is a fraction) and one that permits extension to other explanatory variables, an idea examined further in Chapter 15. The form of the logistic regression model for this example is extremely simple. Let k denote the tree index, from 1 to 42, and let p_k denote the percentage of bitter pit in the apples obtained from each tree, expressed as a fraction between 0 and 1. The explanatory variable considered here is the calcium content in parts per million, x_k. The logistic regression model for this example is then given by:

$$\ln\left(\frac{p_k}{1 - p_k}\right) = \alpha + \beta x_k + e_k. \tag{13.63}$$

The parameter α in this model characterizes the overall incidence of bitter pit in

Figure 13.3: Logistic model prediction errors vs. tree number, for the fraction of bitter pit estimated from calcium concentrations. Trees are grouped by supplemental nitrogen treatment, as indicated at the bottom of the plot

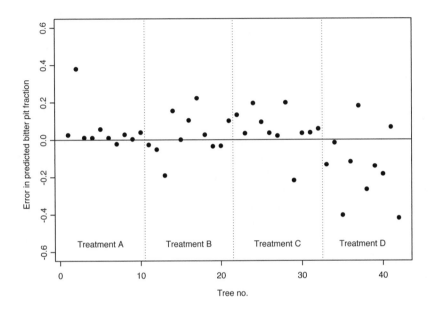

the apples from these 42 trees, while the coefficient β characterizes the influence of calcium content on this incidence. Fitting this model using the generalized linear modeling procedure for logistic regression discussed in Sec. 13.5.1 yields the parameter estimates:

$$\hat{\alpha} = 4.207, \quad \hat{\beta} = -0.032. \tag{13.64}$$

Note that the negative estimate for the coefficient β further supports the idea that the prevalence of bitter pit decreases with increasing calcium concentration.

Fig. 13.3 shows the prediction errors from the logistic regression model defined in Eq. (13.63) with the parameters given in Eq. (13.64). These errors are defined as $e_k = \hat{\pi}_k - p_k$ where $\hat{\pi}_k$ is the logistic model prediction of the bitter pit fraction p_k, computed from the calcium concentration value x_k. Since the trees are grouped by supplemental nitrogen treatment, different patterns of prediction errors in different portions of the plot (i.e., in different treatment groups, separated by vertical lines) suggest the possiblility of additional treatment effects beyond its influence on calcium concentration. In particular, the generally larger scatter seen in the prediction errors associated with the noncontrol treatment groups (i.e., Treatments B, C, and D) and the apparent negative bias seen in the errors for Treatment D suggest that the tentative model "treatment influences calcium, which determines bitter pit" is too simple. Refinements of this model will be examined in Chapter 15.

13.5.4 Application 2: missing data

As an application of logistic regression to ungrouped data, the following example considers the question of whether there is evidence that the missing triceps skinfold thickness (TSF) data in the Pima Indians diabetes dataset is *systematically* missing. Recall that this dataset characterizes 768 female members of the Pima Indian tribe, giving the classification of each patient as either diabetic or non-diabetic, along with recorded values for a number of medical characteristics. Two of these characteristics are body mass index (BMI) and TSF, both representing measures of obesity. In the Pima Indians dataset, BMI is only missing 1.4% of the time, while TSF is missing much more frequently, 29.6% of the time. The following example applies logistic regression to attempt to answer the following question: is there evidence to suggest that patients with missing TSF values exhibit systematic differences in their other recorded characteristics?

For this analysis, we restrict consideration to the $N = 757$ records for which BMI values x_k are available (i.e., recorded as nonzero values), and define $z_k = 0$ if the TSF value is missing (i.e., zero) for record k and $z_k = 1$ if the corresponding TSF value is nonzero. The logistic regression model considered here uses these binary $\{z_k\}$ values as responses, to be predicted from the following seven explanatory covariates in the Pima Indians dataset:

1. NPG, the number of times pregnant;

2. GLC, plasma glucose concentration;

3. DIA, diastolic blood pressure;

4. INS, the serum insulin concentration;

5. BMI, the body mass index;

6. DPF, a computed diabetes pedigree function value;

7. AGE, patient age in years;

8. DIAG, a binary diagnosis (diabetic or nondiabetic).

Using the built-in generalized linear modeling procedure **glm** in the S programming language (either R or S-*Plus*), the logistic regression model obtained by fitting $\{z_k\}$ to these eight candidate explanatory variables is summarized in the analysis of deviance table shown in Table 13.7, in the format described in Sec. 13.4. Specifically, Table 13.7 lists the eight covariates discussed in the preceeding paragraph in order of their inclusion in the model, gives the estimated model coefficient $\hat{\beta}_i$ associated with each variable, its associated standard error

Table 13.7: Analysis of deviance table for the logistic regression model predicting the presence of plausible TSF data observations in the Pima Indian diabetes dataset from the eight covariates listed in the text

No.	Covariate	$\hat{\beta}_i$	$\hat{\sigma}_i$	df	Deviance	P_{χ^2}
1	NPG	-0.018	0.037	1	16.3325	0.0000531
2	GLC	-0.013	0.004	1	1.5028	0.2202358
3	DIA	0.020	0.006	1	19.2852	0.0000113
4	INS	0.802	11.730	1	404.8517	0.0000000
5	BMI	0.027	0.018	1	3.8475	0.0498208
6	DPF	0.676	0.376	1	2.8772	0.0898445
7	AGE	-0.028	0.011	1	6.9092	0.0085752
8	DIAG	0.049	0.280	1	0.0312	0.8597300
—	Null	-0.354	0.751	748	453.2654	—
—	Total	—	—	756	908.9027	—

estimate $\hat{\sigma}_i$, the degrees of freedom (df) associated with each variable, its contribution to the total deviance, and its associated χ^2 probability estimate. The last two rows in the table give the result for the null model containing only the intercept term, along with the total degrees of freedom and deviance.

Recall from the discussion given in Sec. 13.5.2 that while the total deviance is not useful in assessing the overall goodness-of-fit for a model estimated from ungrouped data, the changes in deviance when each term is added do provide a useful basis for assessing their significance. Looking at these deviance changes, it is immediately clear that serum insulin concentration (INS) is the most important covariate to include in this model. In fact, this variable is so strongly associated with the missing variable status z_k for TSF that their odds ratio is infinite. This extremely strong association is closely related to the phenomenon of *separation* in logistic regression models discussed in Sec. 13.5.5, and it causes the model-fitting procedure to issue a warning message that estimated probabilities are near their limiting values of 0 or 1. Also, note the very large standard error associated with this parameter, relative to its estimated value. As discussed in Sec. 13.5.5, this observation is often a warning sign of numerical difficulties in the model fitting procedure, in this case caused by the extremely strong association between INS and z_k.

Refitting the logistic regression model without this term changes the model coefficients estimated for all of the remaining covariates, and it causes the deviance contributions of all terms included after DIA in the original model to increase substantially. For example, the deviance contribution for AGE increases from ~6.91 in the original model to 28.02 in the model without INS. As a con-

Figure 13.4: Boxplot summaries comparing the ranges of four Pima Indian covariates when TSF is missing ($z_k = 0$) vs. when TSF is present ($z_k = 1$)

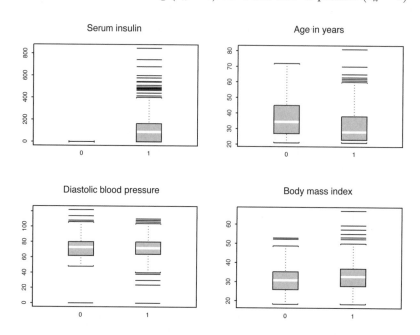

sequence, the χ^2 p-value for this covariate decreases to 1×10^{-7}, making it the most significant variable in the revised model. Overall, all covariates included in this revised model except blood glucose concentration (GLC) and diagnosis (DIAG) are significant, with χ^2 probability estimates less than 0.01.

Fig. 13.4 gives boxplot summaries for four of the eight covariates included in the original model, comparing records with missing TSF values ($z_k = 0$) and nonmissing TSF values ($z_k = 1$). The upper left plot shows the results for serum insulin concentration (INS) and clarifies the strong association between this variable and z_k noted in the preceeding discussion. Specifically, it is clear from this pair of boxplots that INS *always* has the value zero—indicating that it is also missing—whenever TSF is missing. The converse is not true, so these variables are not fully equivalent: INS is sometimes missing when TSF is present. The pair of boxplots in the upper right shows that patients with missing TSF values are generally somewhat older than those for which TSF values are available. A more detailed picture of the differences between patients with TSF missing and those with TSF recorded is given in Table 13.8, which presents the 20% trimmed means for each of the covariates included in the second logistic regression model for both of these patient groups, along with the probability associated with the null hypothesis that these trimmed means are equal, as computed by the

Table 13.8: Comparison of 20% trimmed means for the covariates in the second logistic regression model, for patients with ($z_k = 1$) and without ($z_k = 0$) TSF data. The p-values are from the Yuen-Welch test described in Chapter 9

No.	Covariate	$\bar{x}^{0.2}(z = 0)$	$\bar{x}^{0.2}(z = 1)$	p-Value
1	NPG	4.41	2.77	8.06×10^{-7}
2	GLC	122.30	116.51	1.95×10^{-2}
3	DIA	71.95	71.31	5.60×10^{-1}
4	BMI	30.73	32.56	2.92×10^{-3}
5	DPF	0.33	0.43	3.01×10^{-6}
6	AGE	35.30	29.01	1.79×10^{-8}
7	DIAG	0.33	0.22	1.20×10^{-1}

Yuen-Welch test described in Chapter 9. Consistent with the logistic regression results, the most significant difference seen is that for AGE: the estimated mean age for patients with missing TSF values is 6.3 years greater than that for patients with nonzero TSF values. Conversely, while the logistic regression model results declare the covariate DIA significant ($p = 1.13 \times 10^{-5}$), the Yuen-Welch results do not ($p = 5.60 \times 10^{-1}$), and the boxplot summary given in the lower left in Fig. 13.4 appears consistent with the Yuen-Welch results (i.e., both groups appear to have very similar ranges of DIA values). It is possible that the logistic regression results are more sensitive to the outliers in this covariate, but it is not possible to say without a more detailed analysis.

Overall, the logistic regression results and the comparisons presented in Table 13.8 suggest that patients with missing TSF values differ systematically in several respects from those with nonzero recorded TSF values. First and foremost, for any patient with a missing TSF value, the value for serum insulin (INS) is also missing. In addition, patients with missing TSF values are older on average than those with recorded TSF values (AGE), have been pregnant more frequently (NPG), and have higher blood glucose (GLC) and body mass index (BMI) values. Taken together, these observations suggest that simply omitting patients with missing TSF values from models that incorporate TSF as a covariate is risky. This question is revisited in Chapter 16 in connection with alternatives for handling missing data.

13.5.5 Practical issues: EPV and separation

The following paragraphs briefly discuss two important practical issues that arise in connection with logistic regression. The abbreviation *EPV* stands for "events per variable," and it refers to the number of responses $y_k = 0$ or $y_k = 1$

(whichever number is smaller) relative to the number of covariates included in the model. Often, y_k represents a binary event (e.g., the patient developed a particular disease), which can be relatively rare. In such cases, these rare events actually convey most of the information that determines the model parameters, so it is important to have enough events relative to the number of model covariates to obtain reasonable parameter estimates. Based on the work of Peduzzi *et al.* [232], Hosmer and Lemeshow recommend a minimum EPV value of at least 10 [148, p. 346]. For example, in the missing TSF data example discussed in Sec. 13.5.4, there are 218 records with missing TSF values and 539 records with nonzero recorded TSF values. Applying the "rule of 10" suggests that no logistic regression model should include more than about 22 covariates; the most complex model considered in Sec. 13.5.4 had 8 covariates, implying an EPV value of 27.25, well above this recommended lower limit.

In a detailed analysis based on subsamples of a real dataset, Peduzzi *et al.* were able to vary the EPV for the dataset and compared results obtained for EPV values of 2, 5, 10, 15, 20, and 25 with those obtained from the original dataset with an EPV value of 36 [232]. Specific problems observed at small EPV values included significant biases in the estimated logistic regression model parameters, poor variance estimates, paradoxical associations (i.e., estimated model parameters with the wrong sign), and failure of the iterative regression model fitting procedure to converge. The authors conclude that "for EPV values of 10 or greater, no major problems occurred," but also that "although other factors (such as the total number of events, or sample size) may influence the validity of the logistic model, our findings indicate that low EPV can lead to major problems."

The second practical issue noted at the beginning of this discussion is the problem of *separation*, in which the outcome $\{y_k\}$ can be predicted perfectly from the values of one or more covariates included in the model [148, p. 138]. In such cases, the likelihood function is monotone and therefore exhibits no maximum, implying that maximum likelihood estimates do not exist. Not surprisingly, this situation causes practical difficulties in fitting logistic regression models, which typically give numerical values for the estimated model parameters, even though these values are not meaningful. Also, Hosmer and Lemeshow note [148, p. 139] that complete separation is most likely in cases of small sample sizes and many covariates, as in the small EPV examples just discussed. Some computational procedures detect these problems and provide warning meassages, but a common method of detecting separation is by examining the standard deviation estimates for the parameters. In fact, Hosmer and Lemeshow offer the following advice as a general prescription for detecting numerical difficulties in logistic regression models:

> We believe that the best indicator of a numerical problem in logistic regression is the estimated standard error. In general, any time the the estimated standard error of an estimated coefficient is large relative to the point estimate, we should suspect the presence of one of the data structures described in this section.

The data structures referred to in this quote include collinearity, separation, or the closely related problem of zero cell counts in a binary covariate, like that seen for INS values in the missing TSF logistic regression model considered in Sec. 13.5.4. For that example, the estimated INS coefficient was 0.802 with an estimated standard deviation of 11.730, while the estimated DIAG coefficient was 0.049 with an estimated standard deviation of 0.280. Note that in both cases, the estimated standard deviation is an order of magnitude larger than the point estimate. Removing INS from the model caused the DIAG estimate to increase in magnitude and change sign, from 0.049 to −0.211, reducing the standard error slightly, from 0.280 to 0.215. Also, removing INS from the model eliminated the warning message issued by the logistic regression routine when the model was estimated including INS.

It can be argued that, since the covariates responsible for the separation phenomenon just described are strongly predictive of the response variable, it is undesirable to eliminate them from the logistic regression model, despite the numerical difficulties they cause. It is possible to do this, but it requires special model fitting techniques like those discussed by Heinze and Schemper [137].

13.6 Poisson regression models

In practice, the two most popular applications of the generalized linear modeling strategy described in Sec. 13.4 are logistic regression, described in Sec. 13.5, and Poisson regression, described here. The Poisson regression model is appropriate to cases where the response variable of interest takes the form of integer-valued counts, where the "natural" probability distribution is the Poisson distribution. The notions of *overdispersion* and *underdispersion* were introduced in Chapter 8 in connection with the Poissonness plot, but they are revisited here because they are intimately related to Poisson regression. Specifically, one possible source of overdispersion is an unrecognized dependence of the Poisson mean parameter on an explanatory covariate, and it is precisely this type of dependence that Poisson regression attempts to describe. Consequently, the following discussion of Poisson regression begins with a brief discussion of overdispersion and underdispersion in Sec. 13.6.1, followed by a detailed discussion of the Poisson regression formulation in Sec. 13.6.2. An example illustrating the application of Poisson regression is presented in Sec. 13.6.3.

13.6.1 Over- and underdispersion

Recall from Chapter 3 that the Poisson distribution is a discrete distribution appropriate to counts $\{n_k\}$ defined by the density function:

$$\mathcal{P}\{n_k = n\} = \frac{e^{-\lambda}\lambda^n}{n!}, \tag{13.65}$$

for $n = 0, 1, 2, \ldots$. One of the characteristic features of this distribution is that the mean and the variance are equal, both equal to the single distribution parameter λ. Thus, if the estimated variance for the sequence $\{n_k\}$ is substantially larger than the mean are characterized as *overdispersion*, while those cases where the estimated variance is substantially smaller than the mean are characterized as *underdispersion*. Possible causes of overdispersion include misspecification of the count distribution (e.g., instead of the Poisson distribution, the counts exhibit the negative binomial distribution discussed in Chapter 3, for which the variance is larger than the mean), contamination (e.g., the zero-inflated count models introduced in Chapter 3), or the distributional heterogeneity caused by dependence of the Poisson parameter λ on explanatory covariates. Underdispersion seems to be rarer in practice than overdispersion and requires somewhat different treatment [53, p. 79].

To see how distributional heterogeneity leads to overdispersion, consider the following mixture model between two Poisson distributions, characterized by constants λ_1 and λ_2:

$$\mathcal{P}\{n\} = \pi\mathcal{P}\{n; \lambda_1\} + (1 - \pi)\mathcal{P}\{n; \lambda_2\}, \tag{13.66}$$

where π is a mixing parameter satisfying $0 < \pi < 1$. In practice, this situation could arise if an observed sequence $\{n_k\}$ of counts came from two subgroups, distinguished by some binary explanatory variable (e.g., male vs. female patients, untreated vs. heat-treated material samples, etc.). It follows directly from the mixture model results given in Chapter 4 that the mean of the two-component distribution is:

$$\mu = E\{n\} = \pi\lambda_1 + (1 - \pi)\lambda_2, \tag{13.67}$$

and it is easy to show (Exercise 7) that:

$$\text{var}\{n\} = \mu + \pi(1 - \pi)(\lambda_1 - \lambda_2)^2. \tag{13.68}$$

Note that the second term on the right-hand side of this equation can never be negative, and it is strictly positive for any mixing parameter $0 < \pi < 1$ and distinct distributional parameters $\lambda_1 \neq \lambda_2$. In other words, unrecognized heterogeneity caused by the dependence of the Poisson model parameter on a binary covariate leads to overdispersion, since $\text{var}\{n\} > \mu = E\{n\}$.

While the preceeding example illustrated the fact that dependence on an explanatory covariate leads to overdispersion for a specific example, this result holds more generally [8, p. 8]. Specifically, the following results are direct consequences of the conditional expectation results presented in Chapter 10. If n exhibits a Poisson distribution with a parameter $\lambda(\theta)$ that depends on some explanatory covariate θ, the unconditional mean of n is given by:

$$E\{n\} = E_\theta\{E\{n|\theta\}\} = E\{\lambda(\theta)\}, \tag{13.69}$$

and the unconditional variance is given by:

$$\begin{aligned}
\text{var}\{n\} &= E_\theta\{E\{n|\theta\}\} + \text{var}\{E\{n|\theta\}\}, \\
&= E\{\lambda(\theta)\} + \text{var}\{\lambda(\theta)\}, \\
&= E\{n\} + \text{var}\{\lambda(\theta)\}. \tag{13.70}
\end{aligned}$$

Since the variance term on the right-hand side of the last line is always positive unless θ is known precisely, it follows that the dependence of the Poisson parameter λ on explanatory covariates leads to overdispersion.

As a consequence of the results just presented, Cameron and Trivedi [53, p. 77] note that Poisson regression—which adjusts for the presence of explanatory variables in the Poisson model—reduces overdispersion, but it exacerbates underdispersion since the adjusted variance is reduced, relative to the mean. They also note that if the original overdispersion is sufficiently large, it is generally not fully corrected by Poisson regression. Nevertheless, as the following example illustrates, Poisson regression can lead to the identification of covariates responsible for part of the overdispersion seen in the data, and this can be extremely useful insight.

13.6.2   Poisson regression

Since it is also a special case of the GLM class introduced in Sec. 13.4, the derivation of the Poisson regression modeling strategy is similar to that for logistic regression presented in Sec. 13.5.1. First, note that the Poisson distribution defined in Eq. (13.65) belongs to the exponential family described in Sec. 13.4, corresponding to the following parameter and function definitions:

$$
\begin{aligned}
\theta &= \ln n_k, \\
\phi &= 1, \\
a(\phi) &= 1, \\
b(\theta) &= e^\theta, \\
c(n_k, \phi) &= -\ln n_k!.
\end{aligned}
\tag{13.71}
$$

Consequently, the standard form of the Poisson regression model is:

$$
\ln n_k = \mathbf{x}_k'\beta,
\tag{13.72}
$$

illustrating why this model is also commonly known as the *loglinear model* [8]. The deviance for this model is given by:

$$
D(\{n_k\}; \mathbf{x}_k'\beta) = \sum_{k=1}^{N} w_k \left\{ n_k(\ln n_k - \mathbf{x}_k'\beta) - n_k + e^{\mathbf{x}_k'\beta} \right\}.
\tag{13.73}
$$

As in the case of logistic regression models, the optimal parameter estimate $\hat{\beta}$ is that which maximizes the log likelihood for the data, or equivalently, minimizes the deviance defined in Eq. (13.73). The fitting procedures and numerical concerns are similar for the Poisson and logistic regression models: in the S language implementations (i.e., R and *S-Plus*), both models are fit by the same procedure (procedure **glm**), specifying a different distribution via the **family** option (i.e., "binomial" for logistic regression vs. "poisson" for Poisson regression). Thus, while it was offered for the case of logistic regression, Hosmer and Lemeshow's general cited in Sec. 13.5.5 provides useful guidance for Poisson regression as

Figure 13.5: Poissonness plot for the number of times pregnant (NPG) from the
Pima Indians diabetes dataset. Solid circles represent NPG values occurring
more than once, open circles represent values that occur only once, and open
triangles describe 95% confidence limits for the Poissonness count metameter

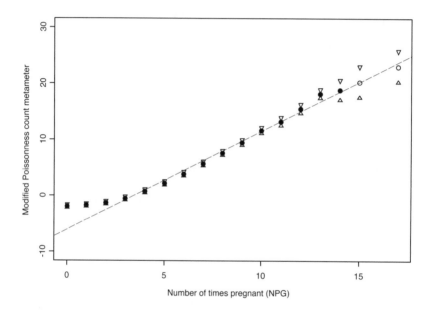

well [148, p. 138]: large estimated standard errors relative to the point estimates
for the model parameters β_i may be indicative of numerical difficulties caused
by data sparsity or collinearity in the explanatory variable set.

13.6.3   Application: NPG in the Pima Indians dataset

Fig. 13.5 shows the Poissonness plot introduced in Chapter 8 for the number
of times pregnant (NPG) from the Pima Indians diabetes dataset described
in Chapter 2. The NPG values that occur only once in the dataset (NPG
= 15 and 17) are marked with open circles, while those that occur two or
more times are marked with solid circles. The open triangles, most evident for
large NPG values, indicate the upper and lower 95% confidence intervals on the
Poissonness count metameter for each NPG value. Overall, it is clear from this
plot that, while the estimated count metameter values are fairly precise, there
is a systematic departure from linearity at small counts, especially zero. One
possible explanation is that the NPG values follow a zero-inflated distribution
like those discussed in Chapter 3; another is that, as with several of the other
variables in the Pima Indians diabetes dataset, zero values are being used to

Figure 13.6: Poissonness plots for NPG in the same format as Fig. 13.5, broken down by age group as indicated above each plot

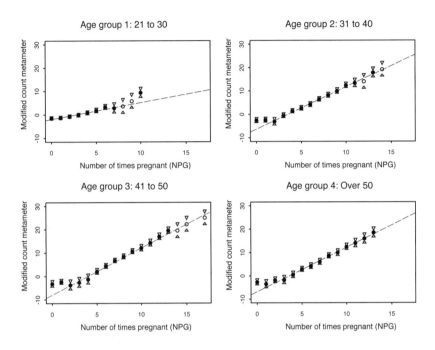

code missing data. The following analysis applies Poisson regression to assess the reasonableness of a third possible explanation: that the NPG values depend systematically on age. Biologically, such a dependence is to be expected, since most births occur in women from about 20 to about 40 years of age [12, p. 222].

A preliminary graphical examination of this idea is presented in Fig. 13.6, which shows Poissonness plots in the same format as Fig. 13.5 for the NPG values in the following four age ranges:

1. Group 1: ages 21 to 30 years;

2. Group 2: ages 31 to 40 years;

3. Group 3: ages 41 to 50 years;

4. Group 4: ages greater than 50 years.

Three conclusions are clear from these plots. First, the range of NPG values varies with age: none of the women in the 21 to 30 age group have been pregnant more than 10 times, while some women in the other age groups have been pregnant between 11 and 17 times. Biologically, this is what we expect: it is not plausible for a 21-year-old to have been pregnant 17 times, while this is feasible (although not common) for a woman over 50. Second, the slopes of the

Table 13.9: Estimated means ($\bar{x}$), variances ($\hat{\sigma}^2$), and overdispersion measures (OD) for the Pima Indians NPG counts, broken down by age group

Age group	$\bar{x}$	$\hat{\sigma}^2$	OD
All ages	3.85	11.35	1.95
21 to 30	2.01	3.22	0.60
31 to 40	5.27	10.23	0.94
41 to 50	7.12	13.95	0.96
50+	5.96	11.29	0.89

reference lines fit to these Poissonness plots are significantly different for the four age groups, supporting the idea of a systematic dependence of NPG on age. Third, while a fit anomaly at low NPG values is evident in three of these four age-group-specific plots, it is less pronounced than that seen in the overall Poissonness plot in Fig. 13.5, and it is not evident in the youngest age group.

As noted in Sec. 13.6.1, dependence on an explanatory covariate like age can be a contributing cause of overdispersion in sequences of count data like the Pima Indians NPG values. Table 13.9 gives a simple assessment of the apparent overdispersion for the NPG data, both overall and broken down by age group. Specifically, Table 13.9 gives the estimated means $\bar{x}$, variances $\hat{\sigma}^2$, and overdispersion estimates (OD) for the indicated age groups:

$$OD = \frac{\hat{\sigma}^2}{\bar{x}} - 1. \tag{13.74}$$

Recall that for the Poisson distribution, the variance is equal to the mean, implying an OD value of zero. It is clear from the results presented in Table 13.9 that, while each age group exhibits some degree of overdispersion, this behavior is much less pronounced within these age groups than it is in the dataset overall. These observations suggest examining a Poisson regression model relating NPG to membership in these four age groups.

Table 13.10 is the analysis of deviance table described at the end of Sec. 13.4 for this Poisson regression model, giving model parameters, their standard errors and the corresponding deviance values obtained from the R procedure **glm**. The p-value associated with the age group coefficient β_1 was computed as the upper tail probability of the χ^2 distribution with one degree of freedom for the deviance change on including this term in the model. The extremely small value computed for this probability strongly supports the conclusion presented above that the distribution of NPG values depends on patient age.

Taken together, the results presented here suggest that the zero values for NPG are probably representative and do not correspond to codes for missing

Table 13.10: Analysis of deviance table for the Possion regression model relating number of times pregnant (NPG) to the four age groups defined in the text for the Pima Indians diabetes dataset

Covariate	$\hat{\beta}_i$	$\hat{\sigma}_i$	df	Deviance	P_{χ^2}
Age group	0.399	0.016	1	615.13	$< 1 \times 10^{-16}$
Null	0.527	0.041	766	1769.42	—
Total	—	—	767	2384.55	—

data, as in the case of diastolic blood pressure and several other variables where zero values are biologically infeasible.

13.7   Exercises

1. Suppose $\{x_k\}$ and $\{y_k\}$ are binary sequences, where $x_k = 1$ with probability p and $y_k = 1$ with probability q. Define the *population odds ratio* as the ratio of the probabilities:

$$O_{xy} = \frac{p_{00}p_{11}}{p_{01}p_{10}},$$

where p_{ij} denotes the joint probability that $x_k = i$ and $y_k = j$ for $i, j = 0$ or 1. Show that $O_{xy} = 1$ when $\{x_k\}$ and $\{y_k\}$ are statistically independent.

2. Derive the sum of squares partition from Eq. (13.11):

$$\sum_{i=1}^{C} \sum_{j=1}^{n} (y_{ij} - \bar{y}_{..})^2 = n \sum_{i=1}^{C} (\bar{y}_{i.} - \bar{y}_{..})^2 + \sum_{i=1}^{C} \sum_{j=1}^{n} (y_{ij} - \bar{y}_{i.})^2.$$

3. Derive the expectations for the between groups and within groups sums of squares, SS_B and SS_W, given in Eq. (13.12):

$$E\{SS_B\} = n \sum_{i=1}^{C} \alpha_i^2 + (C - 1)\sigma_e^2,$$

$$E\{SS_W\} = C(n - 1)\sigma_e^2.$$

4. Show that the C components of the least squares solution $\hat{\theta}$ given in Sec. 13.3.2 are identical with the estimators for μ and $\alpha_1, \alpha_2, \ldots, \alpha_{C-1}$ given in Sec. 13.3 for the ANOVA model.

5. Show that by defining $D_{i,k}$ as in Eq. (13.31), y_k can be expressed as in Eq. (13.20) for the unbalanced one-way ANOVA problem, from which it follows that the regression formulation of this problem has the same form as the balanced one-way ANOVA problem.

6.   a. Show that the Gaussian distribution belongs to the exponential family defined by Eq. (13.46); that is, for $y \sim N(\mu, \sigma)$ determine the constants θ and ϕ and the functions $a(\phi)$, $b(\theta)$, and $c(y, \phi)$ appearing in this expression.

    b. For the Gaussian case with the cannonical link $g(x) = x$, show that the deviance defined in Eq. (13.52) is equal to the residual sum of squares:

$$\text{RSS} = \sum_{k=1}^{N} (y_k - \mathbf{x}_k' \beta)^2.$$

7. Show that the variance of the two-component mixture of Poisson distributions discussed in Sec. 13.6.1 is given by:

$$\text{var}\{n\} = \mu + \pi(1 - \pi)(\lambda_1 - \lambda_2)^2,$$

where $\mu = \pi\lambda_1 + (1 - \pi)\lambda_2$ is the mean of the mixture distribution, π is the mixing parameter, and λ_1 and λ_2 are the parameters associated with the component Poisson distributions on which the mixture is based.

Chapter 14

Characterizing Analysis Results

Sections and topics:

This chapter introduces a variety of useful techniques for characterizing analysis results, all based on making simple modifications to our original dataset and recomputing the results from these modified datasets. Specific examples include both *bootstrap methods* based on sampling from the analysis dataset with replacement, and *subsampling methods*, based on sampling from the analysis dataset without replacement. In either case, the idea is to recompute the analysis results based on a modified dataset and compare the results to obtain an assessment of the variability of the original results. Another important idea introduced in this chapter is the use of *random permutations* to provide a frame of reference for assessing significance, an idea that forms the basis for *computational negative control* strategies. Finally, a fourth specific implementation of this general "make variations and recompute" strategy is that of *deletion diagnostics*, where data subsets are systematically omitted to see how this omission changes the original results. All four of these strategies lead to broadly applicable techniques for characterizing computational results, as examples presented in this chapter illustrate.

Key points of this chapter:

1. The techniques described in this chapter are all based on the idea of replacing working assumptions (e.g., distributional assumptions) with repeated computations. Given a dataset $\mathcal{D}$ and an analysis result R, these methods proceed by first constructing a set $\{\mathcal{D}_i\}$ of modified datasets and then computing the corresponding analysis results $\{\tilde{R}_i\}$ from these datasets.

2. This chapter presents four classes of methods based on the general idea just described. The best known of these is the class of *bootstrap resampling methods*, which construct the modified datasets by sampling records from $\mathcal{D}$ *with replacement,* often providing useful methods for estimating the variability of R (e.g., standard deviations or confidence intervals).

3. A key working assumption on which bootstrap methods are based is a smooth dependence of the characterization R on the data from which it is computed. In cases where these assumptions are not met (e.g., when R is the sample maximum), bootstrap methods can fail. A useful alternative in these cases is the class of *subsampling methods,* which construct the modified datasets $\{\mathcal{D}_i\}$ by sampling from $\mathcal{D}$ *without replacement.*

4. A disadvantage of subsampling methods is that they depend on the size of the subsample chosen, which is necessarily smaller than the size of the original dataset. As results and examples presented here illustrate, the subsampling strategies that are most comparable to bootstrap methods when both methods are applicable are those based on *half-samples* (i.e., random subsamples of size $N/2$ drawn from a dataset of N samples).

5. *Deletion diagnostics* construct the modified datasets $\{\mathcal{D}_i\}$ by deleting subsets and examining the influence of these deletions on R. The most useful deletion diagnostics in practice are *leave-one-out diagnostics* that compute N results from an N-record dataset, each based on a subset of size $N-1$ with one data record omitted from $\mathcal{D}$. As examples presented here illustrate, these methods can be extremely useful in detecting anomalous data observations, which can cause large changes in R on deletion.

6. The three strategies just described all construct modified datasets $\{\mathcal{D}_i\}$ by making what should be "small" changes in the original dataset $\mathcal{D}$. *Computational negative control strategies* construct these datasets by making what should be "large" changes in the original dataset and looking for correspondingly "large" changes in the result R. These techniques are useful in assessing the statistical significance of results that attempt to measure structural relations within a dataset (e.g., correlations between variables).

7. The strategies described here are very general, imposing minimal restrictions on the dataset $\mathcal{D}$ and applicable to a wide range of analysis results R. Significant computational savings can sometimes be achieved by specializing these methods to particular applications (e.g., least squares regression problems), a point discussed further in Chapter 15.

14.1 Analyzing modified datasets

This chapter describes four classes of computational procedures based on the idea of making systematic changes to our original dataset $\mathcal{D}$ and examining the influence of those changes on our computational results. In cases where these systematic changes are small, their consequences tell us something about the sensitivity of our results to small changes in the dataset, giving us a useful measure of variability. In particular, results based on random samples drawn from $\mathcal{D}$ (bootstrap and subsampling methods, described in Secs. 14.4 and 14.5, respectively) can be used to obtain variability estimates and approximate confidence intervals for many different types of computational results. Similarly, results based on what *should be* small changes in $\mathcal{D}$ can be used to detect anomalous data values or data subsets. This is the idea behind deletion diagnostics, introduced in Sec. 14.3 and further specialized to the case of least squares regression models in Chapter 15. Finally, we can make what should be large changes in the original dataset and see whether we obtain correspondingly large changes in our computed result: if so, this provides evidence of the significance of our original result, and if not, it implies that the result could have been due to random chance alone. This idea is the basis for the *computational negative control* strategy described in Sec. 14.2, where *structure-destroying random permutations* are used to create a collection $\{\mathcal{D}_i\}$ of datasets that should be like one another, but lacking the key structure we are interested in characterizing in the original dataset $\mathcal{D}$ (e.g., dependence between two variables in the dataset). Here, if the structure is present in $\mathcal{D}$, the original result should differ substantially from those obtained from the datasets $\{\mathcal{D}_i\}$.

14.1.1 Variation-based analysis procedures

All of the characterization procedures described in this chapter compare a computational result R obtained from a given dataset $\mathcal{D}$ with a collection $\{\tilde{R}_i\}$ of B related results, each obtained from a modified dataset $\mathcal{D}_i$ derived from $\mathcal{D}$. A key feature of these methods is that the modified datasets $\{\mathcal{D}_i\}$ should all be "essentially the same" because of the way they are constructed. This notion of datasets being "essentially the same" corresponds to an assumption of *exchangeability*, an idea discussed in detail in Sec. 14.1.2. Compared with the distributional assumptions imposed in classical hypothesis testing (e.g., the t-test described in Chapter 8 compares the means of two Gaussian random variables), this exchangeability assumption is extremely weak, contributing greatly to the wide applicability of the procedures described here, a point discussed further in Sec. 14.6.

The general structure of all of these procedures is the following:

1. given the original dataset $\mathcal{D}$, generate a collection of B modified datasets $\{\mathcal{D}_i\}$, based on a simple modification mechanism that should make them all exchangeable;

2. compute results $\{\tilde{R}_i\}$ from the datasets $\{\mathcal{D}_i\}$ analogous to the original result R computed from $\mathcal{D}$;

3. interpret the modified results $\{\tilde{R}_i\}$ to provide useful information about the original result R.

The specific methods described here differ in the details of these steps. For example, the compuatational negative control procedures discussed in Sec. 14.2 use random permutations in Step 1, designed to destroy a particular structural feature of $\mathcal{D}$, if it is present. In the specific cases considered here, the results computed in Step 2 are real-valued scalars, since this simplifies the interpretation in Step 3, a point discussed further in Sec. 14.2. This interpretation step considers the question of whether the modified results $\{\tilde{R}_i\}$ are sufficiently different from the original result R to give evidence that the original dataset $\mathcal{D}$ is *not* exchangeable with the modified datasets $\{\mathcal{D}_i\}$. Such exchangeability violations imply that the assumed structure was present in the original dataset, while the absence of these violations suggests that the structure was absent. For comparison, in the bootstrap resampling methods discussed in Sec. 14.4, the modified datasets $\{\mathcal{D}_i\}$ constructed in Step 1 are obtained by sampling the original dataset $\mathcal{D}$ with replacement, while Step 3 assesses the results by constructing confidence intervals for R from the modified results $\{\tilde{R}_i\}$.

14.1.2 The notion of exchangeability

As it is used here, the term *exchangeability* refers to a conceptual equivalence between datasets: informally, two datasets are exchangeable if applying the same analysis procedure to both can be expected to give approximately the same result. This notion of exchangeability corresponds to that given by Draper *et al.* [93], representing a judgment we make about distinct sources of data. What follows is a simple, informal characterization of this notion of exchangeability; for a more detailed discussion in the context of data analysis, refer to the book by Pearson [228, sec. 6.2], or for a complete discussion refer to the original paper by Draper *et al.* [93].

The essential notion is the following. The original dataset $\mathcal{D}$ is assumed to be made up of a collection of N data records, each characterizing a distinct experimental unit of the same type. In medical datasets, these experimental units might be patients or tumor samples, while in physical property datasets, these experimental units might be material samples subjected to tensile strength tests, compositional analysis, or electrical conductivity characterizations. The fundamental working assumption is that these experimental units are similar enough that two distinct data subsets $\mathcal{D}_i$ and $\mathcal{D}_j$, each with the same number (or approximately the same number) of records from $\mathcal{D}$ should give comparable results, $\tilde{R}_i$ and $\tilde{R}_j$ when subjected to the same analysis. This assumption is implicit whenever we consider $\mathcal{D}$ to represent a collection of statistically independent samples, all drawn from the same distribution. In contrast, exchangeability will not hold if the dataset characterizes experimental units drawn from two or more distinct populations whose characterizations R can be expected to differ significantly. *A key point here is that the assumption of exchangeability will be made in the analyses presented in this chapter and then assessed for its reasonableness and/or consequences.* It is also important to note that this exchangeability assumption is analogous to, but much weaker than, the distributional assumptions made in Chapter 9 as the basis for traditional hypothesis tests and confidence interval estimates: there, it is typically assumed that the data observations correspond to statistically independent samples drawn from a fixed distribution (which itself implies exchangeability), and that certain details of this distribution are also known (e.g., its form).

The next paragraph discusses specific examples where exchangeability can be expected to hold, relevant to the specific analytical characterizations considered in this chapter. First, however, it is important to say something about one other working assumption: in all of the characterizations considered in this chapter, it is assumed that R is *permutation-invariant*, meaning that simply re-ordering all of the records in the dataset $\mathcal{D}$ does not alter the value of R. This condition is satisfied by all of the data characterizations considered in this book, but it is *not* satisfied by dynamic data analysis techniques like spectrum estimation, auto- and cross-correlation analysis, or time-series modeling. In particular, note that simply reindexing all of the data values does not change the value of estimated means, variances, medians, skewness measures, etc. Further, this condition also holds for more complex characterizations like Q-Q plots and nonparametric density estimates. It is important to distinguish these simple record reorderings from the *structure-destroying permutations* considered in Sec. 14.2: the permutations considered here simply renumber each record in the dataset $\mathcal{D}$, keeping all of the fields defining each record intact; the structure-destroying permutations considered in Sec. 14.2 randomly scramble the fields between records, introducing deliberate misalignment errors. Permutations of the first kind will have no effect on the results of correlation or regression analysis, but permutations of the second kind can have a very large effect. As a specific example illustrating the difference, consider the estimated correlation coefficient between two sequences, $\{x_k\}$ and $\{y_k\}$, each of length N. The basic estimator given in Chapter 10 is:

$$\hat{\rho}(x, y) = \frac{1}{\sigma_x \sigma_y} \sum_{k=1}^{N} (x_k - \bar{x})(y_k - \bar{y}), \tag{14.1}$$

where $\bar{x}$ and $\bar{y}$ are the sequence means and σ_x and σ_y are their estimated standard deviations. Now, suppose we randomly reorder the records, an operation that corresponds to replacing the index k in the sum in Eq. (14.1) with $\pi(k)$ where π represents a random permutation of the sequence $\{1, 2, \ldots, k\}$. Note that this modification changes the indices on both $\{x_k\}$ and $\{y_k\}$ in this sum, giving the estimator:

$$\tilde{\rho}(x, y) = \frac{1}{\sigma_x \sigma_y} \sum_{k=1}^{N} (x_{\pi(k)} - \bar{x})(y_{\pi(k)} - \bar{y}) = \hat{\rho}(x, y), \tag{14.2}$$

since the replacement $k \rightarrow \pi(k)$ in the sum corresponds to simply renaming the dummy summation index k. In contrast, the structure-destroying random permutations used in the computational negative control method described in Sec. 14.2 corresponds to replacing this sum with:

$$\tilde{\rho}(x, y) = \frac{1}{\hat{\sigma}_x \hat{\sigma}_y} \sum_{k=1}^{N} (x_{\pi(k)} - \bar{x})(y_k - \bar{y}) \neq \hat{\rho}(x, y), \tag{14.3}$$

since x and y now have different indices in the sum. The differences between the results obtained in Eqs. (14.1) and (14.3) are examined in detail in Sec. 14.2.2.

In the computational negative control (CNC) strategy described in Sec. 14.2, the modified datasets $\{\mathcal{D}_i\}$ are obtained by applying structure-destroying random permutations of the type just described, randomly scrambling the fields between records relative to the original dataset $\mathcal{D}$. If the records in $\mathcal{D}$ describe "equivalent" data sources, the effects of two different, statistically independent random permutations should be comparable. Thus, in CNC analyses, it is reasonable to assume the modified datasets $\{\mathcal{D}_i\}$ are exchangeable with each other, but *not* generally with the original dataset $\mathcal{D}$. Indeed, the essence of the CNC strategy is to look for evidence of non-exchangeability between $\mathcal{D}$ and $\{\mathcal{D}_i\}$. In contrast, the deletion diagnostic strategies described in Sec. 14.3 construct modified datasets $\{\mathcal{D}_i\}$ by deleting subsets $\{S_i\}$ from the original dataset $\mathcal{D}$; in the simplest and most commonly used case, these subsets each consist of a single point (i.e., $S_i = \{x_i\}$ for $i = 1, 2, \ldots, N$). In this case, if each data record has the same character, we expect the deletion datasets $\{\mathcal{D}_i\}$ to be both exchangeable with each other and, at least for reasonably large N, exchangeable with the original dataset $\mathcal{D}$. The basis for practical deletion diagnostics is the idea that violations of exchangeability identify data subsets S_i that are *not* exchangeable with the rest of the records in $\mathcal{D}$, thus leading us to identify possible data anomalies. Finally, the modified datasets constructed in the other two methods considered here are obtained by drawing random samples of records from the dataset $\mathcal{D}$. In the bootstrap method discussed in Sec. 14.4, the datasets $\{\mathcal{D}_i\}$ are constructed by sampling *with replacement* from $\mathcal{D}$, while in the subsampling method discussed in Sec. 14.5, these datasets are constructed by sampling *without replacement*. Two important distinctions are first, that sampling with replacement permits bootstrap datasets $\mathcal{D}_i$ to be the same size as the original, while these datasets are necessarily smaller than the original in the subsampling method, but second, that the bootstrap datasets necessarily involve duplicate records while the subsampling datasets do not. In either case, the idea is that if the individual data points are equivalent, the subsets $\{\mathcal{D}_i\}$ obtained by either of these sampling procedures should be exchangeable both with each other and with the original dataset $\mathcal{D}$.

14.2 Computational negative controls

In biological experiments, it is common to include *negative controls:* experimental units that are known not to respond to whatever experimental stimulus is being applied. This practice is useful because it provides a biological frame of reference in judging the responses of other units. In particular, if these units do not respond appreciably more strongly than the negative controls, it suggests that these other units are not responsive to the stimulus, either. This idea has been applied to cluster analysis [231], where the general objective is to partition a collection of objects into k subsets of objects that are similar within subsets and different between subsets. The basis for this partitioning is a dataset that lists, for each object, the values of several different attributes, and objects are judged to be similar or different depending on similarities and differences between their attribute values. Many methods are available for doing cluster analysis [120, 169], and they will generally return a clustering for *any* dataset, including synthetic ones without any inherent cluster structure whose attribute variables are collections of independent random variables. Thus, an important practical question is how significant a particular clustering result is. One way of addressing this question is to compare the clustering obtained from the original dataset with one obtained from modified datasets where any cluster structure that may be present in the original has been destroyed by randomly reassigning the attribute values in the dataset to different objects. Comparing the quality of the original clustering—which can be measured many different ways—with the same quality measure computed from the modified datasets provides a basis for assessing the significance of the original result [231]. Specifically, if the quality of the clustering obtained from the original dataset is not significantly better than that obtained from the randomly permuted datasets, we conclude that the structure suggested by the original result is not significant and could have been observed by random chance alone.

This same approach can be applied to any other data characterization that seeks to describe a structural relationship between variables. In the example presented in Sec. 14.2.2, the question of interest is whether two sequences $\{x_k\}$ and $\{y_k\}$ exhibit a significant association, as measured by one of the correlation measures presented in Chapter 10. There, the general procedure described in Sec. 14.1.1 specializes to the following three steps:

1. Given the N record dataset $\mathcal{D} = \{(x_k, y_k)\}$, construct B modified datasets $\mathcal{D}_i = \{(x_{\pi_i(k)}, y_k)\}$, where $\{\pi_i(k)\}$ is a sequence of B independent random permutations of the index sequence $\{1, 2, \ldots, N\}$.

2. Compute the correlation coefficient $\rho(x, y)$ from the original dataset $\mathcal{D}$ and the corresponding correlation coefficients $\tilde{\rho}_i(x, y)$ from the modified datasets $\{\mathcal{D}_i\}$.

3. Determine whether $\rho(x, y)$ is comparable to the values $\{\tilde{\rho}_i(x, y)\}$ or significantly different from these values, using the significance measures described in Sec. 14.2.1.

A detailed illustration of this approach to assessing the significance of correlation results is presented in Sec. 14.2.2, but it is important to emphasize that the basic strategy is applicable to a variety of other data analysis problems, including any of the regression analysis problems discussed in Chapters 5, 11, or 13, a point illustrated in Chapter 15. The key in all cases is to define suitable structure-destroying random permutations in Step 1 of the procedure and to define suitable real-valued characterizations of the analysis results in Step 2. In cases like multivariate regression analysis that attempt to characterize relations involving more than two variables, the number of available options—and necessary choices—increases in both of these steps, complicating the application of computational negative control (CNC) methods somewhat. In particular, in cases where the result is more complicated than simply a real number like a mean, skewness, or prediction error variance, it is necessary to convert the results into a suitable real-valued summary measure to use the significance assessments described in Sec. 14.2.1 in Step 3. Nevertheless, the general CNC technique remains both relatively simple and potentially very useful in a wide variety of applications.

14.2.1   Empirical probabilities and z-scores

The interpretation of CNC results involves comparing the original result R with the collection $\{\tilde{R}_i\}$ of modified results to determine whether the original dataset $\mathcal{D}$ appears to be exchangeable with the modified datasets $\{\mathcal{D}_i\}$. Since these results are assumed here to be real-valued scalars, there are two simple ways of making this comparison. The first is to compute empirical probabilities and the second is to compute z-scores.

The empirical probability of observing a result as large as R or larger in the set $\{\tilde{R}_i\}$ is given by:

$$\mathcal{E}^+ = \frac{\#\{\tilde{R}_i > R\} + 1}{B + 1}, \tag{14.4}$$

where $\#\mathcal{S}$ denotes the number of elements in the set $\mathcal{S}$ and B is the number of modified results $\{\tilde{R}_i\}$. Similarly, the empirical probability of observing a result as small as R or smaller is given by:

$$\mathcal{E}^- = \frac{\#\{\tilde{R}_i < R\} + 1}{B + 1}. \tag{14.5}$$

For two-sided tests, an empirical estimate of the probability that R lies outside the range of the reference results $\{\tilde{R}_i\}$ is given by:

$$\mathcal{E}^\pm = \begin{cases} \frac{2\#\{\tilde{R}_i > R\} + 1}{B + 1} & R > \text{median } \{\tilde{R}_i\}, \\ 1 & R = \text{median } \{\tilde{R}_i\}, \\ \frac{2\#\{\tilde{R}_i < R\} + 1}{B + 1} & R < \text{median } \{\tilde{R}_i\}. \end{cases} \tag{14.6}$$

The question of which of these empirical probilities is appropriate is application-dependent, a point illustrated in Sec. 14.2.2. In any case, given an empirical probability estimate, we declare the modified datasets $\{\mathcal{D}_i\}$ to be non-exchangeable with the original dataset $\mathcal{D}$ if the appropriate empirical probability is less than a prespecified threshold α. Note that since the smallest empirical probability that can be obtained from a collection of B modified datasets is $1/(B + 1)$, achieved when the original result is more extreme than all of the modified results, it follows that given a threshold probability α, it is necessary to make $B > 1/\alpha - 1$ in constructing our modified datasets.

The primary advantage of empirical probabilities is their direct interpretation, but a limitation is that they make no use of information about how far R lies from the set $\{\tilde{R}_i\}$ when it lies outside the range of these values. That is, $\mathcal{E}^+ = 1/(B + 1)$ whenever $R > \tilde{R}_i$ for all i, regardless of whether R is only slightly larger than the largest $\tilde{R}_i$ value, or orders of magnitude larger. While it is not as directly interpretable, the z-score has the advantage that it does make use of this information. Specifically, the z-score for the result R relative to the modified results $\{\tilde{R}_i\}$ is defined as:

$$z = \frac{R - \mu}{\sigma}, \tag{14.7}$$

where μ is the mean of the modified results $\{\tilde{R}_i\}$ and σ is the standard deviation. In cases where R lies outside the range of $\{\tilde{R}_i\}$, the magnitude of the z-score gives a measure of the associated distance.

A disadvantage of the z-score relative to empirical probabilities is that it is less directly interpretable. Often, Gaussian reference values are used in interpreting z-scores, but the appropriateness of these values depends strongly on both the nature of the computational result R and on the distributional character of the dataset $\mathcal{D}$. A much more conservative but safer alternative is based on the Chebyshev inequality, presented in Chapter 4, which states that for all finite variance distributions:

$$\mathcal{P}\{|x - \mu| > t\sigma\} < \frac{1}{t^2}, \tag{14.8}$$

where μ is the mean of the distribution and σ^2 is the variance. It follows immediately from this result that, for the z-score defined in Eq. (14.7):

$$\mathcal{P}\{|z| > t\} < \frac{1}{t^2}, \tag{14.9}$$

under any finite-variance distribution. Thus, if we observe a z-score of sufficiently large magnitude, we can conclude that the original result is *not* exchangeable with the modified results computed from the perturbed dataset. In particular, note that a z-score larger than 10 exhibits a probability of less than 1% under any finite-variance distribution. Since, as noted in Chapter 4, the Chebyshev bound is extremely conservative, z-scores this large may be taken as strong evidence against exchangeability.

14.2.2   An application: assessing correlations

The brain-weight/body-weight dataset introduced in Chapter 1 provides a nice illustration of the utility of the CNC approach in assessing the significance of

Figure 14.1: Compuatational negative control assessment of brain/body-weight correlations, based on three different representations: the original data sequences, log-transformed data sequences, and rank-ordered data sequences

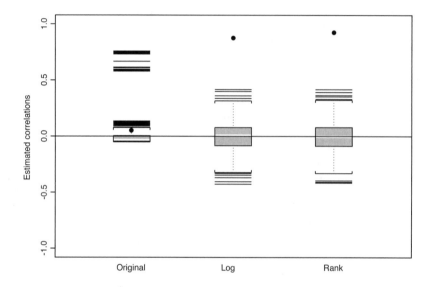

correlation results. Recall that this dataset contained the body weights in kilograms and the brain weights in grams for 65 different animal species. Because of the enormous dynamic range of both of these variables, a scatterplot of one variable against the other gave no suggestion of any relationship between these variables, a conclusion supported by their extremely small product-moment correlation coefficient ($\rho \simeq 0.052$). Conversely, applying logarithmic transformations to both variables greatly reduces this dynamic range and yields a much larger correlation estimate ($\rho \simeq 0.875$), as does rank-ordering both sequences and computing the Spearman rank correlation estimate ($\rho \simeq 0.924$).

Fig. 14.1 shows the results obtained when applying a computational negative control analysis to these three correlation results. Here, the original results are compared with the corresponding correlations computed from $B = 1000$ modified sequences, each obtained by applying an independent random permutation to the $\{x_k\}$ sequence (i.e., the body-weight values). In each case, the estimated correlation value from the unmodified dataset is represented as a solid circle, while the range of random permutation results is represented as a boxplot. Several conclusions are evident from these results. First, the small correlation value computed from the original data sequences is indeed insignificant relative to the

Table 14.1: Computational negative control correlation summary for three representations of the brain/body weight dataset: original, $\log_{10}$-transformed, and rank-ordered (Spearman correlations)

Data sequences	Empirical probability $\mathcal{E}^+$	z-Score	Chebyshev probability bound	5% significant?
Original	0.090	0.414	5.834	No
$\log_{10}$-transformed	0.001	7.070	0.020	Yes
Rank-ordered	0.001	7.518	0.018	Yes

range of random permutation results. This point is both visually evident in Fig. 14.1 and demonstrated quantitatively in Table 14.1, which gives the empirical probabilities, the z-scores, and the Chebyshev probability bounds computed from the z-scores for all three results. For the untransformed data sequences, 89 of the 1000 random permutation values exceed the original correlation value, giving an empirical probability estimate of 9%, meaning that the estimated correlation value is too small to be significant at the 5% level. Similarly, the small z-score for this result yields a Chebyshev probability bound that is larger than 1, both consistent with the postulated insignificance of the result and illustrating the conservativeness of the Chebyshev bound. In contrast, both of the other correlations estimated from the unmodified data sequences exceed all of the random permutation results, as may be seen in Fig. 14.1 and the smallest achievable empirical probabilities shown in Table 14.1 (i.e., $\mathcal{E}^+ = 1/(B+1) \simeq 0.001$). Here, the z-scores are large enough that the Chebyshev probability bounds for both cases are significant at the 5% level.

Before leaving this example, the following point is worth noting. The boxplots shown in Fig. 14.1 are not the simple versions based on Tukey's five-number summary introduced in Chapter 1, but are the *S-Plus* default versions discussed in Chapter 7 that represent outliers with horizontal lines beyond the nominal data range, as determined by the interquartile distance. In typical CNC analyses, the question of interest is whether the original value lies within the range of the random permutation results or lies outside this range, and if so, how far outside this range. Here, however, the alternative boxplots were used because they highlight a cluster of isolated outliers in the correlation results computed from the original data sequences. In fact, these outliers account for many of the 89 random permutation results that exceed the unmodified correlation result for this example. This behavior is unusual and arises here because of the extreme dynamic range of the original data values, responsible for the essentially noninformative appearance of the original scatterplot presented in

Chapter 1 for this example. In particular, the estimated correlation values are highly sensitive to the presence of coincident outliers in the two sequences, a point discussed in Chapter 10 and illustrated further by the example discussed in Sec. 14.3.2. In the original brain/body weight dataset, some of the random permutations induce coincidences between some of the outliers in the two original data sequences that were not present in the raw data, greatly increasing the correlation estimates. We do not see this effect in the other two examples in Fig. 14.1 since both the log transformation and the rank-ordering process greatly reduce the dynamic range of both data variables, reducing the effective magnitude and influence of the outliers.

14.3   Deletion diagnostics

Deletion diagnostics are extremely useful techniques for detecting data anomalies. They are most commonly used in conjunction with least-squares-based linear regression problems where the structure of the exact solution can be exploited to achieve significant computational savings. This point is illustrated in Chapter 15, where several important deletion diagnostics are described. The focus in this chapter is on computationally based deletion diagnostics that can be applied much more broadly: because they do not rely on special problem structures, their applicability is limited only by computational considerations. In some cases, these methods can be impractically slow, but even when a complete deletion-based analysis is not feasible, the underlying idea can be used to assess the influence of any single group of "suspicious" points, however our suspicion may have been aroused. That is, given a suspicious subset $\mathcal{S} \subset \mathcal{D}$, comparing the result computed from $\mathcal{D}$ with and without the subset $\mathcal{S}$ can give us a useful indication of whether these suspicious data points are unduly influential and thus possibly deserving of omission, special treatment, or further study.

14.3.1   The deletion diagnostic framework

As a practical matter, it is important to distinguish between *leave-one-out deletion diagnostics*, constructed by omitting each data point individually and recomputing the results, and more general deletion diagnostics that omit larger data subsets. The reason this distinction is important is that as we increase the size of the subsets we delete, the number of possible modified datasets grows combinatorially. Specifically, the number of possible subsets with k points deleted from a dataset of size N is:

$$\begin{pmatrix} N \\ N - k \end{pmatrix} = \begin{pmatrix} N \\ k \end{pmatrix} = \frac{N!}{(N-k)!k!} = \frac{N \cdot (N-1) \cdots (N-k+1)}{k \cdot (k-1) \cdots 1}. \quad (14.10)$$

This number grows extremely rapidly: for $N = 50$, taking $k = 1$ leads to 50 leave-one-out deletion subsets $\mathcal{D}_i$, while taking $k = 2$ increases this number to 1225 leave-two-out deletion subsets, and taking $k = 25$, corresponding to analyzing all "half subsets" of $\mathcal{D}$, yields approximately 1.26×10^{14} deletion subsets. Thus, while exhaustive leave-one-out deletion diagnostics are practical to both compute and interpret even for fairly large N, more general deletion diagnostics cannot be examined exhaustively, in general.

For leave-one-out deletion diagnostics, the general procedure outlined in Sec. 14.1.1 specializes to the following three steps:

1. From the N point dataset $\mathcal{D}$, construct the N datasets $\mathcal{D}_i$ defined by omitting each successive data point from $\mathcal{D}$ in turn.

2. Compute the analysis result R and the comparison results $\tilde{R}_i$ by applying the same computational procedure to the datasets $\mathcal{D}$ and $\mathcal{D}_i$, for $i = 1, 2, \ldots, N$.

3. For scalar-valued characterizations, plot $\tilde{R}_i$ vs. i, with the value of R indicated on the plot as a horizontal line.

Alternative interpretations in Step 3 are certainly possible, such as the application of outlier detection procedures like those described in Chapter 7 to the deletion sequence $\{\tilde{R}_i\}$, but as the example presented in Sec. 14.3.2 illustrates, graphical representations of the deletion diagnostic results can be extremely informative.

It is also important to note that the restriction to scalar results, while both common and convenient, is not essential. This point is illustrated in the discussion of traditional OLS-based deletion diagnostics in Chapter 15. There, one class of diagnostics is developed on the basis of changes in the model's prediction error variance, a scalar measure, as data points are deleted, while another is developed on the basis of changes in the model parameters as data points are deleted. This second method typically develops a scalar composite measure depending on all of the model parameters, but this raises the question of how the individual terms are to be weighted in forming the composite.

14.3.2 Simulation example: correlation analysis

To illustrate the basic idea of deletion diagnostics in a simple, concrete setting, consider the application of the leave-one-out deletion strategy to correlation analysis. The following example is based on two simulated data sequences, $\{x_k\}$ and $\{y_k\}$, each of length $N = 100$, and the data characterizations of interest are either the product-moment correlation coefficient or the Spearman rank correlation coefficient introduced in Chapter 10. These data sequences are constructed from statistically independent standard Gaussian data sequences (i.e., both are zero mean, unit variance), each contaminated with a single common-mode outlier with the value $+8$ at $k = 50$ (i.e., $x_{50} = y_{50} = 8$).

The upper left plot in Fig. 14.2 shows a scatterplot of the uncontaminated nominal data sequences on which this example is based, while the upper right plot shows the corresponding contaminated data sequences. The common-mode outlier at $(+8, +8)$ is clearly evident in the upper right corner of this plot, and this outlier causes a spurious correlation estimate of $\rho(x, y) = 0.352$, vs. the

Figure 14.2: Graphical summary of data sequences and deletion diagnostics: the upper left plot shows a scatterplot of the uncontaminated nominal data sequences, while the upper right plot is the corresponding scatterplot for the contaminated data sequences; the lower left plot is a leave-one-out deletion diagnostic plot for the product-moment correlation coefficient, and the lower right plot is the same for the Spearman rank correlation coefficient

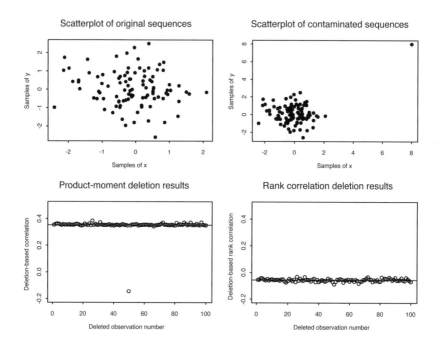

value $\rho(x, y) = -0.142$ for the uncontaminated sequences. The lower-left plot shows the leave-one-out deletion diagnostic plot constructed for this correlation example: the ith point on the plot corresponds to the correlation estimate $\rho(x, y)$ computed from the contaminated data sequence when the ith data point is omitted. For most i, this omission causes only a minor change in the computed correlation coefficient, reflecting the fact that we have made what may be regarded as a 1% change in the dataset. For $i = 50$, however, we remove the common-mode outlier, changing the estimated correlation coefficient from essentially the spurious value computed for the original dataset to a value near that for the uncontaminated data sequence. The key point is that this removal causes a very large change in the result, suggesting—correctly—that the deleted data point is somehow anomalous. In contrast, the lower right plot shows the corresponding deletion diagnostic plot for the Spearman rank correlaiton coefficient between the two contaminated sequences. As discussed in Chapter 7, the Spearman rank correlation coefficient is much less sensitive to outliers than the

Figure 14.3: Leave-one-out deletion diagnostic plot for correlation estimates between two statistically independent sequences contaminated by m common-mode outliers for $m = 0$ (uncontaminated sequence, upper-left), $m = 1$ (outlier at $k = 50$, upper-right), $m = 2$ (outliers at $k = 25, 50$, lower left), and $m = 3$ (outliers at $k = 25, 50, 75$, lower right)

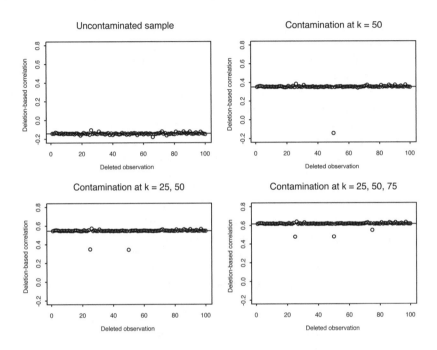

product-moment correlation coefficient is, and this is reflected in the difference between the two deletion diagnostic plots shown here. In particular, the Spearman rank correlation coefficient is essentially insensitive to the deletion of any one point, including the common mode outlier at $k = 50$.

Fig. 14.3 presents leave-one-out deletion diagnostic plots for four related scenarios, all plotted on the same vertical scale to facilitate comparison. In the upper left plot, the results are shown for product-moment correlations computed from the uncontaminated nominal data sequences. Note that this result looks essentially like the Spearman rank correlation diagnostic plot shown in the lower right in Fig. 14.2, illustrating the expected behavior of this deletion diagnostic in the absence of data anomalies. That is, deleting any single point from the nominal data sequence represents a small change in the dataset, and this change causes only a small change in the corresponding product-moment correlation coefficient. The upper right plot in Fig. 14.3 is same as the lower left plot in Fig. 14.2, only plotted on a slightly different scale; as noted in the preceeding discussion, the estimated correlation coefficient is strongly biased by the presence of the common-mode outliers at $k = 50$, decreasing to the level of the upper left plot when this outlier is removed. The lower left plot in Fig. 14.3 shows the corresponding leave-one-out diagnostic for the case where there are two outliers in the dataset: the common-mode pair $x_k = y_k = +8$ at $k = 50$, along with a second common-mode pair $x_k = y_k = -8$ at $k = 25$. The presence of these two outliers causes the bias in the estimated correlation to increase further, and deleting any single point causes this result to decrease to the level of the single-outlier case shown in the upper left plot when the deleted point is one of the outliers (i.e., at $k = 25$ or $k = 50$). Finally, the lower right plot in Fig. 14.3 shows the leave-one-out deletion diagnostic results for a sequence contaminated with three outliers: in addition to those at $k = 25$ and $k = 50$ just described is a third common-mode outlier, $x_k = y_k = 6$ at $k = 75$. As before, this additional outlier causes a further increase in the upward bias of the estimated correlation coefficient, and its removal reduces the result to the two-outlier result shown in the lower left plot.

The key points of this second example are first, that leave-one-out deletion diagnostics can correctly identify the location of data anomalies even when more than one is present in the dataset. Conversely, this example also illustrates that the sensitivity of leave-one-out deletion diagnostics generally declines as the number of anomalies increases. Still, these observations suggest—correctly— that a stepwise leave-one-out deletion strategy can sometimes be effective in identifying multiple anomalies in a dataset. That is, we proceed by first, ten- tatively removing the single data point from a dataset of size N that appears most anomalous and then applying a leave-one-out diagnostic to the remaining dataset of size $N-1$. If one or more points appear anomalous in this dataset, we repeat the process, removing the most anomalous point and applying a leave- one-out diagnostic to the remaining dataset of size $N - 2$. Once we obtain a dataset that appears to have no significant anomalies, the process stops, leav- ing us with a candidate nominal dataset and a collection of possible anomalies. These subsets can be analyzed further to determine the best course of action: omit some or all of the anomalies from the dataset and analyze what remains, include them in the analysis, recognizing them as influential points, or subject them to a separate analysis from the nominal dataset altogether.

14.3.3   Application to the brain/body dataset

As a real data application, again consider the brain weight/body-weight dataset discussed in Sec. 14.2.2. Fig. 14.4 shows the results obtained when the leave- one-out deletion strategy is applied to the product-moment correlations com- puted between the original brain-weight and body-weight sequences. The three dinosaurs included in the dataset are marked as solid circles in this plot and labeled with the type of dinosaur. The brachiosaurus is clearly detected as an anomalous data point by this method, but the other two dinosaurs are not. In addition, the first data point in this sequence is identified as unusual, with its removal causing a substantial decrease in the estimated correlation; this point corresponds to the African elephant.

Figure 14.4: Leave-one-out deletion diagnostic plot for correlation estimates between the original brain weight and body weight sequences. The three dinosaurs included in the dataset correspond to the points marked as solid circles

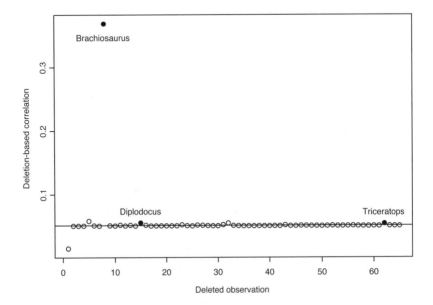

Fig. 14.5 shows the corresponding leave-one-out deletion diagnostic results obtained when the correlations are computed from log-transformed data (lower points, represented as circles), and when they are computed from rank-ordered data (upper points, represented as triangles). Again, data observations corresponding to dinosaurs are marked as solid points and labeled. Here, all three dinosaurs are clearly identified as outliers in the log-transformed data, and only these species are so identified. In contrast, while both the diplodocus and triceratops are identified as outliers in the rank-ordered data, the brachiosaurus is not, but one nondinosaur species is (specifically, the water opossum, one of the smaller animals in the dataset).

The key point of this example is that the utility of a leave-one-out deletion diagnostic depends on both the details of the dataset and the characterization to which it is applied. With the product-moment correlation coefficient applied to the original data sequence, the behavior is quite similar to that seen in the simulation example discussed in Sec. 14.3.2, with a marked change in the estimated correlation when the most extreme outlier is removed, but less of an effect than when all three outliers are removed (the dinosaur-free correlation between the original data sequences is $\rho(x, y) \simeq 0.934$). Similarly, the relatively

Figure 14.5: Leave-one-out deletion diagnostic plot for correlation estimates between the log-transformed (lower points) and the rank-ordered brain/body-weight data sequences. Solid points represent the three dinosaurs

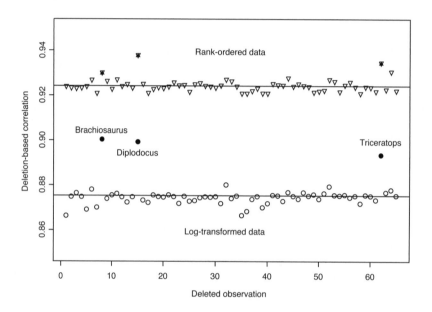

low sensitivity of the rank-based diagnostic to the presence of the dinosaurs in this dataset is not surprising, given the outlier resistance of rank correlations. The most informative result is that based on the log-transformed data, which in this case puts the two variables on a more or less equal footing, as may be seen by comparing the scatter plots for the original and log-transformed data presented in Chapter 1.

14.4   Bootstrap resampling methods

Probably the best known of the subset-based analysis strategies discussed in this chapter is the *bootstrap resampling method*. The basic idea is to use the dataset to estimate the distribution of values for a given data characterization R; from this estimated distribution, we can then construct variability estimates, confidence intervals, or hypothesis tests. The following discussion gives only a brief introduction to this topic, about which much has been written. Very readable introductions with a lot more details are given in the books by Efron and Tibshirani [97] and Davison and Hinkley [85].

14.4.1 The basic bootstrap formulation

In bootstrap resampling methods, the three basic steps outlined in Sec. 14.1.1 specialize to the following:

1. from the original dataset $\mathcal{D}$ of N records, create B modified datasets $\{\mathcal{D}_i\}$ by drawing N records randomly from $\mathcal{D}$ *with replacement*;

2. compute the B results $\{\tilde{R}_i\}$ from the datasets $\{\mathcal{D}_i\}$ using exactly the same procedure as that used to compute the result R from dataset $\mathcal{D}$;

3. interpret the modified results $\{\tilde{R}_i\}$ appropriately to the characterization of interest (e.g., estimate standard deviations for variability assessment, estimate confidence intervals, etc.).

Several points should be noted. First, the sampling with replacement used in Step 1 means that all of the datasets $\{\mathcal{D}_i\}$ necessarily have repeated records. This is one reason that bootstrap analysis is not always appropriate, as in the case of the sample maximum: the bootstrap sample maximum is equal to the largest element in the original dataset approximately 62% of the time [97, p. 81], while the probability of observing this specific value from any continuous probability distribution is zero (see Exercise 1). Second, in contrast to the finite number of samples of size $m < N$ that can be drawn from a dataset of size N without replacement, there is no limit to the number B of samples that can be drawn with replacement. As a practical matter, B is limited primarily by the complexity of evaluating $\tilde{R}$ since this computation is required B times in Step 2. Finally, the question of how many replicates is enough to achieve reasonable results with the bootstrap depends on the characterization desired in Step 3. For example, Efron and Tibshirani suggest, based on their experience, that $B = 50$ is often large enough to give good standard error estimates, and that $B > 200$ is seldom necessary for this case [97, p. 52]. In contrast, values of B for computing confidence intervals typically range from 1,000 to 10,000 [85, 97].

14.4.2 Application to the brain/body dataset

To illustrate the basic idea of the bootstrap, Fig. 14.6 shows the ranges of $B = 1,000$ bootstrap resampling results for three different correlation estimates computed from the brain-weight/body-weight dataset discussed in Secs. 14.2.2 and 14.3.3. Specifically, 1,000 bootstrap resample datasets $\{\mathcal{D}_i\}$ were generated by sampling records from the original dataset $\mathcal{D}$ with replacement. From each of these modified datasets, one of the following three correlation estimates was then computed:

1. the product-moment correlation between the bootstrap body weight values $\{x_k^*\}$ and the bootstrap brain weight values $\{y_k^*\}$;

2. the product-moment correlation between $\{\log_{10} x_k^*\}$ and $\{\log_{10} y_k^*\}$;

3. the Spearman rank correlation between $\{x_k^*\}$ and $\{y_k^*\}$.

Figure 14.6: Ranges of $B = 1,000$ bootstrap correlation estimates for the brain-weight/body-weight dataset, in three representations: the original data, the log-transformed data, and the rank-ordered data

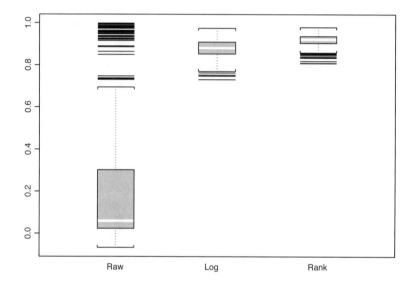

The ranges of these three results are summarized in the boxplots shown in Fig. 14.6, labeled as "Raw," "Log" and "Rank," respectively, and several points are worth noting about these correlation estimates. First, the "Raw" results exhibit the widest range of variation seen for the three methods, significant distributional asymmetry, and a group of very prominent outliers. In fact, these outliers arise from bootstrap samples that do not contain records for any of the three dinosaurs in the original dataset (the probability of such a bootstrap sample is approximately 5%, so we should expect about 50 such samples in 1,000 bootstrap replicates): for these samples, the estimated correlation is ∼0.93, consistent with the cluster of outliers seen here. In contrast, both the log transformed and the rank-ordered bootstrap data sequences exhibit much narrower and more symmetric ranges of variation, reflecting the much reduced influence of the outliers on these correlation estimates. The standard deviations estimated from these bootstrap samples are 0.253 for the original data sequences, 0.041 for the log-transformed sequences, and 0.021 for the rank-ordered sequences, quantifying the range differences seen in Fig. 14.6.

14.5 Subsampling methods

Like the bootstrap methods just described, subsampling methods can be used to assess the variability of a result or to estimate confidence intervals. The difference lies in how the modified datasets are formed. While the bootstrap method constructs modified datasets $\{\mathcal{D}_i\}$ by sampling from the original dataset $\mathcal{D}$ with replacement, subsampling methods construct these datasets by sampling *without* replacement. The results are often similar, although there are important theoretical differences that sometimes manifest themselves in important practical terms.

14.5.1 Subsampling versus the bootstrap

In subsampling methods, the three basic steps outlined in Sec. 14.1.1 specialize to the following:

1. from the original dataset $\mathcal{D}$ of N records, create B modified datasets $\{\mathcal{D}_i\}$ by drawing $m < N$ records randomly from $\mathcal{D}$ *without replacement*;

2. compute the B results $\{\tilde{R}_i\}$ from the datasets $\{\mathcal{D}_i\}$ using the same procedure as that used to compute the result R from dataset $\mathcal{D}$;

3. interpret the modified results $\{\tilde{R}_i\}$ appropriately to the characterization of interest (e.g., estimate standard deviations for variability assessment, estimate confidence intervals, etc.).

The primary practical difference between subsampling and bootstrap methods is that subsampling methods yield a collection $\{\mathcal{D}_i\}$ of modified datasets that are each smaller than the original. This difference arises from the fact that drawing N samples without replacement from a dataset of size N necessarily reconstructs the original dataset, and it raises the practical question of how to choose the subset size m in Step 1 of the above procedure. The effects of varying subsample size are illustrated in the example discussed in Sec. 14.5.2, which compares the subsampling results with bootstrap results and also demonstrates the utility of the following sample-size correction for variance and standard deviation estimates [235, p. 42]. Specifically, if R is any scalar characterization of a dataset $\mathcal{D}$ and $\{\tilde{R}_i\}$ is a collection of B subsampling results, each obtained from subsamples of size m, the following expression gives a useful estimate of the standard deviation of R:

$$S = \sqrt{\left(\frac{m}{N-m}\right)\left[\frac{1}{B-1}\sum_{i=1}^{B}(\tilde{R}_i - \tilde{R})^2\right]}, \qquad (14.11)$$

where $\tilde{R}$ is the mean of the B subsample values $\{\tilde{R}_i\}$. Note that this subsample size correction factor has the value of 1 when $m = N/2$ (i.e., 50% subsamples).

In view of the similarity of subsampling and bootstrap methods, it is reasonable to ask whether one is better than the other and, if so, when. Politis *et al.* [235, sec. 2.3] consider this question, providing five examples where the bootstrap approach is known to fail while subsampling procedures work, including the sample maximum counterexample discussed in Sec. 14.4. Conversely, the bootstrap can give results that are substantially more accurate when the method is applicable, as the authors acknowledge [235, p. 51]:

> We have claimed that subsampling is superior to the bootstrap in a first-order asymptotic sense, since it is more generally valid. However, in many typical situations, the bootstrap is far superior and has some compelling second-order asymptotic properties. These are well studied in Hall [130]. In nice situations, such as when the statistic or root is a smooth function of sample means, a bootstrap approach is often very satisfactory. In other situations, especially those where it is not known that the bootstrap works even in a first-order asymptotic sense, subsampling is preferable.

As a practical matter, the smoothness mentioned in this quote is important as the cases where the bootstrap fails (e.g., the sample maximum example discussed briefly in Sec. 14.4) typically involve nonsmooth behavior.

14.5.2 Application 1: the simulation dataset

As a simple application where the exact answer is known, consider the simulation-based example discussed previously. Fig. 14.7 gives boxplot summaries of the subsample results obtained for the correlations $\hat{\rho}(x, y)$ estimated between two statistically independent standard Gaussian data sequences (i.e., zero-mean, unit variance and each i.i.d.), each of length $N = 100$. These subsample results are computed from $B = 1,000$ random subsamples each of size fN, drawn without replacement from the original dataset, for $f = 0.1$ through 0.9 in steps of size 0.1. The dependence of these results on the subsample fraction f is clearly evident in these boxplots, with the apparent variability decreasing from small values of f where the subsamples are smallest to large values of f where the subsamples are largest. These results are consistent with our expectations based on sample size effects and illustrate the need for the subsample size correction factor introduced in Eq. (14.11). The rightmost boxplot represents the corresponding $B = 1,000$ bootstrap replicates, included for comparison. As noted in the previous discussion, the correction factor $\sqrt{f/(1-f)}$ has the value 1 when $f = 0.5$, and this result is consistent with the fact that the boxplot for $f = 0.5$ is most similar to the bootstrap boxplot.

The estimated means ($\bar{R}_0$) and standard deviations ($\hat{\sigma}_R^0$) for $\rho(x, y)$ computed from these subsampling results are summarized in Table 14.2, along with two other results for comparison. The first is the bootstrap estimate summarized in Fig. 14.7 and the second, labeled "classical," corresponds to the exact mean (i.e., $\rho(x, y) = 0$ for two statistically independent sequences) and a widely used approximate variance result given by Johnson *et al.* [163, p. 556] for Gaussian

Figure 14.7: Comparison of subsample-based correlation estimates for the un-contaminated simulation example, based on a subsampling fraction f between 10% and 90%, with the bootstrap-based correlation estimates for comparison

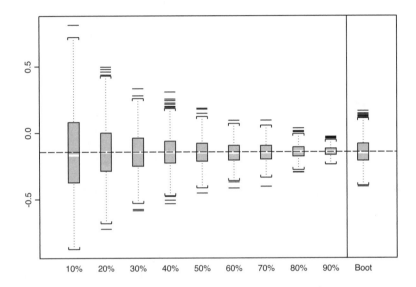

data sequences with correlation ρ (in this case, zero):

$$\text{var}\{\hat{\rho}(x,y)\} \simeq \frac{(1-\rho^2)^2}{N}. \tag{14.12}$$

This approximation gives the classical result that the standard deviation is approximately $1/\sqrt{N}$ for the estimated correlation between two statistically independent Gaussian sequences. These authors also note that the estimated standard deviation has a negative bias, and this observation is consistent with the fact that all of the subsample results and the bootstrap means $\bar{R}_0$ are slightly negative. In fact, all of these values lie between 1.4 and 1.5 standard deviations below the true mean, implying that these negative estimates are not negative enough to be significantly different from zero. The estimated standard deviations $\hat{\sigma}_R^0$ are all within 6% of the bootstrap result, which agrees to three decimal places with the classical estimate $1/\sqrt{N}$ for this example. This result demonstrates the practical utility of the subsample size correction factor in the standard deviation estimate defined in Eq. (14.11).

Table 14.2 also gives the same subsample and bootstrap results computed when the data sequences are contaminated with the pair of common mode outliers considered previously: $x_{50} = y_{50} = 8$. Specifically, the contaminated mean

Table 14.2: Summary of subsample and bootstrap means and standard deviations for the uncontaminated and singly contaminated simulation datasets

f	$\bar{R}_0$	$\bar{R}_c$	$\hat{\sigma}_R^0$	$\hat{\sigma}_R^c$
0.1	-0.142	-0.058	0.106	0.137
0.2	-0.140	0.029	0.105	0.190
0.3	-0.142	0.078	0.103	0.233
0.4	-0.144	0.114	0.103	0.263
0.5	-0.146	0.136	0.099	0.295
0.6	-0.147	0.157	0.098	0.316
0.7	-0.147	0.175	0.106	0.341
0.8	-0.141	0.206	0.103	0.339
0.9	-0.142	0.217	0.102	0.353
Bootstrap	-0.145	0.146	0.100	0.257
Classical	0.000	0.000	0.100	0.100

values are listed in the column labeled $\bar{R}_c$ and the standard deviations in the column labeled $\hat{\sigma}_R^c$. Comparing these values with the corresponding uncontaminated values, $\bar{R}_0$ and $\hat{\sigma}_R^0$, it is clear that both the subsampling and the bootstrap results are quite sensitive to the presence of this pair of common-mode outliers. Further, note the much greater dependence on subsample size m seen in both the mean and the standard deviation of the contaminated subsampling results, relative to the uncontaminated results. In particular, note that although these values differ significantly from the uncontaminated values even for 10% subsamples, the difference between the contaminated and uncontaminated results increases essentially monotonically with increasing subsample size for this example. This behavior is a consequence of the fact that a greater fraction of the subsamples contain the outlier pair as the subsample size increases (see Exercise 2). A graphical illustration of the effects of contamination on the subsampling and bootstrap results may be seen by comparing Figs. 14.7 and 14.8, which give the same boxplot summaries of $B = 1,000$ subsample or bootstrap results for the uncontaminated and contaminated datasets, respectively.

14.5.3 Application 2: the brain/body dataset

As an application of subsampling-based variability assessment to a real dataset, consider the following five correlation estimates for the brain-weight/body-weight dataset: the product-moment correlations computed from the original data sequences, from the log-transformed sequences, and from the rank-ordered sequences, together with the product-moment correlations computed from both

Figure 14.8: Subsample-based correlation estimates for the singly contaminated simulation example, based on a fraction f of the data for f between 10% and 90%, with the bootstrap-based correlation estimates for comparison

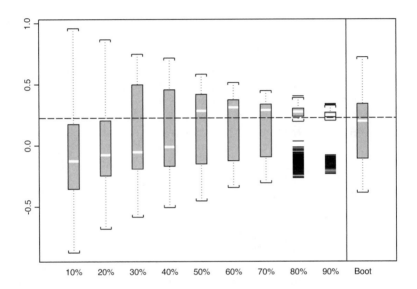

the untransformed and log-transformed data sequences with the dinosaurs removed. Both the subsampling standard deviations for subsamples of size 10% through 90% and the bootstrap standard deviation are given for each of these correlation estimates in Table 14.3. As in the contaminated simulation data sequence example discussed in the previous section, note that the subsample results for the raw correlation estimates vary quite strongly with subsample size, increasing almost monotonically and nearly tripling between the 10% estimate and the 90% estimate. Consistent with both the bootstrap results and the deletion diagnostic results presented previously for this example, this behavior of the subsampling estimates reflects the presence of the outliers (i.e., the dinosaurs) in this dataset. Comparing these results with those computed from the dataset with these outliers removed, it is clear that cleaning the data sequences makes these raw correlation estimates less variable, but these cleaned results are still significantly more variable than either the log-transformed or the rank-ordered results, and this variability is still quite a strong function of the subsampling fraction. The most consistent results are those obtained when the log transformation is applied to the cleaned data sequences, giving the results shown in the

Table 14.3: Comparison of subsample and bootstrap variability estimates for five brain-weight/body-weight correlations: raw data, log-transformed data, rank-ordered data, raw data without dinosaurs, and log-transformed data without dinosaurs

f	$\hat{\sigma}_R^{\text{Raw}}$	$\hat{\sigma}_R^{\text{Log10}}$	$\hat{\sigma}_R^{\text{Rank}}$	$\hat{\sigma}_R^{\text{Clean}}$	$\hat{\sigma}_R^{\text{LgCln}}$
0.1	0.135	0.039	0.029	0.058	0.016
0.2	0.218	0.042	0.023	0.095	0.013
0.3	0.266	0.043	0.023	0.115	0.012
0.4	0.303	0.044	0.023	0.138	0.012
0.5	0.320	0.047	0.024	0.144	0.012
0.6	0.313	0.045	0.022	0.145	0.011
0.7	0.296	0.044	0.022	0.138	0.011
0.8	0.330	0.046	0.023	0.110	0.011
0.9	0.361	0.049	0.024	0.068	0.010
Bootstrap	0.253	0.041	0.021	0.101	0.010

last column. Note that these standard deviation estimates are both the smallest seen in the table and the least dependent on the subsampling fraction.

Overall, these results suggest that the best correlation estimates are those obtained from the log-transformed, cleaned data sequences. This conclusion is further supported by the corresponding mean results: while they are not tabulated here, these mean values decline monotonically from 0.698 for $f = 0.1$ to 0.090 for $f = 0.9$ for the product-moment correlations computed from the raw, untransformed data; in contrast, all of the subsample means are within 1% of the correlation estimate $\hat{\rho} = 0.960$ computed from the cleaned, log-transformed data sequences, as is the bootstrap mean. Fig. 14.9 shows a log-log plot of this dataset with the points corresponding to the three dinosaurs marked as solid circles. This plot is overlaid with a least squares line fit to the log-transformed data with these points removed, and the excellent agreement between this line and the general variation seen in the bulk of the data is consistent with the large correlation estimate computed from these points.

14.6   Applicability of these methods

As noted at the beginning of this chapter, the methods presented here are very broadly applicable, replacing the relatively strong distributional assumptions on which many classical results are based (e.g., expressions for estimator variances or confidence intervals) with simple modifications of a dataset followed by re-

Figure 14.9: Log-log plot of the brain-weight/body-weight dataset, with the three dinosaurs shown as solid circles. The dashed line represents the least squares fit of the log-transformed data without these anomalous points

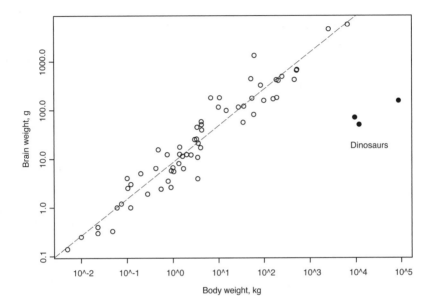

peated computations. Probably the best known method described here is the bootstrap, which provides a basis for estimating both the standard deviation of a result R from a set of B bootstrap replicates $\{\tilde{R}_i\}$ and confidence intervals for it. Since confidence intervals attempt to characterize the behavior of the tail of the distribution of R, it is necessary to make B larger for confidence interval estimation than for variance estimation, as noted in the discussion presented in Sec. 14.4. In fact, bootstrap methods are much more widely applicable than the discussion here illustrates: they can also be used for hypothesis testing, assessing the significance of estimated regression parameters, or combined essentially with deletion diagnostic methods to form the basis for *bootstrap cross-validation methods*, an idea discussed further in Chapter 15. For much more complete discussions of bootstrap methods, refer to the books by Efron and Tibshirani [97] or Davison and Hinkley [85].

While they are not nearly as well known, subsampling methods represent an alternative to bootstrap methods, with the advantage of being more broadly applicable but the disadvantage of usually giving somewhat poorer performance when both methods are applicable. In particular, bootstrap methods require the characterization R to depend sufficiently smoothly on the data, a condition that is often but not always met in practice: it is satisfied by the sample mean, for example, but not for the sample maximum. As noted in Sec. 14.4, the bootstrap fails for this second case, but subsampling methods do not. This point is illustrated in Fig. 14.10, which presents boxplot summaries of four different results. In all cases, the quantity of interest is the maximum value of a sample of length $N = 1,000$ drawn from a uniform distribution on the unit interval $[0, 1]$. The first (leftmost) boxplot was obtained by drawing $B = 1,000$ independent random samples, each of length $N = 1,000$, and computing the corresponding sample maxima. This gives a direct measure of the range of the quantity of interest, based on repeated simulations, something that is possible in simulation examples but not with real datasets. The second boxplot was obtained by applying the bootstrap method to the sample maximum computed from a single random sample. That is, $1,000$ bootstrap replicates were obtained by drawing samples of size N with replacement from this original data sample and computing its maximum. As noted in the discussion in Sec. 14.4, the maximum of the original data sample is included in approximately 62% of these bootstrap replicates, giving a very narrow distribution around this specific value. This behavior is reflected in the extremely narrow range of variation seen for the bootstrap replicates in Fig. 14.10.

The third boxplot in Fig. 14.10 gives the corresponding subsampling results, obtained from $B = 1,000$ half-samples (i.e., $f = 0.5$), and the fourth (rightmost) boxplot summarizes $1,000$ repeated draws from the known distribution of sample maxima for the uniform distribution, presented in Chapter 6. In particular, recall that the order statistics for uniformly distributed random data sequences have beta distributions whose parameters depend on the particular order statistic considered. For the maximum of a sample of size N, the parameters of this beta distribution are $p = N$ and $q = 1$, and the rightmost boxplot in Fig. 14.10 summarizes $1,000$ random samples drawn from this distribution. Comparing these four boxplots, it is clear that the samples drawn from the beta distribution agree well with those obtained from the direct simulations, while the bootstrap replicates do not agree at all well with either of these two results. The subsampling results are intermediate: admittedly, their agreement with the results from the beta distribution and the direct simulations is not spectacular, but it is *much* better than that of the boostrap. This conclusion can also be seen in the standard deviations estimated from these results, which are given in Table 14.4. For comparison, the exact standard deviation may be computed from the beta distribution, and is given by [16, p. 14]:

$$\sigma_{\max} = \frac{1}{N+1}\sqrt{\frac{N}{N+2}} \simeq 0.000998. \tag{14.13}$$

The results presented in Table 14.4 give the standard deviations estimated from each set of results, along with its error relative to the value given in Eq. (14.13).

As a final point in this discussion, note that the half-sample estimator (i.e., subsampling with a sample fraction $f = 0.5$) was chosen for this example because this is the case that *should* give results most similar to the bootstrap when

Figure 14.10: Boxplot summary of $B = 1,000$ estimates of the sample maximum for a uniformly distributed sequence of length $N = 1,000$. The first (leftmost) summarizes the sample maxima of independently drawn random samples; the second summarizes the bootstrap estimates obtained from a single random sample; the third summarizes the subsampling estimates for $f = 0.5$ (half sample estimates) from this same sample; the fourth (rightmost) summarizes repeated draws from the exact distribution of the uniform sample maximum

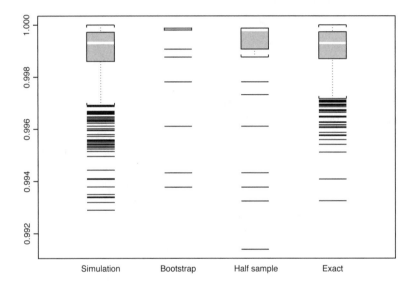

bootstrap methods are applicable. This point was evident both in the boxplot comparisons presented in Sec. 14.5 and in the fact that the correction factor for the standard deviation is equal to 1 for this case. In fact, this expected similarity suggests a possible strategy for evaluating the results of methods that are either new and therefore not yet well characterized, or incompletely known (e.g., results based on a commercial software package whose implementation details may be regarded as proprietary by the supplier). Specifically, the bootstrap and subsample results can be compared: if they are similar, either one can probably be used for variability assessment; since the bootstrap is better known, this is probably the better choice because this assessment will require less explanation than if the subsampling results are used. Conversely, if the results differ strongly, this may mean that the analysis procedure being characterized fails to meet the smoothness conditions required for bootstrap applicability, as in the sample maximum example just presented.

Table 14.4: Standard deviations estimated from the samples summarized in Fig. 14.10

Estimate	$\hat{\sigma}$	Error
Direct simulation	0.001056	5.8%
Bootstrap replicates	0.000471	−52.8%
Half-sample replicates	0.000860	−13.8%
Draws from the exact distribution	0.000895	−10.3%

In contrast to bootstrap and subsample methods, which attempt to describe a data characterization R under the assumption that all of the observations in the original dataset $\mathcal{D}$ are exchangeable, deletion diagnostics are designed to explicitly assess this exchangeability hypothesis by looking for violations. This idea is closely related to the *jacknife*, a technique briefly discussed in Chapter 15 because of its relation to the cross-validation methods introduced there. Also, it is important to note two points about the deletion diagnostic procedures discussed in this chapter. First, they are general purpose, making no use of any problem-specific characteristics: subsamples are systematically removed from the original dataset, the characterization R is recomputed from this modified dataset, and the influence of this subset deletion is assessed. In some problems like least squares regression modeling, specialized methods can be developed that exploit some aspect of the problem structure to significantly improve computational efficiency. This important point is discussed further and illustrated in Chapter 15 where ordinary least-squares-based deletion diagnostics are presented. The second important point about deletion diagnostics is that systematic deletion of all subsets of size k is usually only feasible for $k = 1$, the case of leave-one-out deletion diagnostics considered in this chapter. While more general leave-k-out deletion diagnostics are possible in principle, they are usually impractical since the number of subsets to be deleted increases extremely rapidly with k, a point noted in Sec. 14.3.1.

The three computational assessment strategies just described—the bootstrap, subsampling, and deletion diagnostics—all make what should be "small changes" in our original dataset and examine their influence on computed results, with the expectation that these changes should also be "small." Even in the case of deletion diagnostics, these "small changes in the results" represent our nominal expectations, and we are looking for anomalous data points that may violate these expectations. In contrast, the computational negative control strategy introduced in Sec. 14.2 makes what should be a "large" change in our original dataset and examines the results to see whether they also exhibit a "large" change. Failure to do so is then taken as evidence of insignificance of the original result, which is attempting to characterize some structural feature of the dataset. Like the other three strategies discussed in this chapter, the computational negative control approach is broadly applicable, but it requires a way of making changes to the original dataset that destroy the structure of interest if it is present, and does so in a way that it can be repeated to obtain a large collection of similar but not identical (i.e., exchangeable) modified datasets. In the examples presented here, this modification is accomplished using random permutations that destroy any association that may exist between the different data fields in a data record. The specific examples considered in this chapter applied this idea to correlation estimates, but it can also be applied to regression modeling problems, a point discussed further in Chapter 15, and it has been applied to cluster analysis [231] and to comparing ranked lists [230].

14.7 Exercises

1. Suppose $\mathcal{D}$ is a sequence $\{x_k\}$ of N random variables drawn from any continuous distribution and let R denote the sample maximum: $R = x_{(N)}$. Consider a bootstrap sample $\{x_k^*\}$ and let $\tilde{R} = x_{(N)}^*$. Show that:

$$P\{\tilde{R} = R\} = 1 - (1 - 1/N)^N \to 1 - e^{-1} \simeq 0.632 \text{ as } N \to \infty.$$

(Hint: note that $P\{\tilde{R} \neq R\} = P\{x_{(N)}$ is not drawn from $\mathcal{D}$ in N draws$\}$.) This result illustrates a difficulty that can arise with the bootstrap method since for any continuous distribution the probability that $\tilde{R}$ has *any* specified value is zero.

2. Recall from Chapter 3 that the probability of drawing m black balls in a sample of size p from an urn containing Q white balls and $N - Q$ black balls is given by the hypergeometric distribution. For a dataset of size N containing a single outlier, compute the probability that a subsample of size m contains this outlier. For a data sequence of length $N = 100$, how many of $B = 1000$ independent 10% subsamples can be expected to contain the outlier? Of 50% subsamples? Of 90% subsamples?

Chapter 15

Regression Models III: Diagnostics and Refinements

Sections and topics:

Several previous chapters have dealt with the mechanics of different formulations of the regression problem, including ordinary least squares (OLS), robust alternatives like least trimmed squares (LTS), and generalized linear models for dealing with mixed data types. This chapter takes a broader view of the regression problem, addressing questions like which variables should be included in a regression model, how such a model should be assessed once it has been fit to a dataset, and how these models can be refined if we feel their fit or their form is inadequate, or even if we just want to see how certain alternatives compare. Many of these ideas are presented in greatest detail for the case of OLS linear regression models, both because specialized results are often available for this case and because these models arise most commonly in practice. Conversely, most of the ideas presented here can also be applied in more general settings, though sometimes at the cost of much greater computational effort.

Key points of this chapter:

1. The process of building data models is generally an iterative one, involving tentative decisions about what variables to include in the model, how these variables should be represented, and whether the analysis should be based on all of the available data or only a subset of it.

2. There is no single, universal defining criterion for a "good" data model: the quality of a data model depends on many factors, including obvious ones like *goodness-of-fit* (i.e., how well the data model approximates the data) and model complexity, but also less obvious ones like the consequences of including anomalous data observations (e.g., outliers) or highly correlated explanatory variables in the model (i.e., the problem of *collinearity*). Other factors include the reasonableness of the explanatory variables on which the model is based in the context of the application, and the relationship of the data model developed to other data models that may have significant historical precedent and support in the field.

3. While it is not a completely adequate measure of model quality in and of itself, goodness-of-fit is almost always an important component of model quality, and it can be measured in several different ways. This chapter considers classical examples like *adjusted* R^2 and *F-ratios* that are intimately related to traditional Gaussian data assumptions, along with robust alternatives like the modified R^2 statistic proposed for use with LTS regression, and computational alternatives like cross-validation that replace distributional assumptions with random sampling from the available data.

4. Many different general strategies have been developed for building data models, and this chapter cannot begin to treat all of them, or even provide a comprehensive treatment of a few of them. The primary objective of this chapter is to introduce a number of important and useful ideas, such as univariate screening strategies for preliminary variable selection, statistical significance assessments as an aid in deciding whether to include a particular variable in a model or not, and various diagnostic tools that can be helpful in detecting data subsets or combinations of variables that may cause difficulties in model building.

5. It has been emphasized throughout this book that data modeling is more of an art than a science, in the sense that different datasets, analysis methods, and analysts will generally yield different solutions—sometimes very different solutions—to what can at some level be regarded as the "same" problem. An extremely important practical corollary is that the final decision of whether a data model is "good enough to be adequate" rests with the model developer and *not* with the computational procedures: the model quality measures and diagnostic procedures described in this chapter can be extremely helpful in making this decision, but the decision itself is ultimately the responsibility of the data analyst.

15.1   The model-building process

Previous chapters of this book have been concerned with the mechanics of model-building: what is a regression problem? How is it formulated and solved? What kinds of things do we need to be concerned about that might cause the results not to be representative? These issues are critically important and must be discussed first because they provide the basic understanding necessary before we can formulate and solve regression problems in the context of a real application. Given an understanding of these issues, we are then faced with the larger question of how we use the techniques to formulate and solve specific regression problems for specific purposes. This subject is one of enormous scope, and a single chapter cannot begin to do it justice, but this chapter can and does provide a reasonably comprehensive introduction to the key ideas involved in formulating and solving these problems, and in interpreting the results we obtain.

15.1.1   What is a good data model?

The question of what constitutes a "good" data model is a basic one that provides an important focus for the model-building process: given a dataset $\mathcal{D}$ and a tentative model $\mathcal{M}$, how do we know if the model is any good? In fact, the answer depends strongly on our intended application, but one important general aspect of "model goodness" is how well the model describes its associated dataset. That is, the typical data model $\mathcal{M}$ considered in this book generates a prediction $\hat{C} = \mathcal{M}(\mathcal{D})$ of some characteristic C of a dataset $\mathcal{D}$. Usually, this characteristic is the value of one of the variables included in the dataset, predicted from the values of some of the other variables in the dataset, although more general cases are possible (e.g., the logistic regression models introduced in Chapter 14 estimate probabilities that a given response is 1, not the individual values of the response). *Goodness-of-fit measures* are numerical characterizations of the difference between the characteristic C and its model-based estimate $\hat{C}$, the idea being that this difference should be "small" for a "good" model. Simple goodness-of-fit measures are discussed in more detail in Sec. 15.3, but an extremely important point here is that, by themselves, goodness-of-fit measures are *not* adequate characterizations of data model quality. In particular, we can usually achieve perfect goodness-of-fit (i.e., zero distance between C and $\hat{C}$, however we choose to measure distance) by simply taking our original dataset $\mathcal{D}$ as our data model, giving $\hat{C} = C$. Alternatively, by making the data model complex enough to effectively code all of the details of the original dataset, we can also frequently achieve this same perfect fit by simply reparameterizing the original data as a model of the specified form. A specific example discussed in Chapter 1 was that of fitting a set of N points in the plane to a polynomial of degree $N - 1$: the resulting polynomial "data model" exactly reproduces the original data values, but it is as complex as the original dataset, replacing the original N data values with the N parameters required to define a polynomial of degree $N - 1$.

In typical exploratory data analysis applications, a good model should satisfy the following three criteria:

1. the model predictions $\hat{C} = \mathcal{M}(\mathcal{D})$ should fit the observed values C reasonably well (i.e., $\hat{C} \simeq C$);

2. the model should be simple in form to facilitate subsequent use and interpretation;

3. the explanatory variables included in the model should be meaningful in the context of the application problem that motivated the data analysis.

In the polynomial perfect fit example described above, while the data model provides the best possible fit to the data, it fails with respect to both of the other two criteria. Two key factors that influence all three of these goodness criteria are the form of the model $\mathcal{M}$ and the choice of explanatory variables on which it is based. As noted in Chapters 11, 12, and 13, the regression models considered in this book are those with linear dependence on the unknown model parameters, including the linear regression models for real-valued response variables discussed in Chapters 5 and 11, extensions of these models based on the transformations discussed in Chapter 12, applied to either the response variable or the explanatory variables, and the generalized linear models introduced in Chapter 13 for regression models involving mixed data types. In particular, the general nonlinear regression model fitting problem is not considered here, for two reasons: first, the techniques required to fit and interpret these models are different from (and generally more complicated than) those described here and second, the types of regression models described here are the ones most commonly encountered in practice. In fact, even if a nonlinear regression model is ultimately required, it is generally good practice to start with a simpler (and ideally, closely related) linear regression model to serve as a frame of reference. For a more detailed treatment of nonlinear regression models, refer to the book by Seber and Wild [262].

For either the linear regression models discussed in Chapter 11 or the generalized linear models discussed in Chapter 13, a crucial factor in determining their adequacy is the choice of explanatory variables or *covariates* to include in the model. In practice, this important issue is closely intertwined with the iterative nature of model development, and it is discussed in this chapter from a number of different perspectives. Specifically, this topic is considered from the perspectives of initial variable selection, of assessing the statistical significance of covariates that have been tentatively included in a regression model, of collinearity as a basis for deciding which covariates *not* to include together in a model, and of specific covariate-based iterative model refinement schemes like forward selection. Before going into these detailed discussions, the following section gives a brief overview of the general iterative model-building process as it is typically followed in practice.

15.1.2   The model development cycle

The task of building a data model is an iterative one that begins with a certain preliminary set of working assumptions about what variables to include, how they are to be represented, and how they are likely related. Data models are then built using the appropriate techniques from previous chapters, and the results are examined in light of initial expectations to see how well they conform to those expectations. Sometimes, the agreement is reasonably good, and we may take this as confirmatory evidence in support of our expectations. Indeed, this outcome is precisely what we hope for in a confirmatory data analysis setting, although it is worth emphasizing that confirmatory analyses are typically based on datasets obtained using designed experiments that incorporate a good deal of prior knowledge about the problem at hand. The focus of this book is on exploratory analysis where prior knowledge is generally less abundant, often vague, and sometimes blatantly wrong. In this setting, surprises are more likely and it is important to look for them.

The basic iterative strategy for formulating and solving regression problems advocated in this chapter consists of the following steps:

0. preliminary problem formulation: summarize what is known initially about analysis objectives, the available data, expected results of the analysis, possible sources for additional data and/or information, etc.;

1. detailed problem formulation: specify response and explanatory variables, select the computational procedures to be used in fitting the data models, and decide how the results are to be subsequently used and interpreted;

2. model fitting: apply the selected computational procedure to the available data and generate a detailed data model;

3. model evaluation: assess the overall fit of the model to the data, assess the significance of the explanatory variables included in the model, and look for evidence of data anomalies that may be unduly influencing the results;

4. model refinement: based on the results of Step 3, identify alternative models to be considered, possibly involving fewer explanatory variables (if some of those included initially appear not to contribute significantly), additional or alternative explanatory variables (e.g., data transformations or interaction terms) if the overall goodness-of-fit appears poor, or subsets of the data if there is evidence of anomalous data observations;

5. assessment and possible iteration: based on the results of the previous steps, decide whether the data model obtained so far appears to represent a reasonable fit to the data that offers useful insights into the questions that motivated the analysis in the first place; if so, summarize the results and conclude the analysis; otherwise, go back to Steps 0, 1, 2, 3, or 4, as appropriate, and refine the analysis results further.

All of these steps will be discussed further in connection with the three model-building examples presented in Sec. 15.2, and specific aspects of several of these steps (e.g., preliminary selection of explanatory variables, assessment of the contribution of these variables to the overall model, detection of anomalous and/or influential data points) are discussed in detail in other sections of this chapter.

15.2 Three modeling examples

The following three subsections briefly describe three regression modeling problems that are used to illustrate key ideas and techniques described in subsequent sections of this chapter. All three of these examples are based on datasets described previously, and they include the brain-weight/body-weight simple linear regression example introduced in Chapter 1, a specific multivariate regression problem based on the Pima Indians diabetes dataset introduced in Chapter 2, and a multivariate logistic regression model based on the bitter pit dataset introduced in Chapter 13.

15.2.1 The brain/body dataset

The brain-weight/body-weight dataset has been used extensively in previous chapters to illustrate a variety of points. Since many of these points are directly relevant to the subject of this chapter, it is useful to review some of them in a summary of this example that focuses on fitting and evaluating a simple linear regression model. In particular, since this model involves only a single explanatory variable, specified from the outset, the issues of variable selection and collinearity do not arise, but questions of anomaly detection and treatment are extremely important for this example.

Motivation for examining the relationship between brain weights and body weights of different species has its roots in *allometry* [51], the study of how various characteristics of animal species vary with their size. Further, traditional allometric analysis examines these characteristics one at a time (see, e.g., Calder's comments, "Why not use multiple regression?" [51, pp. 42–43]), motivating the consideration here of a model that predicts one characteristic (brain mass) from one size measure (body mass). Since allometric models are typically power-law relations that are linearized by applying logarithmic transformations to both the response and explanatory variables, the model considered here is a linear regression model relating log of brain weight to log of body weight. Recall from the scatterplots presented for this example in Chapter 1 that the log-log plot of these variables is suggestive of a linear relationship, while the corresponding scatterplot of the untransformed masses is not. Also, recall that the enormous difference between the product-moment correlations and the Spearman rank correlations for this example was strongly suggestive of the need for a transformation.

15.2.2   Predicting triceps skinfold thickness

One of the variables included in the Pima Indians diabetes dataset introduced in Chapter 2 is triceps skinfold thickness (TSF), an alternative measure of obesity to the more popular body mass index (BMI). Unfortunately, as described in Chapter 2, almost 30% of the recorded TSF values are given as zero, a physically impossible value used to code missing data. One of the techniques for handling missing data discussed in Chapter 16 is *regression-based imputation*, in which missing data values are predicted from other variables in the dataset. The purpose of the TSF regression example discussed throughout this chapter is to determine how well we can predict TSF values from other variables in the dataset for the cases where TSF is *not* missing. As discussed further in Chapter 16, our ability to do this is not a sufficient condition for obtaining reasonable estimates for the missing TSF values, but it is a necessary condition for attempting regression-based imputation.

Since it was noted in the discussion of the TSF data given in Chapter 13 that serum insulin data (INS) is *always* missing whenever TSF is missing, it follows that TSF is not a useful covariate to include in a TSF regression model for potential use in regression-based imputation of missing TSF values. Similarly, since the diabetes pedigree function (DPF) is a computed quantity with less clear medical interpretation than the other variables included in the Pima Indians dataset, it is also excluded from consideration. Thus, the TSF models considered here are standard linear regression models that predict the real-valued response variable TSF from one or more of the following six potential explanatory variables:

1. the integer-valued number of times pregnant (NPG);

2. the real-valued plasma glucose concentration (GLC);

3. the real-valued diastolic blood pressure (DIA);

4. the real-valued body mass index (BMI);

5. the integer-valued age in years (AGE);

6. the binary diabetes diagnosis (DIAG).

It is important to note that since the variables GLC, DIA, and BMI all exhibit a few missing data values coded as zeros, all of these observations have been removed from the dataset, leaving a total of $N = 532$ records for which TSF, GLC, DIA, and BMI values are all nonzero. Note also that there were no missing AGE values and that zero is a valid data value for both NPG and DIAG, so no records were omitted on the basis of these variables. The primary questions considered in subsequent discussions of this example are those of variable selection, model evaluation, and iterative model refinement.

15.2.3   Bitter pit and mineral content

The bitter pit example was introduced in Chapter 13, based on a dataset from the book by Andrews and Herzberg [12, no. 59] that gives the measured prevalence of the apple defect bitter pit for 42 apple trees, along with the average apple weight from each tree, a categorical variable indicating which of four possible supplemental nitrogen treatments was given, and the average concentrations of six chemical constituents in the apples. Analyses of parts of this dataset were given in Chapter 13, including both an analysis of variance that examined whether the supplemental nitrogen treatment influence the prevalence of bitter pit, and a univariate logistic regression analysis that examined the influence of calcium concentration on bitter pit. The example presented here is a refinement of both of these previous analyses, examining the question of how the prevalence of bitter pit depends on the following seven potential covariates:

1. supplemental nitrogen treatment, the categorical explanatory variable, assuming the four values A, B, C, and D defined in Chapter 13;

2. the real variable total nitrogen concentration (TN), in parts per million;

3. the real variable plant nitrogen concentration (PN), in parts per million;

4. the real variable phosphorus concentration (P), in parts per million;

5. the real variable potassium concentration (K), in parts per million;

6. the real variable calcium concentration (Ca), in parts per million;

7. the real variable magnesium concentration (Mg), in parts per million.

It was seen in the previous analyses that both supplemental nitrogen treatment and calcium concentration significantly influenced the prevalence of bitter pit, but the analysis presented here examines the broader question of how these influences compare both with each other and with the influences of the other factors listed above. Like the TSF example described in Sec. 15.2.2, the primary emphasis of this example is on variable selection, model evaluation, and iterative refinement, but here for a multivariate logistic regression model instead of a standard linear regression model.

15.3   Assessing goodness-of-fit

While it was noted in the discussion of what constitutes a good data model given in Sec. 15.1.1 that goodness-of-fit is not a *complete* characterization of "model goodness," it is an important component of such characterizations. The following subsections briefly describe some of the most commonly used goodness-of-fit measures for assessing the agreement between a data model and the dataset from which it was derived. Specifically, Sec. 15.3.1 describes the classical R^2 characterization and its associated statistics under classical normality assumptions, while Sec. 15.3.2 briefly describes a robust alternative that is useful in connection with the least trimmed squares (LTS) regression procedure introduced in Chapter 11.

15.3.1   The classical R^2 measure and F-statistics

A popular goodness-of-fit measure for standard linear regression models is the R^2 value, which compares the model's accuracy as a predictor of its real-valued response variable with that of the mean response. More specifically, all of the linear regression models considered in this book predict one response variable y from a collection of one or more explanatory variables $\{x_i\}$. Given the predictions $\{\hat{y}_k\}$ generated by the data model for the N individual response variable observations $\{y_k\}$, the R^2 value is defined as [94, p. 90]:

$$R^2 = \frac{\sum_{k=1}^{N} (\hat{y}_k - \bar{y})^2}{\sum_{k=1}^{N} (y_k - \bar{y})^2}. \tag{15.1}$$

Here, $\bar{y}$ is the mean value of the observed response sequence $\{y_k\}$, and it follows from Eq. (15.1) that the R^2 value is nonnegative and is zero if and only if $\hat{y}_k = \bar{y}$ for all k. Further, if the mean of $\hat{y}_k$ is equal to $\bar{y}$ as in the case of ordinary least squares regression estimators (Exercise 1) and the variance of the predictions $\{\hat{y}_k\}$ is no greater than that of the data values $\{y_k\}$, it follows that $R^2 \leq 1$ (Exercise 2). In fact, note that this upper bound is achievable if $\hat{y}_k = y_k$ for all k, leading to the interpretation of R^2 as a goodness-of-fit measure: $R^2 = 0$ if the model describes the data sequence $\{y_k\}$ no better than the mean does, and $R^2 = 1$ if the model exhibits a perfect fit. For more typical intermediate cases, the larger the R^2 value, the better the model fit.

As noted in Sec. 15.1.1, goodness-of-fit measures like R^2 are not, by themselves, adequate quality measures for regression models. In the case of R^2, this shortcomming is reflected in the fact that R^2 generally increases (in particular, it cannot decrease) as additional covariates are added to a linear regression model, *even if these covariates have negligible explánatory power*. This situation is widely recognized and has motivated extensions like the *adjusted R^2 value*, defined as follows. First, note that R^2 may be written as [252, p. 42]:

$$R^2 = 1 - \frac{\text{SSE}}{\text{SST}_m} = 1 - \frac{\sum_{k=1}^{N} (\hat{y}_k - y_k)^2}{\sum_{k=1}^{N} (y_k - \bar{y})^2}, \tag{15.2}$$

where SSE is the *residual error sum of squares* and SST_m is the *total sum of squares corrected for the mean*. The adjusted R^2 value essentially replaces these sums of squared errors with the corresponding means, obtained by dividing each sum by its corresponding number of degrees of freedom. For the residual error sum of squares, this number is $N - p$, where N is the number of data observations and p is the total number of parameters in the data model (including the intercept term if it is present). Similarly, for the total sum of squares corrected for the mean, the number of degrees of freedom is $N - 1$, leading to the following final expression for the adjusted R^2 value:

$$R_a^2 = 1 - \frac{\text{SSE}/(N-p)}{\text{SST}_m/(N-1)} = 1 - (1 - R^2)\left(\frac{N-1}{N-p}\right). \tag{15.3}$$

Note that if the number of parameters p is small compared with the number of data values N, it follows from Eq. (15.3) that this adjustment has negligible effect, i.e., $R_a^2 \simeq R^2$. Conversely, if R^2 is large as a consequence of gross overfitting—meaning that the number of model parameters is on the order of the number of data points—R_a^2 can be much smaller than R^2.

The arguments leading to this definition of the adjusted R^2 value are closely associated with those leading to the definition of the associated F-statistic that is appropriate to linear regression models when the Gaussian error assumption is approximately valid. Specifically, the F-statistic is defined as the ratio of the *regression mean square (MSR)*:

$$\text{MSR} = \frac{1}{p}\sum_{k=1}^{N}(\hat{y}_k - \bar{y})^2, \tag{15.4}$$

to the *error mean square (MSE)*:

$$\text{MSE} = \frac{1}{N-p-1}\sum_{k=1}^{N}(\hat{y}_k - y_k)^2. \tag{15.5}$$

Under the assumption that the covariates are observed without error and the observed response variable y is subject to independent, identically distributed, zero-mean Gaussian observation errors, it follows that the F-statistic has an F distribution with p and $N - p - 1$ degrees of freedom:

$$F = \frac{\text{MSR}}{\text{MSE}} \sim F_{p,N-p-1}. \tag{15.6}$$

Thus, if the Gaussian error assumption is reasonable, the F-statistic provides a basis for attaching statistical significance to a regression model: if the F statistic is large enough to be significant at a specified level α, the regression model can be said to provide a significantly better fit to the data than the mean does. Conversely, it is important to remember that F-ratios like this one tend to be very non-robust in the face of violations of the Gaussian working assumptions, as discussed in Chapter 9. As a consequence, while F-statistics can provide a useful basis for comparing regression models, it is important not to rely on them exclusively in view of this sensitivity to violations of working assumptions.

To illustrate the ideas just described, Table 15.1 uses the R^2 goodness-of-fit measure, its associated F-statistic, and the corresponding p-value to compare six different prediction models for the Pima Indians triceps skinfold thickness (TSF) data. While all of these p-values are less than 5%, suggesting that each of these variables has some predictive power for TSF, the actual p-values vary by at least 14 orders of magnitude, implying considerable variation in this predictive

Table 15.1: Comparison of regression models predicting TSF from one of six other covariates in the Pima Indians diabetes dataset. The table gives the adjusted R^2 value, its associated F-statistic, and its associated p-value

No.	Covariate	R^2	F	p-Value
1	NPG	0.009	4.84	2.83×10^{-2}
2	GLC	0.051	28.68	1.27×10^{-7}
3	DIA	0.051	28.55	1.36×10^{-7}
4	BMI	0.419	382.50	$< 1 \times 10^{-16}$
5	AGE	0.026	14.16	1.86×10^{-4}
6	DIAG	0.065	36.82	2.47×10^{-9}

power between the different variables. BMI gives by far the best single variable regression model, as seen by comparing the adjusted R^2 values: the BMI model explains just over 40% of the TSF variance, while none of the other single-variable models explain as much as 10%.

15.3.2   Robust goodness-of-fit measures

As emphasized in Chapter 7, outliers arise frequently, and they can adversely influence essentially all classical analysis methods, including OLS-based linear regression. This observation has motivated the development of outlier-resistant alternatives like the least trimmed squares (LTS) approach to linear regression described in Chapter 11. This point is important in connection with the general model-building problem considered here for two reasons. First, the existence of outlier-resistant alternatives to OLS regression means that the two different approaches can be applied to the same regression problem and comparison of the results can give valuable insights into the character of the data. In particular, large differences in estimated model parameters can signal the presence of significant data anomalies. If and when this happens, the appropriate course of action is to identify and study the source of these anomalies, as emphasized by Rousseeuw and Leroy in their book on robust regression [252]. The second important point here is that assessing goodness-of-fit for these outlier-resistant models needs special measures that are also outlier resistant.

As a specific example, the *S-Plus* implementation of the LTS regression procedure provides an alternative to R^2 that is given by the following expression:

$$R_{LTS}^2 = 1 - \frac{\sum_{i=1}^{h} r_{i:n}^2}{\sum_{i=1}^{h} (y_i - \tilde{y})^2}. \tag{15.7}$$

Here, h is the number of data observations used by the LTS estimator (deter-

Table 15.2: OLS vs. LTS regression models predicting TSF from six covariates in the Pima Indians diabetes dataset. Results include the estimated model parameters $\hat{\beta}$ for each method, its R^2 measure, computed as described in the text, and the number (N_o) of outliers declared by the LTS method

No.	Covariate	$\hat{\beta}_{OLS}$	$\hat{\beta}_{LTS}$	R^2_{OLS}	R^2_{LTS}	N_o
1	NPG	0.3021	0.5741	0.009	0.037	9
2	GLC	0.0769	0.0929	0.051	0.056	7
3	DIA	0.1933	0.2643	0.051	0.077	9
4	BMI	0.9902	1.0124	0.419	0.481	6
5	AGE	0.1578	0.2297	0.026	0.033	9
6	DIAG	5.6873	5.8667	0.065	0.085	9

mined by the LTS fitting algorithm), $r_{i:n}$ is the ith rank-ordered LTS prediction residual, and $\tilde{y}$ is the location parameter that minimizes the sum of squares in the denominator. Because the LTS estimator was developed for situations where the OLS Gaussian working assumptions may be badly violated, there is no natural analog of the F-statistic and its associated p-value for R^2_{LTS}. Conversely, since outliers can cause R^2 to either increase or decrease [252, p. 46], an outlier-resistant analog like R^2_{LTS} is useful in comparing LTS models.

Table 15.2 compares LTS and OLS regression results for the six single-predictor models considered in the preceeding discussion, giving the estimated model parameter $\hat{\beta}$ relating each predictor to TSF, with the subscript indicating the fitting method. The table also gives the adjusted R^2 values for the OLS models (labeled R^2_{OLS}) and the corresponding robust R^2 values for the LTS models (labeled R^2_{LTS}). The last column gives the number of outliers identified by the LTS fitting procedure, corresponding to observations whose final regression weights were zero. Although only a small number of outliers were detected in this dataset (between 6 and 9, corresponding to ~1–2% contamination level), their influence on some of the coefficient estimates appears to be substantial. As a specific example, note that the NPG coefficient is nearly twice as large in the LTS results as it is in the OLS results. More generally, note that all of the LTS coefficients are larger in magnitude than the corresponding OLS coefficients, as are the corresponding R^2 goodness-of-fit estimates. Still, the *ordering* of these R^2 values is the same in either case, with both results suggesting the best univariate model is that based on BMI. In fact, the estimated BMI coefficient for these methods only differs by about 2%. The influence of these outliers on variable selection for more complex regression models is examined further in Sec. 15.9.4.

15.4 Initial variable selection

In a large dataset, there may be many potential explanatory covariates that could be included in a data model, and in an exploratory analysis, the choice of which ones are most appropriate to include may not be at all obvious. One possible strategy that has its advocates is to "let the dataset speak for itself" and simply include all possible covariates in the initial model. While this strategy is sometimes possible in practice, it is not always practical (see, e.g., the discussion of EPV limits for logistic regression discussed in Chapter 13), and even in cases it could be implemented, it is often not effective. In particular, it is possible that subsequent model pruning steps will fail to eliminate all covariates that are marginal or even completely irrelevent, especially if outliers or other data anomalies are present. Consequently, it is generally useful to begin by selecting a subset of the possible covariates that appear most appropriate for inclusion in the data model. As a practical matter, this selection process involves two quite distinct considerations: field-specific assessments of relevance, and statistical assessments of association with the response variable of interest. The following discussions consider both of these aspects of preliminary variable selection.

15.4.1 Statistical significance versus purposeful selection

Since the objective of a regression model is to generate the best possible predictions of the response variable from the selected covariates, subject to the constraints imposed by the form of the regression model, one reasonable criterion for initial covariate selection is that they exhibit a significant association with the response variable. Having said this, three details are important. First, it is entirely possible that the most important covariates that should be included in a multivariable model are only effective when included *together*, and not when each one is included by itself. Nevertheless, the largest influences are often individual contributions, and this observation motivates the use of the *univariate screening strategy* presented here: we consider an explanatory variable for inclusion in the model based on its association with the response variable. The second important detail is that we can measure association in many different ways, not all of which will lead to the same results. For the regression problems of interest here, a particularly appropriate way of measuring association is by fitting and comparing univariate regression models, one for each candidate variable, using the model fitting procedure under consideration. Finally, the third key point here is that in the case of real- or integer-valued response variables and real- or integer-valued explanatory variables, it is also useful to compute and compare both the standard product-moment correlations and the Spearman rank correlations. In particular, recall that in the brain-weight/body-weight example discussed in Chapter 1, the large difference between these correlation values provided evidence that a transformation of one or both variables might be useful. The following paragraphs discuss the practical trade-offs between statistical and field-specific variable selection criteria, after which Sec. 15.4.2 briefly considers the question of data transformations.

The goodness-of-fit results discussed in Sec. 15.3 for the Pima Indians TSF regression modeling problem summarized both the ordinary least squares (OLS) results and the least trimmed squares (LTS) results for six regression models, each of which predicted TSF values from one of six explanatory variables. In both cases, the same results for relative goodness-of-fit were obtained when ranking these six models by either the standard R^2 value (R^2_{OLS}) or its robust counterpart (R^2_{LTS}): the single variable with the most explanatory power was BMI, while that with the least explanatory power was NPG. Further, although the R^2_{LTS} values were consistently larger than the R^2_{OLS} values, these values were generally of the same order of magnitude, as were the estimated model coefficients. Thus, the p-values computed from the F-statistics associated with the OLS results appear to offer reasonable guidance in initial variable selection. Since the largest of these p-values is 2.83%, all six of these variables appear to be reasonable candidates for inclusion in the regression model, based on the traditional 5% significance levels typically used in hypothesis testing applications.

While 5% or even 1% significance limits are typical for confirmatory hypothesis testing applications, Hosmer and Lemeshow [148, p. 95] recommend taking $p = 0.25$ in univariate screening applications like the one just described, noting that "use of a more traditional level (such as 0.05) often fails to identify variables known to be important." Conversely, it is also important to note that in cases where a large number of possible covariates are present, statistically based variable selection methods can lead to the inclusion of irrelevant covariates in the model. Jobson emphasizes this point [159, sec. 4.2.3], based on a paper by Freedman [106], who presents the results of a simulation study involving 51 mutually independent random variables, all drawn from a standard normal distribution. Selecting one of these variables as the response and the other 50 as explanatory covariates led to the identification of one covariate as significant at the 5% level and 15 covariates significant at the 25% level.

This last example serves as an important motivation for the adoption of *purposeful variable selection strategies* advocated by Hosmer and Lemeshow [148] that incorporate nonstatistical, field-specific information into the variable selection process. Similar advice is given by Draper and Smith [94, p. 422], who argue that:

> no scientist should be persuaded to abandon his scientific insight and principles in favor of some computerized statistical screening procedure. The use of multiple regression techniques is a powerful tool only if it is applied with intelligence and caution.

15.4.2 Considering variable transformations

Instead of performing univariate screening by fitting one-variable prediction models, a more traditional approach is to compute the product-moment correlations between each candidate explanatory variable and the response variable. For least squares linear regression models involving real explanatory variables, these approaches are equivalent in principle since the single-variable regression

Table 15.3: Comparison of p-values from different ways of measuring univariate association for the Pima Indians TSF data with each of six explanatory variables. The product-moment correlation coefficients are shown ($\hat{\rho}$), along with the probability that the true correlation is zero (p_C) estimated from the asymptotic normality of $\hat{\rho}$ discussed in Chapter 10, the associated z-score relative to 1,000 random permutations (z), its Gaussian probability (p_z), and the probability associated with the covariate computed from the F-statistic associated with the corresponding single-variable regression model (p_F)

No.	Covariate	$\hat{\rho}$	p_C	z	p_z	p_F
1	NPG	0.095	2.83×10^{-2}	2.245	2.48×10^{-2}	2.83×10^{-2}
2	GLC	0.227	1.73×10^{-7}	5.027	4.98×10^{-7}	1.27×10^{-7}
3	DIA	0.226	1.84×10^{-7}	5.167	2.38×10^{-7}	1.36×10^{-7}
4	BMI	0.647	$<1 \times 10^{-16}$	14.961	$<1 \times 10^{-16}$	$<1 \times 10^{-16}$
5	AGE	0.161	1.98×10^{-4}	3.684	2.30×10^{-4}	1.86×10^{-4}
6	DIAG	0.255	4.14×10^{-9}	5.933	2.97×10^{-9}	2.77×10^{-9}

model coefficient is equal to the corresponding product-moment correlation coefficient. In practical terms, this equivalence is illustrated in Table 15.3 for the Pima Indians TSF modeling problem, giving the product-moment correlation estimates $\hat{\rho}$ between TSF and each of the six explanatory variables considered in the model. In addition to the $\hat{\rho}$ value, this table also gives three p-values associated with each covariate: the classical p-value (p_C) computed from the asymptotic normality assumption for $\hat{\rho}$ discussed in Chapter 10, the Gaussian p-value (p_z) associated with the z-score relative to 1,000 random permutations computed from the CNC method discussed in Chapter 14, and the p-value from the OLS linear regression model based on the F-statistic discussed in Sec. 15.3. Note that, for each covariate, all of these p-values are in reasonable agreement, all leading to essentially the same relative assessment of the explanatory power of the six covariates. In particular, note that all of these p-values suggest that BMI is the best covariate to include in a single-variable model.

An advantage of the univariate modeling approach is that it extends to the case of categorical explanatory variables coded as a related set of binary variables as described in Chapter 13. There, the correlation equivalence breaks down since correlations between real-valued and categorical variables are not simply defined. Conversely, an advantage of the correlation-based univariate screening approach is that we can compare both product-moment and Spearman rank correlations to check for evidence of either outliers or the utility of a data transformation. In particular, it has been noted previously that the representation in which a real-valued variable is presented in a dataset is not necessarily the most useful representation for data modeling. This point was demonstrated clearly in Chapter 1 in connection with the brain-weight/body-weight dataset, where a scatterplot of the original variables did not indicate any clear relationship between them, while a log-log plot was strongly suggestive of a linear relationship between the transformed variables. Another point illustrated by that example was that in cases where we need a monotone transformation like the logarithm applied to the response variable, the explanatory variable, or both to bring out the relationship between variables, this can manifest itself as a large difference between the product-moment correlation coefficient and the Spearman rank correlation coefficient.

In connection with the problem of screening candidate explanatory variables considered here, this last observation motivates the following simple comparison approach: compute both of these correlation coefficients and look for large differences in their values. In particular, if the rank correlations are significantly larger than the product-moment correlations for one or a few explanatory variables, this suggests exploring the utility of transformations applied to these explanatory variables, using the ideas described in the following paragraphs. Conversely, if such differences are seen for all or almost all of the explanatory variables, this suggests exploring the utility of transformations applied to the response variable. Of course, it is possible that any large differences seen between these correlation values have other causes, two important ones being outliers and discreteness. Specifically, recall from the discussions presented in Chapter 10 that outliers can cause product-moment correlations to either increase (in the case of coincident outliers in both variables, such as those arising from common-mode failure mechanisms) or decrease (in the more common case of outliers in one variable or statistically independent outliers in both variables), while the rank correlations are largely insensitive to outliers. Similarly, for highly discrete covariates (e.g., binary variables or integer-valued variables involving only a few levels), the presence of a large number of ties in the ranks can cause these two correlation estimates to differ significantly.

An application of this idea to the Pima Indians TSF data and the six explanatory variables described in Sec. 15.2.2 is illustrated in Table 15.4, which gives both the product-moment ($\hat{\rho}$) and rank correlations ($\hat{\rho}_S$) between the nonzero TSF values and each of these six candidate regression variables. Also included in this table are the z-scores z_ρ and z_S for these correlations, each computed relative to 1,000 statistically independent random permutations, along with the p-values computed from these z-scores for a Gaussian distribution. Comparing the values of $\hat{\rho}$ and $\hat{\rho}_S$, we see a decrease of $\sim$15% for NPG, essentially no changes for GLC or DIAG, increases on the order of 10% for DIA and BMI, and an increase of $\sim$40% for AGE.

If differences between product-moment and rank correlations appear large enough to warrant subsequent examination, we are faced with the question of how exactly to proceed. Since there exist an uncountably infinite number of different monotone transformations we could consider, it is obviously not possible to consider all possibilities. In practice, it is often useful—at least initially—to compare the following four: the original, untransformed data; the

Table 15.4: Comparison of product-moment correlations and Spearman rank correlations for the Pima Indians TSF data with each of the six explanatory covariates considered for the linear regression model. Values listed in the table include both correlation estimates, their z-scores relative to 1,000 random permutations, and the Gaussian probabilities associated with these z-scores

No.	Covariate	$\hat{\rho}$	$\hat{\rho}_S$	z_ρ	z_S	p_ρ	p_S
1	NPG	0.095	0.079	2.245	1.874	2.48×10^{-2}	6.98×10^{-2}
2	GLC	0.227	0.228	5.027	5.199	4.98×10^{-7}	1.45×10^{-7}
3	DIA	0.226	0.248	5.167	5.700	2.38×10^{-7}	1.02×10^{-8}
4	BMI	0.647	0.684	14.961	15.600	$<1 \times 10^{-16}$	$<1 \times 10^{-16}$
5	AGE	0.161	0.223	3.684	5.198	2.30×10^{-4}	2.60×10^{-7}
6	DIAG	0.255	0.259	5.933	5.824	2.97×10^{-9}	5.74×10^{-9}

log-transformed data; the square root of the data; and the square of the data. One simple strategy that is particularly appropriate for standard linear regression models is the following, based on normal Q-Q plots. Once we have decided which variable to consider as a candidate for transformation, we construct normal Q-Q plots for the following four data sequences:

1. the original, untransformed data sequence $\{x_k\}$;

2. the natural logarithm $\{\ln x_k\}$ or common logarithm $\{\log_{10} x_k\}$;

3. the square root sequence $\{\sqrt{x_k}\}$;

4. the square sequence $\{x_k^2\}$.

Each of these Q-Q plots should have its own reference line to aid in assessing the linearity of the plot, and ideally these plots should be organized together as a quartet for easy visual comparison. The advantage of constructing and comparing these plots is that they often provide a rapid indication of whether any of the three nonidentity transformations bring the data sequence into better conformity with the nominal Gaussian working assumption. Further, including the square and square root transformations in this collection can give a preliminary indication of whether some other power transformation might be more appropriate. For example, if the square-root transformed plot looks more linear than the original, which in turn looks substantially more linear than either the log-transformed or the square-transformed plots, this is a suggestion that a fractional power transformation like the square root is worth considering. If the square root transformation yields a plot that is better than the original but still fairly nonlinear, it may be useful to examine a more extreme transformation

Figure 15.1: Four normal Q-Q plots for the variable AGE from the Pima Indians diabetes dataset: the original data sequence (upper left), the log-transformed sequence (upper right), the square root sequence (lower left), and the squared sequence (lower right)

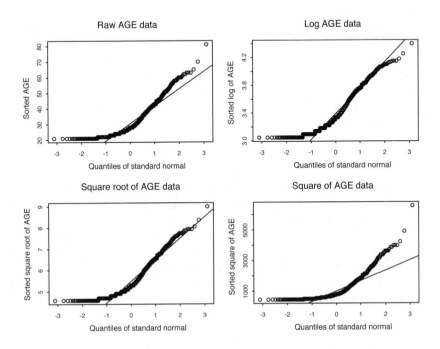

like x^p for some $0 < p < 1/2$. Conversely, if the square root and the original plots are comparably nonlinear but different in shape, it may be reasonable to consider an intermediate fractional transformation like x^p for some $1/2 < p < 1$. Similar reasoning involving the original and square transformations may lead to considering alternative power transformations x^p for $p > 1$.

Fig. 15.1 shows these four plots for AGE, the variable in the Pima Indians diabetes example exhibiting the largest difference between product-moment and Spearman rank correlation values. While both the log and the square root transformations appear somewhat more linear than the other two plots, the dominant feature that is apparent from all four plots is the pronounced lower tail in this age distribution, which none of these transformations affects very much. One possibility—not pursued here—would be to partition the dataset into two subsets, one consisting of those patients of age 21 (the minimum AGE value in the dataset) and the other consisting of those patients whose age is greater than 21. The key point here is to note, first, the possibility of an age anomaly, and second, that simple transformations like those considered here are not likely to be useful in the application considered here. Also, while the analogous plots (not shown) constructed for BMI suggested that log BMI appears to be somewhat better approximated by the normal distribution than the untransformed variable, a model-building exercise similar to the one described in subsequent sections of this chapter for the original variable did not yield a significantly better model. Similarly, the corresponding Q-Q plots (also not shown) for DIA were still less compelling.

A more positive illustration of these ideas is shown in Fig. 15.2 for the response variable TSF, which also demonstrates another advantage of this approach: it can highlight any outlying values that may be present in the data. Specifically, the upper left plot in Fig. 15.2 shows the normal Q-Q plot for the raw TSF data sequence, the upper right plot shows the corresponding log-transformed plot, the lower left shows the results obtained for the square root transformation, and the lower right plot shows the results obtained for the square transformation. Comparing these plots suggests that both the log transformation and the square transformation are too aggressive, making the data distribution less Gaussian than the original. The most linear plot results from the square root transformation, which is indeed quite linear aside from a single apparent outlier. In fact, applying the Hampel identifier described in Chapter 7 to this transformed sequence identifies this point as the only outlier, corresponding to the largest TSF value (99) in the dataset. Taken together, these results suggest, first, omitting this outlying case before modeling, and second, using the square root of TSF rather than TSF itself as the response variable. Both of these ideas are revisited later in this chapter, in connection with the use of LTS regression as a more comprehensive outlier detection strategy.

15.5   Deciding which variables to keep

As discussed further at the end of this chapter, the iterative process of model building can be approached in different ways, involving different levels of effort, and potentially leading to different results. Once we have selected a group of candidate covariates, however, one useful step in this chain is to build the model that includes them all and examine the coefficients for each term. Ideally, if any of these included explanatory variables does not make an important contribution to our ability to predict the outcome, its coefficient will be small enough to be regarded as negligible. In such cases, we lose little in terms of the model's predictive ability by setting the coefficient exactly to zero, thus eliminating that variable from the model. Further, by eliminating this variable, we obtain a simpler data model that is potentially easier to understand and explain to others; in fact, if—as is often the case—we can eliminate several terms from the model, this simplification can be quite substantial. The key practical question is how small the estimated coefficient needs to be for it to be safely regarded as "negligible."

Under the assumption of normally distributed model errors in the case of OLS linear regression, we can formulate this question as a hypothesis test using the ideas introduced in Chapter 9. Specifically, the null hypothesis of interest

Figure 15.2: Four normal Q-Q plots for the variable TSF from the Pima Indians diabetes dataset: the original data sequence (upper left), the log-transformed sequence (upper right), the square root sequence (lower left), and the squared sequence (lower right)

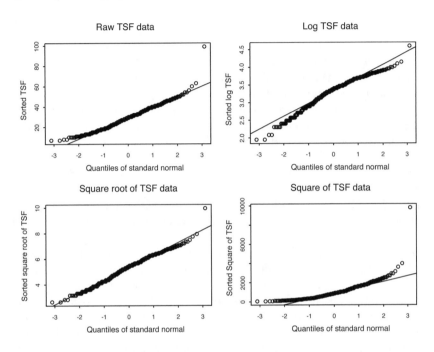

is that we can omit the term from the model, corresponding to the hypothesis that $\beta_j = 0$ where β_j is the model coefficient for the jth explanatory variable included in the model. The *standard error* for the estimated model coefficient $\hat{\beta}_j$ is defined as the square root of its estimated variance [94, p. 25], and it follows from the results presented in Chapter 11 that this value may be written as:

$$\hat{\sigma}_j = \hat{\sigma}_e C_{jj}, \tag{15.8}$$

where $\hat{\sigma}_e$ is the estimated prediction error standard deviation computed from the model:

$$\hat{\sigma}_e = \sqrt{\frac{1}{N-p-1}\sum_{k=1}^{N}(\hat{y}_k - y_k)^2}, \tag{15.9}$$

and C_{jj} is the j,j-diagonal element of the normalized covariance matrix:

$$\mathbf{C} = (\mathbf{X}^T\mathbf{X})^{-1}. \tag{15.10}$$

Here, it is assumed that the original dataset used in building the model contains N data observations, and that the model includes an intercept term (denoted

$\hat{\beta}_0$ in subsequent discussions) and one term for each of p explanatory covariates. These assumptions, together with the assumed normality of the prediction errors, imply that the following quantity should exhibit a Student's t-distribution with $N - p - 1$ degrees of freedom under the null hypothesis:

$$t_j = \frac{\hat{\beta}_j}{\hat{\sigma}_e C_{jj}} \sim t_{N-p-1}. \tag{15.11}$$

This result forms the basis for a specific test of the hypothesis that the estimated parameter $\hat{\beta}_j$ is small enough that its omission should cause no significant degradation of the model's predictive ability.

As a practical matter, several points are worth noting explicitly here. First, note that in the common case where N is large compared with p, the t_{N-p-1} distribution can be approximated reasonably well by the standard normal distribution $N(0, 1)$. Second, note that although the basis for these results is the assumption that the prediction errors $e_k = \hat{y}_k - y_k$ are normally distributed with mean zero and variance σ_e^2, the probabilities associated with the t-statistics t_j defined in Eq. (15.11) are widely used in practice without examining the reasonableness of these working assumptions. While this practice does have its potential pitfalls (e.g., the presence of outliers can cause these probabilities to be highly misleading), it does provide a systematic means of deciding which variables to retain in a data model. *This practice is probably reasonable, so long as the model is examined carefully with regard to the possibility of outliers and other data anomalies, using the techniques described in the subsequent sections of this chapter.* In fact, this same basic strategy is often used to give guidance in developing other types of data models (e.g., logistic regression models or Poisson regression models) that appeal to the asymptotic normality of the estimated model coefficients rather than to the normality of the prediction errors.

The ideas just described are illustrated in Table 15.5, which characterizes the OLS linear regression model that predicts the triceps skinfold thickness (TSF) values from the Pima Indians diabetes dataset, based on the six explanatory variables listed in the table. In addition to listing the explanatory variable names, the table gives their estimated model coefficients $\hat{\beta}$, the standard errors $\hat{\sigma}$ associated with these coefficients, the t-statistics defined in Eq. (15.11), and their associated probabilities under the t-distribution with $N - p - 1 = 525$ degrees of freedom. Consistent with the preliminary variable selection results presented in Sec. 15.4.2, the only explanatory variable whose coefficient is clearly significant is BMI, although with a p-value of approximately 6%, the variable AGE is also a candidate to be retained. Conversely, if we recognize that we are making multiple hypothesis tests and employ any of the corrections for multiple comparison discussed in Chapter 9, we would be less likely to include AGE. For example, the Bonferroni correction would divide the standard acceptance threshold of 5% by the number of covariates, requiring that $p < 0.0083$ for a coefficient to be declared significantly nonzero in this case. Under this more stringent requirement, we would be even less likely to include anything other than BMI in our final model.

Table 15.5: Summary table for the six-variable prediction model for TSF from the Pima Indians diabetes dataset fit via OLS. For each explanatory variable, table entries include the estimated model parameter $\hat{\beta}$, its standard error $\hat{\sigma}$, its associated t-score, and the probability of observing a t-score this large or larger under the Student's t-distribution

No.	Variable	$\hat{\beta}$	$\hat{\sigma}_\beta$	t	$\mathcal{P}(> |t|)$
1	NPG	0.093	0.137	0.675	0.500002
2	GLC	0.013	0.013	1.011	0.312478
3	DIA	-0.013	0.031	-0.410	0.682053
4	BMI	0.966	0.055	17.456	0.000000
5	AGE	0.084	0.045	1.878	0.060901
6	DIAG	0.292	0.895	0.326	0.744534

One way of addressing the possibility that the dataset is not well described by standard normal distributional assumptions is to apply the least trimmed squares (LTS) robust regression approach as a preprocessing step. That is, we fit an LTS model, which provides as a by-product a set of binary weights: data points retained in the model are given a weight of 1, while those dropped from the model are given a weight of 0. Often, we can obtain improved OLS data models by first fitting the LTS model, excluding those points with zero LTS weights from the dataset, and refitting the OLS model. This idea is discussed further in Sec. 15.9.4, where it is applied to the Pima Indians TSF prediction problem considered here; as the results demonstrate, this is a case where omitting a small set of apparently anomalous observations changes some of the OLS model coefficients and their associated significances enough to alter the form of our final data model. Alternatively, the techniques described in the next two sections can also often be used to detect observations or variables that should be omitted from our data model, at least provisionally.

15.6 The problem of collinearity

The term *collinearity* refers to the undesirable effects of including highly correlated explanatory variables together in a data model. In the extreme case where these variables are perfectly correlated—i.e., satisfy an exact linear relationship—it is easy to demonstrate that their presence in a regression model introduces a degeneracy that makes parameter estimation impossible. As a specific example, consider the following regression model between an explanatory variable x and a response variable y:

$$y_k = \alpha_0 + \beta_0 x_k + e_k, \tag{15.12}$$

where α_0 and β_0 are model coefficients estimated from the data observations $\{(x_k, y_k)\}$ that minimize any desired "size measure" for the error sequence $\{e_k\}$ (e.g., the sum of squared errors, sum of absolute errors, etc.). Next, suppose that v is another candidate explanatory variable that is related to x by an exactly linear relationship:

$$x_k = Cv_k + D. \tag{15.13}$$

Now, suppose we include both of these variables in the regression model, considering models of the form:

$$y_k = \alpha_1 + \beta_1 x_k + \gamma_1 v_k + e_k. \tag{15.14}$$

Note that by combining Eqs. (15.13) and (15.14), we can rewrite this model as:

$$y_k = (\alpha_1 + \gamma_1 D) + (\beta_1 + C\gamma_1)x_k + e_k. \tag{15.15}$$

As a consequence, this model is identical with Eq. (15.12) if the following two conditions are satisfied:

$$\begin{aligned} \alpha_1 + \gamma_1 D &= \alpha_0, \\ \beta_1 + \gamma_1 C &= \beta_0. \end{aligned} \tag{15.16}$$

For fixed values of α_0, β_0, C, and D, this set constitutes two simultaneous linear equations in three unknowns (α_1, β_1, and γ_1), which generally has an infinite number of solutions. That is, given any arbitrary value for the coefficient γ_1, we can generally find values for α_1 and α_2 to satisfy Eqs. (15.16). This result means that if an exact linear relationship exists between two explanatory covariates in a model, their coefficients cannot be uniquely estimated from the dataset.

In practice, the problem of collinearity—or *multicollinearity*, as it is sometimes called—typically arises not when two or more variables satisfy an *exact* linear relationship like the one just described, but an *approximate one*. One important case where this problem arises is that were highly correlated explanatory variables are included together. As the following discussion demonstrates, the practical consequences of collinearity can include extreme sensitivity of the parameter estimates to small changes in the data values and very large standard errors for the estimated parameters, supporting the argument of Hosmer and Lemeshow [148] that the observation of very large standard errors is often an indication of a poorly formulated logistic regression problem.

15.6.1 Collinearity and OLS parameter estimates

While it is worth emphasizing that collinearity—the presence of two or more explanatory variables that nearly satisfy a linear relationship—is not only a problem for OLS regression models, but is also an issue for robust alternatives like LTS regression and extensions like generalized linear models. Nevertheless, because the OLS problem has a simple exact solution, we can gain significant insights into the collinearity problem by examining it for the case of OLS linear regression models.

Specifically, consider a linear regression model involving a single response variable y and m explanatory variables x_j for $j = 1, 2, \ldots, m$. In the matrix-vector notation introduced in Chapter 11, this regression model may be written as:

$$\mathbf{y} = \mathbf{X}\theta + \mathbf{e}, \tag{15.17}$$

where $\mathbf{y}$ is the vector of N response variables to be predicted by the regression model, $\mathbf{X}$ is the $N \times m$ matrix formed by including each of the m explanatory vectors as columns, θ is the vector of m regression model parameters to be estimated, and $\mathbf{e}$ is the vector of N model prediction errors. As discussed in Chapter 11, the OLS solution is the parameter vector:

$$\hat{\theta} = (\mathbf{X}^T\mathbf{X})^{-1}\mathbf{X}^T\mathbf{y}. \tag{15.18}$$

Note that if an exact linear relationship exists between two or more explanatory covariates, the matrix $\mathbf{X}^T\mathbf{X}$ will be singular and the inverse in Eq. (15.18) will not exist. The collinearity problem encountered in practice corresponds to the case where the matrix $\mathbf{X}^T\mathbf{X}$ is not singular but is *ill-conditioned*, meaning that it is "nearly singular."

Since the matrix $\mathbf{X}$ contains all of the explanatory variable observations, problems like collinearity manifest themselves as characteristics of this matrix. For this reason, it is instructive to consider its *singular value decomposition (SVD)* [178, p. 35]:

$$\mathbf{X} = \tilde{\mathbf{U}}\mathbf{S}\mathbf{V}^T, \tag{15.19}$$

where $\mathbf{U}$ and $\mathbf{V}$ are *unitary matrices* of dimensions $N \times N$ and $m \times m$, respectively (i.e., their inverse is equal to their transpose, so $\mathbf{U}^T\mathbf{U} = \mathbf{U}\mathbf{U}^T = \mathbf{I}_{N \times N}$ and $\mathbf{V}^T\mathbf{V} = \mathbf{V}\mathbf{V}^T = \mathbf{I}_{m \times m}$), and $\mathbf{S}$ is an $N \times m$ matrix with the following structure:

$$\mathbf{S} = \begin{bmatrix} \mathbf{D} & \mathbf{0} \\ \mathbf{0} & \mathbf{0} \end{bmatrix}. \tag{15.20}$$

Here, $\mathbf{D}$ is a diagonal matrix whose diagonal elements are the *singular values* of $\mathbf{X}$, defined as the positive square roots of the nonzero eigenvalues of the matrix $\mathbf{X}^T\mathbf{X}$. For a rank r matrix, there are r of these numbers so the matrix $\mathbf{D}$ is of dimension $r \times r$, and the entries represented as $\mathbf{0}$ in Eq. (15.20) are compatably dimensioned blocks of zeros so that the matrix $\mathbf{S}$ has the correct dimensions $N \times m$. The case of interest here where $\mathbf{X}^T\mathbf{X}$ is nonsingular but ill-conditioned, implying that the rank of the $\mathbf{X}$ matrix is $r = m$ and some of the diagonal elements of the $m \times m$ matrix $\mathbf{D}$ are very small. In this case, it is convenient to rewrite Eq. (15.19) as:

$$\mathbf{X} = \mathbf{U}\mathbf{D}\mathbf{V}^T, \tag{15.21}$$

where $\mathbf{U}$ is the first m columns of the unitary matrix $\tilde{\mathbf{U}}$. The advantage of this representation is that the OLS parameter estimate can now be expressed as (Exercise 3):

$$\hat{\theta} = \mathbf{V}\mathbf{D}^{-1}\mathbf{U}^T\mathbf{y}. \tag{15.22}$$

Note that since some of the elements of the diagonal matrix $\mathbf{D}$ are very small, the corresponding elements of its inverse are very large. As a consequence, small changes in the response data values $\mathbf{y}$ can manifest themselves in very large changes in the OLS parameter estimate $\hat{\theta}$.

Another common manifestation of collinearity is in very large standard errors for the estimated model parameters. To see how this arises, recall from Chapter 11 that the covariance matrix for the OLS parameter estimate $\hat{\theta}$ is:

$$\mathbf{C} = \sigma_e^2 (\mathbf{X}^T \mathbf{X})^{-1}. \qquad (15.23)$$

It follows from the singular value decomposition of $\mathbf{X}$ that $\mathbf{C}$ can be written as (Exercise 4):

$$\mathbf{C} = \sigma_e^2 \mathbf{V} \mathbf{D}^{-2} \mathbf{V}^T. \qquad (15.24)$$

Since the standard error of the model parameter $\hat{\theta}_i$ is the square root of the i, i diagonal element of $\mathbf{C}$, it follows that, like the influence of changes in data values, the standard errors vary roughly inversely with the singular values of $\mathbf{X}$. Since collinearity can cause these singular values to be extremely small, it can have a very large inflationary effect on the standard errors for the estimated parameters.

15.6.2   Dealing with collinearity

Evidence of possible collinearity often first appears in the form of some type of unpleasant behavior: very large standard errors, extreme sensitivity to small changes in data values, or the wrong sign of the coefficient for one or more explanatory variables. A safer course of action is to first test for the possibility of collinearity, and there are at least two ways of doing this. One is by computing the correlation matrix between all pairs of explanatory variables defining the $\mathbf{X}$ matrix and looking for large values. A particularly well-known example of collinearity is the economic dataset described by Longley [190], consisting of 16 observations of six covariates: GNP, GNP deflator, unemployment, armed forces, population, and year. Of the 15 correlation coefficients between distinct variables, six exceed 0.99: those between GNP and GNP deflator, population or year, those between GNP deflator and population or year, and that between population and year.

A simpler and probably more reliable—though arguably less informative—approach to collinearity detection is to compute the *condition number* of the data matrix $\mathbf{X}$, κ, defined as the ratio of its largest to its smallest singular value. This number is bounded below by 1 and approaches infinity as the matrix becomes singular, so large values of the condition number have been proposed as an indication of collinearity. In particular, Jobson offers the following advice [159, p. 281]:

> If the condition number is less than 100 multicollinearity is not considered to be a problem, while if κ exceeds 1000 the multicollinearity is considered to be severe.

For the Longley economic dataset described above, the condition number is 5751, suggesting severe collinearity under Jobson's criterion and consistent with the known collinearity characteristics of this dataset. Since the singular values of **X** are the diagonal elements of the matrix **D** appearing in the singular value decomposition, the condition number is easily computed if an SVD procedure is available (as it is in the *S* language, with procedure **svd**).

For the Pima Indians diabetes dataset, applying this procedure yields a condition number estimate of 367, large enough to suggest the possibility of collinearity, but not large enough to raise great concern. Computing the correlations between all pairs of explanatory variables yields a largest correlation of approximately 0.5 between the variables DIAG and GLC. Removing DIAG from the dataset reduces the condition number to approximately 60. All in all, these results suggest that collinearity does not appear to be a significant issue for the Pima Indians diabetes dataset.

In cases where collinearity is a significant issue, there are at least two ways to proceed. The simplest and most easily explained is to simply exclude certain explanatory variables, although not everyone likes this approach [159, p. 285], noting that the omission of important variables can cause biased predictions. Still, it is reasonable to at least construct and compare the results obtained with models that omit explanatory variables that are highly correlated with one or more included explanatory variables and examine them in light of the intended application.

The second way of dealing with collinearity is to adopt a specialized regression analysis procedure like *ridge regression* [94, 159, 285]. The basic idea there is to replace the nearly singular matrix $\mathbf{X}^T\mathbf{X}$ with the better-conditioned matrix $k\mathbf{I}_{m\times m} + \mathbf{X}^T\mathbf{X}$, where k is some positive constant. This modification replaces the OLS solution with the *ridge regression solution:*

$$\hat{\theta} = (k\mathbf{I}_{m\times m} + \mathbf{X}^T\mathbf{X})^{-1}\mathbf{X}^T\mathbf{y}. \tag{15.25}$$

Note that when $k = 0$, the ridge regression solution reduces to the OLS solution, while in the limit of large k, the effect of including this term is to "shrink" the OLS estimate toward zero. In terms of the collinearity problem, the reason ridge regression is helpful is that the smallest eigenvalue λ_- and largest eigenvalue λ_+ of a positive-definite $m \times m$ matrix **A** (like, e.g., $\mathbf{X}^T\mathbf{X}$) satisfy the following inequalities:

$$\lambda_-\mathbf{x}^T\mathbf{x} \leq \mathbf{x}^T\mathbf{A}\mathbf{x} \leq \lambda_+\mathbf{x}^T\mathbf{x}, \tag{15.26}$$

for all m-vectors **x**. Thus, the effect of replacing the positive-definite matrix **A** with the modified matrix $k\mathbf{I}_{m\times m}+\mathbf{A}$ is to replace λ_- and λ_+ in these inequalities with $k + \lambda_-$ and $k + \lambda_+$. Since the condition number of the matrix **A** is simply the ratio $\kappa = \lambda_+/\lambda_-$, the effect of this change of the matrix is to modify the condition number as:

$$\tilde{\kappa} = \frac{k + \lambda_+}{k + \lambda_-} \to 1 \text{ as } k \to \infty. \tag{15.27}$$

Further, this effect is monotonic: increasing k always causes $\tilde{\kappa}$ to decrease (Exercise 5). Thus, by choosing k sufficiently large, we can make the ridge regression

solution as well conditioned as we like, at the cost of reducing all of the parameter estimates toward zero. In the limit as $k \to \infty$, our parameter estimates all
go to zero and our model loses all explanatory power. Thus, the art of applying
ridge regression lies in selecting a reasonable value for k. A detailed discussion
of this topic is beyond the scope of this book, but detailed discussions and examples are available; see, for example, Chapter 7 of the book by Vinod and
Ullah [285] or the book by Jobson [159, sec. 4.3.2].

15.7   Finding influential data observations

One of the points emphasized repeatedly in the discussions of outliers given in
earlier chapters (especially Chapter 7) has been the disproportionate influence
of anomalous data observations on modeling results. The following subsections
consider several different approaches that can be taken to detecting highly influential observations—which may or may not be anomalous—in a dataset, in
terms of their influence on regression models. One obvious but unfortunately
often ineffective method is by the examination of model residuals, an approach
discussed in Sec. 15.7.1. A technique that is often more useful in detecting influential observations is based on the *hat matrix*, introduced in Sec. 15.7.2 as a
way of quantifying *leverage* in OLS problems. These ideas are incorporated into
the OLS-specific *deletion diagnostics* introduced in Sec. 15.7.3.

15.7.1   Examining model residuals

Chapter 1 presented the "four R's of exploratory data analysis" as revelation
(i.e., looking at the data), residuals, reexpression, and resistance. Once we have
fit a data model, the examination of the residuals (i.e., the difference between
the predicted and observed values of the response variable) can be extremely
informative, although as the following example illustrates, some care is required
in doing this since significant data anomalies can introduce large enough biases
to completely disguise their presence in the residuals obtained from standard
OLS model fits.

Fig. 15.3 shows a plot of the residuals $\{r_k\}$ obtained from fitting the natural
logarithm of the brain weight (y_k) to the natural logarithm of the body weight
(x_k) via ordinary least squares, as discussed in Chapter 5. Specifically, these
residuals are given by:

$$r_k = y_k - \hat{y}_k = y_k - \alpha x_k - \beta, \qquad (15.28)$$

where α is the slope of the best fit line and β is its y-axis intercept. These
residuals vary over the range $\sim \pm 2$, corresponding to $\sim 20\%$ of the total range
of y variation seen in the original (x_k, y_k) data. One of the motivations for
examining residuals from a model fit is to detect regular patterns, indicative
of additional structure in the dataset that has not been adequately captured
by the model under consideration. No glaringly obvious pattern is seen in
the residuals shown in Fig. 15.3, although careful examination does reveal the

Figure 15.3: OLS residuals, log brain weight vs. log body weight fit

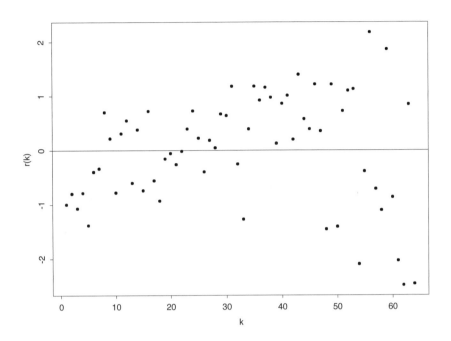

possible existence of two clusters: a main group of residuals with a positive linear trend, and a second group of points in the lower right corner of the plot. These observations turn out to be significant.

Fig. 15.4 shows the normal Q-Q plot of these residuals, which generally appear to conform extremely well to the usual Gaussian error model. In fact, note that the largest apparent deviations from linearity of this plot occur in the upper tail, but as in the preliminary examination of the residuals themselves, nothing appears highly unusual or contrary to the standard working assumptions in linear model building: the observed responses y_k appear to conform reasonably well to a model of the form $y_k = \alpha x_k + \beta + r_k$, where the residuals $\{r_k\}$ look approximately Gaussian. Similarly, neither the traditional 3σ edit rule nor the more effective (but sometimes excessively zealous) Hampel identifier detect any outliers in the sequence $\{r_k\}$. Also, both the traditional product-moment correlation coefficient $\rho \simeq 0.91$ and the Spearman rank correlation $\rho_S \simeq 0.93$ are in reasonable agreement and support the conjecture that a linear relationship exists between x_k and y_k.

In fact, this dataset contains a number of outliers that are responsible for the differences observed in Chapter 5 between the results obtained when fitting y_k on x_k and those obtained when fitting x_k on y_k and transforming back

Figure 15.4: OLS residuals, normal Q-Q plot

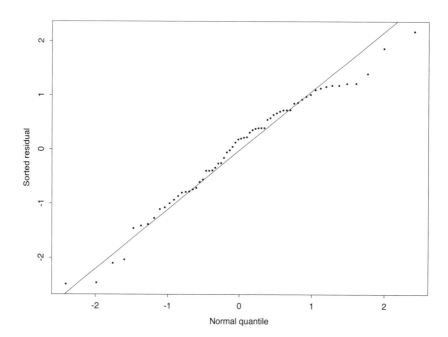

to a relationship of the form considered here. This example is revisited in the following discussions to illustrate the effectiveness (or ineffectiveness) of different techniques designed to detect both *influential* and *anomalous* data points, which may or may not be the same thing.

15.7.2 Leverage and the hat matrix

As the four-point example discussed in Chapter 5 illustrated, regression results obtained from a contaminated dataset can depend strongly on where the contamination occurs within the dataset, a phenomenon called *leverage* discussed briefly in Chapter 5. Although, as those examples demonstrated, leverage effects can be important for other, more robust estimators as well, it is useful to restrict this discussion to the OLS problem because its structure leads to a useful characterization in terms of the so-called *hat matrix*.

Specifically, if $\hat{\theta}$ is the OLS regression estimator defined by Eq. (11.15), it is useful to consider the *prediction* $\hat{\mathbf{y}}$ of the observed response vector $\mathbf{y}$ that is generated by the regression model:

$$\hat{\mathbf{y}} = \mathbf{X}\hat{\theta} = \mathbf{X}(\mathbf{X}^T\mathbf{X})^{-1}\mathbf{X}\mathbf{y} \equiv \mathbf{H}\mathbf{y}. \qquad (15.29)$$

The $N \times N$ matrix $\mathbf{H}$ is called the *hat matrix* in the statistics literature, because it "puts the hat on $\mathbf{y}$," mapping the observed data vector $\mathbf{y}$ into the vector of predictions, $\hat{\mathbf{y}}$. This matrix has a number of interesting properties that are useful in characterizing leverage. In particular, note that the diagonal element H_{ii} expresses the influence of the observed value y_i on its OLS prediction $\hat{y}_i$.

More generally, it is easy to demonstrate that the matrix $\mathbf{H}$ is both symmetric (i.e., $\mathbf{H}^T = \mathbf{H}$) and *idempotent*:

$$\mathbf{HH} = \mathbf{X}(\mathbf{X}^T\mathbf{X})^{-1}\mathbf{X}^T\mathbf{X}(\mathbf{X}^T\mathbf{X})^{-1}\mathbf{X}^T = \mathbf{X}(\mathbf{X}^T\mathbf{X})^{-1}\mathbf{X}^T = \mathbf{H}. \qquad (15.30)$$

It is not difficult to show (Exercise 6) that any $N \times N$ idempotent matrix of rank r has r eigenvalues equal to 1 and $N - r$ eigenvalues equal to 0. Further, the *trace* of a square matrix is the sum of its diagonal elements, and it follows from the definition of matrix multiplication that trace$\{\mathbf{AB}\}$ = trace$\{\mathbf{BA}\}$ [159, p. 551], implying:

$$\begin{aligned} \text{trace } \{\mathbf{H}\} &= \text{trace } \{\mathbf{X}(\mathbf{X}^T\mathbf{X})^{-1}\mathbf{X}^T\} \\ &= \text{trace } \{\mathbf{X}^T\mathbf{X}(\mathbf{X}^T\mathbf{X})^{-1}\} = \text{trace } \{\mathbf{I}_{p \times p}\} = p. \end{aligned} \qquad (15.31)$$

Since the trace is also equal to the sum of the eigenvalues, note that it follows from this result that the rank of the hat matrix is equal to p, the rank of the $p \times p$ matrix $\mathbf{X}^T\mathbf{X}$. An important practical consequence of Eq. (15.31) is that it provides a scale for measuring the influence of individual data points i in the dataset. Specifically, this result implies that the average value for H_{ii} is equal to p/N, and it has been suggested that points i for which $H_{ii} > 2p/N$ may be regarded as *influential points* or *leverage points* [34, p. 17].

Fig. 15.5 shows a practical application of this idea. Specifically, this plot shows the diagonal element H_{ii} of the hat matrix $\mathbf{H}$ vs. the data point index i for the brain-weight/body-weight dataset considered previously. The regression model considered here is the line fit considered in Chapter 5, so the rank of $\mathbf{H}$ is $p = 2$. Because this dataset is of size $N = 64$, it follows that $2p/N = 0.0625$ and a horizontal line is shown in Fig. 15.5 at this threshold value. Three points exceed this threshold value, corresponding to the smallest animal in the dataset ($k = 1$, the lesser short-tailed shrew) and the largest two animals ($k = 63$, the blue whale, and $k = 64$, the brachiosaurus). This example is revisited in Sec. 11.5 in connection with other fitting procedures, but the results presented here provide a first indication of the unusual character of this dataset: the largest animal belongs to the now extinct family of dinosaurs, known for their anomalously large body masses and their anomalously small brains. Conversely, it is important to note that not all leverage points need be outliers, nor does the converse hold: not all outliers are leverage points. In this example, two of the leverage points (the lesser short-tailed shrew and the blue whale) are not outliers, and not all of the outliers (corresponding to 10 dinosaurs) are leverage points. Finally, it is worth emphasizing that leverage depends only on the x variables and not on the y variables; in this example, this observation means that leverage is determined entirely by body mass and is independent of brain mass, so large animals with large brains (e.g., the blue whale) are as likely to be leverage points as large animals with small brains (e.g., the brachiosaurus).

Figure 15.5: H_{ii} vs. i for the brain/body dataset

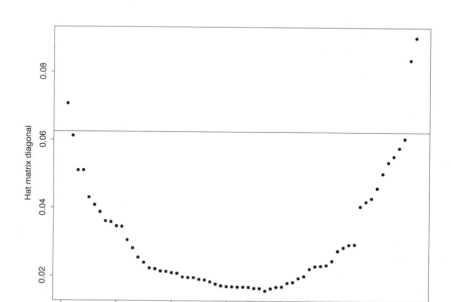

15.7.3 OLS regression diagnostics

One important source of difficulty in regression analysis is the excessive influence of one or a few anomalous data points. An extremely useful general idea is to examine the effects of eliminating data subsets on regression results: large changes indicate highly influential subsets that could be investigated further to determine why they are so influential and whether this influence appears reasonable. This approach has the advantage that it is completely general and can be applied to essentially any problem formulation, but brute force applications of this idea can become computationally prohibitive in unfavorable cases as demonstrated in Chapter 14. Conversely, an extremely practical application of this idea is to examine the influence of each data point individually, especially for the OLS problem where the structure of the OLS solution can be used to great advantage.

To illustrate this idea, consider a dataset $\mathcal{D}$ and denote by $\mathcal{D}_{(i)}$ the modified dataset obtained by deleting data point i. The basic idea behind deletion diagnostics is to compare the results obtained for $\mathcal{D}_{(i)}$ with those obtained for $\mathcal{D}$, noting that large changes suggest that data point i is strongly influential. In the context of the regression problems considered here, there are at least two possible bases for such comparisons: changes in the estimated model parameters, and changes in the prediction errors. The following discussion briefly considers both of these ideas, specialized to the case of OLS regression models where the structure of the exact solution permits computation of these measures without re-solving the regression problem for each modified dataset $\mathcal{D}_{(i)}$. The two diagnostics considered here are *Cook's distance* for parameter changes and the *DFFITS diagnostic* for prediction errors, both of which are closely related to the hat matrix discussed in Sec. 15.7.2 [34].

To obtain these results, first define $\hat{\theta}$ as the OLS parameter estimate computed from $\mathcal{D}$ according to Eq. (11.15) and define $\hat{\theta}_{(i)}$ as the corresponding OLS parameter estimate computed from $\mathcal{D}_{(i)}$. Similarly, define $\mathbf{y}_{(i)}$ as the observation vector $\mathbf{y}$ with y_i omitted and $\mathbf{X}_{(i)}$ as the matrix $\mathbf{X}$ without the row $\mathbf{z}_i^T$ defined by Eq. (11.39); i.e.:

$$\mathbf{y}_{(i)} = \begin{bmatrix} y_1 \\ \vdots \\ y_{i-1} \\ y_{i+1} \\ \vdots \\ y_N \end{bmatrix}, \quad \mathbf{X}_{(i)} = \begin{bmatrix} \mathbf{z}_1^T \\ \vdots \\ \mathbf{z}_{i-1}^T \\ \mathbf{z}_{i+1}^T \\ \vdots \\ \mathbf{z}_N^T \end{bmatrix} \Rightarrow \mathbf{X}_{(i)}^T \mathbf{X}_{(i)} = \mathbf{X}^T \mathbf{X} - \mathbf{z}_i \mathbf{z}_i^T. \quad (15.32)$$

Our ability to avoid re-solving each modified regression problem rests on the following matrix inversion result [159, p. 552]:

$$(\mathbf{A} - \mathbf{u}\mathbf{v}^T)^{-1} = \mathbf{A}^{-1} + \frac{\mathbf{A}^{-1}\mathbf{u}\mathbf{v}^T\mathbf{A}^{-1}}{1 - \mathbf{v}^T\mathbf{A}^{-1}\mathbf{u}}. \quad (15.33)$$

To make use of this result, take $\mathbf{A} = \mathbf{X}^T\mathbf{X}$ and $\mathbf{u} = \mathbf{v} = \mathbf{z}_i$, and invoke Eq. (15.32) to obtain:

$$[\mathbf{X}_{(i)}^T \mathbf{X}_{(i)}]^{-1} = [\mathbf{X}^T\mathbf{X}]^{-1} + \frac{[\mathbf{X}^T\mathbf{X}]^{-1}\mathbf{z}_i\mathbf{z}_i^T[\mathbf{X}^T\mathbf{X}]^{-1}}{1 - H_{ii}}, \quad (15.34)$$

where H_{ii} is the denominator of the hat matrix. This result follows from the observation that:

$$H_{ii} = [\mathbf{X}(\mathbf{X}^T\mathbf{X})^{-1}\mathbf{X}]_{ii} = \mathbf{z}_i^T(\mathbf{X}^T\mathbf{X})^{-1}\mathbf{z}_i. \quad (15.35)$$

Noting that the modified parameter vector is given by $\hat{\theta}_{(i)} = [\mathbf{X}_{(i)}^T\mathbf{X}_{(i)}]^{-1}\mathbf{X}_{(i)}^T\mathbf{y}_{(i)}$, it can be shown (Exercise 7) that:

$$\hat{\theta} - \hat{\theta}_{(i)} = \frac{(\mathbf{X}^T\mathbf{X})^{-1}\mathbf{z}_i r_i}{1 - H_{ii}}, \quad (15.36)$$

where $r_i = y_i - \hat{y}_i$ is the ith residual from the unmodified OLS fit. *The importance of this result is that it permits us to compute the change in the OLS parameter vector $\hat{\theta}$ directly from the OLS residuals and other known components of the OLS problem formulation without having to solve another OLS problem.*

A practical problem with this result is that the change in the parameter vector is itself a vector with p components, complicating the interpretation task: how do we decide what constitutes a "large change in $\hat{\theta}$?" One possible solution is *Cook's squared distance*, which is a scalar measure of the size of this parameter shift [34]. Specifically, this quantity is defined as:

$$\mathrm{CD}^2(i) = \frac{[\hat{\theta} - \hat{\theta}_{(i)}]^T \mathbf{X}^T \mathbf{X}[\hat{\theta} - \hat{\theta}_{(i)}]}{ps^2}, \tag{15.37}$$

where s^2 is the residual variance computed from the OLS solution, defined as:

$$s^2 = \frac{1}{N-p} \sum_{k=1}^{N} r_k^2. \tag{15.38}$$

It follows from Eq. (15.36) that this result may be simplified to:

$$\mathrm{CD}^2(i) = \frac{r_i^2 \mathbf{z}_i^T (\mathbf{X}^T \mathbf{X})^{-1} \mathbf{z}_i}{ps^2(1 - H_{ii})^2} = \frac{r_i^2 H_{ii}}{ps^2(1 - H_{ii})^2}. \tag{15.39}$$

It has been suggested [252, p. 228] that data point i may be regarded as influential if $\mathrm{CD}^2(i) > 1$. A plot of this diagnostic is shown in Fig. 15.6 for the brain-weight/body-weight dataset, with a horizontal line at this suggested threshold. Here, this deletion diagnostic finds nothing suspicious, so we consider a simpler diagnostic based on changes in the model predictions.

This measure is the DFFITS diagnostic discussed by Belsley *et al.* [34], based on a normalized measure of the change δ_i between the prediction $\hat{y}_i = \mathbf{z}_i^T \hat{\theta}$ of the ith response from the OLS model and its prediction $\tilde{y}_i = \mathbf{z}_i^T \hat{\theta}_{(i)}$ by the OLS model *omitting* this point from the dataset; i.e.:

$$\delta_i = \mathbf{z}_i^T [\hat{\theta} - \hat{\theta}_{(i)}] = \frac{\mathbf{z}_i^T (\mathbf{X}^T \mathbf{X})^{-1} \mathbf{z}_i r_i}{1 - H_{ii}} = \frac{H_{ii} r_i}{1 - H_{ii}}. \tag{15.40}$$

It has been suggested [34, p. 15] that a reasonable normalization for this quantity is $s(i)\sqrt{H_{ii}}$, where $s^2(i)$ is the variance of the *other* prediction errors:

$$s^2(i) = \frac{1}{N-p-1} \sum_{k \neq i} [y_k - \mathbf{z}_k^T \hat{\theta}_{(i)}]^2. \tag{15.41}$$

Adopting this normalization leads to the DFFITS deletion diagnostic:

$$\mathrm{DFFITS}_i = \frac{r_i \sqrt{H_{ii}}}{s(i)[1 - H_{ii}]}. \tag{15.42}$$

This diagnostic can assume either positive or negative values, and Belsley *et al.* [34, p. 28] have suggested that DFFITS values larger in magnitude than $2\sqrt{p/N}$ may indicate data points with a large influence on the regression results. The values computed for this deletion diagnostic are shown in Fig. 15.7 for the brain-weight/body-weight dataset. In addition, the threshold values

Figure 15.6: Cook's squared distance for the brain/body dataset

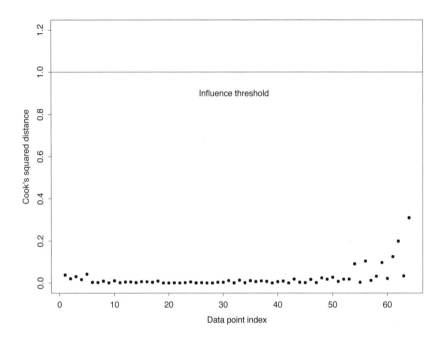

$\pm 2\sqrt{p/N}$ are shown as horizontal lines in this plot, and it may be seen that six data points fall outside these limits. The two values exceeding the upper threshold correspond to data points $k = 56$ and $k = 59$, which are the Asian and African elephant, respectively; the four values falling below the lower threshold correspond to data points $k = 54$ (stegasaurus), $k = 61$ (triceratops), $k = 62$ (diplodocus), and $k = 64$ (brachiosaurus). Like the diagonal elements of the hat matrix, note that this deletion diagnostic has revealed some of the anomalous data points that are present in this dataset, although not all of them. Similarly, this diagnostic has also detected some influential points that do not belong to this anomalous subset. Overall, these results emphasize that OLS influence measures can sometimes be useful in detecting data anomalies, but they are not 100% reliable, and they should not be used blindly for anomaly detection.

15.8 Cross-validation

Classical goodness-of-fit and variable selection criteria for linear regression models are fundamentally based on an assumption that the random variation seen in the response variable is, first, the dominant source of uncertainty, and second, well approximated by a sequence of independent, identically distributed Gaus-

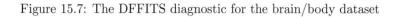

Figure 15.7: The DFFITS diagnostic for the brain/body dataset

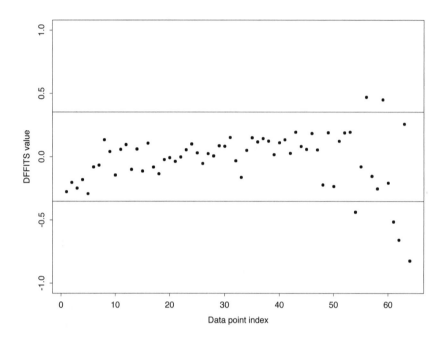

sian observation errors. In cases where these assumptions are badly violated, robust alternatives to ordinary least squares regression have been developed, but there remains the question of how to replace tools like the F-statistic for assessing goodness-of-fit in the absence of distributional knowledge. One approach is to employ bootstrap resampling methods that permit us to base these assessments on the empirical distribution of the data instead of distributional assumptions. One of the most popular of these resampling-based approaches is *cross-validation*, a technique that can be used with a wide variety of different model types. The basic notion is to divide the original dataset into subsets, use some of these subsets to estimate data model parameters, and use the others to assess the quality of these model parameters. Depending on the specific data modeling problem considered (e.g., OLS vs. robust regression methods), the details of these procedures may be quite different; in particular, the availability of an exact solution for OLS problems sometimes makes great simplification possible, as subsequent examples demonstrate. The following subsections describe three versions of this idea: *leave-one-out cross-validation* for goodness-of-fit assessment in Sec. 15.8.1, *K-fold cross-validation* in Sec. 15.8.2, and the use of cross-validation ideas in variable selection in Sec. 15.8.3.

15.8.1 Leave-one-out cross-validation

The basic idea behind leave-one-out cross-validation methods is the systematic omission of each single data point in turn—exactly as in the deletion diagnostics described earlier—followed by a comparison of the predicted and observed values of each omitted data point. A fundamental component of all of the cross-validation methods considered here is a measure of distance $\delta_k = c(y_k, \hat{y}_k)$ between an observed response value y_k and its model-based prediction $\hat{y}_k$. The error measure $c(\cdot, \cdot)$ can be quite general, but two of the most popular choices are the squared error measure $c(y_k, \hat{y}_k) = (y_k - \hat{y}_k)^2$ and the absolute error measure $c(y_k, \hat{y}_k) = |y_k - \hat{y}_k|$. Averaging δ_k values over all observations k then gives different measures of regression model quality, depending on how the predictions $\hat{y}_k$ are defined. The simplest of these averages is the *apparent error* [85, p. 292], given by:

$$\Delta_{app} = \frac{1}{N} \sum_{k=1}^{N} c(y_k, \hat{y}_k), \tag{15.43}$$

where $\hat{y}_k$ is the prediction of y_k from the data model. As a specific example, for the linear OLS model, $\hat{y}_k = \mathbf{x}_k^T \hat{\beta}$, where $\mathbf{x}_k^T$ is the kth row of the explanatory variable matrix $\mathbf{X}$ and $\hat{\beta}$ is the OLS parameter vector estimate obtained from the dataset. Because this error measure is computed from the same data used in building the model, Δ_{app} is sometimes called the *resubstitution error*, and it is known to underestimate the true prediction error [85, p. 292].

To obtain a better prediction error estimate, we need to use different data for error assessment than we do for model building. The simplest way of doing this is to compute the *leave-one-out cross-validation error* [85, p. 293], given by:

$$\Delta_{CV-1} = \frac{1}{N} \sum_{k=1}^{N} c(y_k, \hat{y}_{(k)}), \tag{15.44}$$

where $\hat{y}_{(k)}$ represents the prediction of y_k obtained from the model fit to all observations in the dataset *except* for observation k. That is, let $\mathcal{D}$ denote the complete dataset under consideration and define $\mathcal{D}_{(k)}$ as the dataset with the kth data record (i.e., both the response y_k and all explanatory variables $\mathbf{x}_k^T$) omitted from $\mathcal{D}$. The value $\hat{y}_{(k)}$ is the prediction of y_k obtained when explanatory variables $\mathbf{x}_k^T$ are substituted into the data model constructed from $\mathcal{D}_{(k)}$. In general, computing the N predictions $\hat{y}_{(k)}$ requires fitting N data models, one for each of the N modified datasets $\{\mathcal{D}_{(k)}\}$, but given these results, the error estimate Δ_{CV-1} can be computed for any type of data model and it is generally better than the apparent error estimate Δ_{app} [85, p. 292].

For the case of OLS linear regression models with the squared error estimate, the deletion diagnostic results presented in Sec. 15.7.3 lead directly to the following expression for the leave-one-out cross-validation error:

$$\Delta_{OLS,CV-1} = \frac{1}{N} \sum_{k=1}^{N} \frac{(y_k - \mathbf{x}_k^T \hat{\beta})^2}{(1 - H_{kk})^2}, \tag{15.45}$$

where H_{kk} is the k,k diagonal element of the hat matrix $\mathbf{H}$ defined in Sec. 15.7.2. The great advantage of this result over that for the general case is that Eq. (15.45) does not require any additional model-building steps beyond that required to generate the model of interest: the only additional compuatations required are those necessary to generate the hat matrix diagonal elements, H_{kk}.

15.8.2   K-fold cross-validation

As noted, the leave-one-out cross-validation approach just described is applicable in principle to any data model and any error measure $c(\cdot,\cdot)$, but in the general case it requires N distinct model-building steps, which may be computationally prohibitive if N is large and/or the computational effort required for each model-building step is great. Also, in cases where either the error measure or the model predictions do not depend smoothly on the data values (as in the case of the median discussed in Chapter 7), leave-one-out cross-validation may not work well [85, p. 293]. In both of these situations, alternative cross-validation procedures are desirable, motivating the *K-fold cross-validation strategies* described next.

The simplest implementation of K-fold cross-validation partitions the original dataset into K nonoverlapping subsets, each of approximately equal size. The cross-validation error Δ_{CV-1} defined in Eq. (15.44) is then replaced with:

$$\Delta_{CV,K} = \frac{1}{N} \sum_{k=1}^{N} c(y_k, \hat{y}_{(j(k))}), \tag{15.46}$$

where $\hat{y}_{(j(k))}$ is defined as follows. Let $j(k)$ denote the data subset that contains the kth data observation (i.e., the response y_k and the explanatory variables $\mathbf{x}_k^T$) and let $\mathcal{M}_{(j(k))}$ denote the data model built from the dataset with subset $j(k)$ omitted. The prediction $\hat{y}_{(j(k))}$ is then defined as the prediction of y_k given the explanatory variables $\mathbf{x}_k^T$ and the model parameters defining $\mathcal{M}_{(j(k))}$. Note that, for an arbitrary data model, computing the K-fold cross-validation error $\Delta_{CV,K}$ requires K model-building steps. Davison and Hinkley [85, p. 294] recommend taking:

$$K = \min\{\sqrt{N}, 10\}, \tag{15.47}$$

so that at most 10 model-building steps are required, independent of N.

A potentially significant disadvantage of using the K-fold cross-validation error estimate defined in Eq. (15.46) is that it tends to have a larger bias than leave-one-out cross-validation. To address this problem, Davison and Hinkley describe a bias-corrected variation on K-fold cross-validation called *K-fold adjusted cross-validation* [85, p. 295]. A slight simplification of their procedure consists of the following steps:

1. Select K as defined in Eq. (15.47) and define m_1 through m_{K-1} as $m = \lfloor N/K \rfloor$ and m_K as $N - (K-1)m$, where $\lfloor x \rfloor$ denotes the largest integer that does not exceed x.

2. Apply a random permutation to the record indices k to obtain a reordered dataset $\tilde{\mathcal{D}}$. Let $\mathcal{D}_k$ denote the m_k elements of $\tilde{\mathcal{D}}$ with indices $j = (k-1)m + 1$ through $j = km$ for $k = 1, 2, \ldots, K-1$ and let $\mathcal{D}_K$ denote the last m_K elements of $\tilde{\mathcal{D}}$.

3. Fit the regression model to the complete dataset $\tilde{\mathcal{D}}$; from this model, generate the predictions $\hat{y}_k$ and compute the apparent error Δ_{app} as:

$$\Delta_{app} = \frac{1}{N} \sum_{k=1}^{N} c(y_k, \hat{y}_k).$$

4. For $j = 1, 2, \ldots, K$:

   a. fit the regression model to all records in $\tilde{\mathcal{D}}$ except those in $\mathcal{D}_j$;

   b. from this regression model, compute the predictions $\check{y}_k$ of y_k for all records k from $\tilde{\mathcal{D}}$; from these predictions, compute:

$$\check{\Delta}_j = \frac{1}{N} \sum_{k=1}^{N} c(y_k, \check{y}_k);$$

   c. for all records k from $\mathcal{D}_j$, save $\check{y}_k$ as $\tilde{y}_k$.

5. Combining the $\tilde{y}_k$ values computed in (4c) for all j, compute $\Delta_{CV,K}$ as:

$$\Delta_{CV,K} = \frac{1}{N} \sum_{k=1}^{N} c(y_k, \tilde{y}_k).$$

6. Compute the adjusted K-fold cross-validation error estimate as:

$$\Delta_{adj} = \Delta_{CV,K} + \Delta_{app} - \sum_{j=1}^{K} \left(\frac{m_j}{N}\right) \check{\Delta}_j.$$

Davison and Hinkley also consider a number of other variants of cross-validation, but for the examples they consider, adjusted cross-validation with $K = 10$ seems to represent the most practical strategy in terms of the trade-off between computational efficiency and accuracy. In particular, they note that "adjustment significantly improves cross-validation when group size is not small" [85, p. 301]. Thus, adjusted cross-validation is a broadly applicable strategy for assessing the goodness-of-fit for data models built from large datasets in cases where classical measures are either unavailable or of questionable utility.

15.8.3 Cross-validation for variable selection

The cross-validation strategies described in the previous two subsections are useful in assessing goodness-of-fit for an individual model, but not for comparing models that include different variables. In particular, note that for linear least squares models, the apparent error Δ_{app} is simply the residual sum of squares (RSS) divided by the total number of data observations, N. Since including additional explanatory variables in such a model causes the RSS value to decrease, it follows that minimizing Δ_{app} will cause all candidate explanatory variables to be included in the model. Davison and Hinkley note that the leave-one-out cross-validation error estimate behaves similarly, as do certain bootstrap-based approaches [85, pp. 303–304]. Conversely, they note that the problem can be overcome by using the following bootstrap cross-validation procedure.

In what follows, consider a collection of linear regression models, each based on a different set of explanatory variables. A specific model M from this collection then generates a prediction $\hat{y}_{k,M}$ that is given by:

$$\hat{y}_{k,M} = \mathbf{x}_{k,M}^T \hat{\beta}_M, \qquad (15.48)$$

where $\mathbf{x}_{k,M}^T$ is the vector of explanatory variables for model M used to predict the kth observed response, y_k, and $\hat{\beta}_M$ is the vector of model parameters associated with model M. The bootstrap cross-validation procedure described by Davison and Hinkley proceeds by generating R variants of model M, as follows. For each bootstrap replication r, a fixed number m of records are randomly deleted from the dataset, equivalent to drawing a random sample of size $N - m$ without replacement from the dataset. A model of type M (i.e., involving the same covariates) is then fit to this random subsample to obtain an estimated parameter vector $\hat{\beta}_{M,r}$, where M again indicates the model considered and r specifies the bootstrap sample used. The prediction error associated with this bootstrap replicate model is:

$$\Delta_{B,r} = \frac{1}{N} \sum_{k=1}^{N} (y_k - \mathbf{x}_{k,M}^T \hat{\beta}_{M,r})^2. \qquad (15.49)$$

Note that this error term is computed from the estimated parameter vector $\hat{\beta}_{M,r}$ for *all* N data observations, even though the parameter vector was estimated from a randomly selected subset of the full dataset. This point is important since it means that $\Delta_{B,r}$ is not simply the apparent error for the original model, but a cross-validation error estimate based on predictions of m data samples that were not used in estimating $\hat{\beta}_{M,r}$. Repeating this process R times and averaging the results over all of these bootstrap replications then gives the final error estimate to be used for model selection:

$$\Delta_B = \frac{1}{R} \sum_{r=1}^{R} \Delta_{B,r} = \frac{1}{NR} \sum_{r=1}^{R} \sum_{k=1}^{N} (y_k - \mathbf{x}_{M,k}^T \hat{\beta}_{M,r})^2. \qquad (15.50)$$

To use this measure for model selection, Δ_B would be computed for each model M and the model minimizing Δ_B would be selected as the "best" model. In practice, probably a better approach would be to incorporate Δ_B as part of a broader model selection strategy that also considered total complexity (i.e., the number of explanatory covariates included), the interpretation of the resulting models in terms of the application for which the models are intended, and other criteria like the relation of the models to historical precedent (e.g., is the model that exhibits the second- or third-best Δ_B value a simple variant of another model that is widely adopted in the literature?). These considerations are discussed further in the following section.

Finally, note that to apply the procedure described here, it is only necessary to specify two numbers—R and m—and perform R computations. As a general rule of thumb, since Δ_B is a second moment characterization, it has been recommended that R be at least 100 [85, p. 304]. Since R models of type M must be fit to the dataset, this requirement can make the method impractical in cases where the model-fitting procedure is computationally intensive. Also, based on practical experience with the method, Davison and Hinkley recommend $m \sim (2/3)N$ [85, p. 304], meaning that each bootstrap sample represents approximately one-third of the total dataset, drawn randomly without replacement. A potential disadvantage of this approach, particularly for complex models or small datasets, is that the data matrix $\mathbf{X}$ associated with one or more of these bootstrap samples may become ill-conditioned or singular. If the number of singular models is small relative to R, one possible approach is to simply omit them from the bootstrap average, but this is not feasible in cases where this number is not small relative to R. Also, the need to omit singular cases raises the question of how accurate the bootstrap cross-validation error estimates are. As a practical matter, this method should probably not be used to assess models for which this singularity occurs more than a very few times.

15.9   Iterative refinement strategies

It has been emphasized several times in this chapter that model building is typically an iterative process that involves tentatively fitting one or more models to a dataset, examining the results in various ways, and refining the model based on what we learn from these examinations. Clearly, this strategy can be implemented in several different ways, and the following discussions briefly outline some important approaches that have been described in the regression literature.

One possibility is a user-guided assessment strategy closely related to the *purposeful selection* approach for preliminary variable selection discussed in Sec. 15.4.1. There, model refinement decisions—i.e., the choice of which included variables should be dropped or which nonincluded variables should be added to the model—are made explicitly by the user, guided by a combination of model fit summary data and subject matter expertise. Two alternative possibilities are fully automated ones, based on statistical significance criteria. A point emphasized several other places in this chapter but worth repeating yet again is that statistical inclusion or exclusion criteria should *not* be regarded as either infallible or inviolable: the choices of variables to either include in a data model or exclude from it are ultimately up to the model developer, *not* to a model-fitting package. With this in mind, automated iteration strategies like those described next can be extremely useful in assisting the model builder in this task, especially in cases where there are many possible variables to consider.

15.9.1 All subsets regression

One possible strategy for model building is *all subsets regression*, where all possible combinations of covariates are included in the model and the results are compared. In theory, this approach is quite attractive, since it provides the most complete basis for deciding which covariates to include in a model. Computational complexity considerations limit its practical applicability, however, since it rapidly becomes infeasible as the number of covariates C increases. In particular, since there are 2^C distinct ways of combining C covariates, even for only 10 covariates, all subsets regression would require building and comparing 1,024 models. In the common situation where we wish to fit multivariate regression models to datasets containing many observations and many covariates, both the computational requirements involved in fitting all subset models and the interpretation issues that arise in systematically comparing the results make this approach impractical on the large scale.

Conversely, once other techniques have been used to reduce the field of candidate explanatory variables substantially, all subsets regression may be a reasonable strategy for choosing among restricted covariate subsets. A particularly useful computational tool in evaluating the results of all subsets regression in those cases where it is feasible is *Mallows' C_p measure*, described next. The underlying assumption on which this OLS-based measure rests is that we are considering a family of models, the most complex of which provides an adequate description of the dataset—in general, this model will be more complex than we desire, but its assumed goodness-of-fit characteristics make it a good frame of reference for evaluating other, simpler models. Specifically, assume this most complex model includes M explanatory variables and that all other models under consideration are obtained by deleting one or more explanatory variables from this set. For a candidate model based on $p < M$ explanatory variables, the C_p measure is defined as [94, p. 299]:

$$
\begin{aligned}
C_p &= \frac{\mathrm{RSS}_p}{\mathrm{RSS}_M} - (N - 2[p+1]), \\
&= \frac{\sum_{k=1}^{N} (y_k - \hat{y}_{k,p})^2}{\sum_{k=1}^{N} (y_k - \hat{y}_{k,M})^2} - (N - 2p - 2),
\end{aligned}
\tag{15.51}
$$

where $\hat{y}_{k,p}$ is the prediction of y_k generated by the model with p explanatory variables and $\hat{y}_{k,M}$ is the prediction generated by the model with M variables. (It is important to note that intercept terms are assumed to be included in both models.) Under the assumption that the M-variable model provides an unbiased fit to the data and an adequate basis for estimating the prediction error variance, it follows that the expected value of C_p is $p + 1$ [94, p. 300]. Further, if the p-variable model is good enough to give an unbiased fit to the data, it follows that $C_p \leq p + 1$, while poor models will typically exhibit C_p values that exceed $p + 1$, sometimes by quite a lot. Thus, a useful general approach in comparing a collection of models with $p < M$ variables is to compute C_p and plot it as a function of the number of explanatory variable p: points falling above the line $C_p = p + 1$ are suggestive of poor fits, while points falling on or below this reference line are suggestive of better fits. Again, however, it is worth repeating the point noted earlier in conjunction with bootstrap cross-validation-based model selection procedures: Mallows' C_p is useful as an aid in selecting between candidate regression models, but it should not be used as the sole crieteria for model selection.

15.9.2   Stepwise regression

A more practical systematic iteration strategy for large datasets than all subsets regression is *stepwise regression*, which can be implemented in several different ways. These methods consider one potential explanatory variable at a time, examine its influence on the resulting model fit, and decide whether to include it in the model based on these results. The number of models examined in stepwise regression is of order C^2 for C candidate explanatory variables, an enormous improvement over the 2^C evaluations required for all subsets regression. Even so, this number can be large enough to render stepwise methods infeasible if we attempt to apply them directly to large datasets with many covariates. For this reason, preliminary variable selection procedures like those described in Sec. 15.4 are extremely important when dealing with large datasets with many covariates whose role and importance in the problem at hand is not well understood.

As noted above, stepwise regression can be implemented in several different ways, and the following discussions consider only *forward selection*, probably the simplest of these approaches and one that often gives useful exploratory data models; for discussions of alternatives, refer to the books by Draper and Smith [94] or Jobson [159]. Given a set of C candidate explanatory variables, forward selection first fits the univariate models defined by each of these explanatory variables alone, ranking them in order of goodness-of-fit using measures like those discussed in Sec. 15.1.1. The explanatory variable that results in the best model is then selected for inclusion, provided the improvement in fit for this one-variable model is enough relative to an initial null model (e.g., an intercept-only model) to be deemed significant. Otherwise, the procedure stops: no explanatory model is built, as none of the candidate variables is deemed to have sufficient explanatory power (this situation is rare, but it can happen). Given a first selected explanatory variable, a second set of $C - 1$ two-variable models is then built, each one including the selected variable and one of the remaining variables. These two-variable models are then ranked by goodness-of-fit, and again, the best model is selected for evaluation: if the improvement in goodness-of-fit that results from including it is statistically significant, this second variable is retained and the procedure continues to evaluate three-variable models; otherwise, the process stops and the best-fit two-variable model is accepted as the final result.

The foward stepwise procedure continues until one of two terminating conditions is reached. Typically, termination occurs after some number p of explanatory variables have been selected for inclusion in the model and none of the possible $p + 1$ variable models exhibits a significantly better goodness-of-fit than the p-variable model does. In this case, the p-variable model is selected as the best overall model. Alternatively, it may happen that all of the C explanatory variables yield significant improvements in goodness-of-fit when they are included in the model. In this case, the C variable model is selected as the best overall model. Note that in this last case, the total number of models built and evaluated is $C(C - 1)/2$ (Exercise 1), giving an upper bound on the computational complexity of the forward selection procedure.

15.9.3   Forward selection for the TSF model

To illustrate the practical mechanics of forward selection, consider the problem of building a linear regression model to predict triceps skinfold thickness (TSF) from the six candidate explanatory variables considered in Sec. 15.3.1. It was shown in that discussion that the best univariate predictor was body mass index (BMI), which gave an adjusted R^2 value of 0.4192. The forward selection procedure therefore begins by selecting this variable and first building and comparing the five two-variable models that include BMI together with one of the other explanatory variables. Table 15.6 summarizes the resulting models, fit via ordinary least squares, giving the explanatory variables included, the adjusted R^2 value, the significance of each included variable p_F, based on an F-test comparing the BMI-only model with the two-variable model, the coefficient $\hat{\beta}$ for BMI estimated for each model, and its associated standard error $\hat{\sigma}$. Note that the $\hat{\beta}$ and $\hat{\sigma}$ values are included to provide a reasonableness check on the effects of including the second variable: we expect to see some change in these values when the second variable is added, but very large changes require our attention. In particular, very large changes in $\hat{\beta}$ raise the possibility of an interaction (i.e., including the product of both variables as an additional model term), and very large increases in $\hat{\sigma}$ suggest the possibility of collinearity. The fact that we see neither of these effects here is reassuring. Comparing the adjusted R^2 values suggests that the best variable for inclusion with BMI is patient age (AGE).

The next stage in the forward selection process is to retain the variables BMI and AGE and consider the effects of adding each of the four remaining variables. Carrying out this process yields adjusted R^2 values between 0.4323 and 0.4337, corresponding to F-ratio p-values between 0.229094 and 0.732522. Thus, none of these variables yield significant improvements when included in the model, leading us to adopt the model based on BMI and AGE as best.

15.9.4   An alternative TSF model

Many traditional modeling efforts would stop at this point, using the OLS model based on BMI and AGE to predict TSF. It was noted in Sec. 15.3.2, however, that this dataset appears to contain some outliers: in comparing the univariate

Table 15.6: Summary of two-variable models predicting TSF, listing the explanatory variables, the adjusted R^2 value, the significance of each variable (p_F), the estimated BMI model parameter ($\hat{\beta}$), and its standard error ($\hat{\sigma}$)

No.	Covariates	R^2	p_F	$\hat{\beta}$	$\hat{\sigma}$
0	BMI only	0.4192	—	0.9902	0.0506
1	BMI + NPG	0.4272	0.006725	0.9890	0.0503
2	BMI + GLC	0.4239	0.037685	0.9633	0.0521
3	BMI + DIA	0.4200	0.390459	0.9761	0.0532
4	BMI + AGE	0.4322	0.000538	0.9773	0.0502
5	BMI + DIAG	0.4231	0.057030	0.9598	0.0530

OLS and LTS models, the LTS procedure returned zero weights for a few observations in each case. It is therefore instructive to examine the effects of dropping these observations on the resulting OLS model. Table 15.7 compares the OLS model coefficients and their significances based on t-statistics for the full models containing an intercept term and all six explanatory variables, both for the full dataset and for the dataset with the LTS outliers removed. The results obtained from the original dataset, described by $\hat{\beta}^-$ and $\mathcal{P}^-$, are in reasonable agreement with those obtained in Sec. 15.9.3 using forward selection: the dominant contribution clearly arises from BMI, while a marginal improvement may be obtained by also including AGE. In contrast, removing the nine LTS-detected outliers from this dataset causes AGE to go from marginally significant to completely nonsignificant, while at the same time causing the variable NPG (number of times pregnant) to go from complete nonsignificance to clear significance at the traditional 5% level. Further, dropping the nonsignificant terms from this post-LTS model and refitting yields the following OLS model:

$$\widehat{\text{TSF}} = -4.912 + 0.333 \, \text{NPG} + 0.995 \, \text{BMI}, \qquad (15.52)$$

where the NPG term has an associated p-value of 0.0004. The adjusted R^2 value for this model is 0.4910, $\sim$13% larger than that for the model obtained from the full dataset based on BMI and AGE.

Finally, while the results are not presented in detail, the alternative of predicting the square root of TSF instead of the raw TSF value was also examined. Although the model coefficients were significantly different, applying the LTS regression procedure for outlier detection, removing the anomalous data points and fitting an OLS model gave highly analogous results to the untransformed case. Specifically, the model involved the same two explanatory variables, BMI and NPG, and the adjusted R^2 value was 0.4969, extremely similar to the value of 0.4910 obtained in the untransformed case.

Table 15.7: Comparison of OLS model parameter estimates and their associated
p-values, both before ($\hat{\beta}^-$ and $\mathcal{P}^-$) and after ($\hat{\beta}^+$ and $\mathcal{P}^+$) LTS-defined outliers
have been removed from the dataset

No.	Term	$\hat{\beta}^-$	$\mathcal{P}^-(> |t|)$	$\hat{\beta}^+$	$\mathcal{P}^+(> |t|)$
0	Intercept	-6.390	0.0144	-6.299	0.0079
1	NPG	0.093	0.5000	0.311	0.0130
2	GLC	0.013	0.3125	0.008	0.5294
3	DIA	-0.013	0.6821	0.025	0.3777
4	BMI	0.966	0.0000	0.964	0.0000
5	AGE	0.084	0.0609	-0.012	0.7711
6	DIAG	0.292	0.7445	0.515	0.5197

The key point of this example is to provide a further illustration of the
potential influence of outliers: in addition to changing parameter estimates—
slightly in some cases, and fairly dramatically in others—outliers can also change
the variables selected for inclusion in a multivariable regression model when
using standard variable selection criteria.

15.9.5   Forward selection for the bitter pit model

The basic forward selection strategy can be applied to other data models beyond
standard linear regression models. As a specific example, consider the problem
of predicting the probability of bitter pit in apples via logistic regression, using
the seven explanatory variables defined in Sec. 15.2.3. The first step in this forward selection procedure is to characterize the univariate models summarized in
Table 15.8, which lists each explanatory variable, its associated model param-
eter $\hat{\beta}$ and the parameter standard error $\hat{\sigma}$, the change in deviance relative to
an intercept-only reference model, and the χ^2 probability computed from this
deviance change. An important difference between this example and the others
considered in this chapter is the inclusion of the categorical variable Trt, representing one of four supplemental nitrogen treatments, A, B, C, or D, represented
using three binary indicator variables as discussed in Chapter 13. Thus, while
all other variables have one associated degree of freedom, Trt has three; also,
while each of the three components of Trt has its own parameter estimate and
standard error, the change in deviance and its p-value are associated with the
overall variable Trt, not its individual components.

As noted in Chapter 13, Hosmer and Lemeshow recommend using a relaxed
threshold of 0.25 for initial variable selection, rather than standard significance
levels like 0.05 or 0.01 [148, p. 95]. Under this recommendation, the following

Table 15.8: Summary of the univariate logistic regression models predicting probability of bitter pit from each of seven candidate explanatory variables, including the variable included in the model, the model parameter estimate ($\hat{\beta}$), its associated standard error ($\hat{\sigma}$), the corresponding change in deviance relative to an intercept-only model, and the associated p-value

No.	Covariate	$\hat{\beta}$	$\hat{\sigma}$	δ_{Dev}	$\mathcal{P}_{\chi^2}$
1	Ca	-0.0321	0.015	7.59	0.005857
2	Mg	0.0060	0.010	0.40	0.526765
3	K	0.0002	0.000	0.55	0.454634
4	P	0.0049	0.003	3.75	0.052731
5	PN	0.0024	0.001	5.33	0.020981
6	TN	0.0004	0.000	2.37	0.123779
7	Trt: A	1.2463	1.136	8.21	0.041706
	Trt: B	0.2025	0.485	—	—
	Trt: C	0.6010	0.258	—	—

five variables are considered for model building: Ca, P, PN, TN, and Trt. Since the most significant of these variables is Ca, we select it first, and the forward selection procedure then considers the two-variable models that incorporate this covariate along with the other four remaining candidates. These two-variable models are summarized in Table 15.9, which lists the covariates included in each model, the estimated model parameters $\hat{\beta}$ and their standard errors $\hat{\sigma}$, the change in deviance relative to the one-variable (Ca only) model, and the corresponding χ^2 probability estimate. Since none of these p-values is significant at the standard 5% significance level (indeed, none is even significant at the relaxed level used for univariate selection), the forward selection procedure stops at this stage. The final model is based on the calcium concentration alone, with the negative coefficient suggesting that bitter pit becomes more likely as calcium concentration decreases. This conclusion is consistent with the view presented by the North Carolina Department of Agriculture: "bitter pit can be caused by low soil moisture, low leaf calcium, or an imbalance between calcium and potassium within the leaves" [14].

15.10   Exercises

1. For the OLS linear regression model, the prediction vector $\hat{\mathbf{y}}$ is given by $\hat{\mathbf{y}} = \mathbf{H}\mathbf{y}$ where $\mathbf{H}$ is the hat matrix introduced in Sec. 15.3.1. Prove that

Table 15.9: Summary of the one-variable reference model (0) and the four two-variable logistic regression models predicting the probability of bitter pit from calcium concentration (Ca) and each of four other covariates. Values listed in the table include the logistic model parameters estimated for each variable ($\hat{\beta}$), their standard errors ($\hat{\sigma}$), the change in deviance relative to the one-variable model (δ_{Dev}), and the associated χ^2 probability

Model	Covariate	$\hat{\beta}$	$\hat{\sigma}$	δ_{Dev}	$\mathcal{P}_{\chi^2}$
0	Ca	-0.0321	0.015	—	—
1	Ca	-0.0256	0.017	—	—
	PN	0.0010	0.001	0.63	0.426030
2	Ca	-0.0195	0.016	—	—
	Trt: A	0.7973	1.174	2.56	0.464382
	Trt: B	0.1216	0.492	—	—
	Trt: C	0.4307	0.288	—	—
3	Ca	-0.0276	0.016	—	—
	P	0.0019	0.003	0.44	0.507839
4	Ca	-0.0299	0.015	—	—
	TN	0.0001	0.000	0.22	0.637431

the mean of the predictions is equal to the data mean, i.e.:

$$\frac{1}{N}\sum_{k=1}^{N}\hat{y}_k = \frac{1}{N}\sum_{k=1}^{N}y_k.$$

2. Prove that if the variance of the predictions $\{\hat{y}_k\}$ is no larger than the variance of the data sequence $\{y_k\}$, the maximum achievable R^2 value defined in Eq. (15.1) is 1.

3. Prove that the OLS parameter estimate vector $\hat{\theta}$ is given by:

$$\hat{\theta} = \mathbf{V}\mathbf{D}^{-1}\mathbf{U}^T\mathbf{y},$$

where $\mathbf{U}$, $\mathbf{V}$, and $\mathbf{D}$ are the matrices defined by the singular value decomposition of the data matrix $\mathbf{X}$ given in Eq. (15.21).

4. Show that the matrix $\mathbf{C} = (\mathbf{X}^T\mathbf{X})^{-1}$ can be written as:

$$\mathbf{C} = \sigma_e^2\mathbf{V}\mathbf{D}^{-2}\mathbf{V}^T,$$

where $\mathbf{D}$ and $\mathbf{V}$ are the matrices defined by the singular value decomposition of the data matrix $\mathbf{X}$ given in Eq. (15.21) and σ_e^2 is the variance of the model error term e_k.

5. It was shown that the condition number for the modified matrix $\tilde{\mathbf{C}} = (k\mathbf{I} + \mathbf{X}^T\mathbf{X})^{-1}$ appearing in the ridge regression problem formulation is given by:

$$\tilde{\kappa} = \frac{k + \lambda_+}{k + \lambda_-},$$

where k is the ridge regression tuning parameter. Show that $\tilde{\kappa}$ decreases with increasing k.

6. Prove that an $N \times N$ idempotent matrix with rank r has r eigenvalues equal to 1 and $N - r$ eigenvalues equal to 0.

7. Show that the change in the OLS parameter vector $\hat{\theta}$ on deletion of the ith observation from the dataset is given by:

$$\hat{\theta} - \hat{\theta}_{(i)} = \frac{(\mathbf{X}^T\mathbf{X})^{-1}\mathbf{z}_i r_i}{1 - H_{ii}},$$

where $\mathbf{z}_i$ is the deleted row of the data matrix $\mathbf{X}$, r_i is the residual $r_i = y_i - \hat{y}_i$ associated with the ith response in the full data model, and H_{ii} is the i, i diagonal element of the hat matrix $\mathbf{H}$.

8. Show that the maximum number of models built and evaluated in forward stepwise regression without backward elimination is $C(C - 1)/2$.

Chapter 16

Dealing with Missing Data

Sections and topics:

It has been noted repeatedly in previous chapters that missing data can be a serious problem in practice: observed data values that we would like to have for our analysis are simply not present in the datasets on which we must base the analysis. Chapter 2 emphasized that missing data can arise from a variety of different causes, and the key objective of this chapter is to present a brief, introductory survey of some of the approaches that have been proposed for dealing with missing data in practice. Important observations are, first, that many different methods have been proposed; second, that these different methods can perform very differently in practice; and third, that the relative performance of these methods generally depends on both the *fraction* of missing data and its *nature:* do missing values represent random deletions, or is some systematic omission mechanism at work? This chapter considers all three of these points, illustrating them with simulation-based examples where exact answers are known. Finally, as an example of a real dataset with missing data, the Pima Indians diabetes dataset is reexamined.

Key points of this chapter:

1. Like gross measurement errors, missing data should not arise, but it does. A fairly representative example is the Pima Indians diabetes dataset, containing nine variables: four of these are complete, while the other five exhibit missing data fractions ranging from less than 1% to almost 50%.

2. An extremely important aspect of missing data is the *pattern of missingness:* are the missing values distributed randomly throughout the dataset, or do they exhibit a *systematic pattern?* Three formal models used to describe these patterns are *missing completely at random (MCAR)*, *missing at random (MAR)*, and *missing not at random (MNAR)*.

3. The MCAR and MAR missing data models are sometimes characterized as *ignorable missing data models*, and these cases are generally most amenable to the missing data treatment methods described in this chapter. The MNAR missing data model, also known as *nonignorable missing data*, is both more difficult to detect and more difficult to treat, requiring explicit (and generally untestable) assumptions about the relationship between the observed and missing portions of the dataset.

4. Several different techniques for handling missing data are described in this chapter, including some that are simple but sometimes (although not always) quite effective—complete case analysis is one example—and others that are often *much* more effective, but harder to implement and use (e.g., the *expectation maximization algorithm*).

5. Unfortunately, some simple and intuitively appealing missing data treatment methods (e.g., mean imputation) are often very ineffective, even in the face of relatively benign missing data types (e.g., MCAR missingness). This is particularly the case when repeated data values are generated that may cause analytical difficulties (e.g., when using rank-based methods). A specific example considered here is the use of mean imputation for a univariate MCAR dataset, leading to implosion of the MAD scale estimate (i.e., zero scale estimates).

6. In general, most methods yield similar results when the fraction of missing data is small, but these results can be dramatically different when the missing data fraction is large. Thus, it is probably reasonable to apply complete case analysis (i.e., omission of all incomplete records) if the fraction of omitted records is only a few percent, *unless there is reason to believe the missing records differ systematically from the nonmissing ones.* For larger fractions of missing data, method selection becomes more important, and the results presented here suggest that very simple approaches like mean imputation or hot deck imputation are not adequate: something like the hot deck/regression composite method described in Sec. 16.5.4 or the EM algorithm described in Sec. 16.7 should be considered instead.

16.1   The missing data problem

The problem of missing data was introduced in Chapter 2, where it was noted that there are at least two general strategies for dealing with it: omit incomplete records, or replace the missing values with suitable estimates before proceeding with the analysis. This chapter considers both of these strategies in more detail, along with a third: the *expectation maximization (EM)* algorithm, an iterative approach that alternates between estimating missing values and analyzing the results; a brief overview of the basic ideas behind all three approaches is given in Sec. 16.1.1. One of the key factors in deciding which approach to adopt is the nature of the missing data: can it be characterized as simple random deletions, or is there reason to believe the missing data values are absent due to some *systematic* omission mechanism? This question is considered in Sec. 16.1.2, which introduces three models of missingness: *missing completely at random (MCAR)*, *missing at random (MAR)*, and *missing not at random (MNAR)*.

16.1.1   General missing data strategies

As noted, there are a number of different ways of handling missing data, differing significantly in their characteristics: ease of use, appropriateness in different missing data situations, and accuracy when they are applicable. The general approaches considered here are the following:

1. case deletion, in which missing observations are dropped from the analysis;

2. simple imputation strategies, where missing observations are replaced by estimated values;

3. multiple imputation strategies, where missing observations are replaced by multiple estimated values; multiple analysis results are then generated and combined;

4. the *expectation maximization (EM)* algorithm, an iterative strategy that alternates between imputation and analysis of the imputed results.

Each of these ideas are discussed in detail in subsequent sections of this chapter, but it is useful to briefly introduce them here to give an idea of our range of options in dealing with missing data.

Case deletion corresponds to the simple strategy of omitting missing records, and this can be done in at least two different ways. The simplest is *complete case analysis*, where all incomplete records (i.e., any record with missing values for one or more fields) are eliminated and a standard analysis is applied to the complete data subset that remains. This approach has the advantage of both operational and conceptual simplicity, but it can lead to a substantial reduction of the size of the dataset available for analysis, particularly if the nonmissing portions of several different fields exhibit little overlap. An alternative that is sometimes applicable is *available case analysis*, where records are omitted only as necessary to have a complete data subset available for each portion of the analysis. A specific example considered in Sec. 16.4.1 is the computation of covariance matrices, where the amount of data available for estimating each element of the matrix depends on the specific data fields involved and can be quite different for different matrix elements. The advantage of available case analysis is that it can sometimes make use of a substantially larger portion of the overall dataset, but a significant disadvantage is that it can lead to inconsistent results (e.g., covariance matrices that are not positive-definite). As subsequent examples illustrate, both complete case analysis and available case analysis can lead to significant biases in analysis results if the missing observations are *systematically missing* rather than randomly missing. Finally, a closely related approach that is also sometimes applicable is the omission of one or more variables that exhibit a high fraction of missing data. Obviously, this approach is not applicable in cases where key variables (e.g., the response variable of interest) exhibit high missing data fractions, and it has the disadvantage that we learn nothing about the influence of excluded variables, but there are circumstances where it does represent a reasonable alternative. One such example is the Pima Indians diabetes dataset, discussed in Sec. 16.8.

Simple imputation strategies attempt to replace the missing data values with reasonable alternatives. In principle, the best such strategy would be to simply collect the missing observations from the original data source, but this option is rarely feasible. More practical alternatives include estimating the missing values either from other data sources or from nonmissing values in the same dataset. Specific implementations of this idea can involve either using a fixed replacement value (e.g., mean imputation), random samples drawn from the complete dataset (as in hot deck imputation), more complex approaches like regression-based imputation, or hybrid combinations of these ideas.

A disadvantage of the simple imputation strategies just described is that they tend to reduce the natural variability of the imputed dataset, relative to that of the (unobserved) complete dataset. An alternative strategy that overcomes this limitation is *multiple imputation*, where several different imputations are made for each missing data value, typically by random sampling, and analysis results are repeated for each imputed dataset. Relative to the simple imputation strategies described above, this approach has the advantages of both restoring some of the variability lost with single imputations and providing a basis for assessing the influence of the imputations on the final analysis results. The primary disadvantage of multiple imputation strategies is that they require multiple runs of the complete analysis procedure, and this may be a serious difficulty for either very complex analyses or very large datasets where even simple analyses require long computation times.

Finally, the fourth missing data handling strategy considered here is the expectation maximization (EM) algorithm. As noted, this algorithm is an iterative procedure that alternates between imputation and analysis of the imputed dataset. Different forms of the EM algorithm have been applied in a wide variety of applications, and the discussion given here represents only a brief introduction to the basic ideas; for a more detailed introduction, refer to the book by McLachlan and Krishnan [207].

In general, the choice of which of the above missing data methods to use depends strongly on a number of factors. One key factor is the fraction of missing data: in cases where this fraction is sufficiently small, similar results should be obtained by all methods, so it is logical to choose the one that is simplest to implement, easiest to explain, most intuitive, most convenient (e.g., in terms of available software), or best accepted by researchers in the field. Conversely, there are cases where we are forced to deal with missing data fractions of 50% or more, and in these cases the simpler strategies are often inadequate. A second key factor is the type of analysis we are attempting to perform: in building logistic regression models to characterize rare responses, missing values for explanatory variables in the small set of records exhibiting positive responses are probably not best handled by omission strategies that further reduce the number of these positive responses. Finally, a third key factor in deciding between the different missing data treatment strategies listed above is the nature of the missingness itself, the subject of the following discussion.

16.1.2   Missingness: MCAR, MAR, and MNAR

To describe missing data explicitly, consider a dataset $\mathcal{D}$ that can be viewed as a rectangular table with N rows (records) and M columns (fields per record). Associated with this dataset, assume we have a binary metadata matrix $\mathcal{M}$ with one entry for each entry in $\mathcal{D}$, such that $\mathcal{M}_{ij} = 1$ if the element $\mathcal{D}_{ij}$ is missing, and $\mathcal{M}_{ij} = 0$ if this data element is present. In 1976, Rubin introduced an approach to describing missing data that assumes the metadata $\mathcal{M}$ is random [253]. Imposing different assumptions on the distribution of $\mathcal{M}$ then leads to different missing data models.

The simplest of these models assumes that the data values are *missing completely at random (MCAR)*, corresponding to the following distributional assumption:

$$P(\mathcal{M}_{ij}|\mathcal{D}) = P(\mathcal{M}_{ij}), \qquad (16.1)$$

for $i = 1, 2, \ldots, N$ and $j = 1, 2, \ldots, M$. Here, $P(\mathcal{M}_{ij}|\mathcal{D})$ represents the conditional probability that the missingness metadata value $\mathcal{M}_{ij}$ is 1, given the complete dataset $\mathcal{D}$, and this criterion means that the probability that any data value is missing is independent of both the observed and the unobserved portions of the dataset. In the most general case, the NM probabilities defined by Eq. (16.1) can all be distinct numbers lying between 0 and 1, but in the simplest case, all of these probabilities are equal. This choice represents a particularly useful missing data model, implying that all data entries have an equal probability p of being absent.

Like the popular Gaussian distributional assumption, the MCAR missing data assumption is an extremely strong one, but popular because it sometimes greatly simplifies subsequent analysis. A weaker and thus frequently more realistic assumption is *missing at random (MAR)*, which assumes only that the probability that $\mathcal{M}_{ij} = 1$ does not depend on the *unobserved* data values in $\mathcal{D}$.

That is, the MAR model replaces Eq. (16.1) with:

$$\mathcal{P}(\mathcal{M}_{ij}|\mathcal{D}) = \mathcal{P}(\mathcal{M}_{ij}|\mathcal{D}_{\text{obs}}),$$ (16.2)

where $\mathcal{P}(\mathcal{M}_{ij}|\mathcal{D}_{\text{obs}})$ represents the conditional probability that $\mathcal{M}_{ij} = 1$ given only the *observed* data values $\mathcal{D}_{ij}$. This missing data model is reasonably general, allowing for the possibility of certain types of systematic data omission. For example, the evident differences in age and other characteristics between patients with missing and nonmissing triceps skinfold thickness (TSF) values seen in the Pima Indians diabetes dataset discussed in Chapter 13 are admissible under the MAR model but not under the MCAR model.

Finally, allowing the probability that a data observation is missing to depend on its own value—i.e., the case where Eq. (16.2) *does not* hold—leads to the *missing not at random (MNAR)* model. An alternative terminology for the MNAR model is the *nonignorable missing data model*, since it allows for the possibility of "selective data editing" by some unspecified mechanism, a possibility that can lead to significant biases in analysis results. The MAR model (including the MCAR model as a special case) is often called the *ignorable missing data model* because, if handled appropriately, these biases can be avoided under the MAR model.

These different missing data assumptions and their consequences will be examined further throughout this chapter for the different missing data treatment strategies considered here. The basis for these examinations will be the four simulation-based datasets described in Sec. 16.3, for which the details of the missing data mechanism are known. In practice, of course, these details are rarely known, forcing us to make assumptions and proceed. Given an incomplete dataset, we can attempt to distinguish between the MCAR and more general MAR mechanisms by using techniques like those described in Chapter 13 for the Pima Indians TSF data sequence. In contrast, the issue of nonignorable missing data can only be addressed by invoking, at least tentatively, assumptions that are usually unverifiable, since they postulate a dependence of the probability that data values are not observed on those unobserved values themselves.

16.2 The univariate case

As examples discussed previously have illustrated, missing data is often a multivariate phenomenon. This is particularly true in cases where the fraction of complete records is small, since an important source of incomplete records is the construction of composite datasets from multiple files that do not characterize exactly the same data sources (e.g., medical records compiled from sources where some but not all of the patients appear in all data sources). Conversely, the univariate case is worth considering briefly, since it provides the basis for a number of simple, clear illustrations of the range of treatment strategies available and the consequences of different types of missing data. Also, one of the reasons the univariate case is simpler than the multivariate case is that the range of options for treating missing data is significantly reduced in the univariate case. The following paragraphs illustrate both of these points and provide a useful framework for the discussions of the more general multivariate problems discussed throughout the rest of this chapter.

16.2.1   Univariate issues and strategies

While there are important practical distinctions between the MCAR and MAR missing data models in the multivariate case, these distinctions are less obviously useful in the univariate case. That is, given a univariate data sequence $\{x_k\}$ and its associated missingness indicator sequence $\{w_k\}$, the general situation is characterized by the joint distribution of these two variables, $p(x, w)$. This observation leads us to consider the following two cases:

1. MCAR: $\mathcal{P}\{w_k = 1\}$ is independent of x_k;

2. MNAR: $\mathcal{P}\{w_k = 1\}$ depends on x_k.

In the following discussions, the MCAR model assumes that $\{w_k\}$ is a sequence of Bernoulli trials characterized by a specified probability p. Specification of MNAR models is somewhat more involved, since we must say *how* the distribution of w_k depends on x_k, motivating the consideration of three different MNAR examples here.

Besides simplifying the range of missingness models that need to be considered, restricting attention to the univariate case also reduces the range of available missing data treatment strategies, as noted previously. For example, there is no difference betwee complete case analysis and available case analysis in univariate problems. Similarly, regression-based imputation strategies that predict missing values of one variable from nonmissing values of other variables are not applicable in the univariate case. Consequently, the following discussion considers only two simple imputation strategies: *mean imputation*, where missing x_k values are replaced by the mean of all observed values, and *hot-deck imputation*, where missing x_k values are imputed by sampling with replacement from the set of observed values.

Because the multivariate setting supports a wider variety of multiple imputation strategies, these approaches are deferred until Sec. 16.6, where they are considered for the multivariate simulation examples introduced in Sec. 16.3. Similarly, while the expectation maximization algorithm can be illustrated for simple univariate examples, the approach is much more useful in the multivariate setting. For this reason, two univariate illustrations of the EM algorithm are presented, but these are deferred until Sec. 16.7, where they are included as part of the general discussion of the method.

In view of these considerations, the rest of this section compares three univariate missing data treatment strategies: case deletion, mean imputation, and hot-deck imputation. Note that under the MCAR missingness model, case deletion corresponds to random subsampling: if M of N observations are missing, the MCAR model implies that the observed dataset represents a sample of size

$N - M$ drawn from the complete dataset without replacement. Similarly, the hot-deck imputation strategy described above corresponds to a data augmentation where the M missing observations are estimated by M samples drawn with replacement from the $N - M$ nonmissing observations. Two consequences follow directly from this observation. First, the hot-deck imputation strategy necessarily results in some level of data duplication, which may be significant in cases where discrete distributions introduce analytical complications (e.g., the special treatment required for tied observations in using rank-based methods). Conversely, note that mean imputation generally results in much larger levels of data duplication, since all of the missing observations are imputed using the same value. The second consequence of the equivalence of hot-deck imputation to sampling with replacement is that under MNAR missing data models where the missing observations differ systematically from the available data values, hot-deck imputation can be expected to perform poorly. In fact, all three of the strategies considered here tend to perform poorly for MNAR problems, a point discussed further at the ends of secs. 16.2.3 and 16.2.4.

16.2.2 Four simulation examples

To illustrate the differences between the MCAR and MNAR missingness models in the univariate case, the following discussions compare four different missing data examples. A relatively large missing data fraction of 50% is assumed to make the effects of missing data pronounced, and all of the samples considered are of length $N = 200$. The first example corresponds to the MCAR missingness model with the constant probability $p = 0.5$ that any observation is missing. The other three examples illustrate three different MNAR missingness models, where the probability that x_k is missing is 1 if it satisfies certain simple selection criteria and 0 otherwise. More specifically, MNAR Example 1 omits the largest 50% of the original data observations, MNAR Example 2 omits the 50% most extreme observations (i.e., the 25% largest and the 25% smallest), and MNAR Example 3 omits the central 50% of the observations (i.e., retains only those observations that MNAR Example 2 omits). Representative examples of data sequences generated under each of these models are shown in Fig. 16.1. In each case, missing data observations are represented as open circles while nonmissing observations are represented as solid circles.

16.2.3 Location estimates

The following paragraphs compare the performance of the three missing data handling strategies described in Sec. 16.2.1 for the four univariate simulation examples described in Sec. 16.2.2. These comparisons are presented in the form of boxplot summaries, each one characterizing $r = 1,000$ statistically independent datasets, each generated as described in Sec. 16.2.2. As in the examples presented there, each individual data sample is of size $N = 200$ and the missing data fraction is 50% ($p = 0.5$). To illustrate the consequences of the different missing data mechanisms, boxplot sumaries of the corresponding results

Figure 16.1: The four univariate missing data examples described in the text. Each plot shows a sample of size $N = 200$, with open circles representing missing data observations and solid circles representing nonmissing observations

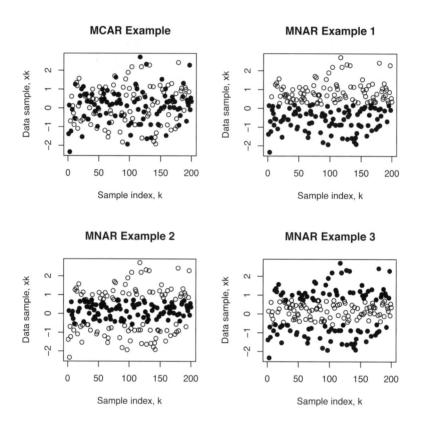

for the complete datasets are also presented. More specifically, the displays presented here summarize the range of computed mean and median values for the r datasets in five cases: the complete dataset (labeled "0"), the MCAR dataset (labeled "C"), and the three MNAR datasets (labeled "N1," "N2," and "N3"). Since they are both unbiased estimates of the population mean, results are shown in each plot for the sample mean and the sample median together.

Fig. 16.2 summarizes these results for the univariate case deletion strategy. The five boxplots to the left of the vertical line summarize the range of the results obtained for the mean for the five datasets, while those to the right of the vertical line summarize the corresponding median results. The horizontal line represents the population mean/median value of zero. Comparing the MCAR results ("C") for both the mean and the median with those for the complete dataset ("0") shows the variability of the MCAR results to be larger than the complete results, consistent with the subsampling interpretation for the MCAR case. Specifically,

Figure 16.2: Comparison of mean and median estimates for the complete dataset (labeled "0") and the four incomplete datasets under the case deletion strategy: MCAR ("C") and MNAR (denoted "N1" through "N3")

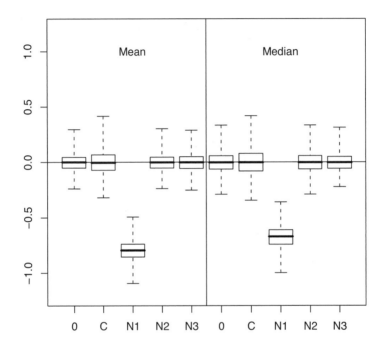

the variability of the MCAR results should be approximately 40% larger than the complete data results, consistent with the differences seen in the figure. Because the first MNAR mechanism considered here systematically eliminates the largest 50% of the data values, case deletion applied to this dataset ("N1") introduces a significant negative bias in both the mean and the median, an effect seen clearly in these boxplots. Conversely, the second and third MNAR missing data mechanisms (results "N2" and "N3") preserve distributional symmetry and these results appear to be—at least for the mean and median—both unbiased and less variable than the MCAR results, comparable with the complete data results. As the corresponding results presented in Sec. 16.2.4 demonstrate, these conclusions do not extend to the scale estimation problem.

The corresponding results obtained under the mean imputation strategy are shown in Fig. 16.3, in the same format as Fig. 16.2. Comparing these two figures, it is clear that both mean imputation and case deletion give quite similar

Figure 16.3: Comparison of mean and median estimates for the five datasets compared in Fig. 16.2 under the mean imputation strategy

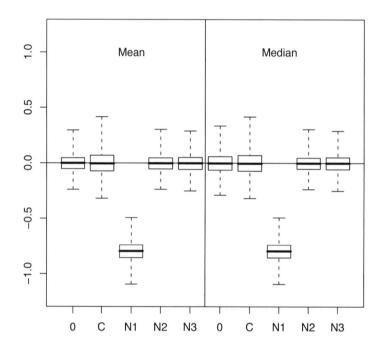

results for these location estimates. On careful comparision, two differences are apparent. First, the mean and median results are in better agreement for the first MNAR dataset ("N1") under mean imputation than under case deletion. Second, the variability of the median results is more comparable to that of the mean results under mean imputation than under case deletion. Both of these observations are a direct consequence of the fact that the mean values are being used to replace the missing 50% of the data, forcing the median results to almost equal the mean results.

Fig. 16.4 summarizes the corresponding results obtained under the univariate hot-deck imputation strategy, again in the same format as Fig. 16.2. Recall that this strategy replaces missing data values with random samples drawn from the available observations with replacement and that, as a consequence, there is significant duplication of individual data values in the imputed dataset. Comparing Fig. 16.4 with both of the previous results, several differences are apparent. The most pronounced of these is the extremely large variability seen

Figure 16.4: Comparison of mean and median estimates for the five datasets compared in Fig. 16.2 under the hot-deck imputation strategy

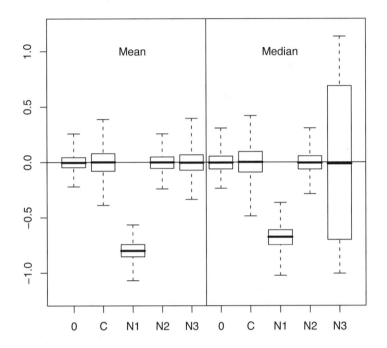

for the median estimates in the third MNAR example ("N3"). Since this dataset excludes the central 50% of the data values, leaving roughly the same number of observations from the upper and lower tails of the distribution, the median exhibits the "swing vote" phenomenon discussed in Chapter 7. That is, the median value will be representative of the upper tail values if more than 50% of the remaining data samples come from the upper tail, and it will be representative of the lower tail values if more than 50% of the remaining data samples come from the lower tail. Under the hot-deck imputation strategy, values for the missing samples are being drawn with replacement from both tails, and the extreme variability of the imputed median is a consequence of the fact that, sometimes a little more than 50% of these imputations are drawn from the upper tail and sometimes a little more than 50% are drawn from the lower tail. While the differences are much less dramatic, note that the hot-deck imputation results for the mean for the "N3" results also exhibit larger variability than that seen for either of the other two missing data strategies considered here.

Overall, the results presented here demonstrate three extremely important points. First, depending on the specifics, the MNAR missing data mechanism can introduce very large biases, as seen in the "N1" results presented here, for all three of the missing data treatment strategies considered. The second key point is that the results obtained by different missing data treatment strategies can be very different, as illustrated by the difference between the median results obtained for the third MNAR dataset ("N3") under hot-deck imputation and those obtained under the other two missing data treatment strategies. Finally, the third key point is that the effectiveness of the missing data treatment strategies considered here also depends on the data characterization considered. Here, these characterizations were restricted to the sample mean and median and the differences were mostly minor, but they can be much more substantial, as the results for scale estimation presented in the following section demonstrate.

16.2.4   Scale estimates

The results presented in Sec. 16.2.3 compared the influence of the four specific missingness mechanisms described in Sec. 16.2.2 and the three different missing data treatment strategies described in Sec. 16.2.1 for the two most popular location estimators: the mean and the median. The following paragraphs present the analogous comparisons for the scale estimation problem, comparing the standard deviation and the MAD scale estimator. As in Sec. 16.2.3, the results are presented as side-by-side boxplot displays, summarizing the results obtained for $r = 1,000$ statistically independent datasets. The same five results are compared for each estimator and each missing data treatment strategy: the complete dataset ("0"), the MCAR dataset ("C"), and the three MNAR datasets ("N1", "N2", and "N3"). Since the underlying distribution for all of these datasets is the standard normal, the target value for all of these estimators is 1, and this value is indicated by a horizontal line in each plot.

Fig. 16.5 summarizes the scale estimation results obtained under the case deletion strategy, in the same format as Fig. 16.2. As in the location estimation problem, note that the results obtained for the MCAR datasets ("C") exhibit greater variability than the complete datasets ("0"), but are again unbiased. In contrast, while the "N2" and "N3" location results were also unbiased—a consequence of the fact that these two missing data mechanisms preserved distributional symmetry—this is no longer true for the scale estimates. In particular, note that both the standard deviation and the MAD scale estimate exhibit substantial biases for all three MNAR mechanisms considered here. Further, the magnitude of this bias can depend strongly on the specific scale estimate considered, a point illustrated clearly by the differences between the standard deviation and MAD scale estimator results for the "N3" dataset.

Fig. 16.6 presents the corresponding scale estimation summary for the results obtained under the mean imputation strategy. The most striking effect seen here is that *all* of the results shown here except the complete data reference cases exhibit nonnegligible biases. Interestingly, the smallest bias is seen for the standard deviations estimated from the "N3" MNAR datasets, but even

Figure 16.5: Comparison of standard deviation and MAD scale estimates for the complete dataset (labeled "0") and the four incomplete datasets under the case deletion strategy. The MCAR dataset is denoted "C," and the three MNAR datasets are denoted "N1" through "N3"

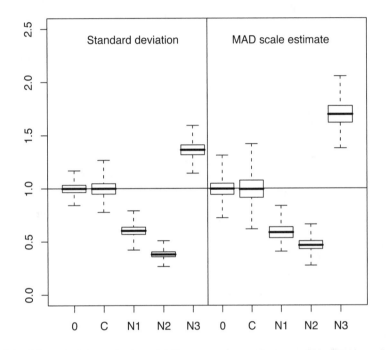

here, systematic underestimation of the true scale is clearly evident. This bias is most severe for the MAD scale estimator, where for all cases except the "N3" datasets, the scale estimate implodes, yielding a scale estimate of exactly or approximately zero. This result is a direct consequence of the large fraction (near 50%) of repeated values introduced by mean imputation, which, coupled with even a few sample values near the mean, is enough to force the imputed sample median to be exactly equal to the observed sample mean and the MAD scale estimate to zero, as discussed in Chapter 7.

Another extremely important point is the significant bias exhibited by both the standard deviation and the MAD scale estimator for the MCAR examples. While this bias is not catastrophic for the standard deviation as it is for the MAD scale estimator, it does illustrate that the MCAR missing data mechanism can induce estimator bias and not just increased variability. In fact,

Figure 16.6: Comparison of standard deviation and MAD scale estimates for the five datasets compared in Fig. 16.5, under the mean imputation strategy

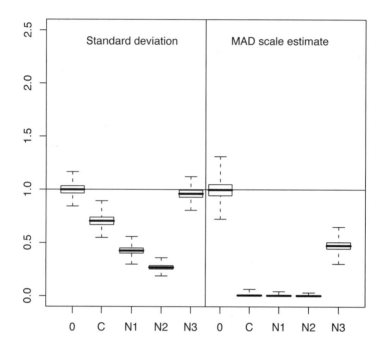

this consequence of mean imputation is well recognized: replacing missing data observations with a single imputed value reduces the overall variability of the dataset, consistent with the smaller standard deviation estimates seen here. What is less widely recognized is that even as "benign" a missing data mechanism as MCAR can induce significant biases for some estimators, depending on the missing data treatment strategy adopted. Finally, before leaving this example, it is worth noting that the unusually "good" behavior of mean imputation for the third MNAR dataset ("N3") is a direct consequence of the unusual missing data pattern in this dataset. Specifically, recall that the third MNAR example excludes the middle 50% of the observed data values. Since this omission preserves distributional symmetry, the mean remains an unbiased estimator and mean imputation here fills in these missing values with a fairly reasonable approximation. Thus, the scale estimate obtained by the standard deviation is almost unbiased for this case, while the MAD scale estimator bias is much less than that seen for the other three incomplete datasets.

Figure 16.7: Comparison of standard deviation and MAD scale estimates for the five datasets compared in Fig. 16.5, under the hot-deck imputation strategy

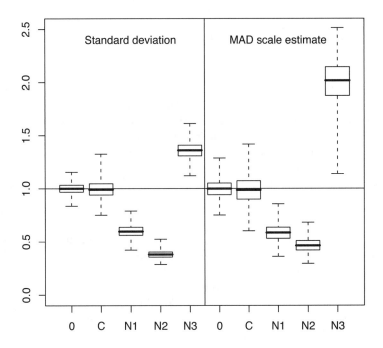

Fig. 16.7 summarizes the scale estimation results obtained under the hot-deck imputation strategy, where missing data values are imputed with samples drawn from the observed portion of the dataset with replacement. Comparing figs. 16.7 and 16.5, it appears that the hot-deck imputation results are comparable to those obtained under case deletion. The two main differences are, first, that the MAD scale estimator exhibits greater bias under hot-deck imputation, and second, that the hot-deck imputation results are somewhat more variable than the case deletion results. This variability is caused by the same factors that were responsible for the extreme variability seen in Fig. 16.4 for the median under hot-deck imputation. In particular, both the median and MAD scale estimator results illustrate the point made earlier that the significant number of repeated values generated by hot-deck imputation can cause rank-based procedures like these to perform poorly.

Overall, both the scale estimation results presented here and the location estimation results presented in Sec. 16.2.3 emphasize that the choice of an appropriate missing data treatment strategy depends on both the analysis objectives (i.e., the estimator considered) and the characteristics of the incomplete dataset. These general conclusions also hold for the more general multivariate problems considered throughout the rest of this chapter, complicated somewhat by both the wider range of practically important missing data mechanisms to consider and the greater variety of missing data treatment strategies available.

16.3 Four multivariate examples

To provide a common set of examples for evaluating the full range of missing data treatment strategies considered in this chapter, the following paragraphs describe four simulation-based bivariate datasets where both the overall characteristics of the data and the details of the missingness mechanism are known. Considering these examples allows us to say what characteristics we *should* obtain for each dataset and compare these expectations with the results we actually obtain for each of the missing data strategies described here.

In all cases, the complete dataset is generated by the nominal data model:

$$y_k = x_k + e_k, \tag{16.3}$$

where $\{x_k\}$ and $\{e_k\}$ are mutually independent, zero-mean, unit-variance i.i.d. Gaussian data sequences. Associated with each of these variables are binary missingness indicators: w_k takes the value 1 if x_k is missing and 0 otherwise, and z_k takes the value 1 if y_k is missing and 0 otherwise. The simulation datasets considered here all contain $N = 200$ records, differing only in the probabilities that data observations are missing and the ways these missingness indicators are defined. For each missing data mechanism, missing data probabilities $p = 0.05$, 0.20, 0.50 and 0.75 are considered, covering a representative range of missing data fractions seen in practice.

The simulated MCAR dataset is generated by randomly omitting $pN/2$ values of both the explanatory variable x_k and the response variable y_k, for the missing data fractions $p = 0.05$, 0.20, 0.50, and 0.75. These datasets are generated by first sampling $pN/2$ index values k from the set $\{1, 2, \ldots, N\}$ without replacement, setting $w_k = 1$ for these values of k, and setting $w_k = 0$ for all other k. The process is then repeated for z_k, drawing an additional $pN/2$ independent samples from this same index set without replacement, setting $z_k = 1$ for these k values and setting all other z_k values to zero.

Scatterplots of the MCAR datasets are shown in Fig. 16.8 for $p = 0.05$ (upper left), $p = 0.20$ (upper right), $p = 0.50$ (lower left), and $p = 0.75$ (lower right). Each plot shows the (x_k, y_k) pairs represented as points in the plane, drawn as open circles when the observation k is incomplete (i.e., $w_k = 1$ and/or $z_k = 1$) and as solid circles when the observation is complete (i.e., $w_k = 0$ and $z_k = 0$). The lack of any evident pattern of missing observations in these plots reflects the MCAR missing data mechanism used to generate the datasets. Also, note that the expected number of incomplete observations is smaller than pN, due to the existence of observations where both x_k and y_k are missing (Exercise 1).

Figure 16.8: Scatterplots of simulated bivariate MCAR datasets, with the probability of missing observations indicated at the top of each plot. Open circles represent incomplete observations (either x_k missing, y_k missing, or both), and solid circles represent complete observations (both x_k and y_k present)

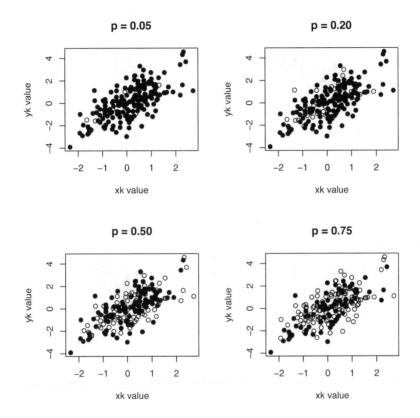

As emphasized in Sec. 16.1.2, the difference between the missing at random (MAR) and missing completely at random (MCAR) mechanisms is that the MAR missingness pattern can depend on the observed data values, while the MCAR missingness pattern cannot. The first of the two simulated MAR datasets considered here exhibits the same fractions of missing x_k and y_k values as the MCAR example, but the probability that y_k is missing is permitted to depend on x_k. Specifically, $pN/2$ observations of x_k are first omitted at random exactly as in the MCAR model (i.e., $w_k = 1$ with probability $p/2$, independent of both x_k and y_k). Then, $pN/2$ observations of y_k are omitted, corresponding to the *smallest nonmissing* x_k values (i.e., $z_k = 1$ for those values of k corresponding to the smallest $pN/2$ values of x_k for which $w_k = 0$).

Analogous to Fig. 16.8, scatterplots of the first MAR simulation dataset are shown in Fig. 16.9 for $p = 0.05$ (upper left), 0.20 (upper right), 0.50 (lower left),

Figure 16.9: Scatterplots of simulated bivariate MAR No. 1 datasets, in the same format as Fig. 16.8

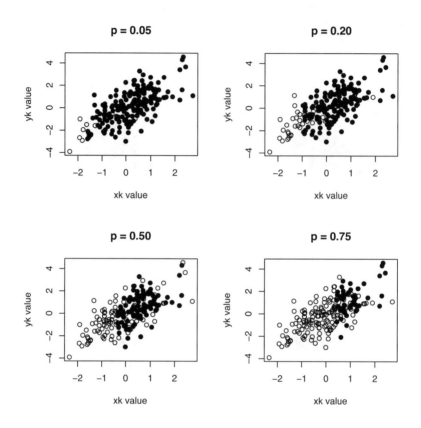

and 0.75 (lower right). In contrast to the MCAR examples presented before, however, a clear structure may be seen in the missing data patterns shown in Fig. 16.9, becoming increasingly evident with increasing p. In particular, note the increasing tendency for points in the lower left portion of the scatterplot to be missing (i.e., shown as open circles). Finally, note that the total number of incomplete records is exactly pN for this dataset, since $pN/2$ observations of the prediction variable x_k are omitted in the first step of the process that generates this dataset, and the $pN/2$ observations of the response variable y_k omitted in the second step of this process are defined by nonmissing x_k values.

The second MAR example shown in Fig. 16.10 is in some sense a dual of the one just described. That is, $pN/2$ observations are omitted from both the explanatory variable x_k and the response variable y_k, but here the y_k observations are omitted at random, while the x_k observations omitted are those corresponding to the $pN/2$ *largest nonmissing* response values y_k. More specifically, this dataset is generated by first selecting $pN/2$ indices k from the set $\{1, 2, \ldots, N\}$

Figure 16.10: Scatterplots of simulated bivariate MAR No. 2 datasets, in the same format as Fig. 16.8

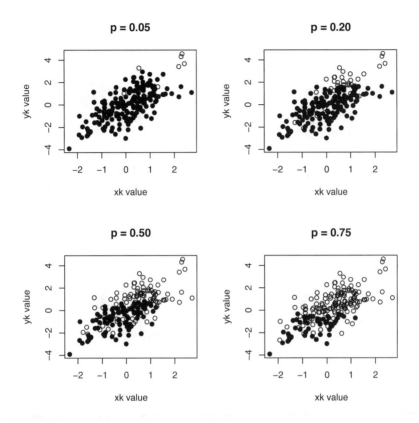

without replacement, setting $z_k = 1$ for these k values and $z_k = 0$ for all other k. Then, w_k is set to 1 for the $pN/2$ values of k for which y_k is largest, among those records with $z_k = 0$. Scatterplots for this case are shown in Fig. 16.10 for $p = 0.05$ (upper left), $p = 0.20$ (upper right), $p = 0.50$ (lower left), and $p = 0.75$ (lower right), and evidence of systematic missingness becomes clearer from these plots as p increases. As before, since records with missing x_k or y_k values are mutually exclusive, the total number of incomplete records is pN.

The fundamental characteristic of the MNAR model is that the probability of data observations being omitted depends on the unobserved portion of the dataset. The specific MNAR example considered here, shown in Fig. 16.11, again has equal numbers of missing observations of x_k and y_k, but now the selection of the missing values depends on the values themselves. Specifically, the MNAR dataset is generated by systematically omitting the smallest $pN/2$ values of x_k and the largest $pN/2$ values of y_k. That is, $w_k = 1$ for the smallest $pN/2$ values of x_k and $z_k = 1$ for the $pN/2$ largest y_k values.

Figure 16.11: Scatterplots of simulated bivariate MNAR datasets, in the same
format as Fig. 16.8

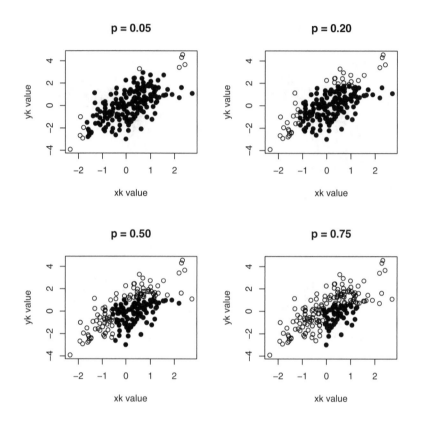

Fig. 16.11 shows scatterplots of these four datasets in the same format as
before: for $p = 0.05$ in the upper left plot, $p = 0.20$ in the upper right, $p = 0.50$
in the lower left, and $p = 0.75$ in the lower right. Here, the systematic pattern of
missing data observations is even more apparent than in the two MAR examples,
since both extreme values in the lower left and those in the upper right tend
to be omitted. Also, note that since x_k and y_k are fairly strongly correlated
in the original dataset, omitting small x_k values tends to cause records with
small y_k values to be incomplete, while omitting large y_k values tends to cause
records with large x_k values to be incomplete. Thus, records with both x_k
and y_k missing tend to be rare, so the actual fraction of incomplete records is
larger than in the MCAR dataset. Specifically, for $p = 0.75$, the MCAR dataset
has 123 incomplete records (61.5%), while the MNAR dataset has 143 (71.5%)
incomplete records.

16.4   Case deletion strategies

Without question, the simplest strategy for dealing with missing data is to omit incomplete data records and base our analysis on the complete portion of the dataset. Despite its apparent simplicity, however, this strategy can be implemented in a number of different ways, each with its own advantages and disadvantages. Further, these advantages and disadvantages can be strongly case dependent, as the following discussions emphasize.

16.4.1   Complete case versus available case analysis

As noted previously, the term *complete case analysis* refers to the practice of simply omitting *all* incomplete data records (i.e., all records in a dataset with one or more missing fields). This omission step is then followed by whatever analysis we wish to perform, applied to the remaining dataset of complete cases. In contrast, *available case analysis* omits only those records containing fields whose values are required for each stage of the analysis. As discussed in Sec. 16.2.1, these approaches are equivalent in the univariate case, but they are not in the multivariate case, where they can lead to significantly different analysis results, depending on circumstances.

To illustrate the distinction between these approaches, consider the problem of estimating the correlation coefficient ρ_{xy} between two data sequences, $\{x_k\}$ and $\{y_k\}$. As discussed in Chapter 10, the standard complete-data estimator for ρ_{xy} is given by:

$$\hat{\rho}_{xy} = \frac{\sum_{k=1}^{N} (x_k - \bar{x})(y_k - \bar{y})}{\left[\sum_{k=1}^{N} (x_k - \bar{x})^2 \sum_{k=1}^{N} (y_k - \bar{y})^2\right]^{1/2}}. \tag{16.4}$$

An important observation in the following discussion is that, since all of the sums appearing in Eq. (16.4) are taken over the same N observations k, exactly the same result would be obtained by replacing these sums with averages, dividing each one by N. In the face of missing data, the complete case estimator simply restricts all of the sums in Eq. (16.4) to those indices k for which neither x_k nor y_k is missing. That is, define K_x as the set of all k for which x_k is nonmissing, K_y as the set of all k for which y_k is nonmissing, and $K_{xy} = K_x \cap K_y$ as the set for which neither x_k nor y_k is missing. Complete case analysis restricts the three sums in Eq. (16.4) to $k \in K_{xy}$. Again, note that the result is exactly the same as that would be obtained if we replaced these sums with averages.

In contrast, the available case analysis replaces each of the three sums in Eq. (16.4) with the corresponding averages *taken over all records k for which each sum is defined*. Specifically, the available case correlation estimator is given by:

$$\tilde{\rho}_{xy} = \frac{(1/N_{xy}) \sum_{k \in K_{xy}} (x_k - \bar{x})(y_k - \bar{y})}{[(1/N_x) \sum_{k \in K_x} (x_k - \bar{x})^2 \cdot (1/N_y) \sum_{k \in K_y} (y_k - \bar{y})^2]^{1/2}}, \tag{16.5}$$

where N_x is the number of records in K_x, N_y is the number of records in K_y, and N_{xy} is the number of records in K_{xy}.

The primary motivation for available case analysis is that it generally uses larger subsets of the dataset in computing some of the quantities that go into a composite analysis result than complete case analysis does. Thus, we can expect that some of these component quantities may be more accurately estimated in available case analysis than they are in complete case analysis, but this does not always lead to a better result overall. As a specific example, the fact that $\hat{\rho}_{xy}$ is bounded between -1 and $+1$ follows from the Cauchy-Schwartz inequality applied directly to Eq. (16.4), a result that extends to complete case analysis. In contrast, the result obtained from available case analysis can violate these constraints [188, p. 54]. By extension, the application of available case analysis to the estimation of covariance matrices can lead to matrices that are not positive-definite, leading to nonsensical subsequent results (e.g., nonelliptical Gaussian probability contour estimates).

16.4.2   Omitting variables and related ideas

Because the fraction of missing observations can be an extremely strong function of the variables nominally included in a dataset, the simplest approach to handling missing data is sometimes to simply exclude one or more variables from the analysis. For example, the fraction of obvious missing data in the Pima Indians diabetes dataset is less than 5% for all but two variables, but for these two variables—triceps skinfold thickness (TSF) and serum insulin concentration (INS)—the missing data fractions are $\sim$25% and $\sim$50%, respectively. Thus, if we are interested in building logistic regression models to estimate the probability of being diagnosed diabetic on the basis of other variables included in the dataset, missing data issues become much less significant if we exclude TSF and INS from our model.

Despite the simplicity of the variable omission strategy just described in cases where it is applicable, it is important to emphasize that this approach does have two potentially serious drawbacks. First, it may happen that the omitted variable or variables are among the best available predictors, or would be if a more complete dataset were available. In such cases, the best data models excluding these variables may be poor enough to be of little use. The second difficulty is closely related: since omitting variables from our data models does not allow us to say anything about the relationship between these variables and the response variable of interest, the decision to omit variables from consideration obviously limits our ability to gain insights from an exploratory data analysis. Since these insights are the primary motivation for doing exploratory data analysis in the first place, the significance of these limitations is clear.

Another missing data strategy that is somewhat analogous to variable omission is the construction of composite replacement variables from a set of closely related variables. This idea is described by Schafer and Graham for survey data under the term *averaging the available items* [258]. There, the idea is that psychological surveys often ask several closely related questions in a survey when attempting to assess a characteristic (e.g., self-esteem, depression, or anxiety) that is not directly quantifiable. For quantitative responses (e.g., "on a scale of 1 to 10, would you say that ..."), it may be reasonable to average all available responses for each individual into a single assessment variable, which is then used as the basis for analysis. This does not address cases where *all* of these related responses are missing, and it may be reasonable to impose a "minimum information restriction," including only those cases where at least some minimum number of these related questions have been answered. Clearly, this approach is not always applicable, and even in cases where it is, Schafer and Graham argue that its theoretical properties are not well understood ("it does not even have a well-recognized name"), but they also note that the method can nevertheless yield reasonable results when it is applicable [258].

16.4.3 Results for the simulation examples

To assess the relative performances of complete case and available case analyses, the following paragraphs compare the product-moment correlation estimates obtained under these two missing data treatment strategies for the four missingness models defined in Sec. 16.3. Like the univariate results given in Sec. 16.2, these results are presented as side-by-side boxplot summaries, with the complete data results included for reference. Specifically, boxplot summaries are presented for simulations of the complete datasets (denoted "FULL"), the MCAR datasets, the MAR datasets (denoted "MAR1" and "MAR2"), and the MNAR datasets.

The results obtained for 5% missing data are shown in Fig. 16.12. In each case, the boxplots summarize $r = 1000$ statistically independent simulated datasets, each defined as in Sec. 16.3. The left-most boxplot (labeled both "0" and "FULL") summarizes the complete data results, which behave as expected: the estimated values appear to be symmetrically distributed about the correct median value, corresponding to the horizontal line in the plot. The results obtained under the MCAR missingness model appear to be slightly more variable, consistent with the effective 5% reduction in sample size, but still unbiased, and there is no obvious difference between the results obtained under complete case analysis ("C") and available case analysis ("A"). In contrast, the MAR1 and MAR2 models exhibit slight negative biases, with the available case analysis results exhibiting consistently larger biases than the complete case results. Finally, for the MNAR missingness model, the biases are consistently worse than for the MAR1 and MAR2 missingness models, and again the available case biases are worse than the complete case biases.

The corresponding results for the case of 75% missing data are shown in Fig. 16.13, in the same format as Fig. 16.12. The increase in variability seen in the MCAR results is more pronounced here than in the case of 5% missing data, but the results remain unbiased, both for complete case analysis and available case analysis. Interestingly, even though available case analysis uses more data than complete case analysis does for the MCAR datasets, the variability of the correlation estimates is greater for the available case analysis. This observation reflects the inconsistency in the data subsets on which the averages of prod-

Figure 16.12: Boxplot comparisons of the correlation coefficients estimated from $r = 1000$ independent simulations of each of the four 5% incomplete datasets defined in Sec. 16.3 under both complete case analysis ("C") and available case analysis ("A"), together with the results for the complete datasets. The horizontal line represents the correct correlation value of $\rho = 1/\sqrt{2}$

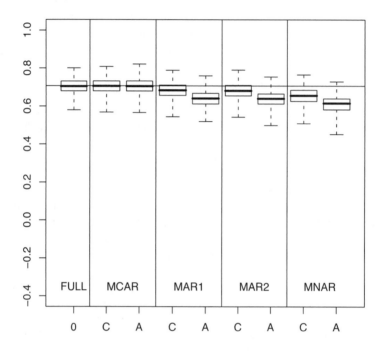

ucts and averages of squares are based in the available case analysis (i.e., the differences between the sets K_x, K_y and K_{xy} defined in Sec. 16.4.1). In fact, this inconsistency is responsible for the biases seen in the available case analysis for the MAR1, MAR2, and MNAR datasets, which are consistently larger than those of the complete case results for these datasets. These results are qualitatively consistent with those seen in the case of 5% missing data, but more pronounced. As expected, results for 20% and 50% missing data (not shown) are intermediate between the 5% and 75% missing data results shown in figs. 16.12 and 16.13.

Overall, these results suggest three things. First, either complete case analysis or available case analysis give unbiased correlation estimates under the MCAR missingness model, although these results are more variable than those

Figure 16.13: Boxplot comparisons of the correlation coefficients estimated from $r = 1000$ independent simulations of each of the four 75% incomplete datasets defined in Sec. 16.3, in the same format as Fig. 16.12

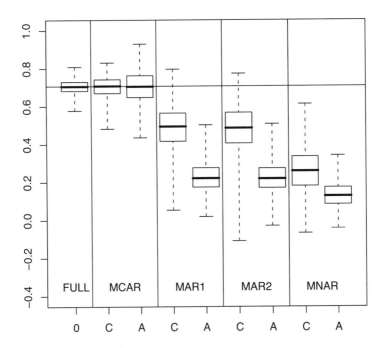

for the complete dataset. Second, systematic missing data mechanisms—either MAR or MNAR—can lead to significant biases under either complete case or available case analysis. Finally, the third conclusion is that available case analysis leads to substantially larger biases in the face of systematic missing data than complete case analysis does. It is important to emphasize that these conclusions are based on the specific cases considered here—i.e., correlation estimates for Gaussian data in the face of specific systematic missing data models—but they are consistent with the observations of others that available case analysis can lead to inconsistencies like estimated correlation matrices that are not positive-definite [188, p. 54]. Since available case analysis is usually harder to implement than complete case analysis, there appears to be little to recommend it in general. In the face of significant missing data, particularly if it is systematically missing, complete case analysis can also perform poorly, motivating the attention given to the other approaches described here.

16.5   Simple imputation strategies

As noted earlier, the term *imputation* refers to the replacement of a missing data value with an estimate, obtained in any one of several different ways. The following subsections describe four specific implementations: *mean imputation, hot-deck imputation, regression-based imputation,* and a hybrid combination of hot-deck and regression-based imputation.

16.5.1   Mean imputation

The term *mean imputation* refers to the use of the mean of the observed values in an incomplete data sequence $\{x_k\}$ as an estimate of missing data values. While this description is unambiguous in the univariate case, the strategy can actually be implemented in more than one way in multivariate settings. Arguably the simplest implementation—and the one considered here—is a componentwise application of this idea to each individual variable separately in a multivariate dataset. To see that alternative implementations are possible, note that we could implement mean imputation by replacing *all components* of any incomplete record with the vector of means computed from the available observations for each individual component. With respect to univariate (i.e., marginal) characterizations of the overall dataset, this alternative strategy is undesirable since it would generally replace some number of *observed values* for each component with a mean imputation. For multivariate characterizations like the correlation coefficient considered here, the consequences of such alternatives are not clear. A detailed examination is not undertaken here because the following results suggest that the performance of mean imputation is, at least for the examples considered here, less effective than complete case analysis. These conclusions are consistent with those presented by Little and Rubin, who note that mean imputation tends to bias both variances and covariances downward [188, p. 61], even for the most favorable case of MCAR data.

Fig. 16.14 shows the results of applying the simple mean imputation strategy described above to representative samples of each of the four incomplete datasets defined in Sec. 16.3. Complete data observations are represented as open circles in each of these plots, while the mean imputations are represented as solid circles. The cross-shaped pattern formed by these mean imputations is a consequence of the fact that one component of each incomplete observation vector is replaced with the mean of the available observations for that component.

Fig. 16.15 gives boxplot summaries for $r = 1000$ correlation estimates constructed from statistically independent bivariate Gaussian datasets, as the fraction of missing data varies from 5% to 75% as indicated above each plot in the figure. Results are presented for the complete dataset (labeled "F" and identical in all four plots), and mean imputations applied to each of the four incomplete datasets described in Sec. 16.3: "C" denotes the MCAR datasets, "R1" and "R2" denote the MAR1 and MAR2 datasets, and "N" denotes the MNAR datasets. As with complete case and available case analysis, the results become systematically worse as the fraction of missing data increases, but there are a

Figure 16.14: Results of mean imputation applied to each of the four incomplete bivariate datasets described in Sec. 16.3, with $p = 0.75$. Open circles represent complete data observations and solid circles represent the mean imputations

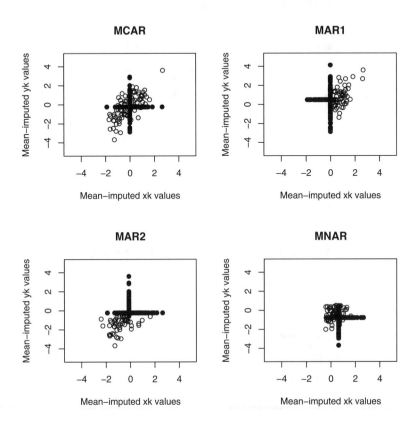

number of important distinctions to note between the mean imputation results presented here and the case deletion results presented earlier. First, note that mean imputation introduces a bias even under the MCAR missingness model, in contrast to both the complete case and available case results presented earlier. Under mean imputation, all of the correlation estimates exhibit a negative bias, becoming more pronounced as we go from the MCAR model to the MAR models MAR1 and MAR2, to the MNAR model. In fact, note that under the MNAR missingness model considered here, by the time we reach 75% missing data, the estimated correlations are slightly but consistently negative, despite the moderately large value of the true correlation, indicated by the solid line in each plot. A detailed comparison of the mean imputation results shown in Fig. 16.15 for $p = 0.75$ with the corresponding case deletion results presented in Fig. 16.13 show that the mean imputation results exhibit consistently worse biases than available case analysis for the MCAR and MNAR scenarios, and

Figure 16.15: Boxplot summaries of the correlations estimated from the full dataset ("F") and under mean imputation for the four incomplete datasets described in Sec. 16.3 ("C" for MCAR, "R1" for MAR1, "R2" for MAR2, and "N" for MNAR), for missing data probabilities $p = 0.05$, 0.20, 0.50, and 0.75

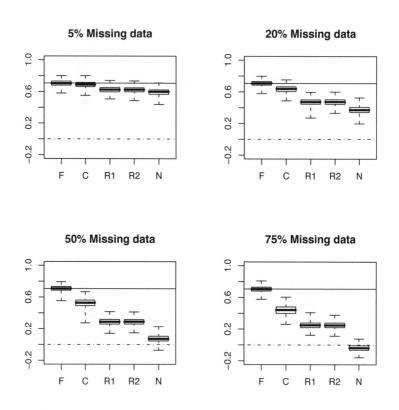

comparable biases for the MAR1 and MAR2 scenarios. For this example, the complete case analysis exhibits consistently smaller bias than mean imputation under all four missing data scenarios considered.

16.5.2 Hot-deck imputation

As Little and Rubin note [188, p. 66], the term *hot-deck imputation* dates back to the days of computer punch cards and refers to a deck of cards with hypothetical response data for survey nonrespondents, approximately matched on the basis of personal characteristics (e.g., similar age, address, etc.). Little and Rubin also note that there is no precise definition of the term in common usage, so they describe several implementations [188, pp. 66–70], including the one considered here. This strategy is based on random sampling with replacement from the

Figure 16.16: Results of hot-deck imputation applied to the four incomplete bivariate datasets described in Sec. 16.3, with $p = 0.75$. Open circles represent complete data observations, and solid circles represent the hot-deck imputations

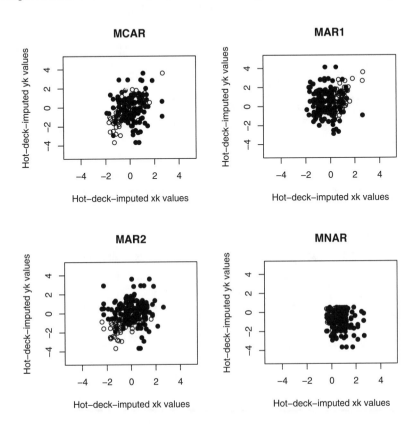

available observations and is implemented separately for each component, like the mean imputation strategy described in Sec. 16.5.1.

Fig. 16.16 shows the hot-deck imputation results for representative samples of the four incomplete datasets defined in Sec. 16.3. As in Fig. 16.14, the probability of incompleteness is $p = 0.75$, with observed data values in each dataset represented as open circles and imputed data values represented as solid circles. In contrast to the pronounced cross-shaped patterns seen in the mean-imputed datasets, the hot-deck imputations exhibit patterns much closer to the circular pattern of an uncorrelated bivariate Gaussian random vector. Indeed, the only one of the three imputed datasets shown in Fig. 16.16 that looks markedly different from an uncorrelated bivariate Gaussian dataset is the MNAR case, where the restriction to the lower right quadrant of the complete dataset imposed by this particular MNAR model is emphasized by the hot-deck imputations.

Figure 16.17: Boxplot summaries of the correlations estimated from the full dataset ("F") and under hot-deck imputation for the four incomplete datasets described in Sec. 16.3 ("C" for MCAR, "R1" for MAR1, "R2" for MAR2, and "N" for MNAR), for missing data probabilities $p = 0.05$, 0.20, 0.50, and 0.75

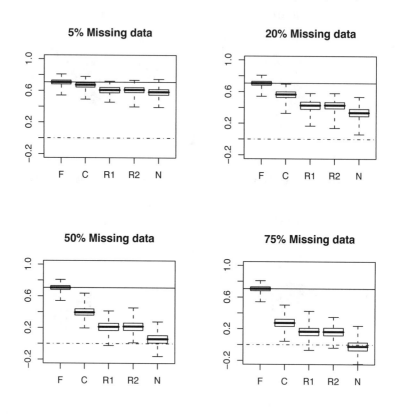

The roughly circular appearance of the imputed data plots in Fig. 16.16 contrasted with the clearly elliptical appearance of the complete dataset suggest that the hot-deck imputation strategy is likely to bias the estimated correlations toward zero in this example. Indeed, this is the case, as may be seen in Fig. 16.17, particularly for large fractions of missing data. As with the mean imputation results summarized in Fig. 16.15, the hot-deck imputation summary shown in Fig. 16.17 consists of four collections of boxplots, one for each of the four missing data levels considered here, with each boxplot summarizing the results for $r = 1000$ statistically independent datasets, generated according to the four missingness models defined in Sec. 16.3. As in all of the examples considered previously, the effects of missing data become much more pronounced as the incompleteness probability p increases. Like the mean imputations described in Sec. 16.5.1 and in contrast to the case deletion results described in

Sec. 16.4.3, note that hot-deck imputation leads to negatively biased correlation estimates even under the MCAR missing data model. As in all of the previous examples, the two MAR models give essentially the same performance, worse than the MCAR case but better than the MNAR case considered here. Overall, the hot-deck imputation strategy gives somewhat more variable results (as expected, in view of the use of random sampling inherent in hot-deck imputation), with biases that appear to be worse for the MCAR case, comparable for the two MAR examples, and *slightly* better for the MNAR example.

16.5.3 Regression-based imputation

Both the mean imputation strategy described in Sec. 16.5.1 and the hot-deck imputation strategy described in Sec. 16.5.2 are componentwise multivariate extensions of the corresponding univariate strategies described in Sec. 16.2. In contrast, regression-based imputation is an inherently multivariable strategy, in which imputed values for one variable are based on regression estimates computed from other, related variables. As a consequence, all of the issues that arise in building regression models are relavent to regression-based imputation: variable selection, regression fitting method selection, model validation, etc. In the bivariate example considered here, these issues simplify greatly, but in dealing with real datasets, they all merit consideration in contemplating and developing a regression-based imputation strategy. More specifically, since the datasets considered here are outlier-free Gaussian data sequences, ordinary least squares regression is appropriate. Also, since the dataset includes two variables, the only possible regression-based imputation strategy for either of these variables uses predictions from the other one. Thus, in what follows, two regression models are built and used for imputing missing data values: one that imputes missing y_k values from observed x_k values, and a second one that imputes missing x_k values from observed y_k values. As discussed in Chapter 5, these problems can be formulated in more than one way, but for this example, the simplest strategy is adopted, based on the following two regression models:

$$\text{Model 1: } \hat{y}_k = ax_k + b,$$
$$\text{Model 2: } \hat{x}_k = cy_k + d. \tag{16.6}$$

Both of these models are fit to the complete portion of the dataset: a regression model of y_k on x_k is fit to all records including both x_k and y_k observations to obtain the coefficients a and b, and the analogous regression model of x_k on y_k is fit to obtain the coefficients c and d. Missing x_k values are then imputed with the corresponding $\hat{x}_k$ values computed from Eq. (16.6), and missing y_k values are imputed with the corresponding $\hat{y}_k$ values.

An important practical corollary to the strategy just described is that records in which *both* x_k and y_k are missing cannot be imputed under the scheme just described. In fact, this problem is a general characteristic of regression-based imputation: if a particular imputation value cannot be computed due to missing values of the covariates on which the regression prediction is based, some other

Figure 16.18: Results of regression-based imputation applied to the four incomplete bivariate datasets described in Sec. 16.3, with $p = 0.75$. Open circles represent complete data observations, and solid circles represent the regression-based imputations

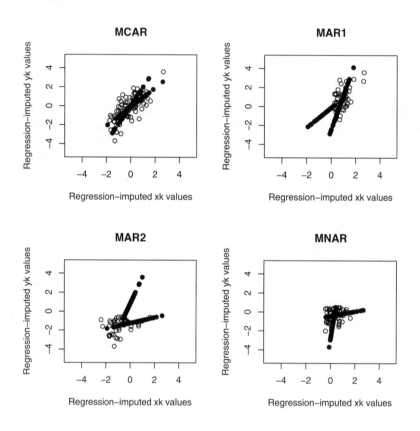

missing data treatment strategy must be adopted for those records. In large multivariate datasets, alternative strategies might include different regression models that do not include the missing covariate. In the bivariate example considered here, however, this alternative is not feasible. Since, of the four imputation strategies considered so far—complete case analysis, available case analysis, mean imputation, and hot deck imputation—complete case analysis has generally given the best results for this example, it is adopted here for the records that cannot be addressed using regression-based imputation. That is, records for which both x_k and y_k are missing (i.e., $w_k = z_k = 1$) are omitted from subsequent analysis and the regression-based imputation strategy described above is applied to those records for which either x_k or y_k is missing, but not both.

Fig. 16.18 shows representative regression-based imputation results obtained under the four missing data scenarios defined in Sec. 16.3. As with the mean

Figure 16.19: Boxplot summaries of the correlations estimated from the full dataset ("F") and under regression-based imputation for the four incomplete datasets described in Sec. 16.3 ("C" for MCAR, "R1" for MAR1, "R2" for MAR2, and "N" for MNAR), for missing data probabilities $p = 0.05, 0.20, 0.50,$ and 0.75

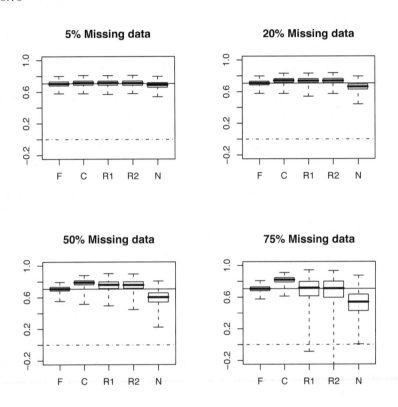

and hot-deck imputation results presented above, open circles represent complete data observations and solid circles represent imputations of incomplete observations. Analogous to the mean imputations, the regression-based imputations form clear linear patterns in the data; in fact, the two lines evident in each of the four plots in Fig. 16.18 correspond to the two regression models defined in Eq. (16.6), one for imputing missing x_k values and the other for imputing missing y_k values. It is clear that these two regression models are different in all cases, and that the extent of this difference depends strongly on the missingness mechanism: these lines are most similar for the MCAR case, least similar for the MNAR case, and intermediate for the two MAR cases.

Fig. 16.19 presents boxplot summaries of the correlations estimated for $r = 1000$ statistically independent datasets, for the four missingness mechanisms defined in Sec. 16.3 and the four missing data probabilities considered here. The format of these plots is the same as that of figs. 16.15 and 16.17 for mean and hot-deck imputation, respectively. Comparing these three sets of figures, several important differences are clear. First, while the biases seen for the mean and hot-deck imputations are consistently negative for all four incomplete datasets, the biases seen under regression imputation are often positive. In particular, for significant fractions of missing data (e.g., 50% or 75%), the MCAR results almost always exhibit a positive bias, while the two MAR models yield results that typically overestimate the true correlation value, but less so than the MCAR model does. Only the MNAR model consistently exhibits a negative bias like that seen in the mean and hot-deck imputation results. Further, the biases exhibited by the regression-based imputation results are consistently smaller in magnitude than those seen for mean imputation and especially those seen for hot-deck imputation. Finally, note that the variability of the regression-based imputation results is, in some cases, much greater than that seen in either the mean imputation or the hot-deck imputation results. This is especially evident in the MAR1 results for $p = 0.75$ where, although the median value is almost exactly correct for the regression-based imputation results (0.711 vs. 0.707, corresponding to a positive bias of 0.5%), the range of these estimated correlations is from -0.448 to 0.934; in contrast, the hot-deck imputation results exhibit a negative bias of approximately 77% (0.162 vs. 0.707), but the total range of variation is much smaller, from -0.040 to 0.350.

16.5.4   Hot-deck/regression composite method

Although it incorporates the relationship between components on which the correlation coefficient is based, regression-based imputation leads to somewhat disappointing results in the preceding example. In particular, the much greater variability seen for the MAR1 and MAR2 cases relative to the hot-deck imputation results raise the question of whether we can do better by combining these ideas. Thus, the following paragraphs examine a composite imputation method that uses both regression model predictions and the notion of sampling with replacement from the hot-deck imputation strategy considered here. The combination yields a composite strategy mentioned in passing in the book on missing data by Little and Rubin [188, p. 61].

As in the regression-based imputation procedure described in Sec. 16.5.3, this composite approach first eliminates all records for which both x_k and y_k values are missing. Next, the two regression models defined in Eq. (16.6) are fit to the complete portion of the remaining dataset as before (i.e., model coefficients a, b, c, and d are estimated from those records for which both x_k and y_k observations are present). Once these models have been obtained, their associated prediction errors are computed for these complete data records:

$$
\begin{aligned}
e_k &= y_k - \hat{y}_k = y_k - ax_k - b, \\
f_k &= x_k - \hat{x}_k = x_k - cy_k - d.
\end{aligned}
\tag{16.7}
$$

Figure 16.20: Results of composite hot-deck/regression imputation applied to each of the four incomplete bivariate datasets described in Sec. 16.3, with $p = 0.75$. Open circles represent complete data observations, and solid circles represent the imputations.

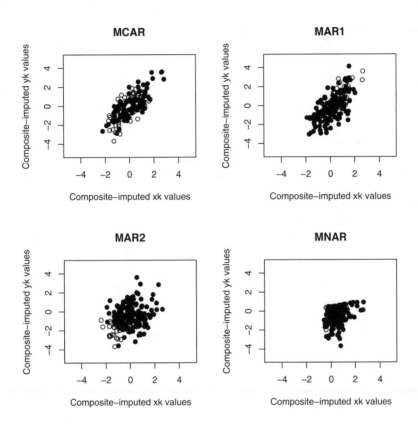

Imputations for missing x_k and y_k values are then computed as:

$$\begin{aligned} \tilde{x}_k &= cy_k + d + \tilde{f}_k, \\ \tilde{y}_k &= ax_k + b + \tilde{e}_k, \end{aligned} \qquad (16.8)$$

where $\tilde{e}_k$ and $\tilde{f}_k$ are obtained by sampling the sets $\{e_k\}$ and $\{f_k\}$ of regression model errors with replacement. The resulting imputations are thus predictions consistent with the regression models fit to the complete dataset rather than exact fits obtained from these regression models.

Fig. 16.20 shows representative imputation results obtained by this composite method for the four missingness models described in Sec. 16.3. As before, open circles in these four plots represent complete data observations while the solid circles represent imputed observations. Comparing these results with those presented earlier for hot-deck and regression-based imputations, it is clear that the composite method imputations are intermediate in character between those of the two methods it combines. In particular, note that for both the MCAR and MAR1 examples, the composite imputations are much more consistent with the elliptical point cloud defined by the complete dataset than are the much more circular point clouds obtained under hot-deck imputation for both of these examples. Also, since the composite method adds estimated errors to the regression model predictions, the imputed values in Fig. 16.20 do not exhibit the dramatic pattern seen in the regression-based imputations shown in Fig. 16.18. Finally, note that a strong dependence of the imputation results on the missingness model is also seen in Fig. 16.20 for the composite method: specifically, note the greater similarity of the composite and hot-deck results for the MAR2 and MNAR datasets compared with the MCAR and MAR1 datasets.

Fig. 16.21 presents boxplot summaries for the correlations computed from $r = 1000$ statistically independent datasets, including both the full datasets "F") and the incomplete datasets generated by the four missingness models defined in Sec. 16.3, labelled "C" for the MCAR results, "R1" and "R2" for the MAR1 and MAR2 results, and "N" for the MNAR results. For the MCAR datasets, the composite imputation strategy gives unbiased correlation estimates for all levels of missing data considered, a results that does not hold for any of the other simple imputation strategies described here. In particular, recall that both mean imputation and hot-deck imputation gave large negative biases for this case, while regression-based imputation gave a smaller but still significant positive bias. Similarly, both mean and hot-deck imputations exhibited even larger negative biases for both the MAR1 and MAR2 datasets, while the regression-based imputations gave unbiased results with much greater variabilities. The composite imputations exhibit exhibit comparably large variabilities to the regression-based imputations for the MAR1 and MAR2 datasets with a clear negative bias, but these biases are much less severe than those for either mean or hot-deck imputations. Finally, the composite imputation results for the MNAR examples also exhibit a comparably large variability to the regression-based imputations, with a worse negative bias, but one that is *much* smaller than those of the mean or hot-deck imputations.

16.6   Multiple imputation

The results presented in secs. 16.4 and 16.5 have demonstrated that different imputation strategies can give very different results for the same dataset. Further, with approaches like hot-deck imputation described in Sec. 16.5.2 and the composite hot-deck/regression procedure described in Sec. 16.5.4 that involve random sampling, we can expect the results to depend on the specific random sample on which the imputations are based. *Multiple imputation strategies* exploit this dependence by constructing a collection of m distinct imputation-based results, a strategy that has two potential advantages. The first is that a comparison of these imputation results can give us an idea of the sensitivity of the computed results to our imputation assumptions. The second potential ad-

Figure 16.21: Boxplot summaries of the correlation estimates for the full dataset
("F") and the composite imputations from the four incomplete datasets de-
scribed in Sec. 16.3 ("C" for MCAR, "R1" for MAR1, "R2" for MAR2, and
"N" for MNAR), for missing data probabilities $p = 0.05, 0.20, 0.50$, and 0.75

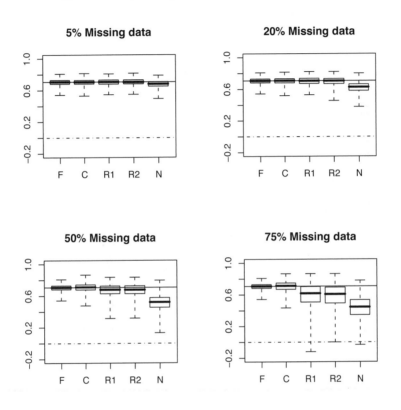

vantage is that, in some cases, we can obtain a better overall result by averaging
over the distinct imputations.

Fig. 16.22 uses multiple imputation to compare both of the imputation
strategies just described for the four missing data examples defined in Sec. 16.3.
Specifically, the figure presents side-by-side boxplot summaries obtained under hot-deck imputation ("HD") and under the composite imputation strategy
("CO") for $r = 1000$ statistically independent imputations, all based on the
same dataset. Two things are clear from these boxplot summaries. First, the
composite strategy is much less sensitive to variations caused by random sampling than the hot-deck imputation strategy. Second, it is also clear from these
results that the hot-deck imputation strategy consistently gives much smaller
correlation estimates than the composite imputation strategy, for all four of the
missingness models considered here. These results further emphasize the point

Figure 16.22: Boxplot summaries comparing the correlation estimates obtained from a single dataset under $r = 1000$ independent hot-deck imputations (HD) with those obtained under $r = 1000$ independent composite imputations (CO), for the case of 75% missing data

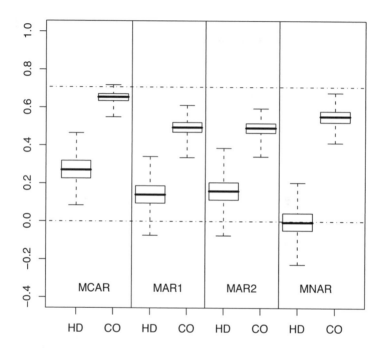

made above that the composite imputation strategy gives much better estimates for this problem than the hot-deck imputation strategy, but this conclusion relies on knowing the correct answer, whereas the first two conclusions are based only on comparisons of the results for an individual dataset.

To be useful in practice, a multiple imputation strategy needs to be based on a collection of distinct imputations, typically based on random samples that are deemed representative under some plausible missing data model. As a consequence, they are more involved to implement than simple single-imputation strategies like mean imputation, although they yield more information (e.g., variability assessments) and may give much better average results than these simpler strategies. A detailed discussion of the development of multiple imputation strategies is beyond the scope of this book, but an excellent treatment is given by Little and Rubin [188].

16.7 The EM algorithm

As noted at the beginning of this chapter, the *expectation maximization (EM) algorithm* is an iterative procedure for handling missing data that alternates between an imputation step and an analysis step. A great deal has been written about the EM algorithm [87, 188, 207], and it has been applied extensively, both to problems involving explicit missing data issues and to problems like fitting mixture distributions that can be cast as a missing data problem, even though this form may not be obvious initially.

16.7.1 General description

Essentially, the EM algorithm is an iterative procedure for solving maximum likelihood (ML) estimation problems with incomplete data where the structure of the complete-data likelihood function is much simpler than that for the incomplete-data problem. Thus, if we had the complete dataset, solving the ML estimation problem of interest would be easy, and this fact is exploited in the EM algorithm. More explicitly, consider a problem where $\{x_k\}$ denotes a random dataset whose underlying probability distribution is $g_c(\{x_k\}; \theta)$, where θ is a vector of unknown distributional parameters. Here, the subscript c stands for "complete," and we first formulate the complete data ML estimation problem, seeking the value (or values) of θ that maximize the complete data log likelihood:

$$L_c(\{x_k\}; \theta) = \ln g_c(\{x_k\}; \theta), \qquad (16.9)$$

where the dataset $\{x_k\}$ is regarded as fixed and θ is regarded as a parameter.

Now, suppose the complete dataset $\{x_k\}$ is not observed, but only a subset, denoted $\{x_k^o\}$ is observed. The probability distribution associated with this observed subset is obtained from the complete distribution $g_c(\{x_k\}; \theta)$ by integrating out the missing observations. To estimate θ from the observed dataset, we need to maximize this modified likelihood, generally a much more difficult problem. The EM algorithm replaces this problem with the following sequence of iterative steps:

0: initialize the procedure by computing a preliminary estimate $\theta^{(0)}$ and setting $j = 0$;

E: compute the expected log likelihood $Q(\theta; \theta^{(j)})$ given the complete data and the prior parameter estimate:

$$Q(\theta; \theta^{(j)}) = E_{\theta^{(j)}}\{L_c(\{x_k\}, \theta) \mid \{x_k^o\}\};$$

M: compute a value $\theta^{(j+1)}$ that maximizes $Q(\theta; \theta^{(j)})$, which is often unique but need not be. That is, $\theta^{(j+1)}$ is any parameter estimate satisfying the condition:

$$Q(\theta^{(j+1)}; \theta^{(j)}) \geq Q(\theta; \theta^{(j)}),$$

for all admissible parameter vectors θ.

724   *Exploring Data*

These ideas are probably easiest to understand clearly in the light of some illuminating examples, like those presented in secs. 16.7.2 through 16.7.6, but a few general comments are in order first.

The procedure just outlined is based on a paper published in 1977 by Dempster *et al.* [87], which has been subsequently applied to an extremely broad range of problems [207]. The utility of the EM algorithm is greatest when the computations involved in the E and M steps are relatively simple, as in cases where the M step problem can be solved in closed form. Essentially, the idea behind the algorithm is that the E step generates imputations of the missing terms in the complete data likelihood $L_c(\{x_k\}; \theta)$, which is similar in spirit but distinct in detail from imputing the missing data values themselves. In particular, it is often possible to express the likelihood in terms of *sufficient statistics:* quantities derived from the data that are themselves sufficient to compute the complete data likelihood. As the following examples illustrate, imputing these sufficient statistics often leads to results that are both simpler and distinct from the results obtained by imputing the missing data values themselves. Consequently, the EM algorithm generally leads to different results than those that would be obtained under data imputation strategies. Finally, it is important to note that the convergence properties of the EM algorithm are well established: Dempster *et al.* have shown that the sequence of likelihood values computed from each iteration of the procedure form a monotonically increasing sequence. Thus, for well-posed problems where the likelihood is bounded so that a maximum likelihood estimate is well defined, the EM algorithm converges.

Despite its widespread applicability and popularity, however, the EM algorithm is not without its limitations. First and foremost, this procedure can converge slowly, particularly in the face of large missing data fractions. This limitation is important enough in practice that a number of extensions have been proposed to speed up the computations; for a discussion of these ideas, refer to the book by McLachlan and Krishnan [207]. Another limitation of the EM algorithm is that, while it often greatly simplifies maximum likelihood estimation problems, there are situations where the required computations are complex enough to severely limit the utility of the approach.

16.7.2   A simple univariate example

To illustrate both the mechanics and the utility of the EM algorithm, consider the simple example of a univariate data sequence under the MCAR missingness model. Specifically, consider a data sequence $\{x_k\}$ consisting of independent draws from a Gaussian distribution with mean μ and variance σ^2. The data characterization objective in this example is to estimate the unknown values of μ and σ^2 in the face of m missing data observations, assuming that these observations are missing completely at random (MCAR). Since the sequence $\{x_k\}$ is i.i.d. and the characterizations of interest here do not depend on the order of the data sequence, there is no loss of generality in assuming that the first $N - m$ elements of $\{x_k\}$ are observed and that the last m elements of this sequence are missing.

Since the observations are statistically independent, the complete data like-lihood for this problem is simply the product of the individual Gaussian proba-bility densities $p(x_k; \mu, \sigma^2)$, so the complete data log-likelihood $L_c(\{x_k\}; \mu, \sigma^2)$ is given by:

$$L_c(\{x_k\}; \mu, \sigma^2) = -\frac{N}{2} \ln 2\pi\sigma^2 - \frac{1}{2} \sum_{k=1}^{N} \frac{(x_k - \mu)^2}{\sigma^2}, \qquad (16.10)$$

$$= -\frac{N}{2} \left[\ln 2\pi\sigma^2 + \frac{\mu^2}{\sigma^2} \right] - \frac{1}{2\sigma^2} \sum_{k=1}^{N} x_k^2 + \frac{\mu}{\sigma^2} \sum_{k=1}^{N} x_k.$$

A key observation here is that the sums $\sum_{k=1}^{N} x_k$ and $\sum_{k=1}^{N} x_k^2$ appearing in the second line of Eq. (16.10) represent sufficient statistics for this problem: the complete data log-likelihood can be computed from these values even if the individual data values $\{x_k\}$ are not known. This observation is important because the E step in the EM algorithm for this problem does not attempt to impute the individual data values, but only these two sufficient statistics.

Specifically, let $\mu_{(j)}$ and $\sigma_{(j)}^2$ denote the estimates of the unknown parameters μ and σ^2 generated at step j of the EM algorithm. The imputations generated by the E step at iteration $j+1$ are the conditional expectations of these sufficient statistics based on the observed data and the parameter estimates from step j. That is, the E step for this problem computes:

$$E\{\sum_{k=1}^{N} x_k \mid \mu_{(j)}, \sigma_{(j)}^2, \{x_k^o\}\} = \sum_{k=1}^{N-m} x_k + m\mu_{(j)},$$

$$E\{\sum_{k=1}^{N} x_k^2 \mid \mu_{(j)}, \sigma_{(j)}^2, \{x_k^o\}\} = \sum_{k=1}^{N-m} x_k^2 + m[\mu_{(j)}^2 + \sigma_{(j)}^2]. \qquad (16.11)$$

As Little and Rubin note in discussing this example [188, p. 168], simple mean imputation applied to this problem would give the same results for the first of these expectations, but not for the second, since the term $\sigma_{(j)}^2$ would be absent.

The M step of the EM algorithm for this problem computes the maximum likelihood estimates for μ and σ^2 from the results of the E step as:

$$\mu_{(j+1)} = \frac{1}{N} E\left\{ \sum_{k=1}^{N} x_k \mid \mu_{(j)}, \sigma_{(j)}^2, \{x_k^o\} \right\},$$

$$= \frac{1}{N} \sum_{k=1}^{N-m} x_k + \frac{m}{N} \mu_{(j)},$$

$$\sigma_{(j+1)}^2 = \frac{1}{N} E\left\{ \sum_{k=1}^{N} x_k^2 \mid \mu_{(j)}, \sigma_{(j)}^2, \{x_k^o\} \right\} - \mu_{(j+1)}^2,$$

$$= \frac{1}{N} \sum_{k=1}^{N-m} x_k^2 + \frac{m}{N} [\mu_{(j)}^2 + \sigma_{(j)}^2] - \mu_{(j+1)}^2. \qquad (16.12)$$

One of the advantages of considering this simple example is that it is easy to compute its limiting results, $\mu_{(\infty)}$ and $\sigma^2_{(\infty)}$, obtained by setting $\mu_{(j)} = \mu_{(j+1)} = \mu_{(\infty)}$ and $\sigma^2_{(j)} = \sigma^2_{(j+1)} = \sigma^2_{(\infty)}$. These conditions lead to the following results, identical to the complete case results for this example:

$$\mu_{(\infty)} = \frac{1}{N-m} \sum_{k=1}^{N-m} x_k,$$

$$\sigma^2_{(\infty)} = \frac{1}{N-m} \sum_{k=1}^{N-m} x_k^2 - \mu^2_{(\infty)}. \tag{16.13}$$

16.7.3   A nonignorable univariate example

Besides illustrating the mechanics of the EM algorithm, the univariate example just considered also provides the basis for a simple example demonstrating the necessity of assumptions in dealing with MNAR (nonignorable) missing data problems. Specifically, consider the following variation of the univariate problem just considered: as before, $\{x_k\}$ is a sequence of N statistically independent, Gaussian random variables, the first $N - m$ of which are observed, and the last m of which are missing. The key difference between these problems is that the distribution of the missing values is different, having mean $\lambda\mu$ and variance $\tau^2\sigma^2$, where λ is any real number and τ is any positive number. Note that if $\lambda = \tau = 1$, this problem reduces to the MCAR example considered in the preceeding example, but for $\lambda \neq 1$ and/or $\tau \neq 1$, the problem becomes MNAR since the distribution of the missing data differs systematically from that of the observed data. *A key characteristic of this problem is that, to apply the EM algorithm, the values of λ and τ must be known.* In fact, this is a fundamental characteristic of dealing with non-ignorable missing data: it is not enough to know (or believe) the missing data values differ systematically from the observed data values, we must know (or be willing to assume) something quantitative about *how* the missing values differ.

To derive an EM algorithm for this problem, it is necessary to first obtain the complete data likelihood, which is given by (Exercise 2):

$$L_c\{\{x_k\} \mid \mu, \sigma^2\} = -\frac{N}{2}\ln(2\pi\sigma^2) - \frac{m}{2}\ln\tau^2 - \frac{1}{2\sigma^2}S_2 + \frac{\mu}{\sigma^2}S_1$$

$$-\frac{\mu^2}{2\sigma^2}[N - m + m\lambda^2/\tau^2], \tag{16.14}$$

where S_1 and S_2 are defined as:

$$S_1 = \sum_{k=1}^{N-m} x_k + \sum_{k=N-m+1}^{N} (\lambda/\tau^2)x_k,$$

$$S_2 = \sum_{k=1}^{N-m} x_k^2 + \sum_{k=N-m+1}^{N} x_k^2/\tau^2. \tag{16.15}$$

It follows from Eq. (16.14) that S_1 and S_2 are the sufficient statistics for this problem, replacing the sum of observations and squared observations that appeared in the MCAR example. Differentiating the complete data log likelihood with respect to μ and σ^2 and setting the results to zero yields the following expressions for the complete data maximum likelihood estimators $\hat{\mu}$ and $\widehat{\sigma^2}$ (Exercise 3):

$$\hat{\mu} = \frac{S_1}{N - m + m\lambda^2/\tau^2},$$

$$\widehat{\sigma^2} = \frac{1}{N}[S_2 - (N - m + m\lambda^2/\tau^2)\hat{\mu}^2]. \tag{16.16}$$

At the jth step of the EM algorithm, the E step computes the expected values of the sufficient statistics S_1 and S_2, based on the observed data and the current estimates of μ and σ^2:

$$E\{S_1 \mid \{x_k^o\}, \mu_{(j)}, \sigma_{(j)}^2\} = \sum_{k=1}^{N-m} x_k + (m\lambda^2/\tau^2)\mu_{(j)},$$

$$E\{S_2 \mid \{x_k^o\}, \mu_{(j)}, \sigma_{(j)}^2\} = \sum_{k=1}^{N-m} x_k^2 + m\sigma_{(j)}^2 + (m\lambda^2/\tau^2)\mu_{(j)}^2. \tag{16.17}$$

The M step of the algorithm then uses these expectations to update the μ and σ^2 estimates as:

$$\mu_{(j+1)} = \frac{\sum_{k=1}^{N-m} x_k + (m\lambda^2/\tau^2)\mu_{(j)}}{N - m + m\lambda^2/\tau^2},$$

$$\sigma_{(j+1)}^2 = \frac{1}{N}\left[\sum_{k=1}^{N-m} x_k^2 - (N - m)\mu_{(j)}^2 + m\sigma_{(j)}^2\right]. \tag{16.18}$$

From these results, it is not difficult to show that the EM algorithm for this problem converges to exactly the same limiting estimators $\mu_{(\infty)}$ and $\sigma_{(\infty)}^2$ as in the MCAR case (Exercise 4). That is, this formulation of the EM algorithm for the specific nonignorable missing data problem considered here converges to the complete case analysis estimators for μ and σ^2, independent of the parameters λ and τ that describe the differences in distribution between the observed and missing data values.

Two aspects of this example are worth emphasizing. First, as noted at the beginning of this discussion, it is necessary to *know* the parameters λ and τ that describe the distributional differences between the observed and missing data values. In practice, this is almost never the case, so what is done instead is to make assumptions about *possible* differences between observed and missing data and then assess the reasonableness of the results. In some cases, we can show that the differences are minor enough to be safely neglected over a reasonable range of assumptions, but in other cases, it may be important to present and compare results obtained under several different missingness models (e.g.,

MCAR vs. specific choices of λ and τ in this example). The second aspect of this example worth emphasizing is that, while the limiting behavior of the algorithm can be shown to correspond to a simple result that is much easier to compute directly, this behavior is not typical, especially for more complex problems. The primary reason for including this example was to provide further illustration of the mechanics of the EM algorithm in a setting where the deails are simple enough to be illuminating.

16.7.4   Specialization to bivariate Gaussian data

Another popular illustrative example of the EM algorithm is its application to an incomplete bivariate Gaussian dataset, discussed by both McLachlan and Krishnan [207, sec. 2.2] and Little and Rubin [188, pp. 170–172]. The following paragraphs describe the steps of the EM algorithm for this problem, and Sec. 16.7.5 applies the procedure to the four simulation examples presented in Sec. 16.3 to provide a comparison of its performance with the other missing data treatment approaches considered here.

As in the discussion of available case analysis given in Sec. 16.4.1, define K_x to be the set of indices k for which x_k is observed, K_y to be the set of indices for which y_k is observed, and the set $K_{xy} = K_x \cap K_y$ to be the indices for which both variables are observed. Similarly, let $|\mathcal{S}|$ denote the number of elements in the set $\mathcal{S}$ and define $N_x = |K_x|$, $N_y = |K_y|$, and $N_{xy} = |K_{xy}|$. Since the EM algorithm effectively replaces missing values of x_k with imputations based on y_k and vice versa, as in the discussion of regression-based imputation strategies discussed in Sec. 16.5.3, records for which both x_k and y_k are missing are omitted from the analysis.

To derive the EM update equations for this example, first note that the joint distribution of the complete dataset is:

$$p(x, y) = p(\mathbf{w}) = \frac{1}{2\pi\sigma_x\sigma_y\sqrt{1 - \rho^2}} \cdot \exp\left\{-\frac{1}{2}(\mathbf{w} - \mu)^T \Sigma^{-1}(\mathbf{w} - \mu)\right\}, \quad (16.19)$$

where $\mu = [\mu_x, \mu_y]^T$ is the vector of component means and Σ is the covariance matrix, given by:

$$\Sigma = \begin{bmatrix} \sigma_x^2 & \rho\sigma_x\sigma_y \\ \rho\sigma_x\sigma_y & \sigma_y^2 \end{bmatrix}. \quad (16.20)$$

Taking the logarithm of the joint density given in Eq. (16.19) leads to the following expression for the complete data log likelihood (Exercise 5):

$$\begin{aligned}
L_c\{\mathcal{D} \mid \mu, \Sigma\} &= -N\ln(2\pi) - \frac{N}{2}\ln\zeta - \frac{1}{2\zeta}\{\sigma_y^2 T_{11} - 2\rho\sigma_x\sigma_y T_{12} + \sigma_x^2 T_{22} \\
&\quad - 2[(\mu_x\sigma_y^2 - \mu_y\rho\sigma_x\sigma_y)T_1 + (\mu_y\sigma_x^2 - \mu_x\rho\sigma_x\sigma_y)T_2] \\
&\quad + N(\mu_x^2\sigma_y^2 + \mu_y^2\sigma_x^2 - 2\rho\mu_x\mu_y\sigma_x\sigma_y)\}.
\end{aligned} \quad (16.21)$$

Here, $\mathcal{D}$ denotes the complete dataset, $\zeta = \sigma_x^2\sigma_y^2(1 - \rho^2)$, and the five sufficient statistics for this problem are the following sums:

$$T_1 = \sum_{k=1}^{N} x_k, \qquad\qquad T_2 = \sum_{k=1}^{N} y_k,$$

$$T_{11} = \sum_{k=1}^{N} x_k^2, \qquad\qquad T_{22} = \sum_{k=1}^{N} y_k^2,$$

$$T_{12} = \sum_{k=1}^{N} x_k y_k. \tag{16.22}$$

The complete data maximum likelihood estimators for these parameters are (Exercise 6):

$$\begin{aligned}
\hat{\mu}_x &= \frac{1}{N} T_1, \\
\hat{\mu}_y &= \frac{1}{N} T_2, \\
\widehat{\sigma^2}_x &= \frac{1}{N} T_{11} - \hat{\mu}_x^2, \\
\widehat{\sigma^2}_y &= \frac{1}{N} T_{22} - \hat{\mu}_y^2, \\
\hat{\sigma}_{xy} &= \frac{1}{N} T_{12} - \hat{\mu}_x \hat{\mu}_y,
\end{aligned} \tag{16.23}$$

where $\sigma_{xy} = \rho \sigma_x \sigma_y$, from which it follows that the estimator of the correlation coefficient ρ is:

$$\hat{\rho} = \frac{\hat{\sigma}_{xy}}{\sqrt{\widehat{\sigma^2}_x \widehat{\sigma^2}_y}}. \tag{16.24}$$

For the E step of the EM algorithm, we exploit the properties of the bivariate Gaussian distribution to obtain improved imputations of the missing portions of the complete data likelihood, conditioned on the observed data. Specifically, recall the following conditional density results from Chapter 10:

$$\begin{aligned}
x_k | y_k &\sim N\left(\mu_x + \frac{\sigma_{xy}(y_k - \mu_y)}{\sigma_y^2}, \sigma_x^2(1 - \rho^2) \right), \\
y_k | x_k &\sim N\left(\mu_y + \frac{\sigma_{xy}(x_k - \mu_x)}{\sigma_x^2}, \sigma_y^2(1 - \rho^2) \right).
\end{aligned} \tag{16.25}$$

Given these results, the E step computes the conditional expectations of the five sufficient statistics, given the observed portion $\mathcal{D}_{obs}$ of the dataset $\mathcal{D}$ and the current estimates of the distribution parameters μ and Σ. For example, the conditional expectation of the first sum is given by:

$$\begin{aligned}
T_1^{(j)} &= E\{T_1 \mid \mathcal{D}_{obs}, \mu^{(j)}, \Sigma^{(j)}\} \tag{16.26} \\
&= \sum_{k \in K_x} x_k + \sum_{k \notin K_x} E\{x_k \mid y_k, \mu^{(j)}, \Sigma^{(j)}\} \\
&= \sum_{k \in K_x} x_k + (N - N_x)\left(\mu_x^{(j)} - \frac{\sigma_{xy}^{(j)} \mu_y^{(j)}}{\sigma_y^{2(j)}} \right) + \left(\frac{\sigma_{xy}^{(j)}}{\sigma_y^{2(j)}} \right) \sum_{k \notin K_x} y_k.
\end{aligned}$$

Applying the same general strategy to the other four sums yields (Exercise 7):

$$T_2^{(j)} = \sum_{k \in K_y} y_k + (N - N_y)\left(\mu_y^{(j)} - \frac{\sigma_{xy}^{(j)}\mu_x^{(j)}}{\sigma_x^{2(j)}}\right) + \left(\frac{\sigma_{xy}^{(j)}}{\sigma_x^{2(j)}}\right)\sum_{k \notin K_y} x_k,$$

$$T_{11}^{(j)} = \sum_{k \in K_x} x_k^2 + \sum_{k \notin K_x}\left[\mu_x^{(j)} + \frac{\sigma_{xy}^{(j)}}{\sigma_y^{2(j)}}(y_k - \mu_y^{(j)})\right]^2 + (N - N_x)\sigma_x^{2(j)}(1 - \rho_{(j)}^2),$$

$$T_{22}^{(j)} = \sum_{k \in K_y} y_k^2 + \sum_{k \notin K_y}\left[\mu_y^{(j)} + \frac{\sigma_{xy}^{(j)}}{\sigma_x^{2(j)}}(x_k - \mu_x^{(j)})\right]^2 + (N - N_y)\sigma_y^{2(j)}(1 - \rho_{(j)}^2),$$

$$T_{12}^{(j)} = \sum_{k \in K_{xy}} x_k y_k + \sum_{k \in K_x \backslash K_y} x_k\left[\mu_y^{(j)} + \frac{\sigma_{xy}^{(j)}}{\sigma_x^{2(j)}}(x_k - \mu_x^{(j)})\right]$$

$$+ \sum_{k \in K_y \backslash K_x} y_k\left[\mu_x^{(j)} + \frac{\sigma_{xy}^{(j)}}{\sigma_y^{2(j)}}(y_k - \mu_y^{(j)})\right], \tag{16.27}$$

where the symbol $\backslash$ denotes set difference: $K_x \backslash K_y$ corresponds to the set of indices k belonging to K_x but not to K_y, with the analogous interpretation for $K_y \backslash K_x$. Given these results, the M step of the EM algorithm for this problem computes the following updated parameter estimates:

$$\mu_x^{(j+1)} = \frac{1}{N}T_1^{(j)},$$

$$\mu_y^{(j+1)} = \frac{1}{N}T_2^{(j)},$$

$$\sigma_x^{2(j+1)} = \frac{1}{N}T_{11}^{(j)} - (\mu_x^{(j+1)})^2,$$

$$\sigma_y^{2(j+1)} = \frac{1}{N}T_{22}^{(j)} - (\mu_y^{(j+1)})^2,$$

$$\sigma_{xy}^{(j+1)} = \frac{1}{N}T_{12}^{(j)} - \mu_x^{(j+1)}\mu_y^{(j+1)},$$

$$\rho^{(j+1)} = \frac{\sigma_{xy}^{(j+1)}}{\sqrt{\sigma_x^{2(j+1)}\sigma_y^{2(j+1)}}}. \tag{16.28}$$

Finally, note that to actually implement this procedure numerically, two additional details must be addressed. First is the issue of how to initialize the iteration sequence. Probably the simplest initialization is to use the complete case analysis results, obtained by replacing N in Eq. (16.23) with N_{xy} and restricting the corresponding sums in Eq. (16.22) to $k \in K_{xy}$. Besides its simplicity, this choice of initialization also has the advantage that the differences between these initial parameter values and those to which the EM algorithm converges provide a direct measure of the benefit of the iterative procedure. The second issue is how to determine when the iterative procedure has converged.

Because it represents a scalar-valued figure of merit, the complete data likelihood defined by Eq. (16.21), computed from the results of each iteration of the EM algorithm, provides a useful measure of progress. Specifically, the iterations terminate when the log likelihood improvement is sufficiently small:

$$\Delta_{(j+1)} = |L_c\{\mathcal{D} \mid \mu^{(j+1)}, \Sigma^{(j+1)}\} - L_c\{\mathcal{D} \mid \mu^{(j)}, \Sigma^{(j)}\}| < \epsilon_T, \qquad (16.29)$$

where ϵ_T is a predefined tolerance limit for convergence.

16.7.5 Application to the simulation examples

To illustrate the practical behavior of the EM algorithm, the following paragraphs examine the results obtained when the formulation described in Sec. 16.7.4 is applied to the four simulation-based examples introduced in Sec. 16.3. In all cases, the EM algorithm is applied to $r = 1000$ independently generated datasets, under each of the four specific missingness models defined in Sec. 16.3, and the results are compared with those obtained from the complete dataset. The EM algorithm is initialized using complete case analysis, and the iterations are stopped when the change in computed log likelihood is less that 1×10^{-6}.

The results obtained for missing data probabilities $p = 0.05$, 0.20, 0.50, and 0.75 are summarized in Fig. 16.23 in essentially the same format as the corresponding figures presented earlier to assess the results of the other imputation strategies considered in this chapter for these simulation datasets. Specifically, Fig. 16.23 gives boxplot summaries for the correlation coefficients computed from the full datasets (labeled "F"), the MCAR datasets (labeled "C"), the MAR datasets (labeled "R1" and "R2"), and the MNAR datasets (labeled "N"). For each of the four missing data fractions considered here, the EM algorithm yields unbiased results for all of the ignorable missing data examples (i.e., the MCAR and MAR datasets). For these cases, the only evident difference between the EM algorithm results and those obtained from the complete dataset is the somewhat greater variability of the EM results, particularly when the missing data fraction is very large (e.g., the results for $p = 0.75$ shown in the lower right plot in Fig. 16.23). These results are in marked contrast to those obtained under any of the single imputation strategies considered here, both with respect to variability and lack of bias for the ignorable cases; compare, e.g., the results for 75% missing data shown here with those shown in Fig. 16.19 in Sec. 16.5.3 for regression-based imputation, or those shown in Fig. 16.21 in Sec. 16.5.4 for the composite hot-deck/regression method applied to the same cases. It is clear from these comparisons that the EM algorithm is capable of yielding dramatically better results, at least under favorable circumstances. Conversely, the same comparisons for the nonignorable case (the MNAR datasets) show that the EM algorithm can give results that are dramatically worse in unfavorable situations. In particular, these same comparisons for the MNAR case show that the EM algorithm yields a bias that is substantially larger than these other two methods, with comparably high variability.

The price paid to obtain the EM algorithm results in the favorable cases is a considerably greater computing time, especially for large fractions of missing

Figure 16.23: Boxplot summaries of the correlations estimated from the full dataset ("F") and from the EM algorithm for the four incomplete datasets described in Sec. 16.3 ("C" for MCAR, "R1" for MAR1, "R2" for MAR2, and "N" for MNAR), for missing data probabilities $p = 0.05$, 0.20, 0.50, and 0.75

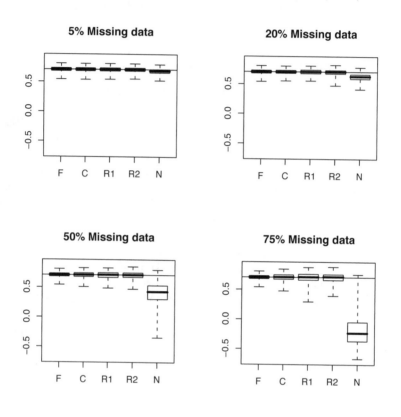

data [188, 207]. In the examples considered here, results for $r = 1000$ independent simulation datasets of size $N = 200$ typically required on the order of a minute for the simple imputation strategies, essentially independent of the fraction of missing data considered. In contrast, the EM algorithm results required approximately twice this long for the case of 5% missing data, and much longer (more than an hour) for the case of 75% missing data. The reason for the difference is that the number of EM algorithm iterations required to reach convergence is a strong function of both the fraction of missing data and the specific dataset considered (e.g., the type of missing data). This point is illustrated in Table 16.1, which gives the number of iterations required to achieve convergence when the EM algorithm is applied to each of the four different missingness models considered here (MCAR, MAR1, MAR2, and MNAR), for the missing data probabilities $p = 0.05$, 0.20, 0.50, and 0.75. All of these datasets

Table 16.1: Numbers of iterations required for the EM algorithm to converge as a function of the missingness pattern and the percentage of missing data

Dataset	$p = 0.05$	$p = 0.20$	$p = 0.50$	$p = 0.75$
MCAR	4	9	17	28
MAR1	7	16	47	80
MAR2	8	16	59	116
MNAR	8	15	58	170

were constructed as described in Sec. 16.3 from the same complete dataset. In all cases, the number of iterations required for EM algorithm convergence increases with increasing p, but the rate of this increase depends strongly on the dataset: from $p = 0.05$ to $p = 0.75$, this increase is a factor of 7 for the MCAR case, about a factor of 10 for the MAR1 dataset, roughly a factor of 15 for the MAR2 dataset, and a factor of 20 for the MNAR dataset. Not surprisingly, the number of iterations required to achieve convergence is consistently the smallest for the MCAR dataset. Also, note that while the differences between the other three cases (MAR1, MAR2, and MNAR) are fairly minor at small and moderate missing data fractions, they become fairly pronounced by the time the missing data fraction reaches 75%.

16.7.6   The Healy-Westmacott regression procedure

The Healy-Westmacott procedure was proposed for regression problems with missing values of the response variable $\{y_k\}$ in 1956 [136], prior to the advent of the EM algorithm, but it has been shown that this procedure can be represented as a special case of the EM algorithm [188, 207]. The essential idea is to alternate between regression-based imputation of the missing values and refinement of the regression model, iterating to convergence. If we make the popular assumption that the regression model errors have a zero-mean Gaussian distribution with unknown variance, this problem may be cast as one of maximum likelihood parameter estimation, leading to the EM algorithm formulation described here.

Specifically, assume that $\{y_k\}$ is a sequence of N response variables, the first $N - m$ of which are observed and the last m are missing. The regression model is based on a collection of p covariates (which may include a dummy vector of all $1's$ to describe a constant intercept term), represented by the p-vector sequence $\{\mathbf{x}_k\}$. An important assumption here is that all N of these covariate vectors are observed since they form the basis for the regression-based imputations of the missing y_k values. The explicit regression model is then given by:

$$y_k = \mathbf{x}_k^T \beta + e_k, \tag{16.30}$$

where $\{e_k\}$ is the sequence of N model prediction errors, assumed to be i.i.d. Gaussian random variables with zero-mean and unknown variance σ^2. Note that since no assumptions are made regarding the relationship between the vectors $\mathbf{x}_k$ for the cases where y_k is observed ($k = 1, 2, \ldots, N - m$) and those where y_k is missing ($k = N - m + 1, \ldots, N$), this general formulation is consistent with either the MCAR or MAR missingness models, but not with the nonignorable MNAR missingness model. In particular, if the covariate vectors $\{\mathbf{x}_k\}$ corresponding to the m missing responses represent a random sample of the complete set of covariate vectors, the missingness model is MCAR; if these covariate vectors differ systematically with respect to one or more components from the non-missing covariate vectors, the missingness model is MAR. Since the conditional distribution of y_k given $\mathbf{x}_k$ and β is the same for the observed and the missing response values, the missingness model cannot be MNAR.

To formulate the EM algorithm for this problem, first note that the complete data likelihood is given by:

$$L_c\{\{y_k\} \mid \{\mathbf{x}_k\}, \beta, \sigma^2\} = -\frac{N}{2}\ln(2\pi\sigma^2) - \frac{1}{2}\sum_{k=1}^{N}\frac{(y_k - \mathbf{x}_k^T\beta)^2}{\sigma^2}$$

$$= -\frac{N}{2}\ln(2\pi\sigma^2) - \frac{1}{2\sigma^2}\sum_{k=1}^{N}y_k^2$$

$$+ \frac{1}{\sigma^2}\sum_{k=1}^{N}y_k\mathbf{x}_k^T\beta - \frac{1}{2\sigma^2}\sum_{k=1}^{N}(\mathbf{x}_k^T\beta)^2. \quad (16.31)$$

Here, since the mean of y_k is not constant but depends on k through its dependence on $\mathbf{x}_k$, the sums of y_k and its square are no longer sufficient statistics, as they were in the univariate case considered in Sec. 16.7.2. Thus, to compute the complete data likelihood for this problem, it is necessary to impute the individual y_k values themselves. This observation forms the basis for the E step of the algorithm, which computes the following conditional expectations:

$$E\{y_k \mid \{\mathbf{x}_k\}, \beta_{(j)}, \sigma_{(j)}^2\} = \mathbf{x}_k^T\beta,$$
$$E\{y_k^2 \mid \{\mathbf{x}_k\}, \beta_{(j)}, \sigma_{(j)}^2\} = (\mathbf{x}_k^T\beta)^2 + \sigma_{(j)}^2. \quad (16.32)$$

The M step updates the estimates $\beta_{(j)}$ and $\sigma_{(j)}^2$ as follows:

$$\beta_{(j+1)} = (\mathbf{X}^T\mathbf{X})^{-1}\mathbf{X}^T\mathbf{y}_{(j)},$$

$$\sigma_{(j+1)}^2 = \frac{1}{N}\left[\sum_{k=1}^{N-m}(y_k - \mathbf{x}_k^T\beta)^2 + m\sigma_{(j)}^2\right], \quad (16.33)$$

where $\mathbf{X}$ is the $N \times p$ matrix whose N rows are the transposed p-vectors $\mathbf{x}_k^T$, and $\mathbf{y}_{(j)}$ is the imputed vector of responses:

$$\mathbf{y}_{(j)} = [y_1, y_2, \ldots, y_{N-m}, \mathbf{x}_{N-m+1}^T\beta_{(j)}, \ldots, \mathbf{x}_N^T\beta_{(j)}]^T. \quad (16.34)$$

An important observation is that the $\beta_{(j+1)}$ update equation above does not depend on $\sigma^2_{(j)}$, but only on $\beta_{(j)}$ and the data matrix $\mathbf{X}$. Consequently, this regression parameter update equation may be iterated to convergence without involving either the E step or the M step for the prediction error variance. Further, given the limiting regression parameter vector $\beta_{(\infty)}$ obtained from these iterations, the variance update equations are simple enough to solve analytically (Exercise 8), giving the final result:

$$\sigma^2_{(\infty)} = \frac{1}{N - m} \sum_{k=1}^{N-m} (y_k - \mathbf{x}_k^T \beta_{(\infty)})^2. \tag{16.35}$$

In fact, it can be shown that the complete case regression model represents a fixed point of the Healy-Westmacott procedure, a result that does not appear to be well known. To establish this result, write the covariate matrix $\mathbf{X}$ and the imputed observation vector $\mathbf{y}_{(j)}$ in partitioned form as:

$$\mathbf{X} = \begin{bmatrix} \mathbf{X}_c \\ \mathbf{X}_m \end{bmatrix} \quad \mathbf{y}_{(j)} = \begin{bmatrix} \mathbf{y}_c \\ \mathbf{X}_m \beta_{(j)} \end{bmatrix}, \tag{16.36}$$

and define $\Delta\beta_{(j)} = \beta_{(j)} - \beta_c$, where β_c is the complete case regression solution:

$$\beta_c = (\mathbf{X}_c^T \mathbf{X}_c)^{-1} \mathbf{X}_c^T \mathbf{y}_c. \tag{16.37}$$

Next, note that the basic iteration for the regression coefficient given in Eq. (16.33) may be rewritten as:

$$
\begin{aligned}
(\mathbf{X}^T \mathbf{X})\beta_{(j+1)} &= \mathbf{X}^T \mathbf{y}_{(j)} \\
\Rightarrow [\mathbf{X}_c^T \mathbf{X}_c + \mathbf{X}_m^T \mathbf{X}_m][\beta_c + \Delta\beta_{(j+1)}] &= \mathbf{X}_c^T \mathbf{y}_c + \mathbf{X}_m^T \mathbf{X}_m[\beta_c + \Delta\beta_{(j)}] \\
\Rightarrow [\mathbf{X}_c^T \mathbf{X}_c + \mathbf{X}_m^T \mathbf{X}_m]\Delta\beta_{(j+1)} &= (\mathbf{X}_m^T \mathbf{X}_m)\Delta\beta_{(j)}.
\end{aligned}
\tag{16.38}
$$

The key observation here is that $\Delta\beta_{(j)} = \mathbf{0}$ for all j is a solution of Eq. (16.38), implying that the complete case regression parameter β_c is a solution of the Healy-Westmacott iteration defined in Eq. (16.33).

16.8 Results for the Pima Indians dataset

The Pima Indians diabetes dataset was introduced in Chapter 2 to illustrate the concept of disguised missing data: although the accompanying metadata indicates the dataset to be complete—and a number of data analysts appear to have treated it as complete—the presence of physiologically infeasible zero values (e.g., for diastolic blood pressure) implies that zero has been used to code missing data. Of the nine variables included in the dataset, five exhibit one or more records with infeasible zeros. Altogether, the dataset includes 768 records and the number of missing values in these five variables ranges from 5 ($\sim$0.7%) to 374 ($\sim$48.7%).

Because this is a real dataset rather than a simulated one, it is not possible to know which—if any—of the standard missing data assumptions are reasonable. The results presented in Chapter 13 for the triceps skinfold thickness variable (TSF, ~29.5% missing) provide strong evidence that the MCAR model is inadequate. Specifically, patients with missing TSF values appear to differ systematically from those with recorded TSF values, particularly with respect to age (these patients appear to be significantly younger) and with respect to the likelihood that the INS variable is also missing (specifically, INS is *always* missing when TSF is missing). It is possible that the MAR model is appropriate to the TSF data (i.e., that the probability of TSF values being missing depend on the other variables in the dataset, but not on the missing TSF values themselves), but it is also possible that the missing TSF data values are MNAR (i.e., that the probability of TSF being missing does depend on the unobserved TSF values). The key point is that we generally cannot know whether the MAR or MNAR models are appropriate since the distinction lies in the missing data values themselves. Thus, to examine the MNAR hypothesis for a real dataset, we must make tentative assumptions about the dependence of missingness on the missing values (e.g., the mean unobserved TSF value is 20% larger than the mean observed TSF value but the variance is the same, corresponding to the slippage model discussed in Chapter 7).

Much of the published analysis of the Pima Indians diabetes dataset is concerned with the development of models to predict the diagnosis of each patient as diabetic or nondiabetic on the basis of the other variables included in the dataset. This problem can be approached in many different ways and a detailed discussion of developing such *binary classification procedures* lies beyond the scope of this book. One applicable method that has been discussed here is logistic regression, introduced in Chapter 14, and the following paragraphs briefly consider the problem of building logistic regression models in the face of the substantial missing data seen in this dataset.

As a first step, consider the results of a complete case analysis. There, all incomplete records are eliminated, reducing the dataset from size $N = 768$ to $N_c = 392$, a reduction of approximately 49%. A logistic regression model is then fit to this complete case dataset, predicting the binary outcome variable DIAG from an intercept term and seven predictors (NPG, PGL, DIA, TSF, INS, BMI, and AGE). A summary of the estimated model coefficients ($\hat{\alpha}_0$), their standard errors (σ_0), and the corresponding p-values for these variables is given in Table 16.2. Taking the standard 5% acceptance threshold for declaring these variables significant in their ability to predict diabetic status, we would retain the intercept term, the plasma glucose concentration (PGL), the body mass index (BMI), and age (AGE), but none of the other terms. In fact, these conclusions do not change even if we adopt the more relaxed preliminary variable selection levels of 15–25% discussed in Chapter 14.

In addition to the model parameters obtained from the logistic regression model fit to all seven candidate prediction variables, the estimated model parameters, their standard errors and their p-values are also listed in Table 16.2 for the reduced model obtained by retaining only those terms showing signifi-

Table 16.2: Complete case logistic regression model summaries for the Pima Indians dataset, predicting diabetic status from the indicated covariates. Columns labeled with the subscript "0" correspond to the full model, and those labeled with the subscript "r" correspond to the reduced model that retains only those terms from the full model that are significant at the 5% level

Term	$\hat{\alpha}_0$	σ_0	p-Value, p_0	$\hat{\alpha}_r$	σ_r	p-Value, p_r
Intercept	-9.459	1.173	7.52×10^{-16}	-9.677	1.042	$< 2 \times 10^{-16}$
NPG	0.067	0.054	2.16×10^{-1}	—	—	—
PGL	0.038	0.006	2.96×10^{-11}	0.036	0.005	1.45×10^{-13}
DIA	-0.003	0.012	7.64×10^{-1}	—	—	—
TSF	0.015	0.017	3.70×10^{-1}	—	—	—
INS	-0.001	0.001	6.16×10^{-1}	—	—	—
BMI	0.070	0.027	8.90×10^{-3}	0.078	0.020	1.09×10^{-4}
AGE	0.038	0.018	3.37×10^{-2}	0.054	0.013	4.40×10^{-5}

cance in the full model. Comparing the original model parameters with those for the reduced model, we see relatively small changes in the coefficients, in general, although the AGE coefficient increases by approximately 40%. In addition, the standard error for AGE also declines by approximately 25%, resulting in a strong increase in significance from about 3% to less than 0.01%.

A key observation here is that neither of the highly missing variables, triceps skinfold thickness (TSF) and serum insulin concentration (INS), appear as significant predictors in the full complete case model. This observation suggests the use of the omission strategy discussed in Sec. 16.4.2 for these two variables. Eliminating these variables and performing a complete case analysis on the rest of the dataset yields an analysis dataset with $N = 724$ observations, corresponding to a missing data fraction of approximately 6%.

The results obtained from this analysis are presented in Table 16.3, in the same format as Table 16.2. That is, parameter estimates $\hat{\alpha}_0$, standard errors σ_0 and the associated p-values p_0 are listed for the intercept term and the five predictor variables NPG, PGL, DIA, BMI, and AGE included in the full logistic regression model that predicts diabetic status from all of these variables. Similarly, the parameter estimates $\hat{\alpha}_r$, standard errors σ_r, and p-values p_r are also given in Table 16.3 for the reduced model that predicts diabetic status only from those variables that appear as significant terms in the full model. Here, these variables include the intercept term, the number of times pregnant (NPG), the plasma glucose concentration (PGL), and the body mass index (BMI).

In all of the logistic regression models examined here, the most significant predictor variable was plasma glucose concentration (PGL), and body mass in-

Table 16.3: Modified complete case logistic regression model summaries for the Pima Indians dataset, predicting diabetic status from the same covariates as Table 16.2, but with TSF and INS excluded from the analysis. The format of this table is the same as that of Table 16.2.

Term	$\hat{\alpha}_0$	σ_0	p-Value, p_0	$\hat{\alpha}_r$	σ_r	p-Value, p_r
Intercept	−8.609	0.802	$< 2 \times 10^{-16}$	−8.689	0.696	$< 2 \times 10^{-16}$
NPG	0.114	0.033	5.85×10^{-4}	0.139	0.028	7.26×10^{-7}
PGL	0.036	0.004	$< 2 \times 10^{-16}$	0.036	0.003	$< 2 \times 10^{-16}$
DIA	−0.010	0.009	2.45×10^{-1}	—	—	—
BMI	0.095	0.016	1.00×10^{-9}	0.088	0.015	2.85×10^{-9}
AGE	0.017	0.010	7.41×10^{-2}	—	—	—

dex (BMI) also appears as a significant predictor variable in all models. The main difference between the models fit by complete case analysis of the dataset with triceps skinfold thickness (TSF) and serum insulin concentration (INS) included and those fit by complete case analysis omitting both of these variables was in the inclusion or exclusion of the variables NPG (number of times pregnant) and AGE. Specifically, the models that included INS and TSF found AGE to be significant but not NPG, while exactly the opposite result was found when these variables were excluded. Since these results were obtained in both cases for reduced models that did not include either INS or TSF as predictors, the difference in these results is apparently due to the much larger size of the dataset that excludes these two highly incomplete variables ($N = 392$ when INS and TSF are included, vs. $N = 724$ when they are excluded).

16.9 General conclusions

The results obtained from the simulation datasets presented here provide the basis for some general conclusions regarding the treatment of missing data. First and foremost, the these results have demonstrated that nonignorable missing data—i.e., the MNAR model, where the probability of an observation being absent depends on its value—can lead to severe biases under any of the missing data treatment strategies considered here. *In general, the problem of nonignorable missing data is difficult to either detect or treat.* As a practical matter, suspected cases of nonignorable missing data are typically handled by making tentative assumptions about how the unobserved part of the dataset differs from the observed part and applying a missing data treatment strategy that takes these differences into account. The example considered in Sec. 16.7.3 represents a case in point: there, our ability to obtain a reasonable model for the complete dataset dependend on our knowing specifically how the missing data characteristics differed from the observed data characteristics. Conversely, the practical corollary is that techniques that work well in the context of ignorable missing data (i.e., the MCAR and MAR models) can work very poorly in the nonignorable case. This point was demonstrated repeatedly in the examples presented here, both in the univariate examples and the bivariate examples considered, where most of the missing data treatment methods considered in this chapter gave badly biased results in the face of nonignorable missing data. In cases where specific nonignorable missingness mechanisms are appropriate (e.g., the case of *censored data*, where observations larger than some maximum value x^* are recorded as x^*), specialized treatment techniques may be available; for a more detailed introduction to these ideas and others useful in dealing with nonignorable missing data, refer to the book by Little and Rubin [188, ch. 15].

Turning to the more tractable problem of ignorable missing data (i.e., the MCAR and MAR missingness models), it is useful to consider the simpler univariate case first before turning to the general multivariate case. Recall that complete case analysis of the univariate MCAR missing data model corresponds to random subsampling, which can be expected to give unbiased estimates of increased variability relative to the analysis of the complete dataset. The four univariate estimators considered here all illustrate this behavior: the mean, median, standard deviation, and MAD scale estimator. Similar behavior is observed for hot-deck imputation, which corresponds to filling in the missing data values by sampling with replacement from the observed data values. In contrast, mean imputation reduces the variability of the imputed dataset relative to the complete dataset and this can cause significant biases even in the case of MCAR data, as seen in the standard deviation estimates considered in Sec. 16.2.4. An even more severe consequence of mean imputation was the catastrophic implosion of the MAD scale estimator for the mean-imputed MCAR datasets, illustrating the point made earlier that strategies like hot-deck imputation or mean imputation that duplicate data values may cause difficulties for rank-based analysis procedures where tied values can be problematic.

As noted earlier in this chapter, the multivariate setting opens a wider range of possibilities for handling missing data than the univariate setting. A case in point is the difference between complete case analysis and available case analysis. The comparisons of the results obtained by these methods presented in Sec. 16.4.1 strongly suggest that complete case analysis is generally to be preferred over available case analysis, despite the fact that available case analysis incorporates more of the data into the results, sometimes substantially more. The key disadvantage of available case analysis is that it does not treat all variables in a multivariate analysis equivalently, and this differential treatment can lead to significant biases, a point illustrated in Sec. 16.4.1 and widely discussed in the missing data literature [188, 258].

As in the univariate case, mean imputation applied to the ignorable missing data bivariate correlation estimation problems considered here tended to give significant biases, even in the MCAR case. In fact, this observation also holds for the hot-deck imputation strategy in the bivariate case, in marked contrast to the univariate case. The clear imputation patterns seen in the plots of the imputed bivariate datasets shown in Sec. 16.5 suggest that part of the difficulty in both cases lies in the inherently "componentwise" character of these imputation strategies. An inherently bivariate strategy is the regression-based imputation method considered in Sec. 16.5.3, but even there, a regular pattern may be seen clearly in the imputed data values that is distinct from that of the rest of the dataset. The MCAR results are still biased in this case, but much less so than in the case of mean or hot-deck imputation. Combining the notion of sampling with replacement on which hot-deck imputation is based with regression imputation leads to the hybrid strategy described in Sec. 16.5.4, which appears to work better than either hot-deck or regression imputation in the case of MCAR data.

Far and away, the best results for the bivariate datasets were obtained using the EM algorithm, which gave unbiased results for all three ignorable missing data examples considered here (MCAR, MAR1, and MAR2). Further, the variability of these results was smaller than that seen for the other missing data treatment methods applied to these cases. Two prices paid for this improved performance were first, the need for a detailed data model (i.e., the EM algorithm considered here assumed knowledge of the bivariate Gaussian nature of the dataset in order to estimate the unknown mean and covariance parameters), and second, the much greater computation times required: cases where hot-deck imputation results were available in about a minute required EM computation times of more than an hour. This aspect of the EM algorithm is well known, particularly in the face of high missing data fractions [207, p. 2].

Overall, the results presented here suggest the following conclusions. First, complete case analysis represents a useful preliminary step, which may be perfectly adequate if the fraction of missing data is small and the missing data mechanism is ignorable. Second, strategies like componentwise mean imputation that cause significant duplication of individual data values can perform very badly, even in the "favorable" case of MCAR data, especially if rank-based analysis is considered. Third, of the simple imputation strategies considered in Sec. 16.5, the hot-deck/regression composite method described in Sec. 16.5.4 gave the best results for the specific examples considered here. Finally, fourth, in problems where the EM algorithm is appropriate and its implementation and application are feasible, this method can give results that are almost as good as those obtained from the complete dataset. The prices paid for this performance are complexity of development when a ready-made implementation is not available, and the potential of long computation times, especially in the face of substantial missing data fractions.

16.10   Exercises

1. Show that the expected number of incomplete observations (i.e., observations with x_k missing, y_k missing, or both) is $p(1 - p/4)N$ for the MCAR simulation dataset.

2. Show that the complete data log likelihood for the univariate example with nonignorable missing data considered in Sec. 16.7.3 is given by Eq. (16.14):

$$
\begin{aligned}
L_c\{\{x_k\} \mid \mu, \sigma^2\} &= -\frac{N}{2}\ln(2\pi\sigma^2) - \frac{m}{2}\ln\tau^2 - \frac{1}{2\sigma^2}S_2 + \frac{\mu}{\sigma^2}S_1 \\
&\quad - \frac{\mu^2}{2\sigma^2}[N - m + m\lambda^2/\tau^2],
\end{aligned}
$$

where S_1 and S_2 are the sufficient statistics defined in Eq. (16.15).

3. Derive the maximum likelihood estimators $\hat{\mu}$ and $\widehat{\sigma^2}$ shown in Eq. (16.16) for the nonignorable missing data example considered in Sec. 16.7.3.

4. Compute the limiting values $\mu_{(\infty)}$ and $\sigma^2_{(\infty)}$ for the EM algorithm described in Sec. 16.7.3 for the univariate MNAR problem considered there.

5. Show that the complete data log likelihood for the bivariate Gaussian example considered in Sec. 16.7.4 is given by Eq. (16.21).

6. Show that the complete data maximum likelihood estimators for the bivariate Gaussian parameters are given by Eq. (16.23).

7. Show that the conditional expectation of the sums T_2, T_{11}, T_{22}, and T_{12} given the observed data and the parameters $\mu^{(j)}$ and $\Sigma^{(j)}$ for the bivariate Gaussian example discussed in Sec. 16.7.4 are given by Eq. (16.27).

8. Show that the limiting variance for the Healy-Westmacot procedure described in Sec. 16.7.6 is given by:

$$
\sigma^2_{(\infty)} = \frac{1}{N-m}\sum_{k=1}^{N-m}(y_k - \mathbf{x}_k^T\beta_{(\infty)})^2,
$$

where $\beta_{(\infty)}$ is the limiting vector of regression coefficients obtained by the procedure.

Appendix: The R data analysis language

All of the examples presented in this book were developed using one of two variants of the S programming language: the commercially-supported implementation *S-plus* [272], or the near-equivalent open-source implementation R [240]. Because it is available at no cost to anyone who wants it, this appendix gives a very brief description of R and how to obtain it. *It is important to emphasize that neither this appendix nor the book overall are intended as an introduction to the* R *software environment.* The intent of this appendix is only to make readers aware of what R is and how to obtain it. A number of comprehensive introductions to the language and its use are available, and the books by Venables and Ripley [283] and Crawley [72] are especially recommended. In addition, much free information is also available from the CRAN Web site discussed below (see the discussion "Obtaining R").

What exactly is R?

R is an open-source programming environment that supports a wide—and constantly expanding—range of statistically-based data analysis procedures. Because it is open-source, R is available at no cost for PC or Apple MAC computers, running under Linux, Mac OS or Windows operating systems, and it is becoming increasingly popular among both academic and commercial user communities. The basic R package ("base R") is an interactive programming language that supports many different statistical and graphical data exploration and analysis procedures. In particular, *all* of the data analysis procedures discussed in this book can be easily implemented in R, and many are available as built-in procedures. The rapid expansion of the capabilities of R is largely due to the support built into the language for optional *packages* that provide additional capabilities beyond those included in base R. A few of these packages are characterized as *recommended* and are included by default in the basic R installation. (One example is the package **MASS** described in Venables and Ripley's book to support many of the more complex analysis procedures discussed there [283].) Other packages are more specialized and are not of interest to all users:

making them available as optional packages that can be freely downloaded provides these capabilities to those who want them without making the overall package enormous and impossible to maintain (at the time of this writing, there are over 2000 optional R packages available).

Obtaining R

The easiest way to obtain R is to go to the Comprehensive R Archive Network (CRAN) at the following Web site:

http://cran.r-project.org/

and follow the instructions given there for downloading the software. In addition, this Web site also provides links to a great deal of online documentation, with the following four being particularly useful:

1. *An Introduction to R* provides a general introduction to the language and its use;

2. *R Installation and Administration* gives detailed instructions on installing and maintaining R on your computer;

3. *R Data Import/Export* describes data import and export features available either in R itself or in various add-on packages;

4. *R: A Language and Environment for Statistical Computing* contains all of the detailed help files for the standard and recommended R packages.

Using R

Once R is installed, simply click on the R icon to run it (or look under "Programs" to find it if the icon is not visible on the desktop). This brings up an interactive session that allows you to enter commands at the ">" prompt. As an extremely simple but illustrative example:

```
> sqrt(pi)
> 1.772454
```

The first line is the command "compute the square root of π," and the second line gives R's result of the computation. As a slightly more representative example, the following pair of commands generates a sample of 100 random numbers with a standard (i.e., zero-mean, unit-variance) Gaussian distribution and plots them:

```
> x <- rnorm(100)
> plot(x)
```

744

The symbol "<-" in the first line assigns the results of the normal random number generator (the "rnorm" function) to the variable "x," and the second line plots this variable, which represents a vector of 100 numbers.

To conclude an *R* session, the command is "q()" (for "quit"), i.e.:

```
> q()
```

When you execute this command, you will be asked whether you wish to save your working session, which includes all data objects or procedures you created during the session; if you do, *R* will restore this working session the next time you invoke the *R* icon.

One important practical point is that any optional add-on package not included in the base *R* distribution must be installed explicitly before it can be used. If your computer is connected to the Internet, this can be done within *R* using the "install.packages" command. For example, to install the optional "beanplot" package, the command is:

```
> install.packages("beanplot")
```

Also, before an available library can be used in an *R* session, it must be loaded using the "library" command, whether it is a "recommended" package included with the base *R* distribution or an "optional" package installed as just described. Thus, to use procedures from the **MASS** library, the following command must be executed:

```
> library(MASS)
```

After you conclude an *R* session, even if you have saved the workspace, any libraries that were loaded during the session must be reloaded in the next session where they are needed (but not reinstalled: any installed libraries are still available, but they must be reloaded using the "library" command before they can be used in a new *R* session).

Finally, it is worth reemphasizing that the discussion presented here is *not* intended as a detailed introduction to *R*, but simply to give a sense of what the package is and what it can do, again noting that all of the results presented in this book can be generated using it. Most of the datasets used in these examples and most of the *R* procedures used to generate the results presented here are available from the Oxford University Press Web site:

http://www.oup.com/us/ExploringData

Bibliography

[1] M. Abramowitz and I.A. Stegun. *Handbook of Mathematical Functions.* Dover, 1972.

[2] F.S. Acton. *Numerical Methods That Work.* Harper and Row, 1970.

[3] J. Aczel and J. Dhombres. *Functional equations in several variables.* Cambridge University Press, 1989.

[4] J. Aczel, F.S. Roberts, and Z. Rosenbaum. On scientific laws without dimensional constants. *J. Math. Anal. Appl.*, 119:389–416, 1986.

[5] L.A. Adamic and B.A. Huberman. Zipf's law and the Internet. *Glottometrics*, 3:143–150, 2002.

[6] P. Adriaans and D. Zantinge. *Data Mining.* Addison-Wesley, 1996.

[7] N. Afari and D. Buchwald. Chronic fatigue syndrome: a review. *Psychiatry*, 60:221–236, 2003.

[8] A. Agresti. *Categorical Data Analysis.* Wiley, 2nd edition, 2002.

[9] S. Albeverio, P. Blanchard, and R. Hoegh-Krohn. A stochastic model for the orbits of planets and satellites: An interpretation of titus-bode law. *Expositiones Mathematicae*, 4:365–373, 1983.

[10] D. Aldous. *Probability Approximations via the Poisson Clumping Heuristic.* Springer-Verlag, 1989.

[11] D.F. Andrews, P.J. Bickel, F.R. Hampel, P.J. Huber, W.H. Rogers, and J.W. Tukey. *Robust Estimates of Location.* Princeton University Press, 1972.

[12] D.F. Andrews and A.M. Herzberg. *Data.* Springer-Verlag, 1985.

[13] Anonymous. The pig in different coordinates. *Chemtech*, October 1986.

[14] Anonymous. Apples. Technical Report 16, North Carolina Department of Agriculture, Agronomic Division, 1994.

[15] F.J. Aranda-Ordaz. On two families of transformations to additivity for binary response data. *Biometrika*, 68:357–363, 1981.

[16] B.C. Arnold, N. Balakrishnan, and H.N. Nagaraja. *A First Course in Order Statistics*. Wiley, 1992.

[17] B.C. Arnonld, E. Castillo, J.M. Sarabia, and L. Gonzalez-Vega. Multiple modes in densities with normal conditionals. *Statistics Probab. Lett.*, 49:353–363, 2000.

[18] M.J. Assael, A. Leipertz, E. MacPherson, Y. Nagasaka, C.A. Nieto de Castro, R.A. Perkins, K. Ström, E. Vogel, and W.A. Wakeham. Transport property measurements on the IUPAC sample of 1,1,2,2-tetrafluoroethane (R134a). *Int. J. Thermophysics*, 21:1–22, 2000.

[19] A.C. Atkinson. *Plots, Transformations, and Regression*. Oxford, 1985.

[20] T.C. Atkinson, K.R. Briffa, and G.R. Coope. Seasonal temperatures in britain during the past 22,000 years, reconstructed using beetle remains. *Nature*, 325:587–592, 1987.

[21] T.C. Atkinson, K.R. Briffa, G.R. Coope, M.J. Joachim, and D.W. Perry. Climatic calibration of coleopteran data. In B.E. Berglund, editor, *Handbook of Holocene Palaeoecology and Palaeohydrology*, chapter 41, pages 851–858. Wiley, 1986.

[22] V.G. Babak. Line tension in thin liquid films and layer thermodynamics. *Rev. Chem. Eng.*, 15:157–221, 1999.

[23] N. Balakrishnan and A.P. Basu, editors. *The Exponential Distribution: Theory, Methods, and Applications*. Gordon and Breach, 1995.

[24] K.P. Balanda and H.L. MacGillivray. Kurtosis and spread. *Can. J. Stat.*, 18:17–30, 1990.

[25] B.C. Barish. The physics of monopole detection. In J.L. Stone, editor, *Monopole '83*, volume 111 of *NATO ASI Series B: Physics*, pages 367–381, 1984.

[26] V. Barnett and T. Lewis. *Outliers in Statistical Data*. Wiley, 2nd edition, 1984.

[27] V. Barnett and T. Lewis. *Outliers in Statistical Data*. Wiley, 3rd edition, 1994.

[28] I. Barrodale and F.D.K. Roberts. An improved algorithm for discrete L_1 linear approximation. *SIAM J. Num. Anal.*, 10:839–848, 1973.

[29] I. Barrodale and F.D.K. Roberts. An efficient algorithm for discrete L_1 approximation with linear constraints. *SIAM J. Num. Anal.*, 15:603–611, 1978.

[30] A.P. Basu. Bivariate exponential distributions. In N. Balakrisnhan and A.P. Basu, editors, *The Exponential Distribution*, chapter 20, pages 327–332. Gordon and Breach Publishers, 1995.

[31] S.L. Beal. Asymptotic confidence intervals for the difference between two binomial parameters for use with small samples. *Biometrics*, 43:941–950, 1987.

[32] G. Belforte, B. Bona, and V. Cerone. Parameter estimation with set membership uncertainty: Nonlinear families of models. In *Proc. 8th IFAC/IFORS Symp. Identification System Parameter Estimation*, pages 399–404, Beijing, August 1988.

[33] G. Belforte, B. Bona, and V. Cerone. Identification, structure selection and validation of uncertain models with set-membership error description. *Math. Comput. Simulation*, 32:561–569, 1990.

[34] D.A. Belsley, E. Kuh, and R.E. Welsch. *Regression Diagnostics: Identifying Influential Data and Sources of Colinearity*. Wiley, 1980.

[35] J.E. Bentley. Metadata: Everyone talks about it, but what is it? Paper 125-26. In *Proc. SAS User's Group Intl. Conf., SUGI26*, Cary, NC, USA, 2001.

[36] R. Bhatia and C. Davis. A better bound on the variance. *Amer. Math. Monthly*, 107:353–357, 2000.

[37] P. Billingsley. *Probability and Measure*. Wiley, 1986.

[38] H.W. Block and A.P. Basu. A continuous bivariate exponential extension. *J. American Statistical Association*, 69:1031–1037, 1974.

[39] P. Borwein and T. Erdelyi. *Polynomials and Polynomial Inequalities*. Springer, 1995.

[40] G.E.P. Box and G.C. Tiao. *Bayesian Inference in Statistical Analysis*. Addison-Wesley, 1973.

[41] R.N. Bracewell. *The Fourier transform and its applications*. McGraw-Hill, 2nd edition, 1978.

[42] J. Breault. Data mining diabetic databases: Are rough sets a useful addition? In *Proc. 33rd Symposium on the Interface, Computing Science and Statistics*, Fairfax, VA, USA, 2001.

[43] P. Bremaud. *An Introduction to Probabilistic Modeling*. Springer-Verlag, 1988.

[44] P.J. Brockwell and R.A. Davis. *Time Series: Theory and Methods*. Springer-Verlag, 2nd edition, 1991.

[45] J.D. Broffitt. Zero correlation, independence, and normality. *Amer. Stat.*, 40:276–277, 1986.

[46] A. Browder. *Mathematical Analysis.* Springer, 1996.

[47] L.D. Brown, T.T. Cai, and A. DasGupta. Confidence intervals for a binomial proportion and asymptotic expansions. *Ann. Statist.*, 30:160–201, 2002.

[48] K.A. Brownlee. *Statistical Theory and Methodology in Science and Engineering.* John Wiley, 1960.

[49] J.A. Burns. Some background about satellites. In *Satellites*, chapter 1. University of Arizona Press, 1986.

[50] R.W. Cahn. Superdiffusion in solid helium. *Nature*, 400:512–513, 1999.

[51] W.A. Calder. *Size, Function, and Life History.* Dover, 1984.

[52] S. Cambanis, S. Huang, and G. Simons. On the theory of elliptically contoured distributions. *J. Multivar. Anal.*, 11:368–385, 1981.

[53] A.C. Cameron and P.K. Trivedi. *Regression Analysis of Count Data.* Cambridge University Press, 1998.

[54] A.D. Caplin, M. Hardiman, and J.C. Schouten. Reply to: Observation of an unexplained event from a magnetic monopole detector. *Nature*, 325:463, 1987.

[55] S. B. Carter, S. S. Gartner, M. R. Haines, A. L. Olmstead, R. Sutch, and G. Wright, editors. *Historical Statistics of the United States, Earliest Times to the Present: Millennial Edition.* Cambridge University Press, 2006.

[56] E. Castillo and J. Galambos. Conditional distributions and the bivariate normal distribution. *Metrika*, 36:209–214, 1989.

[57] A. Cezairliyan. *Specific Heat of Solids*, volume I-2. Hemisphere Publishing Co., 1988.

[58] J. Chandrasekhar. Organic structures with remarkable carbon-carbon distances. *Curr. Sci.*, 63:114–116, 1992.

[59] S.J. Chapman. Drums that sound the same. *Amer. Math. Monthly*, 102:124–138, 1995.

[60] J. Chapuis, C. Eck, and H.P. Geering. Autonomously flying helicopter. In *Proc. 4th IFAC Symposium Advances Control Education*, pages 119–124, Istanbul, 1997.

[61] H. Chernoff, J.L. Gastwirth, and M.V. Johns. Asymtotic distribution of linear combinations of functions of order statistics with applications to estimation. *Ann. Math. Stat.*, 38:52–72, 1967.

[62] U. Cherubini, E. Luciano, and W. Vecchiato. *Copula Methods in Finance.* Wiley, 2004.

[63] Y.S. Chow and H. Teicher. *Probability Theory: Independence, Interchangability, Martingales.* Springer-Verlag, 2nd edition, 1988.

[64] V.J. Clancy. Statistical methods in chemical analysis. *Nature*, 4036:339–340, 1947.

[65] D.G. Clayton. A model for association in bivariate life tables and its application in epidemiological studies of familial tendency in chronic disease incidence. *Biometrika*, 65:141–151, 1978.

[66] D. Collett. *Modelling Binary Data.* Chapman and Hall/CRC, 2nd edition, 2003.

[67] P.L. Combettes. The foundations of set theoretic estimation. *Proc. IEEE*, 81:182–208, 1993.

[68] R.D. Cook and M.E. Johnson. A family of distributions for modelling non-elliptically symmetric multivariate data. *J. Royal Statist. Soc., Ser. B*, 43:210–218, 1981.

[69] F. Copleston. *A History of Philosophy, Volume III: Ockham to Suárez.* Paulist Press, 1953.

[70] J.A. Cordeiro. Logical operators of some statistical computing packages and missing values. *Comp. Stat. and Data Anal.*, 51:2783–2787, 2007.

[71] T.M. Cover and J.M. Thomas. *Information Theory.* Wiley, 2nd edition, 2006.

[72] M.J. Crawley. *The R Book.* Wiley, 2007.

[73] R.A. Cribbie, R.R. Wilcox, C. Bewell, and H.J. Keselman. Tests for treatment group equality when data are nonnormal and heteroscedastic. *J. Modern Applied Statistical Methods*, 6:117–132, 2007.

[74] C. Croux and P.J. Rousseeuw. Time-efficieint algorithms for two highly robust estimators of scale. In Y. Dodge and J. Whittaker, editors, *Computational Statistics*, volume 1, pages 411–428. Physika-Verlag, 1992.

[75] D.J. Currie. Estimating Michaelis-Menten parameters: Bias, variance, and experimental design. *Biometrics*, 38:907–919, 1982.

[76] R.B. D'Agostino and M.A. Stephens, editors. *Goodness-of-fit Techniques.* Marcel Dekker, 1986.

[77] D.J. Daley and D. Vere-Jones. *An Introduction to the Theory of Point Processes*. Springer-Verlag, 1988.

[78] C. Daniel and F.S. Wood. *Fitting Equations to Data*. Wiley, 1971.

[79] C. Daniel and F.S. Wood. *Fitting Equations to Data*. Wiley, 2nd edition, 1980.

[80] G.B. Dantzig. *Linear Programming and Extensions*. Princeton University Press, 1963.

[81] C.J. Date. *An Introduction to Database Systems*. Addison-Wesley, 7th edition, 2000.

[82] L. Davies and U. Gather. The identification of multiple outliers. *J. Amer. Statist. Assoc.*, 88:782–792, 1993.

[83] L. Davies and U. Gather. Rejoinder. *J. Amer. Statist. Assoc.*, 88:797–801, 1993.

[84] N. Davies. *Europe: A History*. Oxford University Press, 1996.

[85] A.C. Davison and D.V. Hinkley. *Bootstrap Methods and their Application*. Cambridge University Press, 1997.

[86] A.P. Dempster. Purposes and limitations of data analysis. In G.E.P. Box, T. Leonard, and C.F. Wu, editors, *Scientific Inference, Data Analysis, and Robustness*. Academic Press, 1983.

[87] A.P. Dempster, N.M. Laird, and D.B. Rubin. Maximum likelihood from incomplete data via the em algorithm (with discussion). *J. Royal Statist. Soc. Ser. B*, 39:1–38, 1977.

[88] D. DesJardins. Outliers, inliers, and just plain liars—new EDA+ (EDA Plus) techniques for understanding data. Paper 169. In *Proc. SAS User's Group Intl. Conf., SUGI26*, Cary, NC, USA, 2001.

[89] P. Diaconis. Theories of data analysis: From magical thinking through classical statistics. In D.C. Hoaglin, F. Mosteller, and J.W. Tukey, editors, *Exploring Data Tables, Trends, and Shapes*, chapter 1, pages 1–36. Wiley, 1985.

[90] P. Diaconis, S. Holmes, and R. Montgomery. Dynamical bias in the coin toss. *SIAM Review*, 49:211–235, 2007.

[91] Y. Dodge. An introduction to statistical data analysis L_1-norm based. In Y. Dodge, editor, *Statistical Data Analysis Based on the L_1-norm and Related Methods*, pages 1–21. North-Holland, 1987.

[92] Y. Dodge. The guinea of multiple regression. In H. Rieder, editor, *Robust Statistics, Data Analysis and Computer Intensive Methods*, pages 91–117. Springer-Verlag, 1996.

[93] D. Draper, J.S. Hodges, C.L. Mallows, and D. Pregibon. Exchangeability and data analysis. *J. Royal Statist. Soc., Ser. A.*, 156, Part I:9–37, 1993.

[94] N.R. Draper and H. Smith. *Applied Regression Analysis.* Wiley, 2nd edition, 1981.

[95] N.R. Draper and H. Smith. *Applied Regression Analysis.* Wiley, 3rd edition, 1998.

[96] D. Dubois and H. Prade. *Fuzzy Sets and Systems.* Academic Press, 1980.

[97] B. Efron and R.J. Tibshirani. *An Introduction to the Bootstrap.* Chapman and Hall, 1993.

[98] R.S. Elliot. *Electromagnetics.* IEEE Press, 1993.

[99] G.J. Evans. A new crystal-growing technique: the reciprocal of the thermodielectric effect. *J. Chem. Soc., Faraday Trans. 1*, 80:2343–2348, 1984.

[100] P. Eyerer. Electric charge separation and charge storage during phase changes in the absence of external electric fields: Thermodielectric effect (Costa Ribeiro effect) and Workman-Reynolds effect. *Adv. Colloid Interface Sci.*, 3:223–273, 1972.

[101] K.-T. Fang, S. Kotz, and K.-W. Ng. *Symmetric Multivariate and Related Distributions.* Chapman and Hall, 1990.

[102] D. Feng and C. Aldrich. Effect of fluid properties on two-phase froth characteristics. *Ind. Eng. Chem. Res.*, 38:4110–4112, 1999.

[103] A. Ferro-Luzzi and W.P.T. James. Adult malnutrition: simple assessment techniques for use in emergencies. *British J. Nutrition*, 75:3–10, 1996.

[104] C.A. Floudas. *Nonlinear and Mixed-Integer Optimization.* Oxford, 1995.

[105] E.B. Fowlkes. *A Folio of Distributions.* Marcel Dekker, 1987.

[106] D.A. Freedman. A note on screening regression equations. *Amer. Stat.*, 37:152–155, 1983.

[107] K. Fukuda, S.E. Strauss, I. Hickie, M.C. Sharpe, J.G. Dobbins, and A. Komaroff. The chronic fatigue syndrome; a comprehensive approach to its definition and study. *Ann. Inst. Med.*, 121:953–959, 1994.

[108] K.Y. Fung and S.R. Paul. Comparisons of outlier detection procedures in weibull or extreme-value distributions. *Commun. Statist. Sim. Comp.*, 14:895–917, 1985.

[109] R.C. Funt and M.A. Ellis. Cork spot and bitter pit of apples. Technical Report HYG-1403-92, Ohio State University Extension, 1992.

[110] W. Gander, G.H. Golub, and R. Strebel. Least-squares fitting of circles and ellipses. *BIT*, 34:558–578, 1994.

[111] M. Gardner. *Fads and Fallacies in the Name of Science*. Dover, 1957.

[112] J. Gastwirth. On robust procedures. *J. Amer. Statist. Assoc.*, 61:929–948, 1966.

[113] C. Geertz. *Local Knowledge*. Basic Books, 1983.

[114] C. Genest, M. Gendron, and M. Bourdeau-Brien. The advent of copulas in finance. *European J. Finance*, 2009. in press.

[115] C. Genest and L.-P. Rivest. Statistical inference procedures for bivariate archimedian copulas. *J. Amer. Statis. Assoc.*, 88:1034–1043, 1993.

[116] G.C. Glatzmaier and W.F. Ramirez. Simultaneous measurement of the thermal conductivity and thermal diffusivity of unconsolidated materials by the transient hot wire method. *Rev. Sci. Instrum.*, 56:1394–1398, 1985.

[117] B. Goethals. Survey on frequent pattern mining. Technical report, Helsinki Institute for Information Technology, 2003.

[118] G.H. Golub and C.F. van Loan. *Matrix Computations*. Johns Hopkins, 2nd edition, 1989.

[119] P. Good. *Permutation Tests*. Springer-Verlag, 1994.

[120] A.D. Gordon. *Classification*. Chapman and Hall, 1999.

[121] C. Gordon, D. Webb, and S. Wolpert. You cannot hear the shape of a drum. *Bull. Amer. Math. Soc.*, 27:134–138, 1992.

[122] S.I. Gradshteyn and I.M. Ryzhik. *Table of Integrals, Series and Products*. Academic Press, 1965.

[123] N.N. Greenwood and A. Earnshaw. *Chemistry of the Elements*. Pergamon Press, 1984.

[124] G.R. Grimmett and D.R. Stirzaker. *Probability and Random Processes*. Oxford, 1992.

[125] J.H. Gundlach, E.G. Adelberger, B.R. Heckel, and H.E. Swanson. New technique for measuring Newton's constant G. *Phys. Rev. D*, 54:R1256–R1259, 1996.

[126] J.H. Gundlach and S.M. Merkowitz. Measurement of Newton's constant using a torsion balance with angular acceleration feedback. *Phys. Rev. Lett.*, 85:2869–2872, 2000.

[127] C.N. Guy. Observation of an unexplained event from a magnetic monopole detector. *Nature*, 325:463, 1987.

[128] S.M. Haile, D.W. Johnson, G.H. Wiseman, and H.K. Bowen. Aqueous precipitation of spherical zinc oxide powders for varistor applications. *J. Amer. Ceram. Soc.*, 72:2004–2008, 1989.

[129] J. Hajek, Z. Sidak, and P.K. Sen. *Theory of Rank Tests.* Academic Press, 1999.

[130] P. Hall. *The Bootstrap and Edgeworth Expansion.* Springer-Verlag, 1992.

[131] D. Halliday, R. Resnick, and J. Walker. *Fundamentals of Physics.* Wiley, 4th edition, 1993.

[132] F.R. Hampel. The influence curve and its role in robust estimation. *J. Amer. Statist. Assoc.*, 69:383–393, 1974.

[133] G.J. Hanh and S.S. Shapiro. *Statistical Models in Engineering.* Wiley, 1967.

[134] W. Härdle. *Applied Nonparametric Regression.* Cambridge University Press, 1990.

[135] T.J. Hastie and R.J. Tibshirani. *Generalized Additive Models.* Chapman and Hall, 1990.

[136] M.J.R. Healy and M. Westmacott. Missing values in experiments analyzed on automatic computers. *App. Stat.*, 5:203–206, 1956.

[137] G. Heinze and M. Schemper. A solution to the problem of separation in logistic regression. *Stat. in Med.*, 21:2409–2419, 2002.

[138] J.C. Hey and W.P. Kram, editors. *Transient Voltage Suppression Manual.* General Electric, 1978.

[139] M.S. High and R.P. Danner. Treatment of gas-solid adsorption data by the error-in-variables method. *AIChE J.*, 32:1138–1145, 1986.

[140] R.J. Hilderman and H.J. Hamilton. Evaluation of interestingness measures for ranking discovered knowledge. In *Proc. 5th Asia-Pacific Conf. Advances Knowledge Discovery*, pages 247–259, Hong Kong, April 2001.

[141] D.C. Hoaglin, F. Mosteller, and J.W. Tukey, editors. *Exploring Data Tables, Trends, and Shapes.* Wiley, 1985.

[142] D.C. Hoaglin, F. Mosteller, and J.W. Tukey, editors. *Fundamentals of Exploratory Analysis of Variance.* Wiley, 1991.

[143] A.E. Hoerl. Fitting curves to data. In J.H. Perry, editor, *Chemical Business Handbook*, pages 55–77. McGraw-Hill, 1954.

[144] M.S. Hoffman, editor. *The World Almanac and Book of Facts.* World Almanac, 1991.

[145] H. Hogan. The 1990 post-enumeration survey: Operations and results. *J. Amer. Statist. Assoc.*, 88:1047–1060, 1993.

[146] R.A. Horn and C.R. Johnson. *Matrix Analysis.* Cambridge University Press, 1985.

[147] P.A. Horvathy. *Introduction to Monopoles.* Bibliopolis, 1988.

[148] D.W. Hosmer and S. Lemeshow. *Applied Logistic Regression.* Wiley, 2nd edition, 2000.

[149] D. Hounshell and J. Smith. *Science and Corporate Technology: DuPont R & D, 1902–1980.* Cambridge University Press, 1989.

[150] J.P. Hoyt. A simple approximation to the standard normal probability density function. *Am. Stat.*, 22:25–26, 1968.

[151] P.J. Huber. *Robust Statistics.* Wiley, 1981.

[152] P.J. Huber. Projection pursuit and robustness. In S. Morgenthaler, E. Ronchetti, and W.A. Stahel, editors, *New Directions in Statistical Data Analysis and Robustness*, pages 139–146. Birkhäuser, 1993.

[153] D. Huff. *How to Lie with Statistics.* W.W. Norton and Co., 1954.

[154] K. Hughes and R. Murphy. Ultrasonic robot localization using Dempster-Shafer theory. In *Proc. SPIE Conf. Neural and Stochastic Methods in Image and Signal Processing*, volume 1766, pages 2–11, San Diego, California, USA, July 1992.

[155] A. Huppert and A.R. Solow. A method for reconstructing climate from fossil beetle assemblages. *Proc. Royal Soc. Lond. B.*, 271:1125–1128, 2004.

[156] H. Jacob. *Using Published Data: Errors and Remedies.* Sage Publications, 1984.

[157] N. Jacobson. *Basic Algebra I.* W.H. Freeman, 2nd edition, 1985.

[158] E.T. Jaynes. On the rationale of maximum-entropy methods. *Proc. IEEE*, 70:939–952, 1982.

[159] J.D. Jobson. *Applied Multivariate Data Analysis*, volume 1. Springer-Verlag, 1991.

[160] N.L. Johnson. Systems of frequency curves generated by methods of translation. *Biometrika*, 36:149–176, 1949.

[161] N.L. Johnson and S. Kotz. *Urn Models and Their Application.* Wiley, 1977.

[162] N.L. Johnson, S. Kotz, and N. Balakrishnan. *Continuous Univariate Distributions*, volume 1. Wiley, 2nd edition, 1994.

[163] N.L. Johnson, S. Kotz, and N. Balakrishnan. *Continuous Univariate Distributions*, volume 2. Wiley, 2nd edition, 1995.

[164] N.L. Johnson, S. Kotz, and A.W. Kemp. *Univariate Discrete Distributions*. Wiley, 2nd edition, 1992.

[165] L.D. Henry Jr. *Zig-Zag-and-Swirl*. University of Iowa Press, 1991.

[166] M. Kac. Can one hear the shape of a drum? *Amer. Math. Monthly*, 73:1–23, 1966.

[167] J. Kacprzyk and M. Fedrizzi, editors. *Fuzzy Regression Analysis*. Physica-Verlag, 1992.

[168] F. Kafka. *The Metamorphasis*. Bantam Books, 1915. S. Corngold translation, published 1972.

[169] L. Kaufman and P.J. Rousseeuw. *Finding Groups in Data*. Wiley, 1990.

[170] D. Kelker. Distribution theory of spherical distributions and a location-scale parameter generalization. *Sankhya*, 32:419–430, 1970.

[171] J.B. Keller. The probability of heads. *Amer. Math. Monthly*, 93:191–197, 1986.

[172] M.G. Kendall. A new measure of rank correlation. *Biometrika*, 30:81–93, 1938.

[173] B.P. Kibble and G.H. Rayner. *Coaxial AC Bridges*. Adam Hilger, 1984.

[174] G. Klambauer. *Mathematical Analysis*. Marcel Dekker, 1975.

[175] P. Kooiman, A. Krose, and R. Ressen. Official statistics: An estimation strategy for the IT-era. In J. Bethlehem and P. van der Heijen, editors, *Proc. 14th Symposium Computational Statistics*, pages 15–26. Physica-Verlag, 2000.

[176] C.T. Kowal. *Asteroids*. Wiley, 2nd edition, 1996.

[177] E. Kreyszig. *Introductory functional analysis with applications*. Wiley, 1978.

[178] A.J. Laub. *Matrix Analysis for Scientists and Engineers*. Society for Industrial and Applied Mathematics, 2005.

[179] M.R. Leadbetter. Extreme value theory under weak mixing conditions. In M. Rosenblatt, editor, *Studies in Probability Theory*, pages 46–110. Mathematical Association of America, 1975.

[180] E.L. Lehmann. *Theory of Point Estimation*. Wiley, 1983.

[181] E.L. Lehmann and J.P. Romano. *Testing Statistical Hypotheses*. Springer-Verlag, 3rd edition, 2005.

[182] S.D. Levitt and S.J. Dubner. *Freakonomics*. William Morrow, 2005.

[183] K.-T. Li and K.-S. Wu. Selective oxidation of hydrogen sulfide to sulfur on vanadium-based catalysts containing tin and antimony. *Ind. Eng. Chem. Res.*, 40:1052–1057, 2001.

[184] H.-P. Liao and J.-H. Hong. Analysis and application of the hierarchical metadata framework of the ISO 19115 standard. Paper gd51-4. In *Proc. 26th Asian Conf. Remote Sensing, ACRS2005*, Hanoi, Vietnam, November 2005.

[185] G.H. Lincoff. *The Audubon Society Field Guide to North American Mushrooms*. Alfred A. Knopf, 1981.

[186] H.C. Ling, M.F. Yan, and W.W. Rhodes. Monolithic device with dual capacitor and varistor functions. *J. Am. Ceramic Soc.*, 72:1274–1276, 1989.

[187] R.J.A. Little and D.B. Rubin. *Statistical Analysis with Missing Data*. Wiley, 1987.

[188] R.J.A. Little and D.B. Rubin. *Statistical Analysis with Missing Data*. Wiley, 2nd edition, 2002.

[189] L. Ljung. *System Identification: Theory for the User*. Prentice-Hall, 2nd edition, 1999.

[190] J.W. Longley. An appraisal of least squares programs for the electronic computer from the point of view of the user. *J. Amer. Stat. Assoc.*, 62:819–841, 1967.

[191] R.D. Luce. On the possible psychophysical laws. *Psych. Rev.*, 66:81–95, 1959.

[192] R.D. Luce. A generalization of a theorem of dimensional analysis. *J. Math. Psych.*, 1:278–284, 1964.

[193] R.D. Luce. Dimensionally invariant numerical laws correspond to meaningful qualitative relations. *Phil. Sci.*, 45:1–16, 1978.

[194] A.L. Mackay. *A Dictionary of Scientific Quotations*. Institute of Physics Publishing, 1991.

[195] A. Madansky. *Prescriptions for Working Statisticians*. Springer-Verlag, 1988.

[196] C.L. Mallows. Robust methods—some examples of their use. *Amer. Stat.*, 33:179–184, 1979.

[197] B.B. Mandelbrot. *The Fractal Geometry of Nature.* W.H. Freeman, 1983.

[198] E.B. Manoukian. *Modern Concepts and Theorems of Mathematical Statistics.* Springer-Verlag, 1986.

[199] R.N. Mantegna, G. Ferrante, and F. Principato. Experimental investigation of theird and fourth moments of $1/f$ noise in microwave devices. In *Proc. 15th Intl. Conf. Noise in Physical Systems and $1/f$ Fluctuations*, pages 439–441, Hong Kong, August 1999.

[200] R.N. Mantegna, G. Ferrante, and F. Principato. Kurtosis experimental detection of electronic noise. In *Proc. 15th Intl. Conf. Noise in Physical Systems and $1/f$ Fluctuations*, pages 442–445, Hong Kong, August 1999.

[201] J. Marron, F. Hernandez-Campos, and F. Smith. Mice and elephant visualization of Internet traffic. In *Proc. 15th Symposium Computational Statistics, COMPSTAT2002*, pages 47–54, 2002.

[202] J.S. Marron and M.P. Wand. Exact mean integrated squared error. *Ann. Statist.*, 20:712–736, 1992.

[203] A.W. Marshall and I. Olkin. A multivariate exponential distribution. *J. American Statistical Association*, 62:30–44, 1967.

[204] P. McCullagh and J.A. Nelder. *Generalized Linear Models.* Chapman and Hall, 1989.

[205] W.A. McGeveran, Jr. et al., editors. *The 2005 World Almanac and Book of Facts.* World Almanac Books, 2005.

[206] I. McHale and P. Scarf. Forecasting international soccer match results using bivariate discrete distributions. Technical Report 332/06, Salford Business School, University of Salford, Manchester, UK, 2006.

[207] G.J. McLachlan and T. Krishnan. *The EM Algorithm and Extensions.* Wiley, 1997.

[208] D.V. Mehrotra and J.F. Heyse. Use of the false discovery rate for evaluating clinical safety data. *Stat. Methods Med. Res.*, 13:227–238, 2004.

[209] M.B. Miles and A.M. Huberman. *Qualitative Data Analysis.* Sage Publications, 1994.

[210] S.H. Mo and J.P. Norton. Fast and robust algorithm to compute exact polytope parameter bounds. *Math. Comput. Simul.*, 32:481–493, 1990.

[211] R. Morris, Jr. *Fraud of the Century: Rutherford B. Hayes, Samuel Tilden, and the Stolen Election of 1876.* Simon and Schuster, 2004.

[212] P.M. Morse and H. Feschbach. *Methods of Theoretical Physics.* McGraw-Hill, 1953.

[213] P.M. Morse and H. Feschbach. *Methods of Theoretical Physics, Part I.* McGraw-Hill, 1953.

[214] F. Mosteller and D.L. Wallace. *Inference and Disputed Authorship: The Federalist.* Addison-Wesley, 1964.

[215] R.B. Nelsen. *An Introduction to Copulas.* Springer-Verlag, 2nd edition, 2006.

[216] S.O. Nelson and P.G. Bartley. Open-ended coaxial-line permittivity measurements on pulverized materials. *IEEE Trans. Instrum. Meas.*, 47:133–137, 1998.

[217] S.O. Nelson, P.G. Bartley, and K.C. Lawrence. Measuring RF and microwave permittivities of adult rice wevils. *IEEE Trans. Instrum. Meas.*, 46:941–946, 1997.

[218] W.G. Nichols and J.D. Gibbons. Parameter measures of skewness. *Commun. Statist. Simul. Comput.*, B8:161–167, 1979.

[219] C.L. Nikias and A.P. Petropulu. *Higher-Order Spectra Analysis.* Prentice-Hall, 1993.

[220] C.L. Nikias and M. Shao. *Signal Processing with Alpha-Stable Distributions and Applications.* Wiley, 1995.

[221] H.W. Ott. *Noise Reduction Techniques in Electronic Systems.* Wiley, 1976.

[222] A. Papoulis. *Probability, Random Variables, and Stochastic Processes.* McGraw-Hill, 1965.

[223] J.K. Patel and C.B. Read. *Handbook of the Normal Distribution.* Marcel Dekker, 1981.

[224] E.S. Pearson, N.L. Johnson, and I.W. Burr. Comparisons of the percentage points of distributions with the first four moments, chosen from eight different systems of frequency curves. *Comm. Statist. Simul. Comput.*, B8:191–229, 1979.

[225] R.K. Pearson. Fitting straight lines to poor quality (x, y) data. *Math. Comput. Modeling*, 16:71–80, 1992.

[226] R.K. Pearson. *Discrete-Time Dynamic Models.* Oxford, 1999.

[227] R.K. Pearson. Exploring process data. *J. Process Control*, 11:179–194, 2001.

[228] R.K. Pearson. *Mining Imperfect Data: Dealing with Contamination and Incomplete Records.* SIAM, 2005.

[229] R.K. Pearson. The problem of disguised missing data. *SIGKDD Explorations*, 8:83–92, 2006.

[230] R.K. Pearson. Reciprocal rank-based comparison of ordered gene lists. In *Proc. GENSIPS'07: IEEE International Conf. Genomic Signal Processing and Statistics*, Tuusula, Finland, June 2007.

[231] R.K. Pearson, T. Zylkin, J.S. Schwaber, and G.E. Gonye. Quantitative evaluation of clustering results using computational negative controls. In *Proc. 4th SIAM International Conf. Data Mining*, pages 188–199, Lake Buena Vista, FL, USA, April 2004.

[232] P.N. Peduzzi, J. Concato, E. Kemper, T.R. Holford, and A. Feinstein. A simulation study of the number of events per variable in logistic regression. *J. Clin. Epidemiol.*, 99:1373–1379, 1996.

[233] B.W. Petley. *The Fundamental Physical Constants and the Frontier of Measurement*. Adam Hilger, Ltd., 1985.

[234] T. Pham-Gia. Value of the beta prior information. *Commun. Statist. Theory Meth.*, 23:2175–2195, 1994.

[235] D.N. Politis, J.P. Romano, and M. Wolf. *Subsampling*. Springer-Verlag, 1999.

[236] G. Polya. *How to Solve It*. Princeton University Press, 2nd edition, 1985.

[237] H. Preston-Thomas and T.J. Quinn, editors. *Techniques for Approximating the International Temperature Scale of 1990*. Bureau International des Poids et Measures, 1990.

[238] M.B. Priestley. *Spectral Analysis and Time Series*. Academic Press, 1981.

[239] M.H. Protter. Can one hear the shape of a drum? revisited. *SIAM Review*, 29:185–197, 1987.

[240] R Development Core Team. *R: A Language and Environment for Statistical Computing*. R Foundation for Statistical Computing, Vienna, Austria, 2009. ISBN 3-900051-07-0, http://www.R-project.org.

[241] D.A. Ratkowsky and D. Martin. The use of multivariate analysis in identifying relationships among disorder and mineral element content in apples. *Austral. J. Agric. Res.*, 25:783–790, 1974.

[242] R.C. Reid, J.M. Prausnitz, and T.K. Sherwood. *The Properties of Gases and Liquids*. McGraw-Hill, 1977.

[243] M. Reyes, R. Nisenbaum, D.C. Hoaglin, E.R. Unger, C. Emmons, B. Randall, J.A. Stewart, S. Abbey, J.F. Jones, N. Gantz, S. Minden, and W.C. Reeves. Prevalence and incidence of chronic fatigue syndrome in witchita, kansas. *Arch. Int. Med.*, 163:1530–1536, 2003.

[244] H. Riemer, S. Mallik, and D. Sudharshan. Market shares follow the Zipf distribution. Technical Report 02-0125, Univ. Illinois Urbana-Champaign College of Business Working Papers, 2002.

[245] F.S. Roberts. *Measurement Theory, with Applications to Decisionmaking, Utility, and the Social Sciences*. Addison-Wesley, 1979.

[246] J.G. Robinson, B. Ahmed, P.D. Gupta, and K.A. Woodrow. Estimation of population coverage in the 1990 United States Census based on demographic analysis. *J. Amer. Statist. Assoc.*, 88:1060–1071, 1993.

[247] R.T. Rockafellar. *Convex Analysis*. Princeton University Press, 1970.

[248] V. Rohatgi and G. Szekely. Sharp inequalities between skewness and kurtosis. *Stat. Probab. Lett.*, 8:297–299, 1989.

[249] P.R. Rosenbaum and J.H. Silber. Matching and thick description in an observational study of mortality after surgery. *Biostatistics*, 2:217–232, 2001.

[250] M. Rosenblatt. Dependence and asymptotic independence for random processes. In M. Rosenblatt, editor, *Studies in Probability Theory*, pages 24–45. Math. Assoc. America, Washington, DC, 1978.

[251] P.J. Rousseeuw and C. Croux. Alternatives to the median absolute deviation. *J. Amer. Statist. Assoc.*, 88:1273–1283, 1993.

[252] P.J. Rousseeuw and A.M. Leroy. *Robust Regression and Outlier Detection*. Wiley, 1987.

[253] D.B. Rubin. Inference and missing data (with discussion). *Biometrika*, 63:581–592, 1976.

[254] H. Sahai and M.I. Ageel. *The Analysis of Variance: Fixed, Random and Mixed Models*. Birkhauser, 2000.

[255] G. Salvadori, C. DeMichele, N.T. Kottegoda, and R. Roiso. *Extremes in Nature: An Approach Using Copulas*. Springer-Verlag, 2007.

[256] G. Samorodnitsky and M.S. Taqqu. *Stable Non-Gaussian Random Processes*. Chapman and Hall, 1994.

[257] N.V. Sastry and M.K. Valand. Densities, viscosities, and relative permittivities for pentane + 1-alcohols (c_1 to c_{12}) at 298.15k. *J. Chem. Eng. Data*, 1998.

[258] J.S. Schafer and J.W. Graham. Missing data: Our view of the state of the art. *Psychol. Methods*, 7:147–177, 2002.

[259] N. Schenker. Undercount in the 1990 census: Special section. *J. Amer. Statist. Assoc.*, 88:1044–1046, 1993.

[260] B. Schweizer and A. Sklar. *Probabilistic Metric Spaces*. North-Holland, 1983.

[261] F.C. Schweppe. *Uncertain Dynamic Systems*. Prentice-Hall, 1973.

[262] G.A.F. Seber and C.J. Wild. *Nonlinear Regression*. Wiley, 1989.

[263] G. Shafer. *A mathematical theory of evidence*. Princeton University Press, 1976.

[264] B.W. Silverman. *Density Estimation for Statistics and Data Analysis*. Chapman and Hall, 1986.

[265] J.S. Simonoff. *Smoothing Methods in Statistics*. Springer-Verlag, 1990.

[266] J.D. Singer. Types of factors and their structural layouts. In D.C. Hoaglin, F. Mosteller, and J.W. Tukey, editors, *Fundamentals of Exploratory Analysis of Variance*, chapter 4, pages 50–71. Wiley, 1991.

[267] J. Sjoberg, J.O. Zhang, L. Ljung, A. Benveniste, B. Delyon, P. Glorennec, H. Hjalmarsson, and A. Juditsky. Nonlinear black-box modeling in system identification: A unified overview. *Automatica*, 31:1691–1724, 1995.

[268] D. Sobel. *Galileo's Daughter*. Fourth Estate, 1999.

[269] C. Spearman. The proof and measurement of association between two things. *Am. J. Psychol.*, 15:72–101, 1904.

[270] Staff. *International Geographic Encyclopedia and Atlas*. Houghton Mifflin, 1979.

[271] S. Stahl. The evolution of the normal distribution. *Mathematics Magazine*, 79:96–113, 2006.

[272] Statsci. *S-PLUS Guide to Statistical and Mathematical Analysis*. Mathsoft Inc., 1995.

[273] D.N. Stivers, J. Wang, G.L. Rosner, and K.R. Coombes. Organ-specific differences in gene expression and UniGene annotations describing source material. In K.F. Johnson and S.M. Lin, editors, *Methods of Microarray Data Analysis III*, chapter 4, pages 59–72. Kluwer Academic Publishers, 2003.

[274] B.E. Storer and C. Kim. Exact properties of some exact test statistics for comparing two binomial proportions. *J. Amer. Stat. Assoc.*, 85:146–155, 1990.

[275] J.M. Stoyanov. *Counterexamples in Probability*. Wiley, 1987.

[276] J.H. Talpe, V.I. Bekeris, and C.E. Acha. Measurement of thermal conductivity and heat capacity in an undergraduate physics laboratory. *Amer. J. Physics*, 58:379–381, 1990.

[277] J.R. Thompson and R.A. Tapia. *Nonparametric Function Estimation, Modeling, and Simulation.* SIAM, Philadelphia, 1990.

[278] D.M. Titterington, A.F.M. Smith, and U.E. Makov. *Statistical Analysis of Finite Mixture Distributions.* Wiley, 1985.

[279] A. Trontell. How the US Food and Drug Administration defines and detects adverse drug events. *Curr. Ther. Res.*, 62:641–649, 2001.

[280] J.W. Tukey. *Exploratory Data Analysis.* Addison-Wesley, 1977.

[281] J.M. Varah. On fitting exponentials by nonlinear least squares. *SIAM J. Stat. Comput.*, 6:30–44, 1985.

[282] P.F. Velleman and D.C. Hoaglin. Data analysis. In D.C. Hoaglin and D.S. Moore, editors, *Perspectives on Contemporary Statistics*, number 21 in MAA Notes, chapter 2. Math. Assoc. America, 1991.

[283] W. Venables and B. Ripley. *Modern Applied Statistics with S.* Springer-Verlag, 2002.

[284] A. Vetere. The Riedel equation. *Ind. Eng. Chem. Res.*, 30:2487–2492, 1991.

[285] H.D. Vinod and A. Ullah. *Recent Advances in Regression Methods.* Marcel Dekker, Inc., 1981.

[286] D. Wagner, R. Nisenbaum, C. Heim, J.F. Jones, E.R. Unger, and W.C. Reeves. Psychometric properties of the CDC Symptom Inventory for assessment of Chronic Fatigue Syndrome. *Popul. Health Metrics*, 3, 2005.

[287] P. Walley. *Statisticsl Reasoning with Imprecise Probabilities.* Chapman and Hall, 1991.

[288] E. Walter and H. Piet-Lahanier. Estimation of parameter bounds from bounded-error data. *Math. Comput. Simul.*, 32:449–468, 1990.

[289] G.A. Watson. *Approximation Theory and Numerical Methods.* Wiley, 1980.

[290] R.J. Webster. Ambient noise statistics. *IEEE Trans. Signal Proc.*, 41:2249–2253, 1993.

[291] J. Wei. Least square fitting of an elephant. *Chemtech*, pages 128–129, February 1975.

[292] D.A. Weintraub. *Is Pluto a Planet?* Princeton University Press, 2007.

[293] P.H. Westfall and S.S. Young. *Resampling-Based Multiple Testing.* Wiley, 1993.

[294] R.L. Wheeden and A. Zygmund. *Measure and Integral.* Marcel Dekker, 1977.

[295] R.R. Wilcox. *Introduction to Robust Estimation and Hypothesis Testing.* Elsevier, 2nd edition, 2005.

[296] W.E. Winkler. Problems with inliers, working paper no. 22. In *Conference of European Statisticians*, Prague, Czech Republic, October 1997.

[297] S. Wolfram. *The Mathematica Book.* Cambridge University Press, 3rd edition, 1996.

[298] A. Wright. *Glut: Mastering Information through the Ages.* Joseph Henry Press, 2007.

[299] J. Yan. Enjoy the joy of copulas: With a package copula. *J. Stat. Softw.*, 21, 2007.

[300] K.K. Yuen. The two-sample trimmed t for unequal population variances. *Biometrika*, 61:165–170, 1974.

[301] L.A. Zadeh. Fuzzy sets. Memo ERL 64-44, Univ. California Berkeley, 1964.

[302] M. Zawadzki and B. Sujak. Behavior of semiconductor low temperature sensors in electromagnetic environments. *Cryogenics*, 23:599–602, 1983.

[303] A. Zerr, G. Miehe, G. Serghiou, M. Schwarz, E. Kroke, R. Riedel, H. Fuess, P. Kroll, and R. Boehler. Synthesis of cubic silicon nitride. *Nature*, 400:340–342, 1999.

[304] L. Zhengyan and L. Chuanrong. *Limit Theory for Mixing Dependent Random Variables.* Kluwer, 1996.

[305] G.M. Ziegler. *Lectures on Polytopes.* Springer-Verlag, 1995.

[306] H.J. Zimmermann. *Fuzzy Set Theory and Its Applications.* Springer-Verlag, 4th edition, 2001.

Index